Regenerative Medicine Using Pregnancy-Specific Biological Substances

Niranjan Bhattacharya • Phillip Stubblefield
(Editors)

Regenerative Medicine Using Pregnancy-Specific Biological Substances

Springer

Editors
Niranjan Bhattacharya
Department of General Surgery,
Obstetrics and Gynaecology and
Clinical Immunology,
Advanced Medical Research Institute,
Gol Park, B.P. Poddar Hospital, and
Vidyasagar State Hospital
Kolkata India

Phillip Stubblefield
Boston University Medical Centre
Deptartment of Obstetrics and Gynaecology
85 E. Concord Street
02118 Boston Massachusetts
USA

ISBN 978-1-4471-5733-5 ISBN 978-1-84882-718-9 (eBook)
DOI 10.1007/978-1-84882-718-9
Springer London Dordrecht Heidelberg New York

British Library Cataloguing in Publication Data
A catalogue record for this book is available from the British Library

Cover design: eStudioCalamar, Figueres/Berlin

Printed on acid-free paper

Springer is part of Springer Science+Business Media (www.springer.com)

Introduction

The legend of Prometheus of Greek mythological fame is well-known, but let it be repeated here with a medical twist. As a punishment for giving fire to humans, Zeus ordered Prometheus to be chained to a rock and sent an eagle to peck at his liver every day. However, Prometheus' liver was able to regenerate itself daily, enabling him to survive. Let us add a story from Hindu mythology to this. The goddess Durga, entreated to save the world from demons, took on Raktabeej, a demon with a unique property: every drop of blood that fell from him was regenerated into another demon, another Raktabeej – and the mother goddess was faced with a difficult task indeed as thousands of demons sprouted from Raktabeej's blood. Interestingly, "Raktabeej" translates as "blood seed."

What appeared to be (symbolic) stories in earlier days may soon become a reality with regenerative medicine. This is a new branch of medicine aimed at the regeneration of diseased or deteriorating organ systems. At the center of this branch of medicine are stem cells and other progenitor cells (cells that differentiate into various organs – a kind of "seed" cell). Interesting work is going on in several centers of excellence focused on different areas of regenerative medicine. For instance, Dr Anthony Atala, a contributor to this book, and his colleagues at the Wake Forest Institute of Regenerative Medicine, North Carolina, have successfully extracted muscle and bladder cells from several patients, cultivated them in petri dishes, and layered them in three-dimensional moulds that resemble the shapes of bladders. Again, Prof Paolo Macchiarini and his associates at the University of Barcelona, Spain, performed the first tissue-engineered trachea transplantation in June 2008.

However, the best examples of the physiological regeneration process can be seen during something that is very natural and common – pregnancy. Scientists, particularly those focusing on reproductive immunology have long sought explanations as to why pregnancies survive in the maternal system, which is essentially hostile. Transplantation biologists have been studying the mechanism of selective upregulation and downregulation of the feto-maternal immune system in order to understand how the fetus evades detection by the mother's hostile immune regulation system. The point here is that pregnancy and neoplasm are two outstanding examples of natural tolerance to homograft.

Pregnancy Is a Unique Phenomenon in the Life of a Female

To many scientists, pregnancy is nothing but an inflammation; to be a little more specific, it is a hormone-initiated chemical inflammation. However, there are many others who do not agree with this concept for different reasons. What is important is that pregnancy is an intricate process, and there are many dimensions to it that are not yet understood. To give an example, reproductive history appears to have a major impact on breast tumorigenesis; it is therefore reasonable to assume that pregnancy and lactation have enduring effects on the cancer susceptibility of multipotent progenitors. These pregnancy-induced mammary epithelial cells (PI-MECs) originate from differentiating cells during the first pregnancy and lactation cycle. They do not undergo apoptosis during post-lactational remodeling, and they persist throughout the remainder of a female's life. (Wagner KU, Smith GH. Pregnancy and stem cell behavior. *J Mammary Gland Biol Neoplasia*. 2005 Jan;10(1):25–36). There are now many new ideas circulating among immunologists, for instance that pregnancy serves as an important vaccination in the life of a female in the prevention of breast cancer, endometrial cancer, etc.

From the inner cell mass of the blastocyst stage, stem cells appear to support pregnancy. Initially, there is the embryonal stem cell, and then fetal and ultimately neonatal stem cells after parturition, and eventually these settle as adult stem cell. Apart from stem cells, the growth and birth of a baby involve the amniotic sac, the placenta, and the umbilical cord. The amniotic sac is a bag of fluid that helps to cushion the fetus from bumps and injury and maintains a constant temperature for its comfort and growth. It is made up of two membranes: the amnion and the chorion. The amniotic fluid protects the fetus from infection. The placenta is the most important organ for birth, linking the blood supply of the fetus to that of the mother, thus facilitating the supply of oxygen through the umbilical cord. Waste products like carbon dioxide are returned along the umbilical cord to be released into the mother's bloodstream. The placenta also protects the growing fetus from infections; moreover, it produces hormones essential for the growth and development of the baby. Toward the end of pregnancy, the placenta also passes antibodies from the mother to the baby, thus giving it immunity in the crucial first 3 months of its life.

Medicinal Uses of Placenta and Amniotic Membrane

Traditional Chinese medicine considers the placenta to be a powerful and sacred medicine full of life force, Qi. New mothers are advised placental capsules in a postpartum course of two capsules at a time with white wine. The wine is supposed to disperse the energy of the placenta throughout the body. This dosage can be taken up to three times a day until the mother feels balanced out. The remaining medicine can be taken homeopathically for the times when one's child may be undergoing separation anxiety, or first steps, weaning a baby, etc.

Some people plant trees or bushes over it, others bury it in a garden to enrich the soil, make placenta prints or membrane art, eat it cooked or raw, and others make medicinal capsules and/or herbal–homeopathic tinctures out of it. Some consider it cannibalism, others find it extremely helpful to ward off "baby blues" experienced in about 80% of women in the first few days or weeks after the birth of a baby. Some situations become more severe and postpartum depression (PPD) may evolve. The placenta medicine is reputed to ward off both the blues and PPD, shorten the post-bleeding time, restore lost hormones, nourish the blood, replenish depleted iron, reduce the overall recovery time from labor and birth for baby and mother after the birth, increase energy, boost the immune systems, and enhance milk production. Placentophagy, or consumption of the placenta, has been around for centuries. In fact, many beauty products contain placental membrane (for instance, the Jodome Organic Placenta Soap).

Modern medicine too became aware of the potential of fetal membranes in the early years of the last century. The first reported use of fetal membranes in skin transplantation was by Davis in 1910 (Davis JW. Skin transplantation with a review of 550 cases at the Johns Hopkins Hospital. *Johns Hopkins Med J.* 1910;15:307–396.) In 1913, Sabella used amniotic membrane on burned and ulcerated skin surfaces and observed lack of infection, marked decrease in pain, and increased rate of reepithelialization of traumatized skin surface (Sabella N. Use of fetal membranes in skin grafting. *Med Records NY.* 1913;83:478–480). Others have demonstrated the use of amniotic membrane as a biological dressing for open wounds including burns and chronic ulceration of the legs (Faulk WP, Mathews R, Stevens PJ, et al. Human amnion as an adjunct in wound healing. *Lancet.* 1980;1:1156–1158).

The first use of amniotic membrane transplantation (AMT) in ophthalmology was by De Rotth in 1940 who reported partial success in the treatment of conjunctival epithelial defects after symblepharon (scarring and adhesions between palpebral and bulbar conjunctiva) (De Rotth A. Plastic repair of conjunctival defects with fetal membranes. *Arch Ophthalmol.* 1940;23:522–525.). Other researchers found its use in caustic burns of the conjunctiva with corneal involvement apart from corneal epithelial defects, neurotrophic corneal ulcers, leaking filtering blebs after glaucoma surgery, pterygium surgery, conjunctival surface reconstruction, bullous keratopathy, chemical or thermal burns, and ocular surface reconstruction for cicatricial pemphigoid or Stevens–Johnson syndrome (Azuara-Blanco A, Pillai CT, Dua HS. Amniotic membrane transplantation for ocular surface reconstruction. *Br J Ophthalmol.* 1999;83:399–402; Baum J. Amniotic membrane transplantation: Why is it effective? *Cornea.* 2002;21:339–341).

The Present Book

The present book is an international attempt to bring researchers working on the potential uses of pregnancy-specific biological substances in regenerative medicine, under one umbrella. More than 72 distinguished authors from five continents have contributed in the 40 chapters of the book. The present President of the Royal College of Obstetrician and Gynaecologists, UK, Prof. Arulkumaran, has highlighted the important aspects of the book in his Foreword and Prof. Elaine Gluckman (Paris), a pioneer in the field of cord blood stem cell transplantation has written the Preamble to the book.

The first four chapters delineate the importance of pregnancy-specific biological substances, particularly the placenta and umbilical cord blood. In the first chapter, Prof. Andrew Burd and Dr. Lin Huang of the Chinese University of Hong Kong calculate the massive global wastage of pregnancy-specific substances and comment that while it is "inevitable that there will be commercial exploitation of some of this source material for extracting specific and defined biological materials … there remains a considerable quantity that is simply going to be discarded and this represents a massive wastage of global resources." Drs Ornella Parolini and Maddalena Soncini explains the significance of the placenta as a source of stem cells and as a key organ for feto-maternal tolerance in the next chapter. They discuss the mechanism of feto-maternal tolerance under certain parameters: (a) expression of nonclassical MHC molecules by trophoblastic cells; (b) expression of the IDO enzyme by placental cells, resulting in tryptophan depletion and kyurenine production; (c) FasL expression by trophoblastic cells; and (d) expression of complement regulator proteins by trophoblastic and decidual cells. The third chapter described the scope and use of the placenta and the umbilical cord in age-old Chinese medicine. According to the author, Prof. P.C. Leung of the Chinese University of Hong Kong, "The human placenta was described as a medicinal material as early as 400BC during Hippocrates' time and in China in 200BC, when it was used as a healing agent after bodily injuries. Legendary figures used it for exclusive reasons. Thus, the great tyrant of the Qin Dynasty in China used human placenta for longevity and the Egyptian Queen Cleopatra used it for cosmetic purposes." Chapter 4 explains the implications of biochemical variations of the umbilical vein and its role in the growth of the fetus in utero. Dr. Bon and Prof. Raudrant examine how regulation of fetal growth involves genetic factors, maternal nutritional factors, circulatory and placental factors, as well as fetal factors, particularly hormonal.

Seven chapters (Chaps. 5–11) of this book deal with the potential use of the discarded placental blood as a true blood substitute. Prof. P. Pranke and Prof. T. Onsten of the Federal University of Rio Grande do Sul, Av Ipiranga, Brazil (Chap. 5) discuss the fundamentals of transfusion strategies with cord blood, citing evidences like higher levels of hemoglobin, hematocrit, mean corpuscular volume, leukocytes, and fetal hemoglobin; and low levels of coagulation factors, diminished expression of erythrocyte antigens, low levels of immunoglobulin, and also an absence of natural antibodies. Drs. T Brune and H Garritsen (Chap. 6) have worked on the problem of autologous transfusion of placental blood in premature babies. They emphasize that adequate blood supply in premature and mature neonates with anemia is a continuous point of discussion in neonatology and transfusion medicine. Dr. Tang-Her Jaing and Dr. Robert Chow discuss (Chap. 7) the utility of cord blood in pediatrics stressing that UCB represents an important new HSC source which has a number of significant advantages over bone marrow.

A clinical experience of cord blood autologous transfusion is described by Dr. Shigeharu Hosono of the Nihon University School of Medicine, Japan in Chap. 8. He is of the view that autologous transfusion prevents the risks of acquiring a transfusion-transmitted infection and releases recipients from allergic reaction. In addition, autologous transfusion of placental blood in premature babies at birth may be considered to be one of the strategies of resuscitation. In a different vein, Prof. Norman Ende et al. from New Jersey, USA, discuss the emergency use of cord blood in disaster scenarios in Chap. 9, where more fresh blood will be needed than available.

In Chap. 10, the struggle to find a blood substitute has been discussed by Prof. E. Moore and his colleagues from the University of Colorado, USA. A legend in the field, Prof. Moore et al. have commented that "The greatest need for blood substitutes worldwide is in patients with unanticipated acute blood loss, and trauma is the most likely scenario. The blood substitutes reaching advanced clinical trials today are red blood cell (RBC) substitutes, derived from hemoglobin. The hemoglobin-based oxygen carriers (HBOCs) tested currently in advanced clinical trials are polymerized hemoglobin solutions…." In the next chapter, Dr. Niranjan Bhattacharya (India), who has been using umbilical cord blood clinically in severe cases of anemia in the background of diseases like leprosy, tuberculosis, thalassemia, malignancies, etc. from the geriatric to the pediatric age group recounts his experiences of using cord blood as a substitute of adult blood. He has emphasized the potential of the impact of donor cytokines on the host system.

Some important questions arise in the context of pregnancy-specific complex phenomena: does pregnancy serve as a vaccination or is it an incomplete vaccination? Does transplacental cell traffic cause inflammation or do they regenerate damaged tissue in certain autoimmune diseases? Is fetal microchimerism a natural occurring phenomenon leading to detectable levels of mononuclear cells in several maternal tissues, such as lungs, heart, spleen, kidney, and bone marrow? These are matters of debate in modern day medical science. In Chap. 12, Prof. Carolyn Troeger, Dr. Olav Lapaire, Dr. Xiao Yan Zhong, and Prof. Wolfgang Holzgreve of the University Women's Hospital, Basel, Switzerland try to explain the implications of feto-maternal cell transfer in normal pregnancy.

In Chap. 13, the immunotherapy potential of cord blood transfusion in cases of advanced breast cancer has been discussed by Dr. Niranjan Bhattacharya (Kolkata, India). He has mentioned that in his clinical experience, there was a rise in hemoglobin concentration after the transfusion of two units of cord blood and a secondary rise was noted during the seventh day assessment. He suggests that this could be the result of the cytokine impact of the fresh cord blood on the hosts' bone marrow. This is a unique phenomenon and never occurs with conventional adult blood transfusions. Moreover, assessment of peripheral blood CD34 level after 72 h of the first two units of cord blood transfusion showed a rise of CD34 from .02% to 79%. This rise appeared to have a good prognostic effect and it raises a serious question about the immunotherapeutic potentialities of cord blood CD34 hematopoietic stem cell.

The four subsequent chapters (Chaps. 14–17) deal with the use and potentialities of cord blood in Neurology. Prof. Martina Vendrame (Temple University, Philadelphia) examines the anti-inflammatory effects of human cord blood and its potential implication in neurological disorders in Chap. 14: "Although initial in vitro evidence pointed to the differentiation of human cord blood cells (HUCBCs) into neuronal and glial lineages, transplantation of these cells never resulted in terminally differentiated neurons. This raised the suspicion that the beneficial effect of HUCBCs in models of central nervous system (CNS) disorders and injury may be attributable to alternative biologic properties. The indication that HUCBCs may have anti-inflammatory and immunoregulatory properties has recently emerged from animal studies. The activation of these systems triggers the production of glucocorticoids and catecholamines which in turn mediates the release of anti-inflammatory interleukins (including IL-10) from CNS resident and infiltrating monocytes, which allow a protective feedback mitigating the initial ischemia-induced pro-inflammatory response. This inhibitory feedback may also induce an immunosuppressive state, which human studies have proven accountable for the infectious complications seen in stroke patients."

Prof. Mariane Secco of Instituto de Biociências Universidade de São Paulo, Brasilia has examined the potential use of HUCB in neuromuscular disorders in Chap. 15, while in the next chapter (Chap. 16) the use of Human Umbilical Cord

Blood Cells in strokes is discussed by Prof. Paul R. Sanberg and his team from Center of Excellence for Aging & Brain Repair, Department of Neurosurgery, University of South Florida College of Medicine, Tampa. According to Prof. Sanberg, "Stroke causes irreversible and permanent damage in the brain immediately adjacent to the region of reduced blood perfusion…Regenerative immediate action is fundamental. Currently, the only effective treatment for stroke, tissue-plasminogen activator, has a very narrow therapeutic window. These disease outcomes should be taken under consideration in developing any therapeutic intervention, especially in cell based therapy. Human umbilical cord blood (HUCB) cells, due to their primitive nature and ability to develop into nonhematopoietic cells of various tissue lineages, including neural cells, may be useful as an alternative cell source for cell-based therapies requiring either the replacement of individual cell types and/or substitution of missing substances."

Chapter 17 is a review article by Dr. Abhijeet Chaudhuri, UK and Dr. Niranjan Bhattacharya, India on the overall use of Placental Umbilical Cord Blood Transfusion for Stem Cell Therapy in Neurological Diseases.

The authors carry this line of thinking further and suggest that placental umbilical cord blood transfusion is potentially an effective therapy for acute ischemic stroke and probably the only treatment that can promote repair of ischemic brain and aid early functional recovery. An added functional advantage of umbilical cord blood in re-perfusing ischemic brain is its high concentration of fetal hemoglobin (Hb F), which has greater oxygen binding capacity than normal adult hemoglobin (HbA). This has been shown to be of considerable therapeutic importance in sickle cell disease and hemoglobinopathy; and in clinical stroke patients, it has the potential for improving oxygenation in the ischemic tissue. Hb F will deliver more oxygen to the surviving neurons in the ischemic core and ischemic penumbra in areas of partial blood flow. The rheological property of term cord blood is also favorable for reperfusion because of lower viscosity.

In the next few chapters, the use of umbilical cord blood, serum, and vein in various disciplines of medicine has been discussed. In Chap. 18, the use of placental umbilical cord blood serum in Ophthalmology has been reviewed by Dr. Kyung-Chul Yoon, of the Department of Ophthalmology, Chonnam National Universtiy Medical School and Hospital, South Korea.

A pioneer in the field, Dr. Kyung-Chul Yoon has commented that umbilical cord blood contains essential tear components, growth factors, and neurotrophic factors such as epidermal growth factor, vitamin A, transforming growth factor-β, substance P, insulin-like growth factor, and nerve growth factor. Umbilical cord serum can provide basic nutrients for epithelial renewal and facilitate the proliferation, migration, and differentiation of the ocular surface epithelium. Serum eye drops made from umbilical cord blood can be used for the treatment of severe dry eye with or without Sjögren's syndrome, ocular complications in graft-versus-host disease, persistent epithelial defects, and neurotrophic keratopathy.

Chapter 19 presents a unique essay on the use of the placental umbilical cord in cardiovascular surgery by Dr. Alan Dardik and Prof. Herbert Dardik of the Yale University School of Medicine. For decades, alternatives to the autologous saphenous vein have been studied. Human umbilical cords are approximately 23 in. long and normally contain one vein and two arteries in a mucopolysaccharide matrix called "Wharton's jelly." At birth, the vessels are collapsed but the vein can easily be dilated up to 7 mm in diameter and the arteries can be dilated up to 4 mm. Roentgenographic studies have shown that the vessels are of uniform diameter. They have no branches, valves, or vasa vasorum. Manometric studies in vitro have shown that these vessels can tolerate pressures in excess of 600 mmHg. In this thought-provoking article, the authors have narrated their experiments with unmodified segments of human umbilical cord veins in the aorta of baboons. They then studied the effects of both dialdehyde starch and glutaraldehyde tanning on umbilical cord vessels prior to their implantation as vascular conduits. Long-term studies have shown that the glutaraldehyde stabilized umbilical vein graft retains its basic architecture. On the basis of improved manufacturing and quality control, this graft has now proved remarkably stable and resistant but certainly not immune to the forces of biodegradation. However, aldehyde cross-linkage of the protein moieties increases tensile strength and masks antigenicity. A polyester (Dacron) mesh is placed about the vein, which is then sterilized and stored in 50% ethanol. Most important, processing with glutaraldehyde sterilizes the tissue of bacteria, viruses, and fungi and renders it nonantigenic.

In Chap. 20, the use of cord blood in Cardiovascular Medicine has been studied by Dr. Peter Hollands, Department of Biomedical Science, University of Westminster, London. According to him, "In order to create angiogenesis in any scenario it is necessary to obtain a cell population which contains a good proportion of endothelial progenitor cells (EPC). Human cord blood has been shown to contain angioblast-like EPC in significantly larger numbers than those found in human peripheral blood . These cord blood EPC were shown to be capable of postnatal neovascularization in the ischaemic hindlimb of rats in vivo. Earlier studies had shown the possible presence of EPC in adult human peripheral blood and it is thought that these cells may reside in the adult bone marrow and are mobilised by tissue ischaemia and associated cytokine release .Similar studies have shown that cord blood progenitors can induce angiogenesis in the in the implanted human thymus in the kidney of NOD/SCID mice .These studies indicate the potential of autologous (and possibly allogeneic) clinical transplantation of cord-blood-derived EPC into ischaemic tissues."

In the subsequent chapter, the concept of Endothelial Progenitor Cells was further supplemented by Dr. Maurizio Pesce et al. from the Laboratorio di Biologia Vascolare e Terapia Genica, Centro Cardiologico Monzino, Milan, Italy. They are of the opinion that EPC antigenic or functional quantification in the peripheral circulation has acquired the value of a diagnostic and prognostic "marker" for cardiovascular disease (CVD) and CVD risk factors. For example, it has been shown that the number of cells expressing EPC markers or showing EPC in vitro clonogenic activity was correlated with the occurrence of acute ischemic events like the vascular trauma. In addition, patients suffering from cardiovascular risk conditions such as old age, male gender, hypertension, diabetes, cigarette smoking, family history of coronary artery disease (CAD), and high LDL cholesterol levels were shown to have significantly reduced levels of circulating EPCs and lower numbers of in vitro clonogenic cells.

The use and potentialities of cord blood in cardiology has been further emphasized in the three following chapters by noted global experts in the field (Chaps. 22–24).

In Chap. 22, the therapeutic potential of placental umbilical cord blood in cardiology was discussed by Dr. Shunichio Miyoshi et al. from the Department of Cardiology, Keio University School of Medicine, Tokyo. According to him, "Cardiomyocytes do not undergo cell division after birth. Once cardiomyocytes become necrotic by myocardial infarction, residual cardiomyocytes do not undergo cell division and cannot restore damaged heart tissue. Therefore, in order to restore severely damaged heart function, heart transplantation from a living donor is performed, which, however, is restricted by a shortage of donors. Embryonic stem cells and somatic stem cells, which have a potential to transdifferentiate into cardiomyocytes, may be able to supply newly generated cardiac muscle cells and restore a severely impaired heart." This group of investigators reported for the first time that murine marrow-derived mesenchymal stem cells (MSCs) can transdifferentiate into cardiomyocytes in vitro by use of 5-azacytidine, which is known to cause nonspecific demethylation of DNA. However, in humans, MSCs could not transdifferentiate into cardiomyocytes by use of 5-azacytidine; cocultivation with murine cardiomyocytes was essential. Moreover, the cardiomyogenic transdifferentiation ratio was extremely low in human marrow-derived MSCs (0.1–0.3%). This result appears reasonable since the human nucleus is protected from spontaneous mutation of gene and neoplasm formation, because human life is longer than that of popular experimental animals.

Prof. Amit Patel and his group from the Center for Cardiac Cell Therapy, University of Pittsburgh Medical Center and McGowan Institute for Regenerative Medicine, USA, next discuss the issue in more general terms. He points out that stem cell therapy such as autologous bone marrow, mobilized peripheral blood, and purified cells have been used clinically since 2001. Over 1,000 patients have received cellular therapy as part of randomized trials till date, with the general consensus being that a moderate but statistically significant benefit occurs. Therefore, one of the important steps in the field is optimizing treatment approaches. They opine that there are three main approaches to optimize stem cell therapy efficacy including: (a) increasing stem cell migration to the heart, (b) optimizing stem cell activity, and (c) combining existing stem cell therapies to recapitulate a "therapeutic niche" and the potential of cord blood in cardiovascular regenerative medicine.

In Chap. 24, the use of cord blood in myocardial infarction has been analyzed by Prof. Robert Henning , University of Florida. His experiments suggest that HUCBC are beneficial in infarcted myocardium and do not require host immunosuppression. HUCBC significantly decrease inflammatory cytokines in hearts with myocardial infarctions and this decrease in inflammatory cytokines is associated with significant decreases in the percentages of myocardial neutrophils and CD3 and CD4 T lymphocytes in the infarcted myocardium. As a consequence, HUCBC can produce a substantial reduction in acute myocardial infarction size in comparison with untreated infarcted hearts when these cells are directly injected into the myocardium, or into the coronary arteries, or given intravenously. Endothelial progenitor cells are normal components of umbilical cord blood that can release pro-angiogenic molecules such as vascular endothelial growth factor. These cells can also express KDR, Tie2/Tek, and VE-cadherin, which are expressed by endothelial cells during new blood vessel formation. In addition, CD34+ HUCBC can integrate into the walls of blood vessels in the periphery of injured tissue and can increase capillary density in ischemic/infarcted muscles.

The use of placental umbilical cord blood in other subspecialties of Regeneration Medicine has been narrated from different centers of repute by noted investigators in the field (Chaps. 25–32).

In Chap. 25, the use of only mesenchymal stem cells from cord blood has been related by Dr. Jose J. Minguell representing the TCA Cellular Therapy, LLC, Covington, LA. Dr. Jose Minguell, a noted expert on mesenchymal stem cells, suggests that hematopoiesis is sustained by a subset of stem cells, which although present as committed progenitors on the yolk sac are unable to reconstitute the entire hematopoietic system. "The multipotent hematopoietic stem cell (HSC) emerges in the AGM region just before the establishment of the hematopoietic liver, where it subsequently expands and colonizes the hepatic tissue and finally the newly formed bone marrow. Concomitantly with the full expression of the self-renewal and differentiation potential of HSC, a fetal liver microenvironment (niche) is formed which plays an instructive role to the HSC. Apparently, one of the first events dealing with the interaction of the duplex stem/progenitor/mature cell and hepatic niche is the expression of cell adhesion proteins (4, 5, and β1 integrins) , on the surface of the primitive erythroid cell, which migrates into the fetal liver (FL) and interact with macrophages within erythroblastic islands in a stage-specific and VCAM-1-dependent process."

In Chap. 26, Jian-Xin Gao and Quansheng Zhou of the Department of Pathology and Comprehensive Cancer Center, Ohio State University, Columbus; Cyrus Tang of the Hematology Research Center, Soochow University, Suzhou, China and Department of Molecular Biology, The Scripps Research Institute, La Jolla, CA write on the current and future status of cord blood stem cell expansion ex vivo. Reconstitution of the immune system with allogeneic hematopoietic stem cells (HSCs) appears to be critical for the cure of hematopoietic malignancies and some autoimmune diseases. The authors feel that cord blood (CB) HSCs are an ideal resource for the reconstitution. The number of HSCs in each CB unit is not sufficient for the patients who require multiple HSC transplantations. Large ex vivo expansion of CB HSCs may overcome the difficulty. Ideally, the expanded CB HSCs should be safe without the risk of cell transformation, preserve the capability of self-renewal and multipotency of differentiation, and be competent in long-term repopulation. In this chapter, the authors have reviewed the recent progress in ex vivo expansion of CB HSCs as well as the current understanding of HSCs, including the cellular and molecular basis which is useful for ex vivo expansion of HSCs.

Prof. Colin McGuckin, University of New Castle Upon Tyne, UK, relates certain advances in cord blood use in regeneration biology in the next chapter. Prof. McGuckin does not require an introduction because of his eminence in the field, but he is a very straightforward person who does not baulk at calling a spade a spade. According to him, "While some governments have and are pouring millions into embryonic stem cell research with no cures, no new drugs and no clinical trials to show for the money, cord blood therapies have already helped over 10,000. Given that the new clinical trials, not least with Type 1 Diabetes, show that there is a growing need, many people alive today could have been treated or supported if a cord blood had been stored for them … 20 years ago cord blood was treating only one or two diseases. 10 years ago only a handful. Now nearly 80 conditions are treatable or supportable with cord blood stem cells. We dream of a day when there will be cord blood banks in every metropolitan city. A dream which is in the making, and a revolution which will continue to grow."

In Chap. 28, Prof. Zygmunt Pojda of the Department of Experimental Hematology, Maria Sklodowska-Curie Memorial Cancer Center and Department of Regenerative Medicine, Warsaw (Poland) discusses the use of non-hematopoietic stem cells of fetal origin from cord blood, umbilical cord, and placenta in Regeneration Medicine. He feels that cord blood, cord, and placenta are a "biological waste" so the cells can be isolated without any medical or ethical contraindications. Further, fetal cells are in many aspects more suitable for clinical purposes than their adult counterparts, having greater proliferation and differentiation potential, lesser cumulation of DNA lesions, lesser risk of pathogen transmission, and reduced host-versus-graft or graft-versus-host reactivity. The chapter offers the characterization of phenotype, expansion capabilities, and in vitro or in vivo differentiation potential of non-hematopoietic stem and progenitor cells present in cord blood, umbilical cord, and placenta, emphasizing the practical aspects of their availability and isolation techniques. Moreover, fetal stem cells are the important tool for the future clinical applications in regenerative medicine, transplantology, oncology, and gene therapy.

Dr. Thomas E. Ichim and Dr Michael P Murphy from Indiana University and Dr. Neil Riordan, from Medistem, USA analyze animal studies involving cord blood and regeneration in the following chapter. They remark that markedly different from use of cord blood for hematopoietic transplants, the use in regenerative medicine does not require ablation of the patient's immune response. This raises the issues of "host versus graft and graft versus host reactions, as well as immunological implications of microchimerism in patients receiving cord blood for regenerative medicine." In this chapter, they discuss the immunology of unrelated cord blood transplants, draw parallels with the similar situation of chimerism in fetal to maternal trafficking of stem cells, and provide a framework for performing future clinical trials that may benefit from the unintended immunological consequences of cord-blood-mediated immune modulation.

Dr. Neil H. Riordan (a coauthor in the last article), Chairman, Medistem Laboratories, Inc., California, next discusses the immune privileges of cord blood in greater detail in Chap. 30. He suggests that T cells from cord blood are known to have a propensity toward an anti-inflammatory phenotype. This is illustrated, for example, in experiments with CD4+ T cells from cord blood which were shown to produce significantly lower IFN-gamma and higher IL-10 upon activation with mature dendritic cells as opposed to control adult blood derived CD4+ T cells. Other experiments have demonstrated hyporesponsiveness to mitogen and MLR stimulation as well as reduced levels of IL-2 production and IL-2 responsiveness as opposed to adult T cells.

In Chap. 31, a basic issue has been discussed: does the trigger for stem cell regeneration start from its environment or niche? Prof. Ian McNiece, Director Regeneration Biology, University of Miami, feels that stem cells should have an appropriate niche for survival and proliferation. He proposes that combination cell therapy will be necessary for optimal tissue repair for heart disease using mesenchymal stem cell (MSC) to repair the microenvironment of ischemic tissue and cardiac stem cells for regeneration of cardiomyocytes.

The use of cord blood in regenerative medicine has also been examined by Prof David T. Harris of the University of Arizona, Tucson, USA. He estimates that up to 128 million individuals may benefit from regenerative medicine therapy, or almost one in three individuals in the USA: "Multipotent stem cells are easily available in large numbers in umbilical cord blood (CB), and may be the best alternative to embryonic stem (ES) cells. CB stem cells are capable of giving rise to hematopoietic, epithelial, endothelial, and neural tissues both in vitro and in vivo. Thus, CB stem cells are amenable to treat a wide variety of diseases including cardiovascular, ophthalmic, orthopedic, neurological and endocrine diseases. Examples of these usages currently in clinical trials include applications that affect the nervous and endocrine system, including cerebral palsy and type I diabetes."

The next three chapters deal with issues in cord blood banking and the authors are experienced cord blood bankers. The problem of cord blood collection variability and banking has been presented in Chap. 33 by Dr. Suzanne Watt, NBS NHS, UK. Citing their work experience in the UK she notes that many of the procedures for processing and storage of UCB in use in England today were established at the New York Cord

Blood Bank in the 1990s and modified in the National Blood Service (now NHS Blood and Transplant or NHSBT). UCB Banking within NHSBT was first instigated in 1995 as part of the South East Regional Blood Transfusion Service. The next chapter too deals with cord blood banking. Here, the donor and collection-related variables affecting product quality in ex-utero cord blood banking have been related by Dr. Sabeen Askari, of the Blood Bank & Transfusion Service, Veterans Affairs Medical Center, Minneapolis. His focus is on the problem of optimizing product quality, which is a current focus in cord blood banking and the effect of various variables. The cell dose is considered as the most important factor compared with HLA in donor choice. A minimum cell dose of >4 × 107 NC/kg at collection and 3 × 107 NC/kg at infusion is recommended. CD34+ cell count correlates with engraftment and a dose of >2 × 105 CD34+ cells/kg is considered optimal; however, it cannot be used for comparative studies between centers due to absence of standardization of counting method. Colony forming units of granulocytes-monocytes colonies (CFU-GM) are also used for measuring the stem cell content of CBUs, but there is significant interlaboratory variability (Barker JN, Davies SM, DeFor T, Ramsey NK, Weisdorf DJ, Wagner JE. Survival after transplantation of unrelated donor umbilical cord blood is comparable to that of human leukocyte antigen-matched unrelated donor bone marrow: results of a matched-pair analysis. *Blood.* 2001;97:2957–2961. Full Text via CrossRef | View Record in Scopus | Cited By in Scopus (167)). The chapter presents a review of selected donor- and collection-related variables and their effect on total volume, nucleated cell count (TNC), and CD34+ cell count of the CBUs that are collected ex-utero.

The problem of collection procedure and variables of cord blood as a transplantation source is discussed next by Dr. Pilar Solves and Dr. Vicente Mirabet of the Tissue Bank, Valencia Transfusion Center, Spain. According to them, "UCB quality is basically defined by three parameters: Total nucleated cells (TNC), CD34+ cells and colony forming units (CFU) content. TNC is a surrogate measure of the stem cell dose in the transplant product and currently the most important factor for donor choice. The most utilized phenotypic marker for stem and progenitor cells is CD34, a glycophosphoprotein. Although it is used as an important clinical marker it is found also on cells that are not stem or progenitor cells. Colony assays (CFU) determine the in vitro functionality of hematopoietic progenitors. These methods use the growth of cells in semi-solid culture media that allow the growth of distinctive colonies. These colonies derive from single cells termed high-proliferative-potential colony-forming cells (HPP-CFCs), multipotential colony forming units (CFU-GEMM for granulocyte, erythroid, macrophage, magakaryocite-containing components), and more lineage-restricted progenitors such as CFU-GM (containing granulocyte and macrophage differentiation capacity), CFU-G (with granulocyte differentiation ability), CFU-M (with macrophage differentiation capacity), CFU-Mega (with megakaryocyte differentiation capacity), and BFU-E (burst-forming-unit-erythroid) .The content of CFU is based on the number of different colonies formed per number of cells plated. Early studies showed that CFU-GM could be grown in vitro from UCB. Further in vitro studies by Broxmeyer et al demonstrated that UCB contains sufficient number of HSC to be used for autologous or allogeneic hematopoietic reconstitution]. Although these three parameters are well correlated, CD34+ cells and CFU content predicts the hematopoietic potential of a UCB unit better than TNC content .UCB characteristics influencing engraftment are total nucleated cells (TNC), CD34+ cell, CFU contents and degree."

The book also has several chapters on the clinical use of amniotic fluid as a ready supplier of stem cells for cell therapy (Chaps. 36–38).

In Chap. 36, Prof. Anthony Atala, Chair, Department of Urology; Director, Institute for Regenerative Medicine. Wake Forest University Baptist Medical Center Winston-Salem, North Carolina presents his perspective on the current and potential uses of the placenta and amniotic fluid. Noting that human amniotic fluid has been used in prenatal diagnosis for more than 70 years, there is now evidence that it may be the source of a powerful therapy for a multitude of congenital and adult disorders since a subset of cells found in the amniotic fluid and placenta appears to be capable of maintaining prolonged undifferentiated proliferation. In addition, these cells can also differentiate into multiple tissue types that encompass the three embryonic germ layers of the embryo, suggesting that they could be used for a myriad of tissue engineering and cell therapeutic applications.

Next, the use of amniotic fluid in nonhealing ulcer dressing is presented on the basis of clinical evidence by Dr. Niranjan Bhattacharya, Kolkata, India. He cites his experience on the use of pregnancy-specific biological substances in burn patients. The development of wound infections is the most common cause of mortality and morbidity among burn patients. A variety of dressings have been used to cover, reduce burn wound sepsis, and promote wound healing. His experiences with placental and amniotic substances indicated that the use of (1) freshly collected placenta at the burn wound site as a dressing material may have a positive cytokine impact on the process of healing; (2) amniotic fluid is a cell therapy source, because of its rich content of epithelial and mesenchymal stem cell component, leaving aside its antibacterial propensity as a helpful adjuvant; (3) amniotic membrane as a temporary biological dressing is an effective method in reducing burn wound sepsis with judicious application mode, that is, chorionic side to augment vasculogenesis and the amniotic side to promote epithelialization. This is an effective step to augment the cell therapy component of the amniotic fluid.

The same investigator suggests that amniotic fluid can play a role (cell therapy) in amelioration in osteoarthritis in the subsequent chapter. A number of different origins have been suggested for amniotic fluid cells. Cells of both embryonic and fetal origins and cells from all three germ layers have been reported to exist in amniotic fluid. Amniotic fluid is a unique fluid made by nature; it is a cocktail of mesenchymal stem cells with antibacterial property, which is used in the study presented by the author as the cell therapy source for the repair of damaged cartilage, synovial membrane, supporting muscles and supporting ligaments, as per the niche provided to these specialized stem cells for regeneration purposes, in advanced and degenerative osteoarthritis with satisfying results. The amniotic fluid, because of its increased viscosity due to protein and other cellular suspension, differs from the steroid treated fluid (normal saline), and may act as a lubricant which diminishes the irritation at the initial phase; and the mesenchymal cells, which do not express HLA antigens, may possibly help in the repair process of the adjacent structures in the joint space as a whole. The study group was divided into a steroid treated group (A) and the amniotic fluid group (B). While 12 out of 26 patients in group B maintained patient satisfaction after 1 year of follow-up, only four patients reported similar satisfaction from group A in the corresponding period.

Pregnancy-specific biological wastage must include the aborted fetal tissue. Chapter 39 deals with the therapeutic potential of the aborted human tissue collected from consenting volunteer mothers donating the fetal tissue for medical research under strict ethical supervision in a free government hospital. This chapter too is contributed by Dr. Bhattacharya and here he narrates his clinical experiences of human fetal neuronal transplantation at a heterotopic site in advanced Parkinsonism. In the present study, both subjective and objective improvement was observed in 83.2% of the patients from the pre-transplant level to the date of assessment, that is, 1 month after the instillation of the fetal cortical graft. Another noteworthy observation was that 75% of the cases in this present study were able to reduce the L-dopa dosage. Similarly, clinical review revealed partial improvement (33%) of the dyskinetic status in one-third of the patients (33%) in the present series in both subjective and objective assessment. Histology and electron microscopy did not reveal any cellular feature of inflammation or immunological reaction of the retrieved fetal tissue at the third month and surprisingly, identical histological changes were noted in the tenth year retrieved tissue. Disability study and cognitive assessment with mini-mental state and mood showed improvement due to the transplant impact ($p < .001$) from its pretransplant value. There were other secondary advantages due to the fetal tissue transplantation, including rise of hemoglobin, weight gain, sense of wellbeing with reduction of aches and pain all over the body, etc. The conclusion of the author is that human fetal neuronal tissue transplantation is an effective remedy in a neuro-degenerative disease like idiopathic Parkinsonism.

Today, the use of stem cells, cord blood, or any pregnancy-related biological waste is subject to controversy. The final chapter of this book deals with ethics-related dilemmas in human research, particularly the global ethical issues surrounding umbilical cord blood donation and banking. The authors are Dr. Gabrielle Samuel, Prof. Ian Kerridge, and Dr. Tracey O'Brien of the Centre for Values, Ethics & the Law in Medicine, University of Sydney, Australia. Since hematopoietic stem cell transplantation (HSCT) is curative therapy for many malignant and nonmalignant conditions, there has been establishment of UCB banks, both not-for-profit "public" banks and private commercial banks, resulting in a large and growing inventory of this type of stem cell. The authors comment that "This has raised a number of important scientific, ethical, legal and political issues. These include: ethical concerns regarding ownership of the blood, the processes for obtaining consent for collection and storage of UCB, issues relating to confidentiality and privacy, questions raised regarding commercial non-altruistic banking, and social justice issues relating to equity of access and equity of care."

We believe that this state-of-the-art book on regenerative medicine where the various uses and potential uses of pregnancy-specific biological substances are detailed will act as a stimulant for senior clinicians and scientists, who may be inspired to further the work of the pioneering medical scientists who have contributed to this volume. Needless to say, the individual authors are responsible for the work discussed or delineated in their respective articles; the editors have only helped to bring their path-breaking ideas and work together within the covers of this single volume.

Foreword

Foreword of the book by the President of the Royal College of Obstetrician and Gynecologist and President (Elect.) FIGO

The editors and authors of the chapters in this book should be congratulated for their phenomenal contribution to knowledge in the area of using cord blood, amniotic fluid, placenta, cord, and its contents for very innovative use in medicine. The book starts with a chapter on the massive wastage of pregnancy-specific biological substances that need to be recognized, as in every country the cord, placenta, and the cord blood are thrown away after delivery of the baby. It would be useful to look at how these tissues could be used in medicine. This is followed by a chapter on the basic science and the role of the placenta and also about the use of cord blood in biochemistry. These chapters on the physiology and the use of cord and the cord blood are followed by cord blood use for therapeutic purposes used as a substitute in transfusion medicine. The use of cord blood in emergency situations especially with the rich hemoglobin concentration in areas where standard blood transfusion is not available is well explained and should be commended. There are centers where cord blood has been used for immunotherapy, especially in special circumstances like advanced breast cancer. This area has not been fully explored and one needs to see whether immunotherapy could be advantageous in cancer and whether the cord blood cells could be used for such a purpose. The possible use in neurological disorders has been explored with some preliminary thoughts. The chapter on identification of stem cells with therapeutic potential in neuromuscular disorders is brilliantly dealt with and holds hope for such conditions.

The book explores the possibility of the cord blood being used in various fields including: orthopedics, ophthalmology, cardiovascular surgery, cardiovascular medicine, and regenerative medicine. Cord-derived mesenchymal stem cells have been studied for a long time and have great potential to be used as a therapeutic agent. The book also covers the area of cord blood collection and procedures in banking and argues the ethics of such banking process. Finally, it ends with some thoughts about the use of amniotic fluid for therapy and the clinical issue of aborted human tissue. It is possible that relevant fetal tissue from aborted fetuses could be used for specific targeted therapy in adult disease. The book covers this area and extends over 400 pages. It will be a good reference source, not only for practicing clinicians, but also for those interested in research in immunotherapy; stem cell therapy; regenerative therapy; and various specialities such as cardiology, neurosurgery, and cardiothoracic surgery. I would highly recommend the book for those who are interested in this area.

Clinical and research organizations should explore the possibility of using pregnancy-specific biological substances in regenerative medicine. One could argue after reading this book whether there should be a separate department within a large maternity unit to collect these tissues and process them in order to give the benefit to those who need it most.

St. George's, University of London Prof. S. Arulkumaran

Preamble

Rationale for cord blood banking: from hematopoietic stem cell transplant to regenerative medicine

Cord blood is an unlimited source of hematopoietic stem cells for allogeneic hematopoietic stem cell transplant. Since the first human cord blood transplant performed 20 years ago, cord blood banks have been established worldwide for collection and cryopreservation of cord blood for allogeneic hematopoietic stem cell transplant. More than 250,000 cord blood units are now available for international exchange of cord blood units. A global network of cord blood banks and transplant centers has been established for a common inventory and study of clinical outcomes. Results of unrelated allogeneic cord blood transplants in malignant and nonmalignant diseases, in adults and children, show that, compared to HLA matched unrelated bone marrow, cord blood transplant has several advantages, including prompt availability of the transplant, decrease of graft-versus-host disease, and better long-term immune recovery resulting in a similar long-term survival. Several studies have shown that the number of cells is the most important factor for engraftment, while some degree of HLA mismatches is acceptable. Progresses are expected to facilitate engraftment including ex vivo expansion of stem cells, intra-bone injection of cord blood cells and double-cord blood transplants. In addition to hematopoietic stem cells, cord blood and placenta contain a high number of non-hematopoietic stem cells including embryonic-like stem cells, mesenchymal cells, endothelial, neuronal, and pancreatic progenitor cells. The absence of ethical concern, the unlimited supply of cells explains the increasing interest of using cord blood for developing regenerative medicine. For this purpose some cord blood banks offer to cryopreserve cord blood from infants for autologous or family purpose in order to use it later in life if needed. This practice has led to considerable controversy but it has also triggered a development of research on criteria of quality for isolation, culture, cryopreservation of stem cell banks facilities responding to ethical and legislative regulations. More recently, it has been shown that new cells could be isolated from the cord, the placenta, and the Wharton jelly. Research should continue for studying the properties of these cells and for the implementation of clinical trials to treat a large variety of degenerative and hereditary disorders.

Paris, France E. Gluckman

Preface

In ancient mythologies, be they from Greece, India, or China, there are stories of kings and emperors seeking the "fountain of youth" or "pearls" that would rejuvenate them. The so-called Philosopher's Stone that medieval alchemists searched for fruitlessly was supposed to not only turn any substance into gold, but also to prolong life and restore youth. Ancient Indian sages practiced "Siddha Vaidya" as well as "tantric" methods for the same reason. In contemporary times, with a better understanding of the human body down to cellular structures and the DNA along with a better knowledge of debilitating diseases and their impact, scientists are looking not at rejuvenation but regeneration.

A natural effect of aging is degeneration; every organ in a human body degenerates as it ages, leading ultimately to, as they say, death due to old age. Congenital defects and damage can also affect organs like the liver, the heart, or the kidney, causing loss of function. Diseases like Parkinsonism or diabetes also cause specific organs to dysfunction. Many of these diseases are also associated with aging and in today's world, improved healthcare has resulted in increasing longevity. Many significant human diseases arising from the loss or dysfunction of specific cell types in the body, such as Parkinson's disease, diabetes, and cancer, are becoming increasingly common. So far, there had been no reprieve from such debilitating diseases or from damage caused by burns or other accidents. Today, however, a new branch of medicine, regenerative medicine, shows much promise.

The term probably comes from a 1992 paper of Leland Kaiser, "The Future of Multihospital Systems,"where in a paragraph subtitled "Regenerative Medicine", the author noted that a "new branch of medicine will develop that attempts to change the course of chronic disease and in many instances will regenerate tired and failing organ systems" (Kaiser L. *Top Health Care Finance*, 1992 Summer; 18:4: 32–45). With work on stem cells getting a new boost in recent years, the process of regenerating dysfunctional and aging organs appears to be no longer a myth but a reality.

Regenerative medicine refers to that branch of medicine which deals with living functional tissues that help to repair or replace damaged or aging tissues, thus regenerating the organ concerned. Research in this field includes cell therapy involving stem cells or progenitor cells, induction of regeneration by biologically active molecules, tissue transplantation, tissue engineering, and the use of cord blood, to mention a few.

Regenerative therapies have been demonstrated (in trials or in the laboratory) to heal broken bones, burns, blindness, deafness, heart damage, nerve damage, etc. It has the potential to cure diseases through repair or replacement of damaged, failing, or aged tissue. Therapies include regeneration of tissues in vitro for future use in vivo as well as direct placement and regeneration of tissue in vivo. However, this branch of medicine is still in its infancy despite strides made in last decade. Much of the work is still confined to animal or laboratory models. The next few years are critical as more and more human trials are undertaken and the true potential of this emerging branch of medicine is expressed.

This is the second effort by the editors to bring together the work of pioneering medical scientists who have ventured into this very exciting field. The first effort resulted in a book, *Frontiers of Cord Blood Science*, which was published by Springer-Verlag in 2009. The focus of the book was on the classical use of stem cells collected from the cord blood; other uses of cord blood and its potentials for use in medicine and bioengineering were also emphasized. This book has broadened the focus to include a variety of pregnancy-induced biological substances that have the potential in healing and regeneration, for instance, the stem cell-rich amniotic fluid, the cytokine rich placenta and its stem cells, the chorionic and amniotic membrane, and the veins of the placental cord. These items that are discarded after birth have been found to have regenerative potential in many diseases and damages to tissues and organs. Scientist from all over the world are researching on pregnancy-specific biological substances on the simple logic that these are the substances which help a zygote to become a full-grown neonate capable of independent survival after birth. This book brings together some of the important work that is being done along with unpublished observations that will help to shape the contours of future therapy in the field of modern regenerative medicine. It promises to be an eye-opener to the enormous potential of hitherto discarded material that had been so far considered as a pure biological waste. The book will have served its purpose if it acts as a stimulant to professionals and clinical scientists who can build on the knowledge and expand the curative potential of pregnancy-specific biological substances.

Acknowledgments

A book of this nature involves the cooperation of many: the contributors, publishers, as well as patients, researchers, and others who have helped the medical scientists with their work. Our thanks go out to all of them although it is not possible to name everyone. However, there are some who need special mention; without them, the book may never have been published.

First, the editors give profuse thanks to Prof. S. Arulkumaran of London University, who is currently the President of the Royal College of Obstetrician and Gynaecology and the Secretary General of FIGO, for writing the Foreword of the book. He is a true clinical scientist and has inspired us time and again with his vision. We are also extremely grateful to Prof. Elaine Gluckman, whose pioneering work in the clinical use of stem cell in modern contemporary medicine is well known, for writing the Preamble to the book.

The editors are particularly grateful to Dr. Clements, Steffan, Editor, Springer-Verlag London Limited, for his keen interest, advice, and support and guidance.

We gratefully acknowledge advise and involvement of Prof. Ian McNiece, Director, Regeneration Biology, University of Miami; Prof. Andrew Burd and Dr. Lin Huang of the Chinese University of Hong Kong; Dr. Neil H. Riordan and Dr. Thomas E. Ichim of Medistem Laboratory, and Dr. Michael P Murphy from Indiana University, USA; Prof. Carolyn Troeger and his team from Switzerland; Prof. Martina Vendrame, Temple University, Philadelphia, USA; Prof. Ernest E. Moore, University of Colorado Health Sciences Center, Denver, Colorado, USA; Prof. Norman Ende, Department of Pathology and Laboratory Medicine, and Prof. Kenneth Swan, Department of Surgery, University of Medicine and Dentistry of New Jersey, USA; and Dr. Alan Dardik, Associate Professor of Surgery, and Prof Herbert Dardik, Yale University School of Medicine, Vascular Biology and Therapeutics, New Haven, CT, USA, in this international project on regenerative medicine.

The editors also gratefully acknowledge the contributions of all the authors who took precious time from their busy schedules in order to help us to complete the book in time. The editors are also grateful to their wives for keeping the home peaceful, creative, and for maintaining a true academic and creative ambiance for research work (Prof. Sanjukta Bhattacharya for Dr. Niranjan Bhattacharya, and Linda Stubblefield, MSW, for Prof. Phillip Stubblefield). We thank them for their encouragement, understanding, and forbearance. Given their own interest in research in their respective fields, it is no surprise that their affection for the book is no less than that of ours.

We were also encouraged and facilitated in our work with creative criticism, comments, suggestions, and guidance from members of our fraternity, students, social activists, and patients, without whose keen interest, advice, and support it would have been difficult to proceed further in this new and vastly unknown field of modern regenerative medicine. May God bless them all for their goodwill and support.

Dr. Niranjan Bhattacharya and Prof. Phillip Stubblefield

Contents

Contributors

Sabeen Askari, MD Department of Pathology and Laboratory Services, Veterans Affairs Medical Center, Minneapolis, MN, USA
Department of Laboratory Medicine and Pathology, University of Minnesota, Minneapolis, MN, USA

Anthony Atala, MD Department of Urology, Wake Forest Institute for Regenerative Medicine, Wake Forest University Health Sciences, Winston-Salem, NC, USA

Niranjan Bhattacharya, DSc, MBBS, MD, MS, FACS (USA) Department of General Surgery, Obstetrics and Gynaecology and Clinical Immunology, Advanced Medical Research Institute, Gol Park, B.P. Poddar Hospital, and Vidyasagore State Hospital, Kolkata, India

Amando Blanquer Umbilical Cord Blood Bank, Valencia, Spain

Chantal Bon Department of Biochemistry, Hôtel Dieu Hospital, Lyon, France

Cesar V. Borlongan Department of Neurology, Medical College of Georgia and Augusta VA Medical Center, Augusta GA, USA

Thomas Brune Children's Hospital, Klinikum Lippe-Detmold, Detmold, Germany

Andrew Burd, MB, ChB, MD, FRCSEd, FHKAM Division of Plastic, Reconstructive and Aesthetic Surgery, Department of Surgery, The Chinese University of Hong Kong, Prince of Wales Hospital, Shatin NT, Hong Kong

Katherine Coldwell: Stem Cell Laboratory, NHS Bood and Transplant, John Radcliffe Hospital, Oxford, UK

Maurizio C. Capogrossi Laboratorio di Patologia Vascolare, Istituto Dermopatico dell' Immacolata, Rome, Italy

Abhijit Chaudhuri, DM, MD, PhD, FACP, FRCPGlasg, FRCP (London) Essex Centre of Neurological Sciences, Queen's Hospital, Romford, UK

Robert Chow, MD, AM, StemCyte International Cord Blood Center, covina, CA, USA

Kathleen M. Coakley, MS Department of Pathology and Laboratory Medicine and Department of Surgery, New Jersey Medical School, University of Medicine and Dentistry, New Jersey, USA

Alan Dardik, MD, PhD, FACS Vascular Biology and Therapeutics, Yale University School of Medicine, New Haven, CT, USA

Herbert Dardik, MD, FACS Vascular Biology and Therapeutics, Yale University School of Medicine, New Haven, CT, USA

L. Eduardo Cruz Cryopraxis and Silvestre Laboratory, Cryopraxis, BioRio, Pólo de Biotecnologia do Rio de Janeiro, Brazil

Norman Ende, MD Department of Pathology and Laboratory Medicine and Department of Surgery, Newark, NJ, USA

David J. Eve Center of Excellence for Aging and Brain Repair, Department of Neurosurgery and Brain Repair, University of South Florida College of Medicine, Tampa, FL, USA

Nicolas Forraz Newcastle Centre for Cord Blood, Stem Cell Institute, Institute of Human Genetics, Newcastle University, UK

Jian-Xin Gao Department of Pathology and Comprehensive Cancer Center, Ohio State University, Columbus, OH, USA

H.S.P Garritsen Department of Transfusion Medicine, Staedtisches Klinikum Braunschweig, Germany

David T. Harris Department of Immunobiology, The University of Arizona, Cord Blood Registry, Tucson, AZ, USA

Robert J. Henning, MD, FACP, FCCP, FACC, FAHA Center for Cardiovascular Research, James A. Haley Hospital/University of South Florida, Tampa, Florida, USA

Naoko Hida Department of Cardiology, Keio University School of Medicine, Tokyo, Japan

Peter Hollands Department of Biomedical Science, University of Westminster, London, UK

Wolfgang Holzgreve Laboratory for Prenatal Medicine, University Women's Hospital, Basel, Switzerland

Shigeharu Hosono, MD, PhD Nihon University Itabashi Hospital, Tokyo, Japan
Department of Pediatrics and Child Health,
Nihon University School of Medicine, Tokyo, Japan

Lin Huang Department of Surgery, The Chinese University of Hong Kong, Prince of Wales Hospital, Shatin, Hong Kong

Thomas E. Ichim Indiana University, Bloomington, IN, USA

Tang-Her Jaing Division of Hematology/Oncology, Department of Pediatrics, Chang Gung Children's Hospital, Chang Gung University, Taoyuan, Taiwan

Jeffrey L. Johnson, MD Department of Surgery, Denver Health Medical Center and University of Colorado Health Sciences Center, Denver, Colorado, USA

Ian Kerridge Centre for Values, Ethics and the Law in Medicine, University of Sydney, NSW, Australia

Olav Lapaire Laboratory for Prenatal Medicine, University Women´s Hospital, Basel, Switzerland

Ping Chung Leung, DSc, MD The Chinese University of Hong Kong, Institute of Chinese Medicine, Prince of Wales Hospital, Shatin, Hong Kong

F. Louwen Women's Health Research Institute, Westfaelische Wilhelms University, Muenster, Germany

Tomohiko Masuno, MD Department of Surgery, Denver Health Medical Center and University of Colorado Health Sciences Center, Denver, Colorado, USA

Colin P. McGuckin University of New Castle Upon Tyne, UK

Ian K. McNiece, PhD Regeneration Biology, Interdisciplinary Stem Cell Institute, University of Miami, Miami, Florida, USA

Jose J. Minguell TCA Cellular Therapy, Covington, LA, USA

Vicente Mirabet, PhD Valencia Transfusion Center, Tissue Bank, Spain

Shunichio Miyoshi, MD, PhD Department of Cardiology, Keio University School of Medicine, Tokyo, Japan

Ernest E. Moore, MD Department of Surgery, Denver Health Medical Center and University of Colorado Health Sciences Center, Denver, Colorado, USA

Hunter B. Moore, BS Department of Surgery, Denver Health Medical Center and University of Colorado Health Sciences Center, Denver, Colorado, USA

Michael P. Murphy Indiana University, Bloomington, IN, USA

Nobuhiro Nishiyama, MD Department of Cardiology, Keio University School of Medicine, Tokyo, Japan

Tracey O'Brien Centre for Values, Ethics and the Law in Medicine, University of Sydney, NSW, Australia

Tor Onsten, MD, PhD Department of Internal Medicine, Federal University of Rio Grande do Sul, Center of Hematology and Transfusion Medicine, Universidade Luterana do, Brasil

Satoshi Ogawa, MD, PhD Department of Cardiology, Keio University School of Medicine, Tokyo, Japan

Dong-Hyuk Park Center of Excellence for Aging and Brain Repair, Department of Neurosurgery and Brain Repair, University of South Florida College of Medicine, Tampa, FL, USA

Ornella Parolini Centro di Ricerca E. Menni, Fondazione Poliambulanza – Istituto Ospedaliero, Brescia, Italy

Amit N. Patel, MD, MS Division of Cardiothoracic Surgery, CTF, Salt Lake City, UT, USA

Maurizio Pesce Laboratorio di Biologia Vascolare e Terapia Genica, Centro Cardiologico Monzino IRCCS, Milan, Italy

Zygmunt Pojda Department of Experimental Hematology, Maria Sklodowska-Curie Memorial Cancer Center, Warsaw, Poland
Department of Regenerative Medicine, WIHiE Institute of Hygiene and Epidemiology, Warsaw, Poland

Giulio Pompilio Laboratorio di Biologia Vascolare e Terapia Genica, Centro Cardiologico Monzino, Milan, Italy

Patricia Pranke, PhD Hematology Laboratory, Federal University of Rio Grande do Sul, Rio Grande do Sul, Brazil

Daniel Raudrant Department of Gynaecology – Obstetrics, Hôtel Dieu Hospital, Lyon, France

Neil H. Riordan, PhD Medistem Panama, Inc., City of Knowledge, Republic of Panama

Ramasamy Sakthivel Case Western Reserve University, Cleveland, Ohio, USA

Gabrielle Samuel Centre for Values, Ethics and the Law in Medicine, University of Sydney, NSW, Australia

Cyndy D. Sanberg Saneron CCEL Therapeutics, Inc., Tampa, FL, USA

Paul R. Sanberg Center of Excellence for Aging and Brain Repair, Department of Neurosurgery and Brain Repair, University of South Florida College of Medicine, Tampa, FL, USA

Mariane Secco Laboratório de Células Tronco, Centro de Estudos do Genoma Humano, Instituto de Biociências, Universidade de São Paulo

Jon Smythe Stem Cell Laboratory, NHS Bood and Transplant, John Radcliffe Hospital, Oxford, UK

Pilar Solves, MD, PhD Valencia Transfusion Center, Cord Blood Bank, Valencia, Spain

Maddalena Soncini Centro di Ricerca E. Menni, Fondazione Poliambulanza – Istituto Ospedaliero, Brescia, Italy

Phillip Stubblefield, MD Department of Obstetrics and Gynecology, University of Boston, Jamaica Plain, MA, USA

Kenneth G. Swan, MD New Jersey Medical School, Newark, New Jersey, USA

Carolyn Troeger Laboratory for Prenatal Medicine, University Women's Hospital, Basel, Switzerland

Akihiro Umezawa, MD, PhD Department of Reproductive Biology and Pathology, National Research Institute for Child Health and Development, Tokyo, Japan

Martina Vendrame, MD, PhD Neurology Department, Temple University, Philadelphia, PA, USA

Natassia Vieira University of São Paulo, Instituto de Biociências Biosciences Institute, São Paulo, Brazil

Suzanne M. Watt, PhD, FRC (Path) NHS Blood and Transplant, University of Oxford, Oxford, UK

Alison E. Willing Center of Excellence for Aging and Brain Repair, Department of Neurosurgery and Brain Repair, University of South Florida College of Medicine, Tampa, FL, USA

Tracy A. Womble Center of Excellence for Aging and Brain Repair, Department of Neurosurgery and Brain Repair, University of South Florida College of Medicine, Tampa, FL, USA

Kyung-Chul Yoon, MD, PhD Department of Ophthalmology, Chonnam National Universtiy Medical School and Hospital, Gwang-Ju, South Korea

Xiao Yan Zhong Laboratory for Prenatal Medicine, University Women´s Hospital, Spitalstrasse, Basel, Switzerland

Mayana Zatz Department of Genetic and Evolutive Biology, Human Genome Research Center, University of São Paulo, São Paulo, Brazil

Quansheng Zhou Cyrus Tang Hematology Research Center, Soochow University, Suzhou, China
Department of Molecular Biology, The Scripps Research Institute, La Jolla, CA, USA

Part I

Massive Wastage of Pregnancy Specific Biological Substances

A Massive Wastage of the Global Resources

1

Andrew Burd and Lin Huang

The United States Census Bureau estimates that in 2008 there will be 135,330,281 human births globally, which means 257 babies are born every minute. The birth of a baby is, and should be, in most cases a wonderful celebration of the process of nature. At the same time, it is an opportunity to reflect upon the complexity of human reproduction with a fertilized ovum initiating a cascade of events that result in the production not just of a new life but also of the "in utero" life-support system. (Fig. 1.1)

The feto-placental unit is a potentially rich resource of biological tissue. It is a human resource, a global resource, a free resource, which, at present, tends to be either wasted or exploited for commercial, rather than humanitarian, gain. This resource comprises:

(a) The placenta
(b) The amniotic membranes
(c) The amniotic fluid
(d) The umbilical cord
(e) The umbilical cord blood

1.1 The Placenta

The placenta develops from the same sperm and egg that gives rise to the fetus and functions as an interface organ between the mother and the fetus having two parts, the fetal part is the Chorion frondosum and the maternal part the Decidua basalis. The chorionic plate or fetal surface of the placenta is typically drawn as round with the umbilical cord emerging from its center. In reality, the shape of the chorionic disc is rarely round and may appear as oval or lobulated depending very much on where it is implanted in the uterus. A study looking at the mean surface area of the human placenta arrived at a figure of just under 286 cm^2.[1] With an average thickness of 2–2.5 cm, this gives a mean volume of just over 600 cc. The typical weight of the placenta is approximately 500 g. Following the birth of the baby, the placenta is delivered and thereafter suffers a variety of fates.

In the western world, it is most often incinerated as biological waste. In other cultures, it is more likely to be revered. Some cultures bury the placenta for various reasons. The ancient Egyptians believed that burial of the placenta was able to protect and ensure the health of the baby and the mother[2]; the Màori of New Zealand traditionally bury the placenta to emphasize the relationship between humans and the earth.[3] Some communities believe that the placenta has power over the life of the baby. In Turkey, the proper disposal of the placenta is believed to promote devoutness in the child later in life.

Human placenta has also been known for its secret power as a medicinal supplement. For the Chinese and Vietnamese, there is a customary practice to prepare the placenta for consumption by the mother. Eating the placenta (placentophagy) is believed to have a variety of potential benefits. For one, the placenta contains high levels of prostaglandin and small amount of oxytocin; this supposedly helps stem bleeding after birth, eases birth stress, and causes the uterus to clean itself out. For another, the placenta is considered rich in vitamins, minerals, iron, protein, and hormones, which would be useful to women recovering from childbirth.[4] More recent research has discovered an active substance in

A. Burd (✉)
Division of Plastic, Reconstructive and Aesthetic Surgery, Department of Surgery, The Chinese University of Hong Kong, Prince of Wales Hospital, Shatin NT, Hong Kong
e-mail: andrewburd@surgery.cuhk.edu.hk

N. Bhattacharya and P. Stubblefield (eds.), *Regenerative Medicine Using Pregnancy-Specific Biological Substances*, DOI: 10.1007/978-1-84882-718-9_1, © Springer-Verlag London Limited 2011

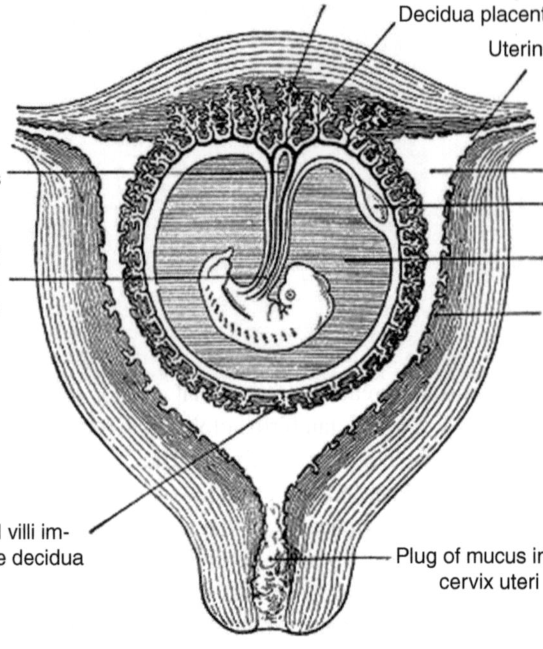

Fig. 1.1 Fetus growing within the womb

placenta, Placental Opioid-Enhancing Factor (POEF) that modifies the activity of endogenous opioids in such a way that produces an enhancement of the natural reduction in pain.[5] Placentophagy could enhance pain tolerance by increasing the opium-like substances activated during childbirth.[6]

The human placenta has also been used as Traditional Chinese Medicine (TCM) for thousands of years.[7] One of the well-known TCM uses dried human placenta, Zi he che, to help with insufficient lactation. A study on 210 women who presented with insufficient milk supply were given dried placenta and 86% of them reported a positive increase in their milk production within a matter of days.[8]

1.2 Amniotic Membranes

At term, the surface area of the human amniotic membrane is approximately 1,300–1,500 cm^2.[9] The membranes actually represent a complex biological structure. The amnion comprises an epithelial layer, which is bathed in amniotic fluid. The epithelial cell layer lies on a basement membrane below which there is a collagen-rich connective tissue matrix forming an interface layer with the chorion. The chorion has a collagen-rich reticular layer, which rests on a basement membrane, which in turn is in continuity with the trophoblasts of the maternal deciduas.

The amniotic membranes have a well-established role in clinical utilization both in the fields of burns and wound care and in ophthalmic surgery. Various preparations of amniotic membrane have been explored including lyophilized, gamma-irradiated amnion,[10] single layer radiation-treated amnion,[11] glycerol-preserved amnion,[12,13] fresh and nonirradiated freeze dried.[14] It is evident that the amniotic membranes are rich in growth factors that can benefit wound healing both of the skin[15] and the corneal epithelium.[16] The use of amniotic membranes transplantation in ophthalmic surgery continues to generate research interest, in particular into ways to sterilize and preserve the membrane while maintaining its biological properties.[17] With regard to the use in burns reports, an application come from areas of great deprivation[18] and centers of considerable affluence[19] with equally promising results. Variations on the amniotic membrane include silver impregnation of amnion[20] and the use of the acellular amniotic membrane matrix for its application in tissue engineering.[21] The use in tissue engineering is, however, not just restricted to the potential as a scaffold but the potential for the amniotic membrane to be a source of stem cells for tissue-engineered constructs.[9]

1.3 Amniotic Fluid

The amniotic fluid surrounds the developing fetus and contains proteins, carbohydrates, lipids, phospholipids, urea, and electrolytes. As the fetus does excrete urine, this forms a significant proportion of the amniotic fluid at the later stages of gestation. The volume of amniotic fluid increases as the fetus grows. It is maximal at about 34 weeks of age with an average volume of 800 mL. This reduces to about 600 mL at term. The therapeutic potential of amniotic fluid is relatively unexplored when compared to the amniotic membranes although a recent paper in Nature Biotechnology has described the isolation of amniotic fluid-derived stem cells.[22]

1.4 The Umbilical Cord

The umbilical cord connects the fetus to the placenta and normally contains two arteries and one vein, which are surrounded by a hyaluronan-rich Wharton's jelly. The umbilical vein supplies the fetus with oxygenated, nutrient-rich blood while the arteries return the nutrient-depleted, deoxygenated blood from the fetus to the placenta. The length of the cord does vary but a study by Malpas published in the *British Medical Journal* reviewed a total of 538 normal cords, and derived an average length of 61 cm.[23]

The human umbilical cord is and has been the source of many clinically applicable products and structures and these can briefly be mentioned under the headings:

- Wharton's jelly
- Vessels
- Epithelium

1.4.1 Wharton's Jelly

Wharton's jelly is named after the seventeenth century English physician and anatomist Thomas Wharton. The predominant molecule is hyaluronan (HA), which is in a higher concentration in the cord (approximately 4 mg/mL) than in any other tissue. HA extracted from the human umbilical cord is commercially available on a potanion salt and is marketed by Sigma chemicals. The extraction fluid has a very high elastoviscosity and the molecular conformation of the extracted HA together with its mechanical properties are most probably not indicative of the biological function.[24] In utero, the Wharton's jelly certainly has a protective effect, surrounding the vessels and maintaining a "cushion" to protect against compression. After birth, however, with a change in temperature after delivery, there will be a conformational change in the Wharton's jelly, which produces a physiological "clamping" of the cord.

There is increasing interest in the potential of Wharton's jelly as a rich source of both tissue and cells. Recent reports have described for the first time the proteoglycan profile of the cord reporting that 1 g of Wharton's jelly contains approximately 2–5 mg of sulfated glycosaminoglycans with decorin strongly predominating over biglycan.[25] Of equal interest is the potential of Wharton's jelly matrix cells to be a source of neurogenic stem cells.[26] Indeed, the value of the mesenchyme as a potential source of a wide variety of stem cells is receiving considerable attention worldwide and this is now being more widely recognized.[27]

1.4.2 Vessels

The vessels in the human umbilical cord have also been identified as a potential source of endothelial stem cells,[28] while the intact vessels have been investigated as potential scaffolds for tissue engineering constructs.[29] It is evident that the potential for laboratory investigation and clinical applications of cells and tissues derived from the umbilical cord vessels is a very pregnant area of research.

1.4.3 Epithelium

The cord lining consists predominantly of a single layer of epithelial cells. Although there is some evidence of selective stratification,[30] when the cord is unraveled and flattened, the surface of the cord would cover an area of approximately 250 cm^2. It is possible to physically remove the vessels from the cord and prepare a "sheet–like" structure that could have potential as temporary biological wound cover. Such applications remain experimental. Of more advanced clinical potential is to use cord lining derived cells for wound cover and this is being explored in particular by the Singapore-based company Cell Research Corporation.[31] The possibility of differentiation into keratinocytes that may have a universal donor potential is driving a considerable interest in this area.[32,33]

1.5 Umbilical Cord Blood

Following birth of the baby, it is possible to collect the blood that has been in the umbilical cord and placenta. The value of blood that may be obtained is influenced by the method of delivery. A Japanese study in 2,000 suggested that Cesarean sections allowed more blood to be collected than after vaginal delivery. The mean volumes were reported as approximated 104 mL for Cesarean sections compared to 85 mL for vaginal deliveries.[34]

The therapeutic use of allogenic human umbilical cord blood (HUCB) is well established for transplantation purposes.[35] What is only now being appreciated is that HUCB can also be used as a transfusion or even an infusion. What is the difference? Simply put, the transplant replaces something that is permanently lost, the transfusion replaces something that is temporarily lost and the infusion can prevent loss.[36] In the developing world where health resources are so limited, the enormous potential of allogenic HUCB is already evident. Indeed, Dr Niranjan Bhattacharya from Calcutta has described his pioneering work using HUCB in conditions as diverse as Malaria, Aids, Leprosy, Advanced Malignancy, and Degenerative disease.[37,38]

The proven utility of cord blood transplants has led to the establishment of cord blood banks, both public and private. In a paper published in 2005, it was reported that at that time there were nearly 100 cord blood banks worldwide with an estimated 200,000 units of cord blood held in the private sector and over 160,000 units registered with the largest public cord blood registry.[39]

In 2007, the World's first public-private cord blood bank was launched in the UK.[40] This is an initiative of Sir Richard Branson and it highlighted the tension in the public-private cord blood bank debate. Essentially, it was reported that the cord blood collections would be split with about 20% of the purified stem cells being set aside for the child's exclusive use and 80% being placed in a public cord blood bank. In the BMJ rapid response to the news of this proposed bank, the question of therapeutic viability of the resulting units was raised by Kenneth Campbell, the Clinical Information Officer of the Leukaemia Research Fund, while we observed that the Royal College of Obstetricians and Gynaecologists (RCOG) of the United Kingdom appeared to be somewhat ambivalent in their position about cord blood banking, to quote, "The cautious comments of the Royal College of Obstetricians and Gynaecologists (RCOG) in their press release of 1 February 2007 (RCOG statement on the setting up of the Virgin Health Bank http://www.rcog.org.uk/index.asp?PageID=1855) contrasts somewhat with their previous comments. On 13th June 2006, they issued a press release to give an authoritative perspective on the hype surrounding commercialized cord blood banking, stating that there is "insufficient evidence" to recommend such banking in low-risk families (Umbilical cord blood banking. Royal College of Obstetricians and Gynaecologists. http://www.rcog.org.uk/index.asp?pageID=545). The reality is that the very last blood a child with leukemia wants is their own, with a teaspoon of stem cells or not. Admittedly, human umbilical cord blood fractions have been shown to modulate the course of such conditions as Alzheimer's disease, prostatic cancer, and type II diabetes (in laboratory mice), but these are hardly diseases of children and nowhere in the world have cryo-preserved stem cells been shown to be clinically effective for longer than the span of childhood".

1.6 Factors Affecting Availability

The delivery of a human baby together with cord and placenta has to be the most fundamental of natural processes. Unfortunately, not all births are healthy and it is estimated that about 90% of HIV-infected children acquire the infection from their mother during pregnancy and child birth.[41,42] In December 2007, the World Health Organization published (WHO) its Global Summary of the AIDS epidemic and indicated that in 2007 approximately 420,000 children under 15 were newly infected with HIV, while 15.4 million women were living with

HIV. Such figures, though tragic, still register that there are potentially over 120 million non-HIV-related births each year. There will be other transmissible diseases that preclude the use of the placenta, cord, and blood such as tuberculosis, malaria, syphilis, but such epidemiology of healthy child birth as exists indicate that there will be over 100 million potential donors each year for the full-term products of conception, excluding the baby.

Another factor that is going to affect the availability of these products is the nature of the birth: vaginal or cesarian. In 1985, the WHO recommended a rate for cesarean sections of about 15% of all live births.[43] While this may reflect the current global average, there are considerable variations depending on cultural and economic factors. In the United States, levels of 30% were reported in 2005 while in Brazil the public healthcare domain reported rates of 35%, whereas in the private hospitals the rate was over double this. Again, as with the estimates from infectious disease, reliable global figures do not exist but indicative figures are available.

What then is the extent of the annual production of this invaluable global resource? Using a comfortable estimate of 100,000,000 healthy live births annually with a 15% rate of cesarean section deliveries gives the following:

Annual global production		
Placenta:	Weight	50 million kilogram
	Volume	60,000 m³
Amniotic membrane:	Area	15 million m²
Amniotic fluid:	Volume	60 million liters
Umbilical cord:	Length	61,000 km
Umbilical cord blood:	Volume (cesarian section)	1,560,000 L
	Volume (vaginal delivery)	7,225,000 L
	Total volume	8,785,000 L

It is evident that there is a considerable amount of a free, readily available, and sustainable human resource. It is inevitable that there will be commercial exploitation of some of this source material for extracting specific and defined biological materials. However, there remains a considerable quantity that is simply going to be discarded and this represents a massive wastage of global resources.

References

1. Yampolsky M, Shlakhter O, Salafia CM, Haas D. Mean surface shape of a human placenta. Available at: http://arxiv.org/abs/0807.2995. Retrieved on 2010-09-14.
2. Buckley SJ. Placenta rituals and folklore from around the World. *Midwifery Today Int Midwife*. 2006;80:58-59.
3. Metge J. Working in/playing with three languages: English, Te Reo Maori, and Maori Bod language. In *Sites* N.S. 2005;2:83-90.
4. Phuapradit W, Chanrachakul B, Thuvasethakul P, Leelaphiwat S, Sassanarakkit S, Chanworachaikul S. Nutrients and hormones in heat-dried human placenta. *J Med Assoc Thai*. 2000;83:690-694.
5. Kristal MB. Enhancement of opioid-mediated analgesia: a solution to the enigma of placentophagia. *Neurosci Biobehav Rev*. 1991;15:425-435.
6. DiPirro JM, Kristal MB. Placenta ingestion by rats enhances delta- and kappa-opioid antinociception, but suppresses mu-opioid antinociception. *Brain Res*. 2004;1014:22-33.
7. Traditional Chinese medicine contains human placenta, Pharmaceutical News, May 8, 2004, http://www.news-medical.net/print_article.asp?id=1333. Retrieved on 2007-12-12
8. Soyková-Pachnerová E, Brutar V, Golová B, Zvolská E. Placenta as a Lactagogon. *Gynacologia*. 1954;138:617-627.
9. Niknejad H, Peirovi H, Jorjani M, Ahmadiani A, Ghanavi J, Seifalian AM. Properties of the amniotic membrane for potential use in tissue engineering. *Eur Cell Mater*. 2008;15:88-99.
10. Gajiwala K, Gajiwala AL. Evaluation of lyophilized, gamma-irradiated amnion as a biological dressing. *Cell Tissue Bank*. 2004;5:73-80.
11. An Y, Zhan HY, Song XH, Liu Y, Sun ZL. Protective effect of single-layer radiation-treated human amnion used as a biological dressing on burn wound in rats. *Chin J Clin Rehab*. 2004;8:256-257.
12. Ravishanker R, Bath AS, Ray R. "Amnion Bank" – the use of long term glycerol preserved amniotic membranes in the management of superficial and superficial partial thickness burns. *Burns*. 2003;29:369-374.
13. Rejzek A, Weyer F, Eichberger R, Gebhart W. Physical changes of amniotic membranes through glycerolization for the use as an epidermal substitute. Light and electron microscopic studies. *Cell Tissue Bank*. 2001;2:95-102.
14. Ganatra MA, Durrani KM. Effect of fresh and freeze dried human amniotic membrane on quantitative bacterial counts in burn wounds. *J Coll Phys Surg Pak*. 1998;8:202-206.
15. Cho DY, Chung BS, Choi KC. The effect of amniotic membrane patch in wound healing of skin defect. *Korean J Dermatol*. 2005;43:926-932.
16. Castillo-Torres F, Lucio-Alva ME, Medina-Zarco A. Conjunctival contraction measurement in pathologies caused by conjunctive lose and amniotic membrane as a therapeutic cover. *Revi Mexi Oftalmol*. 1999;73:251-254.
17. Von Versen-Hoeynck F, Steinfeld AP, Becker J, Hermel M, Rath W, Hesselbarth U. Sterilization and preservation influence the biophysical properties of human amnion grafts. *Biologicals*. 2008;36:248-255.
18. Ramakrishnan KM, Jayaraman V. Management of partial-thickness burn wounds by amniotic membrane:

a cost-effective treatment in developing countries. *Burns*. 1997;23:S33-S36.
19. Branski LK, Herndon DN, Celis MM, Norbury WB, Masters OE, Jeschke MG. Amnion in the treatment of pediatric partial-thickness facial burns. *Burns*. 2008;34:393-399.
20. Singh R, Kumar D, Kumar P, Chacharkar MP. Development and evaluation of silver-impregnated amniotic membrane as an antimicrobial burn dressing. *J Burn Care Res*. 2008;29:64-72.
21. Wilshaw SP, Kearney JN, Fisher J, Ingham E. Production of an acellular amniotic membrane matrix for use in tissue engineering. *Tissue Eng*. 2006;12:2117-2129.
22. De Coppi P, Bartsch G, Siddiqui MM, et al. Isolation of amniotic stem cell lines with potential for therapy. *Nat Biotechnol*. 2007;25:100-101.
23. Malpas P. Length of the human umbilical cord at term. *BMJ*. 1964;1:673-674.
24. Balazs EA. Viscoelastic properties of Hyaluronan and its therapeutic use. In: Garg HG, Hales CA, eds. *Chemistry and Biology of Hyaluronan*. 1st ed. Oxford, UK: Elsevier; 2004.
25. Gogiel T, Bankowski E, Jaworski S. Proteoglycans of Wharton's jelly. *Int J Biochem Cell Biol*. 2003;35:1461-1469.
26. Mitchell KE, Weiss ML, Mitchell BM, et al. Matrix cells from Wharton's jelly form neurons and glia. *Stem Cells*. 2003;21:50-60.
27. Secco M, Zucconi E, Vieira NM, et al. Mesenchymal stem cells from umbilical cord: Do not discard the cord. *Neuromuscul Disord*. 2008;18:17-18.
28. Kestendjieva S, Kyurkchiev D, Tsvetkova G, et al. Characterization of mesenchymal stem cells isolated from the human umbilical cord. *Cell Biol Int*. 2008;32:724-732.
29. Hoenicka M, Jacobs VR, Huber G, Schmid FX, Birnbaum DE. Advantages of human umbilical vein scaffolds derived from cesarean section vs vaginal delivery for vascular tissue engineering. *Biomaterials*. 2008;29:1075-1084.
30. Sanmano B, Mizoguchi M, Suga Y, Ikeda S, Ogawa H. Engraftment of umbilical cord epithelial cells in athymic mice: In an attempt to improve reconstructed skin equivalents used as epithelial composite. *J Dermatol Sci*. 2005;37:29-39.
31. Ruetze M, Gallinat S, Lim IJ, et al. Common features of umbilical cord epithelial cells and epidermal keratinocytes. *J Dermatol Sci*. 2008;50:227-231.
32. Ng W, Nishiyama C, Mizoguchi M, et al. Human umbilical cord epithelial cells express Notch 1: Implications for its epidermal-like differentiation. *J Dermatol Sci*. 2008;49:143-152.
33. Huang L, Wong YP, Gu H, Cai YJ, Ho Y, Wang CC, Leung TY, Burd A. Stem cell-like properties of human umbilical cord lining epithelial cells and the potential for epidermal reconstitution. *Cytotherapy*. 2010 Aug 24. [Epub ahead of print]
34. Yamada T, Okamoto Y, Kasamatsu H, Horie Y, Yamashita N, Matsumoto K. Factors affecting the volume of umbilical cord blood collections. *Acta Obstet Gynecol Scand*. 2005; 79:830-833.
35. Will AM. Umbilical cord blood transplantation. *Arch Dis Child*. 1999;80:3-6.
36. Burd A, Ahmed K, Lam S, Ayyappan T, Huang L. Stem cell strategies in burns care. *Burns*. 2007;33:282-291.
37. Bhattacharya N. A preliminary study of placental umbilical cord whole blood transfusion in under resourced patients with malaria in the background of anaemia. *Malar J*. 2006;5:20.
38. Bhattacharya N. A study of placental umbilical cord whole blood transfusion in 72 patients with anemia and emaciation in the background of cancer. *Eur J Gynaecol Oncol*. 2006;27:155-161.
39. Gunning J. Umbilical cord cell banking – implications for the future. *Toxicol Appl Pharmacol*. 2005;207:S538-S543.
40. Mayor S. World's first public-private cord blood bank launched in UK. *BMJ*. 2007;334:277.
41. Centers for Disease Control and Prevention. HIV/AIDS surveillance report, 2003 (Vol. 15). US Department of Health and Human Services, Centers for Disease Control and Prevention; Atlanta, 2004.
42. Minkoff H. Human immunodeficiency virus in pregnancy. *Obstet Gynecol*. 2003;101:797-810.
43. World Health Organization. Appropriate technology for birth. *Lancet*. 1985;2:436-437.

Part II

Basic Science and the Role of Placenta

2 Placenta as a Source of Stem Cells and as a Key Organ for Fetomaternal Tolerance

Ornella Parolini and Maddalena Soncini

The placenta encloses the very beginnings of the mystery of life, but discloses an ever-increasing amount of information toward our understanding not only of cell development, maturation, and differentiation, but to an even greater extent, the fundamental mechanisms of immunological tolerance.

For many years, the human placenta has attracted the attention of scientists because of the essential role it plays in development of the growing embryo by facilitating gas and nutrient exchange between the mother and fetus, while this tissue has intrigued researchers for an even longer time because of its role in maintaining fetomaternal tolerance. More recently, this tissue has also been investigated as a potential source of stem cells for application in regenerative medicine.

2.1 Placenta Structure

The human term placenta is round or oval in shape with a diameter of 15–20 cm and a thickness of 2–3 cm. The decidua constitutes the maternal portion of the placenta and is derived from the maternal endometrium. The portion of the decidua at which implantation takes place is called the decidua basalis, while the portion adjacent to the chorion leave is termed the decidua capsularis. The decidua parietalis covers the remainder of the endometrium.

O. Parolini (✉)
Centro di Ricerca E. Menni, Fondazione Poliambulanza – Istituto Ospedaliero, Brescia, Italy
e-mail: ornella.parolini@tin.it

The fetal portion of the placenta is composed of the placental disk and the amniotic and chorionic membranes.

The placental disk is composed of the chorionic plate and the basal plate, which form a base and cover, respectively, to enclose the intervillous space. The multilayered chorionic plate faces the amniotic cavity and is composed of a spongy layer, followed by the chorionic mesodermal layer, and a Langans' fibrinoid layer interposed with highly variable amounts of proliferating extravillous cytotrophoblast cells. The amnion covers the face of the chorionic plate, which is closest to the amniotic cavity, while chorionic villi project from the other side of the chorionic plate and either terminate freely in the intervillous space where maternal blood flows, or anchor the placenta through the trophoblast of the basal plate to the endometrium.

Despite the fact that there are different types of villi with different functional specializations, all villi exhibit the same basic structure, consisting of an inner stromal core containing fetal vessels and connective tissue, in which mesenchymal cells, fibroblasts, myofibroblasts, and fetal tissue macrophages (Hofbauer cells) are dispersed. A basement membrane separates the stromal core from an uninterrupted multinucleated outer layer, called syncytiotrophoblast, with single or aggregated cytotrophoblast cells found between the syncytiotrophoblast and its basement membrane. The ramifications of the villous trees differ in their caliber, vessel structure, stromal arrangement, and position within the villous tree itself, and can be distinguished as stem villi, which mechanically support the structure of the villous tree, immature intermediate villi, which act as growth zones and produce new sprouts, and mature intermediate villi and terminal villi, both of which represent the main exchange area in the third trimester placenta. Fetal blood is carried to the villi

N. Bhattacharya and P. Stubblefield (eds.), *Regenerative Medicine Using Pregnancy-Specific Biological Substances*,
DOI: 10.1007/978-1-84882-718-9_2, © Springer-Verlag London Limited 2011

via the branches of the umbilical arteries. After circulating through the capillaries of the villi, the fetal blood absorbs oxygen and nutritional materials from, and transfers waste products to the maternal blood through the villous walls. The purified and nourished fetal blood is then carried back to the fetus via the umbilical vein.

The basal plate is the most intimate and important contact zone between maternal and fetal tissue. It is composed of a superficial stria of Rohr's fibrinoid, which faces the intervillous space, followed by a layer of extravillous cytotrophoblast and connective tissue, and another fibrinoid layer (Nitabuch's fibrinoid layer), which is located next to the compact decidual layer. In term placenta, the basal plate is usually of variable thickness owing to the fact that it loses its typical layering as gestation progresses. Protrusions extending from the basal plate into the intervillous space produce the placental septa, which divide the fetal part of the placenta into the irregular cotyledons.

At the regions of placenta that are in contact with the decidua capsularis during gestation, the intervillous space is obliterated so that the chorionic plate and the basal plate fuse with each other forming the chorionic membrane (commonly called the chorion leave), which consists of a chorionic mesodermal (CM) and chorionic trophoblastic (CT) region. The chorionic mesoderm consists of a network of collagen bundles intermingled with finer fibrils in which fibroblasts and macrophages are usually observed. A basal lamina separates the chorionic mesoderm from the highly variable layer of extravillous trophoblast cells that represent the only residue of the former villi of the chorion frondosum (see section on Embryological Development of the Placenta) intermingled with trophoblastic residues of the primary chorionic plate and basal plate.

The amnion is an uninterrupted membrane, which is in contact with the amniotic fluid on its inner surface, while on the other side it is in contact with the chorion leave, the chorionic plate, and the umbilical cord. The amnion is contiguous over the umbilical cord with the fetal skin.

Structurally, the amniotic membrane is a thin avascular sheet composed of an epithelial layer and connective tissue. The amniotic epithelium (AE), which is in contact with the amniotic fluid, is a single layer of flat, cuboidal to columnar epithelial cells, which is attached firmly to a distinct basal lamina that is in turn connected to the amniotic mesoderm (AM). In the amniotic mesoderm, an acellular compact layer of interstitial collagens I, III, and fibronectin, and a deeper network of widely dispersed fibroblast-like mesenchymal cells and rare macrophages are distinguishable.

The amniotic mesoderm and chorionic mesoderm are loosely connected via a spongy or intermediate layer, which is a reticular zone composed of loosely arranged collagen fibers that results from the incomplete fusion of amniotic and chorionic mesoderm during early pregnancy. Both layers contribute to the mechanical stability of the membranes, but it is the fibers of the compact layer of the AM which confer most of the tensile strength to the fetal membranes.[1,2]

2.2 Embryological Development of the Placenta

Development of the placenta begins as soon as the blastocyst implants in the maternal endometrium (6–7 days after fertilization). At this stage, the blastocyst is a flattened vesicle in which most of the cells form an outer wall (trophoblast), which surrounds the blastocystic cavity (blastocoel). A small group of larger cells, known as the inner cell mass, is apposed to the inner surface of the trophoblastic vesicle. The trophoblast eventually gives rise to the chorion, whereas the embryo, the umbilical cord, and the amnion are derived from the inner cell mass.

As the blastocyst adheres to the endometrial epithelium, the invading trophoblast erodes the deciduas, allowing the embedding of the blastocyst. During implantation, the trophoblastic cells of the implanting pole of the blastocyst show increased proliferation, resulting in a bilayered trophoblast, made up of a multinucleated outer syncytiotrophoblast, which originates from fusion of neighboring trophoblast cells, and an inner, mononucleated cytotrophoblast layer.

By day 8, small intrasyncytial vacuoles appear in the syncytiotrophoblast mass at the implantation pole. These vacuoles grow rapidly and become confluent, forming a system of hematic lacunae separated by lamellae and pillars of syncytiotrophoblast (trabeculae). Primary villi can be observed after invasion of the cytotrophoblast into the trabeculae, while the lacunae form the intervillous space where maternal blood flows.

In early pregnancy, the entire chorionic membrane is covered by villi, which are almost uniform in size, but which soon begin to develop unequally. At the anti-implantation pole, villous degeneration and fibrinoid deposition in the intervillous space give rise to the smooth chorion or chorion leave, while at the implantation pole, villous proliferation forms the leafy chorion or chorion frondosum.

At day 8–9 after fertilization, morphological changes occur in the inner cell mass, which differentiates into two layers, the epiblast and the hypoblast, that together form the bilaminar embryonic disk. From the epiblast, some small cells, that will later constitute the amniotic epithelium, appear between the trophoblast and the embryonic disk and enclose a space that will become the amniotic cavity. The three germ layers of the embryo (endoderm, mesoderm, ectoderm) will also originate from the epiblast.

Once the lining of the amnion has developed, the amniotic cavity surrounds the embryo from all sides and amniotic fluid begins to accumulate within the amniotic cavity. The accumulation of amniotic fluid within the amniotic cavity causes the amnion to expand and ultimately to adhere to the inner surface of the trophoblast (chorion).

From the other side of the bilaminar disk, some cells from the hypoblast migrate along the inner wall of the blastocoel giving rise to the exocoelomic membrane. The exocoelomic membrane and the blastocoel modify to form the yolk sac, while cells of the exocoelomic membrane and the adjacent trophoblast form the extraembryonic reticulum. Some hypoblast cells then migrate along the outer edges of extraembryonic reticulum to form a connective tissue known as the extraembryonic mesoderm, which surrounds the yolk sac and amniotic cavity, and later forms the amniotic mesoderm (AM) and chorionic mesoderm (CM). The amniotic mesoderm and chorionic mesoderm are separated by a cavity called the exocoele, which is compressed during amniotic cavity expansion.[1,3]

All these events occur before gastrulation (third week after fertilization), the process through which the bilaminar disk differentiates into the three germ layers (ectoderm, mesoderm, and endoderm), which leads to the hypothesis that placental tissues themselves may harbor cells that display the potential to differentiate toward different lineages.

2.3 Immunology of the Placenta

Immune evasion by the allogeneic fetus has intrigued immunologists since the beginning of the twentieth century with the observation of Little (1924) that the mother must in some way be able to tolerate the presence and growth of the fetus, leading him to propose that the embryo might have "no definite physiological characteristics which are individual enough to be recognized as foreign by the mother." In 1932, Witebsky and Reich suggested that human trophoblast may be nonantigenic and could be capable of acting as a barrier between the mother and the fetus. However, it was Medawar who identified the truly paradoxical nature of the immunological relationship between the mother and the fetus in 1953, declaring that "the immunological problem of pregnancy may be formulated thus: how does the pregnant mother contrive to nourish within itself, for many weeks or month, a fetus who is an antigenically foreign body?".[4] In what eventually became well known as Medawar's paradox, Medawar proposed that the lack of fetal rejection by the mother might be explained by three mechanisms: (a) that there is an anatomical barrier between the fetus and the mother; (b) that the fetus is antigenically immature; (c) that the maternal immune system might be immunologically inert.[5]

Since the time of Medawar, it has become evident that these mechanisms cannot fully explain why the fetus is not rejected by the mother, and other site-specific immune suppression mechanisms must therefore be considered.

For many years, in accordance with the first mechanism of Medawar's paradox, the trophoblast was considered an impenetrable barrier, which prevents exposure of the fetus to the maternal immune system. More recently, however, bidirectional transfer of fetal and maternal cells through this tissue has been reported by numerous investigators. Fetal cell microchimerism was originally demonstrated in female mice,[6] and long-term persistence of fetal cells in the bone marrow of these animals postpartum has been observed. During human pregnancy, fetal cells enter the maternal circulation from as early as 6 weeks into gestation[7] and can persist in maternal blood and tissues for decades after pregnancy[8] without any signs of graft-versus-host reaction or graft rejection. Data concerning the health consequences of persistent fetal cells in maternal tissues are contradictory. Initially, fetal cells were thought to be implicated in autoimmune diseases, based on the observation that increased levels of fetal microchimerism were detected in women affected by autoimmune diseases.[9-12] However, to date there is no concrete evidence to prove that fetal cells cause autoimmune disease,[13] and increasing scientific evidence now suggests that these cells may actually help to combat disease. Support for this hypothesis comes from studies which show that fetal microchimerism is commonly detected in the peripheral blood of healthy women,[8,14] while the multilineage differentiation potential of fetal cells, which have been transferred to the mother, has also been demonstrated,[15] suggesting that these cells may play a role in tissue regeneration. Furthermore, fetal cell microchimerism may also confer a beneficial effect by performing immune surveillance for malignant cells, as supported by the observation that fetal cell chimerism is reduced in women with breast cancer compared to healthy women.[15-17]

With regard to the second mechanism of Medawar's paradox, it has been shown that fetal cells do in fact express MHC I and MHC II, which are antigenically mature and detectable in maternal circulation.[18] The lack of expression of the classical MHC class I and MHC class II molecules by the trophoblast cells, which are in contact with maternal circulation, was long considered to be a mechanism for evading detection and destruction by maternal cells. However, it was later shown that interstitial trophoblast populations, which are in contact with maternal decidua, do in fact express the MHC class I molecule.[19,20] Furthermore, studies by Shomer and Rogers using transgenic technology showed that expression of allogeneic MHC class I molecules on various trophoblast populations does not increase fetal loss, even in the presence of defects in the Fas/FasL pathway.[21,22]

Finally, concerning the third point of Medawars paradox, it is clear that the maternal immune system is not inert during pregnancy, and is instead able to recognize fetal cells, as proven by the observation that fetal tissues are rejected when transplanted into pregnant rats.[23] Moreover, it has also been shown that the maternal immune system is able to attack the pre-implantation blastocyst when the zona pellucida is removed.[24] Although maternal T cells respond to fetal antigens during normal pregnancy, the nature of the immune response appears to change during gestation, as demonstrated by conflicting data regarding expansion and deletion of maternal T cell subsets at different time points during gestation.[25-28] The production of alloantibodies by maternal B cells to paternally inherited antigens has also been reported, and while alloantibody production increases with subsequent pregnancies, it does not affect the outcome of the pregnancy.[29,30]

2.4 Possible Mechanisms Controlling Fetomaternal Tolerance

Many local mechanisms that contribute to protection of the fetus from the maternal immune system have been identified at the fetomaternal interface, although it is not yet clear how these mechanisms interact with each other.

The most well-known of these mechanisms have been summarized in several reviews[31-36] and those which have been most commonly described include: (a) expression of nonclassical MHC molecules by trophoblastic cells; (b) expression of the IDO enzyme by placental cells, resulting in tryptophan depletion and kyurenine production; (c) FasL expression by trophoblastic cells; (d) expression of complement regulator proteins by trophoblastic and decidual cells.

Regarding the first of the mechanisms listed here, it has been shown that trophoblastic cells express the nonclassical HLA molecules HLA-E, HLA-F, and HLA-G. While the function of HLA-F is unknown, protection of the fetus from allogeneic T-cell responses and NK cell-mediated damage have been attributed to HLA-G,[37] which is supported by the observation that T-cell proliferation is inhibited when these cells are cultured in mixed lymphocyte reactions with HLA-G-transfected cells.[38] In vitro studies have shown that HLA-G can also induce apoptosis of lymphocytes which have been previously activated through the Fas/FasL pathway.[39] Meanwhile, it has been hypothesized that the effect of HLA-G on NK cell activity is not induced directly, but rather, that it requires the expression of HLA-E on trophoblastic cells. It is thought that HLA-G promotes and stabilizes the expression of HLA-E at the cell surface, allowing it to bind the CD94–NKG2 inhibitory receptor on NK cells, which leads to inhibition of NK activity.[40,41] In addition, the interaction of HLA-G with dendritic cells through KIR-related leukocyte Ig-like receptors may have an indirect effect on the immune response by tolerizing dendritic cells and facilitating the generation of regulatory T cells.[42,43]

Regarding the role of Indoleamine 2,3-dioxygenase (IDO) in promoting fetomaternal tolerance, evidence from a study by Munn and Mellor suggests that synthesis of this tryptophan-catabolizing enzyme by placental cells could provide protection of the fetus from maternal T-cells, with the observation that inhibition of this enzyme during murine pregnancy resulted in fetal allograft rejection.[44] IDO is expressed by trophoblast giant cells in mice,[45] and is thought to prevent immune responses to the fetus by inhibiting maternal T cell activation either by depriving T cells of tryptophan[46] or by producing catabolites of tryptophan (kynurenines), which prevent activation and proliferation of T cells, B cells, and NK cells in vitro.[47] However, subsequent studies have shown that IDO-knockout in mice still results in normal litters,[48] suggesting that other mechanisms, such as the presence of another enzyme, tryptophan 2,3-dioxygenase, which also promotes tryptophan catabolism,[49] can compensate for the loss of IDO activity during gestation. It has been reported that IDO may also have indirect effects on immune responses by affecting the function of IDO-expressing dendritic cells, thereby preventing T cell regulation.[50] While tryptophan catabolism appears to be essential in murine pregnancy, its role in human pregnancy is less clear.[32] Although it is known that IDO is expressed by extravillous and villous trophoblast cells in humans, and that its expression increases during the first week of pregnancy and diminishes during the second trimester,[51] IDO deficiency has not been reported as a cause of pathology during human pregnancy.

Support for the hypothesis that apoptosis may be an important determinant in fetomaternal tolerance comes from studies which suggest that maternal tolerance of the fetus may be mediated by the Fas/FasL system, which plays a critical role in promoting apoptosis, and was also identified some years ago as an important pathway for controlling maternal immune responses at the fetomaternal interface.[52-54] The maternal decidua and fetal tissues express FasL on their cell surface and cause apoptosis of activated maternal Fas-expressing lymphocytes,[52,55] with apoptosis detectable at the maternal–fetal interface throughout gestation.[56,57] However, recent studies implicate a more complex role of FasL in fetomaternal tolerance, with the demonstration that this molecule may promote allograft rejection rather than survival.[58,59] Although some mechanisms to explain this have been proposed from studies that report the presence of FasL in trophoblast microvesicles, which can promote fetal rejection,[60,61] a more complete understanding of the role of Fas in fetomaternal tolerance is still required.

A role for the complement system has also been hypothesized in the control of fetomaternal tolerance. This system is a component of natural immunity that can be activated by pathogens, and also after transplantation of allogeneic or xenogeneic cells, resulting in induction of inflammatory cell chemotaxis, enhanced phagocytosis, and promotion of cell lysis by the membrane attack complex. Therefore, the complement system must be tightly regulated in order to protect tissues from damage associated with the inflammatory process, and in the context of fetomaternal tolerance, it has been shown that complement regulatory molecules play an important role in allowing the fetus to regulate maternal processes that would otherwise result in fetal tissue damage. In mice, expression of the complement regulator protein *Crry* prevents deposition of the C3 and C4 complement components, thereby preventing activation of the complement cascade at the fetomaternal interface.[62,63] The role of *Crry* in contributing to fetomaternal tolerance in mice is confirmed by the observation that a deficiency in this protein results in gestational failure.[64] Unlike mice, humans express multiple types of complement regulatory molecules at the fetomaternal interface, such as DAF, MCP, and CD59, and a role for these molecules in regulation of the complement cascade at the C3 level has also been demonstrated.[65,66] The expression of complement regulatory molecules by invading fetal trophoblast cells could be the result of a response to sublytic levels of complement activity, which may be encountered by these cells as they invade the uterine decidua, via a mechanism analogous to that observed during organ transplantation in which increasing levels of antibody and complement activation have been shown to result in increased resistance of the graft to complement-mediated injury.[67]

In trying to understand the mechanisms of fetomaternal tolerance, the possible role of specific leukocyte subtypes that are present at the fetomaternal interface, and which very likely play different and important roles in this process, should also be considered.[31,68] For further reading in this area, we refer readers to comprehensive reviews that have been published describing the characteristics of different leukocyte types which have been identified at the fetomaternal border either at the trophoblastic or decidual level, including NK cells,[69-71] regulatory T cells,[33,72,73] dendritic cells, and macrophages.[74,75] Here, we will focus instead on results which have recently been obtained from studies exploring the immunomodulatory features of cells derived from the amniotic fetal membrane, and their possible roles in fetomaternal tolerance.

Support for the hypothesis that cells derived from the fetal membranes may contribute to fetomaternal tolerance comes from studies which demonstrate that cells isolated from amniotic and chorionic membranes do not induce allogeneic or xenogeneic T-cell responses, and actively suppress T-cell proliferation.[76,77] Furthermore, both human amniotic membrane and human amniotic epithelial cells have been shown to survive for prolonged periods of time after xenogeneic transplantation into immuno-competent animals, including rabbits,[78] rats,[79] guinea pigs,[80] and bonnet monkeys.[81] Additionally, long-term engraftment has been observed after intravenous injection of human amniotic and chorionic cells into newborn swine and rats, with human microchimerism detected in several organs,[76] suggestive of active migration and tolerogenic potential of the xenogeneic cells. In addition, long-term survival of rat amnion-derived cells, with no evidence of immunological rejection or tumor formation, has been observed after allogeneic in utero transplantation of these cells into the developing rodent brain.[82]

Recently, in the stromal layer of the amniotic membrane, two subpopulations have been identified, which differ in their expression of HLA-DR, CD45, CD14, CD86, CD11b, and which possess either T-cell suppressive or stimulatory properties.[83] Even though the roles of these two populations in the amniotic membrane are not yet known, it is tempting to speculate that they may both play a role in controlling fetomaternal tolerance.

In summary, although many mechanisms have been postulated in order to explain maternal acceptance of the fetus, the cause of this phenomenon remains to be clarified and many questions still remain: Is there an initiating mechanism for fetomaternal tolerance, or does it result from the cumulative effect of several mechanisms that interact with each other? If the latter is true, how then are these mechanisms integrated? In any case, it is clear that further studies are needed to gain a complete understanding of the mechanisms of fetomaternal tolerance, which will constitute a fundamental tool for developing strategies of tolerance induction for organ transplantation, cell therapy, and tissue engineering in the future.

2.5 Placenta as a Source of Hematopoietic Stem Cells

Studies performed in mice have proven that in the embryo, hematopoiesis takes place in several anatomical locations, including the yolk sac, the aorta-gonad-mesonephros (AGM), the fetal liver, and the placenta.[84] However, the exact involvement of each of these regions in the processes of emergence, maturation, and expansion of hematopoietic stem cells has not yet been defined.

The mouse placenta is comprised of trophectoderm and two mesodermal components: the chorionic mesoderm, which forms a continuum with the yolk sac (a bilayered organ composed of extraembryonic mesoderm and visceral endoderm) and the allantoic mesoderm, an appendage arising from the posterior primitive streak. The allantois fuses with the chorionic plate and gives rise to the umbilical vasculature and the mesodermal components of the fetal placenta. Interdigitations of the allantoic mesoderm with the trophoblast result in formation of the placental labyrinth, which is the site of oxygen and nutrient exchange between maternal and fetal blood.[84]

The yolk sac, which was long considered to be the only site capable of producing hematopoietic stem cells (HSCs), is the first hematopoietic site to appear in mammals, producing the first primitive blood cells that terminally differentiate after circulating to the fetal liver.[85] The intra-embryonic AGM region, which is composed of the dorsal aorta, its underlying mesenchyme, and the adjacent vitelline and umbilical arteries, can also generate HSCs de novo. Furthermore, a recent study has shown that this region harbors precursors that display high proliferative potential, and the capacity for hematopoietic self-renewal and endothelial cell differentiation.[86]

The fetal liver is the main site of hematopoietic expansion and differentiation during gestation, but unlike the yolk sac and AGM region, it is a site of hematopoietic colonization and not an intrinsic source of hematopoietic cells.[87]

Appreciation of placental contribution to mammalian fetal hematopoiesis was gained after the discovery that the avian allantois retains cells with hematopoietic and endothelial potential.[88] Subsequent studies in mice revealed that the placenta contains multipotential clonogenic progenitors, which are present before liver colonization commences. These cells have the capacity to self-renew and to repopulate the hematopoietic system in irradiated adult hosts.[89,90] It has also been reported that prior to fusion, the allantois and chorion are both potent sources of hematopoietic progenitors, as revealed by their expression of a key transcriptional factor for hematopoiesis (Runx1).[91-94] A recent study has provided further strong evidence that the mouse fetal placenta functions as a hematopoietic organ, with the demonstration that placenta-derived hematopoietic cells are capable of producing both myelo-erythroid and B and T lymphoid progeny, therefore confirming the multipotentiality of HSCs derived from placenta. Interestingly, it has also been demonstrated that HSCs emerge in large vessels within the placenta, leading to the proposal that the small vessels that constitute the placental labyrinth may serve as a niche where HSCs convene for maturation and expansion prior to colonization of the fetal liver.[95]

2.6 Placenta as a Source of Nonhematopoietic Multipotent Stem and/or Progenitor Cells: In Vitro and In Vivo Studies

In addition to playing an essential role in fetal development, nutrition and maintenance of fetal tolerance, and acting as a source of hematopoietic stem cells, placental tissue also draws great interest as a source of other types of progenitor/stem cells, including mesenchymal stem cells.

Since 2002, numerous studies have demonstrated the presence of progenitor cells from different regions of the placenta through in vitro characterization and differentiation experiments. As summarized in recent reviews, various approaches have been reported for isolating cells, which display progenitor cell characteristics from different regions of placental tissues, namely, the mesodermal areas of the amniotic and chorionic fetal membranes, and the amniotic epithelium. Studies exploring the differentiation potential of these cells have yielded promising results indicating that they display plasticity and are capable of in vitro differentiation toward lineages of the three germ layers: ectoderm, mesoderm, and endoderm.[96,97] Here, we will use the nomenclature reported in a recent review when referring to cells derived from the different placental regions: hAEC for human amniotic epithelial cells, hAMSC for human amniotic mesenchymal stromal cells, and hCMSC for human chorionic mesenchymal stromal cells.[96] The review above also sets out a general consensus which has been established regarding the main features of mesenchymal stromal cells from human fetal membranes (hAMSC and hCMSC). Specifically, the minimum criteria for identifying hAMSC and hCMSC include: adherence to plastic; formation of fibroblast colony-forming units; a specific pattern of surface antigen expression whereby mesenchymal markers (CD90, CD73, CD105) are expressed (as shown by greater than 95% positivity for these markers), while hematopoietic markers (CD45, CD34, CD14, HLA-DR) are not expressed (as shown by positivity of less than 2%); fetal origin of the cells and differentiation potential toward one or more lineages including osteogenic, adipogenic, chondrogenic, and vascular/endothelial.[96]

In support of the hypothesis that hAMSC may display some degree of pluripotency, gene expression of octamer binding protein-4 (OCT-4),[77,98-101] SRY-related HMG-box gene 2 (SOX-2), reduced expressin-1 (Rex-1), and Nanog[101] have been reported in these cells, while positivity for the stage-specific embryonic antigens SSEA-3 or SSEA-4 on hAMSC is still debated.[96,102]

A possible association between hAMSC and the neuronal lineage has been demonstrated by studies that show that when freshly isolated, these cells express neuronal (Nestin, Musashi1, neuron-specific enolase, neurofilament medium, MAP2) and glial markers (glial fibrillary acidic protein), with increased expression of some of these observed after differentiation in specific neural induction media.[101,103,104]

The potential of hAMSC to differentiate into hepatocytes was studied by Tamagawa and colleagues, who have shown that these cells express hepatocytic markers such as albumin, α-fetoprotein (α-FP), cytokeratin 18 (CK18), α_1-antitrypsin (α_1-AT), and hepatocyte nuclear factor-4α (HNF-4α). Furthermore, after hepatic induction of these cells, increased expression of the above-mentioned genes was observed, together with production of albumin and α-fetoprotein and storage of glycogen.[105]

Investigation of the cardiomyogenic potential of hAMSC has shown that these cells express the cardiac-specific transcription factor GATA4 and cardiac-specific genes such as atrial myosin light chain (MLC)-2a, ventricular myosin light chain MLC-2v, and the cardiac troponins cTnI and cTnT. hAMSC have also been shown to integrate into cardiac tissue and differentiate into cardiomyocyte-like cells after transplantation into myocardial infarcts in rat hearts.[99]

Enhancement of the cardiomyogenic and vasculogenic differentiation of human amniochorionic-derived cells has been observed after exposure of these cells to a mixed ester of hyaluronan, butyric, and retinoic acid (HBR). In particular, increased expression of cardiomyogenic (GATA4, NKX2.5) and endothelial genes (VEGF, vWF), as well as enhanced expression of the cardiac markers sarcomeric myosin heavy chain, α-sarcomeric actinin and connexin 43, has been observed in HBR-treated amniochorionic cells compared to untreated cells. Meanwhile, injection of both HBR-pretreated and non-pretreated cells into infarcted rat hearts has been shown to result in recovery to essentially normal indices of cardiac function.[106]

In experiments investigating the angiogenic potential of amniotic membrane-derived cells, basal expression of endothelial-specific markers (FLT-1, KDR) and spontaneous differentiation into endothelial cells have been observed, while both of these have been shown to be enhanced by exposure of the cells to vascular endothelial growth factor (VEGF).[100]

Not only do the stromal regions of placenta seem to contain progenitor/stem cells, but interesting data have also been obtained through studies of hAEC. Expression of embryonic stem cell markers such as the stage-specific embryonic antigen SSEA-4, TRA-1–60, and TRA-1–81 has been reported for these cells,[102,107] and in addition, they have also been demonstrated to express molecular markers of pluripotent stem cells, including octamer-binding-protein-4 (OCT-4), SRY-related HMG-box gene 2 (SOX-2), and Nanog.[107,108] The pluripotency of hAEC is further supported by a study of Tamagawa et al., whereby a xenogeneic chimeric embryo was created by mixing amniotic cells with mouse embryonic stem cells, with demonstration that amnion-derived cells were then able to contribute to the formation of all three germ layers.[109]

Interestingly, the mesenchymal marker vimentin, although absent on freshly isolated hAEC, has also been shown appear during culture.[110,111] The significance of the expression of both epithelial and mesenchymal markers by hAEC remains to be elucidated, although it could be due to the spontaneous commencement of differentiation during culture, or perhaps to the so-called epithelial to mesenchymal transition in the amnion, as also suggested by Sakuragawa and co-workers.[104]

To date, numerous studies have been undertaken to explore the differentiation capacity of hAEC, yielding results that confirm the plasticity of these cells.[96]

A neuronal predisposition of hAEC has been demonstrated through pioneeristic studies by Sakuragawa and colleagues, who showed that these cells express neuronal and glial markers,[112] and also perform neuronal functions such as synthesis and release of acetylcholine, catecholamines, neurotrophic factors (brain-derived neurotrophic factor and neurotrophin-3), activin, and noggin.[104,113-117] Furthermore, hAEC conditioned medium has been shown to have neurotrophic effects on rat cortical neurons[116] and can support the survival of chicken neural retinal cells,[118] while more recently, it has also been shown that human amniotic membrane promotes the growth of chicken dorsal root ganglia neurons in the absence of neurotrophic factors.[119]

Preclinical studies in animal models demonstrate that hAEC may be useful for central nervous system regeneration by exhibiting neuroprotective and neuroregenerative effects during acute phases of injury. For example, Sankar and coworkers observed robust regeneration of host axons and enhanced survival of axotomized spinal cord neurons after transplantation of hAEC into lesioned areas of a contusion model of spinal cord injury in monkeys.[81] Improved performance in locomotor tests in cell-treated animals compared to lesion control animals was also observed.[81] Meanwhile, in a rat model of Parkinson's disease, restoration of striatal dopamine levels and behavioral improvement have been observed after transplantation of hAEC,[120-122] while transplantation of these cells into the brains of rats which had undergone middle cerebral artery occlusion resulted in improvement of behavioral dysfunction and reduced infarct volume.[123]

Hepatocyte-like features of hAEC have also been observed in vitro by several groups. These cells have been shown to express liver-enriched transcription factors including hepatocyte nuclear factor (HNF) 3γ and HNF4α, CCAAT/enhancer-binding protein (CEBP) α and β) and CYP450 enzymes, as well as hepatocyte-related genes including α1-antitrypsin (α1AT), cytokeratin 18 (CK18), glutamine synthetase (GS), carbamoyl phosphate synthetase-1 (CPS-1), phosphoenolpyruvate carboxykinase (PEPCK), and drug metabolism-related genes, CYP2D6 and CYP3A4.[107,124] In vitro expression of human serum albumin and α-fetoprotein (AFP) has also been reported for hAEC, as well as typical hepatic functions such as albumin synthesis and production and storage of glycogen.[125,126] Studies in mice suggest that hAEC may also be able to perform hepatic functions in vivo, with human albumin detected in the sera and peritoneal fluid of SCID mice which had received peritoneal implants of human amniotic membrane.[125] Furthermore, a study in which hAEC were transplanted into SCID mice has demonstrated that human α1-antitrypsin could subsequently be detected by Western blot in the sera of these animals,[110] while another study has shown that integrated AFP- and Alb-positive hepatocyte-like cells could be identified in hepatic parenchyma of SCID mice two weeks after hAEC transplantation.[126] Interestingly, the authors of this latter study also showed that hAEC which had been genetically modified to express the *LacZ* gene were able to integrate in liver parenchyma, suggesting that these cells could also be useful as gene carriers for patients with congenital liver disorders.

The ability of hAEC to differentiate toward the pancreatic lineage has also been reported, whereby these cells were induced to produce insulin through culture in the presence of nicotinamide. The insulin-producing hAEC were then able to normalize blood glucose levels after transplantation into streptozotocin-induced diabetic mice.[98] Ultrastructural features characteristic of beta pancreatic cells, as well as expression of the pancreatic marker amylase alpha 2B(AMYB2) and glucagon production have also been observed after culture of hAEC in pancreatic differentiation media.[108]

2.7 Conclusion

To conclude on the possibilities for the future of placenta-derived cells in the clinical setting, it is clear that these cells hold great promise for the reasons that have been discussed in this chapter. The presence of different sources of stem cells in the placenta, from the pluripotent amniotic ectoderm-derived cells to the mesenchymal and hematopoietic stem cells, as well as the plasticity of these cells, which has been shown through in vitro studies, and finally, their promising immunological properties, lead us to hypothesize with confidence that placenta-derived cells, or at least some of their subpopulations, could be applied for the development of new therapeutic strategies. Furthermore, the fact that placental tissue can be procured in nearly unlimited supply without harming the mother or the fetus, as well as the fact that its use raises ethical support rather than objection, and finally, the possibility of collecting and banking these cells at birth, together constitute strong evidence that the placenta indeed represents a potential oasis in the search for new and viable stem cell sources.

Although holding much promise for future applications in regenerative medicine, many questions remain open in the field of placenta-derived stem cell research. Given our current understanding of the cells from placental tissues, perhaps the most important of these is whether it is the plasticity or immunomodulatory properties of placental cells that will make them most useful in clinical applications in the future. Current knowledge leaves open both possibilities, although it appears that ever-increasing attention is being turned toward the effect that these cells have on the surrounding environment. Literature published to date appears to lend stronger support to the hypothesis that placental cells exert the bulk of their actions in vivo by secreting factors which support the growth, survival, or differentiation of other cells, rather than themselves undergoing differentiation to regenerate damaged or diseased tissue. In any case, it is clear that the human placenta still harbors many clues to understanding the processes of tissue development and tolerance, which will no doubt open new doors for the development of therapeutic treatments which can overcome current shortcomings in the field of regenerative medicine.

Acknowledgments The authors express their gratitude to Marco Evangelista for his invaluable help in the revision of this book chapter.

References

1. Benirschke K, Kaufmann P. *Pathology of the Human Placenta*. Berlin: Springer; 2000.
2. Cunningham FG, Macdonald PC, Gant NF, Leveno KJ, Gilstrap LC, Hankins GDF, Clark SL. *Williams Obstetrics*. 20th ed. Stamford, CT: Appleton&Lange; 1997.

3. Moore KL. *The Developing Human*. Philadelphia, PA: W.B. Saunders; 1998.
4. Billington WD. The immunological problem of pregnancy: 50 years with the hope of progress. A tribute to Peter Medawar. *J Reprod Immunol*. 2003;60:1-11.
5. Medawar P. Some immunological and endocrinological problems raised by the evolution of viviparity in vertebrates. *Symp Soc Exp Biol*. 1953;7:320-338.
6. Liegeois A, Escourrou J, Ouvre E, Charreire J. Microchimerism: a stable state of low-ratio proliferation of allogeneic bone marrow. *Transplant Proc*. 1977;9:273-276.
7. Ariga H, Ohto H, Busch MP, et al. Kinetics of fetal cellular and cell-free DNA in the maternal circulation during and after pregnancy: implications for noninvasive prenatal diagnosis. *Transfusion*. 2001;41:1524-1530.
8. Bianchi DW, Zickwolf GK, Weil GJ, Sylvester S, DeMaria MA. Male fetal progenitor cells persist in maternal blood for as long as 27 years postpartum. *Proc Natl Acad Sci USA*. 1996;93:705-708.
9. Kremer Hovinga IC, Koopmans M, Baelde HJ, et al. Chimerism occurs twice as often in lupus nephritis as in normal kidneys. *Arthritis Rheum*. 2006;54:2944-2950.
10. Kuroki M, Okayama A, Nakamura S, et al. Detection of maternal-fetal microchimerism in the inflammatory lesions of patients with Sjogren's syndrome. *Ann Rheum Dis*. 2002; 61:1041-1046.
11. Klintschar M, Schwaiger P, Mannweiler S, Regauer S, Kleiber M. Evidence of fetal microchimerism in Hashimoto's thyroiditis. *J Clin Endocrinol Metab*. 2001;86:2494-2498.
12. Nelson JL, Furst DE, Maloney S, et al. Microchimerism and HLA-compatible relationships of pregnancy in scleroderma. *Lancet*. 1998;351:559-562.
13. Johnson KL, Bianchi DW. Fetal cells in maternal tissue following pregnancy: what are the consequences? *Hum Reprod Update*. 2004;10:497-502.
14. Evans PC, Lambert N, Maloney S, Furst DE, Moore JM, Nelson JL. Long-term fetal microchimerism in peripheral blood mononuclear cell subsets in healthy women and women with scleroderma. *Blood*. 1999;93:2033-2037.
15. Khosrotehrani K, Johnson KL, Cha DH, Salomon RN, Bianchi DW. Transfer of fetal cells with multilineage potential to maternal tissue. *JAMA*. 2004;292:75-80.
16. Gadi VK, Nelson JL. Fetal microchimerism in women with breast cancer. *Cancer Res*. 2007;67:9035-9038.
17. Khosrotehrani K, Stroh H, Bianchi DW, Johnson KL. Combined FISH and immunolabeling on paraffin-embedded tissue sections for the study of microchimerism. *Biotechniques*. 2003;34:242-244.
18. Thellin O, Coumans B, Zorzi W, Igout A, Heinen E. Tolerance to the foeto-placental 'graft': ten ways to support a child for nine months. *Curr Opin Immunol*. 2000;12:731-737.
19. Redline RW, Lu CY. Localization of fetal major histocompatibility complex antigens and maternal leukocytes in murine placenta. Implications for maternal-fetal immunological relationship. *Lab Invest*. 1989;61:27-36.
20. Redman CW, McMichael AJ, Stirrat GM, Sunderland CA, Ting A. Class 1 major histocompatibility complex antigens on human extra-villous trophoblast. *Immunology*. 1984;52:457-468.
21. Shomer B, Toder V, Egorov I, Ehrlich R. Expression of allogeneic MHC class I antigens by transgenic mouse trophoblast does not interfere with the normal course of pregnancy. *Transgenic Res*. 1998;7:343-355.
22. Rogers AM, Boime I, Connolly J, Cook JR, Russell JH. Maternal-fetal tolerance is maintained despite transgene-driven trophoblast expression of MHC class I, and defects in Fas and its ligand. *Eur J Immunol*. 1998;28:3479-3487.
23. Woodruff MF. Transplantation immunity and the immunological problem of pregnancy. *Proc R Soc Lond B Biol Sci*. 1958;148:68-75.
24. Ewoldsen MA, Ostlie NS, Warner CM. Killing of mouse blastocyst stage embryos by cytotoxic T lymphocytes directed to major histocompatibility complex antigens. *J Immunol*. 1987;138:2764-2770.
25. Vacchio MS, Jiang SP. The fetus and the maternal immune system: pregnancy as a model to study peripheral T-cell tolerance. *Crit Rev Immunol*. 1999;19:461-480.
26. Watanabe M, Iwatani Y, Kaneda T, et al. Changes in T, B, and NK lymphocyte subsets during and after normal pregnancy. *Am J Reprod Immunol*. 1997;37:368-377.
27. Tallon DF, Corcoran DJ, O'Dwyer EM, Greally JF. Circulating lymphocyte subpopulations in pregnancy: a longitudinal study. *J Immunol*. 1984;132:1784-1787.
28. Carter J, Newport A, Keeler KD, Dresser DW. FACS analysis of changes in T and B lymphocyte populations in the blood, spleen and lymph nodes of pregnant mice. *Immunology*. 1983;48:791-797.
29. Innes A, Cunningham C, Power DA, Catto GR. Fetus as an allograft: noncytotoxic maternal antibodies to HLA-linked paternal antigens. *Am J Reprod Immunol*. 1989; 19:146-150.
30. Innes A, Power DA, Cunningham C, Dillon D, Catto GR. The alloantibody response to semiallogeneic pregnancy in the rat. I. Alloantibodies in sera and placental eluates directed to RT1A antigens. *Transplantation*. 1988;46:409-413.
31. von Rango U. Fetal tolerance in human pregnancy – a crucial balance between acceptance and limitation of trophoblast invasion. *Immunol Lett*. 2008;115:21-32.
32. Koch CA, Platt JL. T cell recognition and immunity in the fetus and mother. *Cell Immunol*. 2007;248:12-17.
33. Aluvihare VR, Kallikourdis M, Betz AG. Tolerance, suppression and the fetal allograft. *J Mol Med*. 2005;83: 88-96.
34. Petroff MG. Immune interactions at the maternal-fetal interface. *J Reprod Immunol*. 2005;68:1-13.
35. Koch CA, Platt JL. Natural mechanisms for evading graft rejection: the fetus as an allograft. *Springer Semin Immunopathol*. 2003;25:95-117.
36. Mellor AL, Munn DH. Immunology at the maternal-fetal interface: lessons for T cell tolerance and suppression. *Annu Rev Immunol*. 2000;18:367-391.
37. Hunt JS, Petroff MG, McIntire RH, Ober C. HLA-G and immune tolerance in pregnancy. *Faseb J*. 2005;19:681-693.
38. Riteau B, Menier C, Khalil-Daher I, et al. HLA-G inhibits the allogeneic proliferative response. *J Reprod Immunol*. 1999;43:203-211.
39. Fournel S, Aguerre-Girr M, Huc X, et al. Cutting edge: soluble HLA-G1 triggers CD95/CD95 ligand-mediated apoptosis in activated CD8+ cells by interacting with CD8. *J Immunol*. 2000;164:6100-6104.
40. Moffett-King A. Natural killer cells and pregnancy. *Nat Rev Immunol*. 2002;2:656-663.

41. King A, Allan DS, Bowen M, et al. HLA-E is expressed on trophoblast and interacts with CD94/NKG2 receptors on decidual NK cells. *Eur J Immunol.* 2000;30:1623-1631.
42. Shiroishi M, Tsumoto K, Amano K, et al. Human inhibitory receptors Ig-like transcript 2 (ILT2) and ILT4 compete with CD8 for MHC class I binding and bind preferentially to HLA-G. *Proc Natl Acad Sci USA.* 2003;100:8856-8861.
43. Chang CC, Ciubotariu R, Manavalan JS, et al. Tolerization of dendritic cells by T(S) cells: the crucial role of inhibitory receptors ILT3 and ILT4. *Nat Immunol.* 2002;3:237-243.
44. Munn DH, Zhou M, Attwood JT, et al. Prevention of allogeneic fetal rejection by tryptophan catabolism. *Science.* 1998;281:1191-1193.
45. Mellor AL, Munn DH. IDO expression by dendritic cells: tolerance and tryptophan catabolism. *Nat Rev Immunol.* 2004;4:762-774.
46. Munn DH, Shafizadeh E, Attwood JT, Bondarev I, Pashine A, Mellor AL. Inhibition of T cell proliferation by macrophage tryptophan catabolism. *J Exp Med.* 1999;189:1363-1372.
47. Terness P, Bauer TM, Rose L, et al. Inhibition of allogeneic T cell proliferation by indoleamine 2, 3-dioxygenase-expressing dendritic cells: mediation of suppression by tryptophan metabolites. *J Exp Med.* 2002;196:447-457.
48. Baban B, Chandler P, McCool D, Marshall B, Munn DH, Mellor AL. Indoleamine 2, 3-dioxygenase expression is restricted to fetal trophoblast giant cells during murine gestation and is maternal genome specific. *J Reprod Immunol.* 2004;61:67-77.
49. Suzuki S, Tone S, Takikawa O, Kubo T, Kohno I, Minatogawa Y. Expression of indoleamine 2, 3-dioxygenase and tryptophan 2, 3-dioxygenase in early concepti. *Biochem J.* 2001;355:425-429.
50. Munn DH, Sharma MD, Lee JR, et al. Potential regulatory function of human dendritic cells expressing indoleamine 2, 3-dioxygenase. *Science.* 2002;297:1867-1870.
51. von Rango U, Krusche CA, Beier HM, Classen-Linke I. Indoleamine-dioxygenase is expressed in human decidua at the time maternal tolerance is established. *J Reprod Immunol.* 2007;74:34-45.
52. Mor G, Gutierrez LS, Eliza M, Kahyaoglu F, Arici A. Fas-fas ligand system-induced apoptosis in human placenta and gestational trophoblastic disease. *Am J Reprod Immunol.* 1998;40:89-94.
53. Hunt JS, Vassmer D, Ferguson TA, Miller L. Fas ligand is positioned in mouse uterus and placenta to prevent trafficking of activated leukocytes between the mother and the conceptus. *J Immunol.* 1997;158:4122-4128.
54. Runic R, Lockwood CJ, Ma Y, Dipasquale B, Guller S. Expression of Fas ligand by human cytotrophoblasts: implications in placentation and fetal survival. *J Clin Endocrinol Metab.* 1996;81:3119-3122.
55. Coumans B, Thellin O, Zorzi W, et al. Lymphoid cell apoptosis induced by trophoblastic cells: a model of active foeto-placental tolerance. *J Immunol Methods.* 1999; 224:185-196.
56. Smith SC, Leung TN, To KF, Baker PN. Apoptosis is a rare event in first-trimester placental tissue. *Am J Obstet Gynecol.* 2000;183:697-699.
57. Jerzak M, Kasprzycka M, Wierbicki P, Kotarski J, Gorski A. Apoptosis of T cells in the first trimester human decidua. *Am J Reprod Immunol.* 1998;40:130-135.
58. Kang SM, Braat D, Schneider DB, et al. A non-cleavable mutant of Fas ligand does not prevent neutrophilic destruction of islet transplants. *Transplantation.* 2000;69:1813-1817.
59. Allison J, Georgiou HM, Strasser A, Vaux DL. Transgenic expression of CD95 ligand on islet beta cells induces a granulocytic infiltration but does not confer immune privilege upon islet allografts. *Proc Natl Acad Sci USA.* 1997;94:3943-3947.
60. Frangsmyr L, Baranov V, Nagaeva O, Stendahl U, Kjellberg L, Mincheva-Nilsson L. Cytoplasmic microvesicular form of Fas ligand in human early placenta: switching the tissue immune privilege hypothesis from cellular to vesicular level. *Mol Hum Reprod.* 2005;11:35-41.
61. Abrahams VM, Straszewski-Chavez SL, Guller S, Mor G. First trimester trophoblast cells secrete Fas ligand which induces immune cell apoptosis. *Mol Hum Reprod.* 2004; 10:55-63.
62. Miwa T, Zhou L, Hilliard B, Molina H, Song WC. Crry, but not CD59 and DAF, is indispensable for murine erythrocyte protection in vivo from spontaneous complement attack. *Blood.* 2002;99:3707-3716.
63. Matsuo S, Ichida S, Takizawa H, et al. In vivo effects of monoclonal antibodies that functionally inhibit complement regulatory proteins in rats. *J Exp Med.* 1994;180:1619-1627.
64. Xu C, Mao D, Holers VM, Palanca B, Cheng AM, Molina H. A critical role for murine complement regulator crry in fetomaternal tolerance. *Science.* 2000;287:498-501.
65. Jerzak M, Bischof P. Apoptosis in the first trimester human placenta: the role in maintaining immune privilege at the maternal-foetal interface and in the trophoblast remodelling. *Eur J Obstet Gynecol Reprod Biol.* 2002;100:138-142.
66. Holmes CH, Simpson KL, Wainwright SD, et al. Preferential expression of the complement regulatory protein decay accelerating factor at the fetomaternal interface during human pregnancy. *J Immunol.* 1990;144:3099-3105.
67. Dalmasso AP, Benson BA, Johnson JS, Lancto C, Abrahamsen MS. Resistance against the membrane attack complex of complement induced in porcine endothelial cells with a Gal alpha(1-3)Gal binding lectin: up-regulation of CD59 expression. *J Immunol.* 2000;164:3764-3773.
68. Trundley A, Moffett A. Human uterine leukocytes and pregnancy. *Tissue Antigens.* 2004;63:1-12.
69. Manaster I, Mandelboim O. The unique properties of human NK cells in the uterine mucosa. *Placenta.* 2008;29 (Suppl A):S60-S66.
70. Tabiasco J, Rabot M, Aguerre-Girr M, et al. Human decidual NK cells: unique phenotype and functional properties – a review. *Placenta.* 2006;27(Suppl A):S34-S39.
71. Wold AS, Arici A. Natural killer cells and reproductive failure. *Curr Opin Obstet Gynecol.* 2005;17:237-241.
72. Aluvihare VR, Betz AG. The role of regulatory T cells in alloantigen tolerance. *Immunol Rev.* 2006;212:330-343.
73. Terness P, Kallikourdis M, Betz AG, Rabinovich GA, Saito S, Clark DA. Tolerance signaling molecules and pregnancy: IDO, galectins, and the renaissance of regulatory T cells. *Am J Reprod Immunol.* 2007;58:238-254.
74. Laskarin G, Kammerer U, Rukavina D, Thomson AW, Fernandez N, Blois SM. Antigen-presenting cells and materno-fetal tolerance: an emerging role for dendritic cells. *Am J Reprod Immunol.* 2007;58:255-267.
75. Blois SM, Kammerer U, Alba Soto C, et al. Dendritic cells: key to fetal tolerance? *Biol Reprod.* 2007;77:590-598.

76. Bailo M, Soncini M, Vertua E, et al. Engraftment potential of human amnion and chorion cells derived from term placenta. *Transplantation*. 2004;78:1439-1448.
77. Wolbank S, Peterbauer A, Fahrner M, et al. Dose-dependent immunomodulatory effect of human stem cells from amniotic membrane: a comparison with human mesenchymal stem cells from adipose tissue. *Tissue Eng*. 2007;13:1173-1183.
78. Avila M, Espana M, Moreno C, Pena C. Reconstruction of ocular surface with heterologous limbal epithelium and amniotic membrane in a rabbit model. *Cornea*. 2001;20: 414-420.
79. Kubo M, Sonoda Y, Muramatsu R, Usui M. Immunogenicity of human amniotic membrane in experimental xenotransplantation. *Invest Ophthalmol Vis Sci*. 2001;42:1539-1546.
80. Yuge I, Takumi Y, Koyabu K, et al. Transplanted human amniotic epithelial cells express connexin 26 and Na-K-adenosine triphosphatase in the inner ear. *Transplantation*. 2004;77:1452-1454.
81. Sankar V, Muthusamy R. Role of human amniotic epithelial cell transplantation in spinal cord injury repair research. *Neuroscience*. 2003;118:11-17.
82. Marcus AJ, Coyne TM, Black IB, Woodbury D. Fate of amnion-derived stem cells transplanted to the fetal rat brain: migration, survival and differentiation. *J Cell Mol Med*. 2007;12(4):1256-1264.
83. Magatti M, De Munari S, Vertua E, Gibelli L, Wengler GS, Parolini O. Human amnion mesenchyme harbors cells with allogeneic T-cell suppression and stimulation capabilities. *Stem Cells*. 2008;26:182-192.
84. Mikkola HK, Gekas C, Orkin SH, Dieterlen-Lievre F. Placenta as a site for hematopoietic stem cell development. *Exp Hematol*. 2005;33:1048-1054.
85. Palis J, Yoder MC. Yolk-sac hematopoiesis: the first blood cells of mouse and man. *Exp Hematol*. 2001;29:927-936.
86. Yao H, Liu B, Wang X, et al. Identification of high proliferative potential precursors with hemangioblastic activity in the mouse aorta-gonad- mesonephros region. *Stem Cells*. 2007; 25:1423-1430.
87. Houssaint E. Differentiation of the mouse hepatic primordium. II. Extrinsic origin of the haemopoietic cell line. *Cell Differ*. 1981;10:243-252.
88. Caprioli A, Jaffredo T, Gautier R, Dubourg C, Dieterlen-Lievre F. Blood-borne seeding by hematopoietic and endothelial precursors from the allantois. *Proc Natl Acad Sci USA*. 1998;95:1641-1646.
89. Gekas C, Dieterlen-Lievre F, Orkin SH, Mikkola HK. The placenta is a niche for hematopoietic stem cells. *Dev Cell*. 2005;8:365-375.
90. Alvarez-Silva M, Belo-Diabangouaya P, Salaun J, Dieterlen-Lievre F. Mouse placenta is a major hematopoietic organ. *Development*. 2003;130:5437-5444.
91. Zeigler BM, Sugiyama D, Chen M, Guo Y, Downs KM, Speck NA. The allantois and chorion, when isolated before circulation or chorio-allantoic fusion, have hematopoietic potential. *Development*. 2006;133:4183-4192.
92. Ottersbach K, Dzierzak E. The murine placenta contains hematopoietic stem cells within the vascular labyrinth region. *Dev Cell*. 2005;8:377-387.
93. Lacaud G, Gore L, Kennedy M, et al. Runx1 is essential for hematopoietic commitment at the hemangioblast stage of development in vitro. *Blood*. 2002;100:458-466.
94. North T, Gu TL, Stacy T, et al. Cbfa2 is required for the formation of intra-aortic hematopoietic clusters. *Development*. 1999;126:2563-2575.
95. Rhodes KE, Gekas C, Wang Y, et al. The emergence of hematopoietic stem cells is initiated in the placental vasculature in the absence of circulation. *Cell Stem Cell*. 2008;2:252-263.
96. Parolini O, Alviano F, Bagnara GP, et al. Concise review: isolation and characterization of cells from human term placenta: outcome of the first international workshop on placenta derived stem cells. *Stem Cells*. 2008;26:300-311.
97. Parolini O, Soncini M. Human placenta: a source of progenitor/stem cells? *J Reprod Med Endocrinol*. 2006;3: 117-126.
98. Wei JP, Zhang TS, Kawa S, et al. Human amnion-isolated cells normalize blood glucose in streptozotocin-induced diabetic mice. *Cell Transplant*. 2003;12:545-552.
99. Zhao P, Ise H, Hongo M, Ota M, Konishi I, Nikaido T. Human amniotic mesenchymal cells have some characteristics of cardiomyocytes. *Transplantation*. 2005;79: 528-535.
100. Alviano F, Fossati V, Marchionni C, et al. Term Amniotic membrane is a high throughput source for multipotent mesenchymal stem cells with the ability to differentiate into endothelial cells in vitro. *BMC Dev Biol*. 2007;7:11.
101. Tamagawa T, Ishiwata I, Ishikawa H, Nakamura Y. Induced in vitro differentiation of neural-like cells from human amnion-derived fibroblast-like cells. *Hum Cell*. 2008;21:38-45.
102. Miki T, Mitamura K, Ross MA, Stolz DB, Strom SC. Identification of stem cell marker-positive cells by immunofluorescence in term human amnion. *J Reprod Immunol*. 2007;75(2):91-96.
103. Portmann-Lanz CB, Schoeberlein A, Huber A, et al. Placental mesenchymal stem cells as potential autologous graft for pre- and perinatal neuroregeneration. *Am J Obstet Gynecol*. 2006;194:664-673.
104. Sakuragawa N, Kakinuma K, Kikuchi A, et al. Human amnion mesenchyme cells express phenotypes of neuroglial progenitor cells. *J Neurosci Res*. 2004;78:208-214.
105. Tamagawa T, Oi S, Ishiwata I, Ishikawa H, Nakamura Y. Differentiation of mesenchymal cells derived from human amniotic membranes into hepatocyte-like cells in vitro. *Hum Cell*. 2007;20:77-84.
106. Ventura C, Cantoni S, Bianchi F, et al. Hyaluronan mixed esters of butyric and retinoic acid drive cardiac and endothelial fate in term placenta human mesenchymal stem cells and enhance cardiac repair in infarcted rat hearts. *J Biol Chem*. 2007;282:14243-14252.
107. Miki T, Lehmann T, Cai H, Stolz DB, Strom SC. Stem cell characteristics of amniotic epithelial cells. *Stem Cells*. 2005;23:1549-1559.
108. Ilancheran S, Michalska A, Peh G, Wallace EM, Pera M, Manuelpillai U. Stem cells derived from human fetal membranes display multi-lineage differentiation potential. *Biol Reprod*. 2007;77:577-588.
109. Tamagawa T, Ishiwata I, Saito S. Establishment and characterization of a pluripotent stem cell line derived from human amniotic membranes and initiation of germ layers in vitro. *Hum Cell*. 2004;17:125-130.
110. Miki T, Strom SC. Amnion-derived pluripotent/multipotent stem cells. *Stem Cell Rev*. 2006;2:133-142.

111. Toda A, Okabe M, Yoshida T, Nikaido T. The potential of amniotic membrane/amnion-derived cells for regeneration of various tissues. *J Pharmacol Sci*. 2007;105:215-228.
112. Sakuragawa N, Thangavel R, Mizuguchi M, Hirasawa M, Kamo I. Expression of markers for both neuronal and glial cells in human amniotic epithelial cells. *Neurosci Lett*. 1996;209:9-12.
113. Sakuragawa N, Misawa H, Ohsugi K, et al. Evidence for active acetylcholine metabolism in human amniotic epithelial cells: applicable to intracerebral allografting for neurologic disease. *Neurosci Lett*. 1997;232:53-56.
114. Elwan MA, Sakuragawa N. Evidence for synthesis and release of catecholamines by human amniotic epithelial cells. *Neuroreport*. 1997;8:3435-3438.
115. Ishii T, Ohsugi K, Nakamura S, et al. Gene expression of oligodendrocyte markers in human amniotic epithelial cells using neural cell-type-specific expression system. *Neurosci Lett*. 1999;268:131-134.
116. Uchida S, Inanaga Y, Kobayashi M, Hurukawa S, Araie M, Sakuragawa N. Neurotrophic function of conditioned medium from human amniotic epithelial cells. *J Neurosci Res*. 2000;62:585-590.
117. Koyano S, Fukui A, Uchida S, Yamada K, Asashima M, Sakuragawa N. Synthesis and release of activin and noggin by cultured human amniotic epithelial cells. *Dev Growth Differ*. 2002;44:103-112.
118. Tcheng M, Oliver L, Courtois Y, Jeanny JC. Effects of exogenous FGFs on growth, differentiation, and survival of chick neural retina cells. *Exp Cell Res*. 1994;212:30-35.
119. Schroeder A, Theiss C, Steuhl KP, Meller K, Meller D. Effects of the human amniotic membrane on axonal outgrowth of dorsal root ganglia neurons in culture. *Curr Eye Res*. 2007;32:731-738.
120. Kakishita K, Elwan MA, Nakao N, Itakura T, Sakuragawa N. Human amniotic epithelial cells produce dopamine and survive after implantation into the striatum of a rat model of Parkinson's disease: a potential source of donor for transplantation therapy. *Exp Neurol*. 2000;165:27-34.
121. Kakishita K, Nakao N, Sakuragawa N, Itakura T. Implantation of human amniotic epithelial cells prevents the degeneration of nigral dopamine neurons in rats with 6-hydroxydopamine lesions. *Brain Res*. 2003;980:48-56.
122. Kong XY, Cai Z, Pan L, et al. Transplantation of human amniotic cells exerts neuroprotection in MPTP-induced Parkinson disease mice. *Brain Res*. 2008;1205:108-115.
123. Liu T, Wu J, Huang Q, et al. Human amniotic epithelial cells ameliorate behavioral dysfunction and reduce infarct size in the rat middle cerebral artery occlusion model. *Shock*. 2008;29:603-611.
124. Davila JC, Cezar GG, Thiede M, Strom S, Miki T, Trosko J. Use and application of stem cells in toxicology. *Toxicol Sci*. 2004;79:214-223.
125. Takashima S, Ise H, Zhao P, Akaike T, Nikaido T. Human amniotic epithelial cells possess hepatocyte-like characteristics and functions. *Cell Struct Funct*. 2004;29:73-84.
126. Sakuragawa N, Enosawa S, Ishii T, et al. Human amniotic epithelial cells are promising transgene carriers for allogeneic cell transplantation into liver. *J Hum Genet*. 2000;45:171-176.

Placenta and Umbilical Cord in Traditional Chinese Medicine

3

Ping Chung Leung

3.1 History of the Medicinal Use of the Human Placenta

The human placenta was described as a medicinal material as early as 400 BC during Hippocrates' time and in China in 200 BC, when it was used as a healing agent after bodily injuries. Legendary figures used it for exclusive reasons. Thus, the great tyrant of the Qin Dynasty in China used human placenta for longevity and the Egyptian Queen Cleopatra used it for cosmetic purposes.

In the Tang Dynasty (907 AD), human placenta was referred as "human enwrap" in the ancient materia medica.[1] The proper use and description of this entity was started in the Jinn Dynasty when the Taoist paid special attention to this item. The Taoist respected material coming from Nature and believed that what was derived from the parents must be good for the maintenance of health and was likely to have specific indications for certain disease entities. During this period, the Taoist healers gave this item of human tissues the proper terminology of "Purple Turning Lotus Wheel." The human placenta did look like a big round lotus leaf. The Taoist respected structure with mobility, which was correlated with dynamics and efficiency. Therefore, the lotus leaf became a turning wheel.[2] There were placentas with different colors: purple, red, and yellowish. The best-quality product gave a purple color. Hence, the human placenta, when used as a medicinal material, was called "Purple Turning Lotus Wheel" ever since (Fig. 3.1).

Fig. 3.1 Dried human placenta "purple turning lotus wheel"

Ancient Korea and Japan have been under the heavy influence from Imperial China. Records about the "Purple Turning Lotus Wheel" appeared in the medical classics of the two countries in the seventeenth century.

3.2 Preparation of the Human Placenta

"Purple Turning Lotus Wheel" is prepared as a dried entity. It is not used fresh. The dried preparation has a rough, lobulated, and grooved surface, which was attached to the uterus during pregnancy, and a smooth surface, which has given rise to the umbilical cord (Fig. 3.1). In the dried preparation, a varying length of umbilical cord could be defined. During the preparation, special attention was paid on the removal of blood elements from both the vessels and placenta material.

P.C. Leung
Institute of Chinese Medicine, Prince of Wales Hospital,
The Chinese University of Hong Kong,
5/F, School of Public Health Building, Shatin,
NT, Hong Kong SAR
e-mail: pingcleung@cuhk.edu.hk

N. Bhattacharya and P. Stubblefield (eds.), *Regenerative Medicine Using Pregnancy-Specific Biological Substances*,
DOI: 10.1007/978-1-84882-718-9_3,

This procedure could never be thorough, and blood and blood components were always found, although the best quality was described as those cleared of blood and related material.[3]

Probably because of the difficulties involved in the preparation of a neat, clean item and the residual amount of blood seriously affecting the storage, medicinal suppliers have used additional measures like partial steaming, special treatment using rice wine or spice, etc. To simplify storage, the dried preparation is also ground into powder.[4]

3.3 Clinical Use of Human Placenta

From the modern physiological viewpoint, the human placenta is a mysterious source of nutritional supply, which also serves as a filter for the screening out of unwanted metabolites of embryonic activities. Through this filter, complicated exchanges of hormonal substances and cytokines take place. At the stage when the placenta becomes mature, there is no longer direct connection between the maternal and fetal circulation. The placenta works as an efficient transit station for the transfer of oxygen, nutrients, hormones, and cytokines necessary for the growth of the fetus. The human placenta therefore must be rich in this complicated collection of proteins with polypotent functions, which might be related to the steroidal family.

Indeed, the human placenta has been shown to contain various growth factors related to epithelialization, endothelial growth, hemopoiesis, hepatogenesis, and pancreatogenesis.[5]

With this background, the traditional use of the placenta as a polypotent agent for general longevity and for special beauty treatment would gain reasonable justification.

In China, since the practice of "integrated medicine" has become popular, the human placenta has been used in a number of specific situations, either singly or in combination with other herbal items. The main indications include:

1. Anti-infection

The anti-infection property of the human placenta is thought to be due to the presence of globulins with their interferon contents and specific antibodies against the common infections.[6]

2. Immunomodulation

Laboratory experiments have shown that extracts from the human placenta can activate B lymphocytes and promote the secretion of IgM.[7]

3. Hormonal effects

Demonstrable hormonal effects include choriongenic, growth hormone, lactogenic hormone, and testosterone-like activities.[8]

4. Antiasthmatic

Placenta extract has been found to be particularly active against the spasm of bronchial smooth muscles. This has been accompanied by an increase in cAMP, which relaxes smooth muscles.

5. Other indications

Include the treatment of anemia, neurasthenia, and sleep disorders.

3.4 Pharmacological Use of the Human Placenta

Looking through modern Chinese Medicine Pharmacopoeia, the clinical uses of the human placenta could be further understood as follows:

1. Human placenta as sole therapeutic agent
   - Allergic condition: asthma in children and adults[9]
   - Skin diseases: chronic ulcers, resistant eczema; alopecia[10]
   - Degenerative diseases: dementia, coronary heart disease[11]
   - Renal deficiency: renal failure[12]
2. Human placenta as the component of a simple double herb formula
   - Placenta + Ginseng: for the treatment of high-altitude retinal hypoxia[13]
   - Placenta + Cordyceps: for severe gastritis[14]
   - Placenta + Epineurium: for the treatment of menopausal syndrome[15]
3. Placenta used as important component of classic formula
   - Longevity "expert formula" (Tang Dynasty)
   - Rejuvenation formula (Ming Dynasty)
   - Regeneration formula (Qing Dynasty)

4. Use of placenta by current chinese medicine practitioners
   - For gynecological problems, including female infertility, male infertility, and menopausal syndrome[16,17]
   - For respiratory problems, including asthma and tuberculosis[9]
   - For boosting sports performances[7]
   - For local treatment of chronic ulcers[10]

3.5 Adverse Effects of Human Placenta

According to Chinese Medicine Classics, the human placenta has main indications for use as a supplement agent when there is deficiency in the genito-urinary system. The manifestations of such deficiency occur as gynecological syndrome in the female and sexual disorders in the male. Therefore, adverse effects would be expected if the preparations were used otherwise.[18]

In the modern investigations for the safety of traditional Chinese Medicine using rat models, and using different doses of extracts for subcutaneous injection, on short term (14 days) and long term (3–6 months), no adverse effects on body weight, hair integrity, severe allergy, liver, and renal functions were observed.[18]

3.6 Conclusion

Reading through the available literature or the clinical use of the human placenta, one realizes that the efficacy demonstrated is either related to its hormonal contents or hematogenic property.

Using umbilical cord or placenta stem cell directly for the treatment of deficiencies or hemopoietic replacement for hematological malignancies through cell cultures are high-level treatment programs unreachable by the traditional oral use of the human placenta. The oral administration of the placenta could supply some essential components for tissue building like the minerals and some proteins. But the effects after the consumption, through complicated pharmacokinetics and pharmacodynamics processes, are still unknown and unpredictable.

References

1. Yoshida K. *The Mysterious Power of Human Placenta*. Taipei: Shy Mau; 2002.
2. Li Shizhen 1596, Ben Cao Gang Mu, Chinese Classic.
3. Pharmacopoeia of the People's Republic of China. Beijing, china: stationery office; 2005.
4. Zhonghua Ben Cao, Shanghai ke xu ji shu chu ban she, 1999.
5. Liu SL, Tu YX, Liu YS, et al. Clinical use of the human placenta. *China J Chin Mater Med.* 1995;20(1):55-56.
6. Wei LH. Paediatric respiratory infection: treatment using Chinese medicine. *Zhe Jiang J Tradit Chin Med.* 2000; 35(5):16-17.
7. Liu RY, Chen JQ, Liu W, et al. The effect of traditional Chinese medicine "human placenta" on the haematolog parameters of in the track-and field athletes of college. *J Hunan Normal Univ (Med Sci).* 2006;3(3):14-19.
8. Wan CS, Peng YH. Pharmacology and clinical use of Ziheche. *Chin Arch Tradit Chin Med.* 2004;22(10): 1930-1932.
9. OuYang CK, Yang CC. Ziheche and allergic rhinitis. *Zhejiang J Integr Tradit Chin West Med.* 2003;13(1):40-41.
10. OuYang CK, Wu WJ. Ziheche and chromic sore. *Chinese J Tradit Med Sci Technol.* 2001;8(6):357-362.
11. Yuan MS, Chen XL, Chow YP, et al. Ziheche and Senile Dementia. *J Tradit Chin Med Univ Hunan.* 2004; 24(4):39-40.
12. Nie SL, Yao L, Li YC. Ziheche and kidney diseases. *J Med Forum.* 2002;10(14):68-69.
13. Ma Y, Li B, Ha CD, et al. Effect of compound pure dried human placenta and ginseng on retina of healthy young males migrating to highland. *J Clin Rehab Tissue Eng Res.* 2006;10(15):48-52.
14. Wu JY, Di L. Zihche and cordyceps for gastritis xinjiang. *J Tradit Chin Med.* 2005;23(4):48-49.
15. Lor YZ. Ziheche and menopausal syndrome. *New J Tradit Chinese Med.* 1994;12:42.
16. Wu MM. Infertility and Chinese medicine. *J Pract Tradit Chinese Med.* 2006;22(7):419-420.
17. Shi TS, Zhan YY, Xi WX. Ziheche and menstrual disturbances – analysis of 100 cases. *J Pract Tradit Chinese Internal Med.* 2003;17(6):499-500.
18. Dictionary of Chinese Medicine (in Chinese) Shanghai ke xue ji shu chu ban she; 1986.

Part III

Use of Cord Blood in Biochemistry

Use of Umbilical Venous Blood on Assessing the Biochemical Variations of Acid–Base, Nutritional and Metabolic Parameters on Growth-Retarded Fetuses, in Comparison with Gestational Control Cases: A Study

4

Chantal Bon and Daniel Raudrant

4.1 Introduction

Intra-uterine growth retardation (IUGR) affects 5–10% of pregnancies and carries an increased risk of perinatal mortality and morbidity.[1,2] It is suspected when ultrasound examination reveals a fetus who is small for gestational age. However, the general term of IUGR fails to convey the existing heterogeneity of this pathology, whose definition varies according to the length and weight growth curves used, and which etiologies are multiple.[3]

Fetal hypotrophy can be of constitutional origin, but is associated in some cases with fetal distress and vital risks for the infant, requiring close monitoring of pregnancy. The obstetrician needs clinical and/or biological markers to help him identify fetuses who are small but without any particular risk, from those for whom the slowing or stopping of growth indicates a pathological process.[4]

Fetal blood sampling can be performed as early as the second trimester of pregnancy to determine karyotype, and also to measure biological parameters characteristic of both the state of the fetus and the risk of fetal distress.[5,6]

There have been extensive studies on the main biological alterations of fetal blood in case of IUGR[7-16]; however, data are often heterogeneous and vary according to the series under study and the causes of hypotrophy.

C. Bon (✉)
Department of Biochemistry, Hôtel Dieu Hospital, 1, place de l'Hopital, 69288, Lyon Cedex 02, France
e-mail: chantal.bon@chu-lyon.fr

We shall present the results of a study performed on 40 pregnancies complicated with IUGR of various etiologies. On umbilical venous blood (UVB), sampled by cordocentesis, we measured the acid–base balance, the oxygenation level of fetuses and simultaneously the concentrations in several major biochemical components: glucose, pyruvate, lactate, free fatty acids, acetoacetate, beta-hydroxybutyrate, cholesterol. Results were compared with those of a control group of 109 normal fetuses.

The aim of the present study was to investigate about the respiratory and metabolic status of IUGR fetuses, and to identify the distinctive biological disorders associated with growth retardation and fetal distress.

4.2 Materials and Methods

4.2.1 Population Under Study

The pregnant women consulted at the department of Obstetrics of the Hôtel-Dieu hospital in Lyon, France (Professor D. Raudrant). They were informed of the fetal blood sampling and thus gave informed consent. The study was conducted in accordance with the principles of the Helsinki declaration, paragraph II-6, and was approved by the Hospital Ethics Committee.

Gestational age at the time of sampling was calculated from the last menstrual period, and confirmed by early ultrasound examination, performed between 8 and 12 weeks of gestation.

N. Bhattacharya and P. Stubblefield (eds.), *Regenerative Medicine Using Pregnancy-Specific Biological Substances*,
DOI: 10.1007/978-1-84882-718-9_4, © Springer-Verlag London Limited 2011

4.2.1.1 Control Group

It is composed of 109 fetuses, of mean gestational age 26 ± 5.23 weeks of amenorrhea (WA), for whom cordocentesis was performed within the context of prenatal diagnosis; indications were related to some risk incurred by the fetus: suspected infection (toxoplasmosis: 49 cases, rubella: 3 cases, varicella: 18 cases), determination of fetal karyotype (28 cases), risk of thrombocytopenia (11 cases).

These fetuses were not afflicted with the suspected diseases and had a normal karyotype; they all showed normal morphology, growth, and vitality for their gestational age. All infants were born full-term, with a birth weight above the 10th percentile of the department reference curves, and pediatric physical examination confirmed that they were all healthy.

The study group was considered to be a reference population.

4.2.1.2 Pathological Group

It is made up of 40 growth-retarded fetuses with a normal morphology. Fetal blood sampling enabled karyotype analysis, which was found to be normal for all fetuses.

Fetal growth abnormalities were recorded and followed up during successive ultrasound examinations; thanks to the measurement of three characteristic parameters: transverse abdominal diameter (TAD), biparietal diameter (BPD), and femur length; values were compared with those indicated in the department reference tables, according to gestational age.

Fetuses were classified into two groups:

Severe IUGR, 29 cases, of mean gestational age 30.4 ± 4.1 weeks of amenorrhea. BPD and TAD values were below the 10th percentile of the department reference tables (below the 5th percentile in 50% of cases). Diagnosis of hypotrophy was confirmed at birth in reference to Lubchenko's weight curves,[17] and by clinical examination.

Moderate IUGR, 11 cases, of mean gestational age 26.1 ± 4.1 weeks of amenorrhea. Only DAT was below the 10th percentile or nearing the 10th percentile, whereas BPD value ranged within reference limits.

4.2.2 Sampling Procedure

Fetal blood samples were obtained by cordocentesis performed at the umbilical vein under ultrasound guidance, without maternal premedication, and only under local anesthesia at puncture point.[18]

Five hundred microliters of umbilical venous blood were collected in a heparinized syringe for gas and acid–base analyses: pH, partial pressure of CO_2 (pCO_2), bicarbonate and total CO_2 concentrations, partial pressure of oxygen (pO_2), and oxygen saturation (SaO_2). Pyruvate, ketone bodies, free fatty acids (FFA), and cholesterol concentrations were measured on the same sample.

Two hundred microliters of UVB were collected in sodium fluoride for the measurement of glucose and lactate and 200 μL were collected without anticoagulant for the measurement of hCG.

Samples were checked for good quality, particularly the absence of contamination by maternal blood or amniotic fluid.

Immediate checking was made by determining the erythrocyte group iI, then Kleihauer's test, red and white blood cell count, and leukocyte differential count were performed. Serum concentration of hCG in fetal blood was also measured, this test being proposed as an accurate marker of contamination, due to the low hCG level of fetal serum, when compared with maternal blood and/or amniotic fluid.[19]

Fetal blood samples were stored in ice at +4°C and transported without delay to the laboratory. Measurement of pH and blood gases, as well as deproteinization of whole blood with perchloric acid for the quantitative determination of pyruvate and ketone bodies, were performed as soon as the samples were received; deproteinized supernatants were decanted and kept at −20°C.

Blood was centrifuged at 4°C for further analyses. Determination of glucose, lactate, cholesterol, and hCG concentrations was immediately performed; determination of free fatty acids level was performed later on plasma stored at −20°C. Storage conditions of samples were checked for measurements performed after freezing.

4.2.3 Analytical Methods

Analytical methods were regularly used by the laboratory.

Blood gas and acid–base analyses (pH, pO_2, pCO_2, bicarbonate, total CO_2, SaO_2) were performed using the ABL 300 analyzer (Radiometer, Copenhagen, Denmark).

Glucose and lactate concentrations were assayed using the Ektachem 500 automated analyzer (Kodak, New York, USA) with an enzymatic method that uses, respectively, glucose oxidase (*EC 1.1.3.4*) and lactate oxidase (*EC 1.13.12.4*) and reflectometry measurement (reference range in adult blood: 3.6–5.8 mmol/L for glucose and 0.7–2.1 mmol/L for lactate).

Total concentration of free fatty acids was measured with a manual colorimetric enzymatic assay, using an acyl-coenzyme A synthetase (*EC 6.2.1.3*), an acyl-coenzyme A oxidase (*EC 1.3.3.6*), and a peroxidase (*EC 1.11.1.7*) (Biomérieux, Marcy l'Etoile, France).[20]

Pyruvate and ketone bodies assays were performed with an enzymatic fluorimetric micromethod, based on the measure of NADH fluorescence.[21] Fluorescence was quantified in a Kontron SFM spectrofluorometer (Kontron, Zurich, Switzerland). Excitation and emission wavelengths were, respectively, 340 and 460 nm. Reagents (lactate dehydrogenase (*EC 1.1.1.27*), beta-hydroxybutyrate dehydrogenase (*EC 1.1.1.30*), NADH, NAD) were supplied by Boehringer (Mannheim, Germany).

Reference values in adult blood, with the assays used, range from 0.030 to 0.100 mmol/L for pyruvate, 0.018 to 0.078 mmol/L for acetoacetate, and 0.050 to 0.100 mmol/L for beta-hydroxybutyrate.

Cholesterol concentrations were measured with an enzymatic assay using a cholesterol esterase (*EC 3.1.1.13*) and a cholesterol oxidase (*EC 1.1.3.6*) (Biomérieux, Marcy l'Etoile, France); reference range in adult blood is 3.6–7 mmol/L.

Determination of hCG level was performed with the IMx automated analyzer (Abbott Diagnostics, Abbott Park, Chicago, USA) using a microparticle enzyme immunoassay.

4.2.4 *Statistics*

Comparative study of the results obtained in the different groups of fetuses was carried out using Student's t-test (for unpaired samples) and Mann–Whitney U test (nonparametric method).

The possible relationship between the various parameters measured in UVB was investigated by calculating the linear correlation coefficient r and Spearman rank correlation coefficient rs. Fisher's exact test was used to evaluate the statistical signification of correlations.

Linear regression analysis was used in the control group to investigate the possible changes of the parameters according to the gestational age (expressed in weeks of amenorrhea).

A value of p below 0.05 was considered to be statistically significant.

4.3 Results

4.3.1 *Gaseous and Acid–Base Parameters in Umbilical Venous Blood (Table 4.1, Fig. 4.1)*

4.3.1.1 Control Population ($n = 109$)

pH and pO_2 decreased during the gestational period under study (17–41 weeks of gestation), and significant inverse correlation was found between gestational

Table 4.1 Gaseous and acid–base parameters in umbilical venous blood

	Controls	Severe IUGR	Moderate IUGR
n	109	29	11
pH	7.309 0.059	7.199**** 0.147	7.322 0.066
pCO_2 kPa	5.98 0.84	8.32**** 3.23	5.88 0.73
HCO_3^- mmol/L	21.99 1.83	22.60 2.33	22.26 1.46
pO_2 kPa	5.89 1.62	3.35**** 1.53	5.69 1.50
SaO_2	0.73 0.18	0.37**** 0.25	0.74 0.18

Results are expressed in means and standard deviations
n = number of samples
**** $p < 0.0001$

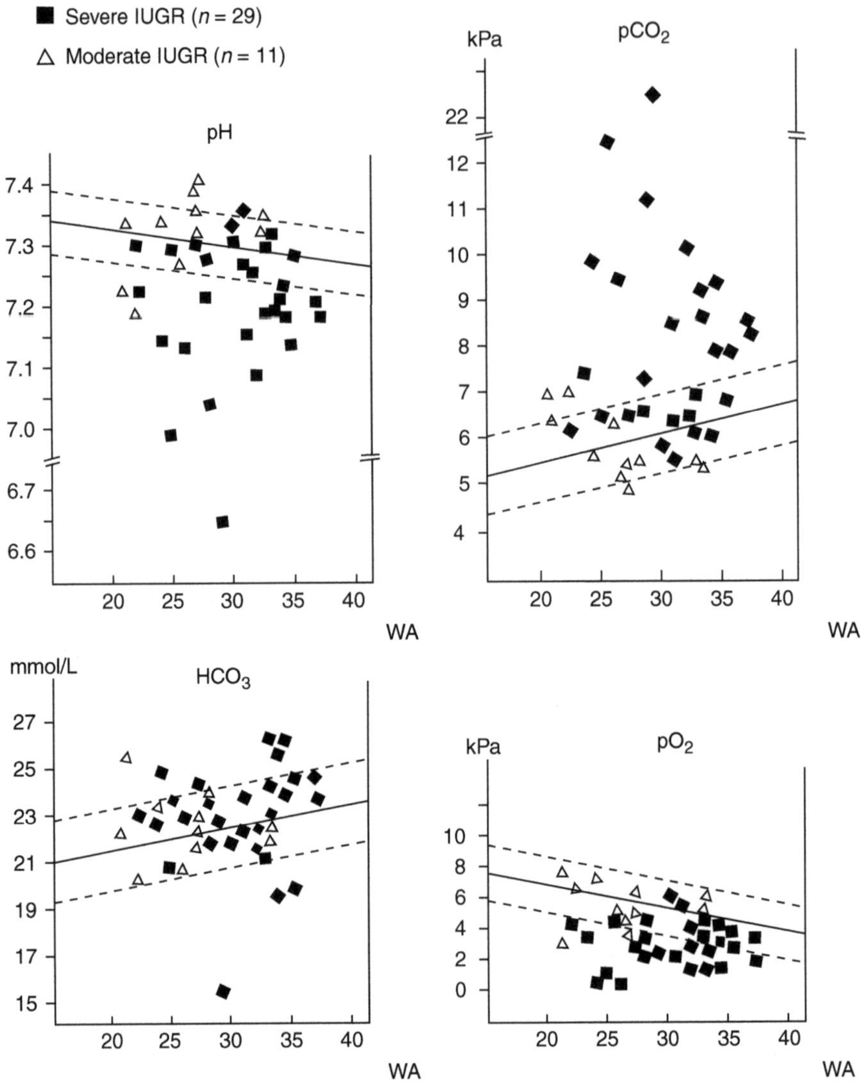

Fig. 4.1 Growth-retarded fetuses. pH, pCO_2, bicarbonate, and pO_2 values in umbilical venous blood, plotted on the reference range for the gestational age, established in control group (mean value ±1 standard deviation)

age and pH on the one hand ($r = -0.301$; $p = 0.0016$), pO_2 on the other hand ($r = -0.374$; $p < 0.0001$).

pCO_2 increased significantly ($r = 0.427$; $p < 0.0001$), as well as bicarbonate concentration, which increased moderately but progressively during gestation ($r = 0.202$; $p = 0.036$).

pH was significantly correlated with pCO_2 ($r = -0.827$; $p < 0.0001$), and with bicarbonate concentration ($r = 0.359$; $p = 0.0001$).

4.3.1.2 Pathological Population

The results obtained in growth-retarded fetuses were compared with those of control fetuses of the same gestational age.

Acidemia and hypoxemia were defined by pH and pO_2 values more than one standard deviation below the mean, and hypercapnia by pCO_2 values greater than one standard deviation above the mean; reference range of bicarbonate concentration was defined by the mean value ±1 standard deviation (means calculated for the gestational age studied).

Severe Growth Retardation ($n = 29$)

pH, pO_2, and SaO_2 for the whole group decreased significantly, and pCO_2 for the whole group increased significantly, when compared with the control group normal values.

Prevalence of hypoxemia is 55%, and that of hypercapnia is identical, with 50% of pCO2 values greater than two standard deviations.

Acidemia occurs in 58% of cases.

Mean bicarbonate concentration was not significantly different from that of the control population; however, bicarbonate level decreased in three cases and increased in five cases.

The transfer of pH and pCO_2 values on a Davenport diagram showed that acidosis was mixed, gaseous, and metabolic, in most cases; pH was significantly correlated with pCO_2 ($r = -0.975$; $p < 0.0001$), and also with plasma bicarbonate concentration ($r = 0.605$; $p = 0.0005$).

Moderate Growth Retardation ($n = 11$)

Mean values of pH, pCO_2, HCO_3^-, pO_2, and SaO_2 were not significantly different from that of the control population. However, the results revealed two cases of acidemia associated with hypercapnia as well as two cases of isolated hypoxemia, pO_2 values being 3.15 and 3.76 kPa.

4.3.2 *Metabolic Parameters in Umbilical Venous Blood* *(Table 4.2, Fig. 4.2)*

4.3.2.1 Control Population ($n = 109$)

Glucose, free fatty acids, acetoacetate, beta-hydroxybutyrate, and cholesterol concentrations were found stable, in UVB within 17 and 41 weeks of gestation (r respectively 0.075; 0.020; 0.018; 0.030; 0.052).

On the other hand, lactatemia and pyruvatemia increased regularly during this gestational period, the coefficients of correlation with gestational age being, respectively, 0.376 ($p < 0.0001$) and 0.430 ($p < 0.0001$).

4.3.2.2 Pathological Population

Results were interpreted in comparison with the normal range established in the control group. The interval of normal values was set within ±1 standard deviation, on each side of the mean, except in the case of free fatty acids, acetoacetate and beta-hydroxybutyrate for which the limits were set to the 10th and 90th percentiles of the group's values.

Reference values for lactate and pyruvate were established according to the gestational age.

Severe Growth Retardation ($n = 29$)

Glucose

Mean UVB glycemia was below that of the control group by 11% and significantly lower. Decrease in blood glucose affects 38% of fetuses, the lower concentration being 2 mmol/L.

A significant correlation was established between glucose and pO_2 in UVB ($r = 0.585$; $p = 0.0009$).

Lactate and Pyruvate

Lactatemia and pyruvatemia were on average significantly higher than that of the control group. Lactate increase affected 45% of fetuses, with a simultaneous pyruvate increase in 41% of cases. There was a wide dispersion of lactate results, with approximately 40% of the values above two standard deviations.

A significant correlation was established between pH and blood lactate level ($r = -0.787$; $p < 0.0001$).

Free Fatty Acids

Mean free fatty acids concentration was found significantly higher than in the control group. The increase in free fatty acids level in UVB affected 50% of the group's fetuses, but remained moderate and did not exceed 1.5 times the normal limit in most cases.

However, it increased all the more so as glycemia was low, and a significant inverse correlation was established between glucose and FFA concentrations ($r = -0.852$; $p < 0.0001$) (Fig. 4.3).

Ketone Bodies

Mean acetoacetate and beta-hydroxybutyrate concentrations were not significantly different from that of the control group.

Table 4.2 Metabolic parameters in umbilical venous blood

		Controls	Severe IUGR	Moderate IUGR
n		109	29	11
Glucose		3.46	3.08****	3.44
mmol/L		0.34	0.44	0.32
n		109	29	11
Lactate		1.41	2.64****	1.65
mmol/L		0.48	1.46	0.43
n		104	29	11
Pyruvate		0.022	0.035****	0.025
mmol/L		0.009	0.014	0.007
n		104	29	11
Lactate	ratio	64.10	71.38	66.17
Pyruvate		17.12	16.10	9.09
n		108	29	11
Free fatty acids		0.124	0.189****	0.135
mmol/L		0.047	0.051	0.039
n		101	29	11
Beta-hydroxybutyrate		0.319	0.320	0.239
mmol/L		0.234	0.126	0.119
n		101	29	11
Aceto-acetate		0.113	0.115	0.083
mmol/L		0.063	0.049	0.031
n		101	29	11
Beta-hydroxybutyrate	ratio	2.82	2.92	2.72
Aceto-acetate		1.18	0.74	0.60
n		104	29	11
Cholesterol		1.65	1.17****	1.87
mmol/L		0.36	0.43	0.23

Results are expressed in means and standard deviations
n = number of samples
**** $p < 0.0001$

Cholesterol

Mean cholesterolemia was below that of the control group by 30% and significantly lower. Decrease in serum cholesterol affected almost 70% of this group's fetuses, the lower concentration being 0.65 mmol/L.

A significant correlation was established between cholesterolemia and umbilical venous pO_2 ($r = 0.823$; $p < 0.0001$) (Fig. 4.4).

Moderate Growth Retardation ($n = 11$)

In cases where growth retardation had been classified as moderate, the mean values of the studied parameters were not significantly different from those of the control population.

One case of moderate hyperlactatemia associated with hyperpyruvatemia has nevertheless been noted.

Fig. 4.2 Growth-retarded fetuses. Glucose, lactate, free fatty acids, and cholesterol values in umbilical venous blood, plotted on the reference range for the gestational age, established in control group (glucose, lactate, cholesterol: mean value ±1 standard deviation; free fatty acids: 10th to 90th percentiles)

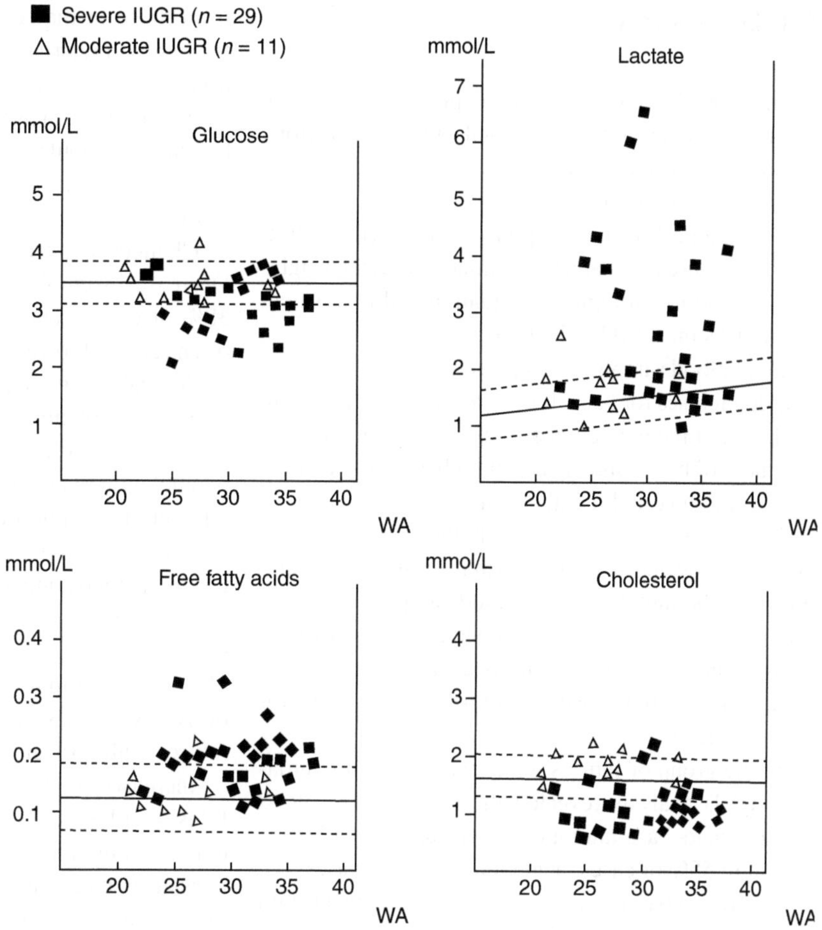

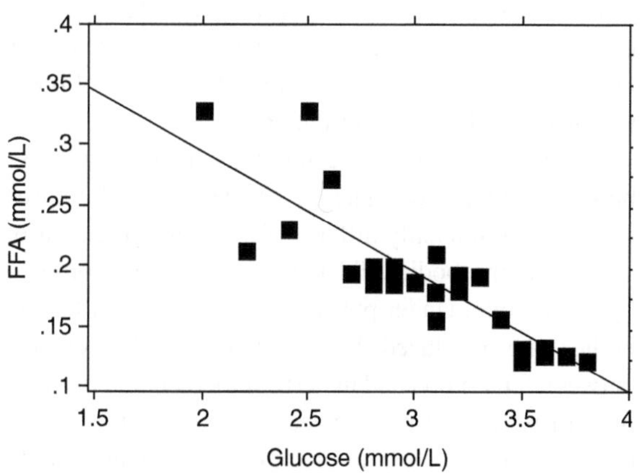

Fig. 4.3 Fetuses with severe growth retardation. Correlation between glucose and free fatty acids concentrations in umbilical venous blood ($n = 29$; $r = -0.852$)

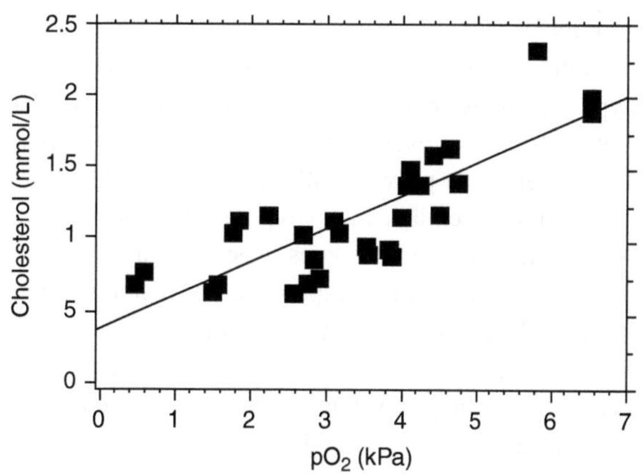

Fig. 4.4 Fetuses with severe growth retardation. Correlation between cholesterol concentrations and pO_2 values in umbilical venous blood ($n = 29$; $r = 0.823$)

4.4 Discussion

Regulation of fetal growth is a complex process, which involves genetic factors, maternal nutritional factors, circulatory and placental factors, as well as fetal factors, particularly hormonal.[3]

The diagnosis of fetal hypotrophy is based on ultrasound measurements and the estimation of fetal weight; it must lead to etiologic investigation and assessment of the risks incurred by the fetus.

Fetal blood sampling was performed on 40 growth-retarded fetuses for karyotype determination; access to the blood of the fetal compartment facilitated the measurement of biochemical variables characteristic of the fetal respiratory and metabolic status.

Results were compared with those of a control group of 109 fetuses with normal growth. The fetal origin of the umbilical cord blood samples was carefully checked in our study protocol, notably with the measurement of hCG serum concentration; reference values in fetal blood were established previously for this parameter.[22]

In the group of fetuses with severe growth retardation, UVB analysis revealed the frequency of blood gas and acid–base alterations: hypoxemia and hypercapnia in 55% of cases, acidemia in 58% of cases.

On the whole, these results are in agreement with those of previous studies;[8,11,14,23] however, the recorded level of acidemia was often more severe than in other series,[9,10,12,24] in relation to increased pCO_2 values. Our pO_2 values interval agrees with the results of other authors.[9-12]

Abnormalities observed in UVB, hypoxemia, and hypercapnia are probably in relation to an alteration of maternal-fetal gas exchange across the placental barrier.

The fetus can provisionally adapt to pO_2 decrease, and then hypoxemia leads to hypoxia. The metabolism of peripheral tissues becomes anaerobic and the production of lactic acid is responsible for metabolic acidosis. Umbilical venous lactatemia was found increased in 45% of fetuses, along with in most cases a concomitant increase in pyruvatemia, which is secondary to hypoxia and to the slowing down of Krebs' cycle.

CO_2 accumulation leads to an accumulation of H^+ ions, resulting in a rapid pH drop and in the installation of respiratory acidosis. It was only rarely compensated by an increase in the bicarbonate plasma concentration, in which mean levels were not significantly different from those of the control group. It is very likely that fetal growth retardation comes with altered renal function,[25] and with a decrease in regulatory capacities, reabsorption of bicarbonate, and secretion of H^+ ions by the renal tubules.

Acidosis was mixed, gaseous, and metabolic, in most cases.

Changes in nutritional parameters were established. Decrease in umbilical venous glycemia, observed in 38% of cases, remained discrete; earlier studies[13,26,27] often reported lower values than ours.

Several causes can be at the origin of blood glucose decrease: depletion of the liver reserves of glycogen, reduction in the transplacental transfer of glucose, in parallel with reduced oxygen diffusion, fetal or placental overconsumption of glucose, due to the anaerobic metabolism, and deficiency in gluconeogenesis capacities in the fetal compartment. Marconi et al.'s findings[28] proved an impaired use of gluconeogenic precursors in growth-retarded fetuses. On the other end, placental glucose consumption was not found altered in a group of IUGR fetuses, when compared with a group of fetuses with normal growth.[29]

Increased FFA levels in UBV, recorded in some fetuses with severe growth retardation, remained moderate; stimulation of lipolysis could constitute a compensatory mechanism for the decrease in glycemia, due to the significant inverse correlation between glucose and FFA concentrations. Our results differ from those of Economides et al.,[15] who did not establish any significant difference between FFA concentrations in growth-retarded fetuses (32 cases) and in fetuses with normal growth (54 cases). However, increased FFA level in the amniotic fluid was reported in case of pregnancies complicated with IUGR.[30]

Umbilical venous concentrations in beta-hydroxybutyrate and acetoacetate in growth-retarded fetuses were not significantly different from that of the control group; ketone bodies did not increase in response to hypoglycemia. Under physiological conditions, ketone bodies are produced by the fetus[31] and exchange between fetal blood and maternal blood operate by simple diffusion across the placenta.[32] Ketone bodies are the preferential substrate of some fetal organs, mainly the brain,[33] and it is likely that they are used as energy substrates by IUGR fetuses, which results in speeding up their turnover and their metabolic clearance.

No study has reported ketone bodies concentrations in the blood of fetuses with growth retardation. A study carried out during the first postnatal week showed the absence of ketogenic response to low blood glucose levels, in a group of 33 hypotrophic children, when compared with 218 children of appropriate weight.[34] De Boissieu et al. suggested the possibility of liver deficiency and the inability for premature infants to convert FFA into ketone bodies.[35]

Umbilical venous cholesterolemia was found significantly decreased in the group of fetuses with severe IUGR, and in almost 30% of cases it was less than two standard deviations below the control group normal values. Most studies on fetal lipids during pathological pregnancies have been carried out at the time of birth and at full term. Spencer et al.[36] showed a decrease in the mean concentrations of total cholesterol and cholesterol esters, in umbilical venous blood sampled beyond 36 weeks of gestation, just after delivery, during 16 pregnancies complicated with growth retardation at the third trimester, compared with 42 normal pregnancies.

Cholesterol is mainly synthesized in the fetal compartment, because maternal cholesterol cannot easily cross the placenta, except during early gestation. Physiologically, cholesterol level is much lower in fetal blood than in maternal blood.[31]

In case of IUGR, insufficient production of cholesterol due to liver deficiency can account for the decrease in fetal cholesterolemia. Roberts et al.[37] suggested the possibility of a fetal liver disorder indicated by an increase in LDH and gamma-glutamyltransferase activities in UVB.

We noticed the frequency of cases of hypoxemia. Cholesterol synthesis requires satisfactory oxygenation conditions, and actually, a significant correlation was established between cholesterolemia and pO_2 in UVB; the most hypoxemic fetuses were those with the lowest cholesterol concentration.

As a result of low cholesterol level, the synthesis of cell membranes may be hindered, and fetal growth slowed down; the fundamental role of cholesterol in embryonic development and particularly in cerebral growth has been emphasized.[38]

Lemery et al.[39] showed a decrease in the fluidity of cell membranes in fetuses with IUGR, in relation to a decrease in cholesterolemia.

Decrease in cholesterol can also be responsible for insufficient steroid production by the fetal-placental unit, particularly estrogens, which play a part in the expansion of maternal plasma volume and facilitate uteroplacental circulation and fetal nutrition.

When growth retardation was classified moderate, UVB biological modifications were less frequent and mainly concerned blood gas abnormalities, only rarely suggesting metabolic disorders. The retarded growth, which had probably appeared late in pregnancy, was not associated with fetal distress.

Etiology of growth retardation could be identified in certain pregnancies. Vascular pathology was diagnosed in 41% of cases: either maternal hypertension with nephropathy and toxemia in ten cases, or placental lesions (infarct and thromboses of villous vessels) without hypertension in six cases. Hypotrophy was probably the consequence of chronic fetal malnutrition related to a deficiency in energy substrates brought by placental transfer.

Infectious fetal cause was identified for three patients, hypotrophy being the result of a viral infection by cytomegalovirus.

Maternal smoking could be the cause in five pregnancies, tobacco being known to have harmful effects on uteroplacental and umbilical circulations, and consequently on fetal nutrition.

In the other pregnancies, etiology of growth retardation could not be explicit, and IUGR was probably idiopathic.

It is also important to emphasize that the series under study included approximately 30% of cases for which growth retardation was revealed on ultrasound examination, whereas the biological tests performed on UBV showed no abnormality. These pregnancies had a favorable outcome but the birth weight was indeed found below or at the limit of normal values for the gestational age under consideration. Under these conditions, hypotrophy is well tolerated, it is probably constitutional, of familial origin, and the term "small-for-gestational-age fetus" would be more appropriate than "growth-retarded fetus."

4.5 Conclusion

This study showed the interest of some biological markers of UVB, during pregnancies complicated with intrauterine growth retardation, a condition distinguished by a wide heterogeneity.

The biological profile provided further information in addition to the ultrasound diagnosis and helped to identify a group with severe abnormalities and at risk of complications sooner or later.

pO_2 is a parameter with an essential interest; the state of hypoxemia, observed in 58% of severe IUGR cases, was responsible for a shift of the oxidative metabolism toward the anaerobic pathway, and an accumulation of lactic acid. In addition to metabolic acidosis, compensated with difficulty by the hypotrophic fetus, respiratory acidosis occurred when placental elimination of CO_2 was insufficient.

The most distinctive acid–base parameters when compared with the control group results were pH, pCO_2, and lactatemia; the frequency and amplitude of their variations accounted for the high risk of mixed acidosis, both gaseous and metabolic.

Energetic parameters were less severely altered, in regard to their mean value, than acid–base parameters. The umbilical venous concentration in glucose, which is an essential energy substrate for the fetus, only decreased in 38% of severe IUGR cases, and it seemed that the fetus retained regulation capacities to protect itself against hypoglycemia. It can also use other energy substrates, as is shown by the increase in the umbilical venous free fatty acids concentration in some pathological pregnancies.

Our results established a significant relationship between hypoxemia and hypocholesterolemia in UVB. Fetal cholesterol is not much correlated with maternal cholesterol and is relevant in that it is a specific marker of the fetus' metabolism. Decrease in cholesterolemia, which occurs frequently in case of growth retardation, indicates a metabolic disorder.

It is very likely that cellular acidosis is unfavorable to the normal course of metabolisms and hinders the synthesis of the normal constituents of cell membranes. The slowing down of growth can also be a protection and survival mechanism, set up by the fetus, in response to the nutrient supply deficiency.

The positive aspect of the study is that it was possible to measure simultaneously acid–base and blood gas parameters, as well as energy substrates and catabolites. However, the information of our work must be balanced in view of the instant nature of our results; fetal blood sampling have proved delicate to perform, not without the risk for the infant, and is difficult to repeat.[40] Our blood gas results reflected an instant measurement and our nutritional results conveyed the balance between fetal production, maternal production, and fetal–maternal exchange.

References

1. Starfield B, McCormick M. Mortality and morbidity in infants with intrauterine growth retardation. *J Pediatr*. 1982; 101:978.
2. Heinomen K, Matilainen R, Koski H, Launiala K. Intrauterine growth retardation (IUGR) in pre-term infants. *J Perinat Med*. 1985;13:171-178.
3. Gluckman PD, Harding JE. The physiology and pathophysiology of intrauterine growth retardation. *Horm Res*. 1997;48(Suppl):11-16.
4. Jones RAK, Roberton NRC. Problems of the small-for-dates baby. *Clin Obstet Gynaecol*. 1984;11:499-524.
5. Meizner I, Glezerman M. Cordocentesis in the evaluation of the growth-retarded fetus. *Clin Obstet Gynecol*. 1992;35: 126-137.
6. Pardi G, Cetin I, Marconi AM, et al. Diagnostic value of blood sampling in fetuses with growth retardation. *N Engl J Med*. 1993;328:692-696.
7. Pardi G, Buscaglia M, Ferrazzi E, et al. Cord sampling for the evaluation of oxygenation and acid-base balance in growth-retarded human fetuses. *Am J Obstet Gynecol*. 1987; 157:1221-1228.
8. Pearce JM, Chamberlain GVP. Ultrasonically guided percutaneous umbilical blood sampling in the management of intra-uterine growth-retardation. *Br J Obstet Gynecol*. 1987; 94:318-321.
9. Weiner CP, Williamson RA. Evaluation of severe growth-retardation using cordocentesis – hematologic and metabolic alteration by etiology. *Obstet Gynecol*. 1989;73(2):225-229.
10. Ribbert LS, Snijders R, Nicolaides KH, Visser GHA. Relationship of fetal biophysical profile and blood gas values at cordocentesis in severely growth-retarded fetuses. *Am J Obstet Gynecol*. 1990;163:569-571.
11. Nicolini U, Nicolaidis P, Fisk NM, Vaughan JF, Fusi L, Gleeson R. Limited role of fetal blood sampling in prediction of outcome in intra-uterine growth retardation. *Lancet*. 1990;336:768-772.
12. Hsieh TT, Kuo DM, Lo LM, Chiu TH. The value of cordocentesis in management of patients with severe preeclampsia. *Asia-Oceania J Obstet Gynecol*. 1991;17(1):89-95.
13. Economides DL, Nicolaides KH. Blood glucose and oxygen tension levels in small for gestational age fetuses. *Am J Obstet Gynecol*. 1989;160:385-389.
14. Nicolini U, Hubinont C, Santolaya J, Fisk NM, Rodeck CH. Effects of fetal intravenous glucose challenge in normal and growth-retarded fetuses. *Horm Metab Res*. 1990;22: 426-430.
15. Economides DL, Crook D, Nicolaides KH. Investigation of hypertriglyceridemia in small for gestational age fetuses. *Fetal Ther*. 1988;3:165-172.
16. Cetin I, Ronzoni S, Marconi AM, Perugino G, Corbetta C, Battaglia FC. Maternal concentrations and fetal-maternal concentration differences of plasma amino-acids in normal and intra-uterine growth-restricted pregnancies. *Am J Obstet Gynecol*. 1996;174:1575-1583.
17. Lubchenko LO, Hansman C, Dressler M, Boyd E. Intra-uterine growth as estimated from liveborn birth-weight data at 24 to 42 weeks of gestation. *Pediatrics*. 1963;32:793-800.

18. Daffos F, Capella-Pavlovsky M, Forestier F. Fetal blood sampling via the umbilical cord using a needle guided by ultrasound. *Prenat Diagn.* 1983;3:271-277.
19. Dommergues M, Bunduki V, Muler F, Mandelbrot L, Morichon-Delvallez N, Dumez Y. Serum hCG assay: a method for detection of contamination of fetal blood samples. *Prenat Diagn.* 1993;13:1043-1046.
20. Okabe H, Uji Y, Nagashima K, Noma A. Enzymatic determination of free fatty acids in serum. *Clin Chem.* 1980;26: 1540-1543.
21. Olsen C. An enzymatic fluorimetric micromethod for the determination of aceto-acetate, beta-hydroxybutyrate, pyruvate and lactate. *Clin Chem Acta.* 1971;33:293-300.
22. Bon C, Gelineau MC, Raudrant D, Pichot J, Revol A. Fœtal blood human chorionic gonadotropin concentrations in normal and abnormal pregnancies. *Immunoanal Biol Spec.* 1999;14:37-46.
23. Rizzo G, Montuschi P, Capponi A, Romanini C. Blood levels of vasoactive intestinal polypeptide in normal and growth retarded fetuses: relationship with acid-base and haemodynamic status. *Early Hum Dev.* 1995;41:69-77.
24. Yoneyama Y, Wakatsuki M, Sawa R, et al. Plasma adenosine concentration in appropriate and small for gestational age fetuses. *Am J Obstet Gynecol.* 1994;170:684-688.
25. Merlet-Benichou C, Leroy B, Gilbert T, Lelièvre-Pegorier M. Retard de croissance intra-utérin et déficit en néphrons. *Med Sci (Paris).* 1993;9:777-780.
26. Nicolini U, Hubinont C, Santolaya J, Fisk NM, Coe AM, Rodeck CH. Maternal-fetal glucose gradient in normal pregnancies and in pregnancies complicated by alloimmunization and fetal growth retardation. *Am J Obstet Gynecol.* 1989;161:924-927.
27. Hubinont C, Nicolini U, Fisk NM, Tanninandorm Y, Rodeck CH. Endocrine-pancreatic function in growth-retarded fetuses. *Obstet Gynecol.* 1991;77(4):541-544.
28. Marconi AM, Cetin I, Davoli E, et al. An evaluation of fetal glucogenesis in intrauterine growth-retarded pregnancies. *Metal Clin Exp.* 1993;42(7):860-864.
29. Magnusson AL, Powell T, Wennergren M, Jansson T. Glucose metabolism in the human preterm and term placenta of IUGR fetuses. *Placenta.* 2004;25:337-346.
30. Urban J, Iwaszkiewicz-Pawlowska A. Concentration of free fatty acids in amniotic fluid and maternal and cord serum in cases of intrauterine growth retardation. *J Perinat Med.* 1986;14:259-262.
31. Bon C, Raudrant D, Golfier F, et al. Feto-maternal metabolism in human normal pregnancies: study of 73 cases. *Ann Biol Clin.* 2007;65(6):609-619.
32. Pere MC. Materno-fœtal exchanges and utilisation of nutrients by the fœtus: comparison between species. *Reprod Nutr Dev.* 2003;43:1-15.
33. Battaglia FC, Meschia G. Fetal nutrition. *Ann Rev Nutr.* 1988;8:43-61.
34. Hawdon JM, Ward Platt MP. Metabolic adaptation in small for gestational age infants. *Arch Dis Child.* 1993;68: 262-268.
35. De Boissieu D, Rocchiccioli F, Kalach N, Bougnères PF. Ketone body turnover at term and in premature newborns in the first two weeks after birth. *Biol Neonate.* 1995;67: 84-93.
36. Spencer JAD, Chang TC, Crook D, et al. Third trimester fetal growth and measures of carbohydrate and lipid metabolism in umbilical venous blood at term. *Arch Dis Child.* 1997;76:21-25.
37. Roberts A, Nava S, Bocconi L, Salmona S, Nicolini U. Liver function tests and glucose and lipid metabolism in growth-retarded fetuses. *Obstet Gynecol.* 1999;94:290-294.
38. Roux C, Wolf C, Mulliez N, Gaona W, Cormier V, Chevy F. Role of cholesterol in embryonic development. *Am J Clin Nutr.* 2000;71(Suppl 5):1270-1279.
39. Lemery DJ, Beal V, Vanlieferinghen P, Motta C. Fetal blood cell membrane fluidity in small for gestational age fetuses. *Biol Neonate.* 1993;64:7-12.
40. Antsaklis A, Daskalaris G, Papantoniou N, Michalas S. Fetal blood sampling-indication-related losses. *Prenat Diagn.* 1998;18:934-940.

Part IV

Use of Cord Blood as Blood Substitute

5 Umbilical Cord Blood Transfusion and Its Therapeutic Potentialities

Patricia Pranke and Tor Onsten

Abbreviations

DM	Diabetes mellitus
GVHD	Graft-versus-host disease
Hb	Hemoglobin
HbA	Adult hemoglobin
HbF	Fetal hemoglobin
Hct	Hematocrit
MCH	Mean corpuscular hemoglobin
MCHC	Mean corpuscular hemoglobin concentration
MCV	Mean corpuscular volume
RBC	Red blood cell
UCB	Umbilical cord blood

Umbilical cord blood (UCB) is being used around the world for stem cell transplants. However, this source could be used in transfusions and its practical use should be encouraged, since the needs of transfusion are increasing considering the possibility of wars, terrorism, natural disasters, and epidemics around the world.

There have been several clinical trials with patients in reference to autologous and allogeneic umbilical cord blood transfusion. Despite the fact that autologous methods are more common throughout the world, the allogeneic use has been studied in order for this transfusion source to be applied to both children and adults.

It is important to consider the hematological particularities of UCB, such as higher levels of hemoglobin, hematocrit, mean corpuscular volume, leukocytes, and fetal hemoglobin, and low levels of coagulation factors. The advantages of using umbilical cord blood in transfusions include diminished expression of erythrocyte antigens, low levels of immunoglobulin, and also an absence of natural antibodies. On the other hand, UCB also has immature nucleated cells with engraftment capacity, which can provoke graft-versus-host disease (GVHD) without leukoreduction. However, blood irradiation before the use of UCB eliminates the risk of GVHD, making the use of allogeneic cord erythrocytes a therapeutically useful option, especially for preterm and lower weight newborns.

5.1 Introduction

Since 1988,[1] UCB has been routinely used in transplants as an alternative to bone marrow transplants, and UCB banks are being built around the world.

To cryopreserve the stem cells, among leukocytes, during the preparation of UCB, erythrocytes, platelets, and plasma are discarded. All attention on UCB use has been given to stem cell transplants only. However, stem cells constitute 0.01% of the nucleated cells of umbilical cord whole blood and the rest of the blood (99.99%) is apparently discarded.[2] Until now, this material has been underestimated as a source of other blood components for autologous and allogeneic transfusion.

Placental vessels contain anything from 75 to 125 mL of blood. Therefore, it has been considered that using this otherwise wasted resource could serve as a means of autologous[3] and, most recently, allogeneic transfusions.[4-6] Taking into consideration that about 100 mL of UCB from each delivery is discarded

P. Pranke (✉)
Hematology Laboratory, Federal University of Rio Grande do Sul, Av Ipiranga, 2752, Porto Alegre, Rio Grande do Sul 90610-000, Brazil
e-mail: patriciapranke@ufrgs.br

N. Bhattacharya and P. Stubblefield (eds.), *Regenerative Medicine Using Pregnancy-Specific Biological Substances*, DOI: 10.1007/978-1-84882-718-9_5, © Springer-Verlag London Limited 2011 and multiplying that by the number of daily deliveries, it is easy to estimate the huge wasted volume, while blood banks are suffering from a lack of donors.

Since UCB volume collected is generally small, initially its use for adult transfusion will be limited. On the other hand, it is sufficient for newborns and low weight children, and it has been successfully used in individual cases around the world. Some estimates indicate that around 80% of infants with birth weights of less than 1,500 g receive at least one red cell transfusion.[7-10]

So, to verify the feasibility of using placental and UCB as a new source of transfusion, it is important to evaluate theoretical advantages and disadvantages, as well as consider published and known experience about the use of UCB in transfusions.

The aim of this chapter is to evaluate the safe application and the therapeutic viability of UCB components for transfusions, based on previous evidence.

5.2 Umbilical Cord Blood as a Source of Components in Transfusional Therapy

There is a rising interest in increasing the therapeutic use of UCB, besides using it as a source of hematopoietic stem cell transplants. One of the alternatives is its use for transfusion goals. This alternative is very interesting, as UCB is abundant and most of the time it is discarded and, consequently, underused.

Autologous blood is widely accepted as a preferred source of red blood cells when blood transfusion is clinically indicated in children and adults because it diminishes problems inherent to allogeneic transfusion, including infectious disease transmission and transfusion reactions. UCB obtained at delivery after cord clamping has been suggested as a source of autologous blood for transfusion in neonates,[11] mainly in preterm and low birth weight newborns, where blood transfusions are often necessary.[6-9,11]

In view of the usual blood volume transfused in neonates being approximately 10–20 mL/kg body weight,[11,12] sufficient UCB for at least one or two autologous transfusions even in extremely low birth weight neonates can therefore be expected to be available.[11,13] Thus, UCB is a feasible alternative source of erythrocytes, as most newborns of 24–27 weeks' gestation will require red blood cell transfusions.[12]

There have been several clinical trials in newborn, pediatric, and adult patients referring to not only autologous but also allogeneic UCB RBC transfusions.[4,6] Notwithstanding the fact that UCB has been considered a feasible alternative source of blood for transfusions, two limitations for its use are its small blood volume, compared to adult blood collected and the higher risk of bacterium contamination. To compensate for the small volume of cord blood collected, it is important to identify the advantage of cord blood as, for example, its immunological particularities.

Features of UCB from a transfusion practice point of view will be analyzed as follows. It is important to compare it to blood obtained from adult donors. The main components used are red blood cell concentrate (RBC), platelet concentrate, and plasma. The most potential and useful component is RBC. The small volume of cord blood probably does not contain enough platelets for transfusion.

The neonate plasma is poor in coagulation factors when compared to adult blood. On the other hand, other features show potential advantages, such as the weak expression of some erythrocyte antigens and the absence of anti-erythrocyte antibodies. The high concentration of progenitor cells brings a theoretical risk of higher implantation of nucleated cells in the patient, mainly in a immunosuppressed receptor leading to chimerism or, even, GVHD. However, this risk could be diminished significantly with a leukoreduction process.

5.3 Human Umbilical Cord Blood Features

5.3.1 Hematologic Parameters of Newborn Blood

Several hematologic parameters are different in neonate blood when compared to adult blood. Among these are the blood volume and erythrocyte mass per kilogram of body weight, as well as hemoglobin concentration, hematocrit, and mean corpuscular volume (MCV), which are higher in newborn than adult blood. Erythrocyte survival in neonate blood is about 60 days, reduced when compared to adult blood. The reduced lifespan of newborn erythrocytes (60–80 days) is most likely explained by the increased osmotic fragility caused by the increased MCV.[14]

The leukocyte number is also higher in newborn blood, mainly mononuclear cells. On the other hand, there is no difference in platelet numbers between newborn and adult blood. The main hematologic parameters from full-term newborn and adult blood are shown in Table 5.1.

5.3.2 Newborn Hemoglobin

At the time of birth, approximately 75% of the hemoglobin is fetal (HbF). As the child grows up, the fetal hemoglobin concentration decreases while the adult hemoglobin (HbA) becomes the main erythrocyte hemoglobin, as shown in Table 5.2. Fetal hemoglobin has an almost 50% larger capacity to transport oxygen than adult hemoglobin. The capacity of the former is to carry 2.08 mL of oxygen per gram of HbF, while the latter has the capacity of 1.39 mL of oxygen per gram of HbA.[16] Fetal hemoglobin also presents higher concentration of 2–3 diphosphoglycerate (2–3-DPG). As 2–3-DPG shifts the oxygen dissociation curve to the right, it increases the oxygen release.[17]

These features are important for transfusional criteria. Theoretically, desired tissue oxygenation can be achieved with smaller increase of hematocrit and, consequently, smaller blood viscosity, due to fetal hemoglobin rich blood. This fact can be interesting in the treatment of anemic patients associated with ischemic disease, or even in patients with sickle cell anemia who need transfusion.

5.3.3 Coagulation Factor Features of Umbilical Cord Blood

The hepatic immaturity of neonates, especially in preterm newborns, and the physiological deficiency of vitamin K, lead to a smaller concentration of pro- and anticoagulant factors in their plasma (Tables 5.3 and 5.4).

Table 5.1 Reference hematologic values in full-term newborns and adults (Adapted from Geaghan[15])

	Newborns		Adults	
	Mean	−2 S.D (or min–max)	Mean	−2 S.D (or min–max)
Blood volume (mL/kg)	86.1		65	(55–75)
Erythrocyte Mass (mL/kg)	43.1		27.5	(25–30)
Hb	16.2	13.5	f:14.0 m:15.5	f:12 m:13.5
Ht%	51	42	f:41 m:47	f:36 m:41
Erythrocytes	4.7	3,9	f:4.6 m: 5.2	f:4 m:5.2
MCV	108	98	90	80
MCH	34	31	30	26
MCHC	33	30	34	31
Reticulocytes (10^6/μL)	0.074	(0.049–0.15)	0.092	(0.058–0.146)
Leukocytes (total)	18.1	(9–30)	7.4	(4.5–11)
Neutrophils	11	(6–26)	4.4	(1.8–7.7)
Lymphocytes	5.5	(2–11)	2.5	(1–4.8)
Monocytes	1.1		0.3	
Eosinophils	0.4		0.2	

Hb hemoglobin, *Hct* hematocrit, *MCH* mean corpuscular hemoglobin, *MCHC* mean corpuscular hemoglobin concentration, *MCV* mean corpuscular volume, *−2 S.D* minus 2 standard deviation, *min* minimum, *max* maximum, *f* female, *m* male

Table 5.2 Erythrocyte hemoglobin concentration from birth to 2 years old, when the concentration remains the same until adulthood (Adapted from Geaghan[15])

Age	HbF%		HbA%	HbA2%	
	Mean	±2 S.D	Mean	Mean	±2 S.D
Newborn	75	61–80	25.0	0	
1 month	60	46–67	39.2	0.8	0.4–1.3
6 months	7	2.7–13	90.5	2.5	2.1–3.1
1 year	2	1.3–5	95.3	2.7	2.0 –3.3
2 years	0.6	0.2–1	96.6	2.8	2.1–3.5

HbA hemoglobin A, *HbA2* hemoglobin A2, *HbF* hemoglobin F, *S.D* standard deviation

Table 5.3 Coagulant factors in full-term and preterm newborn plasma and adult plasma (Adapted from Geaghan[15])

Factor	Full-term newborns		Preterm newborns		Adults	
	Mean	−2 S.D	Mean	−2 S.D	Mean	−2 S.D
Fibrinogen (mg/dL)	246	150	240	150	278	156
F.II (U/mL)	0.45	0.22	0.35	0.21	1.08	0.7
F.VIII (U/mL)	1.68	0.5	1.36	0.21	0.99	0.5
F.IX (U/mL)	0.4	0.2	0.35	0.1	1.09	0.5
F.XII (U/mL)	0.33	0.23	0.22	0.09	0.08	0.52
Antirombin (U/mL)	0.4	0.25	0.35	0.1	–	–
Protein C (U/mL)	0.24	0.18	0.28	0.12	–	–

−2 S.D minus 2 standard deviation

Table 5.4 Coagulation inhibitors in newborn and adult plasma (Adapted from Geaghan[15])

Coagulation inhibitors	Newborns		Adults	
Factor	Mean	Range	Mean	Range
AT.II (antitrombin II)	59.4	42–80	99.8	65–130
Protein C antigen (%)	32.5	21–47	100.8	68–125
Protein C activated (%)	28.2	14–42	98.8	68–129
Protein S (total) (%)	38.5	22–55	99.6	72–128
Protein S (free) (%)	49.3	33–67	98.7	72–128

The smaller volume and reduced concentration of coagulation factors in UCB diminishes the utility of the plasma in correcting coagulation disturbances.

5.3.4 Immunological Features of Umbilical Cord Blood

The placenta barrier protects the fetus against contact with antigens present in maternal circulation and bacterial and viral pathogens very efficiently. The neonate is characterized by a state of true immunological purity. After delivery, the newborn comes into contact with antigenic stimulus of extra-uterus life for the first time. This fact is very important when considering the use of cord blood for transfusion.

Tables 5.5–5.7 show the main immunological features of UCB. It can be observed that IgA levels increase from 4 to 15 times and IgM levels increase from 4 to 30 times, from birth to adult life, whereas total IgG levels increase by just two, and among these,

Table 5.5 Immunoglobulin levels of blood (Adapted from Geaghan[15])

Age	0–30 days	Over 16 years
	Range (95%)	Range (95%)
IgA (mg/dL)	1–20	89–322
IgM (mg/dL)	12–117	59–360
IgG (mg/dL)	221–1,031	632–2,108

Table 5.6 IgG subclasses in preterm and full-term newborn and adult blood (Adapted from Geaghan[15])

IgG subclasses	Preterm	Term	adult
	Range (95%)	Range (95%)	Range (95%)
IgG_1	3.4–9.7	5.8–13.7	4.8–9.5
IgG_2	0.7–1.7	0.6–5.2	1.1–6.9
IgG_3	0.2–0.5	0.2–1.2	0.3–0.8
IgG_4	0.2–0.7	0.2–1.0	0.2–1.1

Table 5.7 Complement levels of blood (Adapted from Geaghan[15])

Complement	0–5 days	Adult
	Range (95%)	Range (95%)
C3	0.26–1.04	0.45–0.83
C4	0.06–0.37	0.11–0.41

IgG_2 is the subclass that increases the most. The level of complement class C3 and C4 does not present differences between neonates and adults. The main difference is in the immunoglobulin level because of the crescent contact with new antigens and pathogens.

5.3.5 Erythrocyte Antigens and Antibodies

Human erythrocytes express polymorphic antigens on their cellular membrane, responsible for hemolytic reactions by incompatibility. The most important antibodies that cause hemolysis are IgM (natural) and IgG (acquired). Notwithstanding, the most important and antigenic blood group is ABH. Natural antibodies against those antigens reach adult levels as early as the third month of extra-uterus life. Anti-A and anti-B antibodies belong to the IgM class and are potent activators of the complement system, causing a severe and potentially lethal intravascular hemolysis.

Healthy neonate blood does not contain acquired antibodies as it has not yet developed natural antibodies against RBC antigens. Newborn erythrocytes do not yet express certain erythrocyte antigens, for example, Kelly, and only express other antigens weakly such as A and B, and are therefore less immunogenic than adult erythrocytes. Table 5.8 shows the most important antigens and antibodies of UCB.

5.4 Hemocomponents from Umbilical Cord Blood

The use of whole blood for transfusion in patients has become an exception and normally whole blood is processed to red cell, platelet, and plasma units before transfusion.[19]

Since blood transfusion in premature or low weight neonates is often necessary,[6-9,11,20] UCB is a good source of hemocomponents for transfusion mainly in newborns

Table 5.8 The main antigens of umbilical cord blood (Adapted from Beutler et al.[18])

Antigen expression	Newborn	1–2 weeks	1 year	Adult
I	Weak	Weak	Strong	Strong after 3 years old
i	Strong	Strong	Undetectable	Undetectable
ABH	Weak	Increasing	Strong	Strong
$Lu^{a\ and\ b}$	Weak	Weak	Weak	Strong after 15 years old
Lewis	Undectable	Detectable	Strong	Strong

as well as premature infants who generally need to receive more transfusions than full-term infants.[9]

Approximately 15–20 mL of UCB per kg of body weight can be harvested.[6,20] Several factors can influence the volume of cord blood collected. It has been shown that there is a direct correlation of volume of UCB collected to newborn[10,11,20-22] and placental weight,[20,21] and gestational age.[6]

Newborn erythrocytes have high concentration of HbF, whose capacity of carrying oxygen is greater than HbA. The main problem of using UCB is its low volume. However, it can be compensated by using more units.

Neonate plasma is deficient in coagulation factors and it does not as yet have natural antibodies against erythrocyte antigens. Newborn plasma is not therapeutically efficient in correcting bleeding due to its factor deficiency. On the other hand, it is less thrombogenic, which is an advantage when a whole UCB transfusion is needed, or when plasma is used to recover blood volume. The lack of antibodies against erythrocyte antigens, mainly natural antibodies, reduces the risk of hemolysis when neonate plasma is transfused. For this reason, iso group transfusion is not needed when whole UCB is used, diminishing the importance of blood fractionation.

Plasma fractionation by centrifugation is necessary with adult whole blood in order to preserve the activity of coagulation factors. When there is no need to preserve coagulation activity, fractionation of whole blood can be done by sedimentation. As sedimentation does not need expensive whole blood centrifuges, it is a cheaper and easier method and therefore well suited for poor and underdeveloped countries.

It is possible to separate erythrocytes and remove leukocytes from UCB by sedimentation without losing quality when stored up to 35 days.[19]

Even if platelet concentration of newborn blood does not differ from adult blood, the total amount per whole UCB unit is smaller, because of its lower volume harvested. Thus, it seems that the use of UCB platelet for transfusion will have little therapeutic importance.

It can be concluded that erythrocytes are the most interesting components in UCB transfusion practice. The potential use of plasma and platelets from UCB in transfusions is small, because of reduced volume and coagulation activity. The reduced erythrocyte volume per cord blood unit can be compensated using more units and by the abundance of the material. The higher oxygen-carrying capacity of HbF, lower thrombogenicity, lower antigenicity, and an absence of natural antibodies make UCB a very attractive source of RBC for transfusion.

The allogeneic UCB transfusion in adults shows an increase in the number of circulating CD34+ cells in the receptor with transient spontaneous engraftment.[23] Therefore, it is a theoretical risk of GVHD due to implantation of viable nucleated cells,[23,24] which can be significantly reduced by using leukocyte filter (7) or eliminated by irradiation before transfusion.[6]

In spite of theoretical GVHD risk, its incidence is rather low[25] and some studies have shown that this engraftment is not enough to provoke such a risk.[23]

5.5 Stored Umbilical Cord Blood Features and Quality

The mean volume of UCB, which can be harvested from term neonates with normal weight, is between 80 and 90 mL.[26] The volume collected from preterm and low weight newborns is lower, achieving volumes of over 15 mL in 60% of harvests.[20] UCB can be stored as whole blood or, after centrifugation or filtration, fractionated blood.

Bacterial contamination may occur during the harvest. However, with an adequate blood collection technique, this contamination rate can be reduced to less than 2%.[6] Thus, the low risk of possible bacterial contamination of placental blood must be carefully balanced against the benefit of avoiding homologous erythrocytes.[7]

UCB can be stored up to 35 days. Compared with adult erythrocyte stored for the same time period, UCB shows a higher hemolysis rate (1.1 ± 8.8 against 0.2 ± 01% from adult blood), higher free hemoglobin (416.9 ± 254 against 82.8 ± 42.4 mg dL from adult blood), and lower pH (6.1 ± 0.1 against 6.8 ± 0.1 from adult blood). Nonetheless, nonleukoreduced cord blood has nucleated cells while in adult blood these cells can be eliminated (4,200 ± 200 and 0.0 ± 0, respectively).[6] After 2–3 weeks of storage, the potassium level in cord blood also starts to increase significantly.[27] The risk of transfusion-related hyperkalemia will therefore limit the secure storage time of UCB to a maximum of 3 weeks to avoid cardiac arythmias.

During storage of nonleukoreduced UCB, TNF* is reduced and TGF-β1 is increased.[27] Alterations in cytokine concentrations during storage of adult allogeneic blood may potentially cause immunomodulation. Why this can happen with UCB is unclear.

5.6 Suggestion for Collection, Preparation, and Storage of Umbilical Cord Blood

UCB use shows some risks when compared with adult blood. Even though there is a higher risk of bacterial contamination at birth collection compared with adult transfusion, this can be reduced by implementing more aseptic collection techniques and testing for bacterial growth.[6]

Scheduled and authorized harvest of full-term and healthy newborn UCB could be a viable suggestion in order to increase and make common practice the use of this material in RBC transfusion.

Sorologic tests can be taken by pregnant women approximately 2 weeks before the delivery in order to avoid harvesting from positive reacting mothers. It is important to establish aseptic collecting techniques and train the obstetricians and staff. Leukoreduction by gravity filtration should be done soon after collection and samples should be sent for microbiological tests. The unit should be stored for no more than 3 weeks and irradiated in the case of transfusion in neonates.

UCB also has immature nucleated cells with engraftment capacity, which can provoke GVHD without leukoreduction, although it has been shown that this risk is minimum.[23,25] Furthermore, it has already been shown that cord blood can be used with safety and at a low risk of immunological and nonimmunological reactions in autologous transfusion in newborns and allogeneic transfusion in children and adults. UCB of healthy full-term neonates with normal weight yields a mean volume of 80 mL of whole blood and from 27 to 30 mL of RBCs after centrifugation. The leukoreduction shows benefits in eliminating nucleated cells and reducing hemolysis and hyperkalemia caused by storage. To diminish transfusional risks caused by hemolysis and hyperkalemia, the period of storage should be reduced from 35 to 21 days.[27] Irradiation before its use eliminates the risk of GVHD, making use of allogeneic cord erythrocytes, a therapeutically useful option especially for preterm and lower weight newborns.

An increase in plasma potassium and a decrease in 2,3-DPG content of erythrocytes during extended storage[6,8] has been shown. Furthermore, morphological changes, including a decreased deformability and an increased osmotic fragility of the erythrocytes, have already been described.[6] Some studies show that 2,3-DPG is totally depleted from erythrocytes after 21 days of storage.[8]

The standard technique for separation of whole blood into plasma and erythrocytes is based on centrifugal force. However, as equipment for blood processing such as centrifuges and the subsequent processing of erythrocytes is expensive and therefore not always available, the use of gravity filter systems have the advantage of removing their necessity. One study showed that placental blood can be separated into its components by gravity with only a hollow-fiber filter system, attaining a quality suitable for later clinical use. One of the advantages of this procedure is that all steps are performed at room temperature. Because no other equipment is necessary and it is possible to use it without electricity, it is our view that this system would be ideal for use in the under resourced world.

5.7 Risk of Infectious Disease due to Allogeneic Umbilical Cord Blood Transfusion

One of the concerns about allogeneic blood transfusion is the risk of viral transmission, although its incidence is rather low. It is estimated that the risk of acquiring human immunodeficiency virus (HIV) is between 1 in 100,000 (0.001%) and 1 in 1 million (0.0001%) per transfusion. For hepatitis B, the risk is 1 in 50,000 (0.002%). Therefore, the risk of viral infections acquired from homologous transfusions does not justify invoking other dangers in an attempt to avoid these rare events.[7]

Despite the small risk of the transmission of infectious diseases through the transfusion of adult blood, the use of UCB diminishes this risk further, because the placenta barrier reduces the chances of vertical maternal-fetal transmission. This is mainly important in places such as Africa, where in some countries more than 50% of the adult population is HIV-positive.

5.8 Therapeutic Use of Umbilical Cord Blood Transfusion

The first autologous UCB erythrocyte transfusion was carried out in 1995 in a neonate.[28] Subsequently, several publications have demonstrated that it is an executable and safe proceeding.[6,10,29,30] Newborns who benefit the most from this proceeding are those with lower weight or preterm neonates, mainly those with cardiopulmonary disease and anemia.[8]

A number of epidemiological and experimental studies have shown that impaired intrauterine growth, resulting in low birth weight, is associated with a variety of adult-onset diseases, including type 2 diabetes, hypertension, hyperlipidemia, cardiovascular disease, stroke, and kidney disease.[25]

A practical limiting factor is that autologous UCB can only fully supply approximately 40% of the transfusional needs of newborns,[20,29] thus in 60% of neonates it is also necessary to use allogeneic blood. UCB use in allogeneic transfusions has been published since 2001. Hundreds of pediatric and adult patients with anemia, associated to several diseases, such as acquired immune deficiency syndrome,[31] ankylosing spondylitis, aplastic anemia,[4,16] benign prostatic hypertrophy,[4] cancer,[16,32] chronic renal failure,[4] diabetes mellitus,[33] leprosy,[24] malaria,[5] rheumatoid arthritis,[4,16,34] systemic lupus erythematosus,[4,16] beta thalassemia,[4,16,35] tuberculosis,[36] and others have already received thousands of allogeneic UCB units, without evidence of immunological or non-immunological reactions.[23,24,31,32,36] Table 5.9 is a resumé of the transfusion clinical trials with RBC of UCB.

Neonates, particularly when extremely preterm, are among the most heavily transfused of all patient groups. It is estimated that 80% of premature neonates with birth weight less than 1.5 kg, and, with rare exception, nearly 100% of extremely preterm infants with birth weight less than 1.0 kg required RBC transfusions every year. A smaller percentage of infants received other blood components such as fresh-frozen plasma, cryoprecipitate, and platelet. Thus, blood component transfusions, particularly erythrocytes, provide a genuine benefit to many preterm infants and are indispensable to the neonatologist.[8]

Many preterm infants who receive blood during the early weeks of life, particularly those with birth weight

Table 5.9 Clinical trials of umbilical cord blood RBC transfusion

Cause of anemia	Transfusion type	Number of units	Number of patients	Age of patients	Year of publication	References
Preterm newborn	Auto	1	1	Newborn	1995	Ballin et al.[28]
Thalassemia, AA, AS, BPH, CRF, RA, and SLE	Alo	174	62	9 - 78	2001	Bhattacharya et al.[4]
Preterm newborn	Auto	52	52	Newborn	2003	Brune et al.[29]
Thalassemia, cancer, AA, AS, RA, and SLE	Alo	413	129	2 - 86	2005	Bhattacharya[16]
Beta thalassemia	Alo	92	14	0.5 - 38	2005	Bhattacharya[35]
Tuberculosis	Alo	106	21	–	2006	Bhattacharya[36]
RA	Alo	78	28	–	2006	Bhattacharya[34]
Cancer	Alo	82	6	–	2006	Bhattacharya[23]
Cancer	Alo	–	72	–	2006	Bhattacharya[32]
DM	Alo	78	39	48 - 74	2006	Bhattacharya[33]
AIDS	Alo	123	16	–	2006	Bhattacharya[31]
Leprosy	Alo	74	16	12 - 72	2006	Bhattacharya[24]
Malaria	Alo	94	39	8 - 72	2006	Bhattacharya[5]

AA aplastic anemia, *AIDS* acquired immune deficiency syndrome, *Alo* allogeneic, *AS* ankylosing spondylitis, *Auto* autologous, *BPH* benign prostatic hypertrophy, *CRF* chronic renal failure, *DM* diabetes mellitus, *RA* rheumatoid arthritis, *SLE* systemic lupus erythematosus lower than 1.0 kg, are given multiple RBC transfusions,[8] which are, generally, correlated to initial hemoglobin value, birth weight, and gestational age.[37] Most RBC transfusions given to neonatal patients are small in volume (10 ± 20 mL/kg).

In neonates with severe respiratory disease, such as those requiring high volumes of oxygen with ventilator support, it is customary to maintain the hematocrit above 40% and hemoglobin concentration above 13 g/dl.[8]

RBC transfusion in newborns has been indicated for: (1) replacement of blood drawn for laboratory tests: (replace if 5–10% of blood volume is removed over a short period); (2) maintenance of optimum hemoglobin and hematocrit in babies with severe respiratory and/or cardiac disease (hemoglobin above 13 g/dL and hematocrit above 40%) evidence that the improvement outcome of transfusion is limited, and (iii) correction of anemia in infants with less severe cardiopulmonary disease or with growth failure (hemoglobin above 10 g/dL and hematocrit above 30%).[9] The risks and benefits of currently used minimal values of hemoglobin and hematocrit to indicate RBC transfusion in newborns have not been tested in randomized controlled clinical trials.

5.8.1 Use of Cord Blood RBCs in Transfusion in Anemia Patients

Anemia in premature newborns with the subsequent need to transfuse allogeneic or autologous red blood cells is a common problem in very low birth weight infants.[9,20,37] Seventy percent of these transfusions are given during the first month of life.[8]

The two most common causes are "physiological" anemia of premature newborns and blood loss due to repeated blood sampling. Anemia of premature newborns results in a lower Hb (6.5–9 g/dL) compared to full-term newborns (10–11 g/dL) and it occurs earlier (4–8 weeks).[9] In extremely low birth weight infants, the causes of anemia and the reasons for RBC transfusions include: the magnitude of blood loss related to the severity and duration of intensive care, erythropoietin treatment failure, and hemodilution caused by rapid weight gain, among others.

Despite limiting the number of donor exposures and transfusion episodes, premature infants still require transfusions of RBC for iatrogenic blood loss and for cardio respiratory instability.[12] Hundreds of infants and adults with anemia have also received allogeneic UCB transfusion such as patients suffering from leprosy,[24] tuberculosis,[36] cancer,[23,32] rheumatoid arthritis,[34] HIV-positive patients,[31] and others.

5.9 Use of Umbilical Cord Blood Transfusion in Sickle Cell Anemia Patients

Most sickle cell anemia patients receive blood during their life. However, one of the potential adverse effects is the high hematocrit and hyperviscosity caused by RBC transfusions,[8] which can cause an increase in the severity of the disease and provoke more sickle cell crisis. To diminish the risks of hyperviscosity due to erythrocytosis, UCB transfusion could be a good approach for these patients.[7] UCB has a high concentration of HbF, which has greater oxygen-binding capacity than normal hemoglobin, and this has been shown to be of considerable therapeutic importance in sickle cell disease or other hemoglobinopathies, since the patient can theoretically receive a smaller volume of blood to receive the same oxygen benefits. HbF will deliver more oxygen to the ischemic core provided there is partial blood flow from subtotal vaso-occlusion or by collateral circulation. The rheological property of term cord blood is also favorable for reperfusion because of lower viscosity.[38]

5.9.1 Use of Umbilical Cord Blood Transfusion in Patients with Malaria

Malaria is an annual killer of over 1 million people mainly in the under resourced world and its essential co-morbidity is anemia, mainly in children.[19,39] The high oxygen affinity and anti-malarial effect of fetal hemoglobin in cord blood are additional advantages for transfusion in malaria patients.[5,30] Without blood transfusions, the patients frequently fail to survive this life-threatening situation.[19] It has been shown that UCB can be used in malaria patients with anemia who need blood transfusions.[39]

5.9.2 Use of Umbilical Cord Blood Transfusion in Patients with Diabetes

Diabetes mellitus (DM) is the most common endocrine disease in all populations and all age groups. Anemia is a common accompaniment of diabetes, particularly in those with albuminuria or reduced renal function, although there are other additional contributory factors. As fetal hemoglobin transport 50% more oxygen than normal hemoglobin, the use of RBC from UCB is theoretically very attractive in patients with DM and anemia since most of them have damaged microcirculation.[33]

Both epidemiological and experimental studies have shown that impaired growth in the uterus due to maternal malnutrition, resulting in low birth weight, is associated with a high incidence of glucose intolerance, insulin resistance, and type 2 diabetes in adult life. Maternal malnutrition is an unavoidable worldwide problem, and therefore, prevention of type 2 diabetes in low birth weight infants who reach adulthood is difficult to achieve. Based on the evidence, it is also proposed that transfusion of human umbilical cord blood to low birth weight infants may offer protection of type 2-DM and other adult onset diseases.[24]

5.9.3 Use of Umbilical Cord Blood Transfusion in Acute Ischemic Stroke Patients

Strokes are a major cause of neurological disability throughout the world. Poststroke functional recovery is limited because of neuronal death and degeneration. Although early reperfusion therapy may improve the outcome, thrombolysis does not reverse ischemic neuronal death and carries the risk of cerebral hemorrhage.[38,40]

Based on some experimental data, human UCB transfusion has been considered possible therapy for ischemic cerebral stroke cases to aid functional recovery. One reason is the higher concentration of HbF in UCB, which has greater oxygen-binding capacity compared with HbA, improving oxygenation in the ischemic tissue. HbF will deliver more oxygen to the surviving neurons in the ischemic penumbra.[38]

Umbilical venous blood also has a high concentration of interleukin-1 receptor antagonist (IL-ra), especially in preterm and in normal term deliveries and is a potent anti-inflammatory cytokine and a target of new clinical stroke trials. Its presence in term newborn UCB suggests that UCB transfusion may potentially attenuate postischemic inflammatory cascade in stroke patients.[38]

Thus, it has been suggested that UCB transfusion could promote better functional recovery in adults with acute ischemic stroke, since UCB transfusion may have the potential to reduce the burden of disability not only in strokes but also in other brain diseases. The collection of cord blood will be parallel with population increase, and as a result, populous countries would be able to use their own resources effectively to treat strokes at a lower cost.[38]

5.10 Conclusions

At present, the placental and the umbilical cord are considered to be biological waste and are usually destroyed. However, UCB is an attractive source of RBC for transfusion for the following reasons: (1) because of its abundance, (2) it can be collected without risks, (3) the fetal hemoglobin has a 50% higher oxygen-carrying capacity, (4) it either does not express or expresses weakly some erythrocyte antigens and is therefore less immunogenic than adult blood, (5) it does not contain or contain very low levels of natural and acquired erythrocyte antibodies.

UCB is easy to collect, filter, and store, which is important in underdeveloped countries or in situations of shortage or war. Maximum time for secure storage should be no more than 3 weeks to avoid the risk of hyperkalemia.

Allogeneic and autologous RBC-UCB has been used in transfusions in a number of clinical situations with very low risk of infection, contamination, or immunological reactions. This makes the use of RBC-UCB in transfusion practice especially interesting in newborns or, for example, in adult patients with ischemic diseases. It is a very viable consideration that the use of UCB transfusion be stimulated in order that many more adult and child patients can benefit from this efficacious clinical approach.

References

1. Gluckman E, Broxmeyer HA, Auerbach AD, et al. Hematopoietic reconstitution in a patient with Fanconi's anemia by means of umbilical-cord blood from an HLA-identical sibling. *N Engl J Med*. 1989;321(17):1174-1178.
2. Bhattacharya N. Placental umbilical cord whole blood transfusion. *J Am Coll Surg*. 2004;4(12):347-348.
3. Roseff SD, Luban NLC, Manno CS. Guidelines for assessing appropriateness pediatric transfusion. *Transfusion*. 2002;42:1398-1413.
4. Bhattacharya N, Mukherijee K, Chettri MK, et al. A study report of 174 units of placental umbilical cord whole blood transfusion in 62 patients as a rich source of fetal hemoglobin supply in different indications of blood transfusion. *Clin Exp Obstet Gynecol*. 2001;28(1):47-52.
5. Bhattacharya N. A preliminary study of placental umbilical cord whole blood transfusion in under resourced patients with malaria in the background of anaemia. *Malar J*. 2006; 5:20.
6. Garritsen HSP, Brune T, Louwen F, et al. Autologous red cells derived from cord blood: collection, preparation, storage and quality controls with optimal additive storage medium (Sag-mannitol). *Transfus Med*. 2003;13:303-310.
7. Strauss RG. Autologous transfusions for neonates using placental blood; a cautionary note. *Am J Dis Child*. 1992;146: 21-22.
8. Strauss RG. Blood banking issues pertaining to neonatal red blood cell transfusions. *Transfus Sci*. 1999;21:7-19.
9. Roberts I. Management of neonatal anaemia: the role of erythropoietin. Rila publications Ltd. *CME Bull Haematol*. 1997;1(1):5-7.
10. Eichler H, Schaible T, Richter E, et al. Cord blood as a source of autologous erythrocytes for transfusion to preterm infants. *Transfusion*. 2000;40:1111-1117.
11. Surbek DV, Glanzmann R, Senn H-P, et al. Can cord blood be used for autologous transfusion in preterm neonates? *Eur J Pediatr*. 2000;159:790-791.
12. Luban NLC. Neonatal red blood cell transfusions. *Vox Sang*. 2004;87(suppl 2):S184-S188.
13. Hosono S, Mugishima H, Fujita H, et al. Umbilical cord milking reduces the need for red cell transfusions and improves neonatal adaptation in infants born at less than 29 weeks' gestation: a randomised controlled trial. *Arch Dis Child Fetal Neonatal Ed*. 2008;93(1):F14-F19.
14. Lurie S, Mamet Y. Red blood cell survival and kinetics during pregnancy. *Eur J Obstet Gynecol Reprod Biol*. 2000; 93(2):185-192.
15. Geaghan SM. Normal blood values: selected reference values for neonatal, pediatric and adult populations. In: Hoffman R, Benz EJ, Shattil SJ, Furie B, Cohen HJ, Silberstein LE, McGlave P, eds. *Hematology, Basic Principles and Practice Elsevier*. 4th ed. Philadelphia: Churchill & Livingstone; 2005.
16. Bhattacharya N. Placental umbilical cord whole blood transfusion: a safe and genuine blood substitute for patients of the under-resourced world at emergency. *J Am Coll Surg*. 2005;200(4):557-563.
17. Walsh TS, Salch E-E-D. Anaemia during critical illness. *Br J Anaesth*. 2006;97:278-291.
18. Segal BG, Palis J. Hematology of the newborn. In: Beutler E, Lichtman MA, Coller BS, Kipps TJ, Seligsohn U, eds. *Williams Hematology*. 6th ed. New York: McGraw-Hill; 2001.
19. Brune T, Fill S, Heim G, et al. Quality and stability of red cells derived from gravity-separated placental blood with a hollow-fiber system. *Transfusion*. 2007;47:2271-2275.
20. Jansen M, Brand A, von Lindern JS, et al. Potential use of autologous umbilical cord blood red blood cells for early transfusion needs of premature infants. *Transfusion*. 2006;46:1049-1056.
21. Canabarro R, Sporleder H, Gomes T, et al. Immunophenotypic evaluation, and physiological and laboratory correlations of hematopoietic stem cells from umbilical cord blood. *Biocell*. 2007;31(3):397-403.
22. Brune T, Garritsen HS, Witteler R, et al. Autologous placental blood transfusion for the therapy of anaemic neonates. *Biol Neonate*. 2002;81:236-243.
23. Bhattacharya N. Spontaneous transient rise of CD34 cells in peripheral blood after 72 hours in patients suffering from advanced malignancy with anemia: effect and prognostic implications of treatment with placental umbilical cord whole blood transfusion. *Eur J Gynaecol Oncol*. 2006; 27(3):286-290.
24. Bhattacharya N. Transient spontaneous engraftment of CD34 hematopoietic cord blood stem cells as seen in peripheral blood: treatment of leprosy patients with anemia by placental umbilical cord whole blood transfusion. *Clin Exp Obstet Gynecol*. 2006;33(3):159-163.
25. Ende N, Reddi AS. Administration of human umbilical cord blood to low birth weight infants may prevent the subsequent development of type 2 diabetes. *Med Hypotheses*. 2006; 66:1157-1160.
26. Lasky LC, Lane TA, Miller JP, et al. In utero or ex utero cord blood collection: which is better? *Transfusion*. 2002; 42(10):1261-1267.
27. Widing L, Bechensteen AG, Mirlashari MR, et al. Evaluation of nonleukoreduced red blood cell transfusion units collected at delivery from the placenta. *Transfusion*. 2007; 47:1481-1487.
28. Ballin A, Arbel E, Kenet G, et al. *Arch Dis Child Fetal Neonatal Ed*. 1995;73(3):181F-183F.
29. Brune T, Garritsen H, Hentschel R, et al. Efficacy, recovery, and safety of RBCs from autologous placental blood: clinical experience in 52 newborns. *Transfusion*. 2003;43(9): 1210-1216.
30. Hassall O, Bedu-Addo G, Adarkwa M, et al. Umbilical cord blood for transfusion in children with severe anaemia in under-resourced countries. *Lancet*. 2003;361:678-679.
31. Bhattacharya N. A preliminary report of 123 units of placental umbilical cord whole blood transfusion in HIV-positive patients with anemia and emaciation. *Clin Exp Obstet Gynecol*. 2006;33(2):117-121.
32. Bhattacharya N. A study of placental umbilical cord whole blood transfusion in 72 patients with anemia and emaciation in the background of cancer. *Eur J Gynaecol Oncol*. 2006; 27(2):155-161.
33. Bhattacharya N. Placental umbilical cord blood transfusion: a new method of treatment of patients with diabetes and microalbuminuria in the background of anemia. *Clin Exp Obstet Gynecol*. 2006;33(3):164-168.

34. Bhattacharya N. Placental umbilical cord whole blood transfusion to combat anemia in the background of advanced rheumatoid arthritis and emaciation and its potential role as immunoadjuvant therapy. *Clin Exp Obstet Gynecol.* 2006; 33(1):28-33.
35. Bhattacharya N. Placental umbilical cord blood transfusion in transfusion-dependent beta thalassemic patients: a preliminary communication. *Clin Exp Obstet Gynecol.* 2005; 32(2):102-106.
36. Bhattacharya N. Placental umbilical cord whole blood transfusion to combat anemia in the background of tuberculosis and emaciation and its potential role as an immuno-adjuvant therapy for the under-resourced people of the world. *Clin Exp Obstet Gynecol.* 2006;33(2):99-104.
37. Hosono S, Mugishima H, Shimada M, et al. Prediction of transfusions in extremely low-birthweight infants in the erythropoietin era. *Pediatr Int.* 2006;48:572-576.
38. Chaudhuri A, Hollands P, Bhattacharya N. Placental umbilical cord blood transfusion in acute. Ischaemic stroke. *Med Hypotheses.* 2007;69:1267-1271.
39. Bhattacharya N. Placental umbilical cord blood transfusion: a novel method of treatment of patients with malaria in the background of anemia. *Clin Exp Obstet Gynecol.* 2006; 33(1):39-43.
40. Chaudhuri A. Treating stroke in the 21st century. *Lancet.* 2007;369(9567):1079-1080.

Autologous Placental Blood Transfusion for the Therapy of Anemic Neonates

6

Thomas Brune, F. Louwen, C. Troeger, W. Holzgreve, and H.S.P. Garritsen

6.1 Background

The majority of neonates will have a hematological uncomplicated adaptation to the new circumstances directly after birth. However, premature neonates and neonates in need of a surgical intervention directly after birth are prone to suffer from a mild to severe anemia, which needs to be corrected.[1,2] Anemia is a common complication of the premature neonate. The etiology is multifactorial and includes especially iatrogenic blood losses due to laboratory examinations,[3-5] a lack of erythropoietin,[4-7] and a lack of nutritive factors.[8,9]

However, the reduction of iatrogenic blood losses,[10,11] prophylactic iron substitution,[12] the use of recombinant human erythropoietin,[13,14] and placento-fetal blood transfusion after delayed cord clamping have reduced but not dispensed with the need for transfusions during the first weeks of life.[15,16] At present, the therapy of choice for anemia is allogeneic blood transfusion.[1,17] Almost 65% of all premature neonates with a birth weight of less than 1,500 g receive at least one erythrocyte transfusion during their first weeks of life.[18] The adequate blood supply of premature and mature neonates with anemia is a continuous point of discussion in neonatology and transfusion medicine. Besides immunological and biohazard considerations,[19-21] parents are subject to important psychological barriers, which cause them to hesitate if neonatal anemia has to be corrected by allogeneic blood. Various alternatives to allogeneic blood transfusions have been discussed. The application of erythropoietin to stimulate autologous erythropoiesis has shown to be of limited success.[1,13,14] Placento-fetal transfusion during delivery by holding the newborn below the level of the uterus and delaying cord clamping might represent an alternative therapy. In prematures with a gestational age of less than 32 weeks, a significantly greater increase in hemoglobin level was determined when compared to infants without placento-fetal transfusion. During a 3-week observation period, this leads to an elevated blood volume and a reduced demand for red cell transfusions in newborns.[15,16] However, this method is not without its risks, because transfusion of an uncontrolled blood volume may lead to a hyperviscosity syndrome, which requires hemodilution in cases of Hct > 70% and can endanger the newborn.[22]

6.2 Placental Blood Collection

For decades, interest in the collection and subsequent transfusion of placental blood has waxed and waned. The feto-placental blood reservoir contains a blood volume of approximately 110 mL/kg fetal weight. Correlated to gestational age, 30–50% of this volume is allocated to the placenta.[23-25] Anderson reported a linear correlation between collected cord blood volume in milliliters and birth weight, but an inverse correlation between relative volume per kilograms birth weight and birth weight.[26] Our group was able to confirm the correlation between total collected blood volume and birth weight, but not the inverse correlation between relative placental blood volume per kg birth weight and birth weight.[18] The average amount of blood collected was approximately 20 mL/kg, irrespective of birth weight. Most research groups

T. Brune (✉)
University Children's Hospital, Universitätsklinikum Magdeburg, Zentrum für Kinderheilkunde, Perinatologie, Gerhard-Hauptmannstr. 35, D-39108 Magdeburg, Germany
e-mail: thomas.brune@med.ovgu.de

N. Bhattacharya and P. Stubblefield (eds.), *Regenerative Medicine Using Pregnancy-Specific Biological Substances*,
DOI: 10.1007/978-1-84882-718-9_6,

punctured the cord vein directly after cord clamping and disinfection, with the placenta still in utero. Using uterine contractions, they obtained placental blood volumes of 80–100 mL.[27,28] Rubinstein reported no difference in the quantities of placental blood obtained before and after delivery of the placenta.[28] Ballin et al. showed that there is no activation of coagulation in the umbilical cord blood within 10 min of cord clamping.[29]

Several studies have shown that the collection volume and cell content depends on various obstetrical factors, e.g., the yield is increased with gestational age, birth weight, early cord clamping, Caucasian or Hispanic ethnicity, male sex, and fetal distress.[30-32] The timing of umbilical cord clamping has been shown to have an important influence on the neonatal blood volume and the subsequent hematological status. If the cord is clamped too soon after birth, the infant may be deprived of a placental blood transfusion, resulting in a lower blood volume and increased risk of anemia in later life, in contrast, if the cords were clamped 3 min or beyond, hypervolemia may result, causing hyperviscosity and delayed or disturbed postnatal cardiorespiratory adaptation in some infants.[23,33]

The results of these and other studies suggest that cord blood collection is most efficient when performed with the placenta still in utero and immediately after cord clamping.[34,35] During cesarean section, this can only be achieved by using a sterilely wrapped collection system that can be used directly on the operation table (e.g., Macopharma MSC1201DU). Most studies on the influence of the mode of delivery on the efficiency of umbilical cord blood collection simply show that the yield is higher in cesarean section compared to operative and spontaneous vaginal deliveries.[30,32] However, in an own study we could show that this is not the case for planned cesarean sections. Compared to secondary cesarean sections, the collected volume was similar, but the cell content was significantly higher in those cases that were delivered for fetal distress[36] (Fig. 6.1). Data on hemoglobin content of umbilical cord blood in regard to obstetrical factors are scarce in the literature. From our results and the fact that the antenatal hemoglobin concentration correlates with the number of total nucleated cells, it can be speculated that the yield in red blood cells is probably similarly best when fetal distress has occurred before delivery.[36,37] Although stem cell collection efficiency is lower in planned cesarean sections and in earlier gestational weeks, this might not have a relevant impact on collection of cord blood for transfusion of neonates, because here in most cases fewer volume and cells are needed compared to stem cell transplantation in children and adults. The collection procedure itself is simple and routinely performed in the course of both vaginal and cesarean deliveries.

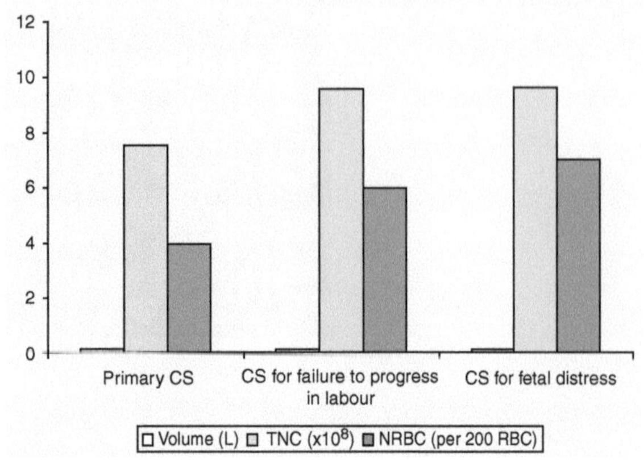

Fig. 6.1 Hematopoietic variables in regard to the indication for a cesarean delivery

6.3 Storage Stability

Several authors have investigated if autologous placental blood could be used as an alternative for allogeneic blood transfusion.[26,38,39] They could show that the quality of stored placental blood is comparable to that of stored adult blood. A decrease in intracellular ATP concentrations, in pH, and in 2,3-DPG concentrations was observed after a storage time of 2–3 weeks, but erythrocytes are capable of regenerating these properties within 24 h of transfusion. Morphological alterations, increased erythrocyte fragility, and an increase in potassium concentration were also reported on levels similar to those observed in adult cells under the same storage conditions.[39-41]

Fetal erythrocytes were found to tolerate storage in CPDA-1 (Citrate-Phosphate-Dextrose-Adenine) better than in CPD (Citrate-Phosphate-Dextrose). The storage-related damage was partially reversible if adenosine was added to the storage medium (as in CPD-A1). Overall, the quality of cord blood stored in CPD or CPDA-1 for several weeks was rated as satisfactory. Bifano et al. reported in 1994 that storage of cord PRC for 28 days in CPDA-1 induced no significant change in hematocrit, ATP content, or erythrocyte morphology. They confirmed, however, an increase of potassium and a decrease in 2,3-DPG-content. These changes were comparable to those found in stored adult Packed Red Cells (PRC).[40] Most other studies were designed to monitor storage conditions within 2 weeks of collection of the (Placental Blood Packed Red Cells) PB-PRC. Most authors set a limited time frame varying from 1 to 14 days for the use of the placental blood[39-42] but showed that this was not sufficient to cover the clinical need. Most prematures and newborns needed transfusions beyond the chosen time frame of 14 days.[2] To cover a more extensive time period for PB-PRC transfusions, we developed and tested a new closed collection system to collect and fractionate cord blood into red cell concentrates containing an extensive storage medium Sag mannitol (Sag-M) for 35 days (Fig. 6.2).[43]

One of the problems inherent in PB-PRCs is that the yield of erythrocytes is variable so that the amount of extended storage medium added must be variable to provide the correct ratio (1:5). Comparison with adult PRCs shows that no significant differences in quality were found between the two products after fractionation (Table 6.1). However, the PB-PRCs displayed a

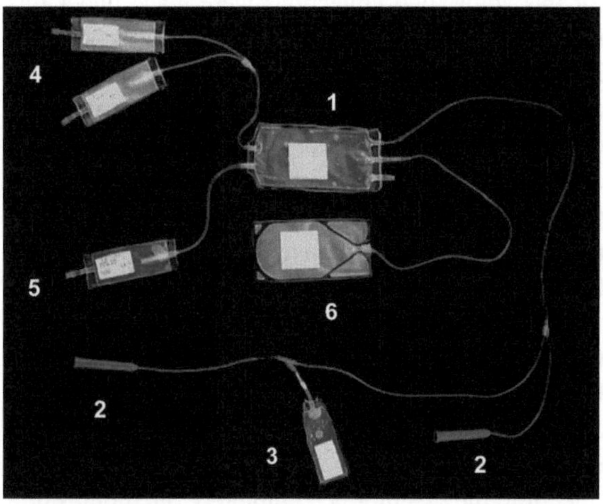

Fig. 6.2 Placental blood collecting system Maco Pharma - Type MXT 2206 DC. (1) Collecting bag for 150 mL blood, containing 21 mL CPD anticoagulant, (2) two cannulas with a diameter of 2.5 mm, (3) an additional reservoir with 8-mL CPD anticoagulant for rinsing the rest of the blood out of the tubes, (4) two mini-blood-bags (20 mL each) for the packed red cells, (5) bag containing 8 mL Sag-M additive solution, (6) plasma bag

higher hemolysis rate than the adult PRCs. This hemolysis rate of 1% after 35 days is still within the range of an adult PRC at 35 days. The Hb-ATP was decreased in the PB-PRC group compared to the adult PRC, a finding also reported by Biffano et al., who showed, however, that this decrease was partly reversible after incubation with adenosine and that the ATP content of the erythrocytes was regenerated within 24 h after retransfusion.[40] Comparison of adult and fetal erythrocytes and hemoglobin revealed that the number of erythrocytes did not differ but that the hematocrit was significantly increased due to the increased MCV of adult red blood cells compared with the initial values.[43]

6.4 Microbial Contamination

The most greatly feared complication in the use of autologous placental blood is the possibility of microbial contamination. The bacterial contamination rates of cord blood collections are reported in several publications and range from 0% to 22%.[26,41,43-47] Important parameters are collection technique, usage of closed collection systems, method of disinfection of the collection site, experience of the person collecting the blood, and the frequency of the bacteremia with an amnion infection syndrome. Neither Paxon nor Brandes detected any contamination of cord blood with bacteria or yeasts in their studies.[38,39] These publications induced RG Strauss to write a cautionary note on the use of placental blood for autologous transfusion.[48] However, it has to be borne in mind that open systems were used for the collection of placental blood in most of these studies, which certainly had a great impact on the contamination rate. The increasing interest in cord blood as a source of stem cells forced development of closed collection systems for cord blood and development of standard operating procedures for collection and processing. More recent examination methods have revealed that iatrogenic bacterial contamination during blood removal deserves less attention than contamination caused by bacteremia already existing in the newborn due to an amnion infection syndrome.[42,44] The measures enabled a reduction of bacterial contamination rate from 10% to <1%.

In an own study from 390 cord blood collection, we found seven cases of bacterial contamination. Two of

Table 6.1 Quality and storage data of RBCs from autologous cord blood compared to adult RBCs

After preparation:			
N = 12	*Placental blood*	*RBC*	
Volume (mL)	23.4 ± 9.1	271.2 ± 7.2	**
Erythroc.(10⁶)/uL	4.6 ± 0.9	6.6 ± 0.3	**
Leukoc./uL	6.300 ± 2.600	2.1 ± 0.9	**
Thromboc./uL	62.685 ± 24.100	3,568 ± 1,193	**
Tot.-Hb (gIdl)	17.1 ± 3.5	18.9 ± 0.4	
HCT (%)	52.6 ± 10.0	56.5 ± 1.3	
Free Hb (gll)	19.4 ± 19.2	14.5 ± 5.4	
Hemolyser. (%)	0.0 ± 0.0	0.0 ± 0.0	
Hb-ATP(umol/gHb)	3.7 ± 0.9	3.3 ± 0.4	
PH	6.4 ± 0.1	7.6 ± 0.2	**
After 35 days storage:			
N = 12	*Placental blood*	*RBC*	
Volume (mL)	23.4 ± 9.1	271.2 ± 7.2	**
Erythroc.(10⁶)uL	4.7 ± 0.9	6.7 ± 0.3	**
Leukoc./10³uL	4.200 ± 200	0.0 ± 0.0*	**
Thromboc./uL	54.6667 ± 11.7851	0.0 ± 0.0*	**
Tot.-Hb(gIdl)	17.1 ± 3.3	19.3 ± 0.5	
HCT (%)	55.2 ± 11.1	63.5 ± 1.4*	
Free Hb(g/L)	416.9 ± 254.5*	82.8 ± 42.4*	**
Hemolyser. (%)	1.1 ± 0.8*	0.2 ± 0.1*	**
Hb-ATP(umol/gHb)	1.2 ± 0.5*	2.3 ± 0.4*	**
PH	6.1 ± 0.1*	6.8 ± 0.1*	**

* $p < 0.05$ compared to after preparation
** $p < 0.05$ compared to cord blood

them were cases of amnionitis, and the diagnosis in both cases was based primarily on the cord blood culture some days before the peripheral blood culture of the neonate itself became positive. According to the inclusion criteria, such patients were to be excluded from the study. The other five cases of contamination were related to normal bacterial skin colonizers with little clinical relevance. In part, they were detected only in the first culture.[43] In our view, a microbial contamination quality control should be performed in every cord blood donation used to prepare PB-PRC, either by means of blood cultures or by the recently introduced molecular testing procedure for detecting microbial contamination.[49] This molecular testing is fast (3–4 h) and can be performed hours before the PB-PRC are transfused, guaranteeing a high level of security in this respect. In both cases of suspected amnionitis, the microbial QC of the collected cord blood provided the first diagnostic proof of an infection. No additional blood had to be drawn from the neonate for diagnostic purposes. Although suspected amnion infection syndrome was an exclusion criterion for this study, two collections were made under this indication. This demonstrates that it is extremely important to have good communication lines in this complex manufacturing system, where gynecologists/obstetricians are responsible for the collection of the cord blood, transfusion medicine is involved in the preparation of the PB-PRC, and neonatologists are responsible for the indication for transfusion and transfusion itself.

The question of safety of blood transfusions is certainly in everyone's mind in view of the risk of transfusion-associated infections such as HIV and HCV. Although transfusion-associated risks can be minimized by stringent diagnostic testing and donor selection, they cannot be completely ruled out. For a neonate, this implies that an infection caused by blood transfusion will affect him/her lifelong, i.e., for many times the period experienced by adults. When PB-PRC transfusions are used, the risk of transfusion-related diseases is confined to vertical transmission from the mother (transplacental infection). In Germany, prenatal infectious disease screening (HIV I+II, HBV, HCV, CMV, EBV) is performed as a routine procedure in all pregnancies. In addition, the PB-PRC product itself is tested in the same as all blood products. If any of these infectious disease markers proves to be positive, the PB-PRC produced is withdrawn and destroyed.

6.5 Maternal Blood Contamination

A number of publications have demonstrated that a regular transfer of maternal cells to the fetus takes place during all stages of the pregnancy,[50,51] especially under pathologic conditions.[52] In the immunocompetent fetus, this initiates no immune reaction. In the case of SCID or Omenn's syndrome, an engraftment of maternal cells and the development of a GvHD was observed.[53,54] In a mixed lymphocyte culture (MLC) between neonatal and maternal cells, it was shown that the proliferation of maternal cells was inhibited by neonatal lymphocytes.[55,56] In our own studies, we eliminated the residual GvHD risk by irradiating the RBC with 30 Gy immediately before transfusion. To prevent alloimmunization against maternal blood group antigens, we determined the blood group, using the Diamed gel system, which offers a simple and sensitive approach to testing for contamination with maternal erythrocytes (Fig. 6.3).[18,57] If contamination with maternal erythrocytes was detected, the PB-PRC was not released but was excluded from transfusion. In an own study, this applied to 3 out of 390 PB-PRC products.[43] In conclusion, the risk of transfusion-associated GvHD can be limited by using the described diagnostics and precautions.

6.6 Pharmacokinetics and Safety

In an own study, we could show that after autologous PB-PRC transfusion, the hemoglobin increase was the same and the hemoglobin per day decrease was slightly but significantly accelerated with 0.32 g/dL compared to 0.24 g/dL for allogeneic blood transfusion (Table 6.2).[58] Based on the calculated hemoglobin decrease, the placental blood group would have used up the transfusion after 9 days, and the allogeneic blood group after 11.5 days. When evaluating the hemoglobin decrease, it has to be taken into account that the blood loss resulting from diagnostic blood sampling accelerated the hemoglobin decrease in both groups. There was a marked difference in the age of the PRCs at the time of transfusion: autologous PB-PRCs were 23 days old on average, but allogeneic adult PRCs only 5 days. There was no difference in the vital parameters like respiratory frequency, heart frequency, and mean arterial blood flow up to 5 h after autologous PB-PRC transfusion compared to allogeneic PRC transfusion. The serum potassium levels also remained stable throughout the observation period in both groups. In conclusion, autologous PB-PRC transfusion proved to be no less effective and no less safe than allogeneic PRC transfusion.[58]

6.7 Efficacy of Autologous Placental Blood in Avoiding Allogenous Blood Transfusions

Initially, it was claimed that the amount of placental blood taken from the umbilical cord would be theoretically sufficient to cover the transfusion demands of anemic neonates. Furthermore, it was reported that

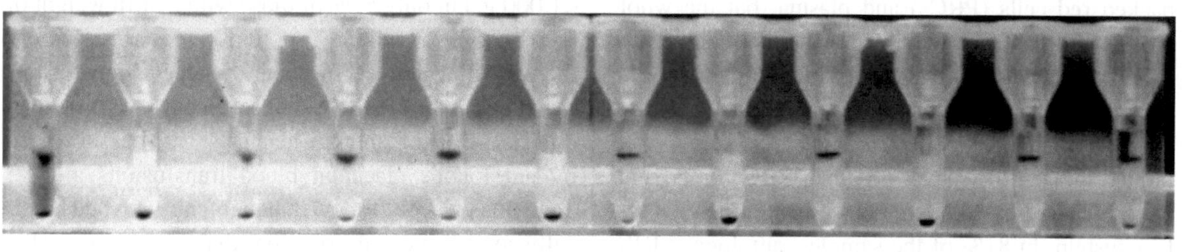

Fig. 6.3 Blood typing of a female twin. Double cell populations were detected in the anti-A, anti-AB, anti-D, anti-C, and anti-Kell fields

Table 6.2 Recovery after 10 mL/kg/BW PRC transfusion

	Autologous placental blood PRCs (Mean ± SD)	Allogeneic blood PRCs (Mean ± SD)
Hb increase directly after transfusion	3.1 ± 1.7 g/dL	3.0 ± 1.3 g/dL
Hb decrease/day after transfusion	0.3 ± 0.2 g/dL*	0.24 ± 0.2 g/dL*

*($p < 0.05$)

Table 6.3 Efficacy in avoiding allogeneic blood transfusions after autologous placental PRC transfusion

	n=	Mean amount of transfused PB-PRCs	Mean amount of transfused allogeneic PRCs	Number of newborns requiring no additional allogeneic blood
Prematures without surgical intervention or bleeding:				
<1,000 g	8	8.5 mL	93 mL	0 (0%)
1,000–2,500 g	22	14 mL	26 mL	9 (41%)
Newborns in need of surgery during the first month of life:				
NEC	11	17.3 mL	121 mL	1 (9.1%)
Open heart surgery	4	Machine filling	Machine filling	0 (0%)
Newborns in need of surgery directly postpartum to correct congenital malformations:				
Gastroschisis	4	24 mL	59 mL	1 (25%)
Meningomyelocele	3	30 mL	15 mL	2 (66.6%)
Total	7	26.5 mL	40 mL	3 (42%)

whole autologous blood derived from cord blood collections could be fractionated into blood components such as PRCs and plasma and stored like homologous blood components.[40-44] However, this knowledge has not yet undergone clinical application on a large scale. The first report of an autologous blood transfusion, carried out on an anemic monozygotic twin, was published in 1977.[59] Two years later, Paxon reported 25 asphyxic prematures who were transfused with autologous cord blood within 24 h of delivery.[38] None of the treated children showed any transfusion-related complications. In the above-mentioned cases, however, the placental blood was not fractionated into packed red cells (PRCs) and plasma, but the whole blood was transfused within a few hours of delivery.

Up to now, four bigger studies investigate the efficacy and the safety of processed autologous placental blood. Eichner et al. collected cord blood from 47 pretermed infants. Whenever possible, RBC components were prepared by standard centrifugation using a six-bag system. In 81% of the samples, autologous RBC components could be processed (vol, 7–87 mL; Hct, 31–82%). But within the group of extremely low birth weight infants (body weight < 1,000 g), a mean cord blood volume of only 37 mL was collected, and the PRC preparation was successful only in exceptional cases. The babies were given either the autologous PB-PRCs or standard allogeneic adult PRC, if autologous blood was not available. Of the 47 infants, 21 were treated with a total of 62 allogeneic and 4 autologous PB-PRC transfusions. Most infants with a body weight over 1,400 g did not need any PRC transfusion.[47] In an own study, we transfuse 52 newborns with PB-PRCs (Table 6.3). The number of newborns requiring no additional allogeneic PRC was calculated. All neonates of the study group with a birth weight of <1,000 g, but only 59% of those with a birth weight of 1,000–2,500 g and 58% of those requiring surgery directly after delivery needed allogeneic transfusions in addition to PB-PRCs. According to well-defined criteria, 40% of anemic neonates can be supported by autologous placental blood transfusions alone.[58] Taguchi et al. could show that in infants who had surgically correctable malformations diagnosed antenatally, 7 of 11 babies (64%) who required transfusion were able to avoid an allotransfusion using autologous placental blood.[60] The group of A. Brand tried to reduce at least 50% allogeneic transfusion needs using autologous placental blood in preterms less than 32 weeks of gestation. After being processed, autologous products (10 mL/kg) were available for only 36% of the total study population and for 27% of the transfused infants and could cover 58% (range, 25–100%) of the transfusion needs within the 21-day product shelf life. She also showed that the availability of autologous products depended most on the gestational age. Infants born between 24 and 28 weeks had the lowest availability (17%). Availability was highest (48%) for the infants born between 28 and 30 weeks. For 42% of the infants with transfusion needs in this group, autologous products were available. For the infants born between 30 and 32 weeks, autologous products were available for 36% of the infants.[61]

All these studies show that the collection and processing of autologous placental blood from neonates is fundamentally possible under clinical conditions. The method seems, however, not to be effective in neonates with a birth weight <1,000 g. Despite the high amount of donated placental blood, the high transfusion requirement of this patient group results in none of these children remaining free from further allogeneic blood transfusion.

6.8 Summary

The recovery, processing, storage, and retransfusion of autologous placental blood is possible in a clinical setting. The advantages offered by this method are of varying nature: although the efficacy is not as high as with other autologous blood-donation programs, there is no risk to the donor. In addition, direct donor screening and laboratory investigations (blood typing, transaminases, as well as screening for viruses), which are necessary for autologous blood as well as for allogeneic blood donations, are not necessary because of the prepartal maternal testing. In addition to many of the present techniques that are utilized to provide blood for neonates, the use of autologous placental blood is a further step towards minimizing donor exposure in this patient population. Compared to allogeneic PRCs, microbial testing (blood culture) as a release criterion involves additional costs. The blood culture of cord blood might also be a fast diagnostic aid in cases of amnionitis without additional blood having to be drawn from the infant.

Although studies with autologous PB-PRCs have failed to meet the originally high expectations concerning therapy of anemia of prematurity, it should be borne in mind that, especially in the case of a premature infant with an uncomplicated course post delivery, allogeneic blood transfusion in the third or fourth week of life involves a risk that can be completely avoided by using autologous placental blood in contrast to high-risk preterms, where the allogeneic blood transfusion represents only one additional treatment risk (e.g., ventilation, multiple transfusions) among many others. Therefore, autologous cord blood storage should be considered in the patients antenatally diagnosed to have surgical malformations.

References

1. Holland BM, Jones JG, Wardrop CA. Lessons from the anemia of prematurity. *Hematol Oncol Clin North Am.* 1987; 1(3):355-366.
2. Shannon KM. Anaemia of prematurity: progress and prospects. *Am J Pediatr Hematol Oncol.* 1990;12(1):14-20.
3. Shannon KM. Anaemia of prematurity: progress and prospects. *Am J Pediatr Hematol Oncol.* 1990;12(1):14-20.
4. Strauss RG. Neonatal anaemia: pathophysiology and treatment. *Immunol-Invest.* 1995;24(1–2):341-351.
5. Obladen M, Sachsenweger M, Stahnke M. Blood sampling in very low birth weight infants receiving different levels of intensive care. *Eur J Pediatr.* 1988;147:399-404.
6. Rhondeau SM, Christensen RD, Ross MP, Rothstein G, Simmons MA. Responsiveness to recombinant human erythropoietin of marrow erythroid progenitors from infants with the "anaemia of prematurity". *J Pediatr.* 1988;112:935-940.
7. Shannon KM, Naylor GS, Torkildson JC, et al. Circulating erythroid progenitors in the anaemia of prematurity. *N Engl J Med.* 1987;317:728-733.
8. Stockman JA. Anaemia of prematurity. *Semin Hematol.* 1975;12(2):163-173.
9. Dallman PR. Anaemia of prematurity. *Ann Rev Med.* 1981; 32:143-160.
10. Bifano EM, Curran TR. Minimizing donor blood exposure in the neonatal intensive care unit. *Clin Pediatr.* 1995; 22(3):657-669.
11. Dallman PR. Anaemia of prematurity: the prospects for avoiding blood transfusions by treatment with recombinant human erythropoietin. *Adv Pediatr.* 1993;40:385-403.
12. Heese HD, Smith S, Watermeyer S, Dempster WS, Jakubiec L. Prevention of iron deficiency in preterm neonates during infancy. *S Afr Med J.* 1990;77:339-345.
13. Shannon KM, Mentzer WC, Abels RI, et al. Recombinant human erythropoietin in the anaemia of prematurity: results of a placebo-controlled pilot study. *J Pediatr.* 1991;118:949-955.

14. Maier RF, Obladen M, Scigalla P, et al. The effect of epoetin beta (recombinant human erythropoietin) on the need for transfusion in very-low-birth-weight infants. European Multicentre Erythropoietin Study Group. *N Engl J Med.* 1994;330(17):1173-1178.
15. Rabe H, Wacker A, Huelskamp G, et al. A randomised controlled trial of delayed cord clamping in very low birth weight preterm infants. *Eur J Pediatr.* 2000;159(10):775-777.
16. Kinmond S, Aitchison TC, Holland BM, Jones JG, Turner TL, Wardrop C. Umbilical cord clamping and preterm infants: a randomized trial. *BMJ.* 1993;306:172-175. Letters: BMJ (1993) 306:578–579.
17. Ramasethu J, Luban NLC. Red blood cell transfusion in the newborn. *Semin Neonatol.* 1999;4:5-16.
18. Brune T, Garritsen H, Witteler R, et al. Autologous placental blood transfusion for the therapy of anaemic neonates. *Biol Neonate.* 2002;81:236-243.
19. Johnson JD, Malachowski NC, Sunshine P, Hafleigh EB, Grumet FC. New transfusion program for an intensive care nursery. *J Pediatr.* 1980;97:806-809.
20. Spanos T, Karageorga M, Ladis V, Peristeri J, Hatziliami A, Kattamis C. Red cell alloantibodies in patients with thalassemia. *Vox Sang.* 1990;58:50-55.
21. Jensen LS, Andersen AJ, Christiansen PM, et al. Postoperative infection and natural killer cell function following blood transfusion in patients undergoing elective colorectal surgery. *Br J Surg.* 1992;79:513-516.
22. Linderkamp O, Nelle M, Kraus M, Zilow EP. The effect of early and late cord-clamping on blood viscosity and other hemorheological parameters in full-term neonates. *Acta Pediatr.* 1992;81:745-750.
23. Usher R, Stephard M, Lind J. The blood volume of the newborn infant and placental transfusion. *Acta Pediatr Scand.* 1963;52:497-512.
24. Yao AC, Moinian M, Lind J. Distribution of blood between infant and placenta after birth. *Lancet.* 1969;2:871-873.
25. Wardrop CAJ, Holland BM. The role and vital importance of placental blood to the newborn infant. *J Perinat Med.* 1995;23:139-143.
26. Anderson S, Fangman J, Wager G, Uden D. Retrieval of placental blood from the umbilical vein to determine volume, sterility, and presence of clot formation. *Am J Dis Child.* 1992;146:36-39.
27. Brossard Y, Van Nifterik J, De Lachaux V, et al. Collection of placental blood with a view to hemopoietic reconstitution. *Nouv Rev Fr Hématol.* 1990;32:427-442.
28. Rubinstein P, Rosenfield RE, Adamson JW, Stevens CE. Stored placental blood for unrelated bone marrow reconstitution. *Blood.* 1993;81(7):1679-1690.
29. Ballin A, Kenet G, Gutman R, Samara Z, Zakut H, Meytes D. Autologous cord blood transfusion. *Acta Paediatr.* 1994; 83:700-703.
30. Cairo MS, Wagner EL, Fraser J, et al. Characterization of banked umbilical cord blood hematopoietic progenitor cells and lymphocyte subsets and correlation with ethnicity, birth weight, sex, and type of delivery: a Cord Blood Transplantation (COBLT) Study report. *Transfusion.* 2005;45(6):856-866.
31. Aroviita P, Teramo K, Hiilesmaa V, Kekomäki R. Cord blood hematopoietic progenitor cell concentration and infant sex. *Transfusion.* 2005;45(4):613-621.
32. Aufderhaar U, Holzgreve W, Danzer E, Tichelli A, Troeger C, Surbek DV. The impact of intrapartum factors on umbilical cord blood stem cell banking. *J Perinat Med.* 2003;31(4):317-322.
33. Yao AC, Lind J. *Placental Transfusion: A Clinical and Physiological Study.* Springfield: Charles C. Thomas; 1982.
34. Surbek DV, Schönfeld B, Tichelli A, Gratwohl A, Holzgreve W. Optimizing cord blood mononuclear cell yield: a randomized comparison of collection before vs after placenta delivery. *Bone Marrow Transplant.* 1998;22(3):311-312.
35. Surbek DV, Aufderhaar U, Holzgreve W. Umbilical cord blood collection for transplantation: which technique should be preferred? *Am J Obstet Gynecol.* 2000;183(6): 1587-1588.
36. Manegold G, Meyer-Monard S, Tichelli A, Pauli D, Holzgreve W, Troeger C. Cesarean section due to fetal distress increases the number of stem cells in umbilical cord blood. *Transfusion.* 2008;48(5):871-876.
37. Shlebak AA, Roberts IA, Stevens TA, Syzdlo RM, Goldman JM, Gordon MY. The impact of antenatal and perinatal variables on cord blood haemopoietic stem/progenitor cell yield available for transplantation. *Br J Haematol.* 1998;103(4): 1167-1171.
38. Paxson CL. Collection and use of autologous foetal blood. *Am J Obstet Gynecol.* 1979;134:708-710.
39. Brandes JM, Roth EF, Berk PD, et al. Collection and preservation of human placental blood. *Transfusion.* 1983;23:325-327.
40. Bifano EM, Dracker RA, Lorah K, Palit A. Collection and 28-day storage of human placental blood. *Pediatr Res.* 1994;36:90-94.
41. Abel M, Elsinger W, Sutor AH, Gregorio G, Peukert W. Untersuchungen zur transfusionsmedizinischen Validität von autologem Plazentablut. *Infusionstherapie.* 1985;12(4): 197-200.
42. Bifano EM, Palit A, Lorah K, Dracker R. Neonatal autologous transfusion: a potential form of hemotherapy. *Pediatr Res.* 1990/1991;29:273A.
43. Garritsen HSP, Brune T, Louwen F, et al. Transfusion of stored autologous red cells from cord blood: collection and preparation with extended storage medium (SAG-M). *Transfusion Med.* 2003;13:303-310.
44. Horn S, Mazor D, Zmora E, Meyerstein N. Storage-induced changes in human newborn red cells. *Transfusion.* 1987; 27:411-414.
45. Golden SM, Petit N, Mapes T, Davis SE, Monaghan WP. Bacteriologic assessment of autologous cord blood for neonatal transfusion. *Am J Obstet Gynecol.* 1984;149: 907-908.
46. Peltonen T. Placental transfusion – advantage and disadvantage. *Eur J Pediatr.* 198 1;13(7):141-146.
47. Eichler E, Schaible T, Richter E, et al. Cord blood as a source of autologous RBCs for transfusion to preterm infants. *Transfusion.* 2000;40:1111-1117.
48. Strauss RG. Autologous transfusions for neonates using placental blood. A cautionary note. *Am J Dis Child.* 1992;146:21-22.
49. Jimenez L, Smalls S, Ignar R. Use of PCR analysis for detecting low levels of bacteria and mold contamination in pharmaceutical samples. *J Microbial Methods.* 2000;41(3): 259-265.

50. Petit T, Gluckman E, Carosella E, Brossard Y, Brison O, Socie G. A highly sensitive polymerase chain reaction method reveals the ubiquitous presence of matemal cells in human umbilical cord blood. *Exp Hematol.* 1995;23:1601-1605.
51. Lo YMD, Lo ESF, Watson N, et al. Two-way cell traffic between mother and foetus: biologic and clinical implications. *Blood.* 1996;88(11):4390-4395.
52. Brune T, Koch HG, Pluempe U, Coenen-Worch V, Harms E, Louwen F. Effect of pathological perinatal conditions on the maternofetal transfer of mononuclear cells. *Fetal Diagn Ther.* 2002;17:110-114.
53. Pollack M, Kapoor N, Sorell M, et al. DR-positive matemal engrafted T cells in a severe combined immunodeficiency patient without graft-versus-host disease. *Transplantation.* 1980;30(5):331-334.
54. Appleton AL, Curtis A, Wilkes J, Cant AJ. Differentiation of materno-foetal GvHD from Omenn's syndrome in pre-BMT patients with severe combined immunodeficiency: case report. *Bone Marrow Transplant.* 1994;14:157-159.
55. Olding LB, Benirschke K, Oldstone MBA. Inhibition of lymphocytes from human adults by lymphocytes from human newborns. *Clin Immunol Immunopathol.* 1974;3(1): 79-89.
56. Adinoifi M. Inhibition of mitosis of maternal lymphocytes by foetal cells. *Lancet.* 1976;10:97.
57. Jamaer D. The use of the double cell population phenomena of the Gel test in posttransfusion complications and after bone marrow transplantation. Present and the Future of The Gel Test, Presentations and Posters from the First International Symposium, Montreux 17–18 March 1993, 1994 by DiaMed AG, 1785 Cressiers/Morat Switzerland, 88–91.
58. Brune T, Garritsen H, Hentschel R, Louwen F, Harms E, Jorch G. Efficacy, recovery, and safety of RBCs from autologous placental blood: clinical experience in 52 newborns. *Transfusion.* 2003;43(9):1210-1216.
59. Brandes JM, Timov-Tritsch I, Suchov P. Autologous heparinized placental blood transfusion for the treatment of the anemic monozygotic twin. Presented to the Haifa Obstetrical Society; 1977; Haifa, Israel.
60. Taguchi T, Suita S, Nakamura M, et al. The efficacy of autologous cord-blood transfusions in neonatal surgical patients. *J Pediatr Surg.* 2003;38(4):604-607.
61. Khodabux CM, von Lindern JS, van Hilten JA, Scherjon S, Walther FJ, Brand A. A clinical study on the feasibility of autologous cord blood transfusion for anemia of prematurity. *Transfusion.* 2008;48(8):1634-1643.

7 Cord Blood: A Massive Waste of a Life-Saving Resource, a Perspective on Its Current and Potential Uses

Tang-Her Jaing and Robert Chow

7.1 Introduction

Although allogeneic stem cell transplantation can cure patients with hematologic malignancies and nonmalignant disorders, limiting factors such as lack of suitable donors and graft-versus-host disease (GVHD) toxicity have led to the exploration of umbilical cord blood (UCB) as an alternative source of hematopoietic stem cells. The unique immunologic properties of UCB likely contribute to a decreased risk of GVHD. Thus, UCB represents a highly convenient hematopoietic stem cell (HSC) source that may significantly expand the HSC donor pool. Cord blood stem cells appear to confer significant advantages over adult stem cells, including ease of procurement, less of a requirement for human leukocyte antigen (HLA) matching, and fever side effects after use in transplant. Because of these characteristics, UCB is now the fastest growing source of stem cells for hematopoietic cell transplantation.

7.2 History of UCB Transplantation

In 1988, the first umbilical cord blood transplantation (UCBT) was performed in Paris, France, on a 5-year-old boy with Fanconi's Syndrome, or Fanconi's Anemia, using his sister's UCB stem cells.[1] To date, he remains disease-free, and is now an adult, free from disease. In 1993, a number of public cord blood banks were set up at Milan, Dusseldorf, and New York. The cord blood bank at the New York Blood Center, founded by Dr. Pablo Rubenstein, is the largest public cord blood bank in the world, with over 30,000 cord blood in their inventory as of March 2008 (NetCord March 2008).

In 1993, the first unrelated donor cord blood transplant was successfully performed by Kurtzberg and colleagues,[2] and by 1995, Wagner et al. published the first series on related donor cord blood recipients.[3] In the Lancet Report, Wagner and his colleagues studied cord blood transplants with "related recipients." The results showed survival and engraftment rates similar to bone marrow transplants.[3] In a 1996 NEJM article, Kurtzberg published data on 25 unrelated cord blood transplant recipients.[2] The following year, in a NEJM report, Gluckman and colleagues published results of the overall survival rates at 1 year, of both related and unrelated cord blood transplant recipients. Survival rates were noted to be 63% for related donors and 29% for unrelated donors.[4]

In 2000, in a retrospective comparison, Rocha and colleagues demonstrate results of lower acute and chronic GVHD in cord blood stem cell transplants as compared to bone marrow transplants.[5] In 2001, the first study of cord blood transplants in adults was published, reporting that 90% of the transplants engrafted. Also in December of 2001, several clinical trials were underway to study the results of multiple mismatched cord blood units in adults.[6]

7.3 Cord Blood Banking

With more than 11 million registered stem cell donors worldwide and limited resources for health systems, a 10/10 or 12/12 HLA-matched donor is still unavailable

T.-H. Jaing (✉)
Division of Hematology/Oncology, Department of Pediatrics, Chang Gung Children's Hospital, Chang Gung University, Taoyuan, Taiwan
e-mail: jaing001@cgmh.org.tw

N. Bhattacharya and P. Stubblefield (eds.), *Regenerative Medicine Using Pregnancy-Specific Biological Substances*,
DOI: 10.1007/978-1-84882-718-9_7,

for most minority recipients and a significant number of Caucasian patients; therefore, it seems reasonable to question if investments in ongoing donor recruitment are useful.[7] Because UCB transplant does not require an exact match, UCB banks are the optimal solution to this donor shortage, especially for minorities.[8] Pubic banks collect UCB for the allogeneic transplantation of any recipient for whom the UCB unit is a suitable HLA-match. In general, obstetric patients are required to sign an informed consent prior to donation and collection of cord blood. Furthermore, specialized collection kits are used for harvesting the cord blood. They contain all the necessary components for collecting the blood, vacutainers for sampling the mother's blood for infectious disease marker, and packaging and shipping materials for sending the collected blood to the cord blood bank.

A further issue in UCB banking is unit quality. Although it is estimated that there are already approximately 400,000 public units banked worldwide (BMDW March 2008), it is not clear as to how many of these are of both sufficient size and quality to be suitable for transplantation.

For private/familial collections, the cord blood samples are obtained by the patient's caregiver, whereas public banks generally use the services of a dedicated technician. In the vast majority of cases, the collections are made after delivery of the infant and ligation of the placenta. In utero collections generally may be easier to perform, take less time to complete, and result in larger volumes of blood than ex utero collection if carried out under ideal situations; however, the disadvantage is the variability in quality and demand on the obstetrician's time.

7.4 Current State of the Art

Over 10,000 unrelated donor UCB transplants have been performed worldwide, with approximately 20% of all pediatric transplants now using this stem cell source.[9-12] In USA and Japan, UCB transplants have now exceeded bone marrow transplant (NMDP 2007). UCB recipients experience a decreased incidence of GVHD but delayed hematopoietic recovery compared to BMT recipients. Based on the extensive experience in UCBT, the total nucleated cell dose infused has emerged as one of the most critical factors in determining speed of engraftment and survival after UCBT.[13-15]

Table 7.1 Potential advantages and disadvantages of UCB

Advantages	Disadvantages
Better minority representation and less HLA restriction	May be insufficient cell dose and increased rates of graft failure in larger pediatric patients and adults
Rapid availability	Usually delayed engraftment
Less severe GVHD	No donor recall for boost
Reduced viral contamination of grafts	Uncertain GVL activity
No donor attrition	Uncertain long-term graft durability
Ability to reschedule easily	Risk of EBV-associated post-transplantation lymphoproliferative disorders (PTLD)

Other surrogate markers of stem cells, such as $CD34^+$ cell dose or colony forming units (CFU) dose may provide better correlation to outcome in single institution studies, but impractical in guiding clinical decisions for unit selection due to the large interlaboratory variances in the performance of these tests among the banks. To improve the blood-forming cells and the usually delayed hematopoietic recovery after cord blood transplantation, certain approaches have been investigated, such as ex vivo expansion of cord blood cells, double cord blood transplantation,[16] co-transplantation of cord blood and haplotype-matched peripheral blood stem cells,[17] plasma depletion processing without red cell reduction,[8] foregoing of post-thaw wash,[18] and reduced-intensity conditioning regimen. While there are many potential strategies to address these problems, umbilical cord blood (UCB) represents an important new HSC source, which has a number of significant advantages over MUD BM (Table 7.1).

7.5 Cord Blood Transplantation for Hemoglobinopathies

Despite improvements in supportive care, patients with beta-thalassemia major or sickle cell disease (SCD) may benefit from the cure that hematopoietic stem cell transplantation offers at some point during their lives. HLA-matched sibling bone marrow donors are not always available and alternative sources of stem cells have been sought, including related and unrelated donor CBT. The outcome of CBT from related and unrelated donors for the treatment of both thalassemia major and SCD is now approaching that for bone marrow transplantation, with around 90% of patients surviving disease-free.[19-22] The main complication is graft rejection, which may be reduced by increasing pre-transplant immune suppression, and by having adequate UCB cell dose.[19,20] While only small series of outcome of CBT for hemoglobinopathies have been reported,[19,20] CBT can be considered as therapy in such patients who are without a suitably matched unrelated volunteer donor. The principal limitation to extending the use of CB stem cells for the cure of hemoglobinopathies is the need to find optimal conditioning regimens used to secure long-term durable engraftment while minimizing morbidity and mortality so that the benefit-risk ratio of this procedure is acceptable to patients and families.[23] Further biological studies and clinical trials are needed to address this aim.

7.6 Double Cord Blood Transplants: Filling a Niche?

A limiting factor to cord blood transplantation has been the tenfold lower cell dose in a UCB unit compared with harvested bone marrow or peripheral blood progenitor cells. Associated with lower doses of infused cells are delayed engraftment and poor immune reconstitution. Double cord blood transplants represent a novel strategy to overcome this limitation. The chimerism data from these studies revealed that typically only one CB "wins the battle" for engraftment.[16,20,24,25] These results may have important scientific implications in terms of understanding the nature of the hematopoietic stem cell niche and how modulation of this niche may impact transplant outcomes.

7.7 Biological Characteristics of UCB

Multiple studies exploring the nature of UCB HSC have shown a higher proportion of primitive hematopoietic progenitors in UCB, with superior in vitro hematopoietic responses and in vivo engraftment capacity compared to adult BM; however, the absolute number of progenitor cells are limited by the collected volume, and appear to be approximately one-tenth of that of bone marrow transplants. Over the past decade, a great deal of interest has been generated on the biological characterization of these cells. In spite of the significant advances in the characterization of these cells, we are still trying to fully understand their biology, both at the cellular and the molecular levels.[26]

A unique characteristic of UCB is the reduced alloreactive response as compared to that with BMT. The precise explanation for this phenomenon is unclear. The median $CD3^+$ cell dose of 8×10^6/kg in UCB units makes it similar to a BM graft after T-cell depletion. Such a T-cell dose is fully capable of reducing significant GVHD, particularly in the setting of HLA mismatch.[27] More likely, cord blood lymphocytes possess phenotypic traits suggestive of T-cell immaturity, including increased co-expression of CD45RA on $CD4^+$ cells, decreased expression of interleukin (IL)-2 receptors on $CD3^+$ T-cells. More than 90% of $CD4^+$ cells co-express CD45RA and CD38, and experiments have shown that these cells serve as suppressor cells with nondetectable helper function.[9,28]

7.8 CB Graft Characteristic, Engraftment, and Outcome

Transplantation of unrelated UCB permits a greater degree of HLA mismatching without an unacceptably high incidence of GVHD.[29] Graft characteristics known to allow rapid donor engraftment in recipients of conventional allografts include nucleated cell dose, CD34 content, colony forming content (CFU) dose, and HLA matching. Cell dose measured by number of nucleated cells is the most important factor; thus, increases in cell dose can partially overcome the presence of HLA incompatibilities.[8,11,15] $CD34^+$ cell quantification in UCB has not been consistently predictive of time to donor hematopoietic engraftment due to the large interlaboratory variabilities. Moreover, the poor correlation between CD34 content of infused UCB grafts and time to hematopoietic engraftment may be confounded by quantification of CD34 in UCB grafts prefreezing versus post-thaw and by reduced surface epitopic density of CD34 on UCB progenitor cells.[30]

7.9 Potential Advantages and Disadvantages of UCB

Despite infusion of HLA class I and II disparate grafts, the incidence and severity of acute GVHD observed in pediatric and adult recipients of UCB grafts is lower when compared to those in recipients of unrelated adult donor grafts.[31-34]

As noted in Table 7.1, there are also several potential disadvantages of UCB SCT. The available cell dose may be insufficient for engraftment, especially in larger pediatric patients and adults. Additional cells from the same donor are not available in case of graft failure, and it is also impossible to obtain donor leukocytes for secondary infusions. Furthermore, the GVHD may also limit graft-versus-leukemia (GVL) activity. Finally, there are limited data on the long-term durability of UCB grafts.

7.10 Clinical Results

7.10.1 Related Donor CB Transplantation

Rocha and colleagues compared the outcomes of 2,052 HLA-identical sibling BMT recipient versus 113 HLA-identical sibling donor UCBT recipients transplanted between 1990 and 1997. The 3-year survival after UCBT of 0.64 versus 0.66 in recipients of BMT was nearly identical.[35] These results were confirmed in multivariate analysis. No difference in relapse-related mortality was observed between the two groups. Furthermore, recipients of UCBT from HLA-identical siblings had a lower incidence of acute and chronic GVHD than recipients of BMT from HLA-identical siblings.

7.10.2 Unrelated Donor CB Transplantation

Clinical series reporting preliminary results of unrelated donor UCBT have demonstrated hematopoietic recovery and sustained engraftment in the vast majority of pediatric patients.[2,4,31,35-39] The results from some of the major pediatric, nonregistry studies are outlined in Table 7.2. In general, the pediatric data indicate that UCBT can be successful, even if the patient and cord blood donor are mismatched at two antigens.[37] Rubinstein's group updated the analysis of results reported to the New York Blood Center and reported that, in addition to cell dose, HLA disparity also has a significant impact on UCBT outcome.[41] Interestingly, there were no significant differences in the engraftment

Table 7.2 Studies of unrelated donor umbilical cord blood (UCB) transplantation in children

Investigator	No. of patients	Diseases	Median follow-up time (months)	Disease-free survival (%)
Kurtzberg et al.[25]	25	ALL, AML, MDS, Fanconi, metabolic disorders, other	13	48
Locatelli et al.[26]	60	ALL, AML	14	34
Gluckman et al.[27]	65	ALL, AML, MDS, CML, lymphoma, Fanconi, metabolic disorders, other	10	29
Wagner et al.[20]	102	ALL, AML, CML, lymphoma, Fanconi, metabolic disorder, other[40]	32	47
Michel et al.[28]	95	AML	31	41
Staba et al.[29]	20	Hurler syndrome	30	85
Kobayashi et al.[30]	15	Wiskott–Aldrich syndrome	13	71
Jiang[41]	32	ALL, AML, AA, CML, lymphoma	18	59

ALL indicates acute lymphoblastic leukemia, *AML* acute myeloid leukemia, *MDS* myelodysplastic syndrome, *CML* chronic myeloid leukemia, *AA* aplastic anemia of recipients of 1 HLA antigen-mismatched versus greater than 1 mismatch, likely explaining why the impact of HLA mismatch had not been previously demonstrated by individual institutions.[24]

The "naive" nature of UCB lymphocytes also permits the use of HLA-mismatched grafts at one to two loci without higher risk for severe GVHD relative to BMT from a fully matched unrelated donor. Although some studies suggest that UCB SCT results in less GVHD, no trial results have suggested that UCB SCT increases the risk of GVHD compared to transplantation of cells from other sources. Some transplantation centers now give UCB the priority as the unrelated HSC source of choice in children and, albeit less frequently, also in adults.[42] Despite HLA mismatching at greater than or equal to two loci in the majority of cases, only 30% of patients experienced Grades III to IV acute GVHD in 122 patients at 36 different transplant centers worldwide.[43] Moreover, preliminary data appeared to show that unwashed HLA-mismatched plasma depleted cord blood result in low cumulative incidence of extensive chronic GVHD when compared with the same products when washed, suggesting a potentially novel method to further reduce severe chronic GVHD.[18] However, most recipients of UCB transplants are young, and younger age is also associated with lower rates of GVHD. Furthermore, UCB is rich in primitive $CD16^{-}CD56^{++}$ NK cells, which possess impressive proliferative and cytotoxic capacities and can be induced to expand using IL-12 or IL-15, so as to mount a substantial GVL effect.[44]

The extent to which HLA disparity between recipient and donor correlates with the frequency and severity of GVHD is still unclear. In an analysis of data from 257 UCB transplantations from Duke and the University of Minnesota, no association was identified between acute or chronic GVHD and the degree of HLA matching.[45]

7.11 Future Potential in Regenerative Medicine and Comparisons to Human Embryonic Stem Cells

Human embryonic stem cells (hES cells) have the potential to form any cell in the body. Despite the initial euphoria, the reality is that hES cells have not been approved by the U.S. FDA to treat any disease. The ethical, legal, moral, and religious objections to this process make the whole technology unacceptable to many. Moreover, many of the scientific and medical hurdles, such as tumorigenicity, require breakthroughs that do not exist yet. Apart from these over-riding objections, there are potential concerns regarding the large number of embryos that may be necessary to obtain sufficient cells for a transplant to just one patient. The technologies to transplant these cells in the various indications have yet to be worked out, and the stability and safety of these hES cells, once transplanted, is completely unknown. Many developed countries have now drawn up legislation or codes, or signed up to Conventions, regulating the creation and use of hES cells. There has been a large political and scientific investment in embryonic stem cells, and so far, patients have yet to benefit from these advances. It is conceivable that even with this level of investment of time and money into embryonic stem cells that no solution to these problems will be available within the near future. In the meantime, there are alternative, readily available sources of stem cells for transplantation and clinical trials that are safe and efficacious.

Umbilical cord blood stem cells are obtained from umbilical cord blood collected at the birth of a baby. This blood will otherwise be discarded along with the placenta. There are no legal, moral, ethical, or religious objections to the collection of the cord blood – it is otherwise a biological waste. The umbilical cord blood stem cells are easily isolated using tried and tested technology and can be stored in liquid nitrogen for many decades, quite possibly for the whole lifetime of the baby.

The stem cells, once stored, are not only available for the baby but also for siblings and immediate family if they are matched at four of six HLA antigens with the donor. There are also an increasing number of cord blood stem cell units being offered for transplantation by public cord blood banks worldwide, thus enabling this technology to benefit even more children and adults suffering from potentially fatal diseases.

Umbilical cord blood stem cells appear to possess the ability to isolate cells from all three lineages as well, and in a form suitable for transplant. Since these stem cells have been transplanted over 10,000 times worldwide for over 70 different diseases including leukemia, lymphoma, and nonmalignant diseases, as well as nonhematological indications such as Krabbe's disease [], the safety profile of UCB stem cells is unquestioned. Current research indicates applications in the areas of heart muscle repair and nerve cell repair may be feasible in the future. These transplants have received very little media attention, despite the clear success stories, and politicians remain largely uninformed on the subject.

7.12 Conclusions

Now that more than 10,000 transplants have been performed worldwide, cord blood should no longer be considered a "waste product" but rather a valuable source of hematopoietic stem cells. Umbilical cord blood stem cells have significant advantages, particularly for transplants involving donors and recipients who are related.

Despite the success of SCT using bone marrow and cytokine-mobilized peripheral blood, several major problems continue to impede progress in the improvement of hematopoietic cell transplantation. UCB, usually discarded, is a largely untapped resource. Cord blood cells possess an enhanced capacity for progenitor cell proliferation and self-renewal in vitro, and clinical experience to date indicates that cord blood is a viable alternative to bone marrow and peripheral blood as a source of stem cells capable of hematopoietic reconstitution.

References

1. Gluckman E, Rocha V. History of the clinical use of umbilical cord blood hematopoietic cells. *Cytotherapy*. 2005;7: 219-227.
2. Kurtzberg J, Laughlin M, Graham M, et al. Placental blood as a source of hematopoietic stem cells for transplantation into unrelated recipients. *N Engl J Med*. 1996;335:373.
3. Wagner JE, Kernan NA, Steinbuch M, Broxmeyer HE, Gluckman E. Allogeneic sibling umbilical-cord-blood transplantation in children with malignant and non-malignant disease. *Lancet*. 1995;346:214-219.
4. Gluckman E, Rocha V, Boyer-Chammard A, et al. Outcome of cord-blood transplantation from related and unrelated donors: Eurocord Transplant Group and the European Blood and Marrow Transplantation Group. *N Engl J Med*. 1997;337: 373-381.
5. Rocha V, Wagner J, Boyer Chammard A, et al. Comparison of graft-versus-host disease in children transplanted with HLA identical sibling umbilical cord blood versus HLA identical sibling bone marrow transplant. *N Engl J Med*. 2000;342:1840.
6. Laughlin M, Barker J, Bambach B, et al. Hematopoietic engraftment and survival in adult recipients of unrelated donor umbilical cord blood. *N Engl J Med*. 2001;344:1815.
7. Schmidt AH, Biesinger L, Baier D, Harf P, Rutt C. Aging of registered stem cell donors: implications for donor recruitment. *Bone Marrow Transplant*. 2008;41:605-612.
8. Chow R, Nademanee A, Rosenthal J, et al. Analysis of hematopoietic cell transplants using plasma-depleted cord blood products that are not red blood cell reduced. *Biol Blood Marrow Transplant*. 2007;13:1346-1357.
9. Tse W, Laughlin MJ. Umbilical cord blood transplantation: a new alternative option. *Hematol Am Soc Hematol Educ Prog*. 2005;377–83.
10. Rocha V, Gluckman E. Clinical use of umbilical cord blood hematopoietic stem cells. *Biol Blood Marrow Transplant*. 2006;12(1 Suppl 1):34-41.
11. Rubinstein P. Why cord blood? *Hum Immunol*. 2006;67: 398-404.
12. Gluckman E, Rocha V. Donor selection for unrelated cord blood transplants. *Curr Opin Immunol*. 2006;18:565-570.
13. Rubinstein P, Carrier C, Scaradavou A, et al. Outcomes among 562 recipients of placental-blood transplants from unrelated donors. *N Engl J Med*. 1997;337:373-381.
14. Rocha V, Cornish J, Sievers EL, et al. Comparison of outcomes of unrelated bone marrow and umbilical cord blood transplantation in children with acute leukemia. *Blood*. 2001;97:2962-2971.
15. Bornstein R, Flores AI, Montalbán MA, del Rey MJ, de la Serna J, Gilsanz F. A modified cord blood collection method achieves sufficient cell levels for transplantation in most adult patients. *Stem Cells*. 2005;23:324-334.
16. Majhail NS, Brunstein CG, Wagner JE. Double umbilical cord blood transplantation. *Curr Opin Immunol*. 2006;18: 571-575.
17. Fernandez M, Regidor C, Cabreera R, et al. Unrelated umbilical cord blood transplants in adults: early recovery of neutrophils by supportive co-transplantation of a low number of highly purified peripheral blood CD34+ cells from an HLA-haploidentical donor. *Exp Hematol*. 2003;31:535-544.
18. Chow R, Wang B, Rosenthal J, et al. A novel method to reduce rates of extensive chronic GVHD (cGvHD) without increased relapse for cord blood transplant. *Biol Blood Marrow Transplant*. 2008;14S:11.
19. Jaing TH, Hung IJ, Yang CP, Chen SH, Sun CF, Chow R. Rapid and complete donor chimerism after unrelated mismatched cord blood transplantation in 5 children with beta-thalassemia major. *Biol Blood Marrow Transplant*. 2005; 11:349-353.
20. Jaing TH, Yang CP, Hung IJ, Chen SH, Sun CF, Chow R. Transplantation of unrelated donor umbilical cord blood utilizing double-unit grafts for five teenagers with transfusion-dependent thalassemia. *Bone Marrow Transplant*. 2007;40:307-311.
21. Locatelli F, Rocha V, Reed W, et al. Related umbilical cord blood transplantation in patients with thalassemia and sickle cell disease. *Blood* 2003;101:2137-2143.
22. Kabbara N, Locatelli F, Rocha V, et al. A multicentric comparative analysis of outcomes of HLA-identical related cord blood and bone marrow transplantation in patients with

beta-thalassemia or sickle cell disease. *Biol Blood Marrow Transplant*. 2008;14S:3.
23. Chow R, Bhat R, Petz L, et al. Unrelated HLA-mismatched cord blood transplantation for transfusion-dependent thalassemia and sickle cell disease – is the benefit risk ratio there yet? *VAK*. 2007;2:57-66.
24. Brunstein CG, Baker KS, Wagner JE. Umbilical cord blood transplantation for myeloid malignancies. *Curr Opin Hematol*. 2007;14:162-169.
25. Haspel RL, Ballen KK. Double cord blood transplants: filling a niche? *Stem Cell Rev*. 2006;2:81-86.
26. Mayani H, Lansdorp PM. Biology of human umbilical cord blood-derived hematopoietic stem/progenitor cells. *Stem Cells*. 1998;16:153-165.
27. Barker JN, Wagner JE. Umbilical cord blood transplantation: current practice and future innovations. *Crit Rev Oncol Hematol*. 2003;48:35-43.
28. Wadlow RC, Porter DL. Umbilical cord blood transplantation: where do we stand? *Biol Blood Marrow Transplant*. 2002;8:637-647.
29. Gluckman E. Current status of umbilical cord blood hematopoietic stem cell transplantation. *Exp Hematol*. 2000; 28:1197-1205.
30. Bender J, Unverzagt K, Walker D, et al. Phenotypic analysis and characterization of CD34+ cells from normal human bone marrow, cord blood, peripheral blood, and mobilized peripheral blood from patients undergoing autologous stem cell transplantation. *Clin Immunol Immunopathol*. 1994;70: 10-18.
31. Wagner JE, Barker JN, DeFor TE, et al. Transplantation of unrelated donor umbilical cord blood in 102 patients with malignant and nonmalignant diseases: influence of CD34 cell dose and HLA disparity on treatment-related mortality and survival. *Blood*. 2002;100:1611-1618.
32. Barker JN, Davies SM, DeFor T, Ramsay NK, Weisdorf DJ, Wagner JE. Survival after transplantation of unrelated donor umbilical cord blood is comparable to that of human leukocyte antigen-matched unrelated donor bone marrow: results of a matched-pair analysis. *Blood*. 2001;97:2957-2961.
33. Laughlin MJ, Eapen M, Rubinstein P, et al. Outcomes after transplantation of cord blood or bone marrow from unrelated donors in adults with leukemia. *N Engl J Med*. 2004;351: 2265-2275.
34. Rocha V, Labopin M, Sanz G, Acute Leukemia Working Party of European Blood and Marrow Transplant Group, et al. Eurocord-netcord registry transplants of umbilical-cord blood or bone marrow from unrelated donors in adults with acute leukemia. *N Engl J Med*. 2004;35:2276-2285.
35. Rocha V, Wagner JE, Sobocinski KA, et al. Graft-versus-host disease in children who have received a cord blood or bone marrow transplant from an HLA-identical sibling. *N Engl J Med*. 2000;342:1846-1854.
36. Locatelli F, Rocha V, Chastang C, et al. Factors associated with outcome after cord blood transplantation in children with acute leukemia: Eurocord-Cord Blood Transplant Group. *Blood*. 1999;93:3662-3671.
37. Michel G, Rocha V, Chevret S, et al. Unrelated cord blood transplantation for childhood acute myeloid leukemia: a Eurocord Group analysis. *Blood*. 2003;102:4290-4297.
38. Staba SL, Escolar ML, Poe M, et al. Cord-blood transplants from unrelated donors in patients with Hurler's syndrome. *N Engl J Med*. 2004;350:1960-1969.
39. Kobayashi R, Ariga T, Nonoyama S, et al. Outcome in patients with Wiskott-Aldrich syndrome following stem cell transplantation: an analysis of 57 patients in Japan. *Br J Haematol*. 2006;135:362-366.
40. Jiang XF, Sheng BJ. Clinical study of unrelated umbilical cord blood transplantation in 32 children patients. *Zhongguo Shi Yan Xue Ye Xue Za Zhi*. 2006;14:829-831.
41. Rubinstein P, Stevens CE. Placental blood for bone marrow replacement: the New York Blood Center's program and clinical results. *Baillières Best Pract Res Clin Haematol*. 2000;13:565-584.
42. Dalle JJH, Duval MM, Moghrabi AA, et al. Results of an unrelated transplant search strategy using partially HLA-mismatched cord blood as an immediate alternative to HLA-matched bone marrow. *Bone Marrow Transplant*. 2004;33:605-611.
43. Ballen KK, Haley NR, Kurtzberg J, et al. Outcomes of 122 diverse adult and pediatric cord blood transplant recipients from a large cord blood bank. *Transfusion*. 2006;46:2063-2070.
44. Cohen Y, Nagler A. Umbilical cord blood transplantation: how, when and for whom? *Blood Rev*. 2004;18:167-179.
45. Ballen K, Broxmeyer HE, McCullough J, et al. Current status of cord blood banking and transplantation in the United States and Europe. *Biol Blood Marrow Transplant*. 2001; 7:635-645.

Clinical Experience of Cord Blood Autologous Transfusion

8

Shigeharu Hosono

8.1 Introduction

With rapid advances in perinatal and neonatal intensive care medicine including surgical intervention over the last 2 decades, mortality rate of infants born prematurely or with surgical disease has changed dramatically. Extremely low birth weight infants remain the population at the greatest risk of repeated red blood cell transfusions after introduction of recombinant human erythropoietin (rHu-Epo) therapy.[1,2] Although blood transfusion is one of the valuable cares in the symptomatic anemic patients in all the fields, the use of homologous blood of adult donor has been reported to cause the risks of infections, graft-versus-host disease (GVHD), and allergic reaction. Irradiation of blood products with 1,500 Gy will prevent post-transfusion GVHD.[3] In spite of the improvement in donor screening and testing implemented recently to increase the safety of homologous blood, the risks of acquiring a transfusion-transmitted infection is still in existence, because nucleic acid amplification test only shortens the window periods.[4]

Autologous transfusion prevents the risks of acquiring a transfusion-transmitted infection. The other benefits of autologous transfusion release recipients from allergic reaction. However, it does not make sense to collect the autologous blood from neonates. On the other hand, the placenta, which is its large reservoir of fetal blood, is a potential source of autologous blood. Several investigators reported storage and use of cord blood as a source of autologous RBCs for transfusion to newborns requiring surgery[5-7] and anemia of prematurity.[8-10]

S. Hosono
Department of Pediatrics and Child Health,
Nihon University School of Medicine, Tokyo, Japan

Moreover, there were several reports on the prevention for anemia of prematurity in developed countries[11-18] or iron deficiency in infancy and childhood in developing countries.[19-22]

In addition, autologous transfusion of placental blood in premature babies at birth might be considered as one of the strategies of resuscitation.

8.2 Concepts of Placental Transfusion

Two methods of placental blood transfusions are performed as stored or unstored placental blood transfusion. Stored blood transfusion has three steps: collection, processing of autologous placental blood, and storage. Another is transfusion from placenta to an infant directly through the umbilical cord at birth. Use of stored placental blood is considered to be effective and safe to treat the anemia including anemia of prematurity and blood loss in surgical patients.

On the other hand, delayed or late clamping of the umbilical cord in full-term and preterm neonates as autologous placental transfusion has been discussed in the last 2 decades. Hosono et al. reported milking of the umbilical cord, as a new concept of placental transfusion reduces also the transfusion in premature infants.[23]

8.3 Concept of Stored Autologous Placental Transfusion

The idea of stored placental blood was first described by Goodall et al.[24] The first clinical use of stored placental blood was not autologous transfusion but allogeneic transfusion.[25] Placental blood, stored up to 15 days, was

N. Bhattacharya and P. Stubblefield (eds.), *Regenerative Medicine Using Pregnancy-Specific Biological Substances*,
DOI: 10.1007/978-1-84882-718-9_8, © Springer-Verlag London Limited 2011

used in 220 transfusions and the results were no less favorable than the fresh adult blood used under similar conditions. Reactions were observed in 11 cases for chills and 1 case for dyspnea and tachypnea.

On the other side, the first report of an autologous blood transfusion was published by Paxson et al. in 1978.[26] Twenty-five asphyxic premature infants received transfusion of autologous cord blood within 24 h after birth. In this case, the placental blood was not fractionated into red blood cells and plasma, but unprocessed whole blood was transfused within a few hours.

In 1995, Ballin et al. first reported usefulness of processed autologous cord blood transfusion for treatment of anemia of prematurity.[8] Imura et al. described that autologous cord blood transfusion has the potential to be a useful alternative to homologous transfusion in newborn requiring surgery.[5]

8.4 The Feasibility of Autologous Blood Collection and Safety

Several authors demonstrated that stored placental blood displayed comparable quality characteristics of stored adult blood.[27-29] Average volume of autologous cord-blood-packed red cells was 27 ± 18 mL/kg[8] and 87% of the specimens provide at least 10 mL/kg of cord-blood-packed red cells, while 53% provided 20 mL/kg of cord-blood-packed red cells.[29] The cord-blood-packed red cells significantly correlate with the birth weight[10,29] and approximately 20 mL placental blood per kilogram body weight can be collected on average, independent of birth weight.[10]

Bifano et al. reported that hematocrit, red cell adenosine triphosphate, and red cell shape of placental blood were maintained for 28 days; however, red cell 2,3-diphosphoglycerate declined from 13.30 ± 1.00 on day 0 to 1.30 ± 0.28 μmol/g Hb on day 28 and potassium levels rose from 8.2 ± 3.4 on day 0 to 32.0 ± 5.9 mmol/L on day 28 significantly, but were not different from the levels reported for adult cells similarly preserved.[30]

8.5 Risk of Autologous Cord Blood Transfusion

The risk of bacterial contamination of stored blood is the most important problem.

The higher bacterial contamination rate of 12.0% was reported in collected placental blood by using an open system for collection of placental blood in 1992[29] compared with banked homologous blood.

The increasing interest in cord blood as a source of stem cell forced development of a closed collection system for cord blood and development of standard operating procedures for collection and processing. In 2002, Brune et al. reported that neither aerobe nor anaerobe contamination was observed in 119 of the stored placental blood products after 35 days of storage.[10] Recently, a cord blood banking system has established for stem cell transplantation and gene therapy. Previous report by Garritsen et al. revealed that seven (1.8%) cases of bacterial contaminations were observed in the 390 cases.[31] On the other hand, Eichler et al. reported 3 (8.6%) of 35 samples as positive by the primary bacterial culture using closed placental collection system.[9] A closed blood bag collection system is a safer method of reducing bacterial contamination than an open syringe collection system. However, bacterial contamination still has room for improvement to establish universal approach.

As regards post-transfusion side effects, there was no intergroup difference in heart rate, mean arterial blood flow, or respiratory frequency before, during, and after transfusion and the vital parameters were not influenced by the autologous placental blood red blood cells transfusion itself.[32] Moreover, the serum potassium levels were stable up to 3 days after transfusions and comparable between the autologous transfusions group and the allogenic transfusion group.[32] Taguchi reported that the blood potassium levels, bilirubin levels, and liver enzymes including aspartate transaminase and alanine transaminase after transfusion were similar between the autologous and allogenic transfusion group.[6]

8.6 Stored Autologous Placental Transfusion for Anemia of Prematurity

In 1995, Ballin et al. first reported usefulness of processed autologous cord blood transfusion for treatment of anemia of prematurity.[8] An infant weighing 1,250 g received two portions of autologous blood on days 5 and 7.

Eichler et al. reported that mean Hb increase of 3.6 g/dL in 4 autologous red blood cell transfusions was comparable with 3.9 g/dL in 62 allogeneic red blood cell transfusions when 15–20 mL per kg of body weight of red blood cells was given.[9]

8.7 Stored Autologous Placental Transfusion for Surgical Newborns

There are two clinical studies and a case report focused on autologous transfusion in surgical newborns with an antenatal diagnosis of congenital anomaly.[5-7] Imura et al. described that autologous cord blood was stored from 50 infants and 12 patients did not undergo surgery within 3 days of birth. Twenty-six infants received autologous cord blood transfusion and 9 infants required no homologous transfusion after autologous cord blood transfusion. Bochdalek hernia, gastroschisis, and omphalocele were three major diseases to require the red blood cell transfusions.[5] Taguchu et al. also showed that 10 of 11 infants antenatally diagnosed to have surgical malformations received autologous cord blood transfusions and 7 of 11 infants could avoid allotransfusions.[6]

8.8 Autologous Cord Blood

The present study demonstrates that the collection, fractionation, and storage of red blood cell concentrates derived from cord blood over 35 days are feasible, and autologous placental red blood cells can replace homologous blood in certain categories of patients or at least reduce the demand for homologous blood. However, we should consider benefit-to-cost ratio. Recent study showed that almost all infants weighting over 750 g are able to avoid red blood transfusion for anemia of prematurity under treating erythropoietin therapy.[2]

8.9 Delayed Cord Clamping as Placental Transfusion

Placental transfusion is generally defined as the following. Placental transfusion refers to the net transfer of blood from the placenta to the infant at birth.[33]

Every obstetrician employs a different technique and usually establishes a pattern that he carries out routinely in his work. The optimal timing of clamping the umbilical cord in preterm neonates and even term neonate at birth is the subject of continuing debate.

During the 1960s, there were several studies of the effects on the neonate of varying the time of umbilical cord clamping. This area of investigation became disinterested, but in the past decade there has been renewed investigation of these topics in both preterm and term infants. The blood volume was, on average, increased in the DCC group after at least a 30-s delay for both vaginal and cesarean deliveries.[34]

8.10 Definition of Delayed and Late Cord Clamping

Early, delayed, and late cord clamping have been defined in a variety of ways.

The minimum time for a delayed or late cord clamping is defined as 30 s. On the other hand, maximum delay is to wait cord clamping until pulsation ceased. Past studies used a delay of cord clamping time varying from 30 to 120 s as the intervention. Early cord clamping is defined as clamping the cord at or before 20 s.[35] However, the type of intervention assessed included a mode of delivery, position of an infant with regard to the placenta, oxytocin treatment of the mother, and milking of the cord.

8.10.1 Aim of Placental Transfusion

Extremely low birth weight infants remain the population at the greatest risk of repeated red blood cell transfusions after introducing the recombinant human erythropoietin (rHu-Epo) therapy.[36,37] Several studies have identified initial hemoglobin value as an important factor related to RBC transfusion.[37-39]

Earlier studies have indicated that a delay in cord clamping enhances the placenta-fetal transfusion, resulting in increase in blood volume after birth.[40,41] In 1967, a review article written by Moss revealed that late clamping results in an increased number of erythrocytes, a rise in hemoglobin, and a higher hematocrit.[42] In the 1970s and 1980s, the discussion about this topic was inactive, but international interest has been revived in the intervention following publication of a trial by Kinmond et al. in 1993.[11] In the developing countries, benefits over ages 2–6 months associated with late cord clamping for at least 2 min in full-term infants include improved hematologic status and a clinical important reduction in the risk of anemia.[43] On the other hand, the conclusions of a Cochrane review are that delayed umbilical cord clamping in premature infants appears to reduce the risk of intraventricular hemorrhage and the need for transfusion, either for anemia or for blood pressure support.[44]

Several studies are now on track to establish the hypothesis that an autologous transfusion of placental blood in premature infants at birth will improve systemic flow, reduce ischemic-reperfusion injury to brain and other organs, reduce need for transfusions with adult blood, and improve growth and disability-free survival.

8.10.2 Magnitude of Placental Transfusion

It is estimated that in full-term infants, the total feto-placental blood volume is roughly 120 mL/kg of fetal weight. After immediate cord clamping, the distribution of blood reflected in the fetus:placenta ratio is approximately 2:1. Allowing placental transfusion to occur for 3 min results in a larger fetal volume (ratio 4:1).[45,46]

In contrast, there were three reports in regard to preterm infants. Sagial et al. found mean blood volume of 79.7 mL/kg in the early cord clamping and 89.6 mL/kg in the 1-min delayed cord clamping (28–36 weeks; mean: 33.5 weeks).[47] A recent report by Aladangady et al. revealed that mean blood volume of 74.4 mL/kg in the delayed cord clamping group was significantly greater than that of 62.7 mL/kg in the early cord clamping group.[48]

8.10.3 Physiology of Placental Transfusion

The increase in cardiac output to the lungs from 8% during the fetal life to 45% immediately after birth necessitates transfer of adequate blood volume,[45] because of the rapid opening up of new vascular beds in the lungs.[49] Expressed in another way, early clamping deprives the infant of a substantial volume of blood. This causes volume depletion and systematic and peripheral hypoperfusion, as pulmonary vascular bed must be filled with blood from systemic circulation instead.

8.10.4 Potential Benefits of Placental Transfusion

8.10.4.1 Preterm Infants

Many reports showed that the decrease in the red blood cell transfusion requirements in the delayed cord clamping may be due to the increase in the initial mean hemoglobin concentration, red blood cell counts, and the abundance of hemopoietic stem cells in the cord blood.[11,13,14] With regard to Apgar scores, Ibrahim et al. showed that 5-min Apgar score in the delayed cord clamping group was higher.[13] Possible explanation may be an increase in blood volume, an increase in mean blood pressure, or both. The mean blood pressure in the delayed cord clamping was higher at 4 h of life.[13,50] Hofmeyr et al. revealed that there was a decrease in the incidence of intraventricular hemorrhage in infants born at less than 35 weeks of gestation after 1 min of delayed cord clamping.[51] A latest systematic review by Rabe et al. showed the higher circulating blood volume during the first 24 h of life, less needs for blood transfusions, and less incidences of intraventricular hemorrhage.[35] By reducing adult blood transfusions, placental transfusion may reduce oxygen toxicity to brain, eye, and lung. Moreover, any reduction in intraventricular hemorrhage is important, because of its association with mortality, later morbidity, and/or developmental delay.

A recent report by Mercer et al. showed less incidences of late-onset sepsis in the delayed cord clamping.[52] It is speculated that sepsis may be a result of immunocompromise as a result of loss of protective primitive hematopoietic progenitor cells along with blood volume at birth, because cord blood of preterm infants born between 24 and 31 weeks of gestation contains the highest concentration of primitive hematopoietic progenitor cells.[53]

8.10.4.2 Term Infants

In developing countries, up to 50% of children become anemic by 12 months of age.[54] Low birth weight and maternal iron deficiency are the risk factors for iron deficiency anemia. Iron deficiency anemia during the infancy is of particular concern because of its potentially detrimental effects on development, some of which might be irreversible even after iron supplement.[55] Generally, it is considered that the most important factors that cause iron deficiency are nutritional in origin. Supplementation and iron fortification have had limited success in reducing global iron deficiency, particularly in developing countries. Benefits for term infants by delayed clamping for at least 2 min over age 2–6 months are improved hematocrit, stored iron, ferritin concentration, and reduction in the risk of anemia.[43] The effect of delayed clamping was significantly greater for infants born to mothers with low ferritin at delivery, breast-fed infants not receiving iron-fortified milk or formula, and infants born with birth weight between 2,500 and 3,000 g.[56]

8.10.5 Potential Harms of Placental Transfusion

8.10.5.1 Preterm Infants

In light-for-dates term infants in the developed countries, there is often an increased incidence of polycythemia and hyperviscosity syndrome, because of chronic hypoxia in utero leading to increase erythropoiesis. However, there was no evidence of polycythemia and hyperviscosity syndrome when delayed cord clamping was done with the premature infants.[35]

Extreme hyperbilirubinemia may lead to bilirubin encephalopathy and may follow polycythemia due to the destruction of an increased red cell mass. Peak bilirubin concentration is higher for infants allocated delayed rather than early clamping[44] and infants needing phototherapy.[17] Many studies showed no significant difference in the mean peak serum bilirubin,[13,16] and duration of phototherapy.[14] Data on exchange transfusion are not reported by any of the previous studies. There was no evidence of any significant harm as measured by the need for phototherapy to treat hyperbilirubinemia.

Lower birth weight infants might be expected to have increased risk of heat loss outside a radiant warmer during the 30–90 s of late clamping. On the other hand, continuing perfusion with warm placental blood might prevent a lowering of body temperature. Three studies show that there was no difference in temperature on admission between the two groups.[14,16,50] If the increase in circulating blood volume can be achieved without compromising on the resuscitation of the infants, there was no statistically significant difference between the groups for cord blood pH[14,57] and 5-min Apgar score.[14,16,50,57] McDonnell reported that 3 of 30 infants born at 28–33 weeks of gestation in the delayed cord clamping group broke the protocol to allow resuscitation to proceed.[12]

8.10.5.2 Term Infants

There was no significant difference in the risk of jaundice within 24–48 h[58-60] and at 3–14 days after birth,[56] and in the proportions of infants who had elevated bilirubin levels (>15 g/dL) that necessitated the use of phototherapy.[58,60]

Risk of polycythemia after birth was more common in neonates allocated to late rather than early clamping at 7 h[56,59] and 24–48 h.[59,60] None of the infants with polycythemia had symptoms of hyperviscosity.[61-63]

No significant difference was observed in the risk of developing either tachypnea or respiratory grunting.[58,59,64]

8.11 Milking of Umbilical Cord

Milking or stripping the umbilical cord towards the baby is the procedure of rapidly transferring extra blood from the placenta to the baby. This procedure is far from entirely new concept. It should be noted that Beck recommended this procedure for premature babies in 1941.[65] Several authors reported that cord stripping was considered the efficacious method of improving the infant's blood volume and offered some technical advantages over delayed clamping.[66-68] An editorial note on the article published in 1950 stated that the recommendation for cord stripping could be accepted as safe and also that the long disregard of cord stripping has been a regrettable lapse in technique.[69] However, active discussion concerning the milking of the umbilical cord has not been considered since 1960.

In 2008, Hosono et al. reasserted that milking the umbilical cord is a safe procedure, reducing the need for red blood cell transfusions, and the need for circulatory and respiratory support in very premature infants.[23]

8.11.1 Concept of Milking of the Umbilical Cord

Concept of milking or stripping the umbilical cord is same as that of delayed cord clamping in the placental transfusion. However, milking of umbilical cord is the procedure of placental transfusions, an alternative to delayed cord clamping so as to not interfere with immediate resuscitation. It has been noted that the hemoglobin values at birth increase with increasing gestational age.[70] Rabe et al. pointed out that attempting to transfuse the infants from the placenta, by leaving the cord unclamped for a longer time at birth, may conflict with a perceived need for immediate resuscitation, which is usually done when the neonate is removed from mother.[44] Furthermore, the more seriously ill infants that have a lower birth weight or a shorter gestational age have a greater need for resuscitation at birth and red blood cell transfusions during the first 3 weeks. Such tiny infants might be expected to increase heat loss outside a radiant warmer during 30–45 s of late clamping. Milking of the umbilical cord might contribute to the shorter gestational age or lower birth weight infants such as extremely premature infants when compared with term infants.

8.12 Procedure of Umbilical Cord Milking

The reported procedure is the following. An infant is placed at or below the level of the placenta, and approximately 20 cm of the umbilical cord is vigorously milked towards the umbilicus two to three times before clamping the cord. The milking speed was approximately 20 cm/2 s.[23] Colozzi et al. suggested that milking of the umbilical cord might be repeated when the cord was no longer distended with blood. Four to eight stripping were usually necessary.[68]

8.13 Potential Benefits of Milking of Umbilical Cord

Effect of umbilical cord milking on the hematological features might be thought to be almost the same as that of delayed cord clamping. Small group study but prospective randomized control study showed that hemoglobin values at the first blood sampling in the milked group was significantly higher when compared with the early cord clamping group in premature infants (16.5 vs. 14.1 g/dL) and that no risk of polycythemia was found between the groups.[23] The solitary report described by Colozzi et al. demonstrated that the milking of the umbilical cord achieved higher hemoglobin values at birth compared with delayed cord clamping in term neonates.[68] Umbilical cord milking was associated with higher initial blood pressure and shorter duration of both ventilation and supplemental oxygen.[23]

8.14 Potential Harms of Milking of Umbilical Cord

The effect of rapid volume expansion by the milking of umbilical cord compared with delayed cord clamping on the hemodynamics is unclear. Placental transfusion by delayed cord clamping is equivalent to giving an adult over 1,600 mL of blood per minute. Placental transfusion by milking of cord is considered to be only more so. Previous report revealed that one important risk factor for intraventricular hemorrhage is thought to be rapid volume overload.[71] However, no significant difference in the risk of intraventricular hemorrhage was found between the milking group and the early clamping group.[23] Several reports have emphasized the importance of initial hypotension as an IVH-inducing factor.[72-74] The blood pressure in the milked group was higher than in the control group. The increase in cardiac output to the lung from 8% during fetal life to the 45% immediately after birth necessitates transfer of adequate blood volume.[45] When the umbilical cord is clamped before an adequate placental transfusion to the infant has occurred, blood volume might be drawn out of the systemic capillary beds resulting in relative hypoperfusion. The additional blood received by milking the umbilical cord before the first breath may serve to expand lung capillaries and alveoli causing more initial capillary erection.[49] Thus, in the milked group, pulmonary capillary erection acts as a "reservoir" as a result of the first breath and may minimize the effect of systemic blood pressure.

Another unresolved problem of umbilical cord milking is the damage of the endothelium of cord vessels leading to coagulopathy because of theoretical concerns that repeated milking may damage endothelium of cord vessels.

8.15 Summary

The procedure of delayed cord clamping time of at least 30 s in the preterm infant in the initial postpartum adaptation period appears to be better than clamping within 30 s and the procedure of milking of umbilical cord might be same as or more useful in the tiny baby.

Tucker and McGuire point out that modern perinatal care and the specific interventions of antenatal steroids and exogenous surfactant have contributed to the improved outcome for very preterm infants.[75] Hutchon expected to add "delayed cord clamping" as specific perinatal interventions that have contributed to trends in improved outcomes for very preterm infants.[76] However, there are not yet any data on the effect of placental transfusion neurodevelopmental outcomes in the longer term. Further large trials are needed to provide these data in order to clarify whether the practice of placental transfusion for very preterm infants should be adopted.

References

1. Maier RF, Obladen M, Müller-Hansen I, et al. Early treatment with erythropoietin beta ameliorates anemia and reduces transfusion requirements in infants with birth weights below 1000 g. *J Pediatr*. 2002;141:8-15.
2. Hosono S, Mugishima H, Shimada M, et al. Prediction of transfusions in extremely low-birthweight infants in the erythropoietin era. *Pediatr Int*. 2006;48:572-576.
3. Luban NL. Prevention of transfusion-associated graft-versus-host disease by inactivation of T cells in platelet components. *Semin Hematol*. 2001;38:34-45.
4. Schreiber GB, Busch MP, Kleinman SH, et al. The risk of trasnsfusion-transmitted viral infections. *N Engl J Med*. 1996;334:1685-1690.
5. Imura K, Kawahara H, Kitayama Y, et al. Usefulness of cord-blood harvesting for autologus transfusion in surgical newborns with antenatal diagnosis of congenital anomalies. *J Pediatr Surg*. 2001;36:851-854.
6. Taguchi T, Suita S, Nakamura M, et al. The efficacy of autologous cord-blood transfusions in neonatal surgical patients. *J Pediatr Surg*. 2003;38(4):604-607.
7. Hosono S, Mugishima H, Nakano Y, et al. Autologous cord blood transfusion in an infant with a huge sacrococcygeal teratoma. *J Perinat Med*. 2004;32(2):187-189.
8. Ballin A, Arbel E, Kenet G, et al. Autologous umbilical cord blood transfusion. *Arch Dis Child Fetal Neonatal Ed*. 1995; 73:F181-F183.
9. Eichler H, Schaible T, Richter E, et al. Cord blood as a source of autologous RBCs for transfusion to preterm infants. *Transfusion*. 2000;40:1111-1117.
10. Brune T, Garritsen H, Witteler R, et al. Autologous placental blood transfusion for the therapy of anaemic neonates. *Biol Neonate*. 2002;81:236-243.
11. Kinmond S, Aitchison TC, Holland BM, Jones JG, Turner TL, Wardrop CA. Umbilical cord clamping and preterm infants: a randomised trial. *BMJ*. 1993;306:172-175.
12. McDonnell M, Henderson-Smart DJ. Delayed umbilical cord clamping in preterm infants: a feasibility study. *J Paediatr Child Health*. 1997;33:308-310.
13. Ibrahim HM, Krouskop RW, Lewis DF, Dhanireddy R. Placental transfusion: umbilical cord clamping and preterm infants. *J Perinatol*. 2000;20(6):351-354.
14. Rabe H, Wacker A, Hülskamp G, et al. A randomised controlled trial of delayed cord clamping in very low birth weight preterm infants. *Eur J Pediatr*. 2000;159(10):775-777.
15. Strauss RG, Mock DM, Johnson K, et al. Circulating RBC volume, measured with biotinylated RBCs, is superior to the Hct to document the hematologic effects of delayed versus immediate umbilical cord clamping in preterm neonates. *Transfusion*. 2003;43(8):1168-1172.
16. Mercer JS, McGrath MM, Hensman A, Silver H, Oh W. Immediate and delayed cord clamping in infants born between 24 and 32 weeks: a pilot randomized controlled trial. *J Perinatol*. 2003;23(6):466-472.
17. Strauss RG, Mock DM, Johnson KJ, et al. A randomized clinical trial comparing immediate versus delayed clamping of the umbilical cord in preterm infants: short-term clinical and laboratory endpoints. *Transfusion*. 2008;48(4):658-665. Epub Jan 10, 2008.
18. Ultee CA, van der Deure J, Swart J, Lasham C, van Baar AL. Delayed cord clamping in preterm infants delivered at 34 36 weeks' gestation: a randomised controlled trial. *Arch Dis Child Fetal Neonatal Ed*. 2008;93(1):F20-F23.
19. Geethanath RM, Ramji S, Thirupuram S, Rao YN. Effect of timing of cord clamping on the iron status of infants at 3 months. *Indian Pediatr*. 1997;34(2):103-106.
20. Gupta R, Ramji S. Effect of delayed cord clamping on iron stores in infants born to anemic mothers: a randomized controlled trial. *Indian Pediatr*. 2002;39(2):130-135.

21. Chaparro CM, Neufeld LM, Tena Alavez G, Eguia-Líz Cedillo R, Dewey KG. Effect of timing of umbilical cord clamping on iron status in Mexican infants: a randomised controlled trial. *Lancet*. 2006;367(9527):1997-2004.
22. Ceriani Cernadas JM, Carroli G, Pellegrini L, et al. The effect of timing of cord clamping on neonatal venous hematocrit values and clinical outcome at term: a randomized, controlled trial. *Pediatrics*. 2006;117(4):e779-e786.
23. Hosono S, Mugishima H, Fujita H, et al. Umbilical cord milking reduces the need for red cell transfusions and improves neonatal adaptation in infants born at less than 29 weeks' gestation: a randomised controlled trial. *Arch Dis Child Fetal Neonatal Ed*. 2008;93:F14-F19.
24. Goodall JR, Andersen FO, Altimas GT, MacPhall FL. An inexhaustible source of blood for transfusion and its preservation. *Surg Gynecol Obstet*. 1938;66:176.
25. Halbrecht J. Fresh and stored placental blood. *Lancet*. 1939;2:1263-1265.
26. Paxson CL Jr. Collection and use of autologous fetal blood. *Am J Obstet Gynecol*. 1979;134(6):708-710.
27. Brandes JM, Roth EF Jr, Berk PD, et al. Collection and preservation of human placental blood. *Transfusion*. 1983; 23(4):325-327.
28. Horn S, Mazor D, Zmora E, Meyerstein N. Storage-induced changes in human newborn red cells. *Transfusion*. 1987; 27(5):411-414.
29. Anderson S, Fangman J, Wager G, Uden D. Retrieval of placental blood from the umbilical vein to determine volume, sterility, and presence of clot formation. *Am J Dis Child*. 1992;146(1):36-39.
30. Bifano EM, Dracker RA, Lorah K, Palit A. Collection and 28-day storage of human placental blood. *Pediatr Res*. 1994; 36(1 Pt 1):90-94.
31. Garritsen HS, Brune T, Louwen F, et al. Autologous red cells derived from cord blood: collection, preparation, storage and quality controls with optimal additive storage medium (Sag-mannitol). *Transfus Med*. 2003;13(5):303-310.
32. Brune T, Garritsen H, Hentschel R, Louwen F, Harms E, Jorch G. Efficacy, recovery, and safety of RBCs from autologous placental blood: clinical experience in 52 newborns. *Transfusion*. 2003;43(9):1210-1216.
33. Yao AC, Lind J. *Placental Transfusion. A Clinical and Physiological Study*. Springfield: Charles C. Thomas; 1982.
34. Aladangady N, McHugh S, Aitchison TC, Wardrop CAJ, Holland BM. Infants' blood volume in a controlled trial of placental transfusion at preterm delivery. *Pediatrics*. 2006;117:93-98.
35. Rabe H, Reynolds G, Diaz-Rossello J. A systematic review and meta-analysis of a brief delay in clamping the umbilical cord of preterm infants. *Neonatology*. 2008;93:138-144.
36. Maier RF, Obladen M, Kattner E, et al. High-versus low-dose erythropoietin in extremely low birth weight infants. The European Multicenter rhEPO Study Group. *J Pediatr*. 1998;132:866-870.
37. Hosono S, Mugishima H, Shimada M, et al. Prediction of transfusions in extremely low-birthweight infants in the erythropoietin era. *Pediatr Int*. 2006;48:572-576.
38. Maier RF, Obladen M, Müller-Hansen I, et al. Early treatment with erythropoietin beta ameliorates anemia and reduces transfusion requirements in infants with birth weights below 1000 g. *J Pediatr*. 2002;141:8-15.
39. Paul DA, Pearlman SA, Leef KH, Stefano JL. Predicting red blood cell transfusions in very low birth weight infants based on clinical risk factors. *Del Med J*. 1997;69:555-561.
40. DeMarsh QB, Windle WF, Alt HL. Blood volume of newborn infant in relation to early and late clamping of umbilical cord. *Am J Dis Child*. 1942;63:1123.
41. Usher R, Shephard M, Lind J. The blood volume of the newborn infant and placental transfusion. *Acta Paediatr*. 1963;52:497-512.
42. Moss AJ, Monset-Couchard M. Placental transfusion: early versus late clamping of the umbilical cord. *Pediatrics*. 1967;40:109-126.
43. Hutton EK, Hassan ES. Late vs early clamping of the umbilical cord in full-term neonates: systematic review and meta-analysis of controlled trials. *JAMA*. 2007;21(297):1241-1252.
44. Rabe H, Reynolds G, Diaz-Rossello J. Early versus delayed umbilical cord clamping in preterm infants. *Cochrane Database Syst Rev*. 2004;18(4):CD003248.
45. Linderkamp O. Placental transfusion: determinants and effects. *Clin Perinatol*. 1982;9:559-592.
46. Yao AC, Moinian M, Lind J. Distribution of blood between infant and placenta after birth. *Lancet*. 1969;2(7626):871-873.
47. Saigal S, O'Neill A, Surainder Y, Chua LB, Usher R. Placental transfusion and hyperbilirubinemia in the premature. *Pediatrics*. 1972;49:406-419.
48. Aladangady N, McHugh S, Aitchison TC, Wardrop CA, Holland BM. Infants' blood volume in a controlled trial of placental transfusion at preterm delivery. *Pediatrics*. 2006; 117:93-98.
49. Lind J. Physiological adaptation to the placental transfusion: the eleventh blackader lecture. *Can Med Assoc J*. 1965;93: 1091-1100.
50. Baenziger O, Stolkin F, Keel M, et al. The influence of the timing of cord clamping on postnatal cerebral oxygenation in preterm neonates: a randomized, controlled trial. *Pediatrics*. 2007;119:455-459.
51. Hofmeyr GJ, Bolton KD, Bowen DC, Govan JJ. Periventricular/intraventricular haemorrhage and umbilical cord clamping. Findings and hypothesis. *S Afr Med J*. 1988; 23(73):104-106.
52. Mercer JS, Vohr BR, McGrath MM, Padbury JF, Wallach M, Oh W. Delayed cord clamping in very preterm infants reduces the incidence of intraventricular hemorrhage and late-onset sepsis: a randomized, controlled trial. *Pediatrics*. 2006;117:1235-1242.
53. Haneline LS, Marshall KP, Clapp DW. The highest concentration of primitive hematopoietic progenitor cells in cord blood is found in extremely premature infants. *Pediatr Res*. 1996;39:820-825.
54. ACC/SCN. Preventing and treating anemia. In: Allen LH, Gillespie SR, eds. *What Works? A Review of the Efficacy and Effectiveness of Nutrition Interventions*. Geneva: ACC/SCN in collaboration with the Asian Development Bank, Manila; 2001:43-54.
55. Grantham-McGregor S, Ani C. A review of studies on the effect of iron deficiency on cognitive development in children. *J Nutr*. 2001;131:649S-666S.
56. Chaparro CM, Neufeld LM, Tena Alavez G, Eguia-Líz Cedillo R, Dewey KG. Effect of timing of umbilical cord clamping on iron status in Mexican infants: a randomised controlled trial. *Lancet*. 2006;367:1997-2004.

57. Hofmeyr GJ, Gobetz L, Bex PJ, Van der Griendt M, Nikodem C, Skapinker R, Delahunt T. Periventricular/intraventricular hemorrhage following early and delayed umbilical cord clamping. A randomized controlled trial. *Online J Curr Clin Trials*. 1993;Doc No 110.
58. Nelson NM, Enkin MW, Saigal S, Bennett KJ, Milner R, Sackett DL. A randomized clinical trial of the Leboyer approach to childbirth. *N Engl J Med*. 1980;302:655-660.
59. Ceriani Cernadas JM, Carroli G, Pellegrini L, et al. The effect of timing of cord clamping on neonatal venous hematocrit values and clinical outcome at term: a randomized, controlled trial. *Pediatrics*. 2006;117:e779-e786.
60. Emhamed MO, van Rheenen P, Brabin BJ. The early effects of delayed cord clamping in term infants born to Libyan mothers. *Trop Doct*. 2004;34:218-222.
61. van Rheenen PF, Gruschke S, Brabin BJ. Delayed umbilical cord clamping for reducing anaemia in low birthweight infants: implications for developing countries. *Ann Trop Paediatr*. 2006;26:157-167.
62. Nelle M, Kraus M, Bastert G, Linderkamp O. Effects of Leboyer childbirth on left- and right systolic time intervals in healthy term neonates. *J Perinat Med*. 1996;24:513-520.
63. Linderkamp O, Nelle M, Kraus M, Zilow EP. The effect of early and late cord-clamping on blood viscosity and other hemorheological parameters in full-term neonates. *Acta Paediatr*. 1992;81:745-750.
64. Yao AC, Lind J, Vuorenkoski V. Expiratory grunting in the late clamped normal neonate. *Pediatrics*. 1971;48:865-870.
65. Beck AC. How can the obstetricians aid in reducing the mortality of prematurely born infants? *Am J Obstet Gynecol*. 1941;42:355-364.
66. Siddall RS, Crissey RR, Knapp WL. Effect on Cesarean section babies of stripping or milking of the umbilical cord. *Am J Obst Gynecol*. 1952;63:1059-1064.
67. Siddall RS, Richardson RP. Milking or stripping the umbilical cord; effect on vaginally delivered babies. *Obstet Gynecol*. 1953;1:230-233.
68. Colozzi AE. Clamping of the umbilical cord – its effect on the placental transfusion. *N Engl J Med*. 1954;250:629-632.
69. McCausland AM, Holmes F, Schumann WR. Management of cord and placental blood and its effect upon the newborn. Part II. *West J Surg Obstet Gynecol*. 1950;58:591-608.
70. Luchtman-Jones L, Schwartz AL, Wilson DV. The blood and hematopoietic system PP1287-1356. In: Marchin RJ, Fanaroff AA, Walsh MC, eds. *Fanaroff and Martin's Neonatal-Perinatal Medicine: Disease of the Fetus and Infant*. 8th ed. Amsterdam: Elsevier Mosby; 2006: Chap. 44.
71. Goldberg RN, Chung D, Goldman SL, Bancalari E. The association of rapid volume expansion and intraventricular hemorrhage in the preterm infant. *J Pediatr*. 1980;96:1060-1063.
72. Fujimura M, Salisbury DM, Robinson RO, et al. Clinical events relating to intraventricular haemorrhage in the newborn. *Arch Dis Child*. 1979;54:409-414.
73. Miall-Allen VM, de Vries LS, Dubowitz LM, Whitelaw AG. Blood pressure fluctuation and intraventricular hemorrhage in the preterm infant of less than 31 weeks' gestation. *Pediatrics*. 1989;83:657-661.
74. Funato M, Tamai H, Noma K, et al. Clinical events in association with timing of intraventricular hemorrhage in preterm infants. *J Pediatr*. 1992;121:614-619.
75. Tucker J, McGuire W. Epidemiology of preterm birth. *BMJ*. 2004;329:675-678.
76. Hutchon DJ. Epidemiology of preterm birth: delayed cord clamping used to be taught and practised. *BMJ*. 2004;329:1287. Author reply.

Emergency Use of Human Cord Blood

9

Norman Ende, Kathleen M. Coakley, and Kenneth Swan

Historically, the use of placental blood was first advocated by George Rubin of New York in 1914. From the nature of the publication, it was utilized in emergency or semi-emergency conditions.[1] In 1934, Soviet surgeon M.S. Malinovski reported on placental blood and 2 years later Bruskin and Farberova reported on 114 transfusions preserved for 6–10 days. Katorovich used blood from the placenta in 1935 for "massive transfusions,"[2,3] which would indicate emergency use of cord blood.

Goodall utilized umbilical cord blood in Canada prior to World War II and some of it may have been used in emergency situations.[4] The amount varied from 100 to 150 cc and was stored up to 60 days. The authors considered fetal blood as an "inexhaustible source" of blood for transfusion and noted "in the many transfusions with cord blood, there has not been one untoward reaction, not a single rise in temperature even to a fraction of a degree."[4] Similarly, Boland utilized cord blood for emergency situations combining adult and placental blood as a "mixture."[5] Cord blood was not utilized in World War II.[6]

In the current literature, beyond Bhattacharya's recommendation for Third World countries to utilize cord blood when adult blood is not available, there is little mention of cord blood for emergency use[7] with one exception, radiation disasters.[8,9]

In 1958, six physicists were exposed to large doses of neutron irradiation following an accident at Veneza, Yugoslavia. They were transported to France and treated with multiple transfusions of homologous bone marrow. Four victims were believed to have had a successful temporary bone marrow graft.[10]

The first temporary human cord blood transplant occurred in Petersburg, Virginia, in 1970.[11] Multiple units of cord blood from different donors were administered to a patient with lymphoblastic leukemia. This case was only matched for A B O blood groups and minimal immunosuppression was used.

In the hope of replacing the bone marrow of the irradiated victims at the Chernobyl disaster in April 1986, HLA matched or partially matched bone marrow was administered to treat a few of the victims. The data obtained during this disaster were limited.[12,13] It was concluded, however, that these transfusions of bone marrow cells were helpful and allowed some patients to recover their own hematopoietic system. Although cells obtained from fetal livers were utilized unsuccessfully in this tragedy, human umbilical cord blood transplants were not attempted.[12,13]

Recent literature, however, has confirmed these earlier studies, that multiple units of cord blood, partially matched for HLA, can be transfused successfully and produce a chimera.[14,15] The current literature indicates that multiple cord blood units can be given, thereby diminishing the necessity of having a close HLA match for a successful transplant.

Previously, there was limited information on immunosuppressed patients receiving greater than two units of cord blood. However, from 1964 to 1974, 139 units of cord blood of various volumes were administered to 15 patients dying from malignancies.[16] The patients were receiving the conventional therapy at that time, which could be considered suppressive, and no adverse events were noted in these patients. A recent review of these cases suggest that at least four transplants occurred and several patients showed a significant rise in hemoglobin concentrations. In addition, on review of their medical records, the only adverse reaction noted was one case of cold antibodies.

N. Ende (✉)
Department of Pathology and Laboratory Medicine
and Department of Surgery, New Jersey Medical School,
University of Medicine and Dentistry of New Jersey, USA

N. Bhattacharya and P. Stubblefield (eds.), *Regenerative Medicine Using Pregnancy-Specific Biological Substances*,
DOI: 10.1007/978-1-84882-718-9_9,

In 1993, Shen reported four patients, transplanted, who received 15 units of cord blood unmatched for HLA with little evidence of Graft-versus-Host disease (GVHD).[17] Partial donor cell engraftment was demonstrated in two of the patients by the presence of Y chromosomes.

Although there were published cases where multiple units of cord blood had been utilized successfully for transplantation, the previous general medical consensus had favored the use of one unit, closely matched to provide the most successful transplant in children. One unit, however, was considered inadequate to provide enough cells for a successful transplant in adults.

Only recently have multiple cord blood units, partly matched, become more frequently utilized to produce successful transplants in adults.[18] In cases where multiple units are utilized, cells from one donor or more may persist. There is one notable incident of severe GVHD; little detail is given concerning this case.[19] In a recent review, two partially HLA-matched cord blood units given after marrow suppression or ablation indicated greater implantation and less GVHD than with bone marrow transplants. When two units of partially matched cord blood were given to marrow-ablated patients, in one publication at least 24% developed a chimera of both donors and one dominated.[14,19,20]

In more current studies (2007), five to seven cord blood units were administered to patients (one unit per 10 kg). Later, HLA typing revealed the presence of a single HLA-type concordant with one of the infused units.[15]

As far as we know, the only patient who received a cord blood transfusion, HLA partially matched, for exposure to lethal levels of irradiation occurred in Tokai, Japan.[21] The patient developed a chimera and rapid autogenous hematopoietic recovery; however, his immune functions were impaired. The patient later developed renal failure and respiratory problems. He did not develop graft-versus-host disease and lived for 210 days.

In 1992 in our laboratory, we first became aware that human cord blood mononuclear cells could produce survivors in lethally irradiated mice.[22] The administration of cells not previously frozen, from multiple units, produced superior clinical results in the animals when compared with blood obtained from a single donor. In addition, radiation survival directly correlated with cord blood mononuclear cell dose.[9,10,22] We have been able to obtain high rates of survival in lethally irradiated mice and in some instances 100% survival. No additional immunosuppression was used. The mice receiving cord blood appear to have two types of responses to the transfusions following irradiation – a chimera with evidence of engraftment lasting up to 12 months,[23] and stimulation of the host's own immune system to recover.[24,25] In these multiple studies, with or without immunosuppressant, no evidence of GVHD was noted in animals.[26,27]

In addition to human cord blood being effective in replacing bone marrow destroyed by irradiation, it might also assist in wound healing. A recent publication indicating human mesenchymal stem cell derived from bone marrow home specifically to radiation-injured tissue in mice.[28] Further dermal-derived multipotent cells promote survival and wound healing in rats with combined radiation and wound injury.[29,30] Recently it has been shown that human cord blood has a similar effect.[29,31]

Although there are multiple examples of treatment of 100 or more victims of radiation disasters, there has been little or no planning for large number (tens of thousands) of casualties. Based on the above considerations, we have developed a theoretical scenario for the treatment of mass casualties from a nuclear explosion (attached). Subsequent plans will certainly be modified by new growth factors, ex vivo expansion of stem cells, protection of the intestinal tract, etc. The essential first requirement for survival, however, following a nuclear explosion, will still *need replacement or recovery of the patient's bone marrow*.

Theoretical Treatment of Mass Casualties Resulting from a Nuclear Explosion in Manhattan: (Scenario based on 10,000 casualties transported across the Hudson river to nearby New Jersey).

Logistics, communication, housing, and transportation of casualties will not be covered in this report, only matters directly related to triage and treatment of radiation and trauma casualties. Emphasis is based more on patient survival than triage.

The report covers:

1. Continuously functioning trauma and burn triage for the first 72 h based on Hospital A (Hospitals B & C are extrapolations of Hospital A). All hospitals are capable of handling trauma.
2. Therapy for mass casualties (radiation and trauma) for the immediate response (72 h) period.
3. Blood supply to support emergency surgery for first 72 h. Three hospitals (A, B, C) available to triage patients and handle trauma, operating at full capacity for 72 h.

(a) Twelve operating rooms
(b) Eighteen operating rooms
(c) Eighteen operating rooms

Assumptions Concerning Triage Assessments of 10,000 casualties

1. Seventy-two hours period of surgical resuscitation
2. Three hours of operating room (OR) time per patient
3. Forty-eight operating rooms × 72 h = 3,456/3 h per patient =1,152 patient capability. All OR personnel, including surgeons, anesthesiologists, nurses, technicians, will be "on" 12 h, "off" 12 h.
4. Two surgeons (primary and secondary, or first assistant surgeon) needed per OR = 96 primary surgeons, 96 first assistant surgeons.
5. Each OR case requires an estimated two units of blood replacement.

If the disaster is the result of conventional explosion or "dirty bomb," casualties will be primarily trauma and secondarily radiation. An estimated 25% will need emergency surgical care. Radiation casualties will be far less than in a nuclear explosion and conventional treatment (as in 100 or less casualties) may be used.

For every 10,000 casualties, an estimated 65% are "walking wounded," 10% are "expectant" and 25% are "priority." All the latter, "…those who probably will survive, with a meaningful survival, if resources (time, personal, and equipment) are used most efficiently…," require operative intervention (surgery OR time).

Therefore:

1. 25% of 10,000 = 2,500 patients need OR time.
2. 3 h per patient = 7,500 h of surgery required.
3. 48 OR × 72 h = 3,456 h available/3 = 1,152 patients.
4. Therefore, if the number of OR cases increases, the number of surgeons and other OR personnel must also increase by 3.
5. 1,448 casualties will need to be relocated to other facilities.

If the disaster is the result of a nuclear explosion, radiation injury will be disproportionately greater than trauma/burn injury; estimating 10% need emergency surgical care.

For every 10,000 casualties, 90% will be "walking wounded" and 10% will require immediate surgical care.

Therefore:

1. 10% of 10,000 = 1,000 patients need OR time
2. 3 h per patient = 3,000 h of surgery required
3. 48 operating rooms between three hospitals × 72 h = 3,456 h (/3 = 1152) patients
4. 1,000 OR cases × 2 units of blood = 2,000

9.1 Radiation Casualties of Fewer Than 100

Based on the existing information on stem cell therapy (American Association of Blood Banks (AABB) Protocol for the Emergency Management of Radiation Victims, March 2006), in radiation casualties fewer than 100, each patient should receive one to two units of cord blood (1–2 billion mononuclear cells, blood type specific, and partially or completely HLA-matched). These patients will receive antibiotics, various cytokines, and colony growth factors as utilized in marrow injury (current, standard of care treatment). Whole blood or packed cell transfusions for trauma patients will be irradiated according to current American Association of Blood Banks (AABB) concepts (AABB Protocol for the emergency Management of Radiation Victims).

9.2 Disproportionately More Radiation Victims

In a scenario where a ground-level nuclear explosion occurs in Manhattan, radiation victims will be disproportionately more frequent than trauma or burn victims. Tall buildings vent burn and blast injury upward; radiation will affect many without their knowledge. First responders will most likely need treatment for radiation exposure (Homeland Security Council and Department of Homeland Security's National Planning Scenario: Radiological Attack, Scenario 11, April 2005). There will be a large number of "walking wounded," unaware of having received irradiation, while those with trauma will be presumed to have received varying degrees of irradiation.

9.3 Blood for the Operating Rooms

Type specific, whole blood is recommended for transfusion in mass disaster patients in need. Based on three hospitals (48 operating units) operating continuously, it is estimated that 2,500 units of blood will be needed for the first 72-h period.

On any given day in New Jersey, approximately 1,000 units of adult blood are available for patient transfusion. Therefore, additional blood will be needed and varying amounts of processing, beyond blood typing, may be performed. For example, irradiation of blood (recommended by AABB) can be logistically difficult and should be utilized only in an emergency consisting of 100 or fewer casualties.

9.4 1,000–10,000 Casualties

All first responders, as well as trauma or burn victims, should be considered to have received a significant level of irradiation and should be treated with one to three cord blood units, depending on availability. Those not treated for trauma or burn but suspected of significant radiation exposure will be treated (paramedics, volunteers, etc.) by one to three cord blood units (frozen or fresh) matched by blood-group, depending on availability. [If there is adequate number of cord blood units and there is personnel and supplies, one to two units (2 billion nucleated cells) will be blood-type and HLA-matched.[19]] *It is highly unlikely that this level of support could be carried out in the foreseeable future.* Therefore, it is recommended that one to three units of cord blood, depending on availability, matched for blood group only, be administered by paramedics to all patients suspected of exposure to lethal levels of irradiation. Three units per patient would be preferred. All patients, with or without trauma, suspected of radiation injury would receive a broad spectrum antibiotic, Levifloxacin (Levaquine) could be given by first responders. Since radiation patients frequently develop vomiting, when they arrive at their destination, a broad spectrum antibiotic could be administered by volunteers or by paramedics. Patient previously treated with cord blood and suspected of irradiation injury should return in 72 h for evaluation when medical emergency support becomes available (FEMA) and treated in a conventional manner.

9.5 Source of Cord Blood

Based on the available information from the existing cord blood banks, 9,000–11,000 frozen, undesignated units are available for emergency use in New Jersey. This does not include the National Cord Blood depository (planned for 150,000 units). On December 20, 2005, Stem Cell therapeutics and Research Act of 2005 became Public Law 109–129 and provides for storage of 150,000 units of cord blood. This National Cord Blood Inventory will be available for emergency use.

9.6 Back Up

All hospitals with facilities to collect cord blood should immediately start collection of umbilical cord blood, with the intention of transferring the units to appropriate treatment centers, processing only for blood type and only the minimal additional tests deemed necessary.[32]

9.7 72 Hours

At the end of 72 h with arrival of national support (FEMA), the medical facilities will begin to revert treatment of casualties to "standard of care" appropriate to the available support.

9.8 Critical Issues

Good Samaritan law in NJ exempts paramedics from malpractice law suits in time of an emergency; however, participating physicians may not be exempt from malpractice suits. This must be corrected on a national basis.

9.9 During the Emergency

Available paramedics and volunteers must be identified in large numbers to administer aid, particularly fluids, to the "walking wounded" (radiation victims without significant visible trauma).

Primary and assistant surgeons must be organized, identified, and available for call.

Many specialized surgeons will not wish to take primary responsibility for the patient but would be first assistants.

9.10 Summary

Currently, there are 9–11,000 undesignated, frozen cord blood units in the state of New Jersey. The three hospitals could conduct emergency life-support surgery on 1,000 (1,152) patients and several thousand "walking wounded" with the adoption of a plan and organization of paramedical support. At this point, the establishment of a logistical plan could potentially save thousands. Currently, no money is needed, simply a plan for basic communication, logistics, and use of the existing facilities.

When disaster occurs and mass casualties result, emphasis is placed on "triage" including radiation disasters. (This is primarily intended to preserve resources and should be exercised on radiation victims with caution.) To exercise triage on overt injuries and burns would be correct. *To do this, however, on mass casualties that have radiation injury and no overt trauma would probably be unwise*. These patients would be difficult to detect and it would be more advisable after the initial 72 h to triage cautiously if conditions and resources warranted it. Survival may vary markedly with different individuals.

Graft-versus-Host disease may occur, particularly in patients who have their marrow completely ablated, but this would be difficult to detect in the first 72 h. When cord blood is utilized, GVHD can usually be adequately treated.

When dealing with mass casualties 10,000 or more, it is critical to coordinate with adjacent cities and states within the first 72 h. Large numbers of volunteers should be appropriately assigned to areas and fully utilized. Dehydration will rapidly claim large numbers of victims due to radiation-induced vomiting. The large numbers of volunteers would be necessary to provide basic needs of fluid and nutrition for the injured as well as medications.

With considerable effort being supported by the government for treatment of radiation victims, the plan described herein will be improved. However, regardless of the final decision on the treatment, *radiation victims basically must have their destroyed marrow replaced.*

Acknowledgment We thank Lynn Baltimore, Senior Research Librarian at the New Jersey Medical School, for assisting in researching the literature for this manuscript and Donald Allegra, M.D., Infectious Disease, for advice.

This chapter is dedicated to the late Dr. Milton Ende and his staff, the physicians and staff of Obstetrical service of Southside Regional Medical Center, Petersburg, Virginia and the Abraham S. Ende Research Foundation that participated in cord blood research since the early 1960s and performed the first temporary cord blood transplant in a patient with lymphoblastic leukemia.

References

1. Rubin G. Placenta blood for transfusion. *N Y Med J*. 1914; 100:421.
2. Podolsky E. *Red Miracle; The Story of Soviet Medicine*. Freeport: Books for Libraries Press; 1972.
3. Drew CR. The role of soviet investigators in the development of the blood bank. *Am Rev Sov Med*. 1943;1(4):360-369.
4. Goodall JR, et al. An inexhaustible source of blood for transfusion and its preservation. *Surg Gynecol Obstet*. 1938;66:176-178.
5. Boland CR, Craig NS, Jacobs AL. Collection and transfusion of preserved blood. *Lancet*. 1939;1:388-391.
6. Coates JB. *Blood Program in World War II*. Washington: Office of the Surgeon General, Dept. of the Army, United States Army Medical Service; 1964.
7. Bhattacharya N. Placental umbilical cord whole blood transfusion: a safe and genuine blood substitute for patients of the under-resourced world at emergency. *J Am Coll Surg*. 2005;200(4):557-563.
8. Ende N, et al. Potential effectiveness of stored cord blood (non-frozen) for emergency use. *J Emerg Med*. 1996;14(6): 673-677.
9. Ende N, et al. Pooled umbilical cord blood as a possible universal donor for marrow reconstitution and use in nuclear accidents. *Life Sci*. 2001;69(13):1531-1539.
10. Mathe G, et al. Transfusions and grafts of homologous bone marrow in humans after accidental high dosage irradiation. *Rev Fr Étud Clin Biol*. 1959;4(3):226-238.
11. Champlin R. The role of bone marrow transplantation for nuclear accidents: implications of the Chernobyl disaster. *Semin Hematol*. 1987;24(3 Suppl 2):1-4.
12. Linnemann RE. Soviet medical response to the Chernobyl nuclear accident. *JAMA*. 1987;258(5):637-643.
13. Barker JN, et al. Transplantation of 2 partially HLA-matched umbilical cord blood units to enhance engraftment in adults with hematologic malignancy. *Blood*. 2005; 105(3):1343-1347.
14. Lister J, et al. Multiple unit HLA-unmatched sex-mismatched umbilical cord blood transplantation for advanced hematological malignancy. *Stem Cells Dev*. 2007;16(1):177-186.
15. Ende M. History of umbilical cord blood transplantation. *Lancet*. 1995;346(8983):1161.
16. Shen BJ, et al. Unrelated, HLA-mismatched multiple human umbilical cord blood transfusion in four cases with advanced solid tumors: initial studies. *Blood Cells*. 1994;20(2–3): 285-292.

17. Weinreb S, et al. Transplantation of unrelated cord blood cells. *Bone Marrow Transplant.* 1998;22(2):193-196.
18. Schoemans H, et al. Adult umbilical cord blood transplantation: a comprehensive review. *Bone Marrow Transplant.* 2006;38(2):83-93.
19. De Lima M, et al. Double-chimaerism after transplantation of two human leucocyte antigen mismatched, unrelated cord blood units. *Br J Haematol.* 2002;119(3):773-776.
20. Nagayama H, et al. Transient hematopoietic stem cell rescue using umbilical cord blood for a lethally irradiated nuclear accident victim. *Bone Marrow Transplant.* 2002;29(3):197-204.
21. Ende N, et al. Murine survival of lethal irradiation with the use of human umbilical cord blood. *Life Sci.* 1992;51(16): 1249-1253.
22. Ende N, et al. The effect of human cord blood on SJL/J mice after chemoablation and irradiation and its possible clinical significance. *Immunol Invest.* 1995;24(6):999-1012.
23. Rameshwar P, et al. Endogenous hematopoietic reconstitution induced by human umbilical cord blood cells in immunocompromised mice: implications for adoptive therapy. *Exp Hematol.* 1999;27(1):176-185.
24. Czarneski J, et al. Effects of cord blood transfer on the hematopoietic recovery following sublethal irradiation in MRL lpr/lpr mice. *Proc Soc Exp Biol Med.* 1999;220(2): 79-87.
25. Ende N. The Berashis cell: a review – is it similar to the embryonic stem cell? *J Med.* 2000;31(3–4):113-130.
26. Ende N. Berashis cells in human umbilical cord blood vs. embryonic stem cells. *J Med.* 2002;33(1–4):167-171.
27. Mouiseddine M, et al. Human mesenchymal stem cells home specifically to radiation-injured tissues in a non-obese diabetes/severe combined immunodeficiency mouse model. *Br J Radiol.* 2007;80(Spec No 1):S49-S55.
28. Shi C, et al. Transplantation of dermal multipotent cells promotes survival and wound healing in rats with combined radiation and wound injury. *Radiat Res.* 2004;162(1):56-63.
29. Burd A, Ahmed K, Lam S, et al. Stem cell strategies in burns care. *Burns.* 2007;33(3):282-291.
30. Ende M, Ende N. Hematopoietic transplantation by means of fetal (cord) blood. A new method. *Va Med Mon (1918).* 1972;99(3):276-280.
31. Valbonesi M, Giannini G, Miglori F, et al. Cord blood (CB) stem cells for wound repair. Preliminary report of 2 cases. Transfus Apher Sci 2004;30:153-156.
32. Ende M. Management of the acute radiation syndrome. *Ann Intern Med.* 2004;141(11):891.

Hemoglobin-Based Oxygen Carriers in Trauma Care: The US Multicenter Prehosptial Trial

10

Ernest E. Moore, Hunter B. Moore, Tomohiko Masuno, and Jeffrey L. Johnson

Supported in part by Northfield Laboratories, Inc. and National Institutes of Health Grants P50GM49222, T32GM08315 and U54GM62119

The current generation of blood substitutes tested in clinical trials are red blood cell (RBC) substitutes, i.e., they are designed primarily to transport oxygen. The products that are now being used in advanced phase clinical trials are derived from hemoglobin and thus are often referred to as hemoglobin-based oxygen carriers (HBOCs). The potential benefits of HBOCs are well known (Table 10.1). The objectives of this brief overview are to outline emerging applications of HBOCs in trauma care and review the scientific background for ongoing patient trials in the USA.

10.1 Potential Role of Hemoglobin-Based Oxygen Carriers in Trauma Care

The US FDA approval of a new product proceeds through phase I, II, and III studies designed to establish safety and efficacy (Table 10.2). FDA regulation defines efficacy as follows: "Effectiveness means a reasonable expectation that … the pharmacologic or other effect of the biologic product … will serve a clinically significant function in the diagnosis, cure, mitigation, treatment or prevention of disease in man."[1] The Center for Biologics Evaluation and Research (CBER) is the review body for the FDA in the arena of biologies and has published a comprehensive listing of "points to consider in the safety evaluation of HBOCs."[2] These points encompass characterization of the product, animal safety testing, and human studies and address the theoretic concerns of Hb solutions raised previously,[3-6] including pulmonary and systemic hypertension, organ dysfunction, oxidative tissue injury, synergy with bacterial pathogens, and immunomodulation. In 1994, CBER convened a workshop with the National Heart, Lung and Blood Institute and the Department of the Army to develop "points to consider in the efficacy evaluation of HBOCs."[7] Documenting a direct clinical endpoint for HBOCs is challenging because this endpoint has never been established for RBCs. Specific recommendations for clinical studies are in three areas: perioperative applications, acute hemorrhagic shock, and regional perfusion. Field trials for postinjury hemorrhagic shock, where RBCs are not available, are difficult because of safety and ethical issues. Decreased perioperative allogeneic RBC transfusion is regarded as a clinical benefit, but the potential risks of HBOCs have to be defined and evaluated as well. Regional perfusion studies include reperfusion following ischemia; for example, as an adjunct during coronary angioplasty (the FDA approved Fluosol DA in 1989 as an O_2-carring drug for this setting).

10.2 Clinical Evaluation of Modified Tetrameric Hemoglobin in Trauma Care: The First Multicenter Trial

Of the modified Hb tetrameric solutions that looked promising in the late 1980s, one formulation was authorized by the FDA for a phase III study in trauma.

E.E. Moore (✉)
Department of Surgery, Denver Health Medical Center,
777 Bannock St, Denver, Colorado, 80204, USA
e-mail: Ernest.moore@dhha.org

N. Bhattacharya and P. Stubblefield (eds.), *Regenerative Medicine Using Pregnancy-Specific Biological Substances*,
DOI: 10.1007/978-1-84882-718-9_10, © Springer-Verlag London Limited 2011

Table 10.1 Potential clinical benefits of hemoglobin-based oxygen carriers in trauma care

Availability: abundant supply
Universally compatible
Prolonged shelf-life
Storage at room temperature
Safety: No disease transmissions
No antigenic reactions
No immunologic effects
Efficacy: enhanced oxygen delivery
Improved rheologic properties

Table 10.2 Potential role of hemoglobin-based oxygen carriers in trauma care

Application	Location
I. Perioperative applications	
Reduce allogeneic RBC transfusions	ED, angiography, OR, ICU
Attenuate transfusion immunomodulation	OR, ICU
II. Acute hemorrhagic shock	
When stored RBCs are unavailable	Field, ED, OR, ICU, remote hospital, civilian disaster, military conflict
More efficient resuscitation	Field, ED, OR, ICU
Low volume resuscitation	Remote hospital, civilian disaster, military conflict
III. Regional perfusion	
Enhance O_2 delivery	
Ischemic reperfusion tissue/organ	OR, ICU
Inflamed tissue	OR, ICU
Ex vivo organ perfusion	Hospital, OR

ED emergency department, *OR* operating room, ICU intensive care unit

HemAssist (Baxter Healthcare, Boulder, CO) consisted of Hb tetramers cross-linked between alpha subunits with bis 3,5-diabromosalicyl fumarate to prevent dissociation into dimers and reduce oxygen affinity. Unfortunately, this clinical trial failed.[8] Regarded by some as a major setback for HBOCs, it is important to emphasize that this US multicenter trial of diaspirin cross-linked Hb (DCLHb) for the treatment of severe traumatic hemorrhagic shock was based on the explicit proposal that "DCLHb was tested not as a substitute for blood but rather as an adjunct to the currently used therapies for enhancing oxygen delivery: fluids, blood, and operative intervention." Although the unexpected outcome raised the issue of comparable study groups, the difference in the primary study endpoint was concerning: the 28-day mortality for the DCLHb group was 46% (24 of 52), compared with 17% (8 of 46) for the control (normal saline) group. Much expert thought and preparation went into the study design of this human trial, but the scientific justification of using a vasoconstricting agent for the initial resuscitation of acute hemorrhagic shock was questionable in retrospect.

The authors rationalized this study design because in preclinical trials "DCLHb has been shown to be effective in enhancing perfusion in small volumes, suggesting a pharmacologic effect that is independent of hemoglobin." But the pharmacologic effect was not always reported as beneficial. In 1993, Hess and coauthors,[9] at the Letterman Army Institute of Research, reported that in a swine model of hemorrhagic shock, DCLHb infusion doubled systemic and pulmonary vascular resistance, and these responses were associated with a fall in cardiac output. In fact, these changes were equivalent to resuscitation with unmodified tetrameric Hb. The authors concluded: "The decrease in cardiac output associated with the vasoconstriction in the Hb-treated animals was equal to the increase in oxygen-carrying capacity – crystalloid or colloid solutions provided equally rapid correction of the elevated whole blood lactate." In a follow-up study,[10] the infusion of low-dose (4 mL/kg = 14 g Hb) DCLHb into swine subjected to hemorrhagic shock prompted the authors to further warn "pulmonary hypertension and low peripheral perfusion may offset benefits for trauma patients." Although the authors of the DCLHb trial cited several animal models that appeared to support their study hypothesis, none of these models replicated their study design – a lesson for future conduct of clinical trials with HBOCs.

The mechanisms responsible for the vasoconstriction resulting from DCLHb administration were investigated before the trauma clinical trial. The increased vascular resistance was believed to be predominantly mediated by the scavenging of NO with an additional component of enhanced endothelin release.[11-13] Subsequently, alternative mechanisms have been proposed, including release-enhanced adrenergic receptor sensitivity and reduced arterial wall shear stress secondary to decreased viscosity.[14,15] Development of DCLHb has been terminated, but the relevance of these basic mechanisms to future trauma care with HBOCs is clear.

10.3 Clinical Safety of Polymerized Hemoglobin in Trauma Care: The New Generation

At this moment, the HBOCs currently tested in Phase III trials are polymerized Hb solutions (Table 10.3). Polymerization addresses several of the problems inherent in tetrameric Hb, i.e., short intravascular retention and reduced colloid osmotic activity. Polymerization also appears to attenuate vasoconstriction associated with the infusion of Hb solutions. A proposed explanation is that tetrameric Hb (65 kDa) extravasates through the endothelium to bind abluminal NO, leading to unopposed vasoconstriction, but polymerized Hb (>130 kDa) remains in the vasculature to bind only luminal NO. Of interest, Hb of the common earthworm, *Lumbricus terrestrics*, is a polymer with a molecular weight of 400 kDa that circulates extracellularly.[16] Mice and rats undergoing exchange transfusion with this naturally occurring polymeric Hb showed no changes in behavior, and nuclear magnetic resonance spectroscopy of the heart confirmed normal O_2-carrying capacity.[17]

Table 10.3 Characteristics of current hemoglobin-based oxygen carriers in Phase III trials

Characteristic	Hemopure	PolyHeme	RBCs
Hemoglobin (g%)	13 g%	10 g%	13 g%
Unit equivalent (g)	30 g	50 g	50 g
Molecular weight (>64 kDa)	≥95%	≥99%	≥100%
P_{50} (mmHg)	38	29	26
Hill coefficient	1.4	1.7	2.7
Oncotic pressure (mmHg)	25	23	25
Viscosity	1.3 cp	2.1 cp	Whole blood = 5–10 cp
Methemoglobin (%)	<10	<8	<1
Half-life	19 h	24 h	31 days
Shelf-life @ 4°C	≥3 years	≥1.5 years	42 days
Shelf-life @ 21°C	≥2 years	≥6 weeks	≥6 h

Cp centipoises, P_{50} tension when hemoglobin-binding sites are 50% saturated

Polymerized HBOCs have undergone extensive preclinical and clinical testing for safety. Hemopure (Biopure Corp, Cambridge, MA), a polymer of bovine Hb, has been used successfully to reduce allogeneic RBC transfusion in elective cardiac,[18] aortic,[19] and hepatic[20] surgery. One study with abdominal aortic reconstruction raised concern about increased systemic vascular resistance,[21] an effect identified in normal volunteers. [22] Recent animal studies designed to replicate prehospital hypotensive resuscitation for hemorrhagic shock have been encouraging.[23-25] Hemopure has been approved for replacement of acute blood loss in South Africa, but there are no published results to date. PolyHeme (Northfield Lab, Evanston, IL) has been evaluated predominantly in acutely injured patients. PolyHeme is derived from outdated human stored blood. After lysis of RBCs, the native tetrameric hemoglobin is polymerized with glutaraldehyde. Pyridoxal phosphate is used to obtain a more physiologic P50. The meticulous, multistep biochemical purification of PolyHeme is believed to eliminate the risk of infection transmission. Under FDA guidance, we initiated clinical trials in trauma to confirm safety with escalating doses of PolyHeme. In the first clinical trial[26], 39 patients received 1 ($n=14$), 2 ($n=2$), 3 ($n=15$), or 6 ($n=8$) units of PolyHeme as their initial resuscitation after acute blood loss. Infusion rates ranged from 1 unit in 175 min to 6 units (300 g) in 20 min. Although the RBC Hb fell to 2.9±0.2 g%, total Hb was maintained at 7.5±0.2 g% with PolyHeme. With respect to safety, the patient's temperature, mean arterial pressure, heart rate, and creatinine clearance did not change during the 72-h study period. Liver function tests and amylase varied substantially because of patient injuries. Cognizant of the vasoconstriction associated with the DCLHb clinical trial, we designed a study to specifically evaluate the vascular response to PolyHeme infusion in acutely injured patients.[27] Patients requiring urgent transfusion were administered either PolyHeme (up to 6 units) or stored RBCs during their initial resuscitation. Systemic arterial pressure, pulmonary arterial pressure, cardiac index, and pulmonary capillary wedge pressure were measured every 4-h postinfusion. There were no significant differences between the groups for these indices or the calculated systemic or pulmonary vascular resistance. Additional issues reported with the clinical use of polymerized Hb solutions include interference of laboratory tests that are based on colorimetric changes from dissolved plasma Hb, inaccuracy of O_2 saturation monitoring because of methemoglobin, mild elevations of serum amylase (but without evidence of pancreatitis), and skin rashes. None of these have been considered clinically significant adverse events to date.

10.4 Clinical Efficacy of Polymerized Hemoglobin in Trauma Care

10.4.1 Perioperative Applications: Reduce Allogeneic RBC Transfusions in Trauma Care

Prompted by the FDA guidelines to demonstrate efficacy, all HBOC companies have pursued what appeared to be the simplest clinically, i.e., to reduce the need for allogeneic RBC transfusions. In collaboration with David B. Hoyt, MD, and the University of California at San Diego, we conducted a randomized trial in patients requiring urgent transfusion.[28] The 44 trauma patients (Injury Severity Score [ISS] $= 21 \pm 1.3$) were allocated to receive stored RBCs or up to 6 units of PolyHeme as their initial blood replacement. The RBC Hb was equivalent preinfusion (10.4 ± 0.4 vs 9.4 ± 0.3 g%); at end infusion, the RBC Hb of the PolyHeme patients fell to 5.8 ± 0.5 vs 10.6 ± 0.3 g% in the control. The PolyHeme group received 4.4 ± 0.3 units, resulting in a plasma Hb of 3.9 ± 0.2 g%. The total number of allogeneic RBC transfusions for the control versus PolyHeme was 10.4 ± 0.9 units versus 6.8 ± 0.9 units ($p < .05$), respectively, through day 1, and 11.3 ± 0.9 units versus 7.8 ± 0.9 units ($p = .06$), respectively, through day 3. After the initial phase, infusion of 4.6 units of stored RBCs in the control group was equivalent to the 5.2 units in the PolyHeme group. Both volumes presumably represented the infused RBCs or PolyHeme lost during acute hemorrhage before operative control. Subsequent replacement volumes were comparable, ultimately sparing the PolyHeme group approximately four units of allogeneic RBC transfusion.

10.5 Perioperative Applications: Reduce Allogeneic RBC Transfusions During Initial

10.5.1 Resuscitation and Thereby Decrease ARDS and MOF

With our long-term interest in the pathogenesis of postinjury MOF[29,30] our working hypothesis extended beyond reduced stored blood use during hospitalization. We proposed that PolyHeme, in lieu of stored RBCs during initial resuscitation, would attenuate the adverse immunoinflammatory effects of allogeneic RBC transfusion and ultimately reduce the incidence of ARDS and MOF. Stored blood is reportedly safer than ever due to comprehensive screening for disease transmission, but the potential adverse effects of packed RBC storage on the immune response to injury and illness are becoming more apparent.[31,32] We have been interested in the proinflammatory effects of stored RBCs and, specifically, their capacity to provoke neutrophil polymorphonuclear (PMN) cytotoxicity. The PMN is a key cellular mediator in the pathogenesis of postinjury MOF. Consequently, PMN functional responses are evaluated as a clinical surrogate for the two-event model of MOF, i.e., inflammatory priming and subsequent activation. The two-event construct of postinjury MOF is based on the fundamental concept that injury primes the innate immune system such that a second insult, during this vulnerable window, provokes unbridled systemic inflammation resulting in organ dysfunction.[33] Priming is defined as an enhanced response to a stimulus that is due to prior exposure of the cell to a different agonist.[34] In our ongoing epidemiologic studies,[35] we have shown that more than 6 units of RBC transfusion within the first 12 h postinjury is an independent risk factor for MOF.[36] Furthermore, the age of transfused blood within the first 6 h postinjury correlates with the incidence of MOF.[37] Previous studies in our center have shown that after severe injury, patients at high risk for MOF have circulating PMNs that are primed for cytotoxicity within the first 6 h postinjury, as marked by the increased surface expression CD11b/CD18, p 38 MAPK activation, release of cytotoxic products in response to fMLP, and delayed apoptosis.[38]

The precise mechanisms linking packed RBC transfusion and PMN priming remain to be established, but many believe that passenger leukocytes accompanying RBCs in storage are important in the generation of proinflammatory agents.[39] Plasma from stored RBCs primes PMNs in vitro, and this effect has been shown to increase progressively from 14 to 42 days of storage.[40] Some investigators have incriminated cytokines (TNF-a, IL_1, IL_6, IL_8, and IL_{18}) generated during storage,[41] while we have focused on proinflammatory lipids presumably generated from the RBC membrane.[42] Metabolites of the arachidonic acid cascade have been strongly implicated in the pathogenesis of transfusion-related acute lung injury.[43] Although prestorage leukoreduction of RBCs decreases the generation of cytokines, this process does not eliminate PMN priming[44] Thus, collectively these studies suggest that a blood substitute devoid of proinflammatory agents will avoid the immunomodulatory consequences of allogeneic RBCs.

In preparation for clinical trials, we conducted in vitro and in vivo studies to test our hypothesis that PolyHeme – free of inflammatory cytokines and lipids – would eliminate the PMN priming previously documented with stored RBCs and translate to reduced ARDS and MOF.[45] Human PMNs were isolated from healthy volunteers and the plasma fraction was separated from packed RBCs at 42 days of storage in our blood bank (the last day stored RBCs can be transfused clinically, but often the first RBCs infused into trauma patients).[46] The isolated PMNs were incubated with either RBC plasma or PolyHeme at concentrations calculated to be equivalent up to 8 units of transfusion. The plasma fraction representing three or more units of stored RBCs primed the human PMNs for enhanced superoxide production and elastase release (Fig. 10.1).

We further tested our hypothesis in an established two-event model of MOF; i.e., trauma/hemorrhagic shock as a priming event followed by TLR4 engagement as an activating event.[47] Our primary study objective was to contrast HBOC versus crystalloid in the prehospital phase, but we expanded the study groups to encompass the possible availability of stored

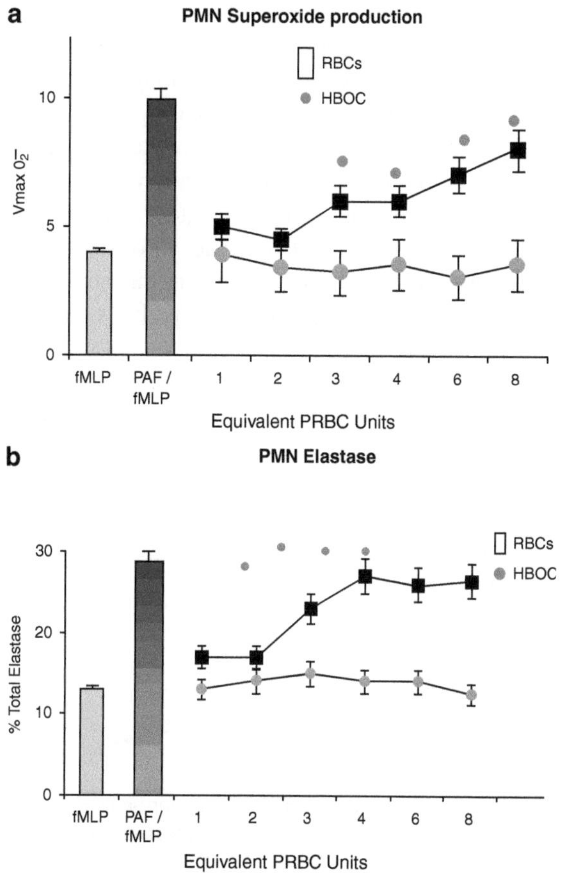

Fig. 10.1 Isolated human neutrophils (PMNs) were incubated with either the plasma fraction from stored RBCs or PolyHeme at concentrations equivalent to 1–8 units of acute transfusion. (**a**) PMN superoxide production; (**b**) PMN elastase release. FMLP is employed as a PMN activator; and PAF (primer) followed by fMLP approximates maximal PMN response. *FMLP* formyl-methionyl-leucyl-phenylalanine, *PAF* platelet-activating factor, $*p<.05$

blood in the field, and our previous inhospital Phase II clinical work with HBOC resuscitation. We selected acute lung injury as our primary study endpoint because ARDS is the first manifestation of postinjury MOF. Rats underwent laparotomy and hemorrhagic shock (30 mmHg × 45 min) and were resuscitated over 2 h in a clinically relevant design, i.e., 2× volume of shed blood (SB) using normal saline (NS) in the first 30 min; ½ volume of SB in the next 30 min; another 2× SB volume with NS over the remaining 60 min. Study groups represented alternative fluid strategies during the first hour of resuscitation: (1) in-hospital SB (standard resuscitation), (2) in-hospital HBOC, (3) prehospital SB, and (4) prehospital HBOC. Global physiologic response was assessed via tissue oxygenation (NIR spectroscopy) and arterial base deficit (BD), and pulmonary response via lung neutrophil (PMN) accumulation and vascular permeability. Prehospital HBOC resuscitation provided the most efficient recovery of tissue oxygenation (Fig. 10.2) and correction of BD, had the greatest reduction in pulmonary PMN accumulation and abrogated acute lung injury (Fig. 10.3). The findings in this controlled in vivo study further supported our hypothesis that initial HBOC resuscitation attenuates the postshock inflammatory response and secondary organ dysfunction.

In our subsequent clinical trial, injured patients requiring urgent transfusion were administered either PolyHeme (up to 20 units = 1,000 g) or stored RBCs for their initial 12 h of resuscitation.[48] PMN priming was determined by the surface expression of CD11b/CD18 in whole blood and superoxide production in isolated PMNs. The study groups (stored RBC [$n = 10$] vs PolyHeme [$n = 9$]) were comparable with respect to

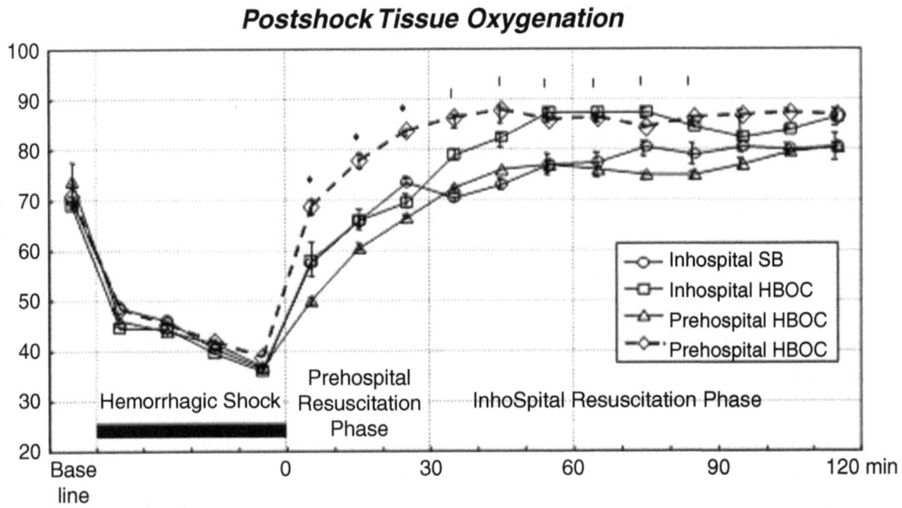

Fig. 10.2 Tissue oxygenation (StO_2 = tissue oxygen saturation) was monitored continuously with an NIRS device placed on the animal's hind limb *$p < .05$ vs other groups + < .05 prehospital HBOC and in-hospital hemoglobin-based oxygen carrier (*HBOC*) vs. prehospital SB and in-hospital SB

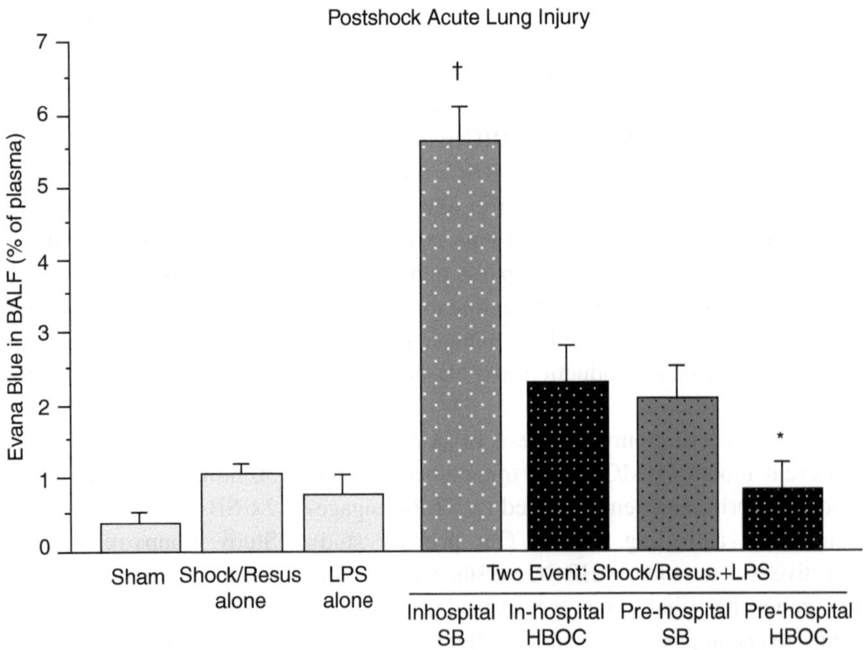

Fig. 10.3 Acute lung injury, determined by Evans blue alveolar extravasation, was evaluated at the end of the study (8 h postinsult). Simulated prehospital hemoglobin-based oxygen carriers (*HBOC*) resuscitated abrogated early ALI * $p < .05$ vs in-hospital SB, †$p < .05$ vs all other Two-Event groups

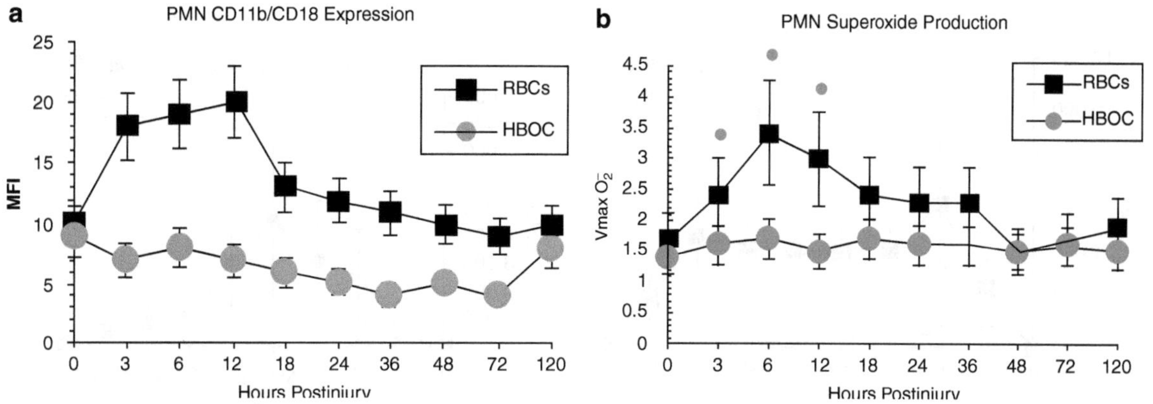

Fig. 10.4 Circulating neutrophils (PMNs) from injured patients who underwent initial resuscitation with either stored RBCs or PolyHeme. (**a**) PMN CD11b/CD18 receptor expression in whole blood; (**b**) PMN superoxide production in isolated cells. *$p < .05$

injury severity (ISS = 27.9 ± 4.5 vs 21.9 ± 2.7), physiologic compromise (emergency department pH = 7.22 ± 0.04 vs 7.19 ± 0.08), and Hb transfusion in the first 24 h (units = 14.1 ± 2.0 vs 14.5 ± 1). Circulating PMNs from patients resuscitated with stored RBCs manifested evidence of priming through increased CD11b/CD18 expression and enhanced superoxide production (Fig. 10.4). Three patients (30%) in the stored RBC group died of MOF, while all patients in the PolyHeme group survived.

To further investigate the impact of early resuscitation with PolyHeme in lieu of stored RBCs, we extended our clinical trial to evaluate the systemic levels of proinflammatory cytokines (IL_6, IL_8), counterregulatory cytokines (IL_{10}, IL_{11}), and markers of endothelial activation (sICAM, sE-selectin).[49] The study groups (stored RBC [$n = 7$] vs PolyHeme [$n = 18$]) were comparable with respect to injury severity. Patients resuscitated with stored RBCs had higher levels of the proinflammatory cytokines IL_6 and IL_8, and higher levels of the counterregulatory cytokine IL_{10} (Fig. 10.5), with a trend toward higher sICAM, and sE-selectin levels. We have not enrolled a sufficient number of injured patients to definitively address the ultimate study objective – reduction of postinjury MOF. But the incidence of MOF in the acutely injured patients given PolyHeme during their initial resuscitation for whom we had complete data ($n = 20$) was 15%, contrasted with a predicted incidence of 37% ($p < 0.05$) based on our MOF prediction model.[45] In sum, these clinical trials in trauma patients suggest that PolyHeme, used in the early resuscitation of patients with hemorrhagic shock, attenuates the immunodysfunction associated with stored RBC transfusion and thereby reduces the incidence of postinjury MOF.

10.6 Acute Hemorrhagic Shock: When Stored RBCs are Unavailable in Trauma Care

The most compelling indication for HBOC is the scenario in which stored RBCs are unavailable. This potential benefit for military use has largely driven the development of HBOCs, but there are also a number of key applications in civilian trauma care. Most conspicuous is the role in prehospital care, particularly for extended transport times. But there are also remote hospitals throughout the country in which stored blood is simply not available or is rapidly depleted when multiple casualties are encountered. There have been well-designed animal models, which strongly suggest that prehospital low-volume resuscitation with HBOCs can save lives.[23-25] Despite the evidence, the scientific design and ethical conduct of clinical trials to establish efficacy of HBOCs when RBCs are unavailable remain a challenge.[50,51] To best approximate this scenario, we compared the 30-day mortality in 171 trauma patients given up to 20 units (1,000 g) of PolyHeme, compared with a historic control of 300 surgical patients who refused stored RBCs on religious grounds.[52] The trauma patients received rapid infusion of 1 to 2 units ($n = 45$), 3–4 units ($n = 45$), 5–9 units ($n = 47$), or 10–20 units ($n = 34$) of PolyHeme; 40 patients had a nadir

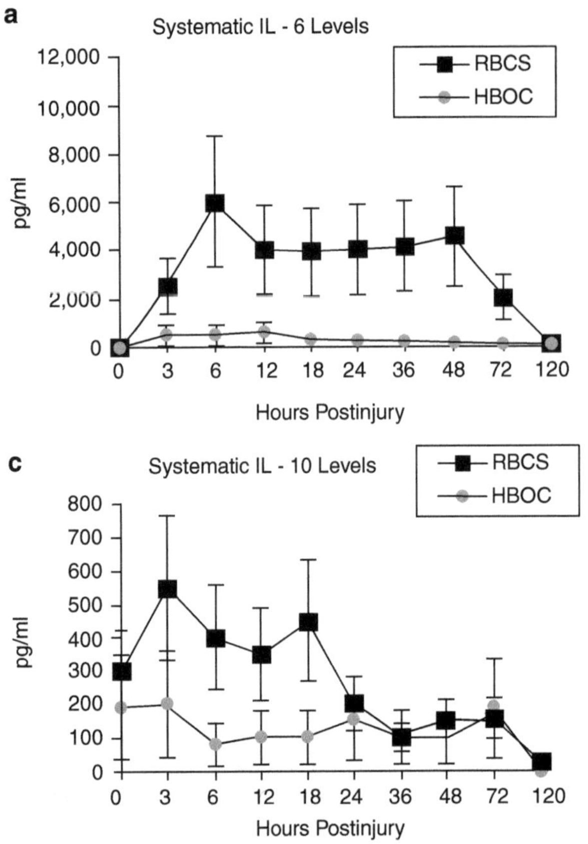

Fig. 10.5 Systemic interleukin IL_6, IL_8, and IL_{10} from injured patients who underwent initial resuscitation with either stored RBCs or PolyHeme. (**a**) IL_6; (**b**) IL_8; and (**c**) IL_{10}. *$p < .05$

RBC Hb≤3 g% (mean = 1.5 ± 0.7 g%). Total Hb was adequately maintained (mean = 6.8 ± 1.2 g%) via plasma Hb added by PolyHeme. The 30-day mortality was 25.0% (10 of 40 patients), compared with 64.5% (20 of 31 patients) in the control patients (Fig. 10.6).

A personal experience with PolyHeme during our in-hospital FDA-approved phase II studies has convinced us that the time has arrived for licensing of HBOCs for trauma care.[45] An 18-year-old man arrived by ground ambulance at our emergency department in extremis after a gunshot wound to the abdomen from a high-velocity elk-hunting rifle (30.06, hollow soft point 220 gr, muzzle energy 2,840 ft/lb). Because of immediate availability, 10 units of PolyHeme (maximum dose permitted at the time) were administered during the first 14 min of in-hospital resuscitation, representing greater than 91% of total circulating Hb at end infusion (RBC Hb = 0.7 g%). The missile entered the left midabdomen and exited posteriorly. At laparotomy, we encountered an avulsed shattered left kidney with secondary aortic and vena caval perforations, a partially transected superior mesenteric vein, and destructive injuries to his distal duodenum, proximal jejunum, midileum, and descending and sigmoid colon. In addition, he had massive soft tissue loss in the retroperitoneum, including the psoas and paraspinous muscles, and suffered a concussive spinal cord lesion with resultant paraplegia. The patient received an additional 40 units of packed RBCs during initial laparotomy, but ultimately this gentleman survived to discharge without organ failure. We believe that the immediate infusion of this HBOC was pivotal in maintaining sufficient O_2 delivery during the critical period of massive blood loss to save this man's life.

10.7 Current Phase III US Multicenter Prehospital HBOC Trial

The optimal resuscitation fluid for acute blood loss remains unclear, and the practical options for prehospital care have been limited to expansion of the

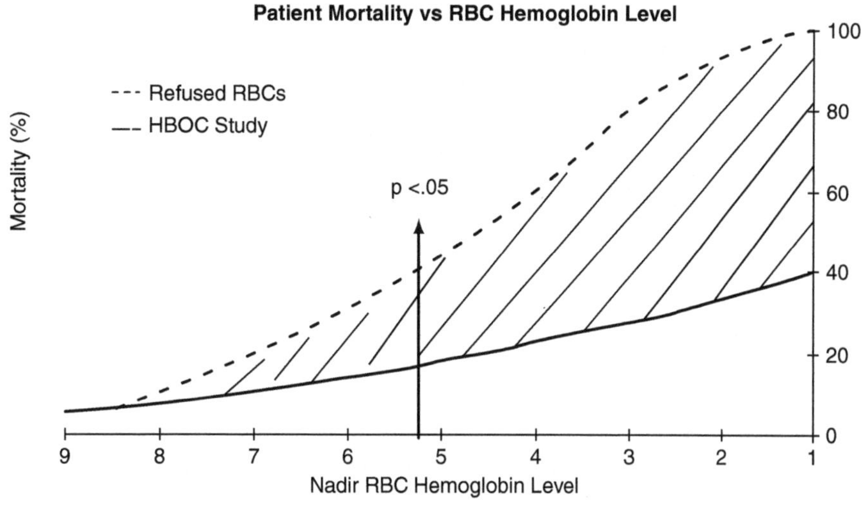

Fig. 10.6 The 30-day mortality is compared in surgical patients who refused stored RBC transfusion versus injured patients who were initially resuscitated with PolyHeme. The computer-generated curves are based on nadir Hb levels. Mortality was significantly less ($p<.05$) in the PolyHeme group when RBC Hb≤5.3 g%; i.e., critical anemia

circulating blood volume. The issue is magnified in the combat scenario where access to blood transfusion is further delayed.[53] Resurgent interest in defining optimal field resuscitation has challenged the long-standing practice of unbridled crystalloid loading,[54] citing the potential risk of exacerbating hemorrhage via dislodging hemostatic clots[55] and diluting plasma coagulation factors. Conversely, the magnitude of oxygen debt following hemorrhagic shock correlates directly with adverse outcome.[56-57] The availability of HBOCs offers a new strategy for this clinical "catch 22." Consequently, with this background and preliminary data, we initiated a multicenter prehospital trial in the USA in January 2003. Severely injured patients, blunt or penetrating, with an SBP≤90 mmHg due to acute blood loss are randomized at the scene to receive either the standard crystalloid resuscitation or PolyHeme. In the hospital, for the initial 12 h postinjury, the control group receives stored RBCs as needed while the study group is administered PolyHeme up to 6 units and then stored RBCs as needed. The primary study endpoint is 30-day mortality; the secondary endpoints include amount of stored RBC transfusion uncrossmatched RBC administration, incidence of circulation Hb<5 g% and incidence of ARDS and MOF. The study is conducted, by necessity, with exception to informed consent.[58] By definition, critically injured patients are unable to provide informed consent and, due to the exigent circumstance, legal guardians or next of kin are often not accessible or appropriate for proxy consent during the narrow therapeutic window. One of the recent major advances in trauma research is the FDA codification of "Exception from Informed Consent Requirements for Emergency Research" detailed in the Code of Federal Regulation, Title 21, part 50, section 24 (21CFR 50.24), which became effective from November 1, 1996 (DHHS/FDA, 1996). The seven fundamental components of this regulation are outlined in Table 10.4. Research protocols using this Exception from Informed Consent

Table 10.4 Exception from informed consent requirements for emergency research in the USA

Human subjects are in life-threatening situation; available treatments are unsatisfactory
Obtaining informed consent is not feasible
Participation in the research holds out the prospect of direct benefit to the subjects
The clinical investigation could not practicably be carried out without the waiver
The investigational plan defines the length of the potential therapeutic window; investigator has committed to attempting to contact a legally authorized representative during that window
The institutional research board has approved the informed consent document and procedures
Additional protection of the rights and welfare of the subjects include: community consultation, public disclosure, establishment of an independent data-monitoring committee, and consent to continue the study is obtained from the patient as soon as possible

must be conducted under a separate investigational new drug (IND) application to the FDA. The study population in the multicenter prehospital trial was targeted for 720 patients, and the study remained unchanged after four planned interim analyses (60–120–250–500–720 patients) by an independent data safety monitoring committee.

References

1. Department of Health and Human Services: 21 CFR601.25(d) (2)
2. Center for Biologics Evaluation and Research. Points to consider in the safety evaluation of hemoglobin-based oxygen carriers. *Transfusion*. 1991;31:369-371.
3. Alayash Al. Oxygen therapeutics – can we tame hemoglobin? *Nat Rev Drug Discov*. 2004;3:152-159.
4. Creteur J, Sibbald W, Vincent JL. hemoglobin solutions – not just red blood cell substitutes. *Crit Care Med*. 2000;28:3025-3034.
5. McFaul SJ, Bowman PD, Villa VM. hemoglobin stimulates the release of proinflammatory cytokines from leukocytes in whole blood. *J Lab Clin Med*. 2000;135:263-269.
6. Winslow RM. Current status of blood substitute research – towards a new paradigm. *J Intern Med*. 2003;253:508-517.
7. Center for Biologics Evaluation and Research. Points to consider on efficacy evaluation of Hemoglobin and perfluorocarbon based oxygen carriers. *Transfusion*. 1994;34:712-713.
8. Sloan EP, Koenigsberg M, Gens D, et al. Diaspirin cross-linked hemoglobin (DCLHb) in the treatment of severe traumatic hemorrhagic shock – a randomized controlled efficacy trial. *JAMA*. 1999;282:1857-1864.
9. Hess JR, MacDonald VW, Brinkley WW. Systemic and pulmonary hypertension after resuscitation with cell-free hemoglobin. *J Appl Physiol*. 1993;74:1769-1778.
10. Poli de Figueiredo LF, Mathru M, et al. Pulmonary hypertension and systemic vasoconstriction may offset the benefits of acellular hemoglobin blood substitutes. *J Trauma*. 1997; 42:847-856.
11. Gulati A, Sen AP, Sharma AC, et al. Role of ET and NO in resuscitative effect of diaspirin cross-linked hemoglobin after hemorrhage in rat. *Am J Physiol*. 1997;273:H827-836.
12. Schultz SC, Grady B, Cole F, Hamilton I, Burhop K, Malcolm DS. A role for endothelin and nitric oxide in the pressor response to diaspirin cross-linked hemoglobin. *J Lab Clin Med*. 1993;122:301-308.
13. Rohlfs RJ, Bruner E, Chiu A, et al. Arterial blood pressure responses to cell-free hemoglobin solutions and the reaction with nitric oxide. *J Biol Chem*. 1998;273:12128-12134.
14. Boura C, Caron A, Longrois D, et al. Volume expansion with modified hemoglobin solution, colloids, or crystalloid after hemorrhagic shock in rabbits: effects in skeletal muscle oxygen pressure and use versus arterial blood velocity and resistance. *Shock*. 2003;19:176-182.
15. Wettstein R, Cabrales P, Erni D, et al. Resuscitation from hemorrhagic shock with MalPEG-albumin: comparison with MalPEG-hemoglobin. *Shock*. 2004;22:351-357.
16. Fushitan K, Imai K, Riggs AF. Oxygen properties of hemoglobin form the earthworm Lumbricus terrestric. *J Biol Chem*. 1986;261:8414-8423.
17. Hirsch RE, Jelicks LA, Wittenberg BA, et al. A first evaluation of the natural high molecular weight polymeric lumbricus terrestric hemoglobin as an oxygen carrier. *Artif Cells Blood Substit Immobil Biotechnol*. 1997;25:429-444.
18. Levy JH, Goodnough LT, Greilich PE, et al. Polymerized bovine hemoglobin solution as a replacement for allogeneic red blood cell transfusion after cardiac surgery: results of a randomized, double-blind trial. *J Thorac Cardiovasc Surg*. 2002;124:35-42.
19. LaMuraglia GM, O'Hara PJ, Baker WH, et al. The reduction of the allogenic transfusion requirement in aortic surgery with hemoglobin-based solution. *J Vasc Surg*. 2000;31: 299-308.
20. Standl T, Burmeister MA, Horn EP. Bovine haemoglobin-based oxygen carrier for patients undergoing haemodilution before liver resection. *Br J Anaesth*. 1998;80:189-194.
21. Kasper SM, Walter M, Grune F, et al. Effects of a hemoglobin-based oxygen carrier (HBOC-201) on hemodynamics and oxygen transport in patients undergoing preoperative hemodilution for elective abdominal aortic surgery. *Anesth Analg*. 1996;83:921-927.
22. Hughes GS Jr, Antal EJ, Locker PK, et al. Physiology and pharmacokinetics of a novel hemoglobin-based oxygen carrier in humans. *Crit Care Med*. 1996;24:756-764.
23. Handrigan MT, Bentley TB, Oliver JD, et al. Choice of fluid influences outcome in prolonged hypotensive resuscitation after hemorrhage in awake rats. *Shock*. 2005;23:337-343.
24. Manning JE, Katz LM, Brownstein MR, et al. Bovine hemoglobin-based oxygen carrier (HBOC-201) for resuscitation of uncontrolled, exsanguinating liver injury in swine. *Shock*. 2000;13:152-159.
25. McNeil JD, Smith DL, Jenkins DH, et al. Hypotensive resuscitation using an polymerized bovine-based oxygen carrying solution leads to reversal of anaerobic metabolism. *J Trauma*. 2001;50:1063-1075.
26. Gould SA, Moore EE, Moore FA, et al. Clinical utility of human polymerized hemoglobin as a blood substitute after acute trauma and urgent surgery. *J Trauma*. 1997;43:325-332.
27. Johnson JL, Moore EE, Offner PJ, et al. Resuscitation of the injured patient with polymerized stroma-free hemoglobin does not produce systemic or pulmonary hypertension. *Am J Surg*. 1998;176:612-617.
28. Gould SA, Moore EE, Hoyt DB, et al. The first randomized trial of human polymerized hemoglobin as a blood substitute in acute trauma and emergent surgery. *J Am Coll Surg*. 1998;187:113-122.
29. Botha AJ, Moore FA, Moore EE, et al. Postinjury neutrophil priming and activation – an early vulnerable window. *Surgery*. 1995;118:35-365.
30. Ciesla DJ, Moore EE, Johnson JL, et al. A 12 year prospective study of postinjury multiple organ failure. *Arch Surg*. 2005;140:432-440.
31. Aiboshi J, Moore EE, Ciesla DJ, et al. Blood transfusion and the two-insult model of postinjury multiple organ failure. *Shock*. 2001;15:302-306.
32. Silliman CC, Moore EE, Johnson JL, et al. Transfusion of the injured patient: proceed with caution. *Shock*. 2004;21: 291-299.

33. Moore EE, Moore FA, Harken HA, et al. The two-event construct of postinjury multiple organ failure. *Shock*. 2005;24s: 71-75.
34. Ingraham LM, Allen JM, Higgins CP, et al. Metabolic membrane and functional responses of human polymorphonuclear leukocytes to platelet activating factor. *Blood*. 1982;59:1259-1266.
35. Sauaia A, Moore FA, Moore EE, et al. Early predictors of postinjury multiple organ failure. *Arch Surg*. 1994;129:38-45.
36. Moore FA, Moore EE, Sauaia A. Blood transfusion – an independent risk factor for postinjury multiple organ failure. *Arch Surg*. 1997;132:620-625.
37. Zallen G, Offner PJ, Moore EE, et al. Age of transfused blood is an independent risk factor for post-injury multiple organ failure. *Am J Surg*. 1999;178:570-572.
38. Biffl WL, Moore EE, Zallen G, et al. Neutrophils are primed for cytotoxicity and resist apoptosis in injured patients at risk for multiple organ failure. *Surgery*. 1999;126:198-202.
39. Bordin JO, Heddle NM, Blajchman MA. Biologic effects of leukocytes present in transfused cellular blood products. *Blood*. 1994;84:1703-1721.
40. Nielsen HJ, Reimert CM, Pedersen AM, et al. Time-dependent spontaneous release of white cell and platelet derived bioactive substances from stored human blood. *Transfusion*. 1996;36:960-965.
41. Shanwell A, Dristiansson M, Remberger M, et al. Generation of cytokines in red cell concentrates during storage is prevented by pre-storage white cell reduction. *Transfusion*. 1997;36:678-684.
42. Silliman CC, Paterson AJ, Dickey WO, et al. The association of biologically active lipids with the development of transfusion-related acute lung injury. *Transfusion*. 1997;37:719-726.
43. Silliman CC, Voelkel NF, Allard JD, et al. Plasma and lipids from stored packed red blood cells cause acute lung injury in an animal model. *J Clin Invest*. 1998;101:1458-1467.
44. Biffl WL, Moore EE, Offner PJ, et al. Plasma from aged stored red blood cells delays neutrophil apoptosis and primes for cytotoxicity-abrogation by post storage washing but not prestorage leukoreduction. *J Trauma*. 2001;50:426-1.0.
45. Moore EE. Blood substitutes: the future is now. *J Am Coll Surg*. 2003;196:1-17.
46. Partrick DA, Moore EE, Barnett CC, et al. Human polymerized hemoglobin as a blood substitute avoids transfusion-induced PMN priming for superoxide and elastase release. *Shock*. 1997;7:24.
47. Johnson JL, Moore EE, Offner PJ, et al. Resuscitation with a blood substitute abrogates pathologic postinjury neutrophil cytotoxic function. *J Trauma*. 2004;50:449-456.
48. Masuno T, Moore EE, Cheng AM, et al. Prehospital hemoglobin based oxygen carrier (HBOC) resuscitation attenuates acute lung injury. *Surgery*. 2005;138:335-341.
49. Huston P, Peterson R. Withholding proven treatment in clinical research. *N Engl J Med*. 2001;345:912-913.
50. McRae AD, Weijer C. Lessons from everyday lives-amoral justification for acute care research. *Crit Care Med*. 2002; 30:1146-1151.
51. Gould SA, Moore EE, Hoyt DB, et al. The life-sustaining capacity of human polymerized hemoglobin when red cells may be available. *J Am Coll Surg*. 2002;195:445-455.
52. Gawande A. Casualties of war-military care for the wounded from Iraq and Afghanistan. *N Engl J Med*. 2004;351: 2471-2478.
53. Bickell WH, Wall MJ, Pepe PE, et al. Immediate versus delayed fluid resuscitation for hypotensive patients with penetrating torso injuries. *N Engl J Med*. 1994;331:2206-1109.
54. Sandeen JL, Coopes VG, Holcomb JB. Blood pressure at which rebleeding occurs after resuscitation in swine and aortic injury. *J Trauma*. 2004;54:5440-5117.
55. Claridge JA, Schulman AM, Young JS. Improved resuscitation minimizes respiratory dysfunction and blunts interleukin – 6 and nuclear factor – kB activation after traumatic shock. *Crit Care Med*. 2002;30:1815-1819.
56. Dunham CM, Siegel JH, Weitreter L, et al. Oxygen debt and metabolic academia as quantitative predictors of mortality and severity of the ischemic insult in hemorrhagic shock. *Crit Care Med*. 1991;19:231-241.
57. Shoemaker WC, Appel PL, Kram HB. Tissue oxygen debt as a determinant of lethal and nonlethal postoperative organ failure. *Crit Care Med*. 1988;16:1117-1120.
58. Department of Health and Human Services. 21 CFR 50.24.

Placental Umbilical Cord Blood as a True Blood Substitute with an Edge

11

Niranjan Bhattacharya

11.1 Background

It was William Harvey (1628) who first discovered that blood flows through the vessels in one direction. Previously, it was believed that blood flows through the vessels forward and backward, like tides in the sea. In the past 381 years, scientists have attempted blood transfusion from animals like dogs, lambs, goats, and calves. Dead soldiers' blood or horse's blood were also used in an attempt to treat patients; the resulting horrific consequences led the church and/or the state to intervene and stop further research. However, there can be no denying the genuine need of blood or a suitable blood substitute for transfusion in humans during war and peace, given the enormous demand because of its life-saving potential.

The ideal blood substitute should not have any group specificity, that is, it should be a universal group, which can be transfused to a person of any blood type without tests. It should also be free of infectious agents, allergens, and should not deteriorate with storage. It should be ideal for disaster situations and should be freely available, even in remote areas. It must have a large-scale, economical manufacturing system for global use in times of need.

There have been many genuine attempts at isolating such a product that will fulfill all the required criteria as stated. Two major attempts included structural or genetic change in the hemoglobin molecule. They incorporated genetic or chemically modified hemoglobin and the isolation of a new oxygen carrier, perflurochemicals.

N. Bhattacharya
Department of General Surgery, Obstertrics and Gynaecology and Clinical Immunology, Advanced Medical Research Institute, Gol Park, B.P. Poddar Hospital and Vidyasagore Hospital, Kolkata 700034, India

Hemoglobin can be extracted from red cells by removing the cell membranes to form stroma-free hemoglobin. However, this cannot be used as a blood substitute because it causes renal toxicity and there is insufficient release of oxygen. Using biotechnological criteria, encapsulated hemoglobin or crosslink hemoglobin was prepared, but these products did not clear the US FDA criteria. The other alternative was the isolation of perflurochemicals, which can carry oxygen and carbon dioxide, the two gases handled byhemoglobin. Perfluorocarbons dissolve more oxygen than any other known inert liquid. Phospholipid-stabilized emulsions of this compound were prepared and it was found that there was no complement activation activity. Though manufactured with formidable cost, this product range also did not pass the stringent FDA criteria for wide unrestricted global use, because of human toxicity at an unacceptable rate. There is a very interesting chapter in this book by Ernest E. Moore et al. on this area of research; as such, it will be redundant to give further details here.

The attempt to find a safe blood substitute for transfusion is continuing, particularly after the onset of the pandemic, HIV/AIDS, and other diseases, which require special screening of blood for transfusion to rule out infections. The steep rise in the cost of screening for HIV (1 & 2), Hepatitis B & C, as well as other viruses like the Western Nile virus, added to the cost of development of the infrastructure to cope with and incorporate newer advances in molecular screening techniques for the bacterial or the viral antigen load of the transfused blood, have led to renewed thinking on safe blood. The need and emphasis on the development of a safe blood substitute has become a focal point of research.[1]

All over the world, millions of people are saved every year as a result of blood transfusions. At the same time, many still die because of an inadequate supply of

N. Bhattacharya and P. Stubblefield (eds.), *Regenerative Medicine Using Pregnancy-Specific Biological Substances*,
DOI: 10.1007/978-1-84882-718-9_11,

safe blood and blood products, particularly in developing countries. A reliable supply of safe blood is essential to improve health standards at several levels, especially for women and children, and particularly in the poorer sections of society anywhere in the world. Half a million women still die of complications related to pregnancy and childbirth, and 99% of these are in developing countries. Hemorrhage accounts for 25% of the complications and is the most common cause of maternal death. Malnutrition, thalassemia, and severe anemia are prevalent diseases in children, which require blood transfusion, apart from other complicated diseases. Over 80 million units of blood are collected every year, but the tragedy is that only 39% of this is collected in the developing world, which contains 82% of the global population.[2]

There are about 100 million births in the world every year. In India alone, there are more than 20 million births, which mean the production of that many placentas. One of the products of the placenta is cord blood and this has immense potential. The placenta is a complex organ that regulates feto-maternal interactions. Many cytokines that influence the lymphohematopoietic environment are produced in the placenta in abundance.

The contention of this paper is that cord blood, collected after the birth of a child, can be safe for transfusion. The objective of this chapter is to bring together various studies on the safe use of placental umbilical cord whole blood in various diseases, many of them conducted by the author.

These studies show that cord blood can be used as a safe alternative to adult blood. Further studies indicate that the special components of cord blood, which has sustained, protected, and developed the fetus in the mother's womb until it has grown into a regular baby, may play a role in immunotherapy. These components are passed on to the system of the recipient when umbilical cord whole blood is transfused.

The CD34 cell, for instance, has a possible nonspecific killing potential for cancer cells. These CD34 cells move throughout the body before ultimately settling at specific sites in the bone marrow and other specific zones in the hematopoietic system. There is a possibility that in case of cancer victims who were treated with cord blood transfusions, these CD34 cells of the cord blood kill cancer cells directly during their meeting, or indirectly assist the intrinsic cancer killing process of the host system by interacting with NK (natural killer) cells, which are key mediators of innate immunity contributing to immunosurveillance by recognizing and killing tumor and virus-infected cells. They are cytolytic and produce inflammatory cytokines.

There are now evidences that there is a rise in the peripheral blood CD34 base level after transfusion in some cases, which may have positive prognostic or immunotherapeutic implications previously not dreamt of. It was observed in studies conducted by the present author that where the rise was substantial and steep followed by a slow decrease over the next few months, there were good prognostic implications, and similarly no rise of CD34 level even after multiple transfusions of cord blood appeared to be a very bad prognostic sign. As such, though it is too early to state anything positively, there is a distinct possibility that cord blood transfusion may provide some extra benefits of cell therapy or cytokine therapy with eventual regeneration impact that is absent in adult blood transfusion, because of its intrinsic antigenic immaturity and the cytokine support system as discussed in the subsequent paragraph.

11.2 Antigenic Difference of Adult RBC and Cord Blood RBC

The red cell collected from the newborn's cord blood differs from the adult RBC in many ways, viz., there is an increase in the immunoreactive myosin in red cell membrane,[3] and the total value of lipid, phospholipid, and cholesterol is more in cord blood red cells than in the adult equivalent.[4] Even the antigen expression of cord blood RBC differs from the adult RBC. A, B, S, and Lutheran antigens are expressed in lesser amount in cord blood than in adult blood; in addition, there is a complete absence of Lewis antigen in the cord blood.[5] There are also fundamental metabolic differences between cord blood and adult blood, for example, the activities of phosphoglycerate kinase, enolase, glyceraldehyde-3-phosphate dehydrogenase, glucose phosphate isomerase, etc., of the Embden–Meyerhof pathway are definitely increased in cord blood,[6] and even nonglycolytic enzymes such as carbonic anhydrase and acetylcholine esterase are distinctly different from adult blood.[7] However, this antigenic difference does not have an adverse clinical impact on transfusion of cord blood in the host system.

11.3 Basic Immunological Characters of Cord Blood Constituents

Cord blood is the blood collected aseptically from the placenta after the birth of a healthy baby. It is globally known for its role as a source of hematopoietic cells, but it also contains potent angiogenesis stimulating cells: $CD34^+$, $CD11b^+$ fraction, which is approximately less than half of the $CD34^+$ fraction of cord blood. This was demonstrated to possess the ability to differentiate into functional endothelial cells in vitro and in vivo.[8] Mesenchymal stem cells are classically defined as adhering to plastic and expressing a nonhematopoietic cell surface phenotype, consisting of $CD34^-$, $CD45^-$, $HLA\text{-}DR^-$, while possessing markers such as STRO-1, VCAM, CD13, CD29, CD44, CD90, CD105, SH-3, and STRO-1.[9] These mesenchymal stem cells are also found in cord blood.

In addition, cells with markers and activities resembling embryonic stem cells have been found in cord blood. Zhao et al. identified a population of $CD34^-$ cells expressing OCT-4, Nanog, SSEA-3, and SSEA-4, which could differentiate into cells of the mesoderm, ectoderm, and endoderm lineage.[10]

The first widespread utilization of cord blood as a stem cell source was in the treatment of pediatric hematological malignancies after myeloablative conditioning. Outside oncology, the clinical use of cord blood stem cells has expanded into various areas that range from reconstituting a defective immune system to correcting congenital hematological abnormalities, as well as for inducing angiogenesis. In addition to current clinical use, cord blood is under intense experimental investigation in preclinical models of pathophysiologies that range from myocardial ischemia, to stroke and muscle regeneration.[11-13]

Hematopoietic stem cells from cord blood are now harvested in many laboratories all over the world and stored in cord blood banks, but they constitute only 0.01% of the nucleated cells of the cord blood. The rest, that is 99.99% of the blood, which is wasted, is rich in fetal hemoglobin, growth factors, and cytokine-filled plasma. Moreover, in the womb, the fetus benefits from the mother's in-built defenses against diseases and the placental environment is basically infection-free in the case of a healthy newborn. This in itself indicates safety, and it may have benefits yet unknown.

The blood volume of a fetus at term is 80–85 mL/kg. The placental vessel at term contains approximately 150 mL of cord blood. Cord blood contains three types of hemoglobin, HbF, HbA, and HbA2, of which HbF constitutes the major fraction. HbA accounts for 15–40% and HbA2 is present only in trace amounts at birth. HbF, which is the major component, has a greater oxygen-binding affinity than HbA.

11.4 Contemporary Experience with Allogeneic Cord Blood Transfusion as an Adult Blood Substitute

The discussion so far has indicated the worldwide demand for safe blood or blood substitutes; it has also mentioned the potential benefits of cord blood constituents. The clinical research of the present group of researchers over the last 10 years has clearly shown that cord blood can be used safely as a blood substitute. Moreover, its higher hemoglobin content and growth factors have the potential to benefit patients in varying diseases.

There have been a few attempts over the past decades to use it as a blood substitute, one early endeavor being that of Halbrecht in 1939, which was later reported in *Lancet*. Medical science was still at a relatively primitive stage at the time: acid citrate dextrose was not available as an anticoagulant, and neither was there any screening against bacteria or viruses prior to blood transfusion.[14]

In recent years, there has been deep interest in the clinical use of cord blood by Bhattacharya et al. who have done extensive work in the field. It should be noted here that in all cases necessary consent was obtained and the precautions of standard blood transfusion protocol were followed. The consent of the institutional ethical committee was sought and obtained for all cord blood transfusions. Their first publication was in 2001 when they reported on the transfusion of 174 units of umbilical cord whole blood to patients with various diseases. The blood was collected aseptically from the umbilical vein after cesarean section in standard pediatric blood transfusion bags, after the removal of the baby from the operative field and after confirming the stable condition of the mother. The work was conducted from April 1, 1999 to August 11, 2000. According to the study, the volume of cord blood varied from 50 to 140 mL with a mean of 86 ± 16 mL.

The cord blood was transfused immediately (within 3 days of collection) to 62 patients with age varying from 9 to 78 years. 32 patients were suffering from different stages and grades of malignancy. The remaining 30 included patients suffering from thalassemia major, aplastic anemia, systemic lupus erythematosus (SLE), chronic renal failure, rheumatoid arthritis, and ankylosing spondylitis, while a geriatric group of patients had benign prostatic hypertrophy. All tolerated the procedure of cord blood transfusion well, without any immunological or nonimmunological reaction.[15]

In another study in sub-Sharan Africa on a pediatric population, a different group of investigators reported (2003) that shortage of blood for transfusion contributes substantially to mortality of children with severe anemia in sub-Saharan Africa and suggested that umbilical cord blood could be an additional and readily available source of blood. The authors of this study tried to demonstrate the possibility of gathering cord blood from a busy Ghanaian labor ward. The mean volume of each blood sample obtained from the umbilical cord was 85 mL (SD 28.0). This amount of blood was sufficient to raise the hemoglobin concentrations of 28 (21%) of 131 children needing transfusions in the same hospital, by 30 g/L.[16]

In another report published in the *Journal of the American College of Surgeons* (2004), the experience of transfusing 413 U of placental umbilical cord whole blood (collected after lower uterine cesarean section from consenting mothers and subjected to the ethical committee protocol) to 129 informed consenting patients (cord blood range: 50–146 mL; mean 86±7.6 mL SD; median 80 mL; mean packed cell volume 48±4.1% SD; mean hemoglobin concentration 16.2±1.8 g/dL SD) was narrated. The list of consenting patients included 54 men and 75 women. Patient age varied from 2 to 86 years. Seventy-three patients (56.58%) suffered from advanced cancer and 56 (43.42%) had other diseases such as ankylosing spondylitis, lupus erythematosus, rheumatoid arthritis, aplastic anemia, and thalassemia major. The investigator did not encounter a single case of immunologic or nonimmunologic reaction in the follow-up.[17]

Bhattacharya et al. have published more reports on their work with cord blood as a safe blood substitute in various diseases. In a study of its impact in beta thalassemia, reported in 2005, 92 units of cord blood were transfused in 14 patients with beta thalassemia with severe anemia (hemoglobin concentration varying from 3.5 to 5.9 g/dL with mean hemoglobin 4.6 g/dL). The results showed that the transfusion was extremely effective in all 14 patients (male to female ratio 1:1, age varying from 6 months to 38 years). In that series, the collection of the blood varied from 57 to 136 mL mean 84±7.2 mL SD, median 87 mL, mean packed cell volume 45±3.1 SD, mean hemoglobin concentration 16.4±1.6 g/dL SD. Here, too, the authors did not encounter a single case of immunological or nonimmunological reaction.[18]

Another case study focused on the use of cord blood in diabetes with varying degrees of nephropathy and severe anemia. In this series, 39 informed consenting patients (22 males+17 females, aged 48–74 years, mean 59.6 years) were randomized into two groups: Group A (control cases $N=15$, males=8 and females=7) and Group B (study group $N=24$, males=14 and females = 10). In Group A, the rise of hemoglobin (Hgb) after two units of adult blood transfusion was 1.5–1.8 g/dL, as seen after a 72-h blood sample assessment. In Group B, patients received two of four units of freshly collected cord blood transfusion each (two units at a time), depending on availability and compatibility. The rise of Hgb after 72 h, after a transfusion of two units, was 0.6–1.5 g/dL. Microalbuminuria was assessed in both groups after 1 month of treatment with transfusion and other identical support. The mean result was 152±18 mg SD of albumin per gram of creatinine excreted through 24-h urine (pretransfusion mean excretion was 189±16 mg) in Group A and 103±16 mg SD of albumin excretion per gram of creatinine in 24-h excretion of urine in Group B (pretransfusion mean excretion was 193±21 mg). Univariate analysis using Fisher's exact test was performed for the results of Groups A and B. The difference between Group A and B values and its comparison with the pretransfusion microalbuminuria appeared to be statistically significant ($p < 0.003$).[19]

A fourth study concentrated on malaria patients suffering from anemia. Cord blood, because of its rich mix of fetal and adult hemoglobin, and the anti-malarial effect of cord blood, is an ideal choice in case of malaria with anemia necessitating blood transfusion. Here, cord blood was safely transfused to 39 informed, consenting patients (age varying from 8 to 72 years) suffering from anemia in the background of malaria, 22 of whom tested positive to Plasmodium falciparum, 17 to Plasmodium vivax. The collected volume of cord blood from each placenta (Unit) in this series varied from 52 to 143 mL, with a mean packed cell volume of 48.9 ± 4.1 SD and a mean hemoglobin concentration of 16.4 ± 1.6 g% SD. The blood was generally transfused immediately after collection, following the standard adult blood transfusion protocol of screening and cross-matching between the donor and the recipient, but if a volunteer was not immediately available, it was refrigerated and transfused within 72 h. For inclusion in this study, the patient's plasma hemoglobin had to be 8 g% or less (the pretransfusion hemoglobin in the malaria-infected patients in this series varied from 5.4 to 7.9 g/dL).[20] The rise of hemoglobin within 72 h of two units of freshly collected cord blood transfusion was 0.5–1.6 g/dL. Each patient received two to six units of freshly collected cord blood transfusion (two units at a time), depending on availability and compatibility. No clinical immunological or nonimmunological reaction was encountered in this series as well. The author concluded that properly screened cord blood is safe for transfusion, in victims of severe malarial anemia who need transfusion support.

In a subsequent report published in *Malaria Journal,* Bhattacharya expanded on the efficacy of this type of transfusion, reporting on further tests that were undertaken, which justified the clinical safety of the protocol of cord whole blood transfusion as there was no rise of urea, creatine, bilirubin, sugar from the pretransfusion level in the posttransfusion period.[21]

Blood transfusion or erythropoietin injection with adequate hematinic reserve is effective in normal situations (if the hemoglobin is 8 g/dL or less), but these are not that effective in anemia with a chronic disease background like rheumatoid arthritis. The next study took up this issue: 78 units (42–136 mL mean 80.6 ± 3.6 mL SD, median 82.4 mL, mean packed cell volume 48.2 ± 2.1 SD, mean percent hemoglobin concentration 16.4 ± 1.5 g/dL SD) of placental umbilical cord whole blood from consenting mothers undergoing lower uterine cesarean section (LUCS) were transfused (from April 1, 1999 to April 2005) to 28 informed consenting patients with advanced rheumatoid arthritis. The patients received 2–6 units of freshly collected placental umbilical cord blood without encountering any clinical, immunological, or nonimmunological reactions. Three days after completion of the transfusion of placental umbilical cord blood, a peripheral blood hematopoietic stem cell (CD34) estimation revealed a rise from the pretransfusion base level (0.09%), varying from 2.03% to 23%, which returned to base level in most of the cases at the 3-month CD34 re-estimation, without provoking any clinical graft-vs-host reaction in any of the patients.[22]

When it comes to immunocompromised cancer patients, many hospitals find it difficult to cope with the specialized requirements of blood transfusion, for instance, leucoreduction, selective apheresis, irradiation of the blood, viral inactivation of the blood by solvent and/or detergent treatment, or photochemical inactivation using psoralen or long wavelength ultraviolet light and cytomegalovirus-safe blood. The exorbitant cost of red blood cell (RBC) substitutes like hemoglobin-based oxygen carriers or perflurocarbon emulsions, or liposome-encapsulated hemoglobin, is simply unacceptable for an average oncological patient in the developing world. Bhattacharya et al. also transfused umbilical cord blood to cancer patients with hemoglobin levels at 8 g/dL or less, requiring special care, the objective being to find out whether it was a safe substitute for adult blood. In this series, the collection of cord blood varied from 54 to 128 mL, mean 82 ± 7.6 mL SD; mean packed cell volume 48 ± 4.1% SD; mean percent hemoglobin concentration 16.4 1 ± 1.6 g/dL SD. Not a single case of immunological or nonimmunological reaction was encountered after transfusion of the cord blood in this series as well.[23]

Tuberculosis causes approximately 1.5 billion latent infections, eight million new clinical cases, and three million deaths annually, making it the most prevalent infectious disease in the world. Anemia and malnutrition are essential comorbidities with tuberculosis. The investigators transfused 106 units (48–148 mL mean 81 ± 6.6 mL SD, median 82 mL, mean packed cell volume 49.4 ± 3.1 SD, mean percent hemoglobin concentration 16.3 ± 1.7 g/dL SD) of freshly collected placental umbilical cord whole blood (from April 1, 1999 to 2005) to 21 informed consenting patients. Cord blood transfusion was done in tuberculosis patients with less than 8 g/dL hemoglobin. The patients received 2–21 units each without encountering any clinical, immunological, or nonimmunological reactions. Three days after completion of the placental umbilical cord blood transfusion, the peripheral blood hematopoietic stem cell (CD34) estimation revealed a rise from the pretransfusion base level (0.09%), varying from 2.99% to 33%, which returned to base level in 66.66% of the patients at the 3-month CD34 re-estimation, without provoking any clinical graft-vs-host reaction in any of them.[24]

Bhattacharya et al. have also reported on the transfusion of cord blood to 16 consenting HIV-positive patients (12 cases had full-blown AIDS) with anemia and emaciation between April 1, 1999 and July 1, 2005. 123 units of placental umbilical cord whole blood (62–154 mL mean 85 ± 8.4 mL SD, median 82 mL, mean packed cell volume 48.8 ± 4.2 SD, mean percent hemoglobin concentration 16.3 ± 1.6 g/dL SD) were transfused, and the transfusion resulted in a sense of well-being, weight gain, and a transient rise of CD34 cell count.[25]

11.5 Other Impacts of Cord Blood Transfusion, i.e., Cell Therapy Potential on the Host

While placental umbilical cord blood has been shown again and again to be a safe alternative to adult blood in the various studies conducted by Bhattacharya et al., which indicate that no immunological or nonimmunological reaction took place in any of the cases in different studies, in the follow-up posttransfusion period, the reports of which are summarized earlier. These studies demonstrated that there is more to cord blood than being just a safe substitute for adult blood. That cord blood can play a role in cell therapy is indicated by the rise of CD 34 from the pretransfusion base level in some of the cases studied along with improvement of the bone marrow cellularity. A series on leprosy patients, mentioned below, is a case in point.

Fifteen males and one female, aged 12–72 years (mean 48.4 years) participated in this series: five cases were pausibacillary type (PB) and 11 cases were multibacillary type (MB). The clinical spectrum of the cases varied widely from the tuberculoid to the lepromatous type and one patient presented with gangrene, preceding an autoamputation of the leg, which was infested with maggots. All the patients were suffering from anemia with a hemoglobin conc of less than 8 g/dL, because of poor nutrition or the effect of dapsone, which can cause anemia. Seven days after completion of the placental umbilical cord blood transfusion, apart from the rise of hemoglobin, the peripheral blood hematopoietic stem cell (CD34) estimation revealed a rise from the pretransfusion base level (.09%), varying from 3.6% to 16.2%, in 75% of the cases, without provoking any clinical graft-vs-host reaction in any of the leprosy victims. This value returned to normal within 3 months in most cases.[26]

The immunotherapy potential of cord blood transfusion and the prognostic implications of such a therapy were assessed in another study.[27] Six cases of advanced cancer with severe anemia were treated with cord blood transfusion after informed consent. A patient with breast sarcoma received the lowest amount of cord blood (6 units), while another patient with breast cancer received the largest amount (32 units). The youngest patient, suffering from non-Hodgkin's lymphoma, was a 16-year-old boy who received eight units of cord blood to combat anemia. The fourth patient with metachronous lymph node metastasis received 15 units, while yet another patient with breast cancer received 14 units and the sixth patient with lung cancer received seven units. Studies of CD34 levels showed an initial rise followed by a fall in two cases, two cases registered very little effect on the CD34 level, i.e., no change from the baseline, and one case demonstrated a very slow rise from the baseline. However, one case showed a frequent steep rise up to 99% and a sustained high CD34 level. After initiation of treatment for stage IV breast cancer in 2001, this patient is still alive with clinical remission of the disease in 2009.

Another published report tried to explain the scientific feasibility of cord blood transfusion for its anti-inflammatory cytokine content and a high fetal hemoglobin content in case of ischemic stroke. The authors concluded that cord blood transfusion could be very effective in cerebral stroke to aid functional recovery.[28,29]

11.6 Special Advantages of Cord Blood Transfusion

Everything under the sun has been tried in the struggle to find an effective, safe, and viable alternative source of readily available blood for transfusion purposes in the last 300 years, viz., from animal to cadaver transfusion as mentioned earlier. The search intensified after 1914, when the shortage of adult blood for transfusion during World War I highlighted the need for a safe substitute. Pharmacological means like erythropoietin, different growth factors, hemoglobin-based oxygen carriers utilizing chemical or genetic changes in hemoglobin, drawing from the concepts of latest biotechnological marvels, i.e., liposomes, and also from nanotechnology concepts, have been attempted by many worthy individuals from different centers of excellence. However, in the words of the hematological legend, Prof Wintrobe, "The alternative of blood transfusion is the transfusion of blood." Nature itself appears to have provided an abundant and safe substitute, and while man was apparently in the dark about it, the animal world appears to have been aware of the significance of placental substances, of which cord blood is but one, because even herbivorous animals like the cow swallow their own placentas after the birth of their babies.

Cytokine production of the cord blood cell is less than that of adult blood, which could be due to antigen naïveté, (especially in case of T cells and NK) cells, lower cytokine gene expression in pretranscription level, mRNA accumulation or the intrinsic instability.[30] There may be a difference in the signaling pathway also. Cord blood cells may downregulate immune and autoimmune functions by maturing into high IL10 and IL2 producing TH1 cells.[31] In healthy individuals, alternatively activated macrophages are found in the placenta and the lung to protect them from unwanted inflammation and immune reactions. They produce anti-inflammatory cytokines like IL1 antagonist; in addition, they lack the expression of pro-inflammatory cytokines IL1, TNF alpha, IL6, IL12, and other macrophage inflammatory protein. The overall effect is the suppression of inflammation. These activated macrophages eventually go to the cord blood to impart their unique characteristics.[32]

If To sum up, blood, because of its rich mix of fetal and adult hemoglobin, high platelet and WBC counts, and a plasma filled with cytokine and growth factors, as well as its hypo-antigenic nature and altered metabolic profile, has all the potentials of a real and safe alternative to adult blood transfusion. In case of anemic patients of different etiologies (with hemoglobin 8 g/dL or less), as mentioned earlier, viz., malarial anemia,[33] geriatric anemia,[34-36] thalassemia,[37,38] HIV,[39] leprosy,[40] only to name a few, leaving aside a true need of blood in emergencies as seen in Ireland,[41] or by accidental collateral victims of allied bombing in the Iraq war,[42] there is an emergency need for a true blood substitute that can be met by cord blood.

Scientists have searched meticulously through all the available resources; there was great hope when *Nature* published an early news that scientists have found a blood substitute in the sea worm, which has the potentialities to replace RBC functions,[43] but the massive waste of true mothers' blood has never been looked into seriously as a blood substitute.

Our friend Prof. Andrew Burd, who is an important contributor to this book, has roughly estimated the total wastage of the cord blood per year:

Umbilical cord blood	Volume (cesarian section)	1,560,000 L
	Volume (vaginal delivery)	7,225,000 L
	Total volume	8,785,000 L

In fine, placental umbilical cord whole blood is a unique cocktail made by nature, which has enormous potentialities in transfusion, transplantation, and molecular therapy impact of its cytokine content.

Acknowledgment The Department of Science and Technology, Govt of West Bengal, supported the investigator with research grant during his tenure at Bijoygarh State Hospital from 1999 to 2006. The author gratefully acknowledges the support of the patients who volunteered for this research work. Guidance of Prof. K. L. Mukherjee of Biochemistry and Prof. M. K. Chhetri, former Director of Health Services, are also acknowledged.

References

1. Refer to http://members.rediff.com/bloodbank/History.htm. Accessed on April 29, 2009.
2. Goodnough LT et al. Blood transfusion. *N Engl J Med*. 1999;340(6):438-447.
3. Matovcik LM, Groschel-Stewart U, Schrier SL. Myosin in adult and human erythrocyte membrane. *Blood*. 1986;67:1668.
4. Tuan D, Feingold E, Newman M, Weissman SM, Forget BG. Different 3' end points of deletions causing delta beta-thalassemia and hereditary persistence of fetal hemoglobin: implications for the control of gamma-globin gene expression in man. *Proc Natl Acad Sci USA*. 1983;80(22):6937-6941.
5. Marsh WL. Erythrocytes' blood groups in human. In: Nathan DG, Oski FA, eds. *Hematology of Infancy and Childhood*. 3rd ed. Philadelphia, PA: W.B. Saunders; 1987.
6. Travis SF, Kumar SP, Paez PC, Delivoria-Papadopoulos M. Red cell metabolic alterations in postnatal life in term infants: glycolytic enzymes and glucose-6-phosphate dehydrogenase. *Pediatr Res*. 1980;14(12):1349-1352.
7. Stevenson SS. Carbonic anhydrase in new born infants. *J Clin Invest*. 1943;22:403.
8. Hildbrand P, Cirulli V, Prinsen RC, et al. The role of angiopoietins in the development of endothelial cells from cord blood CD34+ progenitors. *Blood*. 2004;104:2010-2019. doi:10.1182/blood-2003-12-4219.

9. De Ugarte DA, Alfonso Z, Zuk PA, et al. Differential expression of stem cell mobilization-associated molecules on multi-lineage cells from adipose tissue and bone marrow. *Immunol Lett*. 2003;89:267-270.
10. Zhao Y, Wang H, Mazzone T. Identification of stem cells from human umbilical cord blood with embryonic and hematopoietic characteristics. *Exp Cell Res*. 2006;312:2454-2464.
11. Brzoska E, Grabowska I, Hoser G, et al. Participation of stem cells from human cord blood in skeletal muscle regeneration of SCID mice. *Exp Hematol*. 2006;34:1262-1270.
12. Hu CH, Wu GF, Wang XQ, et al. Transplanted human umbilical cord blood mononuclear cells improve left ventricular function through angiogenesis in myocardial infarction. *Chin Med J (Engl)*. 2006;119:1499-1506.
13. Leor J, Guetta E, Feinberg MS, et al. Human umbilical cord blood-derived CD133+ cells enhance function and repair of the infarcted myocardium. *Stem Cells*. 2006;24:772-780. doi:10.1634/stemcells.2005-0212.
14. Halbrecht J. Fresh and stored placental blood. *Lancet*. 1939;2:1263. doi:10.1016/S0140-6736(00)74023-2.
15. Bhattacharya N, Mukherijee K, Chettri MK, Banerjee T, Mani U, Bhattacharya S. A study report of 174 units of placental umbilical cord whole blood transfusion in 62 patients as a rich source of fetal hemoglobin supply in different indications of blood transfusion. *Clin Exp Obstet Gynecol*. 2001;28(1):47-52.
16. Hassall O, Bedu-Addo G, Adarkwa M, Danso K, Bates I. Umbilical-cord blood for transfusion in children with severe anaemia in under-resourced countries. *Lancet*. 2003; 361(9358):678-679.
17. Bhattacharya N. Placental umbilical cord whole blood transfusion: a safe and genuine blood substitute for patients of the under-resourced world at emergency. *J Am Coll Surg*. 2005; 200(4):557-563.
18. Bhattacharya N. Placental umbilical cord blood transfusion in transfusion-dependent beta thalassemic patients: a preliminary communication. *Clin Exp Obstet Gynecol*. 2005; 32(2):102-106.
19. Bhattacharya N. Placental umbilical cord blood transfusion: a new method of treatment of patients with diabetes and microalbuminuria in the background of anemia. *Clin Exp Obstet Gynecol*. 2006;33(3):164-168.
20. Bhattacharya N. Placental umbilical cord blood transfusion: a novel method of treatment of patients with malaria in the background of anemia. *Clin Exp Obstet Gynecol*. 2006; 33(1):39-43.
21. Bhattacharya N. A preliminary study of placental umbilical cord whole blood transfusion in under resourced patients with malaria in the background of anaemia. *Malar J*. 2006; 5:20.
22. Bhattacharya N. Placental umbilical cord whole blood transfusion to combat anemia in the background of advanced rheumatoid arthritis and emaciation and its potential role as immunoadjuvant therapy. *Clin Exp Obstet Gynecol*. 2006; 33(1):28-33.
23. Bhattacharya N. A study of placental umbilical cord whole blood transfusion in 72 patients with anemia and emaciation in the background of cancer. *Eur J Gynaecol Oncol*. 2006; 27(2):155-161.
24. Bhattacharya N. Placental umbilical cord whole blood transfusion to combat anemia in the background of tuberculosis and emaciation and its potential role as an immuno-adjuvant therapy for the under-resourced people of the world. *Clin Exp Obstet Gynecol*. 2006;33(2):99-104.
25. Bhattacharya N. A preliminary report of 123 units of placental umbilical cord whole blood transfusion in HIV-positive patients with anemia and emaciation. *Clin Exp Obstet Gynecol*. 2006;33(2):117-121.
26. Bhattacharya N. Transient spontaneous engraftment of CD34 hematopoietic cord blood stem cells as seen in peripheral blood: treatment of leprosy patients with anemia by placental umbilical cord whole blood transfusion. *Clin Exp Obstet Gynecol*. 2006;33(3):159-163.
27. Bhattacharya N. Spontaneous transient rise of CD34 cells in peripheral blood after 72 hours in patients suffering from advanced malignancy with anemia: effect and prognostic implications of treatment with placental umbilical cord whole blood transfusion. *Eur J Gynaecol Oncol*. 2006; 27(3):286-290.
28. Chaudhuri A, Hollands P, Bhattacharya N. Placental umbilical cord blood transfusion in acute ischaemic stroke. *Med Hypotheses*. 2007;69(6):1267-1271.
29. Chaudhuri A. Treating stroke in the 21st century. *Lancet*. 2007;369(9567):1079.
30. Riordan NH, Chan K, Marleau AM, Ichim TE. Cord blood in regenerative medicine: do we need immune suppression? *J Transl Med*. 2007;5:8.
31. Madrigal JA, Cohen SBA, Gluckman E, Charron DJ. Does cord blood transplantation result in lower graft vs host disease. It takes more than two to tango. *Hum Immunol*. 1997;56:1-5.
32. Chang MD, Polard JW, Khalile H, et al. Mouse placental macrophages have a decreased ability to present antigen. *Proc Natl Acad Sci USA*. 1993;90:462-466.
33. Clarke T. Newborns might help malaria kids – Blood from umbilical cords could treat anemia caused by tropical disease. Available at: http://www.nature.com/nsu/021111/021111-11html. Accessed November 15, 2002.
34. Bhattacharya N. Placental umbilical cord whole blood transfusion [letter]. *J Am Coll Surg*. 2004;1992:347-348.
35. Bhattacharya N, Bandopadhyay T, Bhattacharya M, Bhattacharya S. Do not discard 99.99% of the human placental umbilical cord blood for the sake of stem cells only. Available at: http://bmj.com/cgi/eletters/323/7304/60#16874. Accessed October 5, 2001.
36. Bhattacharya N, Bandyopadhyay T, Bhattacharya M, Bhattacharya S. Immunization and fetal cell /tissue transplant: a new strategy for geriatric treatment. Available at: http://bmj.com/cgi/eletters/323/7320/1025/b#21055. Accessed April 6, 2002.
37. Bhattacharya N. Study of the utility of placental cord blood in meeting the transfusion needs of beta-thalassaemic patients. *Reg Health Forum*. 2009;12(2):16-27.
38. Bhattacharya N. The safe use of placental umbilical cord whole blood transfusion in patients suffering with anemia and thalassemia in under-resourced regions of the world. Available at: http://bmj.com/cgi/eletters/321/7269/1117#62372. Accessed June 9, 2004.
39. Bhattacharya N. Umbilical cord whole blood transfusion in HIV patients with anemia and emaciation. Available at: http://bmj.com/cgi/eletters/327/7414/562-a#59738. Accessed May 17, 2004.

40. Bhattacharya N. A preliminary study report on placental umbilical cord blood transfusion in victims of anemia with leprosy in under-resourced regions of the world. Available at: http://bmj.comcgi/eletters/328/7454/1447/=63828. Accessed June 22, 2004.
41. Bhattacharya N. Umbilical cord whole blood transfusion: A suggested strategy to combat blood scarcity in Ireland. Availableat:http://bmj.com/cgi/eletters/324/7330/134/c#19096. Accessed January 27, 2002.
42. Bhattacharya N. Utilization of a genuine blood substitute: a suggestion to the medical fraternity in Iraqi hospital. Available at: http://bmj.com/cgi/eletters/326/7391/675#30850. Accessed March 30, 2003.
43. Hoag H. Blood substitute from worm shows promise – hemoglobin from sea creature could replace red cells. Available at: http://www.nature.com/nsu/030602/030602-7html. Accessed June 4, 2003.

Part V

Immunotherapy Potential of Fetal Cell in Maternal System

Implications of Feto-maternal Cell Transfer in Normal Pregnancy

12

Carolyn Troeger, Olav Lapaire,
Xiao Yan Zhong, and Wolfgang Holzgreve

12.1 Introduction

Traditionally, the placenta has been thought to be a well-built barrier that separates the genetically different mother and offspring. This has been studied and discussed by K.E. von Baer already in 1828 when he found separate vascular beds in dogs.[1] In the following years, it has been suggested that there is communication between the placenta and the host mother, because syncytiotrophoblasts have been found in the lungs of women who died of eclampsia.[2] Transplacental fetal "bleeding" was recognized only in cases with a significant placental trauma, e.g., car accident, amniocentesis, chorionic villous sampling, termination of pregnancy, and external version done for breech presentation.[3-7] Fetal red blood cell levels in maternal blood were at that time best identified with the Kleihauer–Betke technique, which makes use of the different hemoglobin elution of fetal and maternal red blood cells in an acidic milieu of pH 3.[8] Since the sensitivity of this assay is limited, feto-maternal red cell traffic has long been related to abnormal courses of pregnancy only. Besides, this test fails to detect those 10% of fetal erythrocytes that already contain adult hemoglobin A and false-positively identifies those maternal red blood cells that during a normal pregnancy contain fetal hemoglobin. Additionally, elevated levels of HbF-containing cells are found in thalassemia further limiting the use of this test.[9-11]

The passage of maternal cells into the fetal circulation has been suggested in cases where rhesus D positive cells were found in genetically rhesus negative fetuses, the so-called "grandmother" effect.[12] The transfer and persistence of maternal leukocytes and stem cells in the fetal organs and blood might have implications for the use of umbilical cord blood for stem cell transplantation and for the vertical transmission of infectious agents.[13,14] Vice versa, the traffic of fetal cells into the maternal circulation and their persistence over decades can be used for noninvasive prenatal diagnosis during an ongoing pregnancy, but might also have implications on the later maternal health status in terms of auto-immune disorders, such as scleroderma.[15]

12.2 Materno-fetal Cell Transfer and Its Implications on Immunity and Tolerance

Although massive transplacental transfer of red blood cells has probably the most serious consequences in regard to anemia and immunization, traffic of other blood particles is also of interest. Maternal neoplasm during pregnancy is a rare event, but might have serious consequences not only for the mother, but also for the fetus. Metastases of several cancers and lymphomas have been found in the placenta. In leukemias, however, the intervillous space might be crowded with leukemic cells without passing through. Owing to fetal intrauterine death or elective termination of pregnancy, information on the actual rate of vertical transmission is sparse. Obviously, neoplastic cells do not always cross the placental barrier.[16,17] Whereas placental metastases of the maternal malignant disease appear quite often, the risk of fetal involvement seems to be different between the entities (see Table 12.1).[18] A possible defense mechanism has been recently brought up as a tendency of the malignant cells to be transformed

C. Troeger (✉)
Laboratory for Prenatal Medicine, University Women´s Hospital, Spitalstrasse 21, CH-4031, Basel, Switzerland
e-mail: ctroeger@uhbs.ch

N. Bhattacharya and P. Stubblefield (eds.), *Regenerative Medicine Using Pregnancy-Specific Biological Substances*,
DOI: 10.1007/978-1-84882-718-9_12, © Springer-Verlag London Limited 2011

Table 12.1 Rates of metastatic vertical transmission in the most common maternal malignancies during pregnancy[18,91]

Disease	N	Placental metastasis	Fetal metastasis
Malignant melanoma	27	24	6
Breast	15	15	0
Lung	10	8	2
Leukemia	9	6	3
Lymphoma	6	4	3

back to benign cells in the fetal environment.[19,20] In most cases of malignant maternal tumors, the intervillous tumor colonies fail to exhibit vascularization, like in the so-called pseudometastases from tumors of the breast, pancreas, lungs, ovary, rectum, squamous cell carcinoma of the skin, and medulloblastoma.[21] From these findings in solid organ malignancies and from the missing correlation between the numbers of leukemic cells in the fetal circulation and massive leukemic cell infiltration of the intervillous space, a selective transport mechanism of the placenta instead of a passive influx gradient must be postulated. This mechanism might be insufficient to prevent the fetus from certain infections by cellular organisms, such as trypanosoma or toxoplasma. In such cases, in the infected placentas a destructive villitis and phagocytosis by Hofbauer cells supports the transfer of infectious cells into the fetal circulation.[22,23]

Also, for normal pregnancies the findings suggest that the materno-fetal transplacental cell traffic does not appear in a dose-dependent manner, since an increased maternal white blood cell count does not correspond to the WBC count in the neonate.[24] Additionally, the "placental barrier" must be able to differentiate between normal and abnormal leucocytes, most probably through a receptor–ligand interaction, since it is known for long that normal leukocytes traffic, whereas obviously neoplastic leukocytes only rarely cross into the fetal blood.[25] Maternal multipotent stromal cells for instance use a VEGF-A- and integrin-dependent pathway.[26] A further indication that the transfer is not dose- and size-dependent comes from the observation that the amount of materno-fetal traffic is lower than the cell transfer from the fetus into the mother, and both ways do not correlate with each other.[27,28] However, both phenomena have implications on health status of the chimeras in terms of immunization and tolerance.[29] It has been noted that occasional transfer of maternal lymphocytes into the fetus might lead to chimerism and possible graft-versus-host disease.[30,31] It seems that microchimerism within the fetus might have different implications than similar cell traffic into the mother due to differences in the maturity of the immune system. Opposite outcomes of microchimerism in terms of tolerance or alloreactivity are also observed after solid organ transplants. Here, microchimerism in hosts with immature T cells resulted in tolerance, whereas in immunologically mature hosts, immunity and graft rejection occurred.[32] This might explain why maternal cells can persist in the offspring until adult life in cases with a normal alloreactive immune system.[33] These cells might have an active role in the development of the fetal immune system. Maternal activated T-cells cross the placenta and can induce antigen-specific tolerance in the developing fetuses.[34] This might explain why certain individuals are to a certain degree tolerant to noninherited maternal human leukocyte antigens (NIMA).[35] Most probably, this NIMA-induced allo-tolerance is due to a profound inhibition of allospecific T and B cell responses in the offspring.[36] Additionally, levels and tissue distribution of maternal microchimerism depends on the MHC zygocity, indicating that the fetal immune system, though immature, recognizes the maternal cells.[37] Whether microchimerism leads to beneficial regenerative effects or to autoimmunity might be related to MHC zygocity since it has been shown that low levels of microchimerism are associated with MHC heterozygocity, which tend to protect from autoimmunity.[38,39] Whether persistent maternal microchimerism leads to autoimmune diseases such as scleroderma is a matter of ongoing and conflicting debate.[33] Though one study revealed HLA-compatibility as a risk factor for systemic sclerosis, another study did not.[40,41] There is also some evidence that maternal microchimerism is associated with myositis, since two independent groups described increased levels of maternal cells in offspring´s blood and tissue compared to unaffected siblings as a control group.[42,43] A possible other candidate would be neonatal lupus, which is associated with maternal autoantibodies and maternal HLA genotype.[44]

In summary, a physiological maternal microchimerism exists in adult life and might have an impact on the normal development of the fetal immune system. Though immature, the fetal immune system might very well recognize the maternal microchimeric cells, which can be concluded from different transplantation outcomes. Whether this microchimerism might also contribute to autoimmune diseases requires further research. The materno-fetal interface obviously is capable of differentiating between different cell types, thus allowing transfer of some but not every cell type. The adhesion and transport mechanisms are so far not well understood.

12.3 Feto-maternal Cell Traffic and Its Use for Noninvasive Prenatal Diagnosis

Fetal lymphocytes frequently pass into the maternal blood and bone marrow and persist for many years post partum.[45] Originally, Selypes and Lorencz described a method of noninvasive determination of fetal sex and karyotype using these fetal lymphocytes.[46] This led to a new development in prenatal diagnosis using fetal cells enriched from maternal blood and their analysis by fluorescence in situ hybridization[47] (see Fig. 12.1). However, only one to six fetal cells are present in 1 mL of maternal blood during the second trimester of pregnancy making the approach very challenging.[48] All technical aspects such as enrichment and detection techniques as well as reproducibility in different laboratories have been extensively evaluated by a prospective, multicenter clinical so-called NIFTY trial.[49-51]

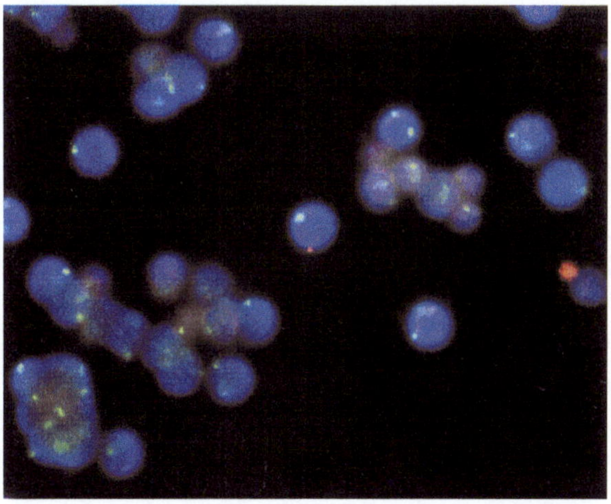

Fig. 12.1 Fetal male cell (*arrow*) enriched from maternal blood detected by fluorescence in situ hybridization using DNA-probes specific for Y-(red) and X-chromosome (green signal) and DAPI-counterstaining

Here, it has been shown that the sensitivity of detecting a fetal cell in maternal blood was 41% with a false-positive rate of 11% in normal pregnancies, whereas validity of this method increased in cases with aneuploidies to a sensitivity of 74% and a false-positive rate of 1–4%. These low rates of sensitivity are due to the fact that in normal pregnancies, only half of the currently enriched erythroblasts have been shown to be of fetal origin.[52] This lack in specificity can be explained by a lack of fetal-specific antigens that are expressed by the target erythroblasts and of suitable antibodies.[53,54]

In contrast to erythroblasts, other cell types such as erythroid progenitors would potentially overcome the technical limitations in that they could be expanded in culture and therewith enable metaphase analyses of the chromosomes and increase their numbers in vitro. The problem with fetal progenitors in maternal blood is that they are outnumbered by maternal progenitors if simply a co-culture is performed. This might explain why promising results of some groups could not be repeated by others.[55-57] Making use of flow cytometry, Bohmer et al. first sorted the cells using gamma and beta globin as markers to differentiate between presumably fetal and maternal cells.[58] Using optimal culture conditions with charcoal-adsorbed human umbilical cord serum and omitting IL-3 followed by FACS-sorting, however, did not allow detecting any fetal progenitor in maternal blood samples from mid-trimester blood of normal and abnormal pregnancies, whereas in the controls (maternal blood drawn after elective termination of pregnancy), a similar protocol worked.[48] These results show that the number of fetal progenitors in maternal blood is even lower than that of fetal erythroblasts (two fetal CD34+ cells compared to 20 erythroblasts in 20 mL of maternal blood) and therewith might be under the detection limit and so far clinical application for noninvasive prenatal diagnosis is still waiting.[59]

An alternative method of noninvasive prenatal diagnosis focused on the polymerase chain reaction (PCR) for the detection of Y-chromosome bearing cells.[60,61] For both methods of single-cell analysis, FISH and PCR, accuracy is limited due to the apoptotic character and dense nucleus of the enriched fetal erythroblasts leading to false-negative results.[62,63]

New advances in noninvasive prenatal diagnosis, therefore, deal with cell-free fetal DNA from maternal plasma or serum (see Fig. 12.2). This material is easier to handle. Additionally, concentrations of cell-free fetal DNA seem to be higher than the amount of fetal

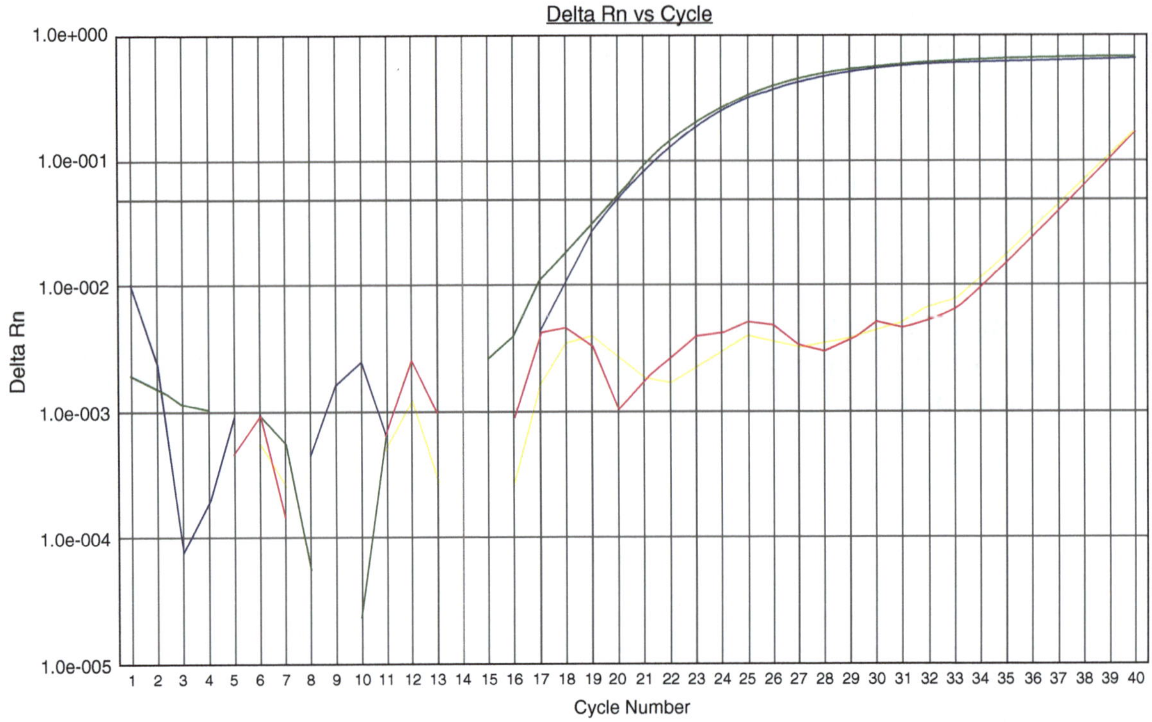

Fig. 12.2 Example of a very sensitive quantitative polymerase chain reaction (*PCR*) assay to detect trace amounts of male DNA in maternal blood using primers and probes for DYS (*yellow and red curve*) and GAPDH (*blue and green curve*)

cells in maternal blood since the cell-free fetal DNA is not the result of transplacental cell traffic, but stems from the shedding of trophoblasts into the maternal circulation.[64-68]

In contrast to fetal cell-free DNA, isolated fetal cells offer a pure source of the entire genomic DNA, without any contamination by maternal DNA. However, efforts need to be expended on the refinement of the specific isolation and rapid automated detection and analysis of the rare fetal cells in maternal blood and also on the evaluation of other putative fetal target cells, such as transferred fetal stem cells and deported trophoblasts.

12.4 Feto-maternal Cell Traffic and Its Long-Term Consequences

Whether any transplacental cell traffic is a cause of inflammation or is just a bystander to regenerate the damaged tissue in cases of certain autoimmune diseases is a matter of ongoing debate.[69-71] Analyzing murine syngeneic and allogeneic pregnancies, Khoshrotehrani and colleagues showed that fetal microchimerism is a naturally occurring phenomenon leading to detectable levels of mononuclear cells in several maternal tissues, such as lungs, heart, spleen, kidney, and bone marrow.[72] The lower levels of fetal microchimerism in allogeneic compared to syngeneic pregnancies indicate that the haplo-identical fetal cells are very well recognized by the maternal immune system. It has been shown for human and murine pregnancies that levels decrease quite after delivery indicating that the maternal immune effector cells shift back to their normal alloresponse.[65,72] Although this group was unable to detect microchimeric fetal cells in maternal mouse brain during pregnancy, others showed a relevant proportion of fetal progenitors that obviously were able to cross the blood-brain barrier during pregnancy.[73] The cells even increased in number during a period of 4 weeks post partum and adopted a local phenotype like perivascular macrophage-, neuron-, astrocyte- , and oligodendrocyte-like cell type. The levels of fetal microchimerism were increased in brain injuries suggesting an active role in tissue regeneration. Whether this is just associated with inflammation and phagocytosis of damaged tissue or formation of truly new neurons needs further attention. Similar uncertainties exist for an animal model on murine maternal hepatic injury.[74] Also here, higher levels of microchimerism were detected after chemical compared to surgical laceration, with increasing levels between 4 and 8 weeks after injury.

It has been shown that fetal progenitors could persist within the maternal blood and tissue over decades.[75] Since these CD34+ and CD34+/CD38+ fetal progenitors are capable of differentiating into functional T and B cells, the increased levels of fetal cell traffic under certain instances such as preeclampsia, surgical termination of pregnancy, and aneuploidies might have a long-term impact on maternal health status.[48,76-78] It is unclear yet why certain women with fetal microchimerism develop autoimmune diseases, and others do not (see Table 12.2). Whether HLA compatibility is an issue is still controversial. Animal models suggest that similarities in MHC lead to higher levels of microchimerism.[29] However, in humans with and without scleroderma, HLA compatibility had no influence on the development of the disease,[79] although some HLA haplotypes seem to be more associated with fetal microchimerism, such as HLA-DQA1*0501.[80] However, this finding is currently controversially discussed.[81] Conflicting results are also reported for fetal cell microchimerism levels in primary biliary cirrhosis.[82,83] It can be discussed whether in some autoimmune conditions such as lupus, fetal microchimerism seems to be correlated with the severity of the disease, since fetal cells are detected in a patient who died of the complications, but not in the cutaneous lesions of patients who obviously did not have such a bad outcome.[84,85] Similar results are found in cases with systemic sclerosis, where fetal cells were detected in patients suffering from the disease. Analyzing their skin, increased levels of microchimerism were found in the lesion, whereas levels were lower in adjacent uninvolved skin of these patients.[86] Still, these findings do not support a causal relationship with the development of an autoimmune disease. This comes more from results of animal models of systemic sclerosis, where vinyl chloride led to typical sclerosis-like lesions in parous mice with increased levels of microchimeric cells in the lesions, whereas similar toxic treatment did not lead to skin involvement in virgin mice.[87] To evaluate a probable causal relationship between feto-maternal cell traffic and the development of an autoimmune disease, those women with known increased levels of fetal microchimerism such as in preeclampsia[77] must be followed prospectively. So far, prospective population-based studies on a correlation between elevated levels of microchimerism during pregnancy and a later development into an autoimmune disease are missing. However, recently, it is discussed whether preeclampsia and HELLP syndrome can be regarded as autoimmune diseases, since autoantibodies against the angiotensin receptor AT1, vascular endothelium, thyroid and nuclear antigens are described in preeclampsia, eclampsia, and HELLP.[88-90]

Table 12.2 Fetal microchimerism (FMC) in maternal diseases

Disease	Controls	Peripheral blood of affected patients	Tissue of affected patients	References
Systemic sclerosis (SSc)	Positive FMC, but lower level	FMC in CD4+ higher than CD8+	FMC in skin	Nelson,[69] Artlett,[92] and Ohtsuka et al.[93]
Polymorphic eruption of pregnancy (PEPP)	Negative	n.d.	FMC in skin	Aractingi et al.[94]
Primary biliary cirrhosis (PBC)	Positive FMC at similar level	Positive FMC	FMC in liver	Fanning et al.[82] and Schöniger-Hekele et al.[83]
Thyroiditis	FMC also in adenoma, goiter, carcinoma, grave's disease	Negative	FMC in Hashimoto	Srivasta et al.[95] Klintschar et al.[96] and Ando et al.[97]
Sjögren's syndrome	Positive FMC	Negative	Negative	Toda et al.[98] and Aractingi et al.[99]
Multiple sclerosis (MS)	Negative FMC in unaffected twins	Positive FMC in affected twins	n.d.	Willer et al.[100]
Lupus erythematosus	Positive FMC at similar level	Positive FMC similar to controls, higher in nephritis	FMC in several organs higher than in controls	Mosca et al.[101] and Kremer et al.[102]

n.d. not done

References

1. Baer KEv. *Untersuchungen über die Gefässverbindungen zwischen Mutter und Frucht in den Säugetieren*. Leipzig: Voss; 1828.
2. Schmorl G. *Pathologisch-anatomische Untersuchungen über puerperale Eklampsie*. Leipzig: FCW Vogel; 1893.
3. Pearlman MD, Tintinalli JE, Lorenz RP. Blunt trauma during pregnancy. *N Engl J Med*. 1990;323:1609-1613.
4. Goodlin RC, Clewell WH. Sudden fetal death following diagnostic amniocentesis. *Am J Obstet Gynecol*. 1974;118: 285-288.
5. Warren RC, Butler J, Morsman JM, et al. Does chorionic villus sampling cause feto-maternal haemorrhage? *Lancet*. 1985;1:691.
6. Voigt JC, Britt RP. Feto-maternal haemorrhage in therapeutic abortion. *BMJ*. 1969;2:395-396.
7. Pollack M, Montague ACW. Transplacental hemorrhage in postterm pregnancies. *Am J Obstet Gynecol*. 1968;102: 383-387.
8. Kleihauer E, Braun H, Betke K. Demonstration von fetalem Hämoglobin in den Erythrocyten eines Blutausstriches. *Klin Wochenschr*. 1957;35:637-638.
9. Mollison PL. Quantification of transplacental haemorrhage. *BMJ*. 1972;3:31-34. correction p. 115.
10. Woodrow JC, Finn R. Transplacental haemorrhage. *Br J Haematol*. 1966;12:297-309.
11. Leiberman JR, Mazor M, Cohen A. Detection of fetal blood. *Am J Obstet Gynecol*. 1989;60:60-64.
12. Taylor JF. Sensitization of Rh-negative daughters by their Rh-positive mothers. *N Engl J Med*. 1967;276:547-551.
13. Rubinstein A, Goldstein H, Calvelli T, et al. Maternofetal transmission of human immunodeficiency virus-1: the role of antibodies to the V3 primary neutralizing domain. *Pediatr Res*. 1993;33:76-78.
14. Bucher C, Stern M, Buser A, et al. Role of primacy of birth in HLA-identical sibling transplantation. *Blood*. 2007;110: 468-469.
15. Holzgreve W, Hahn S, Zhong XY, et al. Genetic communication between fetus and mother: short- and long-term consequences. *Am J Obstet Gynecol*. 2007;196:372-381.
16. Cavell B. Transplacental metastasis of malignant melanoma. Report of a case. *Acta Paediatr Suppl*. 1963;146:37-40.
17. Brodsky I, Baren M, Kahn SB, et al. Metastatic malignant melanoma from mother to fetus. *Cancer*. 1965;18: 1048-1054.
18. Jackisch C, Louwen F, Schwenkhagen A, et al. Lung cancer during pregnancy involving the products of conception and a review of the literature. *Arch Gynecol Obstet*. 2002;268:69-77.
19. Astigiano S, Damonte P, Fossati S, et al. Fate of embryonal carcinoma cells injected into postimplantation mouse embryos. *Differentiation*. 2005;73:484-490.
20. Ma F. The benign transformation tendency of malignant tumor cells for intrauterine transplantation. *Swiss Med Wkly*. 2007;137:561.
21. Lemtis H, Hörmann G. Das Verhalten der Placentaschranke gegenüber Metastasen maligner Tumoren der Mutter. *Arch Gynakol*. 1965;202:471-473.
22. Altshuler G. Toxoplasmosis as a cause of hydranencaphaly. *Am J Dis Child*. 1973;127:427-429.
23. Bittencourt AL. Congenital chagas disease. *Am J Dis Child*. 1976;130:97-103.
24. Bierman HR, Kelly K, Cordes F, et al. The influence of histamine upon the circulating leukocyte level in patients with the leukemias. *Blood*. 1956;11:709-719.
25. Schröder J. Transplacental passage of blood cells. *J Med Genet*. 1975;12:230-242.
26. Chen CP, Lee MY, Huang JP, et al. Trafficking of multipotent mesenchymal stromal cells from maternal circulation through the placenta involves vascular endothelial growth factor receptor-1 and integrins. *Stem cells*. 2008; 26:550-561.
27. Lo YM, Lo ES, Watson N, et al. Two-way cell traffic between mother and fetus: biologic and clinical implications. *Blood*. 1996;88:4390-4395.
28. Lo YM, Lau TK, Chan LY, et al. Quantitative analysis of the bidirectional fetomaternal transfer of nucleated cells and plasma DNA. *Clin Chem*. 2000;46:1301-1309.
29. Bonney EA, Matzinger P. The maternal immune system's interaction with circulating fetal cells. *J Immunol*. 1997;158: 40-47.
30. Kadowaki J, Thompson RI, Zuelzer WW. XX-XY lymphoid chimaerism in congenital immunological deficiency syndrome with thymic alymphoplasia. *Lancet*. 1965;2: 1152-1156.
31. Githens JH, Muschenheim F, Fulginiti VA, et al. Thymic alymphoplasia with XX-XY lymphoid chimerism secondary to probable maternal-fetal transfusion. *J Pediatr*. 1969;75: 87-94.
32. Anderson CC, Matzinger P. Immunity or tolerance: opposite outcomes of microchimerism from skin grafts. *Nat Med*. 2001;7:80-87.
33. Maloney S, Smith A, Furst DE, et al. Microchimerism of maternal origin persists into adult life. *J Clin Invest*. 1999; 104:41-47.
34. Wan W, Shimizu S, Ikawa H, et al. Maternal cell traffic bounds for immune modulation: tracking maternal H-2 alleles in spleens of baby mice by DNA fingerprinting. *Immunology*. 2002;107:261-267.
35. Claas FH, Gijbels Y, van der Velden-de MJ, et al. Induction of B cell unresponsiveness to noninherited maternal HLA antigens during fetal life. *Science*. 1988;241:1815-1817.
36. Andrassy J, Kusaka S, Jankowska-Gan E, et al. Tolerance to noninherited maternal MHC antigens in mice. *J Immunol*. 2003;171:5554-5561.
37. Vernochet C, Caucheteux SM, Kanellopoulos-Langevin C. Bi-directional cell trafficking between mother and fetus in mouse placenta. *Placenta*. 2007;28:639-649.
38. Nelson GW, Martin MP, Gladman D, et al. Cutting edge: heterozygote advantage in autoimmune disease: hierarchy of protection/susceptibility conferred by HLA and killer Ig-like receptor combinations in psoriatic arthritis. *J Immunol*. 2004;173:4273-4276.
39. Kaplan J, Land S. Influence of maternal-fetal histocompatibility and MHC zygosity on maternal microchimerism. *J Immunol*. 2005;174:7123-7128.
40. Artlett CM, Welsh KI, Black CM, et al. Fetal-maternal HLA compatibility confers susceptibility to systemic sclerosis. *Immunogenetics*. 1997;47:17-22.
41. Lambert NC, Evans PC, Hashizumi TL, et al. Cutting edge: persistent fetal microchimerism in T lymphocytes is associated

with HLA-DQA1*0501: implications in autoimmunity. *J Immunol.* 2000;164:5545-5548.
42. Reed AM, Picornell YJ, Harwood A, et al. Chimerism in children with juvenile dermatomyositis. *Lancet.* 2000;356: 2156-2157.
43. Artlett CM, Ramos R, Jiminez SA, Childhood Myositis Heterogeneity Collaborative Group, et al. Chimeric cells of maternal origin in juvenile idiopathic inflammatory myopathies. *Lancet.* 2000;356:2155-2156.
44. Buyon JP. Neonatal lupus and autoantibodies reactive with SSA/Ro-SSB/La. *Scand J Rheumatol Suppl.* 1998;107:23-30.
45. Schröder J, Schröder E, Cann HM. Fetal cells in the maternal blood. Lack of response of fetal cells in maternal blood to mitogens and mixed leukocyte culture. *Hum Genet.* 1977;38:91-97.
46. Selypes A, Lorencz R. A noninvasive method for determination of the sex and karyotype of the fetus from the maternal blood. *Hum Genet.* 1988;79:357-359.
47. Gänshirt-Ahlert D, Burschyk M, Garritsen HS, et al. Magnetic cell sorting and the transferrin receptor as potential means of prenatal diagnosis from maternal blood. *Am J Obstet Gynecol.* 1992;166:1350-1355.
48. Bianchi DW, Farina A, Weber W, et al. Significant fetal-maternal hemorrhage after termination of pregnancy: implications for development of fetal cell microchimerism. *Am J Obstet Gynecol.* 2001;184:703-706.
49. de la Cruz F, Shifrin H, Elias S, et al. Prenatal diagnosis by use of fetal cells isolated from maternal blood. *Am J Obstet Gynecol.* 1995;173:1354-1355.
50. de la Cruz F, Shifrin H, Elias S, et al. Low false-positive rate of aneuploidy detection using fetal cells isolated from maternal blood. *Fetal Diagn Ther.* 1998;13:380.
51. Bianchi DW, Simpson JL, Jackson LG, National Institute of Child Health and Development Fetal Cell Isolation Study, et al. Fetal gender and aneuploidy detection using fetal cells in maternal blood: analysis of NIFTY I data. *Prenat Diagn.* 2002;22:609-615.
52. Troeger C, Zhong XY, Burgemeister R, et al. Approximately half of the erythroblasts in maternal blood are of fetal origin. *Mol Hum Reprod.* 1999;5:1162-1165.
53. Troeger C, Holzgreve W, Hahn S. A comparison of different density gradients and antibodies for enrichment of fetal erythroblasts by MACS. *Prenat Diagn.* 1999;19:521-526.
54. Prieto B, Cándenas M, Venta R, et al. Isolation of fetal nucleated red blood cells from maternal blood in normal and aneuploid pregnancies. *Clin Chem Lab Med.* 2002;40:667-672.
55. Tutschek B, Reinhard J, Kögler G, et al. Clonal culture of fetal cells from maternal blood. *Lancet.* 2000;356: 1736-1737.
56. Campagnoli C, Roberts I, Kumar S, et al. Clonal culture of fetal cells from maternal blood. *Lancet.* 2001;357:962.
57. Zimmermann B, Holzgreve W, Zhong XY, et al. Inability to clonally expand fetal progenitors from maternal blood. *Fetal Diagn Ther.* 2002;17:97-100.
58. Bohmer RM, Zhen D, Bianchi DW. Differential development of fetal and adult haemoglobin profiles in colony culture: isolation of fetal nucleated red cells by two-colour fluorescence labelling. *Br J Haemotol.* 1998;103:351-360.
59. Bianchi DW. Fetomaternal cell traffic, pregnancy-associated progenitor cells, and autoimmune disease. *Best Pract Res Clin Obstet Gynaecol.* 2004;18:959-975.
60. Adinolfi M, Camporese C, Carr T. Gene amplification to detect fetal nucleated cells in pregnant women. *Lancet.* 1989;2:328-329.
61. Lo YM, Patel P, Wainscoat JS, et al. Prenatal sex determination by DNA amplification from maternal peripheral blood. *Lancet.* 1989;2:1363-1365.
62. Babochkina T, Mergenthaler S, De Napoli G, et al. Numerous erythroblasts in maternal blood are impervious to fluorescent in situ hybridization analysis, a feature related to a dense compact nucleus with apoptotic character. *Haematologica.* 2005;90:740-745.
63. Hahn S, Garvin AM, Di Naro E, et al. Allele drop-out can occur in alleles differing by a single nucleotide and is not alleviated by preamplification or minor template increments. *Genet Test.* 1998;2:351-355.
64. Lo YM, Corbetta N, Chamberlain PF, et al. Presence of fetal DNA in maternal plasma and serum. *Lancet.* 1997;350: 485-487.
65. Ariga H, Ohto H, Busch MP, et al. Kinetics of fetal cellular and cell-free DNA in the maternal circulation during and after pregnancy: implications for noninvasive prenatal diagnosis. *Transfusion.* 2001;41:1524-1530.
66. Zhong XY, Holzgreve W, Hahn S. Cell-free fetal DNA in the maternal circulation does not stem from the transplacental passage of fetal erythroblasts. *Mol Hum Reprod.* 2002;8:864-870.
67. Hahn S, Holzgreve W. Prenatal diagnosis using fetal cells and cell-free fetal DNA in maternal blood: what is currently feasible? *Clin Obstet Gynecol.* 2002;45:649-656. discussion 730–732.
68. Alberry M, Maddocks D, Jones M, et al. Free fetal DNA in maternal plasma in anembryonic pregnancies: confirmation that the origin is the trophoblast. *Prenat Diagn.* 2007;27: 415-418.
69. Nelson JL. Microchimerism and autoimmune disease. *N Engl J Med.* 1998;338:1224-1225.
70. Khosrotehrani K, Bianchi DW. Fetal cell microchimerism: helpful or harmful to the parous woman? *Curr Opin Obstet Gynecol.* 2003;15:195-199.
71. Khosrotehrani K, Bianchi DW. Multi-lineage potential of fetal cells in maternal tissue: a legacy in reverse. *J Cell Sci.* 2005;118:1559-1563.
72. Khosrotehrani K, Johnson KL, Guégan S, et al. Natural history of fetal cell microchimerism during and following murine pregnancy. *J Reprod Immunol.* 2005;66:1-12.
73. Tan XW, Liao H, Sun L, et al. Fetal microchimerism in the maternal mouse brain: a novel population of fetal progenitor or stem cells able to cross the blood-brain barrier? *Stem Cells.* 2005;23:1443-1452.
74. Khosrotehrani K, Reyes RR, Johnson KL, et al. Fetal cells participate over time in the response to specific types of murine maternal hepatic injury. *Hum Reprod.* 2007;22:654-661.
75. Bianchi DW, Zickwolf GK, Weil GJ, et al. Male fetal progenitor cells persist in maternal blood for as long as 27 years postpartum. *Proc Natl Acad Sci USA.* 1996;93:705-708.
76. Khosrotehrani K, Leduc M, Bachy V, et al. Pregnancy allows the transfer and differentiation of fetal lymphoid progenitors into functional T and B cells in mothers. *J Immunol.* 2008; 180:889-897.
77. Holzgreve W, Ghezzi F, Di Naro E, et al. Disturbed feto-maternal cell traffic in preeclampsia. *Obstet Gynecol.* 1998; 91:669-672.

78. Bianchi DW, Williams JM, Sullivan LM, et al. PCR quantitation of fetal cells in maternal blood in normal and aneuploid pregnancies. *Am J Hum Genet*. 1997;61:822-829.
79. Evans PC, Lambert N, Maloney S, et al. Long-term fetal microchimerism in peripheral blood mononuclear cell subsets in healthy women and women with scleroderma. *Blood*. 1999;93:2033-2037.
80. Lambert NC, Distler O, Müller-Ladner U, et al. HLA-DQA1*0501 is associated with diffuse systemic sclerosis in Caucasian men. *Arthritis Rheum*. 2000;43:2005-2010.
81. Artlett CM, O'Hanlon TP, Lopez AM, et al. HLA-DQA1 is not an apparent risk factor for microchimerism in patients with various autoimmune diseases and in healthy individuals. *Arthritis Rheum*. 2003;48:2567-2572.
82. Fanning PA, Jonsson JR, Clouston AD, et al. Detection of male DNA in the liver of female patients with primary biliary cirrhosis. *J Hepatol*. 2000;33:690-695.
83. Schöniger-Hekele M, Müller C, Ackermann J, et al. Lack of evidence for involvement of fetal microchimerism in pathogenesis of primary biliary cirrhosis. *Dig Dis Sci*. 2002; 47:1909-1914.
84. Johnson KL, McAlindon TE, Mulcahy E, et al. Microchimerism in a female patient with systemic lupus erythematosus. *Arthritis Rheum*. 2001;44:2107-2111.
85. Khosrotehrani K, Mery L, Aractingi S, et al. Absence of fetal cell microchimerism in cutaneous lesions of lupus erythematosus. *Ann Rheum Dis*. 2005;64:159-160.
86. Sawaya HH, Jimenez SA, Artlett CM. Quantification of fetal microchimeric cells in clinically affected and unaffected skin of patients with systemic sclerosis. *Rheumatology (Oxford)*. 2004;43:965-968.
87. Christner PJ, Artlett CM, Conway RF, et al. Increased numbers of microchimeric cells of fetal origin are associated with dermal fibrosis in mice following injection of vinyl chloride. *Arthritis Rheum*. 2000;43:2598-2605.
88. Gleicher N. Why much of the pathophysiology of preeclampsia-eclampsia must be of an autoimmune nature. *Am J Obstet Gynecol*. 2007;196(5):e1-7.
89. Xia Y, Zhou CC, Ramin SM, et al. Angiotensin receptors, autoimmunity, and preeclampsia. *J Immunol*. 2007;179: 3391-3395.
90. Weitgasser R, Spitzer D, Kartnig I, et al. Association of HELLP syndrome with autoimmune antibodies and glucose intolerance. *Diabet Care*. 2000;23:786-790.
91. Alexander A, Samlowski WE, Grossman D, et al. Metastatic melanoma in pregnancy: risk of transplacental metastases in the infant. *J Clin Oncol*. 2003;21:2179-2186.
92. Artlett CM, Cox LA, Ramos RC, et al. Increased microchimeric CD4+ T lymphocytes in peripheral blood from women with systemic sclerosis. *Clin Immunol*. 2002;103:303-308.
93. Ohtsuka T, Miyamoto Y, Yamakage A, et al. Quantitative analysis of microchimerism in systemic sclerosis skin tissue. *Arch Dermatol Res*. 2001;293:387-391.
94. Aractingi S, Berkane N, Bertheau P, et al. Fetal DNA in skin of polymorphic eruptions of pregnancy. *Lancet*. 1998;352:1898-1901.
95. Srivatsa B, Srivatsa S, Johnson KL, et al. Microchimerism of presumed fetal origin in thyroid specimens from women: a case-control study. *Lancet*. 2001;358:2034-2038.
96. Klintschar M, Schwaiger P, Mannweiler S, et al. Evidence of fetal microchimerism in Hashimoto's thyroiditis. *J Clin Endocrinol Metab*. 2001;86:2494-2498.
97. Ando T, Imaizumi M, Graves PN, et al. Intrathyroidal fetal microchimerism in Graves' disease. *J Clin Endocrinol Metab*. 2002;87:3315-3320.
98. Toda I, Kuwana M, Tsubota K, et al. Lack of evidence for an increased microchimerism in the circulation of patients with Sjögren's syndrome. *Ann Rheum Dis*. 2001;60:248-253.
99. Aractingi S, Sibilia J, Meignin V, et al. Presence of microchimerism in labial salivary glands in systemic sclerosis but not in Sjögren's syndrome. *Arthritis Rheum*. 2002;46: 1039-1043.
100. Willer CJ, Dyment DA, Risch NJ, et al. Twin concordance and sibling recurrence rates in multiple sclerosis. *Proc Natl Acad Sci USA*. 2006;100:12877-12882.
101. Mosca M, Curcio M, Lapi S, et al. Correlations of Y chromosome microchimerism with disease activity in patients with SLE: analysis of preliminary data. *Ann Rheum Dis*. 2003;62:651-654.
102. Kremer Hovinga IC, Koopmans M, Baelde HJ, et al. Tissue chimerism in systemic lupus erythematosus is related to injury. *Ann Rheum Dis*. 2007;66:1568-1573.

Early Reports on the Prognostic Implications and Immunotherapeutic Potentials of Cd34 Rich Cord Whole Blood Transfusion in Advanced Breast Cancer with Severe Anemia

13

Niranjan Bhattacharya

13.1 Introduction

Anemia is the commonest comorbidity, which prevents an aggressive and effective treatment initiation in case of advanced cancer. It increases with the progression of the disease.[1] Severe anemia can cause subsequent tumor cell hypoxia, which can reduce the tumorocidal effect of each modality of treatment including radiation.[2-8] Correction of anemia often improves the quality of life of cancer patients. Corrective options include supplementation of different erythropoietin preparations and dietary enrichment and supplementation, and finally, red cell transfusion.

Advanced cancer patients, by virtue of their frequent exposure to transfusion, develop HLA alloantibodies, which can adversely affect therapy, for example, the refractoriness of platelet functions. Thus, cancer patients should ideally receive specially processed blood products, for example, leucoreduced, irradiated, cytomegalovirus seronegative blood products. Leucoreduction can prevent febrile nonhematological reactions including HLA alloimmunization. Blood components are irradiated to prevent potentially lethal transfusion induced graft-vs-host disease. Irradiation interferes with the ability of the lymphocytes to proliferate. A minimum dosage of 2,500 cGi radiation is recommended for blood products before transfusion to a cancer patient to make the cells hypoantigenic and prevent alloantibodies and platelet refractoriness.[9] Owing to disease load or treatment, cancer patients are often immune-compromised and thus become predisposed to a wide variety of bacterial, viral, and fungal infections and allied cellular-mediated immune responses.[9]

In the present study, viable readily available alternatives were examined in the search for a solution to the problem of severe anemia in patients with advanced breast cancer where the patients were unable to arrange fresh whole blood or packed cell (RBC) for treatment. This was required to make the patient suitable for chemotherapy or palliative surgery to decrease the tumor load.

It has been noted that in the animal kingdom, swallowing the afterbirth by the mother is a general norm. Even herbivorous animals swallow the placenta after the birth of their babies (for example, the cow). Nature appears to have provided a precious wisdom to some of its creatures. But humans seem to be unaware of the positive properties of the womb.

There is up to 150 mL blood in the placenta, which has higher hemoglobin content than adult blood. It has a high fetal hemoglobin content, which is a normal stress response in pregnancy anemia, thyrotoxicosis, etc., and it can also carry more oxygen. This blood is hypoantigenic. The placental barrier is formidable in preventing the entry of bacteria, viruses, and other offensive substances into the fetal circulation. Even in cases of HIV infection, transmission may occur at the end of gestation through alternative routes, such as chorioamnionitis with leakage of the virus into the amniotic cavity or through trophoblast damage.[10]

N. Bhattacharya
Department of General Surgery, Obstertrics and Gynaecology and Clinical Immunology, Advanced Medical Research Institute, Gol Park, B.P. Poddar Hospital and Vidyasagore Hospital, Kolkata 700034, India

N. Bhattacharya and P. Stubblefield (eds.), *Regenerative Medicine Using Pregnancy-Specific Biological Substances*,
DOI: 10.1007/978-1-84882-718-9_13,

If collected aseptically from the umbilical cord of healthy babies, after lower uterine cesarian section (LUCS) of mothers screened for hepatitis B, C and HIV1 and 2 during their antenatal period, this blood can be used as an emergency source of fresh blood for transfusion purposes.[11]

The present work is based on the premise that placental umbilical cord blood can serve as a replacement for adult blood in cancer patients with anemia, and may have other multifaceted advantages. The present group of investigators have reported earlier on their experience with the transfusion of 413 units of freshly collected placental umbilical cord blood, in which it was noted that not a single case of immunological or nonimmunological reaction was encountered.[11,12] What is also significant is that placental blood has a very high concentration of hematopoietic stem cell CD34.

In this study, on patients of advanced breast cancer who mostly received the randomized multimodality treatment of surgery, radiation, antiestrogen drug, and chemotherapy with CMF, the rise of hemoglobin and the fate and implications of the CD34 in the peripheral blood following the transfusion of placental umbilical cord whole blood have been examined. This cord blood project was sponsored by the Department of Science and Technology, Government of West Bengal, Calcutta, from April 1999.

13.2 Materials and Methods

Between 1999 and 2009, 44 patients with advanced breast cancer who were negative for expression of estrogen receptor [ER], progesterone receptor [PgR], Her-2/neu, epidermal growth factor receptor [EGF-R], p53, Bcl-2, etc., as noted through immunohistochemistry tests, were enrolled for the cord blood transfusion regime to combat anemia (hemoglobin less than 8 g/100 mL of blood) and emaciation in order to make them fit for subsequent palliative surgery, radiation, hormone therapy, and chemotherapy.

Human placental umbilical cord blood was collected from consenting mothers aseptically after lower uterine cesarean section under general or regional anesthesia. If there was gross prematurity or dysmaturity or the projected weight of the fetus was less than 2 kg, or if there was any specific disease that the mother was suffering from like hepatitis or HIV, etc., the cord blood collection was abandoned. Cord blood was collected from only informed, healthy mothers after the birth of their healthy babies.

The methodological details of the cord blood transfusion protocol have been reported by the present investigators earlier.[11] Flow analysis cytometry was done routinely for estimating the CD34 level of the peripheral blood 3 days after the transfusion of the cord blood in sex and HLA-randomized patients from Ranbaxy Laboratories. No patient received any growth factor or specific immunosuppressive drug during the cord blood transfusion.

13.3 Result and Analysis

Breast cancer represents a major health problem, with more than a million new cases and 370,000 deaths worldwide yearly. It is one of the commonest female cancers in India. Treatment of breast cancer includes surgery, drugs (hormone therapy and chemotherapy), and radiation. Options and understanding of how to use cytotoxic chemotherapy in both advanced and early-stage breast cancer have made substantial progress in the past 10 years, with numerous landmark studies identifying clear survival benefits for newer approaches.[13]

Principle of treatment regimen for Stage IV disease in this study: 44 patients were ultimately enrolled for the present study after receiving clearance from the hospital-based ethical committee. Basically, we offered cosmetic surgery to prevent fungation or debulking surgery to reduce the tumor cell load followed by anti-estrogen hormone therapy, chemotherapy with six cycles of CMF (cyclophosphomide, methotrexate, 5-flurouracil) and fresh whole screened cord blood transfusion depending on its availability. We received the informed consent of all concerned parties, that is, donor and the recipient of the cord blood transfusion before the enrollment for the present protocol.

Of the 44 patients of biopsy-proven advanced breast cancer in this series, 28 cases were suffering from infiltrating ductal carcinoma; there was invasive lobular carcinoma in eight cases, medullary carcinoma in four cases, adenocarcinoma in two cases followed by mucinous and schirrous carcinoma in one case each (vide Table 13.1). Eighteen patients had blood group B followed by blood group A in 16 cases. Six cases had blood group O followed by AB in four cases only.

Table 13.1 Change in Hemoglobin level and CD34 level of the peripheral blood 72 h after first 2 units of cord blood transfusion

Sl. No.	Histology	Primary Treatment, i.e., definite surgery or debulking, or fungation area plastic reconstruction attempt	Transfusion of placental umbilical cord blood in units	Peripheral cord blood CD34 level after 72 h of Cord blood transfusion	Hemoglobin level before cord blood transfusion in g%	Hemoglobin level after 72 h of cord blood transfusion in g%	Hemoglobin level after 7 days of 2 units cord blood transfusion in g%
(1)	Infiltrating ductal carcinoma	Palliative surgery + 6 cycle chemotherapy with CMF + radiotherapy + antiestrogen	19 units	Pretransfusion: 0.02% and posttransfusion: 0.07%	5.4	6.6	7.3
(2)	Infiltrating ductal carcinoma	Palliative surgery + 6 cycle + radiotherapy + antiestrogen chemotherapy with CMF	5 units	Pretransfusion: 0.02% and posttransfusion: 0.08%	5.8	6.4	6.9
(3)	Infiltrating ductal carcinoma	Palliative surgery + 6 cycle chemotherapy with CMF + radiotherapy + antiestrogen	2 units	Pretransfusion: 0.02% and posttransfusion: 0.03%	6.1	6.7	7.2
(4)	Infiltrating ductal carcinoma	Palliative l surgery + 6 cycle chemotherapy with CMF + antiestrogen	4 units	Pretransfusion: 0.02% and posttransfusion: 0.05%	5.9	6.9	7.4
(5)	Invasive lobular carcinoma	Palliative surgery + 6 cycle chemotherapy with CMF + radiotherapy + antiestrogen	7 units	Pretransfusion: 0.02% and posttransfusion: 0.13%	6.3	7.1	7.5
(6)	Infiltrating ductal carcinoma	Palliative surgery + 6 cycle chemotherapy with CMF + Radiotherapy + antiestrogen	6 units	Pretransfusion: 0.02% and posttransfusion: 0.11%	6.1	6.8	7.2
(7)	Infiltrating ductal carcinoma	Palliative surgery + 2cycle chemotherapy with CMF + radiotherapy + antiestrogen	5 units	Pretransfusion: 0.02% and posttransfusion: 1.1%	6.9	7.6	8.1

(continued)

Table 13.1 (continued)

Sl. No.	Histology	Primary Treatment, i.e., definite surgery or debulking, or fungation area plastic reconstruction attempt	Transfusion of placental umbilical cord blood in units	Peripheral cord blood CD34 level after 72 h of Cord blood transfusion	Hemoglobin level before cord blood transfusion in g%	Hemoglobin level after 72 h of cord blood transfusion in g%	Hemoglobin level after 7 days of 2 units cord blood transfusion in g%
(8)	Adenocarcinoma	Palliative surgery + 5cycle chemotherapy with CMF + antiestrogen	2 unit	Pretransfusion: 0.02% and posttransfusion: 0.23%	6.4	7.0	7.8
(9)	Invasive lobular carcinoma	Palliative surgery + 6 cycle chemotherapy with CMF	32 units	Pretransfusion: 0.02% and posttransfusion: 79%	5.9	6.6	7.2
(10)	Invasive lobular carcinoma	Palliative surgery + 6 cycle chemotherapy with CMF + radiotherapy + antiestrogen	2 units	Pretransfusion: 0.02% and posttransfusion: 0.04%	6.2	6.9	7.3
(11)	Infiltrating ductal carcinoma	Palliative surgery + 6 cycle chemotherapy with CMF + radiotherapy	4 units	Pretransfusion: 0.02% and posttransfusion: 0.02%	7.1	7.7	8.2
(12)	Infiltrating ductal carcinoma	Palliative surgery + 6 cycle chemotherapy with CMF + antiestrogen	5 units	Pretransfusion: 0.01% and posttransfusion: 0.03%	7.6	8.4	8.8
(13)	Infiltrating ductal carcinoma	Palliative surgery + 6 cycle chemotherapy with CMF + radiotherapy + antiestrogen	5 units	Pretransfusion: 0.08% and posttransfusion: 0.03%	6.9	7.7	8.2
(14)	Invasive lobular carcinoma	Palliative surgery + 6 cycle chemotherapy with CMF + antiestrogen	4 units	Pretransfusion: 0.08% and posttransfusion: 0.19%	5.6	6.4	7.9
(15)	Infiltrating ductal carcinoma	Palliative surgery + 6 cycle chemotherapy with CMF + radiotherapy + antiestrogen	5 units	Pretransfusion: 0.02% and posttransfusion: 0.38%	5.4	6.2	7.9

(16)	Invasive lobular carcinoma	Palliative surgery + 3cycle chemotherapy with CMF + radiotherapy + antiestrogen	4 units	Pretransfusion: 0.01% and posttransfusion: 0.23%	6.2	6.8	7.9
(17)	Infiltrating ductal carcinoma	Palliative surgery + 6 cycle chemotherapy with CMF + antiestrogen	5 units	Pretransfusion: 0.01% and posttransfusion: 41%	7.3	7.9	8.7
(18)	Infiltrating ductal carcinoma	Palliative surgery + 6 cycle chemotherapy with CMF + radiotherapy + antiestrogen	3 unit	Pretransfusion: 0.02% and posttransfusion: 0.17%	7.9	8.7	8.9
(19)	Infiltrating ductal carcinoma	Palliative surgery + 6 cycle chemotherapy with CMF + antiestrogen	2 unit	Pretransfusion: 0.02% and posttransfusion: 0.15%	6.3	7.2	8.1
(20)	Invasive lobular carcinoma	Palliative surgery + 6 cycle chemotherapy with CMF + radiotherapy + antiestrogen	3 unit	Pretransfusion: 0.03% and posttransfusion: .19%	6.7	7.3	8.1
(21)	Infiltrating ductal carcinoma	Palliative surgery + 6 cycle chemotherapy with CMF + antiestrogen	7 units	Pretransfusion: 0.02% and posttransfusion: 28%	6.1	6.9	7.6
(22)	Invasive lobular carcinoma	Palliative surgery + 2cycle chemotherapy with CMF + antiestrogen	2 units	Pretransfusion: 0.07% and posttransfusion: 33%	5.9	6.8	7.4
(23)	Infiltrating ductal carcinoma	Palliative surgery + 6 cycle chemotherapy with CMF + radiotherapy + antiestrogen	14 units	Pretransfusion: 0.01% and posttransfusion: 0.24%	5.8	6.8	7.3
(24)	Infiltrating ductal carcinoma	Palliative surgery + 6 cycle chemotherapy with CMF + antiestrogen	5 units	Pretransfusion: 0.02% and posttransfusion: 54%	5.9	6.7	7.4

(continued)

Table 13.1 (continued)

Sl. No.	Histology	Primary Treatment, i.e., definite surgery or debulking, or fungation area plastic reconstruction attempt	Transfusion of placental umbilical cord blood in units	Peripheral cord blood CD34 level after 72 h of Cord blood transfusion	Hemoglobin level before cord blood transfusion in g%	Hemoglobin level after 72 h of cord blood transfusion in g%	Hemoglobin level after 7 days of 2 units cord blood transfusion in g%
(25)	Infiltrating ductal carcinoma	Palliative surgery + 4cycle chemotherapy with CMF + radiotherapy + antiestrogen	7 units	Pretransfusion: 0.02% and posttransfusion: 0.03%	5.6	6.4	7.1
(26)	Lobular carcinoma	Palliative surgery + 6 cycle chemotherapy with CMF + radiotherapy + antiestrogen	2 unit	Pretransfusion: 0.02% and posttransfusion: 0.01%	6.2	7.1	7.7
(27)	Infiltrating ductal carcinoma	Palliative surgery + 5cycle chemotherapy with CMF + radiotherapy	7 units	Pretransfusion: 0.02% and posttransfusion: 0.08%	5.7	6.7	7.3
(28)	Infiltrating ductal carcinoma	Palliative surgery + 6 cycle chemotherapy with CMF + antiestrogen	6 units	Pretransfusion: 0.08% and posttransfusion: 0.27%	5.8	6.9	7.4
(29)	Medullary carcinoma	Palliative surgery + 6 cycle chemotherapy with CMF + radiotherapy + antiestrogen	5 units	Pretransfusion: 0.04% and posttransfusion: 67%	6.1	6.9	7.7
(30)	Adenocarcinoma	Palliative surgery + 2cycle chemotherapy with CMF + radiotherapy + antiestrogen	2 unit	Pretransfusion: 0.01% and posttransfusion: 0.03%	7.9	8.6	9.2
(31)	Infiltrating ductal carcinoma	Palliative surgery + 6 cycle chemotherapy with CMF + radiotherapy + antiestrogen	33 units	Pretransfusion: 0.02% and posttransfusion: 69%	7.1	7.9	8.6
(32)	Infiltrating ductal carcinoma	Palliative surgery + 3cycle chemotherapy with CMF + radiotherapy + antiestrogen	2 units	Pretransfusion: 0.08% and posttransfusion: 0.26%	6.2	7.2	7.9

(33)	Infiltrating ductal carcinoma	Palliative surgery + 6 cycle chemotherapy with CMF + radiotherapy + antiestrogen	5 units	Pretransfusion: 0.01% and posttransfusion: 0.32%	6.5	7.3	7.9
(34)	Medullary carcinoma	Palliative surgery + 6 cycle chemotherapy with CMF + radiotherapy + antiestrogen	3 units	Pretransfusion: 0.01% and posttransfusion: 54%	6.3	7.2	7.8
(35)	Infiltrating ductal carcinoma	Palliative surgery + 6 cycle chemotherapy with CMF + antiestrogen	7 units	Pretransfusion: 0.02% and posttransfusion: 0.19%	6.9	7.7	8.4
(36)	Infiltrating ductal carcinoma	Palliative surgery + 6 cycle chemotherapy with CMF + antiestrogen	2 units	Pretransfusion: 0.02% and posttransfusion: 0.29%	6.8	7.6	8.2
(37)	Mucinous carcinoma	Palliative l surgery + 6 cycle chemotherapy with CMF + antiestrogen	5 units	Pretransfusion: 0.04% and posttransfusion: 0.06%	7.2	8.1	8.9
(38)	Infiltrating ductal carcinoma	Palliative surgery + 6 cycle chemotherapy with CMF + radiotherapy + antiestrogen	10 units	Pretransfusion: 0.05% and posttransfusion: 47%	6.1	6.9	7.6
(39)	Medullary carcinoma	Palliative l surgery + 6 cycle chemotherapy with CMF	6 units	Pretransfusion: 0.02% and posttransfusion: 0.03%	5.8	6.8	7.2
(40)	Infiltrating ductal carcinoma	Palliative surgery + 2 cycle chemotherapy with CMF + radiotherapy + antiestrogen	2 unit	Pretransfusion: 0.02% and posttransfusion: 0.02%	6.2	6.9	7.6

(continued)

Table 13.1 (continued)

Sl. No.	Histology	Primary Treatment, i.e., definite surgery or debulking, or fungation area plastic reconstruction attempt	Transfusion of placental umbilical cord blood in units	Peripheral cord blood CD34 level after 72 h of Cord blood transfusion	Hemoglobin level before cord blood transfusion in g%	Hemoglobin level after 72 h of cord blood transfusion in g%	Hemoglobin level after 7 days of 2 units cord blood transfusion in g%
(41)	Infiltrating ductal carcinoma	Palliative surgery + 6 cycle chemotherapy with CMF + radiotherapy + antiestrogen	3 unit	Pretransfusion: 0.02% and posttransfusion: 0.03%	7.1	7.9	8.4
(42)	infiltrating ductal carcinoma	Palliative surgery + 5cycle chemotherapy with CMF + radiotherapy + antiestrogen	3 unit	Pretransfusion: 0.01% and posttransfusion: .18%	5.9	6.7	7.2
(43)	Medullary carcinoma	Palliative surgery + 6 cycle chemotherapy with CMF + radiotherapy + antiestrogen	7 units	Pretransfusion: 0.02% and posttransfusion: 0.36%	5.2	6.9	7.8
(44)	Schirrous carcinoma	Palliative surgery + 6 cycle chemotherapy with CMF + radiotherapy	2 units	Pretransfusion: 0.02% and posttransfusion: 0.05%	6.9	7.7	8.4

Thirty-four of the 44 patients belonged to the age group of 31–40, followed by four patients in the 41–50 age group. The age groups of 51–60, 71–80, and 81–90 were represented by two patients each. Forty-two patients of the present series were women and two patients were men. Diabetes was associated with 18 patients while hypothyroid was detected in eight patients; other medical illnesses included hypertension in 19, ischemic heart disease in four, rheumatoid arthritis in four, and tuberculosis in three patients.

In the present study, all cases belonged to the Stage IV level of the disease at the time of presentation at the hospital. Of these, 40 cases received antiestrogen therapy, 34 cases received six cycles of CMF, and the rest[10] of the cases received two to five cycles of CMF. Twenty-eight cases received radiotherapy for locoregional control; another six cases received radiotherapy at the metastasis site (vide Table 13.1). All patients enrolled in the present study irrespective of their age and background were negative for estrogen and progesterone receptor status.

In the analysis of the overall survival rate of this group of advanced breast cancer with multimodality of treatment offered to them, it was noted that first year survival was 90.9%, second year survival was 68.18%, third year survival was 54.54%, fourth year survival was 27.27%, fifth year survival was 11.36%, and the sixth year survival was 6.81% (Table 13.2 and Fig. 13.2).

To combat the coexisting anemia, the patients received 2–33 units of cord blood each depending on availability and priority; this was given on the basis of the severity of the anemic condition. One patient received 33 units, followed by another who received 32 units of fresh cord blood transfusion. Both received 10–12 units in a row without any transfusion-related problem. Another patient received 19 units followed by transfusion of 14 units in another patient and 10 units in yet another patient. Six patients received 7 units of blood each followed by 6 units each in three cases, 5 units in 10 cases, while the remaining patients received 2–4 units each (vide Table 13.1). There was no transfusion-related problem such as acute intravascular hemolysis or delayed extravascular hemolytic reaction. Other clinical reactions like febrile, allergic, anaphylactoid, alloimmunization, graft-vs-host disease, or acute lung injury were not observed in any of the cases. Problems of bacterial, viral transmission, hypothermia, thrombocytopenia, cardiac overload, or failure were also not encountered in the present study involving the transfusion of 263 units of freshly collected cord blood made to advanced breast cancer patients with severe anemia.

Some interesting observations were made when analyzing the data and calculating the survival statistics for the advanced breast cancer patients. A rise in the hemoglobin level was noted after the transfusion of 2 units of freshly collected and screened cord blood, from 0.6 to 1.7 g/100 mL, mean 0.8 ± 0.19 g%. There was, surprisingly, another boost in hemoglobin concentration, as assessed on the seventh day assessment, which could probably be due to the cytokine impact of the fresh cord blood on the host's bone marrow. The secondary rise of hemoglobin as assessed on the seventh day's peripheral blood level varied between

Table 13.2 Cumulative survival of patients with advanced cancer breast treated with cord blood as a part of multimodality treatment

Year of detection and the number of cases in ()	1st year survival	2nd year survival	3rd year survival	4th year survival	5th year survival	6th year survival
1999 (5)	3	3	1	1	Nil	Nil
2000 (7)	7	5	4	2	Nil	Nil
2001(8)	7	5	5	4	2	1
2002 (6)	6	4	3	1	1	1
2003(5)	5	3	3	1	1	1
2004(7)	6	5	5	2	1	Nil
2005(6)	6	5	3	1	Nil	Nil
Cumulative survival with multimodality treatment	1st year: 90.9%	2nd year: 68.18%	3rd year: 54.54%	4th year: 27.27%	5th year: 11.36%	6th year: 6.81%

0.2 and 1.7 g% with a mean rise of 0.6±0.12 g% (Fig. 13.1, Table 13.1). This is a unique phenomenon and is not observed in conventional adult blood transfusion.

Moreover, assessment of peripheral blood CD34 level after 72 h of the first two units of cord blood transfusion showed a rise of CD34 from 0.02% to 79%. The rise was very steep in 9 out of the 44 Stage IV cases (Fig. 13.3). This steep rise in the peripheral blood CD34 appears to have had an overall good prognostic index (Figs. 13.2 and 13.7). Out of the nine cases who registered an abrupt rise, eight cases were still living even after the fourth year. The one patient who died after

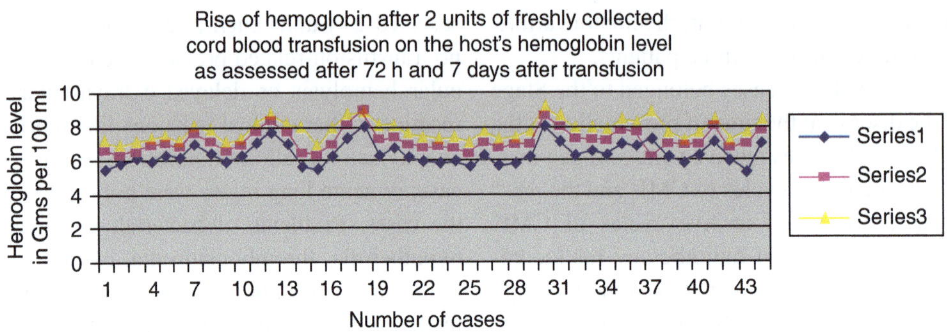

Fig. 13.1 Series 1: pretransfusion hemoglobin in g/100 mL of blood. Series 2: posttransfusion hemoglobin in g/100 mL of blood as seen after 72 h. Series 3: posttransfusion hemoglobin in g/100 mL of blood seen after 7 days

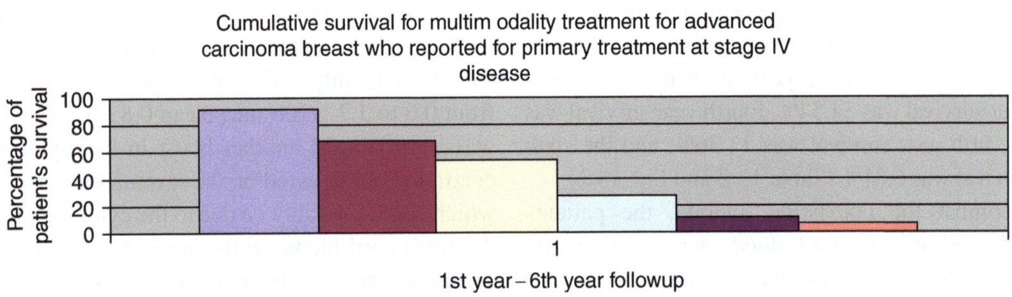

Fig. 13.2 The cumulative survival with multimodality treatment in case of advanced breast cancer. These anemic patients were treated with cord blood support only to combat their anemia

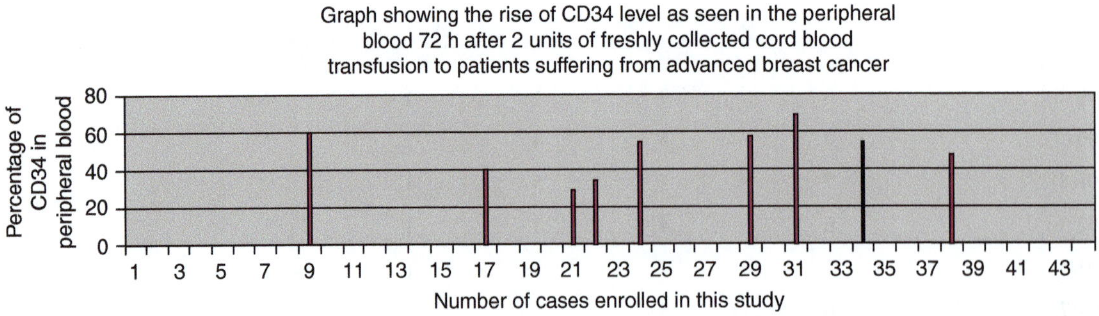

Fig. 13.3 The substantial rise of CD34 level after transfusion of 2 units of freshly collected and screened cord whole blood to advanced cancer breast patients. What is interesting is the fact that 8 out of 9 patients were seen living at the end of the fourth year of follow-up and screening, and the patient who showed the highest rise is living today and is healthy. Predictability of prognosis of breast cancer on the peripheral blood CD34 rise after 72 h from different graph patterns is noted here. This is a bad prognostic graph

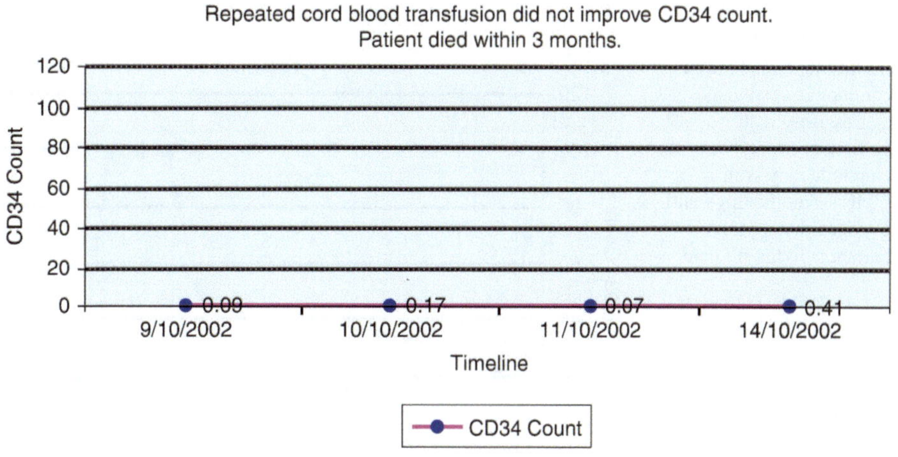

Fig. 13.4 This graph notes the prognosis of a patient (serial no 16) with Stage IV breast carcinoma who was aggressively treated with palliative surgery + 3 cycles of chemotherapy with CMF + radiotherapy + antiestrogen support; however, the patient died within 3 months. There was no rise of CD34 after repeated transfusions of cord blood

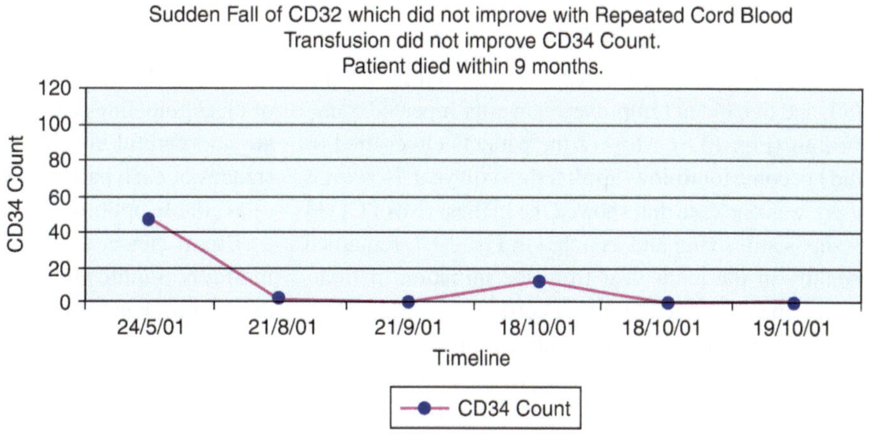

Fig. 13.5 The patient in Serial no 27, suffering from Stage IV infiltrating ductal carcinoma, was treated with aggressive palliative surgery + 5 cycles of chemotherapy with CMF + radiotherapy protocol, but the CD34 level showed a sudden fall, which did not improve in spite of recurrent transfusions

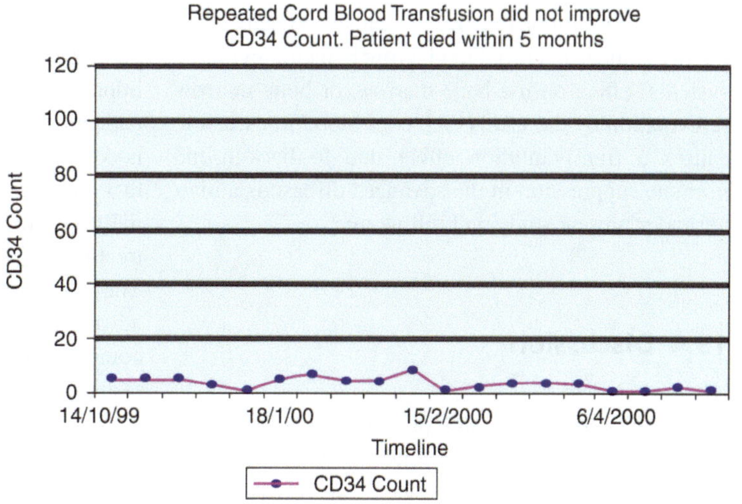

Fig. 13.6 The patient in Serial no 1, suffering from Stage IV infiltrating ductal breast carcinoma, was treated aggressively with palliative surgery + 6 cycle chemotherapy with CMF + radiotherapy + antiestrogen therapy; however, the patient died within 5 months

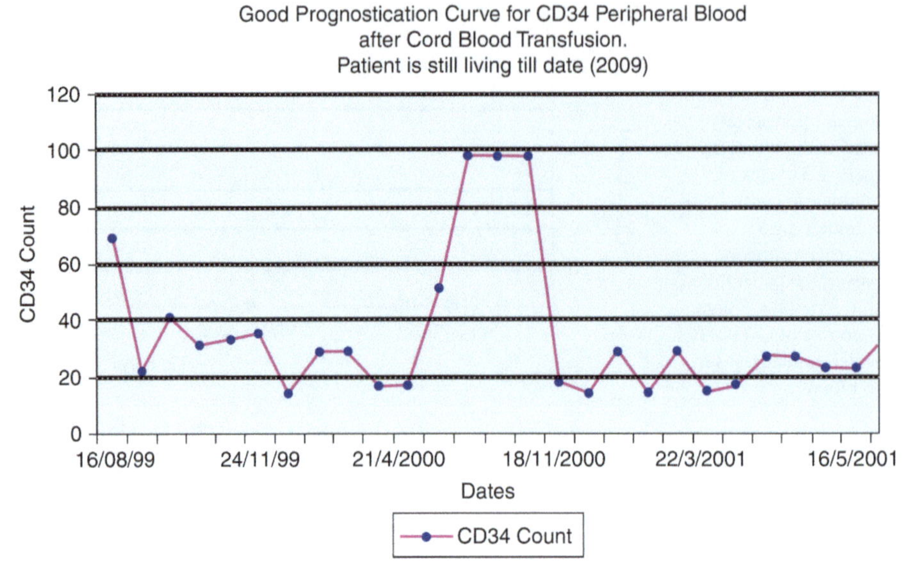

Fig. 13.7 Serial no 31 suffering from Stage IV infiltrating ductal breast carcinoma was treated aggressively with palliative surgery + 6 cycle chemotherapy with CMF + radiotherapy + anti-estrogen therapy; the patient is living till date

9 months registered an initial rise followed by a steep fall, and this did not improve again after repeated transfusions (Fig. 13.5). Most of the patients either died or did not come for follow-up after the sixth year. However, there was one case that showed the highest rise of CD34 – she is still living and as noted in Fig. 13.7, remained healthy in the tenth year from the initiation of treatment. She received 33 units of cord blood.

New approaches for the treatment of advanced cancer include immunotherapeutic strategies, but the type and extent of spontaneous immune responses against tumor antigens remain unclear. A dominance of TH2 cytokines in the patients' sera, reported previously, suggests the possibility of systemic tumor-induced immunosuppression, which potentially inhibits the induction of tumor-reactive T cells.[14] Whether the freshly collected cord blood growth factor cytokine systems' effect on the bone marrow, or bone marrow rejuvination by the CD34-rich cord blood transfusion, causes a transplantation effect due to background immune suppression in the advanced disease is a matter under present study and follow-up.

13.4 Discussion

Treatment choices are still predominantly based on practice determined by individual or collective experience and the historic development of treatment within a locality. Blanket application of international, national, or local guidelines is usually impractical or inappropriate and careful consideration of the detailed circumstances of each patient is required to make optimal use of available options.

Breast cancer is widely prevalent in India and the mortality is quite high. One reason is the late reportage of any problem by the patients due to factors that are beyond the scope of this study. In this series, all patients were treated in a free government hospital.

The problem of treatment of breast cancer in a free state government hospital in Calcutta are distinctly different from the treatment offered at specialized private hospitals and clinics devoted to senological problems only. At government hospitals, most patients report at an advanced stage of the disease (Stage IV). Some photographs have been given here to serve as visual impressions of the state of the disease when the relevant patients reported to the government hospital. One thing is common in all the cases. While they were all aware that they were suffering from breast cancer, they all initially opted for nonsurgical, nonallopathic forms of treatment with the sole intention of preserving the breast. They reported to the government hospital at a very advanced stage of the disease with anemia and other comorbidities, after exhausting all their financial resources in search of a painless nonsurgical treatment.

When the work began in 1999, the initial purpose was not to assess the implication of cord blood transfusion on the prognosis, or to examine the immunotherapeutic potentialities of cord blood transfusion. The idea was to transfuse freshly collected umbilical cord whole blood as an alternative to adult blood to treat their anemic condition. Patients who were admitted to the government hospital for free treatment with advanced cancer burden and severely anemic state were included in the study and readily available and screened cord blood as fresh adult blood substitute was provided to them for transfusion, after getting necessary permission from all concerned. The safety of cord blood transfusion is now a globally acknowledged phenomenon as per published reports from different groups reporting from different centers of excellence in the world, from the pediatric to the adult age group. Here too, there was no immunological or nonimmunological reaction as a result of cord blood transfusion.

While analyzing the data relating to the cord blood transfusion, the rise in hemoglobin etc., was noted with positive surprise since the rise in hemoglobin was two-fold. The first rise, as mentioned, was noted after 72 h of 2 units of cord blood transfusion (0.6–1.7 g/100 mL of blood, mean 0.8 ± 0.19 g SD) and the second on the seventh day (0.2–1.7 g/100 mL of blood with the mean rise of 0.6 ± 0.12 g SD) as seen in Fig. 13.1 and Table 13.1. It was postulated that this secondary rise could be due to the associated cytokine impact of the donated cord blood on the host's bone marrow. In Fig. 13.2 and Table 13.2, the cumulative survival of cancer patients undergoing multimodality treatment is presented.

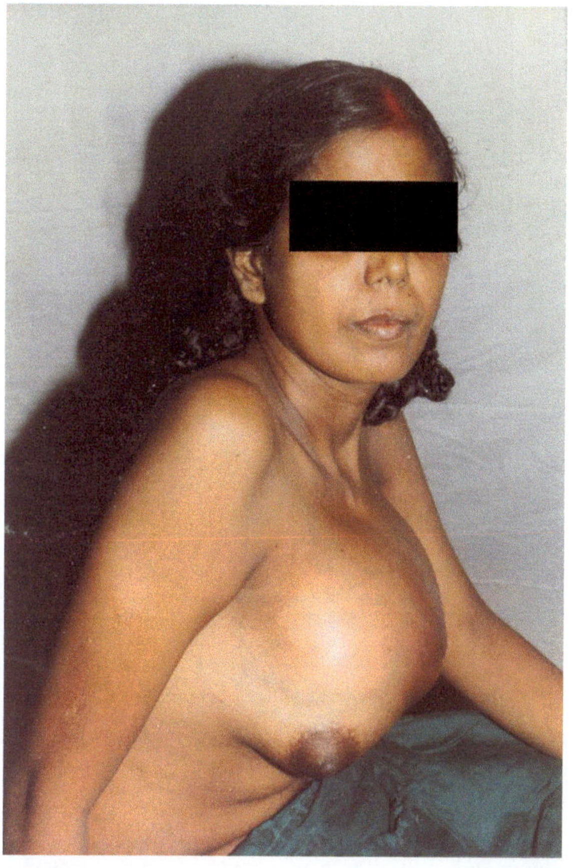

Fig. 13.8 Serial no 1 (Table 13.1) suffering from stage IV infiltrating ductal carcinoma of the breast

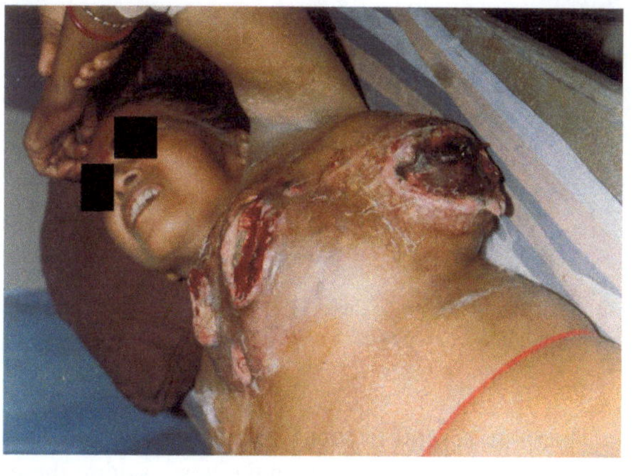

Fig. 13.9 This is a patient (serial no 16, Table 13.1) of stage IV breast carcinoma

The CD34 assessment of the peripheral blood too showed a rise from its base level of 0.02% to as high as 79%, 72 h after 2 units of cord blood transfusion. In nine cases, the rise was very steep, while in other cases there was no rise or only a marginal rise (Fig. 13.3). The most significant observation was the survival of eight of the nine cases at the end of the fourth year, and particularly that of serial number 31 (Figs. 13.7 and 13.10) where the rise was very high. What is remarkable is that Case No. 31, a Stage IV breast cancer patient, was living in the tenth year after the initiation of therapy. This is the phenomenon that the present investigator wishes to highlight for the scientific community. The real question is why is there a rise as noted in Fig. 13.7, which retrospectively suggests good prognosis. The second question is why did repeated cord blood transfusions not improve the peripheral CD34 count in some patients (Figs. 13.4–13.6, 13.8, 13.9, and 13.11); all of them died within a short period, i.e., within a year, thus retrospectively justifying the framing of the concept of a bad prognostic graph. This is the main issue resulting from the present study. The question asked here is why is this so?

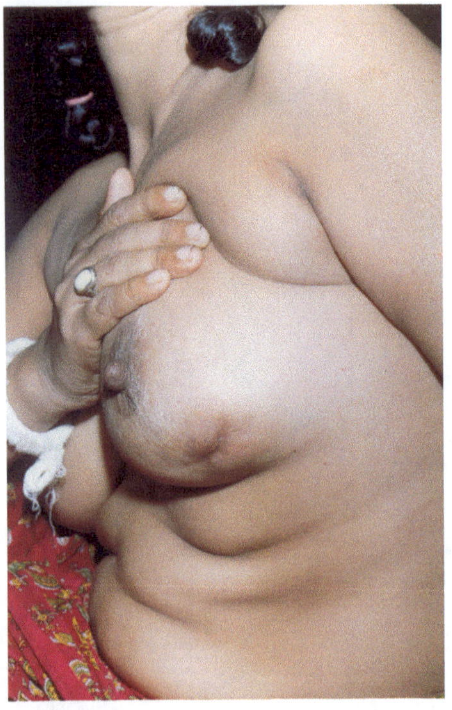

Fig. 13.10 Serial no 31 (Table 13.1) suffering from stage IV infiltrating ductal carcinoma of the breast

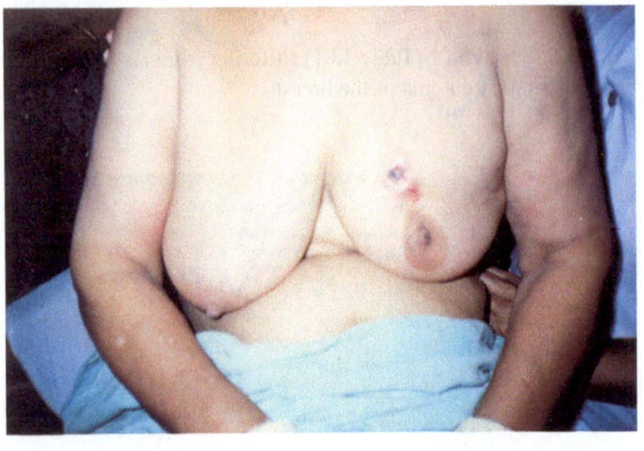

Fig. 13.11 Serial no 27 (Table 13.1) suffering from Stage IV infiltrating ductal carcinoma

There are several factors that may lead to the survival of the donor leucocytes in the host system: there could be a structural or functional immunosuppressive condition persisting in the host system due to the disease itself; or it could be an effect of immunosuppression due to anticancer drugs; the role of nutrition, opportunistic clinical or subclinical infection, etc., should also not be discounted.

The persistence of donor leukocytes in the transfusion recipient is termed microchimerism (MC). It is likely that microchimerism reflects engraftment in the recipient of donor hematopoietic stem cells. This is very uncommon in transfusion for elective surgery, sickle cell anemia, thalassemia, and HIV.[14,15] Long-term white blood cell (WBC) microchimerism of at least 2 years has been reported in trauma patients receiving fresh nonleukoreduced (non-LR) blood.[16] A better understanding of factors determining clearance versus chimerism of transfused leukocytes is critical to the prevention of alloimmunization and transfusion-induced graft-vs.-host disease, and potentially, to the induction of tolerance for transplantation.[17]

It should be pointed out here that pregnancy and neoplasm represent the most interesting examples of immune accommodation seen in mammalian biology. Cytokines of maternal origin act on placental development. At the same time, antigen expression on the placenta determines maternal cytokine patterns.[15]

In case of tumors, the expression of (human leucocytic antigen–G) HLA-G protein on the surface of primitive melanoma and metastatic cells confers protection from natural killer (NK) cells and cytopathic T lymphocyte (CTL) lytic activity.[18,19]

The placenta has a unique microenvironment and its sensitization impact on cord blood cells may have a role in transient transplantation impact on the host system. Trophoblast cells of the placenta invade deep into the maternal uterine tissue to establish a life-giving connection with the maternal blood supply.[20,21] The placenta is a complex organ that regulates maternal-fetal interactions.[22,23] Cord blood is known for its lack of activator factors for inflammation. The decreased cytokine production of the cord blood serum compared to that of adult blood could be due to less cytokine gene and mRNA accumulation. It could also be due to the antigen naïveté of T cells or the NK (natural killer) cells. Moreover, there may be an overall difference in the signaling pathway for the cord blood in comparison with adult blood. In addition, cord blood cells may downregulate the immune and the autoimmune response by maturing into high IL10 and low IL2-producing TH1 cells.[24] In healthy individuals, alternatively activated macrophages are found in the placenta and the lungs to protect them from unwanted inflammatory and immune reaction. They produce anti-inflammatory cytokine like IL1 antagonist; further, they lack the expression of pro-inflammatory cytokines such as the IL1, tumor necrotic factor alpha (TNFα), IL6, IL12, and other macrophage inflammatory proteins. The overall effect is suppression of inflammation. These activated macrophages eventually go to the cord blood.[25]

As already mentioned, the patient described in serial No 31 (Figs. 13.7 and 13.10) showed a sustained and steep rise in peripheral blood CD34 level after 2 units of cord blood transfusion. There was no clinical graft-versus-host disease in any of the cases. The preliminary bone marrow study also suggested a positive impact on the host bone marrow with cord blood transfusion with improved cellularity in that patient. However, the pertinent question still remains as to why in nearly identical clinical states of the disease load, there was no rise of CD34 in some patients (Figs. 13.4–13.6, 13.8, 13.9, and 13.11 bearing the serial no 16, 27, and 1) despite recurrent transfusions of CD34-rich cord blood, which have been categorized here in the bad prognostic graphs.

There is a possibility that in case of the cancer patients who were treated with cord blood transfusions, these CD34 cells of the cord blood killed cancer cells directly during their meeting, or indirectly assisted the intrinsic cancer killing process of the host system by interacting with NK cells, which are the key mediators of innate immunity contributing to immunosurveillance by recognizing and killing tumor and virus-infected cells. They are cytolytic and produce inflammatory cytokines. This increases the scope and potentialities of the CD34 cells of the cord blood as immunotherapeutic modulators, which is the key concern of the present paper. For continuation of the tolerance state, a certain degree of chimerism (coexistence of cells of genetically different individuals) is needed. This is best achieved if the inoculation contains cells capable of self-renewal, i.e., stem cells.[26]

The positive prognostic significance of this unique phenomenon may be due to (a) nonspecific killing of the cancer cells by the CD34 cells of the donated cord blood, as mentioned earlier, or (b) through induction of the dendritic cells (DC) of the cord blood, which are important accessory cells that are capable of initiating an immune response. The generation of functional DC from mononuclear cells isolated from human umbilical cord blood cells has already been reported. It has been shown that the cord blood-derived antigen-specific CTL can cause killing of human leukemic cells (K562) and breast cancer cells (MDA-231).[27,28] Further, it is a well-documented phenomenon that apart from hematopoietic stem cells, the cord whole blood also contains mesenchymal stem cells (MSC), which do not express HLA and are hence not attacked by the host's HLA-dependent cytotoxic reactions. This could theoretically be exploited as a new therapeutic tool in cancer therapy in order to amplify immune responses against tumor-specific antigens.

A recent report has suggested that mononuclear cells (MNCs) from UCB (umbilical cord blood) were used to generate both interleukin-2 (IL-2)-activated killer (LAK) cells and tumor-specific cytotoxic T lymphocytes (CTLs). UCB-derived LAK cells showed a significant in vitro cytotoxicity against IMR-32, SK-NMC, and U-87 human neuroblastoma and glioblastoma.[29] Another report has suggested that umbilical cord blood mononuclear cells generate CD45RA naïve T lymphocytes when cultured under serum-deprived conditions with appropriate combinations of growth factors. These *ex vivo* generated T cells resemble precursors for the lymphoid lineage present in adult bone marrow in terms of active transcription and have all the potentialities of cancer immunotherapy.[30,31]

13.5 Conclusion

Global research has allowed us to refine breast cancers further into prognostic groups based on a gene expression profile. Generally, young age, a large primary tumor, a high grade tumor, the presence of HER2 protein (HER2/neu is also known as ErbB-2, ERBB2, which stands for "Human Epidermal growth factor Receptor 2" and is a protein giving higher aggressiveness), or the BRCA (breast cancer susceptibility) gene have bad prognosis.

In the present set of experience with cord blood transfusion, it is important to note that there was a transient rise in the CD34 cells of the peripheral blood in the bone marrow, up to 79% in one case. This phenomenon has visible prognostic significance as can be seen from patients who are living today, i.e., October 20, 2009. On the other hand, no rise in the peripheral CD34 cell level, even after repeated cord blood transfusions, justifies bad prognosis as seen in Figs 13.4–13.6. The pathophysiology and molecular mechanism of this phenomenon is currently under the scientific scrutiny of the present investigator. The sudden rise in CD34 with its positive prognostic correlation may hint at the future immunotherapeutic potential of this form of therapy.

Acknowledgment The Department of Science and Technology, Government of West Bengal, supported the investigator with a research grant during his tenure at Bijoygarh State Hospital from 1999 to 2006. The author gratefully acknowledges the support of the patients who volunteered for this study. The guidance of Prof. K. L. Mukherjee of Biochemistry, and Prof. M. K. Chhetri, the former Director of Health Services, West Bengal, is also acknowledged.

References

1. Crouch Z, DeSantis ER. Use of erythropoietin-stimulating agents in breast cancer patients: a risk review. *Am J Health Syst Pharm*. 2009;66(13):1180-1185. Review.
2. Mith RE Jr, Tchekmedyian S. Practitioners'practical model for managing cancer elated anemia. *Oncology (Huntingt)*. 2002;16(9 Suppl 10):55-63.
3. Pirker R, Wiesenberger K, Pohl G, Minar W. Anemia in lung cancer: clinical impact and management. *Clin Lung Cancer*. 2003;5(2):90-97.
4. Tchekmedyian NS. Anemia in cancer patients: significance, epidemiology, and current therapy. *Oncology (Huntingt)*. 2002;16(9 Suppl 10):17-24.
5. Steensma DP. Management of anemia in patients with cancer. *Curr Oncol Rep*. 2004;6(4):297-304.
6. Kolesar JM. Novel approaches to anemia associated with cancer and chemotherapy. *Am J Health Syst Pharm*. 2002;59(15 Suppl 4):S8-S11.
7. Harrison LB, Shasha D, Homel P. Prevalence of anemia in cancer patients undergoing radiotherapy: prognostic significance and treatment. *Oncology*. 2002;63(Suppl 2):11-18.
8. Ludwig H, Fritz E. Anemia of cancer patients: patient selection and patient stratification for epoetin treatment. *Semin Oncol*. 1998;25(3 Suppl 7):35-38.
9. BCSH Blood Transfusion Task Force. Guideline for the gamma irradiation of the blood components for the prevention of the transfusion associated graft vs host disease. *Transfus Med*. 1996;6:261.
10. Goodnough T. Transfusion medicine – blood conservation – second of two parts. *N Engl J Med*. 1999;340(7):525-533.
11. Tscherning-Casper C, Papadogiannakis N, Anvret M, et al. The trophoblastic epithelial barrier is not infected in full-term placentae of human immunodeficiency virus-seropositive mothers undergoing antiretroviral therapy. *J Virol*. 1999;73(11):9673-9678.
12. Bhattacharya N. Placental umbilical cord whole blood transfusion: a safe and genuine blood substitute for patients of the under-resourced world at emergency. *JAM Coll Surg*. 2005;200(4):557-563.
13. Bhattacharya N, Mukherjee KL, Chettri MK, Banerjee T, Mani U, Bhattacharya S. A study report of 174 units of placental umbilical cord whole blood transfusion in 62 patients as a rich source of fetal hemoglobin supply in different indications of blood transfusion. *Clin Exp Obstet Gynecol*. 2001; 28(1):47-52.
14. Hussain SA, Palmer DH, Stevens A, Spooner D, Poole CJ, Rea DW. Role of chemotherapy in breast cancer. *Expert Rev Anticancer Ther*. 2005;5(6):1095-1110.
15. Lee Tzong-Hae, Paglieroni Teresa, Ohto Hitoshi, Holland Paul V, Busch Michael P. Survival of donor leukocyte subpopulations in immunocompetent transfusion recipients: frequent long-term microchimerism in severe trauma patients. *Blood*. 1999;93(9):3127-3139.
16. Utter GH, Owings JT, Lee TH, et al. Blood transfusion is associated with donor leukocyte microchimerism in trauma patients. *J Trauma*. 2004;57(4):702-707. discussion 707–708.
17. Lee TH, Paglieroni T, Utter GH, et al. High-level long-term white blood cell microchimerism after transfusion of leukoreduced blood components to patients resuscitated after severe traumatic injury. *Transfusion*. 2005;45(8): 1280-1290.
18. Szekeres-Bartho J. Immunological relationship between the mother and the fetus. *Int Rev Immunol*. 2002;21(6): 471-495.
19. Carosella ED. HLA-G: fetomaternal tolerance. *C R Acad Sci III*. 2000;323(8):675-680.
20. Ishitani A, Sageshima N, Lee N, et al. Protein expression and peptide binding suggest unique and interacting functional roles for HLA-E, F, and G in maternal-placental immune recognition. *J Immunol*. 2003;171:1376-1384.
21. Sargent IL. Maternal and fetal immune responses during pregnancy. *Exp Clin Immunogenet*. 1993;10:85.
22. Le Bouteiller P, Rodriguez AM, Mallet V, Girr M, Guillaudeux T, Lenfant F. Placental expression of HLA class I genes. *Am J Reprod Immunol*. 1996;35:216.
23. McCracken S, Layton JE, Shorter SC, Starkey PM, Barlow DH, Mardon HJ. Expression of granulocyte-colony stimulating factor and its receptor is regulated during the development of the human placenta. *J Endocrinol*. 1996;149: 249-258.
24. Lee T-H, Donegan EA, Slichter S, Busch MP. Transient increase in circulating donor leukocytes following allogeneic transfusions in immunocompetent recipients compatible with donor cell proliferation. *Blood*. 1995;85:1207.
25. Madrigal JA, Cohen SBA, Gluckman E, Charron DJ. Does cord blood transplantation result in lower graft vs host disease.It takes more than two to tango. *Hum Immunol*. 1997;56:1-5.
26. Chang M-DY, Polar JW, Khalili H, et al. Mouse placental macrophages have a decreased ability to present an antigen. *Proc Natl Acad Sci USA*. 1993;90:452-456.
27. Roitt I, Brostoff J, Male D. *Immunology*. 6th ed. London, UK: Mosby; 2001:205-206: Chap. 12.
28. Joshi SS, Vu UE, Lovgren TR, et al. Comparison of phenotypic and functional dendritic cells derived from human umbilical cord blood and peripheral blood mononuclear cells. *J Hematother Stem Cell Res*. 2002;11(2):337-347.
29. Schmitz-Winnenthal FH, Volk C, Z'graggen K, et al. High frequencies of functional tumor-reactive T cells in bone marrow and blood of pancreatic cancer patients. *Cancer Res*. 2005;65(21):10079-10087.
30. Joshi AD, Clark EM, Wang P, et al. Immunotherapy of human neuroblastoma using umbilical cord blood-derived effector cells. *J Neuroimmune Pharmacol*. 2007;2(2):202-212. Epub 2006 Oct 10.
31. Migliaccio AR, Alfani E, Di Giacomo V, Cieri M, Migliaccio G. Ex vivo amplification of T cells from human cord blood. *Pathol Biol (Paris)*. 2005;53(3):151-158. Review.

Part VI

Use of Placental Umbilical Cord Blood in Neurology

Anti-inflammatory Effects of Human Cord Blood and Its Potential Implication in Neurological Disorders

14

Martina Vendrame

14.1 Introduction

Although initial in vitro evidence pointed to the differentiation of human cord blood cells (HUCBCs) into neuronal and glial lineages, transplantation of these cells never resulted in terminally differentiated neurons. This raised the suspicion that the beneficial effect of HUCBCs in models of central nervous system (CNS) disorders and injury may be attributable to alternative biologic properties. The indication that HUCBCs may have anti-inflammatory and immunoregulatory properties has recently emerged from animal studies.

14.2 Inflammatory Response in Neurological Disorders and Brain Injury

CNS ischemia induces a local inflammatory response characterized by activation of brain inflammatory cells (such as resident microglia), infiltrating monocytes, and lymphocytes, which produce a variety of pro-inflammatory molecules implicated in the mediation of neuronal injury.[13] Alterations in the peripheral immune status have been demonstrated in studies in humans and in animal models of stroke; this cascade of events occurring outside the CNS appears to follow the initial brain pro-inflammatory events.[21,42] Brain and immune system contribute to this pro-inflammatory state with a complex molecular and cellular interplay, also involving the hypothalamic–pituitary–adrenal axis and the sympathetic nervous system.[54] The activation of these systems triggers the production of glucocorticoids and catecholamines, which in turn mediates the release of anti-inflammatory interleukins (including IL-10) from CNS resident and infiltrating monocytes, which allow a protective feedback mitigating the initial ischemia-induced pro-inflammatory response. This inhibitory feedback may also induce an immunosuppressive state, which human studies have proven accountable for the infectious complications seen in stroke patients.[36]

The changes in the immune system that follow brain injury have also been characterized as alterations of function and cellular composition of peripheral lymphoid organs, such as spleen, thymus, and lymph nodes.[36] For instance, in the middle cerebral artery occlusion (MCAO) rat model of stroke, a reduction in spleen size and function has been observed.[1,19] Additionally, the cellular and functional alterations seen in these peripheral lymphoid organs are mirrored in the injured brain by a peak in infiltrating lymphocytes as well as activated microglia. The significance of the effect of these invading leukocytes in ischemia has been shown by the reduction of ischemic volume in mice deficient in the leukocyte adhesion molecule, CD11a/CD18.[43]

14.3 Immunomodulatory Strategies for Therapy of CNS Disorders

The extensive literature indicating that inflammatory responses can contribute to the pathophysiology and clinical outcome of stroke[12,21,36,54] has suggested that

M. Vendrame
Neurology Department, Temple University,
3401 North Broad Street, Parkinson Pavilion Suite 558,
Philadelphia, PA 19140
e-mail: vendrame.martina@gmail.com

N. Bhattacharya and P. Stubblefield (eds.), *Regenerative Medicine Using Pregnancy-Specific Biological Substances*,
DOI: 10.1007/978-1-84882-718-9_14,

the immune system may also be harnessed therapeutically. Studies in the MCAO model of stroke have shown that the induction of tolerogenic responses to E-selectin[11] or myelin basic protein (MBP)[3] by their mucosal instillation can reduce the extent of ischemic injury. Moreover, adoptive transfer of splenocytes-derived rats tolerized with these antigen can provide protection from ischemia to naïve animals.[3] Additional evidence has suggested that lymphocytes can mediate neuroprotection and that spontaneous or exogenously boosted T-cell-mediated neuroprotection correlated with early activation of microglia as antigen-presenting cells.[41,55] These findings contributed the "proof of principle" that cell-mediated immunomodulatory therapy could allow protection of the injured and/or ischemic brain.

14.4 Immunological Properties of HUCBCs

14.4.1 Phenotypical Characteristics of HUCBCs

Stem cells isolated from cord blood are part of a subpopulation of cells expressing CD34, a common marker of human hematopoietic stem cells. This population is about the 1% of the mononuclear fraction of the HUCBCs. This population has been identified to contain a set of subtypes that when differentiated bear different CD markers depending on the type of cell they will become. Initial, now historical, evidence supported the idea that when in specific culture conditions, HUCBCs undergo a phenotypic conversion into neurons and/or glial cells.[7,20,23,39] Later, a CD133+ cell population and a CD33– population with distinct differentiation potential were isolated. The CD133+ cell population was shown to express neuronal markers including nestin and Musashi as well as neural filament, class III-beta-tubulin, neuron-specific enolase, MAP2, and NeuN in response to retinoic acid expression; the CD133– population did not express neuronal markers and only became burst forming unit-erythrocyte colonies.[23]

14.4.2 Immunomodulatory Properties of HUCBCs

In vitro experiments have shown that HUCBCs can produce several growth factors and cytokines. CD34 positive cells can express nerve growth factor (NGF) and its receptor TrkA.[6] HUCBCs are also able to selectively produce large amounts of IL-10 after stimulation with an anti-CD3 antibody and other interleukins such as IL-2, thus promoting a Th1/Th2 switch response.[37] Others show that HUCBCs can produce IL-8, MCP-1, and IL-1alpha, without requiring specific culture conditions.[31] Moreover, others have shown that CD34+/CD133+ HUCBCs produce angiopoietin-1 (Ang-1), angiopoietic-2 (Ang-2) factors and vascular endothelial growth factor (VEGF) and their receptors, indicating a possible role of HUCBCs in regulation of both angiopoiesis and hematopoiesis.[34] However, so far there is no in vivo evidence supporting that HUCBCs can produce cytokines once transplanted in the animals.

HUCBCs are well known for their tolerogenicity as demonstrated by the decreased severity and less frequent incidence of graft-versus-host disease observed after cord blood transplants compared to bone marrow transplants for the treatment of hematological malignancies.[25] These unique immunological properties of cord blood made possible to utilize allogeneic cells for regenerative applications without fully compromising the recipient immune system.[38] The mechanism of this cord blood tolerance is not fully understood but it has been related to the higher presence of immature T lymphocytes and a deficient production of IFN-γ in the HUCBCs compared to bone marrow cells.[24] The mononuclear fraction of cord blood is typically a heterogeneous population composed mostly of lymphocytes (about two thirds),[35] which includes high numbers of CNS-antigen-specific tolerogenic T lymphocytes.[16] Additionally, HUCBCs can reconstitute the entire adaptive immune system of transplanted immuno-deprived mice, as intrahepatic injection of CD34+ human cord blood cells into conditioned newborn Rag2–/–gammac–/– mice leads to de novo development of B, T, and dendritic cells, formation of structured primary and secondary lymphoid organs, and production of functional immune responses.[48]

14.5 Experimental Evidence of Anti-inflammatory Properties of HUCBCs in Models of CNS Disorders

14.5.1 Site of Migration and Engraftment of Intravenously Transplanted HUCBCs

HUCBCs can successfully induce functional recovery in animal models of ischemic and hemorrhagic stroke,[9,30] traumatic brain injury (TBI) and spinal cord injury,[28,33] and amyotrophic lateral sclerosis (ALS).[18]

The majority of the studies use intravenous injection as the route of cell transplantation. One important question about experiments involving the parenteral delivery of cells is whether they can reach and engraft into the CNS. Table 14.1 presents a summary of the main site of localization of HUCBCs after intravenous delivery in models of CNS disease. Some studies have shown the presence of HUCBCs in the brain after intravenous delivery.[9,18,28] In the studies by Chen et al.[9] and Lu et al.,[28] which use a model of stroke and a model of TBI, respectively, HUCBCs were localized within the brain parenchyma 2–4 weeks after transplant through positive immunoreactivity for human nuclei (HuNu). Approximately 10^6 cells were transplanted and quantification of cells revealed the presence of about 30,000 cells in the injured hemisphere. Another study employing an animal model of ALS localized transplanted HUCBCs to a variety of brain regions 10–12 weeks after transplantation.

Other studies failed to identify intravenously delivered HUCBCs within the brain. Ende et al.[10,15] showed that after intravenous injections of approximately 3×10^7 cells and a higher dose of approximately 7×10^7 cells in a model of ALS, there was no human DNA in the brain while human DNA was detected in spleen and lymph nodes by Reverse Transcriptase-PCR. The same authors show that the intravenous delivery of HUCBC in the nonobese diabetic (NOD) mouse model of autoimmune type I diabetes was found by RT-PCR for human growth hormone exclusively in the spleen.[14] Later, localization of HUCB in lymphoid organs such as spleen, bone marrow, lymph node s, and thymus was shown by others.[18,49] A study by SPECT imaging data indicated a high tracer uptake in the lung, liver,

Table 14.1 Localization of HUCBCs after IV delivery

Animal model	Number of cells delivered	Localization of cells in vivo	Method for cell localization	Reference
SOD1 mice (ALS)	3.4×10^7–3.5×10^7	Spleen, liver, lung	RT-PCR	Chen and Ende[10]
SOD1 mice (ALS)	7.0×10^7–7.3×10^7	Spleen, liver, lung	RT-PCR	Ende et al.[15]
G93A SOD1 mice (ALS)	1×10^6	Brain, spleen, liver, lung, kidneys, and heart	Immunohistochemistry for HuNu	Garbuzova-Davis et al.[18]
MCAO rats	2.3×10^6–5.3×10^6	Brain	*Immunohistochemistry for HuNu*	Chen et al.[9]
T.B.I. rats	2×10^6	Brain	Immunohistochemistry for HuNu	Lu et al.[28]
MCAO rats	1–5×10^7	Lung, liver, spleen, and kidney	SPECT imaging for (111) In-oxine-prelabelled cells	Makinen et al.[29]
MCAO rats	2×10^5	No cells detected in the brain	Imunofluorescence for green fluorescent protein on prelabelled cells	Borlongan et al.[5]
MCAO rats	5×10^7	Brain and spleen	PCR for Human HG3PDH	Vendrame et al.[49]

spleen, and kidney, but not in the brain immediately after administration or 24 h postadministration.[29] Of note, one study reported that HUCBC protection from MCAO induced neurological deficits could be attained without CNS entry of the HUCBC.[5]

14.5.2 Modulation of Splenocyte Phenotype and Function by HUCBCs

In bone marrow and spleen, HUCBCs have been shown to effect de novo development of lymphocytes as well as provide adjuvant factors to promote endogenous hematopoietic reconstitution.[48] In different rat models of brain injury, CNS-specific autoreactive T-lymphocytes could either rescue neurons from inflammation-related damage[11] or enhance neurotoxicity depending on the subtype of T-lymphocytes.[55] Splenocytes derived from animals tolerized with CNS myelin antigens such as MBP were able to mediate protection against ischemia when adoptively transplanted into naïve rats.[3,4] Based on these reports, and given the ability of HUCBC to mediate endogenous hematopoietic reconstitution during adoptive therapy,[25] investigators suspected that HUCBC treatment of stroke and brain injury rats involved immunomodulatory events. Intravenous delivery of cord blood cells could prevent the stroke-induced reduction in spleen size, which has been previously reported to occur following stroke in rats.[19] Moreover, the decrement of the CD8+/CD4+ splenocytes T-lymphocytes ratios following stroke (due either to the induction of CD8+ T cell death mechanisms in the spleen and/or their mobilization to the damaged brain[19]) was also rescued by HUCBC therapy.[51]

The capacity of HUCBC to modulate in vivo the proliferative response of splenocytes in MCAO animals was further investigated; HUCBC were found to inhibit the stroke-induced proliferative response of T-lymphocytes to Concavalin-A.[51] Furthermore, the evidence that the supernatants from the spleen cells culture assays of the HUCBC-treated animals contain increased IL-10 and decreased IFNgamma amounts (when compared with those from nontreated MCAO rats) was another finding indicating the regulatory tolerogenic T cell function of HUCBC.[51] This functional phenotyping of the splenocytes have suggested that following ischemic brain injury, the peripheral immune response may be skewed toward a Th1-type response and HUCBCs alters this response to Th2/3-type responses (meaning greater ratio of CD8+/CD4+ cells that release IL-10 while demonstrating only minimal proliferation). This type of immune response has been thought to be more amenable to recovery from ischemic brain injury.[3,4] Interleukin-10 has been shown to directly mediate infarct size decrease when delivered intravenously to stroked animals,[44] and when secreted endogenously by mobilized CD4+ T-lymphocytes.[17] Of note, HUCBCs have also been shown to have large numbers of T-lymphocytes recognizing CNS myelin proteins[16] indicating a mechanistic parallel between the effects described for HUCBCs and those reports of CNS antigen-specific tolerogenic T-lymphocytes-mediated benefit of ischemic brain injury.

One hypothesis is that the CD8+ cells that are normally activated and mobilized out of the spleen after stroke may be deactivated by the HUCBC treatment and their traffic out of the spleen may be subsequently prevented. Additionally, the intercellular mechanisms controlling this response could be associated to the production of a particular cytokine repertoire since, for instance, reduced IFNgamma and elevated IL-10 levels are associated with HUCBC treatment. Given that the great part of the HUCBCs that are generally transplanted in these experiments is composed of immature lymphocytes, HUCBCs may directly repopulate cells within the spleen that may have been mobilized after the ischemic insult.

Further studies have confirmed the role of the spleen as an ancillary organ that may be indirectly implicated in brain damage/repair. In MCAO rats, removal of the spleen significantly reduced neurodegeneration after ischemic insult, as shown by reduction of ischemic injured brain parenchyma and decreased numbers of activated microglia, macrophages, and neutrophils in the brain.[1] HUCBCs may exert their protective effect by modulating the peripheral immune response as mediated by the spleen, which is a major contributor to the inflammation that enhances neurodegeneration after stroke.

14.5.3 Modulation of Brain Inflammatory Cells and Cytokines by HUCBCs

Following stroke, microglia are activated mainly to scavenge dead cells. During this phase, microglia produce a number of substances including the pro-inflammatory cytokines tumor necrosis factor-alpha (TNFalpha) and interleukin-1 beta (IL-1beta), which can subsequently harm the surviving brain, by extending the brain injury beyond the initial infarcted area into the surrounding healthy parenchyma.[13] At the same time, a number of blood-derived inflammatory cells (neutrophils, macrophages, as well as B- and T-cells) cross the blood–brain barrier (BBB) and reach the infarcted brain tissue, further potentiating the CNS injury.[13,22,26,27] With the understanding of this cellular and molecular pathogenic mechanism of brain ischemia, several studies have employed anti-inflammatory agents in attempts to limit the severity of postischemia damage. For instance, agents that block the leukocytes adhesion to endothelia (which cross the BBB), such as antibodies to ICAM-1, have been reported to decrease ischemic damage in animal models.[58] Additionally, another approach to control the ischemia-induced inflammation has been the modulation of pro-inflammatory cytokine production in the brain.[53,57]

The effect of HUCBC treatment on the expression of inflammatory cells in the brain of rats subjected to stroke has been explored in vivo.[50] Flow cytometry analysis (FACS) revealed a population of CD45 and CD11b positive cells, which were significantly increased after MCAO in rats and dramatically diminished after intravenous delivery of HUCBCs.[50] CD11b and CD45 are both expressed by microglia and macrophages, but resting microglia (CD11b+/CD45low+) can be distinguished from macrophages (CD11b+/CD45high+) by their previously characterized FACS phenotype.[8] The finding that HUCBCs can control microgliosis is of particular interest, given that chronic microgliosis is thought to mediate neuronal damage not only in stroke injury but also in other neurodegenerative diseases.[46,47] In light of these results, some investigators have experimented HUCBC therapy in a mouse model of Alzheimer's disease and successfully shown that HUCBCs can decrease the amount of beta-amyloid plaques and associated astrocytosis, and this is most likely occurring through regulation of microglial phagocytosis.[32] Additional evidence has demonstrated that HUCBCs can facilitate the proliferative activity of the aged neural stem/progenitor cells and increase neurogenesis.[2] Interestingly, the increase in neurogenesis as a result of cord blood cells treatment seemed to be due to a decrease in inflammation, as a decrease in the number of activated microglia was found.[2] These authors used optical fractionator method of design-based stereology to identify and count the number of OX-6 positive cells in the dentate gyrus 15 days after a single injection of HUCBCs. OX-6 is a marker for MHC-II positive cells and therefore a marker for activated microglia. In animals treated with HUCBCs, the number of activated microglia in the dentate gyrus was nearly half than that of the controls (average OX-6 positive cells in the HUCBC group was 678.7 ± 155.3, and $1{,}217 \pm 128$ in the nontreated animals, $p < 0.05$). Additionally, this significant decrement correlated with the increase in number of proliferating cells (i.e. increase in neurogenesis).

The modulation of other lymphocytic cell lineages after CNS injury by HUCBCs has also been explored. In the previous mentioned study,[50] FACS showed a twofold increase in B220/CD45 positive cell infiltrate (B cells) into MCAO animal brains, and significantly fewer B220/CD45 cells after HUCBC infusion.[50] The role of B cells in the pathogenesis of stroke is not fully clear, but it is generally accepted that B cells also possess phagocytic and antigen-presenting cell functions reminiscent of microglia. Interestingly, it has been shown that myelin basic protein tolerized stroked rats have decreased infarct volume and a prominent brain B-cell population.[3] The tolerogenic phenomena may favor the phagocytic phenotype of the B-lymphocytes over that of antigen presenting and the phagocytic phenotype may help with the removal of neuronal and other cell debris without the production of pro-inflammatory cytokines. The decreased brain infiltration by B-lymphocytes by HUCBC might be associated with the prevention of antigen primed B cells or mature plasma cells from entering the infarcted CNS.

T-lymphocyte infiltration into the brain infarction after stroke was also examined by FACS gating of CD45/CD3 positive cell populations. Elevated levels of T-cells in stroked brain were found but no effect of HUCBCs in preventing the trafficking of these CD45/CD3 positive cells in the stroke brain was observed.[40,50]

After injection in the animal, HUCBCs can also play a role in the production of endogenous cytokines, most likely exerting an immunomodulatory response. Ischemic brain injury in the stroked rat results in significantly elevated brain TNFalpha, IL-1beta, and IL-2 mRNA levels, relative to those of sham rats.[27] Interestingly, HUCBC intravenous delivery to stroked rats was found to decrease the mRNA levels of TNFalpha and IL-2.[50] As activated microglia are a major source of brain TNFalpha, the decrease in cerebral TNFalpha mRNA is consistent with the decreased microgliosis in HUCBC-treated MCAO animals. Additionally, the decreased brain levels of IL-2 mRNA levels following HUCBC treatment of stroked rats would also be in concert with the decreased microglial activation afforded by HUCBC treatment.

When protein levels of these cytokines were measured, levels of TNFalpha, IL-2, IL-1beta, and IL-10 were significantly higher in MCAO rats than sham animals.[50] However, the HUCBC treatment-associated blockade of brain ischemia induced IL-2 mRNA was not mirrored at the protein level.[50] The increased brain levels of IL-10 following ischemic injury have been reported to be part of an endogenous neuroprotective response to control the extent of neuronal injury.[44] HUCBC treatment of MCAO rats did not modulate brain IL-10 levels; it did significantly decrease both TNFalpha, IL-1beta brain protein levels when compared with nontreated MCAO rats.[50] TNF-α is a pro-inflammatory cytokine released by microglia cells and it has been associated with neuronal apoptosis.[45] Experimental studies have shown that blockage of its expression can induce histological and pathological benefits.[56] The importance of the pathogenetic role of TNFalpha has also been suggested by clinical studies showing that serum levels of TNF-α directly correlate with poor prognostic outcomes in stroke patients.[21] Furthermore, stroke patients with low plasma levels of IL-10 have been observed to undergo a more rapid clinical deterioration.[52] These findings highlighted the importance of HUCBC effect in shifting the balance between pro-inflammatory and anti-inflammatory cytokines after cerebral ischemic injury as seen in animal stroke models.

14.6 Conclusion

A growing body of evidence indicates that HUCBCs may ameliorate brain ischemic damage and protect from cell death through mechanisms beyond neural cell replacement and/or neurotrophic support. Immunomodulatory therapy has been proposed to be effective for the treatment of stroke and other neurodegenerative diseases. The underlined molecular mechanisms regulating the modulation of the immune response by HUCBCs are not fully understood. These may relate to the production of cytokines by HUCBCs or the stimulation of endogenous cytokines, control of microgliosis, and/or infiltration of T-cells and B-cells into the injured brain. Moreover, a peripheral mechanism implicating the spleen as harboring organ for T cells that may be phenotypically changed and mobilized by HUCBCs could also be involved.

References

1. Ajmo CT Jr, Vernon DO, Collier L, et al. The spleen contributes to stroke-induced neurodegeneration. *J Neurosci Res*. 2008;86:2227-2234.
2. Bachstetter AD, Pabon MM, Cole MJ, et al. Peripheral injection of human umbilical cord blood stimulates neurogenesis in the aged rat brain. *BMC Neurosci*. 2008;9:22.
3. Becker K, Kindrick D, Mccarron R, Hallenbeck J, Winn R. Adoptive transfer of myelin basic protein-tolerized splenocytes to naive animals reduces infarct size: a role for lymphocytes in ischemic brain injury? *Stroke*. 2003;34:1809-1815.
4. Becker KJ, Mccarron RM, Ruetzler C, et al. Immunologic tolerance to myelin basic protein decreases stroke size after transient focal cerebral ischemia. *Proc Natl Acad Sci USA*. 1997;94:10873-10878.
5. Borlongan CV, Hadman M, Davis Sanberg C, Sanberg PR. Central nervous system entry of peripherally injected umbilical cord blood cells is not required for neuroprotection in stroke. *Stroke*. 2004;35(10):2385.
6. Bracci-Laudiero L, Celestino D, Starace G, et al. CD34-positive cells in human umbilical cord blood express nerve growth factor and its specific receptor TrkA. *J Neuroimmunol*. 2003;136:130-139.
7. Buzanska L, Machaj EK, Zablocka B, Pojda Z, Domanska-Janik K. Human cord blood-derived cells attain neuronal and glial features in vitro. *J Cell Sci*. 2002;115:2131-2138.
8. Campanella M, Sciorati C, Tarozzo G, Beltramo M. Flow cytometric analysis of inflammatory cells in ischemic rat brain. *Stroke*. 2002;33:586-592.
9. Chen J, Sanberg PR, Li Y, et al. Intravenous administration of human umbilical cord blood reduces behavioral deficits after stroke in rats. *Stroke*. 2001;32:2682-2688.
10. Chen R, Ende N. The potential for the use of mononuclear cells from human umbilical cord blood in the treatment of amyotrophic lateral sclerosis in SOD1 mice. *J Med*. 2000;31:21-30.
11. Chen Y, Ruetzler C, Pandipati S, et al. Mucosal tolerance to E-selectin provides cell-mediated protection against ischemic brain injury. *Proc Natl Acad Sci USA*. 2003;100:15107-15112.
12. Del Zoppo GJ, Becker KJ, Hallenbeck JM. Inflammation after stroke: is it harmful? *Arch Neurol*. 2001;58:669-672.
13. Dirnagl U, Iadecola C, Moskowitz MA. Pathobiology of ischaemic stroke: an integrated view. *Trends Neurosci*. 1999;22:391-397.
14. Ende N, Chen R, Reddi AS. Transplantation of human umbilical cord blood cells improves glycemia and glomerular

hypertrophy in type 2 diabetic mice. *Biochem Biophys Res Commun*. 2004;321:168-171.

15. Ende N, Weinstein F, Chen R, Ende M. Human umbilical cord blood effect on sod mice (amyotrophic lateral sclerosis). *Life Sci*. 2000;67:53-9.
16. Fredrikson S, Sun JB, Huang WX, Li BL, Olsson T, Link H. Cord blood contains high numbers of autoimmune T cells recognizing multiple myelin proteins and acetylcholine receptor. *J Immunol*. 1993;151:2217-2224.
17. Frenkel D, Huang Z, Maron R, et al. Nasal vaccination with myelin oligodendrocyte glycoprotein reduces stroke size by inducing IL-10-producing CD4+ T cells. *J Immunol*. 2003; 171:6549-6555.
18. Garbuzova-Davis S, Willing AE, Zigova T, et al. Intravenous administration of human umbilical cord blood cells in a mouse model of amyotrophic lateral sclerosis: distribution, migration, and differentiation. *J Hematother Stem Cell Res*. 2003;12:255-270.
19. Gendron A, Teitelbaum J, Cossette C, et al. Temporal effects of left versus right middle cerebral artery occlusion on spleen lymphocyte subsets and mitogenic response in Wistar rats. *Brain Res*. 2002;955:85-97.
20. Ha Y, Choi JU, Yoon DH, et al. Neural phenotype expression of cultured human cord blood cells in vitro. *Neuroreport*. 2001;12:3523-3527.
21. Intiso D, Zarrelli MM, Lagioia G, et al. Tumor necrosis factor alpha serum levels and inflammatory response in acute ischemic stroke patients. *Neurol Sci*. 2004;24:390-396.
22. Jander S, Kraemer M, Schroeter M, Witte OW, Stoll G. Lymphocytic infiltration and expression of intercellular adhesion molecule-1 in photochemically induced ischemia of the rat cortex. *J Cereb Blood Flow Metab*. 1995;15:42-51.
23. Jang YK, Park JJ, Lee MC, et al. Retinoic acid-mediated induction of neurons and glial cells from human umbilical cord-derived hematopoietic stem cells. *J Neurosci Res*. 2004; 75:573-584.
24. Kilpatrick TJ, Butzkueven H, Emery B, Marriott M, Taylor BV, Tubridy N. Neuroglial responses to CNS injury: prospects for novel therapeutics. *Expert Rev Neurother*. 2004; 4:869-878.
25. Lewis ID. Clinical and experimental uses of umbilical cord blood. *Intern Med J*. 2002;32:601-9.
26. Liu PK, Grossman RG, Hsu CY, Robertson CS. Ischemic injury and faulty gene transcripts in the brain. *Trends Neurosci*. 2001;24:581-588.
27. Liu T, Clark RK, Mcdonnell PC, et al. Tumor necrosis factor-alpha expression in ischemic neurons. *Stroke*. 1994;25: 1481-1488.
28. Lu D, Sanberg PR, Mahmood A, et al. Intravenous administration of human umbilical cord blood reduces neurological deficit in the rat after traumatic brain injury. *Cell Transplant*. 2002;11:275-281.
29. Makinen S, Kekarainen T, Nystedt J, et al. Human umbilical cord blood cells do not improve sensorimotor or cognitive outcome following transient middle cerebral artery occlusion in rats. *Brain Res*. 2006;1123:207-215.
30. Nan Z, Grande A, Sanberg CD, Sanberg PR, Low WC. Infusion of human umbilical cord blood ameliorates neurologic deficits in rats with hemorrhagic brain injury. *Ann N Y Acad Sci*. 2005;1049:84-96.
31. Newman MB, Willing AE, Manresa JJ, Sanberg CD, Sanberg PR. Cytokines produced by cultured human umbilical cord blood (HUCB) cells: implications for brain repair. *Exp Neurol*. 2006;199:201-208.
32. Nikolic WV, Hou H, Town T, et al. Peripherally administered human umbilical cord blood cells reduce parenchymal and vascular beta-amyloid deposits in Alzheimer mice. *Stem Cells Dev*. 2008;17(3):423.
33. Nishio Y, Koda M, Kamada T, et al. The use of hemopoietic stem cells derived from human umbilical cord blood to promote restoration of spinal cord tissue and recovery of hindlimb function in adult rats. *J Neurosurg Spine*. 2006;5: 424-433.
34. Pomyje J, Zivny J, Sefc L, Plasilova M, Pytlik R, Necas E. Expression of genes regulating angiogenesis in human circulating hematopoietic cord blood CD34+/CD133+ cells. *Eur J Haematol*. 2003;70:143-150.
35. Pranke P, Failace RR, Allebrandt WF, Steibel G, Schmidt F, Nardi NB. Hematologic and immunophenotypic characterization of human umbilical cord blood. *Acta Haematol*. 2001;105:71-76.
36. Prass K, Meisel C, Hoflich C, et al. Stroke-induced immunodeficiency promotes spontaneous bacterial infections and is mediated by sympathetic activation reversal by poststroke T helper cell type 1-like immunostimulation. *J Exp Med*. 2003;198:725-736.
37. Rainsford E, Reen DJ. Interleukin 10, produced in abundance by human newborn T cells, may be the regulator of increased tolerance associated with cord blood stem cell transplantation. *Br J Haematol*. 2002;116: 702-709.
38. Riordan NH, Chan K, Marleau AM, Ichim TE. Cord blood in regenerative medicine: do we need immune suppression? *J Transl Med*. 2007;5:8.
39. Sanchez-Ramos JR, Song S, Kamath SG, et al. Expression of neural markers in human umbilical cord blood. *Exp Neurol*. 2001;171:109-115.
40. Schroeter M, Jander S, Witte OW, Stoll G. Local immune responses in the rat cerebral cortex after middle cerebral artery occlusion. *J Neuroimmunol*. 1994;55:195-203.
41. Shaked I, Porat Z, Gersner R, Kipnis J, Schwartz M. Early activation of microglia as antigen-presenting cells correlates with T cell-mediated protection and repair of the injured central nervous system. *J Neuroimmunol*. 2004; 146:84-93.
42. Smith CJ, Emsley HC, Gavin CM, et al. Peak plasma interleukin-6 and other peripheral markers of inflammation in the first week of ischaemic stroke correlate with brain infarct volume, stroke severity and long-term outcome. *BMC Neurol*. 2004;4:2.
43. Soriano SG, Coxon A, Wang YF, et al. Mice deficient in Mac-1 (CD11b/CD18) are less susceptible to cerebral ischemia/reperfusion injury. *Stroke*. 1999;30:134-139.
44. Spera PA, Ellison JA, Feuerstein GZ, Barone FC. IL-10 reduces rat brain injury following focal stroke. *Neurosci Lett*. 1998;251:189-192.
45. Sredni-Kenigsbuch D. TH1/TH2 cytokines in the central nervous system. *Int J Neurosci*. 2002;112:665-703.
46. Streit WJ, Walter SA, Pennell NA. Reactive microgliosis. *Prog Neurobiol*. 1999;57:563-581.

47. Tan J, Town T, Paris D, et al. Microglial activation resulting from CD40-CD40L interaction after beta-amyloid stimulation. *Science*. 1999;286:2352-2355.
48. Traggiai E, Chicha L, Mazzucchelli L, et al. Development of a human adaptive immune system in cord blood cell-transplanted mice. *Science*. 2004;304:104-107.
49. Vendrame M, Cassady J, Newcomb J, et al. Infusion of human umbilical cord blood cells in a rat model of stroke dose-dependently rescues behavioral deficits and reduces infarct volume. *Stroke*. 2004;35(10):2390-2395.
50. Vendrame M, Gemma C, De Mesquita D, et al. Anti-inflammatory effects of human cord blood cells in a rat model of stroke. *Stem Cells Dev*. 2005;14:595-604.
51. Vendrame M, Gemma C, Pennypacker KR, et al. Cord blood rescues stroke-induced changes in splenocyte phenotype and function. *Exp Neurol*. 2006;199:191-200.
52. Vila N, Castillo J, Davalos A, Esteve A, Planas AM, Chamorro A. Levels of anti-inflammatory cytokines and neurological worsening in acute ischemic stroke. *Stroke*. 2003;34:671-675.
53. Wang X, Feuerstein GZ, Xu L, et al. Inhibition of tumor necrosis factor-alpha-converting enzyme by a selective antagonist protects brain from focal ischemic injury in rats. *Mol Pharmacol*. 2004;65:890-6.
54. Woiciechowsky C, Schoning B, Lanksch WR, Volk HD, Docke WD. Mechanisms of brain-mediated systemic anti-inflammatory syndrome causing immunodepression. *J Mol Med*. 1999;77:769-780.
55. Wolf SA, Fisher J, Bechmann I, Steiner B, Kwidzinski E, Nitsch R. Neuroprotection by T-cells depends on their subtype and activation state. *J Neuroimmunol*. 2002;133:72-80.
56. Yang GY, Gong C, Qin Z, Ye W, Mao Y, Bertz AL. Inhibition of TNFalpha attenuates infarct volume and ICAM-1 expression in ischemic mouse brain. *Neuroreport*. 1998;9:2131-2134.
57. Yang GY, Liu XH, Kadoya C, et al. Attenuation of ischemic inflammatory response in mouse brain using an adenoviral vector to induce overexpression of interleukin-1 receptor antagonist. *J Cereb Blood Flow Metab*. 1998;18:840-847.
58. Zhang RL, Chopp M, Li Y, et al. Anti-ICAM-1 antibody reduces ischemic cell damage after transient middle cerebral artery occlusion in the rat. *Neurology*. 1994;44:1747-1751.

15 Transforming "Waste" into Gold: Identification of Novel Stem Cells Resources with Therapeutic Potential in Neuromuscular Disorders

Mariane Secco, Mayana Zatz, and Natassia Vieira

15.1 Introduction

Once considered a biological waste product, umbilical cord blood (UCB) has emerged as a viable source of hematopoietic stem cells for transplantation. At the time of World War II, stored placental blood was explored as a source of blood for transfusion, and found to be similar in effect to fresh adult blood. During the 1970s, it was recognized that UCB contains hematopoietic progenitor cells. The suggestion that cryopreserved UCB could be used as a source of stem cells in much the same way as transplantation of bone marrow paved the way for the first successful human UCB transplant, performed in France in 1988. The recipient remains alive and well 18 years later. Since then, more than 2,000 transplants have been performed worldwide, most using stored units in private or public cord blood banks.

Advantages of UCB include no risk to the donor, reduced risk of viral contamination of the graft, rapid donor identification, and faster availability. Considering the high-risk malignancies and congenital neurodegenerative disorders, the rate is extremely important, since it can affect survival and neurological outcome. Compared with the use of unrelated bone marrow, which can take on average 4.5 months, the time from search of a compatible donor to transplant using UCB can be as little as 3 weeks. Moreover, preliminary clinical experience suggests that cord blood stem cells produce a less severe graft-versus-host disease (GVHD) when compared with bone marrow cells from adult donors. This means that cord blood donors and recipients do not need to be completely HLA-matched as it is required for successful bone marrow transplants.

The field of stem cell research is increasing exponentially and encompasses a wide range of topics – from enhancing our understanding of cellular development to applying these findings to repair and the future creation of organs. This is illustrated by the fast-increasing number of publications describing and characterizing new noncontroversial stem cells resources that are or may be in the future routinely used in medical treatments, including umbilical cord and adipose tissue.

The ability of multipotent stem cells to divide into several cell types (referred to as "plasticity") has led to speculation about their potential for therapeutic application across a range of inherited and acquired diseases. For example, transplantation of isolated endothelial and mesenchymal-like progenitor cells from adipose tissue prolongs survival in animal models of diabetes, Parkinson's disease, and Alzheimer's disease. Similarly, infusion of a subpopulation of stem cells derived from UCB enhances morphological and functional recovery in an animal model of stroke. Although clinical usefulness of this line of research is still to be established in human trials, it represents an exciting potential growth area for the use of these "waste" stem cells resources as a therapeutic approach.

15.2 Umbilical Cord

Human umbilical cord blood (UCB) has been explored in recent decades as an alternative source to bone marrow (BM) for cell transplantation and therapy because of its hematopoietic and nonhematopoietic (mesenchymal) components. In contrast to BM aspiration, human

M. Zatz (✉)
Department of Genetic and Evolutive Biology,
Human Genome Research Center, University of São Paulo,
Rua do Matão, n. 106, São Paulo, Brazil
e-mail: mayazatz@usp.br

N. Bhattacharya and P. Stubblefield (eds.), *Regenerative Medicine Using Pregnancy-Specific Biological Substances*,
DOI: 10.1007/978-1-84882-718-9_15,

UCB is obtained by a simple, safe, and painless procedure when the baby is delivered.[26]

Since the late 1980s, UCB has become an unreplaceable source of progenitor cells for transplantation of hematopoietic stem cells to treat some hematological disorders. Furthermore, other advantages of UC as a source of stem cells were reported. Preclinical studies showed a higher percentage of progenitor cells CD34+ in UCB compared with BM, suggesting that more primitive progenitors may be abundant in neonatal blood. In addition, UCB can be cryopreserved and stored in banks, eliminating delays and uncertainties that may exist regarding marrow collection from unrelated donors.

Despite these advantages, other points should be taken into account. Future development of potential abnormalities of the newborn donor's HSC into adult life and their effects on the recipient is unknown at the time of transplant. Moreover, the limited number of hematopoietic progenitor cells contained in collected UCB units may contribute to failed and delayed kinetics of donor hematologic engraftment and restrict their use in adult recipients. The minimum numbers of HSC in UCB units required to provide durable engraftment in ablated adult recipients is not firmly established. Usually, it is considered that a unit of cryopreserved UCB cell is not sufficient for an adult weighing more than 50 kg.

Published reports have summarized transplant outcomes for more than a thousand patients undergoing UCB-related and UCB-unrelated allogeneic transplantation. Owing to the limited amounts of stem cells in UCB, these treatments focus primarily on pediatric recipients with high risk or recurrent hematologic malignancies and a smaller proportion with nonmalignant hematologic disorders. Furthermore, previous studies have shown that the UCB therapeutic potential extends beyond the hematopoietic component suggesting a regenerative potential in solid organs as well. However, the clinical usefulness of hematopoietic cell transplantation for other diseases depends on the expansion, homing, and division of transplanted cells. Several innovative strategies aimed at increasing the cell dose, including bridging the engraftment period using either an additional allogeneic carrier cell population or by using host hematopoiesis, improving cell homing to the recipient bone marrow space, and ex vivo expansion are currently being investigated.

Regarding neuromuscular disorders, the focus of this chapter, other studies have shown that subpopulations of HUCB cells can be into muscle cells.[10,18] Additionally, CD34, transmembrane glycophosphoprotein known to be expressed by human hematopoietic progenitor cells has been associated with both the quiescent and activated states of myogenic progenitor cells.[3] More recently, in vivo differentiation of human umbilical cord blood cells into muscle cells was shown in a dysferlin mouse model (*sjl Muscular Dystrophy Mice*), suggesting that human umbilical cord blood has myogenic precursors.[14]

However, there is evidence that other stem or progenitor populations are of great interest. The mesenchymal stem cells (MSC), for example, comprise a rare population of multipotent precursors that are capable of supporting hematopoiesis in BM stroma. Previous studies suggest that MSC could be into several cell types, including chondrocytes, osteocytes, adipocytes, and myocytes. Moreover, phenotypic and genetic evidence suggests that MSC are an immature cell type, being a potentially useful model for developmental biology studies and therapeutic application.

Therefore, the search for alternative sources of MSC is of significant value. It has been reported that MSC could be isolated from various tissues, including BM, periosteum, trabecular bone, adipose tissue, synovium, skeletal muscle, deciduous teeth, fetal pancreas, lung, liver, amniotic fluid, UCB, and umbilical cord tissues (UC). Among those, UCB and UC may be ideal sources due to their accessibility, painless procedures to donors, promising sources for autologous cell therapy and lower risk of viral contamination. However, the process of isolation of MSC is at the expense of losing hematopoietic stem cells in cord blood. In addition, the data on the isolation of cord blood-derived MSC are controversial. Some researchers succeeded in isolating these cells, whereas others failed or obtained a low yield. Based on our experience, the efficiency in isolating MSC from blood in approximately 100 umbilical cord units stands around 10%.

Very recently, we compared, for the first time, the efficiency in obtaining MSCs from match-paired UCB and UC samples harvested from the same donors, which were processed simultaneously and under the same culture conditions.[22] Although MSCs from blood were obtained from only one of the ten samples, we were able to generate primary MSCs cultures from all cord samples with a 100% yield.

An important issue in cellular therapy studies is the availability of alternative stem cell sources and the efficacy of isolation techniques to yield a reasonable amount of viable cells that could be successfully expanded. Despite the advantages of HSC from UCB in hematopoietic reconstitution, results from our study demonstrated that UC, and not UCB, is the best choice for isolating MSCs for future applications.

The UC did not gain enough attention in the 1970s and 1980s, and was a discarded material after delivery. Two main questions drove scientists to re-examine its stromal cells and extracellular matrix (ECM) composition in the 1990s. One was the search for a possible reason and consecutive structural alterations in pre-eclampsia cases.[2] A series of ECM components were found altered in pre-eclamptic patients associated with the "premature aging" of this tissue. The second reason was the cellular identification of UC stromal cells, which share some common characteristics with the MSC, including the cell-surface markers expression and the potential to divide into several tissues.[11,13]

As the UC stromal cells are originated from extra-embryonic mesoderm, adipogenic, chondrogenic, osteogenic, cardiomyogenic, and skeletal myogenic inductions have been the most studied cell lineages.[4]

It was demonstrated that MSC from UC are capable of forming premature adipocytes bearing smaller intracytoplasmic lipid droplets. Compared to bone marrow MSCs, some authors found that MSC from UC generated significantly more fat-containing cells than bone marrow MSCs by day 21 after induced differentiation.[13] On the other hand, previous researchers did not report any significant adipogenic differences between these two cell types. Adipogenically induced cells specifically express adipocyte-specific genes, as lipoprotein lipase and plasminogen activator inhibitor-1 (PAI-1), which suggest that the culture conditions used could mimic the in vivo differentiation and initiate the adipogenic pathways.[11,15]

Chondrogenic induction was investigated in a few studies using pellet cultures in conditioned media to form three-dimensional cell spheres or polyglycolic (PGA) scaffolds, in which collagen fiber formation, GAG accumulation, and chondrocyte differentiation was observed. One to 2 mm cell spheres resembling articular cartilage were formed within 3 weeks containing many chondrocytes embedded in a mucopolysaccharide-rich stroma. The functional differentiation was confirmed by the de novo synthesis of type-II collagen fibers, which are normally synthesized by chondroblasts.[11]

Osteogenic potency is demonstrated as the formation of alkaline phosphatase-positive aggregates and von Kossa stained nodules with an associated expression of osteopontin, an osteo-specific matricellular protein. Other studies have showed the functional propriety of these cells. Recently, the use of MSC from UC was shown to build mineralized bone nodules of 300–800 μm in diameter containing a collagenous ECM in their inner structure.[21] Additionally, other authors have suggested that MSC from UC generate a greater extent of mineralization than bone marrow-derived MSC.[1]

Differentiation into cardiomyocytes or skeletal myocytes of MSC from UC was less examined. 5-Azacytidine, a chemical analogue of the cytosine nucleoside in the DNA and RNA helix, is currently used as a key chemical initiator of myogenic differentiation in vitro. Other protocols describe the use of conditioned medium of rat cardiomyocyte to differentiate MSC from cardiomyocytes. The induced cells synthesized several cardiac muscle-specific proteins, but they did not form any sarcomeres as naturally found in mature cardiomyocytes.[27] In many studies that examined the regenerative potency of transplanted MSCs, it has been directly or indirectly demonstrated that undifferentiated cells could incorporate into the myocardium and could somehow repair the injured tissue. Therefore, these partly differentiated cells may hold the potency to divide into cardiomyocytes when appropriate in vivo signals are received. Recently, differentiation of MSC from UC into skeletal myocytes has been successfully achieved both in vitro and in vivo.[5,9]

In many lineages tested so far, the MSC from UC seem promising for regenerative therapeutic applications, especially in orthopedic and cartilaginous interventions. However, compared to other sources of stem cells, especially hematopoietic and nonhematopoietic ones in bone marrow, in vivo studies concerning UC cells are very few in number, though all can be regarded as potentially valuable.

15.3 Adipose Tissue

As aforementioned, future cell-based therapies for non-genetic disorders require a source of autologous pluripotent stem cells that can be easily obtainable. In cases where the umbilical cord was not stored, another adult stem cell source may be used. Human bones contain an MSC population that has the potential to divide into adipocytes, chondrocytes, myoblasts, and osteoblasts. These cells represent a promising source for tissue-engineering strategies. However, traditional bone marrow procedures may be painful, requiring general or spinal anesthesia and may yield low numbers of MSCs upon processing (approximately 1 MSC per 10^5 adherent stromal cells). A low stem cell number requires an *ex vivo* expansion step to obtain clinically significant cell numbers. Such a step is time-consuming, expensive, and risks cell contamination and loss. An ideal source of stem cells would, therefore, should be both easy to obtain, result in minimal patient discomfort, and yet be capable of yielding cell numbers substantial enough to obviate extensive expansion in culture.

Adipose-derived stem cells (ASCs) are multipotent and hold promise for a range of therapeutic applications. Approximately 300,000 liposuction surgeries are performed in the United States each year. These procedures yield anywhere from 100 mL to 3 L of lipoaspirate tissue and this material is routinely discarded.[12] A variety of names have been used to describe the plastic adherent, fibroblast-like, multipotent cell population isolated from collagenase digests. The International Fat Applied Technology Society reached a consensus to adopt the term "adipose-derived stem cells" (ASCs) to identify this cell population. However, it is accepted that some investigators will use the acronym to mean "adipose-derived stromal cells."[8]

ASCs can be maintained in vitro for extended periods with stable population doubling and low levels of senescence. Characterization of ASCs shows that the majority of cells are of mesodermal or mesenchymal origin with low levels of contaminating pericytes, endothelial cells, and smooth muscle cells. Finally, ASCs divide in vitro into adipogenic, chondrogenic, myogenic, neurogenic, and osteogenic cells in the presence of lineage-specific induction factors. ASCs display a population doubling time of 2–4 days, depending on the culture medium and passage number.[28] The ASCs maintain their telomere length with progressive passage in culture.[16] ASCs can be cryopreserved and expanded easily in vitro. Under the conditions commonly used, these cells develop a fibroblast-like morphology. The greatest number of adipocytes can be obtained from cultures plated at low density.

ASCs can be isolated from human subcutaneous adipose tissue and divided into cells of the adipocyte lineage. These ASCs-derived adipocytes develop important features known from mature adipocytes, such as lipolytic capacity upon catecholamine stimulation, anti-lipolytic activity, and the secretion of typic aladipokines, such as adiponectin and leptin. Furthermore, ASCs retain their adipocyte differentiation capacity through multiple passages.[7] ASCs have the same ability for osteogenic differentiation, and this ability is maintained with increasing donor age.[23] In a functional study, cultured ASC cells have the potential for dividing into a cardiomyocyte-like phenotype with specific cardiac marker gene expression and pacemaker activity when cultured in a semisolid methylcellulose medium containing interleukin (IL)-3, IL-6, and stem cell factor.[19] Moreover, the differentiated cells were capable of responding to adrenergic and cholinergic stimuli. The transplantation of monolayered ASCs onto the scarred myocardium in murine myocardial injury models results in cardiomyocyte differentiation, angiogenesis, expression of cardiomyocyte-specific markers, and improvement of cardiac function. Using ASCs isolated from mouse brown adipose tissue, infarction area could be reduced and left ventricular function could be improved after transplantation of these cells in a mouse model of myocardial infarction. However, these data were obtained exclusively from animal models of murine origin and cannot be transferred into the human system.[24] In mice, adipose tissue-derived stromal vascular cells have a considerable proangiogenic potential regarding vessel incorporation, postischemic neovascularization, and vessel-like structure formation.[17] Incubation of ASCs under neurogenic conditions can create a cell population of PLA cells into early neuronal and/or glial progenitors. There was no expressing expression of mature neuronal or glial markers. However, the fact that human PLA cells express NeuN, in addition to the increased expression of several early neuronal and glial markers, is not proof that these cells will ultimately divide into mature neurons that are capable of undertaking complex electrophysiologic and synaptic functions.[25] were successful in differentiating human ASCs into cells with a pancreatic endocrine phenotype using the differentiation factors activin-A, exendin-4, HGF, and pentagastrin. ASCs treated with HGF, oncostatin M (OSM), and dimethylsulfoxide have the potential to develop a hepatocyte-like phenotype expressing albumin and alpha-fetoprotein. ASCs cannot acquire the potential to undergo a complete hematopoietic differentiation program as do BM. However, ASCs might support hematopoiesis in some way. Lethally irradiated mice could have their bone marrow reconstituted by cells isolated from adipose tissue. Using this experimental approach, ASCs from subcutaneous adipose tissue were reported to support the complete differentiation of hematopoietic progenitors into myeloid and B lymphoid cells.[6] However, these cells were unable to maintain the survival and self-renewal of hematopoietic stem cells. In vitro, human ASCs can express alpha-smooth muscle actin, calponin, and SM22, consistent with a smooth muscle phenotype.[20]

ASCs can be used in cell therapy by engraftment and differentiation in the host damaged tissue. ASCs may also be delivered into an injured tissue and stimulate the improvement in a paracrine manner and/or stimulating the endogenous stem cells to the site. In all contexts, it is important to consider that ASCs can be used for both autologous and allogeneic transplantation. Autologous ASCs offer advantages from regulatory, histocompatibility, and infectious perspectives. Independent studies have reported that passaged human ASCs lack the expression of MHC-II, which is maintained even after differentiation. This indicates that the ASCs may not elicit a T-cell response in vivo. Such findings may have a profound impact on the application of ASCs in therapeutic purposes. The allogeneic transplantation of ASCs will reduce the cost of cell therapies.

15.4 Conclusion

In recent years, in parallel to the enormous effort to explore novel and alternative sources of stem cells in the human body, the UC and adipose-tissue appeared as a promising reservoir of "waste" cells that could be readily used as multipotent stem cells. It is very likely that many more will come in the near future since *in vivo* tests of *in vitro* differentiated or undifferentiated stem cells are just starting to be examined in several disease models and in regenerative medicine, including neuromuscular disorders.

References

1. Baksh D, Yao R, Tuan RS. Comparison of proliferative and multilineage differentiation potential of human mesenchymal stem cells derived from umbilical cord and bone marrow. *Stem Cells*. 2007;25:1384-1392.
2. Bankowski E, Sobolewski K, Palka J, et al. Decreased expression of the insulin-like growth factor-I-binding protein-1 (IGFBP-1) phosphoisoform in preeclamptic Wharton's jelly and its role in the regulation of collagen biosynthesis. *Clin Chem Lab Med*. 2004;42:175-181.
3. Beauchamp JR, Heslop L, Yu DS, Tajbakhsh S, et al. Expression of CD34 and Myf5 defines the majority of quiescent adult skeletal muscle satellite cells. *J Cell Biol*. 2000;151(6):1221-1234.
4. Can A, Karahuseyinoglu S. Concise review: human umbilical cord stroma with regard to the source of fetus-derived stem cells. *Stem Cells*. 2007;25(11):2886-2895.
5. Conconi MT, Burra P, Di Liddo R, et al. CD105(+) cells from Wharton's jelly show in vitro and in vivo myogenic differentiative potential. *Int J Mol Med*. 2006;18:1089-1096.
6. Corre J, Barreau C, Cousin B, et al. Human subcutaneous adipose cells support complete differentiation but not self-renewal of hematopoietic progenitors. *J Cell Physiol*. 2006; 208:282-288.
7. Dicker A, Le Blanc K, Astrom G, et al. Functional studies of mesenchymal stem cells derived from adult human adipose tissue. *Exp Cell Res*. 2005;308:283-290.
8. Gimble JM, Katz AJ, Bunnell BA. Adipose-derived stem cells for regenerative medicine. *Circ Res*. 2007;100(9): 1249-1260.
9. Giordano A, Galderisi U, Marino IR. From the laboratory bench to the patient's bedside: An update on clinical trials with mesenchymal stem cells. *J Cell Physiol*. 2007;211: 27-35.
10. Ishikawa F, Drake CJ, Yang S, et al. Transplanted human cord blood cells give rise to hepatocytes in engrafted mice. *Ann NY Acad Sci*. 2003;996:174-185.
11. Karahuseyinoglu S, Cinar O, Kilic E, et al. Biology of the stem cells in human umbilical cord stroma: in situ and in vitro surveys. *Stem Cells*. 2007;25:319-331.
12. Katz AJ, Llull R, Hedrick MH, Futrell JW. Emerging approaches to the tissue engineering of fat. *Clin Plast Surg*. 1999;26:587-603.
13. Kobayashi K, Kubota T, Aso T. Study on myofibroblast differentiation in the stromal cells of Wharton's jelly: expression and localization of alphasmooth muscle actin. *Early Hum Dev*. 1998;51:223-233.
14. Kong KY, Ren J, Kraus M, et al. Human umbilical cord blood cells differentiate into muscle in *sjl* muscular dystrophy mice. *Stem Cells*. 2004;22:981-993.
15. Lu LL, Liu YJ, Yang SG, et al. Isolation and characterization of human umbilical cord mesenchymal stem cells with hematopoiesis-supportive function and other potentials. *Haematologica*. 2006;91:1017-1026.
16. Madonna R, Willerson JT, Geng YJ. Myocardin a enhances telomerase activities in adipose tissue mesenchymal cells and embryonic stem cells undergoing cardiovascular myogenic differentiation. *Stem Cells*. 2008;26:202-211.

17. Moon MH, Kim SY, Kim YJ, et al. Human adipose tissue-derived mesenchymal stem cells improve postnatal neovascularization in a mouse model of hindlimb ischemia. *Cell Physiol Biochem*. 2006;17:279-290.
18. Pesce M, Orlandi A, Iachininoto MG, et al. Myoendothelial differentiation of human umbilical cord blood-derived stem cells in ischemic limb tissues. *Circ Res*. 2003;93:51-62.
19. Planat-Benard V, Menard C, Andre M, et al. Spontaneous cardiomyocyte differentiation from adipose tissue stroma cells. *Circ Res*. 2004;94:223-229.
20. Rodríguez LV, Alfonso Z, Zhang R, Leung J, Wu B, Ignarro LJ. Clonogenic multipotent stem cells in human adipose tissue differentiate into functional smooth muscle cells. *Proc Natl Acad Sci USA*. 2006;103(32):12167-12172.
21. Sarugaser R, Lickorish D, Baksh D, et al. Human umbilical cord perivascular (HUCPV) cells: a source of mesenchymal progenitors. *Stem Cells*. 2005;23:220-229.
22. Secco M, Zucconi E, Vieira NM, et al. Multipotent stem cells from umbilical cord: cord is richer than blood! *Stem Cells*. 2008;1:146-150.
23. Shi YY, Nacamuli RP, Salim A, et al. The osteogenic potential of adipose-derived mesenchymal cells is maintained with aging. *Plast Reconstr Surg*. 2005;116:1686-1696.
24. Strem BM, Zhu M, Alfonso Z, et al. Expression of cardiomyocytic markers on adipose tissue-derived cells in a murine model of acute myocardial injury. *Cytotherapy*. 2005;7:282-291.
25. Timper K, Seboek D, Eberhardt M, et al. Human adipose tissue-derived mesenchymal stem cells differentiate into insulin, somatostatin, and glucagon expressing cells. *Biochem Biophys Res Commun*. 2006;341:1135-1140.
26. Tse W, Laughlin MJ. Umbilical cord blood transplantation: a new alternative option. *Hematol Am Soc Hematol Educ Prog*. 2005;2005:377-383.
27. Wang HS, Hung SC, Peng ST, et al. Mesenchymal stem cells in the Wharton's jelly of the human umbilical cord. *Stem Cells*. 2004;22:1330-1337.
28. Zuk PA, Zhu M, Mizuno H, et al. Multilineage cells from human adipose tissue: implications for cell-based therapies. *Tissue Eng*. 2001;7:211-228.

Human Umbilical Cord Blood Cells for Stroke

16

Dong-Hyuk Park, Alison E. Willing, Cesar V. Borlongan, Tracy A. Womble, L. Eduardo Cruz, Cyndy D. Sanberg, David J. Eve, and Paul R. Sanberg

16.1 Introduction

In the United States each year, about 780,000 people experience a new or recurrent stroke. About 600,000 of these are first attacks, and 180,000 are recurrent attacks. A report from the American Heart Association Statistics Committee and Stroke Statistics Subcommittee said that every 40 s someone in the United States suffers from stroke, and every 3–4 min someone dies from stroke.[84] When considered separately from other cerebrovascular diseases, stroke ranks third among all other causes of death behind heart diseases and cancer. It is also reported that among ischemic stroke survivors who were at least 65 years of age, 50% suffered from hemiparesis and 30% were unable to walk without some assistance 6 months after the stroke. Moreover, 26% were dependent on others to assist in their daily activities.[84] Although recently the number of stroke survivors has increased in virtue of the development of intensive care, it is still necessary to explore more efficient treatments including rehabilitative therapies that will allow for greater recovery. Currently, there are very few treatment options for an ischemic event in the brain. The tissue-plasminogen activator (t-PA), the only US Food and Drug Administration (FDA)-approved thrombolytic medicine for acute ischemic stroke, acts by dissolving the blood clot in the vessel and restoring proper blood flow. However, even when t-PA is administered within 3 h after the initial onset of an embolic stroke, there is only a 30–35% success rate for recovery,[65,74] and this effectiveness decreases with time. There are also significant risks such as recurrent stroke and intracranial hemorrhage.[74] Unfortunately, the therapeutic window is so narrow that only 3–5% of stroke victims can arrive at the hospital within this time frame and be diagnosed. Of those, less than 30% are eligible for t-PA.[70] In fact, only a few patients with stroke have the opportunity of complete restoration with rapid management. The remainder can only anticipate supportive and rehabilitative care with permanent disability.[70]

There are two areas, the so-called "core and penumbra," that comprise the injury site at an ischemic insult in the stroke brain. Cell loss occurs in the core site of stroke, that area adjacent to the vessel in which blood flow has been obstructed or decreased, mainly through excitotoxicity-induced necrosis.[70] In fact, so far it has been a major focus of stroke research to prevent the neuronal death from spreading over into the penumbra, the area in which cells can survive after proper reperfusion.[70] Many studies have focused on new treatments that act by anti-inflammatory action or neuroprotection, which have so far turned out to be ineffective.[113]

There are two events that lead up to neurodegeneration in stroke. The primary reaction is glutamate excitotoxicity owing to the lack of oxygen and nutrients that occurs right after the blood flow decreases below a threshold.[93] The secondary event of neurodegeneration is the inflammatory reaction, such as oxygen free radical and nitric oxide production,[23,60] activation of microglia (macrophages in the central nervous system [CNS]),[56,94] and recruitment of other inflammatory cells through the leaky blood–brain barrier.[29,35,122] Therefore, an ideal management regime for stroke must be multifaceted, responding to all insults simultaneously with both anti-inflammatory and neuroprotective abilities.[70] It must also have a longer window of

P.R. Sanberg (✉)
Center of Excellence for Aging and Brain Repair, Department of Neurosurgery and Brain Repair, University of South Florida College of Medicine, MDC 78, 12901 Bruce B. Downs Boulevard, Tampa, FL 33612, USA
e-mail: psanberg@health.usf.edu

N. Bhattacharya and P. Stubblefield (eds.), *Regenerative Medicine Using Pregnancy-Specific Biological Substances*,
DOI: 10.1007/978-1-84882-718-9_16,

opportunity that increases the survival rate and improves subsequent outcome. These requirements lead to exploration of new therapeutic strategies, such as cell-based therapy to replace the lost cells and release anti-inflammatory and/or neuroprotective factors in stroke or other neurodegenerative diseases.

The new theory that mature neurons can be regenerated has been accepted, since neurogenesis in the adult CNS of mammals including humans has been demonstrated.[7,13,42,61] Neural stem cells (NSCs), existing in specific brain regions, are capable of producing new neurons throughout an organism's lifetime. Adult NSCs appear to be responsible for nervous tissue homeostasis and repair throughout adulthood because they can bring about several differentiated cell types of the CNS, including neurons and glial cells, even though the number of newborn neurons tends to decrease with age.[7]

Increased neurogenesis in the subventricular zone and the dentate gyrus of the hippocampus has been shown in both the adult rodent experimental epilepsy and stroke models.[75] These results suggest that the adult brain has regenerative potential even though this compensation does not provide complete recovery, and the changes in the microenvironment of the CNS, as a response to insult, may affect the endogenous repair system. With exogenous threats, environmental alteration and their byproducts including cytokines and various trophic factors regulate the destiny of adult NSCs and control their proliferation and differentiation.[5,42] Furthermore, similar to embryonic stem cells, tissue-specific adult stem cells, such as NSCs or hematopoietic progenitor cells possess the ability for self-renewal and can transdifferentiate into other cell lineages. It has been shown that adult NSCs can differentiate into non-CNS derivatives, such as blood[9] or skeletal muscle cells and vice versa.[32] This plasticity of adult stem cells opens new approaches for their application in the treatment of various degenerative disorders.

However, currently the application of human NSCs as well as embryonic and fetal stem cells for cell-based therapies is hampered by their availability due to ethical and moral issues.[26,46] Other technical difficulties exist, such as purification and isolation of the cell type needed during in vitro cell culture for transplant.[80,88] These limitations drive the search for new sources of stem cells. Many researchers have been focusing on the potential of the hematopoietic stem cell as a target for cell-based therapies in various diseases, including those unrelated to hematopoiesis such as neurodegenerative diseases as well as hematologic disorders.

Since bone marrow (BM) transplantation was first performed in 1968, it has become the gold standard source for the hematopoietic stem cells used to repopulate blood lineages after myeloablation in a number of malignant and nonmalignant hematological diseases.[79] Although many researchers have demonstrated that BM stem cells can give rise to nonblood lineage cells including nervous tissue,[30,76,86] BM transplantation, especially allogeneic, has several weaknesses that limit cell-based therapy. The most notable disadvantages are as follows: first, the time from donor searching to transplantation is relatively long (average 135 days).[53] Second, even the cost of cell preparation is considerable (at least more than $25,000).[53] Third, it is difficult to find human leukocyte antigen (HLA)-matched donors with BM, which is a critical cause of graft-versus-host disease (GvHD) in allograft transplantation. Only 30% of patients find an eligible matched donor to keep them from succumbing to GvHD.[1] Fourth, BM recipients are very easily infected by viruses (90%).[6] These disadvantages can be avoided by using hematopoietic stem cells from autologous BM or autologous peripheral blood with granulocyte colony-stimulating factor. However, preexisting malignancies may preclude the patient from receiving BM autografts owing to the possibility of recontamination with tumor cells.[71] Fifth, in patients with stroke, "timing" is important, and it may not be feasible to recover, manipulate, and process quality control cells, in a current good manufacturing practices (GMP) environment from the patient within the therapeutic window.

Human umbilical cord blood (HUCB) has already been successfully transplanted for the repopulation of blood cell lineages in children with hematological malignant and nonmalignant diseases.[92] Since the first clinical HUCB cell transplantation appeared in 1988 in which a 5-year-old patient was successfully treated for Fanconi anemia, by implantation of UCB from an HLA-identical sibling,[38] more than 6,000 patients have undergone HUCB transplantation worldwide, many of them with unrelated donors.[57,102,109] So far, HUCB use has been limited to hematological disorders, but current research suggests that it could also be a promising treatment for nonhematological diseases.[31,71] Numerous reports reveal the many advantages of using HUCB cells for cellular therapies. Most of all, HUCB cells, even from an allogeneic origin, are immature and therefore result in a lower incidence of immune rejection, GvHD, and posttransplant infections compared with BM transplants.[54,102] Additionally, there are higher numbers of immature stem cells in HUCB compared with those in adult BM, as demonstrated by telomere length.[104] The immaturity of cord blood cells and their immune naïveté might be related to the optimal effects of these cells for hematopoietic and somatic cell-based therapies. Now, we describe recent findings, including our own, showing the promise of HUCB cells, and identifying their specific properties as a new therapeutic paradigm in the treatment of stroke.

16.2 The Characteristics of Human Umbilical Cord Blood Cells

Since the first transplant was performed in a patient with Fanconi anemia,[38] HUCB transplantations have increased worldwide. Moreover, HUCB from unrelated donors has been successfully transplanted for children and adult patients.[40,58,83,85,108] Hematopoietic progenitor cells, the most primitive stem cells in HUCB, are able to reconstitute blood lineages over a long period of time.[10,69,99] The number of myeloid progenitors in HUCB is similar to BM,[10] but HUCB cells have a greater colony-forming ability,[68] can proliferate in long-term cultures in vitro with stimulation of different growth factors, and possess longer telomeres compared with the adult BM cells.[104] Moreover, HUCB transplants, compared with adult BM stem cell transplants, are better at repairing the host hematopoietic progenitor reservoir.[31] It has been shown that even a single HUCB cell can provide enough hematopoietic stem cells for both short- and long-term engraftment.[92]

The mononuclear fraction of HUCB is primarily composed of lymphocytes and monocytes.[77] Electron microscopic comparison between cord blood, peripheral blood, and BM cells shows that HUCB has more immature myelomonocytic cells and a small number of mature neutrophils with unique ultrastructure components, such as nuclear pockets in the neutrophils, which augmented the delivery of RNA to the cytoplasm.[67]

Lymphocytes found within HUCB can be divided into a comparable B-cell population and a lower absolute number of T-cells ($CD3^+$) with a higher $CD4^+$/$CD8^+$ ratio compared with adult peripheral blood.[45,77] Comparing the characteristics of B-cell differentiation in vitro between $CD34^+$ cord blood and adult peripheral blood, the B-cell progenitors from cord blood are found to be more immature.[47] This may delay mature B-cell production. The relative immunological immaturity of HUCB cells, compared with adult cells, reflects a higher proportion of immature T-cells ($CB45RA^+$) and lower numbers of mature memory T-cells ($CD45RO^+$).[45] Moreover, cord blood lymphocytes expressed fewer proinflammatory cytokine receptors such as interleukin (IL)-2, IL-6, IL-7, tumor necrosis factor (TNF)-α, and interferon (IFN)-γ than in adult blood cells,[124] whereas, they produced more anti-inflammatory cytokine IL-10, which down-regulates the expression of CD86 on dendritic cells (DCs).[39,78] Consequently, less CD86 expression may prevent initiation of T-cell mediated-inflammation.[11] By contrast, increased IL-10 may activate regulatory T-cells, which subsequently inhibits antigen-mediated immune responses.[3]

Other cells also differ between adult and cord blood. HUCB has fewer $CD56^+$ cytotoxic T-cells, while higher numbers of natural killer (NK) cells[25] which can inhibit T-cell proliferation and TNF- α production.[28] DCs are the sentinels of the body, which begin the immune reaction within the lymph nodes. The DCs in cord blood take on lymphoid characteristics that are more likely to be involved with colonizing neonatal tissue, whereas there are more myeloid DCs in adult blood.[114] The lymphoid DCs activate the anti-inflammatory action of T- helper cell 2, which along with the naïve T-cells, may down-regulate immune responses.[2,114]

The immature immunological properties of HUCB cells appear to result in a prolonged immunodeficient status following transplantation,[36,98] which may explain the low incidence of GvHD and viral transmission. Consequently, it allows for less strict HLA-matched donor requirements and a shorter waiting period for treatment.[71] Rocha and colleagues showed that GvHD occurred remarkably less in children with HUCB transplantation compared with BM recipients when the source was from an HLA-identical sibling.[83] Additionally, they demonstrated that GvHD incidence even in unrelated HLA-mismatched UCB recipients was considerably lower than in HLA-identical BM recipients.[82]

Recently, in in vitro cell culture, we have found two different subpopulations of mononuclear HUCB cells; adherent and floating.[18] A significant number of progenitor/stem and neural cell antigens are expressed by the floating cells whereas lymphocytes expressing hematopoietic antigens comprised about 53% of the adherent cell population. These results indicate that nonhematopoietic cells exist within the mononuclear HUCB population and appear to have the potential to become neural-like cells. On the basis of the results in our study, HUCB includes a primitive stem cell population that may give rise to both hematopoietic and neural cells.[33]

However, because HUCB collection can only be performed at a single time-point (i.e., during birth) and the number of mononuclear cells extracted from one donor is very limited,[71] current investigation for clinical applications has focused on the *ex vivo* expansion of the stem cells to yield a sufficient amount for adult transplantation.[62] Unfortunately, the current methods for expanding HUCB cells can neither maintain the quality of the hematopoietic progenitors through to the end product[49,81] nor make up for the cells lost in the storage process, let alone expand them to a suitable quantity for implantation.[48,63,101,120] In terms of cell-based therapies, expansion of the target cells is a critical factor for providing appropriate dosages for both children and adults, as well as decreasing the time to engraftment, consequently elevating the likelihood of treatment success, as well as for recurrent therapy in the case of graft failure.[71] However, instead of simply global multiplication of all cells, selective amplification of relative proportions of cell subpopulations that home to marrow, form colonies, and repopulate specific blood lineages are required[71] because blanket expansion does not reduce the time to engraftment.[49]

Recently, several studies about *ex vivo* expansion of different cell subpopulations of HUCB have been reported. Wei et al.[112] reported that T, NK cells as well as $CD34^+$ cells could be expanded from UCB in the same medium using a combination of different cytokines. The expanded mononuclear cells could reconstitute hematopoiesis as well as eliminate minimal residue leukemia in transplanted mice. Shirvaikar et al.[91] revealed that the combination of thrombopoietin with IL-3 used for expansion not only provided adequate proliferation of megakaryocytic progenitors from cord blood but also prevented these cells from undergoing apoptosis. They also found that these megakaryocytic progenitors had homing potential. Therefore, these cells could be applied to complement cord blood grafts and escalate platelet recovery in transplant recipients.

Selection of an optimal stem cell population and characterization of the desired stem cells should be decided before expansion for engraftment.[90] First, expanded cells need to preserve telomerase length to keep their pluripotent ability.[120] Telomerase activity is high in hematopoietic progenitor cells and is most active during proliferation.[19] Second, identifying the target cells for expansion among multiple subpopulations is critical. The stemness or maturity of a cell can be determined by the cell's presence of, or lack of, a combination of cell surface antigens. These are also potential markers for the selection of progenitor cells for expansion and implantation. For example, the $CD34^+$ population, a marker for hematopoietic stem cells, in HUCB can be defined as more primitive than those in BM because CD38, a marker for pre-lymphoid cells,[14,24] is absent in a higher proportion of them. $CD34^+$ cells in HUCB comprised roughly 1% of the mononuclear fraction and appeared to be more immature than those found in BM.[71] Another hematopoietic stem cell marker CD 133 has also been identified, and it could replace CD34 as an indicator for the selection and expansion of hematopoietic cells for transplantation.[55,121] About 80% of $CD34^+$ cells have been shown to express CD 133, and more than 97% of CD 133^+ cells are positive for CD34 in fresh cord blood.[44] Although CD 133^+ cells comprised 0.67% of the total mononuclear cells of HUCB,[64] expansion of CD 133^+ and $CD133^+CD34^+$ cells was more prevalent than those from $CD34^+$ cells.[44] These results indicate the more primitive properties of $CD133^+$ hematopoietic stem cells compared with $CD34^+$ cells. Furthermore, $CD133^+$ cells, which have been detected in the fetal brain, are considered to be NSCs,[96,103] even though it is still unclear whether the $CD133^+$ cells in HUCB are phenotypically and functionally the same as the NSCs found in fetal brain.

The nonhematopoietic stem cells, such as the mesenchymal stem cell (MSC), have also been detected in UCB.[41,119] MSCs and MSC-like progenitors can be isolated from amniotic fluid and placenta as well as HUCB. MSCs derived from HUCB show impressive plasticity as they can differentiate into cells that comprise all three germ-lines.[50,59,119] However, unlike MSCs from BM, the definitive phenotype, as well as identification of the surface antigens of HUCB-derived MSC, has remained controversial. Yang et al.[119] identified MSCs as being positive for CD13, CD29, CD44, and CD90 and negative to CD14, CD31, CD34, CD45, CD51/61, CD64, CD106, and HLA-DR, whereas Robinson et al.[81] defined the MSC as being positive for CD16, CD73, CD90, and CD105 and negative for

CD31, CD34, CD45, CD80, and HLA-DR. In support, we have demonstrated that HUCB MSCs were negative for CD3, CD11b, CD19, CD34, and CD45 in culture.[27]

Although HUCB MSCs failed to produce macrophage (CFU-M), granulocyte-erythroid-macrophage-megakaryocyte (CFU-GEMM), or granulocyte-macrophage (CFU-GM) hematopoietic colonies in methylcellulose, supernatants from cultured HUCB MSCs promoted survival of NT2N neural cells and peripheral blood mononuclear cells that were cultured in a hostile environment that induced cell stress and limited protein synthesis.[27] Furthermore, HUCB MSCs expressed the neural cell surface antigen (A2B5), the neurofilament polypeptide (NF 200), the oligodendrocyte precursor marker 04, intermediate filament proteins characteristic of neural differentiation (nestin and vimentin), as well as glial fibrillary acidic protein (GFAP) and the neural progenitor marker (TuJ1) following incubation in neural differentiation medium.[27] We also showed that HUCB MSCs could modulate the immune reaction, using coculture with murine splenocytes.[27] These results suggest that HUCB MSCs have multiple functions that allow them to exert therapeutic potency in the treatment of neurological diseases.

16.3 From the Beginning to the Current Niche for HUCB in Neuroscience; Focusing on In vitro Studies

Several in vitro reports have been published showing the multipotent nature of HUCB cells, including differentiation into neural cells, thus helping to advance their promising usefulness in the treatment of neurodegenerative diseases. Our group revealed that with exposure to retinoic acid (RA) and nerve growth factor (NGF), mononuclear HUCB cells usually expressed neuronal and/or glial markers such as those for early neural precursors (musashi-1, nestin, TuJ1), mature neurons [NeuN, microtubule associated protein (MAP2)], and astrocytes (GFAP). Moreover, cells treated with a combination of both factors increased TuJ1 and GFAP expression by approximately two times.[87] Zigova et al.[123] also showed that with RA and NGF, TuJ1 and GFAP positive cells derived from UCB mononuclear cells survived in the subventricular zone of the rat neonatal forebrain. By contrast, under standard growth media conditions (Dulbecco's modified Eagle media), mononuclear HUCB cells also expressed neural markers, such as nestin, TuJ1, MAP2, and GFAP.[34] Of interest, in this study, at 2 weeks after plating, colocalization of nestin and MAP2, as well as expression of TuJ1 was observed. These findings may be dependent on the cell cycle or development.[34] In addition, the increased number of cells expressing the primitive hematopoietic stem cell marker, CD133, in cells cultured for 7 days probably gives rise to cells characterized by both immature and mature properties at the same time.

Other researchers have also determined that HUCB cells can express typical neural antigens within the CNS. Bicknese et al.[8] and Buzanska et al.[12] demonstrated HUCB cells expressing class βIII-tubulin, GFAP, and galactosylceramide (GalC), neuronal, astrocytic, and oligodendrocyte markers, respectively. Interestingly, by using the $CD34^-/CD45^-$ nonhematopoietic mononuclear cell fraction, Buzanska et al.[12] derived a clonogenic line of HUCB-NSCs expressing nestin and GFAP. With exposure to neuromorphogen/RA, 40% of the HUCB-NSCs expressed βIII-tubulin and MAP2, 30% expressed the astrocytic markers (GFAP and S100β), and 11% expressed the oligodendrocytic phenotype (GalC). A combination of RA and brain-derived neurotrophic factor appears to induce neural differentiation in HUCB-NSC cultures.[52] During 7 days of coculture with this combination, rat astrocytes, or hippocampal slices, 80% of cells expressed β-III-tubulin and 64% coexpressed MAP2.[52]

A number of studies have shown the nonneuronal plasticity of HUCB cells. Goodwin et al.[41] showed that a subset of mononuclear cells from HUCB, which had been preserved in continuous culture for more than 6 months did not express antigens for hematopoietic differentiation. After exposure to osteogenic, adipogenic agents, or basic fibroblast and epidermal growth factors, HUCB mononuclear cells expressed bone, fat, and neural markers, respectively. These data suggest that HUCB has a specific cell population, which is able to express multilineage antigens, showing the outstanding plasticity that is integral to cell-based therapies even though it is unclear whether these cells are a stem cell population, multiple differentiated progenitors, or cells with transdifferentiation capacity.[41] To confirm the presence of multipotent stem cells in HUCB, McGuckin and colleagues[66] developed a negative immunomagnetic selection technique that depletes HUCB from hematopoietic antigen-expressing cells, thereby isolating a discrete lineage-negative stem cell population. These selective lineage-negative cells from mononuclear HUCB cells gave rise to adherent cells with neuroglial progenitor cell morphology as well as primitive nonadherent hematopoietic progenitors simultaneously over 8 weeks. An analysis of gene expression revealed upregulation of primitive neuroglial progenitor cell antigens for GFAP, nestin, musashi-1, and necdin.[66]

Recently, we determined the capability of mononuclear HUCB cells to express neural antigens after in vitro exposure to defined media supplemented with a cocktail of growth and neurotrophic factors.[19] Embryonic stem cells have high proliferative potential and resistance to rejection, which could result in deleterious side effects following transplantation, such as fibrosis and malignancies, whereas cord blood stem cells are relatively quiescent under normal circumstances.[100] The intrinsic quiescent potential of these cells needs to be activated by exogenous factors within the environment,[22] without which the cells do not proliferate uncontrolled. Floating mononuclear HUCB cells treated with growth factors and neurotrophins, survived longer in vivo and exhibited an increase in their proliferation and expression of neural antigens.[19]

Cells that are to be considered for transplantation into humans are preferred to be of human origin. This presents a dilemma in HUCB in vivo research. Xenografting of human cells into the rodent is widely accepted in the bioresearch field and often produces encouraging outcomes.[110] As mentioned, HUCB cells have a high immune tolerance, in virtue of their inability to generate cytotoxic T-cells that respond to allogeneic antigens, and to produce the proinflammatory cytokines such as IFN-γ and TNF-α. Despite the immunological naïveté of HUCB cells, graft rejection is so vigorous after transplantation into adult or aged rat brains that strong immunosuppression is often required to protect the transplant.[111] The limited survival of the grafts in animal models is probably a result of an immune reaction. Therefore, we recently performed in vitro and in vivo studies to determine the ability of HUCB cells to survive and the effect of the immune response on their survival by implantation into the normal striatum of nonobese diabetic severe combined immunodeficient (NOD SCID) mice.[110] In vitro, long-term culture of HUCB cells resulted in several different populations of cells including neuron-like cells. These cells were positive for both TuJ1 (early neuronal marker) and nestin (primitive neural marker). Five days after transplantation of HUCB-derived neuronal-like cells, we found many differentiated cells expressing characteristic neuronal proteins. However, at 1 month postgrafting, HUCB cells were no longer detected. There were no signs of T-cell mediated rejection, such as CD4 and CD8 lymphocytes and minimal changes in microglia and astrocytes, suggesting that cell loss may result from an alternative cause than a T-cell mediated immune response. This study indicates that HUCB cells can survive for a long time and display neural characteristics dependent on the environmental conditions in vitro.[110] It is unclear why the same does not appear to be true in vivo. However, cell survival following transplantation might be dependent on a combination of several donor and recipient characteristics rather than a single cause.

Regardless, on the basis of several previous in vitro studies, HUCB cells in virtue of their primitive nature and ability to differentiate into nonhematopoietic lineage cells may be promising for cell-based therapies requiring either the replacement of individual cell types and/or substitution of missing substances (e.g., dopamine in Parkinson's disease).

16.4 The Application of HUCB in Stroke Research Focusing on In vivo Studies

Management for stroke is complicated because it can affect multiple anatomical brain structures, different neuronal cell populations, and various neuroanatomical pathways.[33] Additionally, the timing of treatment is crucial because of the progressive ischemic process with cumulative cell damage.[33] However, the restrictive therapeutic 3-h window of the currently approved treatment (tPA) means that many patients that fall outside of this window suffer from irreversible sequela. This disappointing reality fuels the search for the development of any therapeutic modality, especially, a cell-based therapy.

With respect to adult stem cells, Chopp and colleagues used BM stromal cells as a target cell in transplantation therapy for stroke and demonstrated neurological recovery in rats after focal cerebral ischemia as well as migration of the grafts to ischemic sites and differentiation into both neuronal and glial phenotypes.[15,17,21] Interestingly, the authors suggested that the recovery mechanisms are more likely to be the secretion of trophic factors from transplanted cells, which may promote endogenous neurogenesis and angiogenesis rather than direct cell replacement.

According to the discussion by Savitz et al.[89] in ongoing clinical trials, patients afflicted with stroke showed considerable improvement in neurological outcomes on several scales following intrastriatal implantation of the immortalized NT2N cell line, compared to baseline measurements before treatment. Some patients treated with these cells improved on a series of memory, recall, and visuospatial/constructional ability tests on repeat testing 6 months after transplantation. Another preliminary study showed that intrastriatal implantation of fetal neural cells (from the lateral ganglionic eminence) of the pig prevented further neurological deficits during the acute phase in patients with basal ganglia infarction even though the study was terminated by the FDA after 2 years because of the absence of any significant improvement on the modified Rankin scale.[89] Although current clinical data are too preliminary and insufficient to assess efficacy, the results suggest that cell transplantation is technically feasible and can be performed safely.

HUCB is another promising source of multipotent stem cells that has shown affirmative effects in in vivo studies for the treatment of stroke. Chen et al.[16] showed that transplantation of mononuclear HUCB cells via the intravenous route at 24 h or 7 days after middle cerebral artery occlusion (MCAO) significantly improved functional deficits in a rat model. A histological examination revealed that grafted mononuclear HUCB cells existed mainly in the cortex and striatum of the injured hemisphere in the ischemic boundary zone whereas few cells were detected in the contralateral side. Some of these mononuclear HUCB cells were positive for the endothelial cell marker FVIII (8%), GFAP (6%), MAP2 (3%), and NeuN (2%) in an immunohistochemistry study.[16] Xiao et al.[118] have isolated a nonhematopoietic stem cell line from HUCB. Intravenous administration of these cells reduced infarct volume in rats with ischemic brain injury. Some of the transplanted cells were double labeled for human nuclei and NeuN, though it was unlikely that they contributed to the recovery.

In one of our studies, the efficiency of intravenous versus intrastriatal transplantation of mononuclear HUCB cells was determined to assess which route provided the greatest behavioral improvement in rats with permanent MCAO.[115] Behavioral recovery was not different between the two delivery routes, as spontaneous hyperactivity was significantly less in both cell-treated groups regardless of the delivery route, 24 h after stroke compared with the control MCAO group. Also, transplanted animals learned the task faster compared with control rats in the passive avoidance test. However, in the step test at 2 months after treatment with the mononuclear HUCB cells, only the rats treated via intravenous delivery showed significant functional improvements. These findings suggest that the intravenous route of administration for mononuclear HUCB cell transplantation may be more effective compared with direct striatal implantation in exerting long-term benefits to the stroke animal.

We also examined the dose effect of mononuclear HUCB cells after MCAO on the behavioral recovery and stroke volume in rats.[105] Rats were intravenously injected with between 10^4 and 5×10^7 mononuclear HUCB cells 24 h after MCAO. At 4 weeks after infusion, behavioral deficits improved significantly when 10^6 or more mononuclear HUCB cells were transplanted. The inverse relationship between cell dose and infarction volume was significant at the higher cell doses. Moreover, transplanted cells were detected only in the lesioned hemisphere and spleen by immunofluorescence for human nuclei and polymerase chain reaction analysis.[105] These findings showed a dose relationship between introduced cells, functional improvement, and spared neuronal volume using mononuclear HUCB cell transplantation in the MCAO rat stroke model.

Despite several reports of HUCB cells expressing neural phenotypes both in vitro and in vivo,[34,43,87,123] few cord blood cells are detected in the ischemic region compared with the number of transplanted cells,[20,105,106,115,116] indicating that cell replacement is not the mechanism responsible for the functional recovery seen in these animal studies. Recently, we found that HUCB may act by not only cell replacement but also the production of neurotrophic, neuroprotective, or anti-inflammatory factors (Fig. 16.1). The multifaceted therapeutic effects of HUCB have been extensively studied using the MCAO model of embolic stroke.

Neuroinflammation is a major pathogenesis of neurodegeneration including stroke. The number of $CD45^+/CD11b^+$ cells (microglia) and $CD45^+/B220^+$ cells (B cell) increases in the brain of rats following permanent MCAO.[106] We demonstrated that mononuclear HUCB cell transplantation significantly decreased the

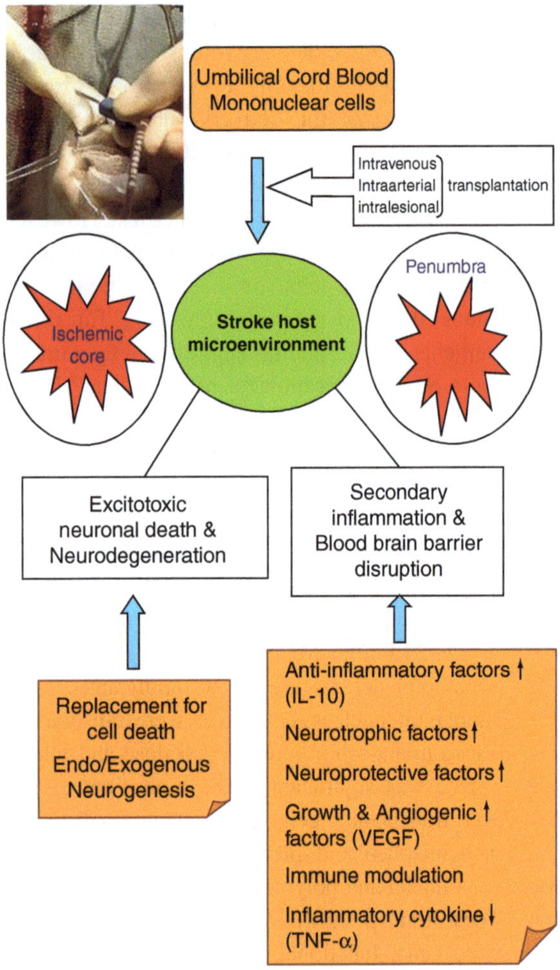

Fig. 16.1 Schematic diagram depicting how umbilical cord blood can beneficially influence the stroke microenvironment. UCB cells may exert their beneficial effects with respect to stroke either directly on cell survival (*left side*) or indirectly via release of factors that affect the secondary inflammation and BBB disruption (*right side*)

number of $CD45^+/CD11b^+$ and $CD45^+/B220^+$ cells in the injured rat's brain.[106] The reduction of $CD45^+/CD11b^+$ cells after HUCB cell treatment is of particular interest, given that chronic microgliosis is believed to mediate neuronal injury, not only in ischemic stroke but also in other neurodegenerative diseases.[94,97] These cellular changes were accompanied by a reduction in the mRNA and protein expression of proinflammatory cytokines, and in nuclear factor kappa B (NF-κB) DNA binding activity in the brain of stroke animals following treatment with mononuclear HUCB cells.[106] HUCB transplantation exerts dual anti-inflammatory effects as follows: reduction of the proinflammatory cytokines such as TNF-α and IL-1β[106] and inhibition of both activated microglia and astrocytes[70] in the brain following stroke. Although the $CD34^+$ component of the transplant may promote angiogenesis and subsequently bring about a neurotrophic effect, the potential anti-inflammatory effects of UCB therapy also may play a significant role in protection against neuronal death.[95]

With respect to neuroinflammation, we also found that, following MCAO, rat spleen size was decreased concomitant with their $CD8^+$ T-cell counts.[107] Spleen size reduction following MCAO was found to correlate with the extent of ischemic damage, whereas HUCB cell treatment rescued the spleen weight and splenic $CD8^+$ T-cell counts and reduced the amount of brain injury. Additionally, splenocyte proliferation assays revealed that HUCB cell transplantation inhibited MCAO-associated T-cell proliferation by increasing the production of IL-10 while decreasing IFN-γ.

The secretory function of HUCB cells is also noteworthy. Recently, in vitro we investigated the expression of cytokines and chemokines released from HUCB cells under a number of culturing conditions.[73] The heterogeneous cells from HUCB mononuclear fractions consistently secreted a variety of chemokines and cytokines under different culture conditions, but in particular, IL-8, monocyte chemoattractant protein-1 (MCP-1), and IL-1α. These chemokines are extensively produced throughout the human body, especially in the brain, and play a key role in the inflammatory reaction as the first "line of defense" mechanism. These results suggest that these factors may be partially responsible for the functional gains observed in stroke animal models investigating the therapeutic use of HUCB cells.

Recently, we demonstrated in vivo that some chemokines were increased in the ischemic brain area.[51] MCP-1 and macrophage inflammatory protein (MIP-1α) are involved in monocyte accumulation into the CNS under pathological conditions.[4,37] Thus, we hypothesized that MCP-1 and MIP-1 α may also participate in the mobilization of HUCB toward the lesioned area. In this study, MCP-1 and MIP-1α expression were significantly increased in the ischemic side of the rat's brain and significantly promoted HUCB cell migration compared with the contralateral side.[51] This cell migration dwindled in the presence of polyclonal antibodies against MCP-1 or MIP-1α. Chemokine receptors were also constitutively displayed on the surface of HUCB cells. These findings indicate that the increased chemokines in the ischemic site can bind surface receptors of HUCB cells and may induce migration of systemically delivered cells into the CNS.

As mentioned earlier, the HUCB cells are composed of mixed subpopulations such as immature T cells, B cells, monocytes/macrophages, and stem cells. Thus, it is necessary to reveal whether the beneficial effects of HUCB treatment can be attributed to a specific cell population. Recently, we demonstrated that the whole mononuclear fraction of cord blood, as well as stem cell-, T-cell-, and B-cell-depleted preparations improved the function of the impaired left forelimb to a similar extent that exceeded the performance of MCAO-only animals, whereas animals that received the monocyte/macrophage-depleted HUCB preparation performed more poorly than those receiving the other HUCB cells.[117] Additionally, HUCB administration significantly reduced MCAO-induced hyperactivity, while depletion of stem cells, monocytes ($CD14^+$), and B cells prevented this recovery following stroke.[117] Although an assessment of the infarct size and inflammatory responses await determination, monocytes/macrophages should prove to be critical to HUCB-induced recovery following stroke.

Determining the therapeutic window of cell transplantation for treatment of stroke is a major concern in research for the clinical application. Therefore, our group determined the optimal time to administer HUCB cells after stroke. Initially, using ischemic tissue extracts in an in vitro assay system, we studied the migration ability of mononuclear HUCB cells.[72] This assay revealed that the migratory activity of these cells toward both the hippocampal and striatal extracts increased when harvested 24–72 h after the stroke. Furthermore, the extracts had elevated levels of cytokine-induced neutrophil chemoattractant-1 (CINC-1) and MCP-1 at 48 h after MCAO, suggesting that these materials may be implicated in the cell migration. We also found that growth-regulated oncogene/CINC-1 (the rat equivalent of human IL-8) and MCP-1 were expressed in a time-dependent pattern similar to that observed in the migration assays on analysis of the ischemic extracts. The existence of chemokines in the supernatant may also lead to further exploration of the mechanisms responsible for the in vivo migration of mononuclear HUCB cells after stroke.[72] These results gave us hope that we may be able to extend to 24–72 h post stroke, the current 3 h therapeutic window available for approved thrombolytic treatment for stroke, by using mononuclear HUCB cell transplantation.

In a recent in vivo study, we revealed that the HUCB treatment window is also relatively narrow. When we intravenously administered HUCB cells at times ranging from 3 h to 30 days after MCAO, we observed maximal improvements with treatment at 48 h.[70] This means that cell transplantation in stroke might only be useful over a relatively short time frame to either sufficiently modulate the immune system or produce secretory factors to act on cells within the brain to exert therapeutic effects.[114]

In conclusion, stroke is a series of complicated cascades of inflammatory events that eventually lead to pronounced cell death adjacent to the blocked vasculature. This cascade is time-dependent, and when HUCB cells are administered intravenously 48 h following stroke onset, they may be able to reverse impending cell death. The administration of HUCB stem cells more likely provides neuroprotection, inhibits spreading of apoptosis, and modulates the immune/inflammatory response to injury both peripherally and locally, rather than acts to directly replace lost cells. Furthermore, neovascularization is a major process in inflammation and tissue repair. According to Savitz et al.,[89] generally transplantation does not succeed if there is a severe arterial occlusion without collateral circulation, which indicates that an adequate blood supply is integral to graft survival and tissue regeneration. With respect to the role of angiogenesis in transplant engraftment and tissue repair, there is an additional advantage in using HUCB cells for stroke since cord blood contains many $CD34^+$ cells including endothelial progenitor cells, which may be useful in proangiogenic neovascularization therapy.[33] Essentially, if present obstacles that apply to the use of HUCB cells in human stroke patients, such as uncertain safety, imperfect expansion processes, and limited graft survival in the recipient, can be overcome in the future, the transplantation of HUCB cells may become a promising new treatment modality with multiple therapeutic effects and a longer effective time frame in a single transplant that no other currently available pharmacological agent could mimic.

16.5 Disclosure statement

AEW and CVB are consultants for and PRS is cofounder and Chairman of Saneron CCEL Therapeutics, Inc. (SCTI, Tampa, FL, USA). SCTI is a University of South Florida start-up company, which is developing UCB-derived treatments for neurodegenerative and cardiovascular disorders. LEC is chairman of Cryopraxis Cryobiology Ltd, which is developing UCB-derived treatments.

References

1. Armitage JO. Bone marrow transplantation. *N Engl J Med.* 1994;330(12):827-838.
2. Arpinati M, Green CL, Heimfeld S, Heuser JE, Anasetti C. Granulocyte-colony stimulating factor mobilizes T helper 2-inducing dendritic cells. *Blood.* 2000;95(8):2484-2490.
3. Asseman C, Powrie F. Interleukin 10 is a growth factor for a population of regulatory T cells. *Gut.* 1998;42(2):157-158.
4. Babcock AA, Kuziel WA, Rivest S, Owens T. Chemokine expression by glial cells directs leukocytes to sites of axonal injury in the CNS. *J Neurosci.* 2003;23(21):7922-7930.
5. Baizabal JM, Furlan-Magaril M, Santa-Olalla J, Covarrubias L. Neural stem cells in development and regenerative medicine. *Arch Med Res.* 2003;34(6):572-588.
6. Behzad-Behbahani A, Pouransari R, Tabei SZ, et al. Risk of viral transmission via bone marrow progenitor cells versus umbilical cord blood hematopoietic stem cells in bone marrow transplantation. *Transplant Proc.* 2005;37(7):3211-3212.
7. Bernal GM, Peterson DA. Neural stem cells as therapeutic agents for age-related brain repair. *Aging Cell.* 2004;3(6): 345-351.
8. Bicknese AR, Goodwin HS, Quinn CO, Henderson VC, Chien SN, Wall DA. Human umbilical cord blood cells can be induced to express markers for neurons and glia. *Cell Transplant.* 2002;11(3):261-264.
9. Bjornson CR, Rietze RL, Reynolds BA, Magli MC, Vescovi AL. Turning brain into blood: a hematopoietic fate adopted by adult neural stem cells in vivo. *Science.* 1999;283(5401): 534-537.
10. Broxmeyer HE, Hangoc G, Cooper S, et al. Growth characteristics and expansion of human umbilical cord blood and estimation of its potential for transplantation in adults. *Proc Natl Acad Sci USA.* 1992;89(9):4109-4113.
11. Buelens C, Willems F, Delvaux A, et al. Interleukin-10 differentially regulates B7–1 (CD80) and B7–2 (CD86) expression on human peripheral blood dendritic cells. *Eur J Immunol.* 1995;25(9):2668-2672.
12. Buzanska L, Machaj EK, Zablocka B, Pojda Z, Domanska-Janik K. Human cord blood-derived cells attain neuronal and glial features in vitro. *J Cell Sci.* 2002;115(Pt 10):2131-2138.
13. Cameron HA, McKay R. Stem cells and neurogenesis in the adult brain. *Curr Opin Neurobiol.* 1998;8(5):677-680.
14. Cardoso AA, Li ML, Batard P, et al. Release from quiescence of CD34+ CD38– human umbilical cord blood cells reveals their potentiality to engraft adults. *Proc Natl Acad Sci USA.* 1993;90(18):8707-8711.
15. Chen J, Li Y, Wang L, et al. Therapeutic benefit of intravenous administration of bone marrow stromal cells after cerebral ischemia in rats. *Stroke.* 2001;32(4):1005-1011.
16. Chen J, Sanberg PR, Li Y, et al. Intravenous administration of human umbilical cord blood reduces behavioral deficits after stroke in rats. *Stroke.* 2001;32(11):2682-2688.
17. Chen J, Zhang ZG, Li Y, et al. Intravenous administration of human bone marrow stromal cells induces angiogenesis in the ischemic boundary zone after stroke in rats. *Circ Res.* 2003;92(6):692-699.
18. Chen N, Hudson JE, Walczak P, et al. Human umbilical cord blood progenitors: the potential of these hematopoietic cells to become neural. *Stem Cells.* 2005;23(10):1560-1570.
19. Chen N, Kamath S, Newcomb J, et al. Trophic factor induction of human umbilical cord blood cells in vitro and in vivo. *J Neural Eng.* 2007;4(2):130-145.
20. Chen SH, Chang FM, Tsai YC, Huang KF, Lin CL, Lin MT. Infusion of human umbilical cord blood cells protect against cerebral ischemia and damage during heatstroke in the rat. *Exp Neurol.* 2006;199(1):67-76.
21. Chen X, Li Y, Wang L, et al. Ischemic rat brain extracts induce human marrow stromal cell growth factor production. *Neuropathology.* 2002;22(4):275-279.
22. Chivu M, Diaconu CC, Bleotu C, Alexiu I, Brasoveanu L, Cernescu C. The comparison of different protocols for expansion of umbilical-cord blood hematopoietic stem cells. *J Cell Mol Med.* 2004;8(2):223-231.
23. Clemens JA. Cerebral ischemia: gene activation, neuronal injury, and the protective role of antioxidants. *Free Radic Biol Med.* 2000;28(10):1526-1531.
24. Conrad PD, Emerson SG. Ex vivo expansion of hematopoietic cells from umbilical cord blood for clinical transplantation. *J Leukoc Biol.* 1998;64(2):147-155.
25. D'Arena G, Musto P, Cascavilla N, et al. Flow cytometric characterization of human umbilical cord blood lymphocytes: immunophenotypic features. *Haematologica.* 1998;83(3):197-203.
26. Daar AS, Bhatt A, Court E, Singer PA. Stem cell research and transplantation: science leading ethics. *Transplant Proc.* 2004;36(8):2504-2506.
27. El-Badri NS, Hakki A, Saporta S, et al. Cord blood mesenchymal stem cells: potential use in neurological disorders. *Stem Cells Dev.* 2006;15(4):497-506.
28. El Marsafy S, Dosquet C, Coudert MC, Bensussan A, Carosella E, Gluckman E. Study of cord blood natural killer cell suppressor activity. *Eur J Haematol.* 2001;66(4):215-220.
29. Emerich DF, Dean III RL, Bartus RT. The role of leukocytes following cerebral ischemia: pathogenic variable or bystander reaction to emerging infarct? *Exp Neurol.* 2002; 173(1):168-181.
30. Ferrari G, Cusella-De Angelis G, Coletta M, et al. Muscle regeneration by bone marrow-derived myogenic progenitors. *Science.* 1998;279(5356):1528-1530.
31. Frassoni F, Podesta M, Maccario R, et al. Cord blood transplantation provides better reconstitution of hematopoietic reservoir compared with bone marrow transplantation. *Blood.* 2003;102(3):1138-1141.
32. Galli R, Borello U, Gritti A, et al. Skeletal myogenic potential of human and mouse neural stem cells. *Nat Neurosci.* 2000;3(10):986-991.
33. Garbuzova-Davis S, Willing AE, Saporta S, et al. Novel cell therapy approaches for brain repair. *Prog Brain Res.* 2006;157:207-222.
34. Garbuzova-Davis S, Willing AE, Zigova T, et al. Intravenous administration of human umbilical cord blood cells in a mouse model of amyotrophic lateral sclerosis: distribution, migration, and differentiation. *J Hematother Stem Cell Res.* 2003;12(3):255-270.

35. Garcia JH, Liu KF, Yoshida Y, Lian J, Chen S, del Zoppo GJ. Influx of leukocytes and platelets in an evolving brain infarct (Wistar rat). *Am J Pathol*. 1994;144(1):188-199.
36. Garderet L, Dulphy N, Douay C, et al. The umbilical cord blood alphabeta T-cell repertoire: characteristics of a polyclonal and naive but completely formed repertoire. *Blood*. 1998;91(1):340-346.
37. Glabinski AR, Tuohy VK, Ransohoff RM. Expression of chemokines RANTES, MIP-1alpha and GRO-alpha correlates with inflammation in acute experimental autoimmune encephalomyelitis. *Neuroimmunomodulation*. 1998; 5(3–4):166-171.
38. Gluckman E, Broxmeyer HA, Auerbach AD, et al. Hematopoietic reconstitution in a patient with Fanconi's anemia by means of umbilical-cord blood from an HLA-identical sibling. *N Engl J Med*. 1989;321(17):1174-1178.
39. Gluckman E, Rocha V. History of the clinical use of umbilical cord blood hematopoietic cells. *Cytotherapy*. 2005;7(3): 219-227.
40. Gluckman E, Rocha V, Boyer-Chammard A, et al. Outcome of cord-blood transplantation from related and unrelated donors. Eurocord Transplant Group and the European Blood and Marrow Transplantation Group. *N Engl J Med*. 1997; 337(6):373-381.
41. Goodwin HS, Bicknese AR, Chien SN, Bogucki BD, Quinn CO, Wall DA. Multilineage differentiation activity by cells isolated from umbilical cord blood: expression of bone, fat, and neural markers. *Biol Blood Marrow Transplant*. 2001;7(11):581-588.
42. Gritti A, Vescovi AL, Galli R. Adult neural stem cells: plasticity and developmental potential. *J Physiol Paris*. 2002; 96(1–2):81-90.
43. Ha Y, Choi JU, Yoon DH, et al. Neural phenotype expression of cultured human cord blood cells in vitro. *Neuroreport*. 2001;12(16):3523-3527.
44. Hao SG, Sun GL, Wu WL, Wu YL. Studies on the dynamics of biological characteristics of CD133+ cells from human umbilical cord blood during short-term culture. *Zhongguo Shi Yan Xue Ye Xue Za Zhi*. 2003;11(6):569-575.
45. Harris DT, Schumacher MJ, Locascio J, et al. Phenotypic and functional immaturity of human umbilical cord blood T lymphocytes. *Proc Natl Acad Sci USA*. 1992;89(21): 10006-10010.
46. Henon PR. Human embryonic or adult stem cells: an overview on ethics and perspectives for tissue engineering. *Adv Exp Med Biol*. 2003;534:27-45.
47. Hirose Y, Kiyoi H, Itoh K, Kato K, Saito H, Naoe T. B-cell precursors differentiated from cord blood CD34+ cells are more immature than those derived from granulocyte colony-stimulating factor-mobilized peripheral blood CD34+ cells. *Immunology*. 2001;104(4):410-417.
48. Hows JM, Bradley BA, Marsh JC, et al. Growth of human umbilical-cord blood in longterm haemopoietic cultures. *Lancet*. 1992;340(8811):73-76.
49. Jaroscak J, Goltry K, Smith A, et al. Augmentation of umbilical cord blood (UCB) transplantation with ex vivo-expanded UCB cells: results of a phase 1 trial using the AastromReplicell System. *Blood*. 2003;101(12):5061-5067.
50. Jeong JA, Gang EJ, Hong SH, et al. Rapid neural differentiation of human cord blood-derived mesenchymal stem cells. *Neuroreport*. 2004;15(11):1731-1734.
51. Jiang L, Newman M, Saporta S, et al. MIP-1alpha and MCP-1 induce migration of human umbilical cord blood cells in models of stroke. *Curr Neurovasc Res*. 2008;5(2):118-124.
52. Jurga M, Markiewicz I, Sarnowska A, et al. Neurogenic potential of human umbilical cord blood: neural-like stem cells depend on previous long-term culture conditions. *J Neurosci Res*. 2006;83(4):627-637.
53. Kernan NA, Bartsch G, Ash RC, et al. Analysis of 462 transplantations from unrelated donors facilitated by the National Marrow Donor Program. *N Engl J Med*. 1993;328(9): 593-602.
54. Knutsen AP, Wall DA. Kinetics of T-cell development of umbilical cord blood transplantation in severe T-cell immunodeficiency disorders. *J Allergy Clin Immunol*. 1999;103 (5 Pt 1):823-832.
55. Kobari L, Giarratana MC, Pflumio F, Izac B, Coulombel L, Douay L. CD133+ cell selection is an alternative to CD34+ cell selection for ex vivo expansion of hematopoietic stem cells. *J Hematother Stem Cell Res*. 2001;10(2):273-281.
56. Ladeby R, Wirenfeldt M, Garcia-Ovejero D, et al. Microglial cell population dynamics in the injured adult central nervous system. *Brain Res Brain Res Rev*. 2005;48(2):196-206.
57. Laughlin MJ, Barker J, Bambach B, et al. Hematopoietic engraftment and survival in adult recipients of umbilical-cord blood from unrelated donors. *N Engl J Med*. 2001; 344(24):1815-1822.
58. Laughlin MJ, Eapen M, Rubinstein P, et al. Outcomes after transplantation of cord blood or bone marrow from unrelated donors in adults with leukemia. *N Engl J Med*. 2004;351(22): 2265-2275.
59. Lee KD, Kuo TK, Whang-Peng J, et al. In vitro hepatic differentiation of human mesenchymal stem cells. *Hepatology*. 2004;40(6):1275-1284.
60. Liao SL, Chen WY, Raung SL, Kuo JS, Chen CJ. Association of immune responses and ischemic brain infarction in rat. *Neuroreport*. 2001;12(9):1943-1947.
61. Lie DC, Song H, Colamarino SA, Ming GL, Gage FH. Neurogenesis in the adult brain: new strategies for central nervous system diseases. *Annu Rev Pharmacol Toxicol*. 2004;44:399-421.
62. Liesveld JL. When what you have is not enough: optimizing cord blood transplantation in adults. *Leuk Res*. 2003;27(3): 197-199.
63. Liu Y, Liu T, Fan X, Ma X, Cui Z. Ex vivo expansion of hematopoietic stem cells derived from umbilical cord blood in rotating wall vessel. *J Biotechnol*. 2006;124(3):592-601.
64. Ma Y, Zou P, Xiao J, Huang S. The expression and functional characteristics of AC133 antigen in cord blood hematopoietic cells. *Zhonghua Nei Ke Za Zhi*. 2002;41(12):798-800.
65. Marler JR, Goldstein LB. Medicine. Stroke--tPA and the clinic. *Science*. 2003;301(5640):1677.
66. McGuckin CP, Forraz N, Allouard Q, Pettengell R. Umbilical cord blood stem cells can expand hematopoietic and neuroglial progenitors in vitro. *Exp Cell Res*. 2004;295(2):350-359.
67. Mikami T, Eguchi M, Kurosawa H, et al. Ultrastructural and cytochemical characterization of human cord blood cells. *Med Electron Microsc*. 2002;35(2):96-101.
68. Nakahata T, Ogawa M. Hemopoietic colony-forming cells in umbilical cord blood with extensive capability to generate mono- and multipotential hemopoietic progenitors. *J Clin Invest*. 1982;70(6):1324-1328.

69. Nayar B, Raju GM, Deka D. Hematopoietic stem/progenitor cell harvesting from umbilical cord blood. *Int J Gynaecol Obstet*. 2002;79(1):31-32.
70. Newcomb JD, Ajmo CT Jr, Sanberg CD, Sanberg PR, Pennypacker KR, Willing AE. Timing of cord blood treatment after experimental stroke determines therapeutic efficacy. *Cell Transplant*. 2006;15(3):213-223.
71. Newcomb JD, Sanberg PR, Klasko SK, Willing AE. Umbilical cord blood research: current and future perspectives. *Cell Transplant*. 2007;16(2):151-158.
72. Newman MB, Willing AE, Manresa JJ, Davis-Sanberg C, Sanberg PR. Stroke-induced migration of human umbilical cord blood cells: time course and cytokines. *Stem Cells Dev*. 2005;14(5):576-586.
73. Newman MB, Willing AE, Manresa JJ, Sanberg CD, Sanberg PR. Cytokines produced by cultured human umbilical cord blood (HUCB) cells: implications for brain repair. *Exp Neurol*. 2006;199(1):201-208.
74. NINDS. Tissue plasminogen activator for acute ischemic stroke. *N Engl J Med*. 1995;333(24):1581-1587.
75. Parent JM. Injury-induced neurogenesis in the adult mammalian brain. *Neuroscientist*. 2003;9(4):261-272.
76. Petersen BE, Bowen WC, Patrene KD, et al. Bone marrow as a potential source of hepatic oval cells. *Science*. 1999; 284(5417):1168-1170.
77. Pranke P, Failace RR, Allebrandt WF, Steibel G, Schmidt F, Nardi NB. Hematologic and immunophenotypic characterization of human umbilical cord blood. *Acta Haematol*. 2001;105(2):71-76.
78. Rainsford E, Reen DJ. Interleukin 10, produced in abundance by human newborn T cells, may be the regulator of increased tolerance associated with cord blood stem cell transplantation. *Br J Haematol*. 2002;116(3):702-709.
79. Ramsay NK, Davies S, Wagner J, McGough E, McGlave PB. Bone marrow transplantation. New strategies for treating malignant disease. *Minn Med*. 1996;79(4):23-28.
80. Riaz SS, Jauniaux E, Stern GM, Bradford HF. The controlled conversion of human neural progenitor cells derived from foetal ventral mesencephalon into dopaminergic neurons in vitro. *Brain Res Dev Brain Res*. 2002;136(1):27-34.
81. Robinson S, Niu T, de Lima M, et al. Ex vivo expansion of umbilical cord blood. *Cytotherapy*. 2005;7(3):243-250.
82. Rocha V, Cornish J, Sievers EL, et al. Comparison of outcomes of unrelated bone marrow and umbilical cord blood transplants in children with acute leukemia. *Blood*. 2001;97(10):2962-2971.
83. Rocha V, Wagner JE Jr, Sobocinski KA, et al. Graft-versus-host disease in children who have received a cord-blood or bone marrow transplant from an HLA-identical sibling. Eurocord and International Bone Marrow Transplant Registry Working Committee on Alternative Donor and Stem Cell Sources. *N Engl J Med*. 2000;342(25):1846-1854.
84. Rosamond W, Flegal K, Furie K, et al. Heart disease and stroke statistics – 2008 update: a report from the American Heart Association Statistics Committee and Stroke Statistics Subcommittee. *Circulation*. 2008;117(4):e25-e146.
85. Rubinstein P, Carrier C, Scaradavou A, et al. Outcomes among 562 recipients of placental-blood transplants from unrelated donors. *N Engl J Med*. 1998;339(22):1565-1577.
86. Sanchez-Ramos J, Song S, Cardozo-Pelaez F, et al. Adult bone marrow stromal cells differentiate into neural cells in vitro. *Exp Neurol*. 2000;164(2):247-256.
87. Sanchez-Ramos JR, Song S, Kamath SG, et al. Expression of neural markers in human umbilical cord blood. *Exp Neurol*. 2001;171(1):109-115.
88. Sathananthan AH, Trounson A. Human embryonic stem cells and their spontaneous differentiation. *Ital J Anat Embryol*. 2005;110(2 Suppl 1):151-157.
89. Savitz SI, Dinsmore JH, Wechsler LR, Rosenbaum DM, Caplan LR. Cell therapy for stroke. *NeuroRx*. 2004; 1(4):406-414.
90. Shih CC, DiGiusto D, Forman SJ. Ex vivo expansion of transplantable human hematopoietic stem cells: where do we stand in the year 2000? *J Hematother Stem Cell Res*. 2000;9(5):621-628.
91. Shirvaikar N, Reca R, Jalili A, et al. CFU-megakaryocytic progenitors expanded ex vivo from cord blood maintain their in vitro homing potential and express matrix metalloproteinases. *Cytotherapy*. 2008;10(2):182-192.
92. Sirchia G, Rebulla P. Placental/umbilical cord blood transplantation. *Haematologica*. 1999;84(8):738-747.
93. Slusher BS, Vornov JJ, Thomas AG, et al. Selective inhibition of NAALADase, which converts NAAG to glutamate, reduces ischemic brain injury. *Nat Med*. 1999;5(12):1396-1402.
94. Streit WJ, Walter SA, Pennell NA. Reactive microgliosis. *Prog Neurobiol*. 1999;57(6):563-581.
95. Taguchi A, Soma T, Tanaka H, et al. Administration of CD34+ cells after stroke enhances neurogenesis via angiogenesis in a mouse model. *J Clin Invest*. 2004;114(3): 330-338.
96. Tamaki S, Eckert K, He D, et al. Engraftment of sorted/expanded human central nervous system stem cells from fetal brain. *J Neurosci Res*. 2002;69(6):976-986.
97. Tan J, Town T, Paris D, et al. Microglial activation resulting from CD40-CD40L interaction after beta-amyloid stimulation. *Science*. 1999;286(5448):2352-2355.
98. Thomson BG, Robertson KA, Gowan D, et al. Analysis of engraftment, graft-versus-host disease, and immune recovery following unrelated donor cord blood transplantation. *Blood*. 2000;96(8):2703-2711.
99. Todaro AM, Pafumi C, Pernicone G, et al. Haematopoietic progenitors from umbilical cord blood. *Blood Purif*. 2000; 18(2):144-147.
100. Traycoff CM, Abboud MR, Laver J, Clapp DW, Srour EF. Rapid exit from G0/G1 phases of cell cycle in response to stem cell factor confers on umbilical cord blood CD34+ cells an enhanced ex vivo expansion potential. *Exp Hematol*. 1994;22(13):1264-1272.
101. Traycoff CM, Kosak ST, Grigsby S, Srour EF. Evaluation of ex vivo expansion potential of cord blood and bone marrow hematopoietic progenitor cells using cell tracking and limiting dilution analysis. *Blood*. 1995;85(8):2059-2068.
102. Tse W, Laughlin MJ. Umbilical cord blood transplantation: a new alternative option. *Hematol Am Soc Hematol Educ Prog*. 2005;2005:377-383.
103. Uchida N, Buck DW, He D, et al. Direct isolation of human central nervous system stem cells. *Proc Natl Acad Sci USA*. 2000;97(26):14720-14725.

104. Vaziri H, Dragowska W, Allsopp RC, Thomas TE, Harley CB, Lansdorp PM. Evidence for a mitotic clock in human hematopoietic stem cells: loss of telomeric DNA with age. *Proc Natl Acad Sci USA*. 1994;91(21):9857-9860.
105. Vendrame M, Cassady J, Newcomb J, et al. Infusion of human umbilical cord blood cells in a rat model of stroke dose-dependently rescues behavioral deficits and reduces infarct volume. *Stroke*. 2004;35(10):2390-2395.
106. Vendrame M, Gemma C, de Mesquita D, et al. Anti-inflammatory effects of human cord blood cells in a rat model of stroke. *Stem Cells Dev*. 2005;14(5):595-604.
107. Vendrame M, Gemma C, Pennypacker KR, et al. Cord blood rescues stroke-induced changes in splenocyte phenotype and function. *Exp Neurol*. 2006;199(1):191-200.
108. Wagner JE, Barker JN, DeFor TE, et al. Transplantation of unrelated donor umbilical cord blood in 102 patients with malignant and nonmalignant diseases: influence of CD34 cell dose and HLA disparity on treatment-related mortality and survival. *Blood*. 2002;100(5):1611-1618.
109. Wagner JE, Broxmeyer HE, Byrd RL, et al. Transplantation of umbilical cord blood after myeloablative therapy: analysis of engraftment. *Blood*. 1992;79(7):1874-1881.
110. Walczak P, Chen N, Eve D, et al. Long-term cultured human umbilical cord neural-like cells transplanted into the striatum of NOD SCID mice. *Brain Res Bull*. 2007; 74(1–3):155-163.
111. Walczak P, Chen N, Hudson JE, et al. Do hematopoietic cells exposed to a neurogenic environment mimic properties of endogenous neural precursors? *J Neurosci Res*. 2004;76(2):244-254.
112. Wei Y, Huang Y, Zhang Y, et al. Ex vivo expansion of CD34(+) and T and NK cells from umbilical cord blood for leukemic BALB/C nude mouse transplantation. *Int J Hematol*. 2008;87(2):217-224.
113. Weiner HL, Selkoe DJ. Inflammation and therapeutic vaccination in CNS diseases. *Nature*. 2002;420(6917): 879-884.
114. Willing AE, Eve DJ, Sanberg PR. Umbilical cord blood transfusions for prevention of progressive brain injury and induction of neural recovery: an immunological perspective. *Regen Med*. 2007;2(4):457-464.
115. Willing AE, Lixian J, Milliken M, et al. Intravenous versus intrastriatal cord blood administration in a rodent model of stroke. *J Neurosci Res*. 2003;73(3):296-307.
116. Willing AE, Vendrame M, Mallery J, et al. Mobilized peripheral blood cells administered intravenously produce functional recovery in stroke. *Cell Transplant*. 2003; 12(4):449-454.
117. Womble TA, Green S, Sanberg PR, Pennypacker KR, Willing AE. $CD14^+$ Human umbilical cord blood cells are essential for neurological recovery following MCAO. *Cell Transplant*. 2008;17(4):485-486.
118. Xiao J, Nan Z, Motooka Y, Low WC. Transplantation of a novel cell line population of umbilical cord blood stem cells ameliorates neurological deficits associated with ischemic brain injury. *Stem Cells Dev*. 2005;14(6):722-733.
119. Yang SE, Ha CW, Jung M, et al. Mesenchymal stem/progenitor cells developed in cultures from UC blood. *Cytotherapy*. 2004;6(5):476-486.
120. Yao CL, Feng YH, Lin XZ, Chu IM, Hsieh TB, Hwang SM. Characterization of serum-free ex vivo-expanded hematopoietic stem cells derived from human umbilical cord blood CD133(+) cells. *Stem Cells Dev*. 2006;15(1):70-78.
121. Yin AH, Miraglia S, Zanjani ED, et al. AC133, a novel marker for human hematopoietic stem and progenitor cells. *Blood*. 1997;90(12):5002-5012.
122. Zhang RL, Chopp M, Chen H, Garcia JH. Temporal profile of ischemic tissue damage, neutrophil response, and vascular plugging following permanent and transient (2H) middle cerebral artery occlusion in the rat. *J Neurol Sci*. 1994; 125(1):3-10.
123. Zigova T, Song S, Willing AE, et al. Human umbilical cord blood cells express neural antigens after transplantation into the developing rat brain. *Cell Transplant*. 2002;11(3):265-274.
124. Zola H, Fusco M, Macardle PJ, Flego L, Roberton D. Expression of cytokine receptors by human cord blood lymphocytes: comparison with adult blood lymphocytes. *Pediatr Res*. 1995;38(3):397-403.

17 Placental Umbilical Cord Blood Transfusion for Stem Cell Therapy in Neurological Diseases

Abhijit Chaudhuri and Niranjan Bhattacharya

17.1 Introduction

Human nervous system is a highly differentiated and specialized tissue that minimizes its potential for regeneration and self-renewal in response to injury. Conventional attempt at the repair of adult brain and spinal cord by pharmacological therapy has been largely unsuccessful and had very limited clinical impact. Yet developmental and acquired neurological disabilities contribute to significant socio-economic burden in any community.[1] In the past decade, possible use of stem cells and neural progenitor cells as a potential therapy in neurological diseases has been explored (Table 17.1), with variable success reported in a very few patients in the early studies.

17.2 Stem Cell Therapy in Neurological Diseases

Because of the functionally restricted capacity of the developed adult nervous system for self-renewal and regeneration, stem cells can potentially repair and replace injured nervous system and restore function. From experimental studies, it is known that stem cells can migrate to the area of injury after transplantation into human brain. Engrafted stem cells have been shown to function at least at a physiological level by releasing or responding to neurohormones and neurochemicals. There is also evidence that stem cells may have possible tropism ("homing") for the site of injury in the nervous system even when administered at a different area or intravenously.

Embryonic stem cells have traditionally been considered to be the best candidates for therapy in neurodegenerative diseases. These cells, which are derived from the embryonic blastula, are highly pluripotent but, because of their ability to proliferate rapidly, carry the risk of teratoma in the transplanted tissue. Human embryonic stem cells have also been isolated and grown in cultures and, when transplanted, can successfully migrate and transdifferentiate in a site-specific fashion. However, there have been few successes in clinical trials so far. An alternative source of stem cells is bone marrow, which has the advantage of autologous transplantation. Bone marrow contains both hematopoietic stem cells, which can differentiate into microglia, astrocytes and possibly neurons[2]; and mesenchymal cells, which can migrate, and in vitro studies express cell surface markers for astrocytes, oligodendrocytes and neurons.[3]

For disease-specific, cell replacement therapy (Table 17.2), cultured embryonic neural stem cells have been evaluated as possible candidates. Neural stem cells are derived from the embryonic neuroepithelium, but isolation of specific cells from in vitro culture is often difficult because of the mixed population of expanded neural precursor cells. The task is further complicated by the fact that expanded neural precursor cells do not proliferate indefinitely, and their differentiation into candidate replacement cells requires careful manipulation of culture conditions, often yielding relatively small number of neurons making this approach highly time consuming, expensive and limited in availability for therapeutic use even if found to

A. Chaudhuri (✉)
Essex Centre for Neurological Sciences, Romford,
Queen's Hospital, Rom Valley Way,
Romford RM7 0AG, UK
e-mail: chaudhuria@gmail.com

N. Bhattacharya and P. Stubblefield (eds.), *Regenerative Medicine Using Pregnancy-Specific Biological Substances*,
DOI: 10.1007/978-1-84882-718-9_17,

Table 17.1 Current approach on stem cell-based therapy in neurological diseases

Source of stem cells	Neurological conditions	Present use	Comments
Adult neural precursor cells	Parkinson's disease Ischemic stroke	Human Experimental	Unlikely to be effective on its own
Olfactory ensheathing cells	Brachial plexus injury	Human	Limited in application
Embryonic stem cells	Parkinson's disease Spinal cord injury Motor neuron disease	Human Experimental	Future trials likely
Fetal stem cells	Parkinson's disease Huntington's disease	Human	Likely to be limited in applicability due to ethical reasons
Cultured stem cells	Retinal disease Parkinson's disease	Human	Limited in application
Autologous hematopoietic or mesenchymal stem cells	Multiple sclerosis Ischemic stroke Motor neurone disease	Human	Future trials likely

Table 17.2 Cell Replacement Therapy (CRT) in neurological diseases

Neurological diseases	Assumed disease mechanism	Candidates for CRT
Parkinson's disease	Dopamine deficiency in nigrostriatal neurons	Dopaminergic cells
Multiple sclerosis	Inflammatory injury to Oligodendrocytes	Oligodendrocyte progenitor cells, Schwann cells
Motor neuron disease	Loss of spinal anterior horn cells	Alpha motor neurons
Huntington's disease	Loss of striatal spiny neurons	GAB Aergic striatal projection neurons
Hemispheric stroke	Cortical neurons	Neuronal cells

be clinically effective. A novel attempt has been made to immortalize neuronally restricted progenitor cells derived from fetal tissue with telomerase by inducing overexpression of human telomerase reverse transcriptase through a viral vector[4] to maintain long-term proliferative capacity without replicative senescence restricting cell division (the so-called Hayflick limit).

In the developed brain, a small population of new neurons can differentiate from adult neural precursor cells that are primarily found in the subependymal layer of the ventricular zone and dentate gyrus of hippocampus. However, mobilization of adult neural precursor cells into sites of brain injury is difficult. However, autologous grafts of adult neural precursor cells are a therapeutic possibility. The success of this approach, previously attempted in Parkinson's disease and experimental models of stroke, very much depends on the assumption that the neural precursor cells are themselves not involved in the disease process, which is not likely to be the case in neurodegenerative conditions such as Alzheimer's disease or HIV-encephalopathy. Consequently, stem cell therapy with adult neural precursor cells has limited potential of success. Although neurogenesis may occur from neural stem cells in the subventricular zone,[5] adequate blood supply or angiogenesis will be a prerequisite for the survival and development of migrating cells into new neurons and establish functional connection. Failure of differentiation of the transplanted cells and of resident precursor cells at the site of injury in the human brain is probably also contributed by unfavorable microenvironment at the site of injury.

17.3 Umbilical Cord Blood: An Alternative Source of Stem Cells

Placental umbilical cord blood provides an alternative source of stem cells, which have been discussed in detail elsewhere in this book. Broadly, there have been three approaches to the use of placental umbilical cord blood in clinical practice. In the first approach, HLA-matched, hematopoietic stem cells have been isolated from umbilical cord blood and invested in patients. This has been well established as a treatment for several disorders during past 20 years since the first report of using HLA-matched, sibling umbilical cord blood cells to reconstitute a child with severe Fanconi anemia. Indeed, using cryopreserved, HLA-compatible, umbilical cord blood in preference to bone marrow may be justified in cases with metabolic diseases, severe combined immunodeficiency, aplastic anemia and high risk leukemia or myelodysplastic syndromes.[6] Human umbilical cord blood is an alternative source of hematopoietic stem cell transplantation and would be suitable as a therapy in certain neurological disorders where autologous bone marrow transplantation is considered a suitable therapy.

The second approach uses umbilical cord blood derived non-hematopoietic stem cells for neural transplants. Placental umbilical cord blood offers a viable source of stem cells for neurogenesis comparable to embryonic and fetal tissue-based cell therapy. In addition, umbilical cord blood has a more undifferentiated stem cell population compared with adult hematopoietic tissue in the bone marrow. These primitive stem cells, which include mesenchymal stem cells, are multi-potent and have the ability to transdifferentiate into multiple lineages, including neural tissue. Exposure of cord blood cells to basic fibroblast growth factor and human epidermal growth factor in culture was found to induce expression of neural and glial markers.[7] Glial fibrillary acidic protein and neuron-specific neural protein were expressed by cord blood cells 10 days after culture with brain-derived neurotrophic factor.[8] Mesenchymal stem cells in human umbilical cord blood rapidly transdifferentiate into neural cell lines.[9] Indeed, the morphology of cells from umbilical cord blood resembles that of mesenchymal stem cells from bone marrow.[10] The cell renewal capacity of these cells is remarkable. Under conditions favorable for neurogenesis, umbilical cord blood cells are known to exhibit the morphology and express genes specific for neuroglial cells. In addition, these differentiated cells blood express functional voltage-gated ion channels and exhibit resting membrane potential characteristic of neuronal cells.[10] In another study, human umbilical cord derived hematopoietic stem cells transdifferentiated into neurons and glia after induction with retinoic acid.[11]

Another approach is to use human placental umbilical cord whole blood as a transfusion therapy. Human placental umbilical cord transfusion has already been shown to be therapeutically effective in a range of medical diseases[12,13] and considered to be a suitable candidate for the treatment of acute ischemic stroke.[14] An important property of placental umbilical cord blood is its anti-inflammatory effect. Inflammatory changes may contribute to clinical deterioration in acute ischemic stroke and multiple sclerosis and probably plays a role in rapidly progressive neurodegenerative disorders. Umbilical venous blood has a high concentration of interleukin-1 receptor antagonist (IL-ra), especially in pre-term and in normal-term deliveries.[15] IL-ra is a potent anti-inflammatory cytokine, and there is experimental evidence that intravenous administration of umbilical cord blood cells in animals with middle cerebral artery reduced mRNA and protein expression of pro-inflammatory cytokines.[16] There is also early evidence that umbilical cord blood therapy may promote neovascularization and reduce excitotoxic cell damage, both of which are of therapeutic advantage in acute ischemic stroke and neurodegenerative disorders. Transplantation of endothelial progenitor cells from cord blood was associated with an increased number of microvessels in a rat model of diabetic neuropathy[17] and infusion of human umbilical cord blood cells protected against cerebral ischemia and excitotoxic injury during heat stroke in rats.[18] Human placental umbilical cord blood transfusion is likely to be more cost-effective as a therapy than cord blood or stem cell transplant and reduces the need of myeloablative therapy and the risk of graft-versus-host disease that may occasionally complicate cord blood transplantation.

In summary, human umbilical cord blood transfusion offers a unique therapeutic option in neurology because of its cellular constituents of hematopoietic and mesenchymal stem cells with potential to migrate, transdifferentiate and repair injured nervous system, and its chemical constituents with potential to modify several major pathophysiological pathways (ischemia, inflammation, excitotoxicity and loss of neuronal function) that have been implicated in neurological disorders. Unlike site-specific stem cell therapy from embryonic, fetal or cultured neuronal transplants, umbilical cord blood will be therapeutically effective after peripheral, intravenous administration. Selective stem cell infusions from pooled blood do not appear to have any significant additional benefit because progenitor stem cells from the transfused umbilical cord blood can enter the central nervous system relatively easily in areas of breakdown of blood–brain barrier.

17.4 Therapeutic Uses of Human Placental Umbilical Cord Blood Transfusion in Neurology

Human placental umbilical cord blood transfusion may be a potentially effective therapy for a number of common and disabling neurological diseases (Table 17.3). Placental umbilical cord blood transfusion is potentially an effective therapy for acute ischemic stroke and probably the only treatment that can promote repair of ischemic brain and aid early functional recovery.[14] An added functional advantage of umbilical cord blood in reperfusing ischemic brain is its high concentration of fetal hemoglobin (Hb F), which has greater oxygen binding capacity than normal adult hemoglobin (Hb A). This has been shown to be of considerable therapeutic importance in sickle cell disease and hemoglobinopathy and, in patients with clinical stroke, has the potential for improving oxygenation in ischemic tissue. Hb F will deliver more oxygen to the surviving neurons in the ischemic core and ischemic penumbra in areas of partial blood flow. The rheological property of term cord blood is also favorable for reperfusion because of lower viscosity.

Table 17.3 Potential uses of placental umbilical cord blood transfusion in neurological diseases

Neurodevelopmental diseases
Optic nerve hypoplasia
Genetic diseases, e.g., neurometabolic disorders
Acute
Hemispheric ischemic stroke
Disabling relapses of multiple sclerosis
Traumatic spinal cord injury
Traumatic brachial plexus injury
Chronic
Alzheimer's disease
Parkinson's disease
HIV encephalopathy
Motor neuron disease
Multiple sclerosis
Multiple system atrophy
Myotonic dystrophy
Traumatic brain injury
Hereditary and diabetic neuropathies

Actively relapsing multiple sclerosis is characterized by acute perivenous demyelination with breakdown of blood–brain barrier. In a recently published study, autologous, non-myeloablative hematopoietic stem cell transplantation was considered to be effective in reversing neurological deficit in relapsing-remitting multiple sclerosis.[19] Human umbilical cord blood transfusion is likely to be an effective option in relapsing-remitting multiple sclerosis because of its anti-inflammatory effect and its population of hematopoietic and mesenchymal stem cells with the potential to repair oligodendroglial and neuronal injury and without risking the side effects of immunosuppressive therapy, which is required as a conditioning treatment for autologous bone marrow transplantation. Acute traumatic injuries to brachial plexus and spinal cord are likely examples where transfusion of placental umbilical cord blood would be of potential value because of its anti-inflammatory and regenerative properties.

HIV encephalopathy (HIV-associated dementia or AIDS-dementia complex) is the most common neurological complication of HIV infection and is not fully reversible by antiretroviral therapy. One of the postulated mechanisms of neurodegeneration that contributes to progressive motor and cognitive deficits in HIV encephalopathy is the inhibition of adult neural precursor cells in the dentate gyrus of hippocampus by HIV-envelope glycoprotein, gp-120.[20] Placental umbilical cord blood transfusion in combination with antiretroviral therapy has the potential to reverse the pathological process and promote brain repair.

Human umbilical placental cord blood transfusion may have a potential role in developmental brain disorders presenting early in early life, and the widely reported success of umbilical cord stem cell transplant in a young girl with blindness due to septo-optic dysplasia[21] offers hope to many others.

For chronic neurological diseases such as progressive multiple sclerosis, Parkinson's disease or motor neuron disease, neurodegeneration due to excitotoxic injury is considered to be the final common pathogenic pathway contributing to functional deficit. For reasons already discussed, long-term treatment with placental umbilical cord blood transfusions (once every few weeks or so) may have the potential to reduce the rate of neurodegeneration and disease progression; clearly, ethically approved controlled clinical trials are necessary to confirm the hypothesis. However, when compared with alternative methods of stem cell delivery, human placental umbilical cord blood transfusion is relative safe, uncomplicated and cost-effective. In addition, hematopoietic and mesenchymal stem cell population in the umbilical cord blood are stable, likely to migrate easily at the site of injury through blood–brain barrier and engraft early. In addition, transfusion therapy requires no additional immunosuppressive and/or myeloablative treatment indicated for stem cell transplants, and the side-effects (similar to mismatched transfusion) are well known and avoidable (Table 17.4).

From a logistic point of view, provision of a national bank of human placental umbilical cord blood suitable for transfusion in neurological patients will require voluntary donation from mothers, and the collaboration of the National Blood Service for human placental umbilical cord blood processing and storage. Cord blood will require being HLA and ABO/Rh blood group typed and stored before transfusion to the national standards already in place for adult donor blood units. Once a suitable recipient is identified after ABO/Rh blood group match and 50% or greater HLA haplotype match, the cord blood unit will be released for transfusion by the processing and storage laboratory. As any of the proposed clinical trials is effectively an HLA and fully cross-matched blood transfusion study, regulatory requirements for Phase I and Phase II clinical trials could easily be met by a suitably designed pilot study. Infection, anaphylaxis and possible transfusion reactions from ABO-incompatibility are likely to be the most common risks of umbilical cord blood transfusion, but these are similar to the adverse events associated with adult blood products in current practice.

Table 17.4 Potential advantages of human umbilical cord blood transfusion as a vehicle for stem cell therapy

• Contains both hematopoietic and mesenchymal stem cells capable of differentiation into multiple lineages
• Stability of constituents in appropriate storage condition
• Non-invasive
• Easy to administer
• Treatment can be repeated as often as needed
• No additional need for immunosuppression or myeloablative therapy
• Safe
• Known complications (mismatched transfusion) that are predictable and can be prevented
• High cost-effectiveness
• Likelihood of functional benefit

17.5 Conclusion

Chronic neurological disorders, developmental as well as acquired, impose significant disability on sufferers and carries high socio-economic burden. For many of these conditions, the only available therapy is rehabilitative or supportive, and the quality of life experienced by severely disabled patients is often rated worse than death. There is a growing debate in many countries on the right of such patients to choose assisted suicide (euthanasia) as the preferred option, which is a sad reflection on the current state of disease-specific therapy. At the same time, there is emerging consensus among most clinicians and researchers that stem cell therapy is an alternative that requires to be explored seriously in neurological diseases. Yet the opportunity to test widely available and relatively easily usable stem cells from placental umbilical cord blood is being missed and will continue to be wasted in favor of more technology driven and industry sponsored research, which may benefit only a favorable few. Human placental umbilical cord blood, which is still regarded as medical waste, can be collected, stored and transfused as part of the established protocol of national blood bank system and transfusion service in any country without any major resource implications. Placental umbilical cord blood transfusion can be used without risking the ethical objections of embryonic and fetal stem cell therapy, and it has the potential to reduce the burden of disability in many brain diseases. Clearly, human placental umbilical cord blood transfusion in selected neurological diseases should become an urgent priority for clinical research.

References

1. Olesen J, Leonardi M. The burden of brain disease in Europe. *Eur J Neurol.* 2003;10:471-477.
2. Mezey E, Chandross KJ. Bone marrow: a possible alternative source of cells in the adult nervous system. *Eur J Pharmacol.* 2000;405:297-302.

3. Priller J, Persons DA, Kleitt FF, et al. Neurogenesis of cerebellar Purkinje neurons from gene-marked bone marrow cells in vivo. *J Cell Biol*. 2001;155:733-738.
4. Roy NS, Nakano T, Keyoung HM, et al. Telomerase immortalization of neuronally restricted progenitor cells derived from the human fetal spinal cord. *Nat Biotechnol*. 2004;22:297-305.
5. Sanai N, Tramontin AD, Quinones-Hinojosa A, et al. Unique astrocyte ribbon in adult human brain contains neural stem cells but lack chain migration. *Nature*. 2004;427:479-494.
6. Cairo MS, Wagner JE. Placental and/or umbilical cord blood: an alternative source of haematopoietic stem cells for transplantation. *Blood*. 1997;90:4665-4678.
7. Bicknese AR, Goodwin HS, Quinn CO, et al. Human umbilical cord blood cells can be induced to express markers for neurons and glia. *Cell Transplant*. 2002;11:261-264.
8. Zhao ZM, Lu SH, Zhang QJ, et al. The preliminary study on in vitro differentiation of human umbilical cord blood cells into neural cells. *Zhonghua Xue Ye Xue Za Zhi*. 2003;24:484-487.
9. Fu YS, Shih YT, Cheng YC, Min MY. Transformation of human umbilical mesenchymal cells into neurons in vitro. *J Biomed Sci*. 2004;11:652-660.
10. Lee OK, Kuo TK, Wei-Ming Chen, et al. Isolation of multipotent mesenchymal stem cells from umbilical cord blood. *Blood*. 2004;103:1669-1675.
11. Jang YK, Park JJ, Lee MC, et al. Retinoic acid-mediated induction of neurons and glia from human umbilical cord-derived haematopoietic stem cells. *J Neurosci Res*. 2004;75:573-584.
12. Bhattacharya N, Mukherjee K, Chettri MK, et al. A study report of 174 units of placental umbilical whole cord blood transfusion in 62 patients as a rich source of fetal haemoglobin therapy in different indications of blood transfusion. *Clin Exp Obstet Gynaecol*. 2001;28:47-52.
13. Bhattacharya N. Placental umbilical cord blood transfusion: a new method for treatment of patients with diabetes and microalbuminuria in the background of anemia. *Clin Exp Obstet Gynaecol*. 2006;33:164-168.
14. Chaudhuri A, Hollands P, Bhattacharya N. Placental umbilical cord blood transfusion in acute ischaemic stroke. *Med Hypotheses*. 2007;69:1267-1271.
15. Fukuda H, Masuzaki H, Ishimaru T. Interleukin-6 and interleukin-1 receptor antagonist in amniotic fluid and cord blood in patients with pre-term, premature rupture of the membranes. *Int J Gynaecol Obstet*. 2002;77:123-129.
16. Vendrame M, Gemma C, de Mesquita D, et al. Anti-inflammatory effects of human cord blood cells in a rat model of stroke. *Stem Cells Dev*. 2005;14:595-604.
17. Naruse K, Hamada Y, Nakashima E, et al. Therapeutic neovascularization using cord blood-derived endothelial progenitor cells for diabetic neuropathy. *Diabetes*. 2005;54:1823-1828.
18. Chen SH, Chang FM, Tsai YC, et al. Infusion of human umbilical cord blood protects against cerebral ischaemia and damage during heatstroke in the rat. *Exp Neurol*. 2006;199:67-76.
19. Burt RK, Loh Y, Cohen B, et al. Autologous non-myeloablative haemopoietic stem cell transplantation in relapsing remitting multiple sclerosis: a phase I/II study. *Lancet Neurol*. 2009;8:244-253.
20. Okamoto S, Kang Y-J, Brechtel CW, et al. HIV/gp120 decreases adult neural progenitor cell proliferation via checkpoint kinase-mediated cell cycle withdrawal and G1 arrest. *Cell Stem Cell*. 2007;1:230-236.
21. Britten N. Stem cell doctors bring a blind girl into the light. The Daily Telegraph March 5, 2009;7.

Part VII

Use of Placental Umbilical Cord Blood Serum in Ophthalmology

18 Umbilical Cord and Its Blood: A Perspective on Its Current and Potential Use in Ophthalmology

Kyung-Chul Yoon

18.1 Introduction

Tears play an essential role in maintaining the health of ocular surface including corneal and conjunctival epithelia. Tear components such as epidermal growth factor (EGF) and vitamin A have an important effect on the regulation of proliferation, differentiation, and maturation of the ocular surface epithelium. Severe dry eye syndrome or keratoconjunctivitis sicca, the major ocular surface disease, results in surface changes characterized by squamous metaplasia and by changes in the quantity and quality of the tear film. Conventional treatments for ocular surface diseases include the application of artificial tears, topical corticosteroids or cyclosporin A, therapeutic contact lenses, protective goggles or anterior chamber glasses, punctal occlusion, and tarsorrhaphy. However, despite these therapies, many patients complain of persistent symptoms and continue to show signs associated with ocular surface changes. The additive application of substances that enhance epithelial proliferation and differentiation can be helpful in the treatment of these patients. Traditionally, autologous blood had been used in the treatment of various ocular surface disorders, because it harbors growth factors and essential tear components. Recently, we first found that umbilical cord blood also contains essential tear component, growth factors, and neurotrophic factors and can be used in many ocular surface diseases, including dry eye syndrome, persistent epithelial defects, and neurotrophic keratitis.

K.-C. Yoon
Department of Ophthalmology, Chonnam National University Medical School and Hospital, 8 Hak-Dong, Dong-Gu, Gwang-Ju 501-757, South Korea
e-mail: kcyoon@jnu.ac.kr

18.2 Rationale of Serum Therapy in Ophthalmology

Serum has been known to be effective in many ocular surface disorders. Biomechanical and biochemical properties of serum are similar to normal tears. The rationale of serum therapy is that serum can supply important tear components and provide an ocular surface with basic nutrients for epithelial renewal, which are lacking in artificial tears. Serum contains many growth factors and tear components which are EGF, vitamin A, transforming growth factor (TGF)-β, acidic and basic fibroblast growth factor (aFGF, bFGF), platelet-derived growth factor (PDGF), hepatocyte growth factor (HGF), fibronectin, and serum antiproteases such as α_2 macroglobulin, and can facilitate the proliferation, migration, and differentiation of the ocular surface epithelium. EGF suppresses apoptosis and may accelerate the proliferation of corneal epithelial cells, vitamin A can suppress the progression to squamous metaplasia in keratoconjunctivitis sicca, while TGF-β controls the proliferation of corneal epithelial cells and maintains cells in undifferentiated conditions. Fibronectin also accelerates epithelial cell migration and anchorage, and α_2 macroglobulin plays a role in the suppression of collagenase in the cornea. Serum also harbors neurotrophic substances such as substance P, insulin-like growth factor (IGF-1), and nerve growth factor (NGF) and can be useful for the restoration of the ocular surface integrity in patients with neurotrophic keratitis. The oil existing in serum can act as a substitute for the lipid components produced by the meibomian glands, and prealbumin contributes to the stability of the tear film. Therefore, eye drops produced by serum may be able to provide both lubrication and nutrition for ocular surface epithelium.

N. Bhattacharya and P. Stubblefield (eds.), *Regenerative Medicine Using Pregnancy-Specific Biological Substances*,
DOI: 10.1007/978-1-84882-718-9_18,

In an animal model of post-LASIK (laser in situ keratomileusis) corneal epithelial defects, 20% serum eye drops induced faster epithelial healing than artificial tears, resulting in (1) a decrease in keratocyte apoptosis and migration of fibroblasts and myofibroblasts in the wound site, (2) a decrease in the migration of inflammatory cells, and (3) the consequent inhibition of cytokine release.[1]

18.3 History of Serum Use in Ophthalmology

Peripheral blood serum has higher vitamin A, TGF-β1, IGF-1, NGF, fibronectin and lysozyme concentrations and lower immunoglobulin A, EFG, and vitamin C concentrations than tears.[2] Clinically, autologous serum eye drops have been effectively applied to treat severe dry eye in Sjögren's syndrome or graft-versus-host disease (GVHD), persistent epithelial defects, neurotrophic keratopathy, superior limbic keratoconjunctivitis, and recurrent corneal erosion.

Several studies have been reported on the efficacy, stability, and safety of autologous serum eye drops for the treatment of dry eye syndrome. Fox et al.[3] initially reported the beneficial effect of autologous serum application to dry eye in Sjögren's syndrome and the advantage of serum over artificial tears. Tsubota et al.[4,5] used 20% autologous serum in severe dry eye associated with Sjögren's syndrome and in the reconstruction of the ocular surface in ocular pemphigoid or Stevens-Johnson syndrome. Rose bengal and fluorescein staining scores (ocular surface epitheliopathy score) were improved after 4 weeks of serum treatment in 12 patients, and mucin expression in cultured conjunctival epithelial cells was significantly upregulated. Ogawa et al.[6] used 20% autologous serum to treat severe dry eye in patients with chronic GVHD. Controlled studies in dry eye or severe ocular surface disease showed that autologous serum treatment led to better improvement in symptoms and signs of dry eye, compared with conventional treatment.[7-9] Noble et al.[7] reported that symptom scores and impression cytologic findings showed significant improvement with autologous serum treatment. Kojima et al.[8] reported that symp tom scores, break-up time, and rose bengal and fluorescein staining scores significantly improved with autologous serum compared with preservative-free artificial tears.

Autologous serum eye drops have been used in the treatment of persistent corneal epithelial defects and neurotrophic keratitis. Tsubota et al.[10] reported that, following application of 20% autologous serum in 16 patients, healing was observed within 2 weeks in 43.8%, and within 1 month in 62.5%. Poon et al.[11] used 50% or 100% autologous serum and reported that healing was achieved in an average of 29 days in 60.0%, and that after serum use was discontinued, epithelial defect recurred in 55.6%. They also used autologous serum eye drops in four patients with neurotrophic keratitis and achieved a successful result in two patients. Young et al.[12] reported that a corneal epithelial defect healed within 1 month in two of three patients treated with 20% autologous serum eye drops. Matsumoto et al.[13] also used 20% autologous serum in 11 patients with neurotrophic keratitis and reported that the epithelial disorders healed completely in all eyes within an average of 17.1 days after treatment. Both visual acuity and corneal sensitivity improved significantly after treatment.

Autologous serum have been used as an adjuvant to macular hole surgery.[14] Serum treatment can improve ocular surface epitheliopathy and subjective symptoms in superior limbic keratoconjunctivitis, and reduce the number of recurrence in recurrent corneal erosion.[15,16] It can be also used for corneal epithelial abrasions in diabetic patients undergoing vitrectomy and late-onset aqueous oozing through filtering bleb after trabeculectomy.[17,18]

18.4 Introduction of Umbilical Cord Serum in Ophthalmology

Although autologous serum therapy is beneficial for many ocular disorders, it has several disadvantages. Autologous serum therapy requires repeated blood collection from patients to obtain fresh serum, which may lead to discomfort or treatment refusal. Furthermore, blood sampling is difficult to be taken in patients with poor general condition or blood dyscrasia, especially hemotologic malignancy.

Fetal bovine serum is widely used in the laboratory to promote cell growth in culture. It is known to accelerate the migration of corneal cells in vitro and to modulate the clonal growth and differentiation of cultured limbal and corneal epithelium. An initial study showed that umbilical cord serum led to faster epithelial healing in the cornea compared to autologous serum.[19] If umbilical cord serum also contains essential tear components and growth factors, the use of umbilical cord serum in place of autologous serum may show greater promise in the treatment of various ophthalmologic diseases. Under the assumption, we analyzed many tear factors in umbilical cord serum and tried to use umbilical cord serum eye drops in patients with many ocular surface diseases. We have verified that umbilical cord serum contains a high concentration of many growth and neural factors and essential tear components, such as EGF, vitamin A, TGF-β, substance P, IGF-1, and NGF, and that umbilical cord serum eye drops can be used safely and effectively for the treatment of severe dry eye with or without Sjögren's syndrome, GVHD, persistent epithelial defects, and neurotrophic keratopathy.[20-24] The levels of EGF, vitamin A, and TGF-β in umbilical cord serum and normal peripheral blood serum were investigated.[21] EGF and TGF-β concentrations in umbilical cord serum were three and two times higher than those in peripheral blood serum, respectively. Although vitamin A concentration in umbilical cord serum was lower than that in peripheral blood serum, it was higher compared with that in tears and may be sufficient to prevent squamous metaplasia. We also analyzed and compared the concentrations of substance P, IGF-1, and NGF in umbilical cord serum with those in normal peripheral blood serum and tears.[23] Substance P, IGF-1, and NGF concentrations in umbilical cord serum were much higher than those in tears. Umbilical cord serum contained higher NGF and lower IGF-1 levels compared with levels in peripheral blood serum. Although statistically insignificant, substance P concentrations were also higher in umbilical cord serum than peripheral blood serum (Table 18.1).

Umbilical cord serum therapy has several advantages compared with autologous serum therapy. Because a large amount of serum can be obtained from the umbilical vein at one time, many patients can benefit from this sampling without waiting for additional preparations. In addition, umbilical cord serum therapy is also feasible in patients with a poor general condition or blood dyscrasia.

18.5 Preparation of Umbilical Cord Serum Eye Drops

Umbilical cord blood has been obtained from mothers with vaginal or cesarean section delivery after obtaining informed consent. Laboratory examination for hepatitis B and C virus and human immunodeficiency virus (HIV) in the donor should be performed twice at the first and third trimesters. A volume of 200–250 ml of the umbilical cord blood can be collected from the umbilical vein after fetal delivery. Umbilical cord blood should be kept for 2 h at room temperature to be clotted. After centrifugation at 3,000 g for 15 min, the serum is isolated carefully under sterile conditions in a laminar air flow hood. Then, the serum is diluted to a 20% concentration with balanced salt solution (BSS). The process of the serum preparation, such as clotting time, centrifugation, and diluents, has an effect on the composition and epitheliotrophic effects of serum. A long clotting time, extensive and high-speed centrifugation, and dilution with BSS can improve the ability of serum eye drops, supporting proliferation, migration, and differentiation of the ocular surface epithelium.[28] The diluted serum is aliquoted into sterile 5-ml bottles with ultraviolet light protection. Patients are instructed to keep an opened bottle in a refrigerator at 4°C and to store unopened eye drop preparation bottles in

Table 18.1 Comparison of components in umbilical cord serum, peripheral blood serum, and tears[20,22,25–27]

	Umbilical cord serum	Peripheral blood serum	Tears
EGF	0.48 ± 0.09 ng/mL	0.14 ± 0.03 ng/mL	1.9–9.7 ng/mL
TGF-β	57.14 ± 18.98 ng/mL	31.30 ± 12.86 ng/mL	2–10 ng/mL
Vitamin A	230.85 ± 13.39 ng/mL	372.34 ± 22.32 ng/mL	0.4–10.6 ng/mL
Substance P	245.3 ± 53.9 pg/mL	169.5 ± 81.0 pg/mL	69.8 ± 24.9 pg/mL
IGF-1	239.0 ± 77.1 ng/mL	375.5 ± 51.3 ng/mL	75.7 ± 50.5 ng/mL
NGF	729.7 ± 72.0 pg/mL	401.7 ± 98.1 pg/mL	107.5 ± 70.9 pg/mL

EGF epidermal growth factor, *TGF-β* transforming growth factor β, *IGF-1* insulinlike growth factor 1, *NFG* nerve growth factor a freezer at −20°C. Each opened bottle is discontinued after 1 week of use. The duration of maximum storage is 3–6 months.

Umbilical cord serum eye drops are applied six to ten times a day as required in addition to the preservative-free artificial tears. Antibiotics eye drops are instilled two times a day simultaneously with the serum at 5-min intervals.

18.6 Safety and Stability

In peripheral blood serum, EGF, vitamin A, and TGF-β are well preserved for up to 1 month in a refrigerator at 4°C and 3 months in a freezer at −20°C.[10] Serum eye drops can be safely used under a strict protocol for preparation and storage. Serum has a bacteriostatic effect because it contains antibacterial agents such as Ig G, lysozyme, and complement. In addition, serum contains no preservatives, and thus serum therapy can avoid the risk of preservative toxicity. In vitro toxicity testing demonstrated that serum eye drops have reduced toxicity compared with unpreserved hypromellose.[11]

18.7 Application of Umbilical Cord Serum Eye Drops in Ophthalmology

18.7.1 Persistent Corneal Epithelial Defects

Persistent corneal epithelial defects are defined as a failure of the epithelium to regrow over a defect within the expected time course. They are caused by medications, chemical or thermal injury, postherpetic infection, autoimmune diseases such as rheumatoid arthritis, ocular cicatrical pemphigoid, and erythema multiforme, neurologic diseases, tear film anomalies, and metabolic disturbances. Persistent epithelial defects may progress through the anterior stroma and lead to subsequent stromal ulceration, resulting in significant ocular morbidity and visual loss.

The use of umbilical cord serum eye drops is effective for the treatment of persistent corneal epithelial defects. In a randomized controlled clinical trial, 30 eyes were treated with umbilical cord serum and 29 eyes with autologous serum.[19] 58.1% in the cord serum group and 37.9% in the autologous serum group healed by therapy. The median percentage decrease in the size of the epithelial defect was significantly greater in the cord serum group. In our previous study, 14 patients (14 eyes) with persistent epithelial defects that had persisted for at least 2 weeks despite conventional treatment were treated with 20% umbilical cord serum eye drops six times a day.[20] The mean duration of epithelial defects before treatment was 7.2 ± 6.3 weeks, and the mean defect area was 7.86 ± 7.32 mm^2. The epithelial defects healed within 2 weeks in 42.9% and between 2 and 4 weeks in 42.9%. Two eyes (14.2%), one of which had chemical burn and the other had herpetic neurotrophic keratopathy, healed between 4 and 8 weeks. Mean healing time in effective or partially effective cases was 2.75 ± 1.06 weeks.

18.7.2 Dry Eye Syndrome

We demonstrated the effect of umbilical cord serum eye drops in 31 patients (55 eyes) with severe dry eye syndrome who were refractory to conventional treatments and had symptoms of dry eye for more than 3 months, low tear film break up time (BUT, <5 s), low Schirmer test (5 mm), and positive fluoresecin or rose bengal vital staining (≥3).[21] Twenty patients (38 eyes) of them had Sjögren's syndrome. Among 11 patients (17 eyes) who didn't have Sjögren's syndrome, 6 patients (8 eyes) had chemical burn, 3 patients (6 eyes) had Stevens-Johnson syndrome, 1 patient (2 eyes) had cicatrical pemphigoid, and 1 patient (1 eye) had herpetic keratitis. Symptom score, tear film BUT, and keratoepitheliopathy score improved significantly at 1 and 2 months after treatment. There was no statistically significant change in Schirmer and corneal sensitivity test results. In impression cytologic analysis, the grade of conjunctival squamous metaplasia and goblet cell density improved significantly at 2 months after treatment (Fig. 18.1). Moreover, most patients were satisfied with umbilical cord serum therapy and wanted to continue use of the serum. There was no growth in bacterial and fungal cultures of the serum. No significant complications associated with the use of umbilical cord serum were observed.

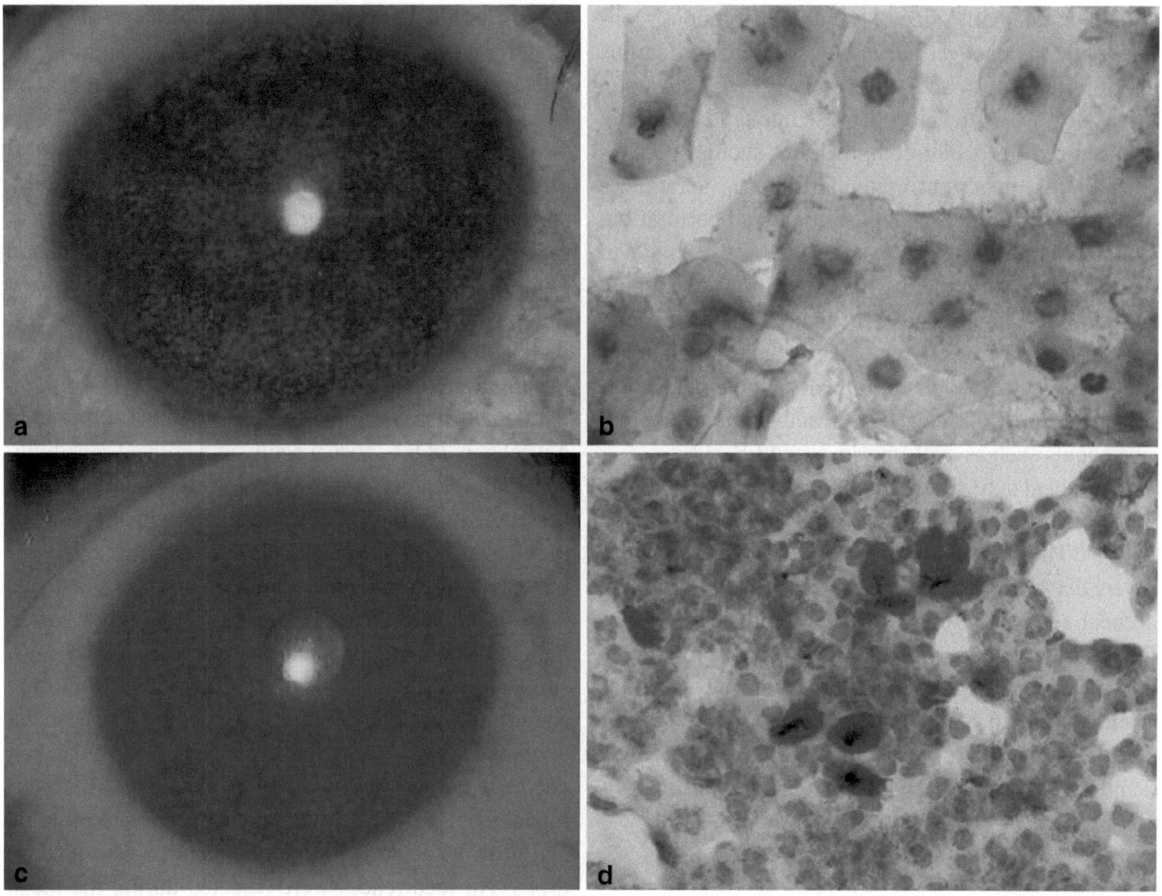

Fig. 18.1 Fluorescein staining and impression cytologic finding in dry eye syndrome. (**a**) Before umbilical cord serum treatment, the corneal epithelium was diffusely stained by fluorescein. (**b**) Conjunctival impression cytology in this patient showed a loss of goblet cells and large, polygonal epithelial cells with a nucleocytoplasmic ratio of 1:6 (PAS, × 400). (**c**) Two months after umbilical cord serum therapy, the corneal epithelium was markedly improved. (**d**) Conjunctival impression cytology after treatment showed many PAS positive goblet cells and small, round epithelial cells with a nucleocytoplasmic ratio of 1:2 (PAS, × 400)

A prospective case control study comparing the therapeutic effect between autologous serum and umbilical cord serum eye drops in the treatment of severe dry eye syndrome was performed.[24] Ninety-two eyes of 48 patients with severe dry eye syndrome (34 eyes of 17 patients with Sjögren's syndrome and 58 eyes of 31 patients with non-Sjögren's syndrome) were treated with either 20% autologous serum (41 eyes of 21 patients) or umbilical cord serum eye drops (51 eyes of 27 patients). Both autologous serum and umbilical serum treatments led to improvement in the symptom score, tear film BUT, keratoepitheliopathy score, and impression cytologic findings. However, symptom and keratoepitheliopathy scores were lower after1 and 2 months of umbilical cord serum treatment compared with autologous serum treatment. In patients with Sjögren's syndrome, goblet cell density was higher after 2 months of umbilical cord serum treatment compared with autologous serum treatment. Therefore, umbilical cord serum eye drops were found to be more effective in decreasing symptoms and keratoepitheliopathy in severe dry eye syndrome and increasing goblet cell density in Sjögren's syndrome compared with autologous serum eye drops.

18.7.3 Ocular Complications of GVHD

GVHD is one of the major complications after allogeneic hematopoietic stem cell transplantation. Sixty percent to 90% of patients with acute or chronic GVHD suffer from various ocular diseases, such as dry eye or keratoconjunctivitis sicca, pseudomembranous conjunctivitis, cataract, retinal vasculopathy, central serous chorioretinopathy, and posterior scleritis. Among these, dry eye is developed most frequently and can result in serious ocular complications such as punctuate keratitis, persistent epithelial defects, keratinization of the cornea, corneal ulceration, and perforation. Ocular complications secondary to GVHD is known to be difficult to be treated with conventional treatment including the application of artificial tears, therapeutic contact lenses, protective goggles or anterior chamber glasses, punctal occlusion, topical retinoic acid, topical or systemic corticosteroids, and immunosuppressive agents such as cyclosporin A and FK 506.

Umbilical cord serum treatment is an effective way to treat severe ocular surface manifestations associated with GVHD. In a study of 12 patients (24 eyes) with severe dry eye associated with GVHD, symptom score, corneal sensitivity, tear film BUT, and keratoepitheliopathy score improved significantly after 2 months of umbilical cord treatment, and the improvement was maintained by 6 months after treatment.[22] No significant differences were existed in the parameters between 2 and 6 months after treatment. No significant complications associated with serum use were observed.

18.7.4 Neurotrophic Keratitis

Neurotrophic keratitis is a degenerative corneal disease characterized by impaired healing of the corneal epithelium due to damage of trigeminal corneal innervation. The depletion of trophic mediators such as acetylcholine and substance P as well as an aqueous tear deficiency has been reported to contribute to the pathogenesis of this disease. It is caused by herpes simplex and herpes zoster infection of the ocular surface, chemical, physical, and surgical injuries, neurosurgical procedures such as acoustic neuroma, meningioma, and aneurysm, and systemic disease such as diabetes, multiple sclerosis, and leprosy. The progression of this disease can lead to a corneal ulcer, melting, and perforation. The treatment of neurotrophic keraitis is challenging. Conventional treatments, including the application of therapeutic contact lenses, artificial tears, cyanoacrylate glue, conjunctival flap, tarsorrhaphy, and amniotic membrane transplantation, have therapeutic limitation. The topical application of neurotrophic substances such as substance P, IGF-1, NGF has been tried to promote corneal wound healing.[29-32] However, these substances may induce side effects such as contact hypersensitivity, conjunctival hyperemia, photophobia, and ocular pain.

We have recently found that umbilical cord serum is effective to treat neurotrophic keratitis that was unresponsive to conventional treatment.[23] Twenty-eight eyes (28 patients) with neurotrophic keratitis were treated with 20% umbilical cord serum eye drops six to ten times a day, after 2 weeks of a washout period. The epithelial defect healed completely in all eyes with a mean healing time of 4.4 ± 4.0 weeks (Fig. 18.2). The epithelial defect healed within 2 weeks in 28.6% and within 4 weeks in 78.6%. Visual acuity and corneal sensitivity were significantly improved after treatment. Corneal sensitivity was improved in all patients with herpes simplex keratitis, 71.4% of patients with diabetes, 66.7% of patients with trigeminal neuralgia, 40.0% of patients with stroke, and 20.0% of patients with herpes zoster. Patients who had leprosy and who underwent a neurosurgical procedure, irradiation, or keratoplasty did not show improvement in corneal sensitivity. The corneal epithelial disease resolved even in patients who had very low corneal sensitivity after treatment. This finding may be due to the continuous supply of neurotrophic and growth factors with the umbilical cord serum eye drops. No recurrences were detected during the mean follow-up duration of 7.9 months.

18.7.5 Miscellaneous

Eye drops from umbilical cord serum can be also used in ocular complications associated with Stevens-Johnson syndrome and ocular cicatrical pemphigoid, ocular surface keratinizatin with extensive meibomian gland loss associated with lid tattooing, recurrent corneal erosions, Mooren's ulceration, and epithelial maintenance after refractive surgery or ocular surface reconstruction procedures.

18.8 Complications and Considerations

Although no significant complications have been reported after the use of umbilical cord serum eye drops, one should keep in mind the possibility of

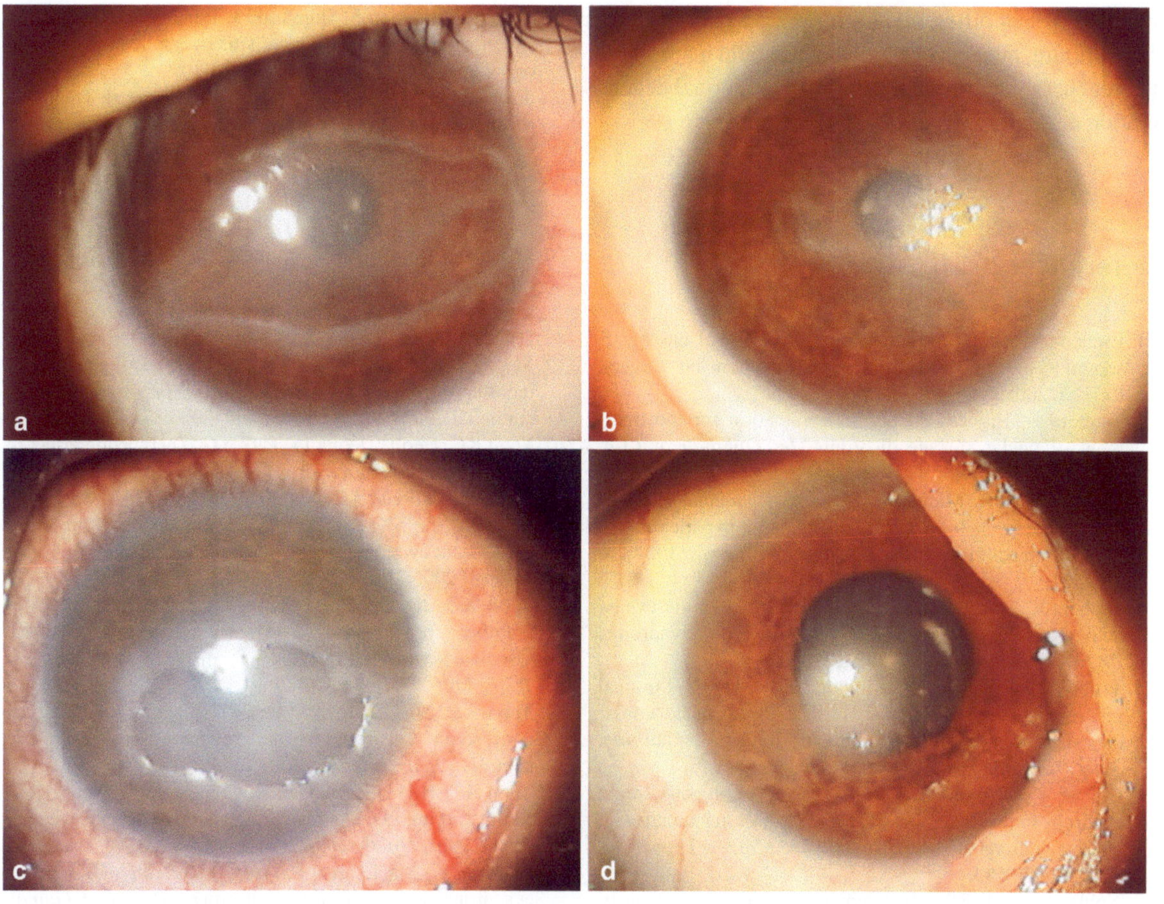

Fig. 18.2 Slit lamp photographs in patients with neurotrophic keratitis. (**a** and **b**) Moderate-staged neutrophic keratitis before (**a**) and after (**b**) treatment with umbilical cord serum eye drops. (**c** and **d**) Advanced-staged neutrophic keratitis before (**c**) and after (**d**) treatment with umbilical cord serum eye drops. The corneal epithelial defect healed with a significant decrease in corneal haze

several problems. Despite two laboratory examinations in pregnant donors, the possibility of transmission of blood-borne infectious disease or blood-borne diseases which at present cannot be detected cannot be absolutely ruled out. Additional HIV testing, with a shortened window period, with the p24 antigen detection method as well as routine laboratory examination for the virus should be performed on all pregnant donors. Additional possible problems include freezer storage, risk of allergies, and the potential risk of bacterial contamination. In a study of patients with hematopoietic progenitor cell transplantation, 6 of 11 autologous serum tears samples were contaminated after 30 days of use.[33] In consideration of possibility of serum contamination after longer periods of use, the serum eye drops should be replaced frequently. As the preparation of umbilical cord serum eye drops is time consuming and labor intensive, the costs should be considered. Complex political, legal, and regulatory issues should be also considered in advance of the serum use.[25]

18.9 Future Application

Mesenchymal stromal cells can be isolated not only from bone marrow, but also from umbilical cord blood, adipose tissue, and amniotic fluid. Human umbilical cord blood contains mesenchymal progenitor cells as well as hematopoietic progenitor cells.[34] Corneal stromal cells have mesenchymal stromal cell-like characteristics.[35] Therefore, umbilical cord blood is expected to be used in corneal tissue engineering and regeneration. In addition, human umbilical cord blood cells can differentiate into retinal nerve cells.[36] In the future, human cord blood cells could be used for the regeneration of retinal nerve cells in retinal degeneration or dystrophy.

References

1. Esquenazi S, He J, Bazab HEP, Bazan NG. Use of autologous serum in corneal epithelial defects post-lamellar surgery. *Cornea*. 2005;24:992-997.
2. Tsubota K, Higuchi A. Serum application for the treatment of ocular surface disorders. *Int Ophthalmol Clin*. 2000;40:113-122.
3. Fox RI, Chan R, Michelson JB, et al. Beneficial effect of artificial tears made with autologous serum in patients with keratoconjunctivitis sicca. *Arthritis Rheum*. 1984;27: 459-461.
4. Tsubota K, Goto E, Fujita H, et al. Treatment of dry eye by autologous serum application in Sjogren's syndrome. *Br J Ophthalmol*. 1999;83:390-395.
5. Tsubota K, Satake Y, Ohyama M, et al. Surgical reconstruction of the ocular surface in advanced ocular cicatrical pemphigoid and Steven-Johnson syndrome. *Am J Ophthalmol*. 1996;122:38-52.
6. Ogawa Y, Okamoto S, Mori T, et al. Autologous serum eyedrops for the treatment of severe dry eye in patients with chronic graft-versus-host disease. *Bone Marrow Transplant*. 2003;31:579-583.
7. Noble BA, Loh RS, MacLennan S, et al. Comparison of autologous serum eye drops with conventional therapy in a randomized controlled crossover trial for ocular surface disease. *Br J Ophthalmol*. 2004;88:647-652.
8. Kojima T, Ishida R, Dogru M, et al. The effect of autologous serum eyedrops in the treatment of severe dry eye disease: a prospective randomized case-control study. *Am J Ophthalmol*. 2005;139:242-246.
9. Tananuvat N, Daniell M, Suillivan LJ, et al. Controlled study of the use of autologous serum in dry eye patients. *Cornea*. 2001;20:802-806.
10. Tsubota K, Goto E, Shimmura S, et al. Treatment of persistent epithelial defect by autologous serum application. *Ophthalmology*. 1999;106:1984-1989.
11. Poon AC, Geerling G, Dart JK, et al. Autologous serum eyedrops for dry eyes and epithelial defects: clinical and in vitro toxicity studies. *Br J Ophthalmol*. 2001;85: 1188-1197.
12. Young AL, Cheng AC, Ng HK, et al. The use of autologous serum tears in persistent epithelial defect. *Eye*. 2004;18:609-614.
13. Matsumoto Y, Dogru M, Goto E, et al. Autologous serum application in the treatment of neurotrophic keratopathy. *Ophthalmology*. 2004;111:1115-1120.
14. Banker AS, Freeman WR, Azen SP, et al. A multicentered clinical study of serum as adjuvant therapy for surgical treatment of macular holes. *Arch Ophthalmol*. 1999;117: 1499-1502.
15. Goto E, Shimmura S, Shimazaki J, Tsubota K. Treatment of superior limbic keratoconjunctivitis by application of autologous serum. *Cornea*. 2001;20:807-810.
16. del Castillo JM, de la Casa JM, Sardina RC, et al. Treatment of recurrent corneal erosions using autologous serum. *Cornea*. 2002;21:781-783.
17. Schulze S, Sekundo W, Kroll P. Autologous serum for the treatment of corneal epithelial abrasions in diabetic patients undergoing vitrectomy. *Am J Ophthalmol*. 2006;142:207-211.
18. Matsuo H, Tomidokoro A, Tomita G, Araie M. Topical application of autologous serum for the treatment of late-onset aqueous oozing or point-leak through filtering bleb. *Eye*. 2005;19:23-28.
19. Vajpayee RB, Mukerji N, Tandon R, et al. Evaluation of umbilical cord serum therapy for persistent corneal epithelial defects. *Br J Ophthalmol*. 2003;87:1312-1316.
20. Yoon KC, Heo H, Jeong IY, Park YG. Therapeutic effect of umbilical cord serum eydrops for persistent corneal epithelial defect. *Korean J Ophthalmol*. 2005;19:174-178.
21. Yoon KC, Im SK, Park YG, et al. Application of umbilical cord serum eyedrops for the treatment of dry eye syndrome. *Cornea*. 2006;25:268-272.
22. Yoon KC, Jeong IY, Im SK, et al. Therapeutic effect of umbilical cord serum eyedrops for the treatment of dry eye associated with graft-versus-host disease. *Bone Marrow Transplant*. 2007;39:231-235.
23. Yoon KC, You IC, Im SK, et al. Application of umbilical cord serum eyedrops for the treatment of neurotrophic keratitis. *Ophthalmology*. 2007;114:1637-1642.
24. Yoon KC, Heo H, Im SK, et al. Comparison of autologous serum and umbilical cord serum eye drops for dry eye syndrome. *Am J Ophthalmol*. 2007;144:86-92.
25. Geerling G, MacLennan S, Hartwig D. Autologous serum eye drops for ocular surface disorders. *Br J Ophthalmol*. 2004;88:1467-1474.
26. Speek AJ, van Agtmaal EJ, Saowakontha S, et al. Fluorometric determination of retinol in human tear fluid using high-performance liquid chromatography. *Curr Eye Res*. 1986;5:841-845.
27. Ohashi Y, Motokura M, Kinoshita Y, et al. Presence of epidermal growth factor in human tears. *Invest Ophthalmol Vis Sci*. 1989;30:1879-1882.
28. Liu L, Hartwig D, Harloff S, et al. An optimised protocol for the production of autologous serum eyedrops. *Graefes Arch Clin Exp Ophthalmol*. 2005;243:706-714.
29. Brown SM, Lamberts DW, Reid TW, et al. Neurotrophic and anhidrotic keratopathy treated with substance P and insulin-like growth factor 1 [letter]. *Arch Ophthalmol*. 1997;115: 926-927.
30. Chikama T, Fukuda K, Morishige N, Nishida T. Treatment of neurotrophic keratopathy with substance-P-derived peptide (FGLM) and insulin-like growth factor I [letter]. *Lancet*. 1998;351:1783-1784.
31. Tan MH, Bryars J, Moore J. Use of nerve growth factor to treat congenital neurotrophic corneal ulceration. *Cornea*. 2006;25:352-355.
32. Lambiase A, Rama P, Bonini S, et al. Topical treatment with nerve growth factor for corneal neurotrophic ulcers. *New Eng J Med*. 1998;338:1174-1180.
33. Leite SC, de Castro RS, Alves M, et al. Risk factors and characteristics of ocular complications, and efficacy of

autologous serum tears after haematopoietic progenitor cell transplantation. *Bone Marrow Transplant*. 2006;38: 223-227.

34. Erices A, Conget P, Minguell JJ. Mesenchymal progenitor cells in human umbilical cord blood. *Br J Haematol*. 2000;109:235-242.
35. Choong PF, Mok PL, Cheong SK, Then KY. Mesenchymal stem cell-like characteristics of corneal keratocytes. *Cytotherapy*. 2007;9:252-258.
36. Kioke-Kiriyama N, Adachi Y, Minamino K, et al. Human cord blood cells can differentiate into retinal nerve cells. *Acta Neurobiol Exp (Wars)*. 2007;67:359-365.

Part VIII

Use of Placental Umbilical Cord in Cardiovascular Surgery

Umbilical Vein Grafts for Lower Limb Revascularization

19

Alan Dardik and Herbert Dardik

The *autologous saphenous* vein is the standard against which the performance of all other vascular grafts is measured. However, even autologous saphenous vein is far from ideal. It may be absent or diseased or too small or too short, and time is required for procurement and preparation. Finally, after implantation as an arterial conduit, it may undergo various forms of biodegradation. Therefore, an alternative vascular conduit is still needed when revascularization is mandatory but the saphenous vein is absent or inadequate, for any reason.

For decades, alternatives to the autologous saphenous vein have been studied by surgeons, engineers, and textile personnel. Some of these alternatives have proved more useful than others, although still others have been discarded. For example, unmodified heterografts are no longer used because of accelerated biodegradation. Unmodified allograft saphenous veins enjoyed brief popularity, but despite apparent low immunogenicity, most were ultimately rejected. Perhaps, pretreatment of allografts or even xenografts with chemical agents or freezing techniques might result in biologic materials that retain function. Clinical experience with aldehyde-processing has, in fact, provided unique insights into the usefulness and challenges of biologic materials deployed as vascular conduits.

A. Dardik (✉)
From the Sections of Vascular Surgery, Yale University School of Medicine, New Haven, CT and
Englewood Hospital and Medical Center, Englewood, NJ and Vascular Biology and Therapeutics, Yale University School of Medicine, 10 Amistad Street, Room 437, PO Box 208089, New Haven, CT 06520-8089, USA
e-mail: alan.dardik@yale.edu

Human umbilical cords are approximately 50 cm. in length and normally contain one vein and two arteries in a mucopolysaccharide matrix called "Wharton's jelly" (Fig. 19.1). At birth, the vessels are collapsed, but the vein can easily be dilated up to 7 mm in diameter and the arteries can be dilated up to 4 mm. Roentgenographic studies have shown that the vessels are of uniform diameter. They have no branches, valves, or vasa vasorum. Manometric studies in vitro have shown that these vessels can tolerate pressures in excess of 600 mmHg.

Initially, we implanted unmodified segments of human umbilical cord veins in the aorta of baboons.[1] Although early patency was achieved, predictable rejection occurred within several weeks of implantation (Fig. 19.2a, b). On gross histologic examination, aneurysm formation and thrombosis were present; on microscopy, necrosis, microabscesses, and macrophage and plasma cell infiltration were seen. Previous attempts to use unmodified umbilical cord vessels had met with similar failure.[2-4]

Inspired by the pioneering work of Carpentier[5] and Rosenberg[6] and their coworkers, we studied the effects of both dialdehyde starch (DS) and glutaraldehyde tanning on umbilical cord vessels prior to their implantation as vascular conduits. Unlike the previous failed attempts that occurred predictably without aldehyde processing, success was now routinely achieved, first in the laboratory (Fig. 19.2c) and later in a small pilot clinical study.[7,8] The superiority of glutaraldehyde as a tanning agent compared with DS was apparent histologically and was confirmed by immunologic studies (Fig. 19.3).[9]

In addition, chemical and physical analyzes of the interface, including internal reflection spectroscopy and contact angle determinations, were used to characterize the intimal surface and to compare it with that of

N. Bhattacharya and P. Stubblefield (eds.), *Regenerative Medicine Using Pregnancy-Specific Biological Substances*,
DOI: 10.1007/978-1-84882-718-9_19,

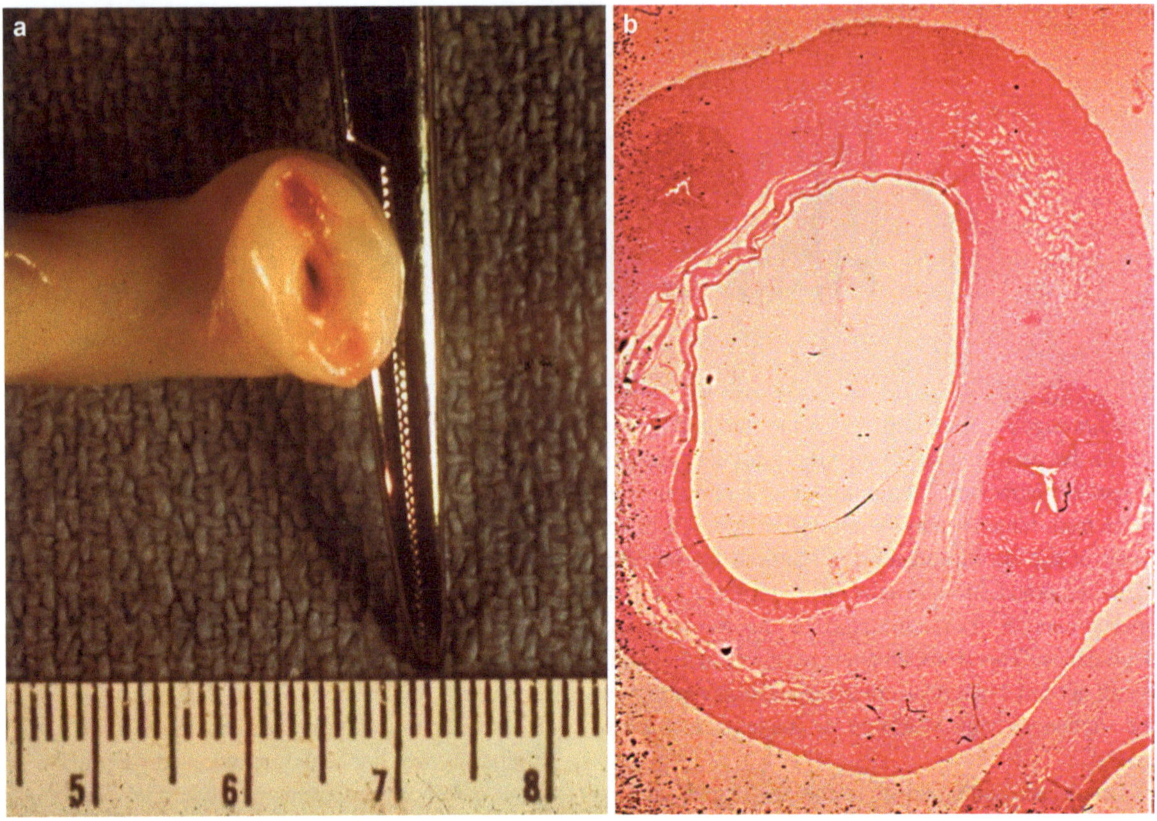

Fig. 19.1 (**a**) Umbilical cord obtained at delivery. (**b**) Larger umbilical vein at center and two smaller umbilical arteries are visualized clearly by microscopy

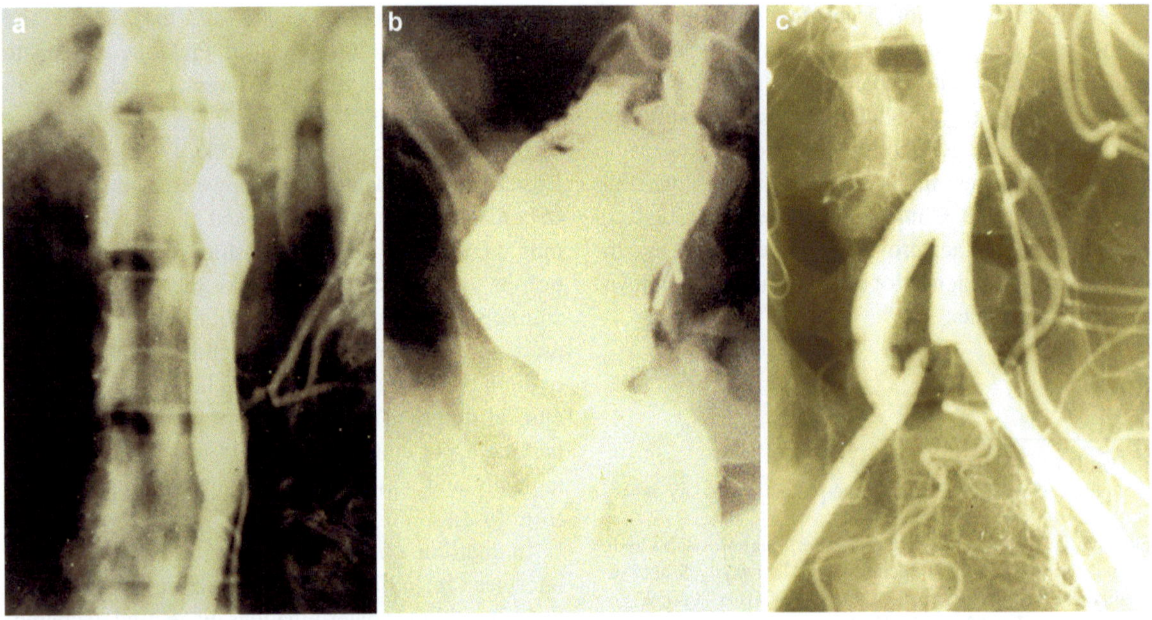

Fig. 19.2 (**a**) Unmodified umbilical cord implanted as an interposition graft in the baboon. (**b**) Aneurysm of untreated umbilical vein graft at 6 weeks. (**c**) Aldehyde-treated umbilical vein graft (aorto-iliac bypass) at 11 months showing retention of structure and continued function as a graft

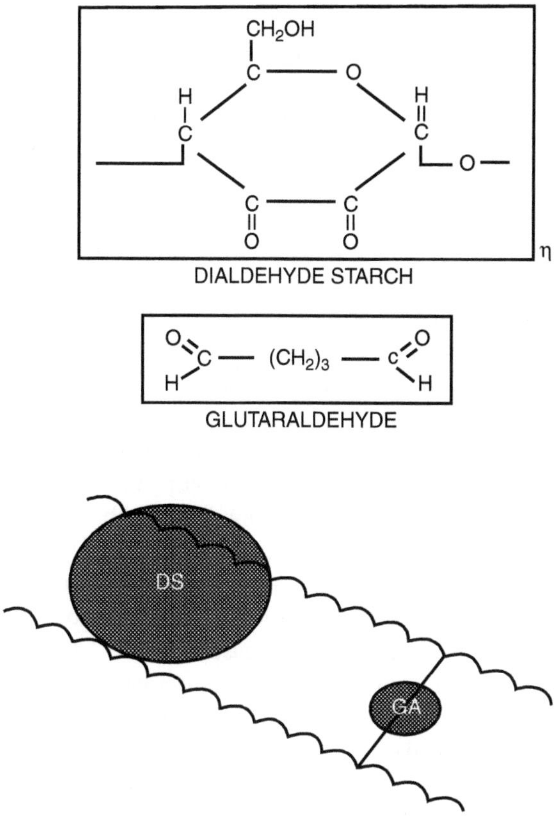

Fig. 19.3 Glutaraldehyde (GA) is superior to dialdehyde starch (DS) as a tanning agent since it is a small molecule and can be more effective in establishing cross-links between the amine groups of the substrate collagen

natural blood vessels, bovine heterografts, and synthetic materials.[10,11] The former method "fingerprinted" the flow surface and provided important information regarding the presence and amount of lipid deposition; the latter measured critical surface tension and, therefore, surface energy as a marker for thrombogenicity (Fig. 19.4).

Mechanical tests have been employed to ensure the adequacy of the cross-links produced by aldehyde tanning. Light, scanning, and transmission electron microscopy have yielded much information regarding structure. Long-term studies have shown that the glutaraldehyde-stabilized umbilical vein graft retains its basic architecture. On the basis of improved manufacturing and quality control, this graft has now proved remarkably stable and resistant but certainly not immune, to the forces of biodegradation.

Current umbilical grafts consist of manually stripped and prepared veins that have been stabilized with glutaraldehyde under optimal pH and temperative control (Fig. 19.5). Aldehyde cross-linkage of the protein moieties increases tensile strength and masks antigenicity. A polyester (Dacron) mesh is placed about the vein, which is then sterilized and stored in 50% ethanol. Most important, processing with glutaraldehyde sterilizes the tissue of bacteria, viruses, and fungi and renders it nonantigenic (Table 19.1).

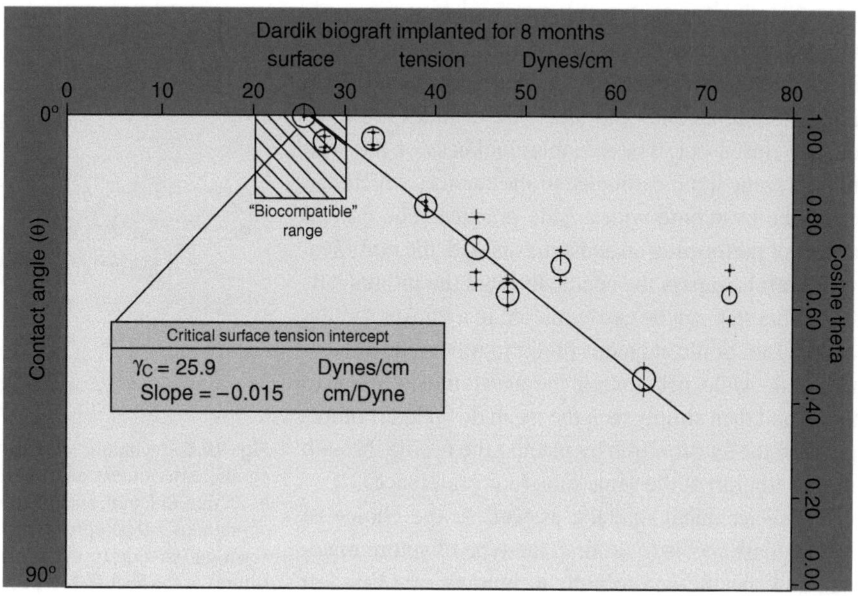

Fig. 19.4 Contact angle data plot (Zisman Plot) characterizing the blood flow surface of a patent, glutaraldehyde-stabilized human umbilical cord vein harvested after 8 months of human arterial service. Values of 25 ± 5 dynes/cm reflect thromboresistance

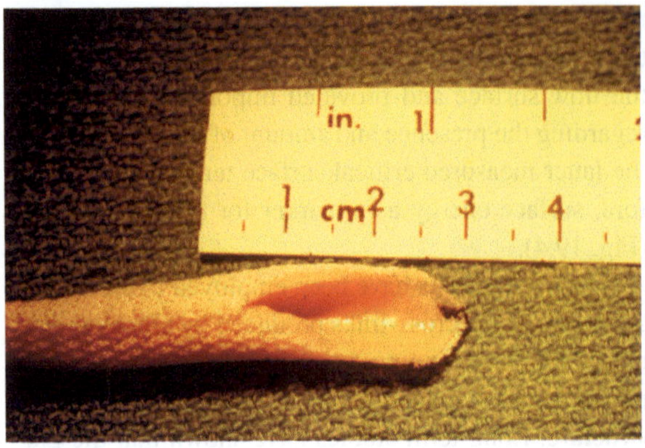

Fig. 19.5 Umbilical vein graft employed for lower extremity revascularization. The inner diameter is 5 mm, and the outer surface is covered with an elastic polyester mesh. Recently manufactured grafts have a uniform thin wall with a dense mesh on the outer surface, including a guideline

Table 19.1 Effects of glutaraldehyde on biological material

Reduces antigenicity
Increases strength via cross-linking
Inhibits degradation
Sterilizes tissue

19.1 Operative Technique

With the glutaraldehyde-stabilized human umbilical vein graft, several important technical maneuvers must be observed. Foremost is the gentleness with which the graft must be handled. Intimal fracture and extensive mural dissection of blood can occur if the grant is handled roughly or if standard clamps are applied. Before implantation, alcohol and aldehyde residues are thoroughly rinsed out. The variable thickness of the wall may present some difficulty to the novice, which can easily be overcome with a little practice. The critical aspect of performing an anastomosis with the umbilical vein graft is to pass the needle through the intimal surface. This step can be easily missed in a thicker section of the graft. Some surgeons prefer to reflect the polyester mesh while performing the anastomosis tissue to tissue and then simply tack the mesh down later; others perform the anastomosis by placing the needle through mesh and graft at the same time (our preference).

These technical matters, as well as the choice of which anastomosis to do first, the type of suture material, and specific anastomotic techniques, are best left to the preference of the experienced surgeon. *Interrupted suture technique* is advocated at the toe and heel of distal anastomosis, particularly at tibial and peroneal sites. During tunneling of the graft, it is essential to pass the graft through a metallic or plastic tube. If the unprotected graft is pulled through tissue planes, damage may occur because of the high friction between the polyester mesh and the host tissues. We prefer to place all grafts to the anterior tibial and peroneal arteries in the lateral subcutaneous position. Similarly, a *medial subcutaneous position* is employed for distal posterior tibial bypasses. The *anatomic position* is employed for all popliteal and proximal posterior tibial bypasses. Systemic heparinization is employed and monitored intraoperatively by the activated clotting time test.

Patients threatened with imminent limb loss and requiring reconstruction to a tibial or peroneal artery and where a prosthetic graft is necessary should be considered for an adjunctive, distal arteriovenous fistula, which produces an increased velocity of blood flow through the graft above the thrombotic threshold level.[12,13] Although most of the augmented graft flow is diverted into the low-resistance venous circuit, distal arterial flow is maintained, albeit at low pressure and decreased velocity.

Most fistulae are constructed by the common ostium technique. Parallel arteriotomy and venotomy incisions (~25 mm long) are anastomosed to create a posterior suture line. This permits the bypass graft to be anastomosed end to side to a common ostium (Fig. 19.6).

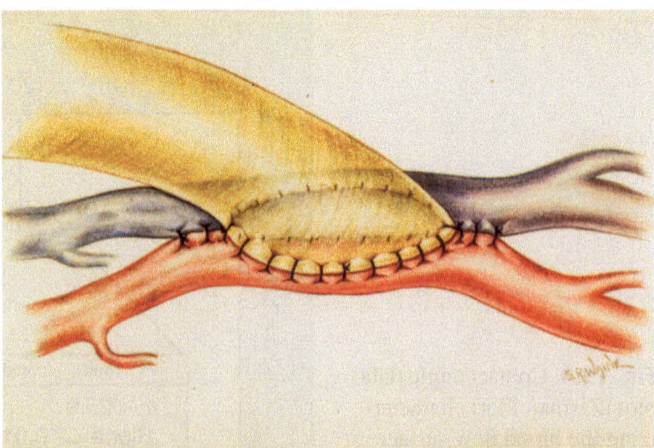

Fig. 19.6 Technique of a distal arteriovenous fistula involves parallel arteriotomy and venotomy incisions, approximately 20 to 25 mm in length and anastomosing the apposing walls with a continuous 7-0 polypropylene suture. The umbilical vein graft is sewn end-to-side to the arteriovenous ostium with interrupted sutures at the heel and toe, continuous along the lateral margins

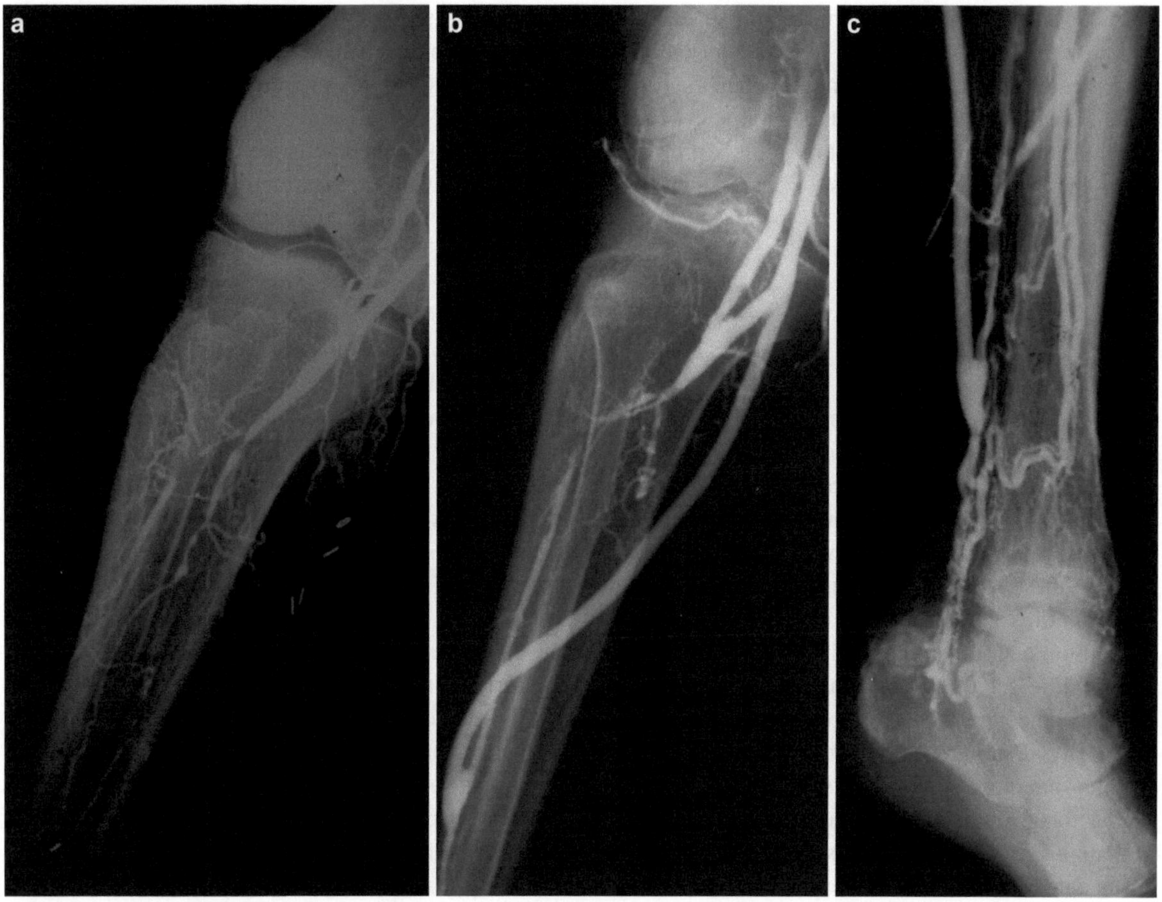

Fig. 19.7 Completion angiograms depicting a femoropopliteal below knee bypass into a blind segment (**a**), a femoropopliteal sequential anterior tibial bypass (**b**), and a femoroposterior tibial bypass with an adjunctive distal arteriovenous fistula (**c**)

Several variations of this technique are possible. The success of this operation depends not only on the quality of the venous circuit but also on the skill, patience, and commitment by the vascular surgeon in performing technically demanding procedures. Using a tourniquet in this type of operation helps to simplify the procedure and to decrease the time required for its performance. This method has become routine in our practice.

Completion intraoperative arteriography has proved successful, and we believe that it should be routine along with the performance of completion duplex sonography for all lower extremity bypass operations. With these techniques, technical errors or unsuspected pathology is often detected, and, just as important, it is possible to obtain a clear record that provides an accurate short-term prognosis and guide to future direction of care (Fig. 19.7).

During the operation, the patient receives therapeutic doses of heparin. If the patient does not have significant cardiac or renal disease, we also employ low-molecular-weight dextran during surgery and for an additional 2 days postoperatively (500 mL/day), for a total of 3 days.[14] Postoperatively, the patients continue to receive heparin until they can be switched to sodium warfarin (Coumadin). The dose of warfarin is usually adjusted to maintain the patient within an approximately therapeutic range, but on occasion we employ an empirical low-dose regimen (2.5 mg/day) for patients who are noncompliant or when the anticoagulation level is difficult to control. Aspirin is generally not used for umbilical vein grafts, although in exceptional circumstances we have combined aspirin, low-dose Coumadin, and subcutaneous heparin. Clopidogrel (Plavix) can also be employed as an alternative antiplatelet agent.

Patients are generally immobilized for several days with a foam knee immobilizer. This is kept on while the patient is in bed, but it is removed when the patient starts to walk. Length of hospitalization is dependent on the status of the foot; hospitalization is prolonged when the presence of lesions or gangrenous sites necessitates care.

19.2 Results

Since the clinical trials with umbilical vein grafts for lower-limb revascularization were initiated in 1974, doubt and skepticism with regard to their efficacy have coexisted with reports demonstrating superior performance as a prosthetic alternative to autologous vein. Most of these studies comparing the performance of vascular grafts are flawed by (1) short-term to medium-term follow-up, (2) varying indications for operations, and (3) differences in underlying pathology. Few prospective randomized studies exist; of those that do, practically all are flawed by poor case selection, bias, or unfair interpretation of results.[15-17]

Our total experience with the glutaraldehyde-stabilized umbilical vein graft now exceeds 1,300 cases. We documented our first decade of experience (1975–1985) to show half-life patencies for popliteal, tibial, and peroneal bypasses of 6.5, 2.3, and 1.7 years, respectively.[18] Some authors have expressed concern regarding the latter two numbers and also with the finding of a 36% incidence of aneurysm and 21% incidence of dilation after 5 years of implantation. Despite these concerns about graft degeneration, only 6% of patients have actually required surgical intervention because of aneurysmal graft dilation within 5 years after implantation.

A follow-up report during the second decade of experience with the umbilical vein graft (1985–1995) confirmed improving secondary patency rates at 5 years: 65% and 45% for popliteals and crurals, respectively, during this latter period compared with 57% and 33%, as reported during the first decade of experience.[19] Furthermore, only two graft aneurysms were discovered during the 1985–1995 period, suggesting an improvement in cord vein selection and processing.

Our current data in the third decade of umbilical vein graft use confirm the trend toward better primary and secondary patency rates as well as better control of biodegradation. Interestingly, surveillance studies of autologous vein grafts are also showing a 30% rate of morphologic change during the first year after implantation and lesser percentages during subsequent years. When added, the total percentage of duplex-demonstrated changes in morphology in autologous vein reconstruction approaches 50%! Admittedly, most of these changes do not require intervention, but this is also true of all biologic grafts, a fact not acknowledged by critics of the umbilical vein graft.

For the years 1990–2000, data for primary and secondary cumulative graft patency rates are shown in Fig. 19.8. There were no statistically significant differences between any of the primary and secondary patency rates of a particular type of reconstruction. Secondary patency rates for popliteal and tibial/peroneal reconstruction at 6 years were 67% and 47%, respectively. These figures were 11% and 4% better than their respective primary patency rates. Cumulative limb salvage rates at 5 years were 80% and 65%, respectively, for popliteal and tibial/peroneal reconstructions. We believe that the marked improvement in the tibial/peroneal data from our prior experience is due to our increased use of the adjunctive distal arteriovenous fistula.[13,20] Figure 19.9 shows our meta-analysis of reported studies comparing umbilical vein grafts to autologous vein and polytetrafluoroethylene (PTFE) grafts.

Factors relating to graft failure include (1) quality of the artery at the site of distal anastomosis and, more important, (2) distal runoff. Calcification, reduced diameter, presence of thrombus, and absence of the pedal arch are additional factors associated with early graft thrombosis, but none is absolute. In fact, durable patency can be achieved in some of these cases, a reflection of multiple factors and especially technical skill. Case selection is therefore critical, but currently available evaluation modalities, both angiographic and noninvasive, can be misleading.

Intraoperative angiography may be helpful, but arterial exploration with direct manual injection of heparinized saline provides the simplest and fastest answer regarding the functional status of the distal circulation. High resistance, as measured by this method, albeit crude and subjective, is associated with predictable early failure. Most of the causes of high resistance can be confirmed by arteriography. Under these circumstances, one should consider abandoning the bypass procedure particularly if, in the case of a crural bypass, an adjunctive distal arteriovenous fistula

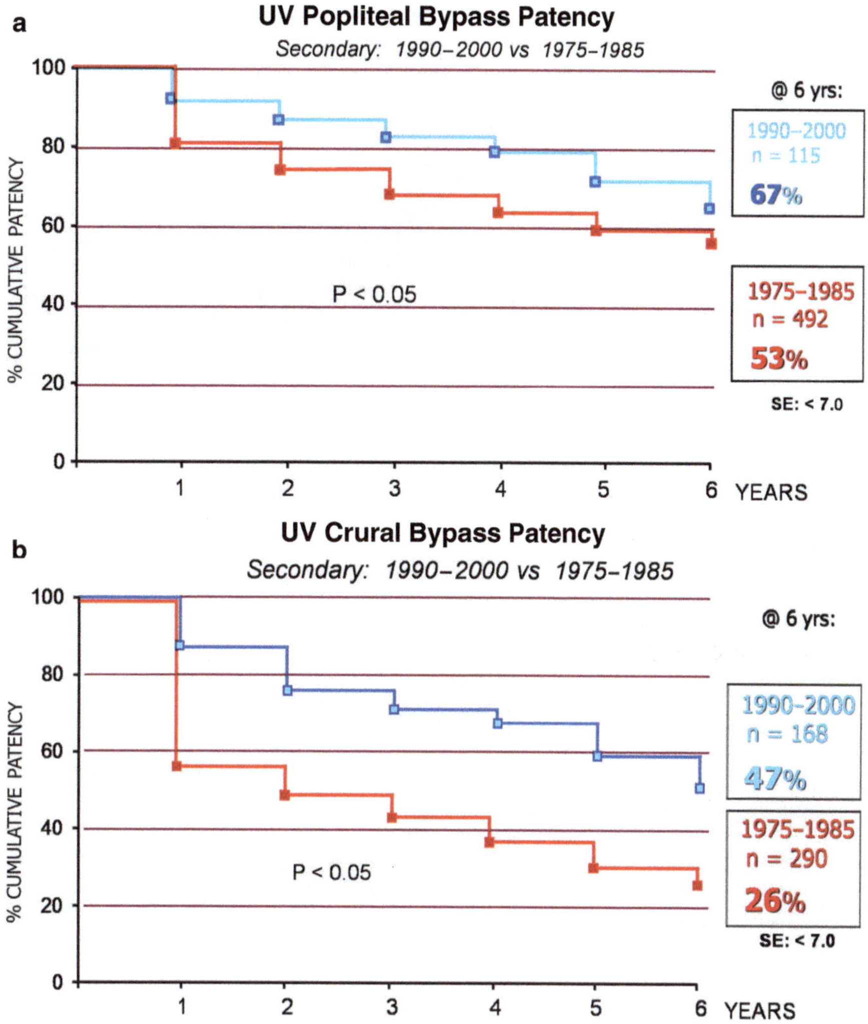

Fig. 19.8 Cumulative primary and secondary patency rates for umbilical vein grafts placed distally into the popliteal artery (predominantly below the knee) from 1990 to 2000 (**a**). Results for crural bypasses performed during this same period (**b**)

cannot be constructed.[13] These are obviously desperate circumstances, and we must consider the general clinical status of the patient before proceeding with reconstructive rather than ablative, surgery.

19.3 Complications

19.3.1 Thrombosis

Thrombosis is the most common complication associated with any vascular graft. With the umbilical vein, thrombosis can result from faulty technique, unfamiliarity with the graft's unique handing characteristics, or poor case selection. Thrombosis resulting from the graft itself occurs where the flow surface is damaged, thereby permitting blood to dissect intramurally. Defective grafts are possible, of course, but these are almost always detected during the manufacturing process by a number of quality control tests.

If the intraoperative arteriogram shows poor runoff or absence of runoff and if a more distal anastomosis would be impossible to perform, early postoperative thrombosis is accepted as a failure and reoperation is not considered. In all other situations, early and unanticipated thrombosis should be immediately reexplored and its cause should be searched for and corrected. Because a distal arteriovenous fistula is routinely constructed with all of our crural bypasses employing an umbilical vein graft, reexploration for early thrombosis

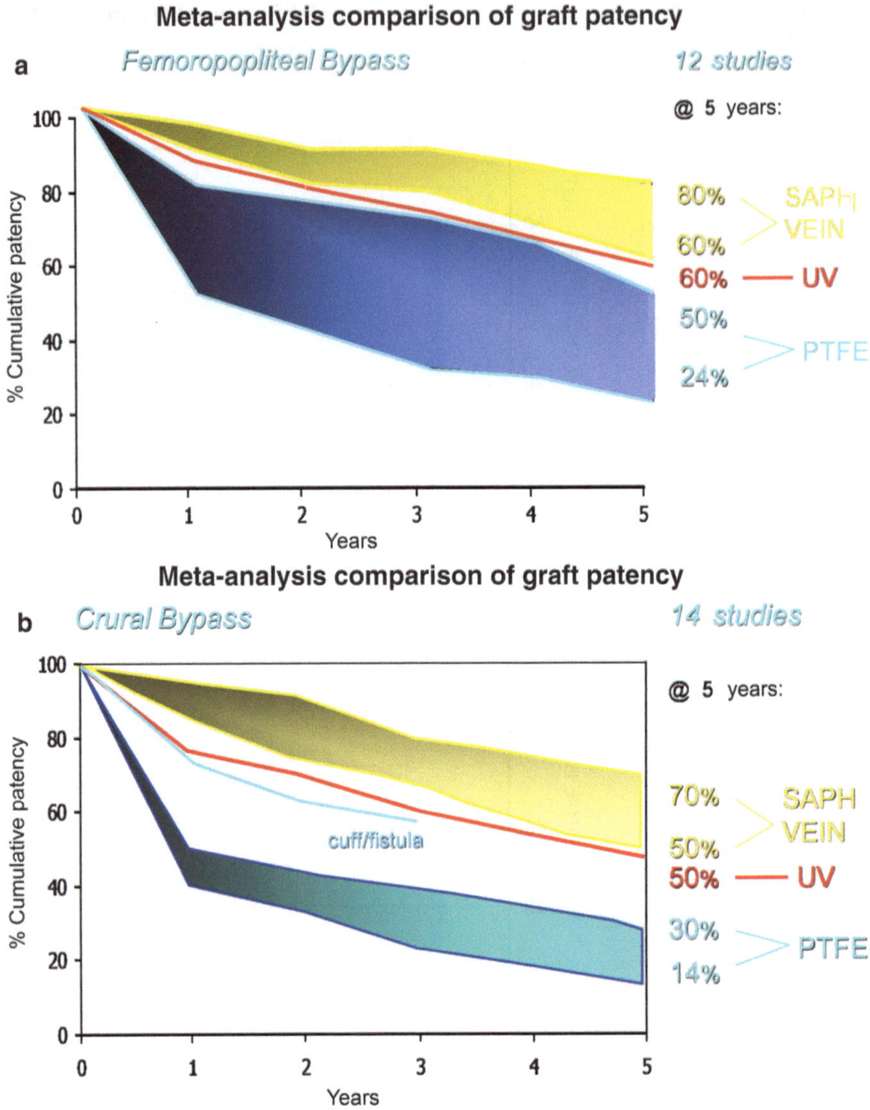

Fig. 19.9 Meta-analysis comparing graft patency of umbilical vein popliteal (**a**) and tibial/peroneal (**b**) bypasses to both saphenous vein and PTFE

is performed if the venous outflow is good (size and quality), regardless of the state of the arterial runoff.

The most important aspect of successful thrombectomy of an umbilical vein graft is extreme gentleness in its performance.[22] Overinflation of a balloon catheter can easily disrupt the graft. It is preferable to evacuate most of the thrombus by gentle external massage or by direct saline irrigation through proximal and distal arteriotomies prior to passage of a balloon catheter. In fact, we omit the use of the balloon catheter for the graft itself if the other maneuvers are clearly effective. Late graft thrombosis can be similarly managed, but preoperative arteriography should be performed to evaluate proximal and distal disease progression. Thrombolysis is an option and can be an effective alternative to surgery. Lesions responsible for thrombolysis may be discovered with clot lysis and treated by endovascular or direct means. Placement of a new graft even to areas remote from the original bypass may be preferable, particularly when an intense desmoplastic response has occurred at the anastomosis and an endovascular approach is not feasible or predictably successful.

Late graft thrombosis does not inevitably lead to amputation, particularly when digital or forefoot amputations or ulcerations have already been healed. The rate of limb salvage following tibial bypass is similar to that obtained with popliteal reconstructions.

Although peroneal reconstruction is associated with the lowest salvage rate, the preservation of more than half of such limbs for 2–4 years emphasizes the importance of continuing peroneal artery reconstruction in appropriately selected cases.[21]

19.3.2 Infection

Infection represents one of the most serious complications of any vascular prosthesis. Although the umbilical vein graft is not immune to infection, the glutaraldehyde tanning employed during its manufacture seems to impart some degree of increased resistance. This is unlike the bovine heterograft which, tanned with DS, would often dissolve when infected, resulting in life-threatening hemorrhage. The umbilical vein graft is able to maintain structural integrity under these adverse conditions to permit appropriate preparation and revision.

The incidence of infection for umbilical vein grafts is similar to the incidence with saphenous vein grafts (<3%). Although most infected grafts must be removed to cure the sepsis and prevent bleeding, some grafts may be salvaged by intensive antibiotic therapy combined with wide local drainage, muscle transposition, or flap advancement to cover the graft.

19.3.3 Aneurysms

True aneurysm of the umbilical vein graft is extremely uncommon today. This is in contradistinction to the bovine heterograft. Superior tanning achieved with glutaraldehyde, use of an outer polyester mesh, and preservation of the umbilical vein's intrinsic anatomy are all believed to contribute to greater resistance to biodegradation. The development of an aneurysm probably reflects a combination of host metabolic factors and inadequate or reversal of aldehyde cross-links. Aneurysms involving the umbilical vein graft should be resected, if they are symptomatic or larger than 2–3 cm, and replaced with an interposition new bypass graft.

Because almost all false aneurysms that occur at anastomotic sites are due to faulty technique, the incidence of this complication has fallen dramatically with increasing experience. Some false aneurysms still occur because of excessive mural thinning following endarterectomy of the recipient vessel. Correction of false aneurysms may require patch angioplasty or interposition grafting. If infection is present, wide drainage, graft excision, and extra-anatomic bypass may all be necessary.

19.3.4 Intimal Hyperplasia

Because the glutaraldehyde-stabilized umbilical vein graft is not viable, it is intrinsically incapable of developing intimal hyperplasia. Hyperplasia may, however, originate in the host vessels near the anastomosis and may extend into the graft lumen. This is believed to be due in part to local turbulence, which in turn, is due to residual disease, compliance mismatch, wide angulation of the anastomosis, and size disparity between the graft and host vessels. Surgical trauma is also a key ingredient. Correction of these factors, when possible, may decrease the incidence of intimal hyperplasia but does not eliminate it; this problem must be resolved at the molecular level.

19.4 Summary

The glutaraldehyde-stabilized umbilical vein graft is a musculocollagenous tube lined by a thromboresistant basement membrane and covered with a polyester mesh. Its use for lower extremity vascular reconstruction has been clinically assessed since 1974. On the basis of long-term studies of patency and durability, this conduit has an important role in revascularization of the lower limb (Fig. 19.10). Experience, judgment, and meticulous technique are prerequisites for securing long-term graft patency. No graft material can compensate for failure to adhere to these principles. The use of glutaraldehyde-stabilized umbilical vein facilitates these operations by providing a reliable, nonantigenic material that is mechanically equivalent to normal arteries and physically and chemically biocompatible. By studying the umbilical vein graft, we have gained new insights into the use of biologic material as a conduit for lower limb revascularization. An immediate advantage has been the observation of the efficacy of the

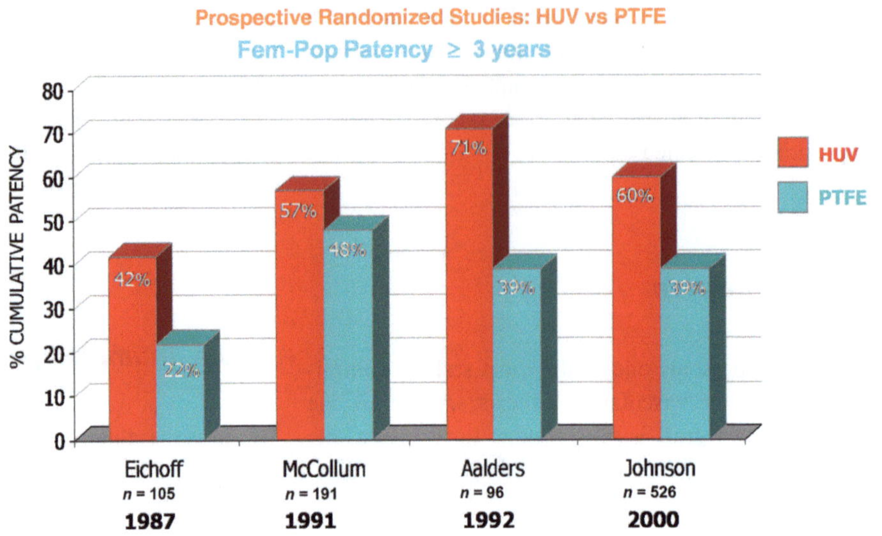

Fig. 19.10 Comparison of graft patency between umbilical vein and PTFE bypasses as reported in randomized clinical trials

umbilical vein graft in reversing ischemia and achieving limb salvage. Nonetheless, prospective randomized studies are necessary to access the comparative role of this graft with others currently available.

References

1. Dardik H, Dardik I. The fate of human umbilical cord vessels used as interposition arterial grafts in the baboon. *Surg Gynecol Obstet*. 1975;140:567.
2. Anzola J, Palmer TH, Welch S. Long femoral and iliofemoral grafts. *Surg Forum*. 1951;2:223.
3. Nabseth DC, Wilson JT, Tan B, et al. Fetal arterial heterografts. *Arch Surg*. 1960;81:929.
4. Yong NK, Eiseman B. The experimental use of heterologous umbilical vein grafts as aortic substitutes. *Singapore Med J*. 1962;3:52.
5. Carpentier A, Blondeau P, Laurens P. Replacement des valvules mitales et triscuspides pare des heterogrefes. *Ann Chir Thorac Cardiovasc*. 1968;7:33.
6. Rosenberg N, Martinez A, Sawyer PN, et al. Tanned collagen arterial prosthesis of bovine carotid origin in man: Preliminary studies of enzyme-treated heterografts. *Ann Surg*. 1966;164:247.
7. Dardik H, Dardik I. Successsful arterial substitution with modified human umbilical vein. *Ann Surg*. 1976;183:252.
8. Dardik H, Ibrahim IM, Dardik I. Modified and unmodified umbilical vein allografts and xenografts employed as arterial substitutes: a morphologic assessment. *Surg Forum*. 1976; 26:286.
9. Perloff LJ, Christie BA, Ketharanathan V, et al. A new replacement for small vessels. *Surgery*. 1981;89:31.
10. Baier RE, Abbott WM. Comparative biophysical properties of the flow surfaces of contemporary vascular grafts. In: Dardik H, ed. *Grafts Materials in Vascular Surgery*. Miami, FL: Symposia Specialists; 1978:79.
11. Baier RE, Akers CH, Perlmutter S, et al. Processed human umbilical cord veins for vascular reconstructive surgery. *Trans Am Soc Artif Intern Organs*. 1976;22:514.
12. Dardik H. The use of an adjunctive arteriovenous fistula in distal extremity bypass grafts with outflow obstruction. In: Kempczinski RF, ed. *The Ischemic Leg*. Chicago: Year Book Medical Publishers; 1985.
13. Dardik H, Silvestri R, Alasio T, et al. Improved method to create the common ostium variant of the distal arteriovenous fistula for enhancing crural prosthetic graft patency. *J Vasc Surg*. 1996;24:240.
14. Rutherford R, Jones DN, Bergentz SE, et al. The efficacy of Dextran 40 in preventing early postoperative thrombosis following difficult lower extremity bypass. *J Vasc Surg*. 1984;1:767.
15. Aalders GJ, Van Vroonhoven JMV, Lobach JHC, et al. PTFE versus human umbilical vein in clinical trial. *J Cardiovasc Surg*. 1988;29:186.
16. Eickhoff JH, Broome A, Ericsson BF, et al. Four years' results of a prospective, randomized clinical trial comparing polytetrafluoroethylene and modified human umbilical vein for below-knee femoropopliteal bypass. *J Vasc Surg*. 1987;6:506.
17. McCollum C, Kenchington G, Alexander C, et al. PTFE or HUV for femoro-popliteal bypass: A multi-centre trial. *Eur J Vasc Surg*. 1990;5:435.
18. Dardik H, Miller N, Dardik A, et al. A decade of experience with the glutaraldehyde-tanned human umbilical cord vein graft for revascularization of the lower limb. *J Vasc Surg*. 1988;7:336.
19. Dardik H. The second decade of experience with the umbilical vein graft for lower-limb revascularization. *Cardiovasc Surg*. 1995;3:265.
20. Dardik H. The threatened limb. *Sci Med*. 1997;4:44.
21. Dardik H, Ibrahim IM, Dardik I. The role of the peroneal artery for limb salvage. Ann Surg. 1979;189:189.
22. Dardik H. Technical aspects of umbilical bypass to the tibial vessels. *J Vasc Surg*. 1984;1:916.

Part IX

Use of Cord Blood in Cardiovascular Medicne

Cord Blood Stem Cells in Angiogenesis

20

Peter Hollands

Human umbilical cord blood contains stem cells with very different characteristics to those found in adult bone marrow and mobilized adult peripheral blood stem cells.[1,2] Cord blood stem cells, compared with adult stem cells, have been found to expand longer in culture, have longer telomeres and produce larger hemopoietic clones in vitro.[3,4] Perhaps, most interesting was the observation that cord blood contains mesenchymal progenitor cells, thus opening up a whole new area of research in cord blood stem cell technology.[5]

20.1 Endothelial Progenitor Cells (EPC)

To create angiogenesis in any scenario, it is necessary to obtain a cell population that contains a good proportion of endothelial progenitor cells (EPC). Human cord blood has been shown to contain angioblast-like EPC in significantly larger numbers than those found in human peripheral blood.[6] These cord blood EPC were shown to be capable of postnatal neovascularization in the ischemic hindlimb of rats in vivo. Earlier studies had shown the possible presence of EPC in adult human peripheral blood,[7] and it is thought that these cells may reside in the adult bone marrow and are mobilized by tissue ischemia and associated cytokine release.[8] Similar studies have shown that cord blood progenitors can induce angiogenesis in the implanted human thymus in the kidney of NOD/SCID mice.[9] These studies indicate the potential of autologous (and possibly allogeneic) clinical transplantation of cord-blood derived EPC into ischemic tissues.

As the hemangioblast is thought to be the common precursor of both EPC and hemopoietic stem cells, it is not surprising that cord blood is a good source of EPC. In the early stages of differentiation of EPC, the cells express the surface antigens CD34, KDR and Tie-2.[10,11] This is very similar to hemopoietic stem cells, which also express the surface antigens CD34, KDR and Tie-2.[12] However, as hemopoietic stem cells differentiate, CD34 expression is lost.[13] The CD34 antigen is therefore generally considered to be an appropriate marker for the isolation of EPC from umbilical cord blood.[14] It has also been reported that cord blood derived EPC can differentiate faster than peripheral blood derived EPC.[15] These data support the concept that cord blood derived EPC have increased proliferative capacity possibly related to their high cell-cycle rate and increased telomere length.[16,17] More recent studies show that the angiopoietin-1&2 and Tie-2 expressing EPC belong to the CD34+/CD133+/CD45+ immature hemopoietic population.[18] It is therefore clear that human umbilical cord blood contains EPC, which may have massive potential in angiogenesis and tissue engineering, and perhaps the two most exciting areas in this field are cardiology and ophthalmology.

20.2 Angiogenesis in Acute Myocardial Infarction (MI)

The possible therapeutic role of stem cells in acute MI has created much interest recently, especially the use of autologous bone marrow stem cells.[19,20]

P. Hollands
Department of Biomedical Science, University of Westminster, 115, Cavendish Street, London, W1W 6UW, UK
e-mail: p.hollands@wmin.ac.uk

N. Bhattacharya and P. Stubblefield (eds.), *Regenerative Medicine Using Pregnancy-Specific Biological Substances*,
DOI: 10.1007/978-1-84882-718-9_20, © Springer-Verlag London Limited 2011

Recipients of autologous bone marrow did show increased left ventricular ejection fraction, reduced ventricular dilation and improved cardiac wall motion when compared with the control group. It still remains unclear as to the role of autologous bone marrow in the context, but it seems unlikely that there is any de novo formation of cardiomyocytes, but there may be beneficial paracrine and angiogenic effects.[21] In addition, carrying out a bone marrow aspiration on a patient who has recently undergone an MI is an extra procedure that, unless absolutely necessary, should be avoided.

Umbilical cord blood represents a source of stem cells that are potentially very valuable in the treatment of acute MI and do not require invasive procedures on an already sick patient. Studies have shown that umbilical cord blood CD34+, cells when cultured with specific cytokines, express endothelial markers such as KDR, CD31 and CD26E.[22] In addition, it has been shown that human umbilical cord blood derived mesenchymal stem cells can differentiate into myocardium-like cells in vitro,[23] and it was subsequently demonstrated that human umbilical cord blood mononuclear cells, when injected into the periphery of an infarct, not only survive in the host heart but also improve left ventricular function.[24] It is also interesting to note that xenogeneic human cord blood stem cells can be transplanted into rat myocardium without rejection, and that such transplants show significant left ventricular remodeling and high levels of angiogenesis.[25]

Perhaps the most clinically useful experimental observation in this field is that cord blood cells can be administered intravenously and will migrate preferentially to the site of MI in NOD/scid mice.[26] The cell migration mechanism is not fully understood, but one possible important molecule may be stromal cell derived factor-1 (SDF-1), which is upregulated in MI and has been suggested as a possible stem cell homing factor.[27,28] There is no doubt that further experimental work is required to fully evaluate the therapeutic potential of cord blood cells in MI. Nevertheless, it is clear that cord blood cells have a significant experimental effect in MI, and that further work will bring these concepts to clinical trial and eventually to the benefit of patients in need.

20.3 Angiogenesis in Retinal and Chordial Abnormalities

There are many ophthalmic pathologies that are associated with or result in vascular abnormalities such as age-related macular degeneration, diabetic retinopathy, retinopathy of prematurity, neovascular glaucoma and even inherited retinal degenerations, such as retinitis pigmentosa. Adult stem cells have already shown some potential in the treatment of such conditions.[29] Cord blood has been shown to contain EPC,[30] and these EPC express CD34 and CD11b, and when they differentiate, there is an up-regulation expression of Tie-1 and secretion of Ang-1.[31] Adult bone marrow cells, comparable to this found in cord blood, have also been shown to promote retinal angiogenesis,[32] especially as hemangioblasts during hypoxia-stimulated retinal angiogenesis.[33] In addition, bone marrow cells can contribute to angiogenesis in murine laser-stimulated choroidal angiogenesis.[34-36]

Myeloid progenitor cells, such as those found in cord blood, have been shown to promote angiogenesis by specifically targeting sites of VEGF expression.[37] On analysis, the donor cells involved in angiogenesis were shown to be CD11b+ with a complete absence of all cells of the lymphoid lineage. A similar study has shown that SDF-1 released from platelets can mobilize and recruit CD11b+ myeloid cells to a critical role in angiogenesis.[38] These same myeloid progenitor cells were subsequently shown to be capable of differentiating into microglia and promoting angiogenesis in ischemic retinopathy.[39] The importance of microglia in retinal angiogenesis is now generally accepted, and if myeloid progenitor cells from cord blood can provide a reliable source of these cells, then therapeutic applications in ischemic retinopathies are the obvious next steps.[40]

20.4 Conclusion

The current scientific literature and ongoing research show the massive potential of cord blood stem cell technology in regenerative medicine and especially in angiogenesis. Endothelial progenitor cells derived from cord blood have been shown to have clear potential angiogenic roles in both myocardial infarction and degenerative ocular pathologies. The next step is to take these observations to clinical trial to confirm or refute this potential and if clinical trials prove positive to bring this technology to the patients waiting for help.

References

1. Nakahata T, Ogawa M. Hemopoietic colony-forming cells in umbilical cord blood with extensive capability to generate mono- and multipotential hemopoietic progenitors. *J Clin Invest.* 1982;70:1324-1328.
2. Broxmeyer HE, Douglas GW, Hangoc G, Cooper S, et al. Human umbilical cord blood as a potential source of transplantable hematopoietic stem/progenitor cells. *Proc Natl Acad Sci USA.* 1989;86:3828-3834.
3. Smith S, Broxmeyer HE. The influence of oxygen tension on the long term growth in vitro of hematopoietic progenitor cells from human cord blood. *Br J Haematol.* 1986;63:29-34.
4. Salahuddin SZ, Markham PD, et al. Long term suspension cultures of human cord blood myeloid cells. *Blood.* 1981; 58:931-938.
5. Erices A, Congnet P, Minguell JJ. Mesenchymal progenitor cells in human umbilical cord blood. *Br J Haematol.* 2000; 109:235-242.
6. Toyoaki M, Ikeda H, et al. Transplanted cord blood-derived endothelial precursor cells augment postnatal neovascularization. *J Clin Invest.* 2000;105:1527-1536.
7. Asahara T, et al. Isolation of putative progenitor endothelial cells for angiogenesis. *Science.* 1997;275:965-967.
8. Takahashi T, et al. Ischemia and cytokine-induced mobilization of bone marrow-derived endothelial progenitor cells for neovascularization. *Nat Med.* 1999;5:434-438.
9. Crisa L et al. Human cord blood progenitors sustain thymic T-cell development and a novel form of angiogenesis. *Blood.* 1999;94:3928-3940.
10. Millauer B, Wizigmann-Voos S, et al. High affinity VEGF binding and developmental expression suggests Flk-1 as a major regulator of vasculogenesis and angiogenesis. *Cell.* 1993;72:835-846.
11. Sato TN, Tozawa Y, et al. Distinct roles of the receptor tyrosine kinase Tie-1 and Tie-2 in blood vessel formation. *Nature.* 1995;376:70-74.
12. Kato O, Tauchi H, et al. Expression of the vascular endothelial growth factor (VEGF) receptor gene, KDR, in haematopoietic cells and inhibitory effect of VEGF on apoptotic cell death caused by ionizing radiation. *Cancer Res.* 1995;55:5687-5692.
13. Civin CI, Strauss LC, et al. Antigenic analysis of hematopoiesis. III. A hematopoietic progenitor cell surface antigen defined by a monoclonal antibody raised against KG-1a. *J Immunol.* 1984;133:157-165.
14. Murohara T, Ikeda H, et al. Transplanted cord blood-derived endothelial precursor cells augment postnatal neovascularization. *J Clin Invest.* 2000;105:1527-1536.
15. Kalka C, Iwaguro H, et al. Generation of differentiated endothelial cells from mononuclear cells of human umbilical cord blood. *Circulation.* 1999;100:I-749.
16. Mayani H, Landsdorp PM. Thy-1 expression is linked to functional properties of primitive hematopoietic progenitor cells from human cord blood. *Blood.* 1994;83:2410-2417.
17. Vaziri H, Dragowska W, et al. Evidence for a mitotic clock in human hematopoietic stem cells: loss of telomeric DNA with age. *Proc Natl Acad Sci USA.* 1994;91:9857-9860.
18. Pomyje J, Zivny J, et al. Expression of genes regulating angiogenesis in human circulating hematopoietic cord blood CD34+/CD133+ cells. *Eur J Haem.* 2003;70:142-150.
19. Wollert KC, Meyer GP, et al. Intracoronary autologous bone-marrow cell transfer after myocardial infarction: the BOOST randomised controlled clinical trial. *Lancet.* 2004; 364:141-148.
20. Chen SL, Fang WW, et al. Effect on left ventricular function of intracoronary transplantation of autologous bone marrow mesenchymal stem cells in patients with acute myocardial infarction. *Am J Cardiol.* 2004;94:92-95.
21. Kamihata H, Matsubara H, et al. Implantation of bone marrow mononuclear cells into ischemic myocardium enhances collateral perfusion and regional function via side supply of angioblasts, angiogenic ligands and cytokines. *Circulation.* 2001;104:1046-1052.
22. Shin JW, Lee DW, et al. Isolation of endothelial progenitor cells from cord blood and induction of differentiation by ex vivo expansion. *Yonsei Med J.* 2005;46:260-267.
23. Cheng F, Zou P, et al. Induced differentiation of human cord blood mesenchymal stem/progenitor cells into cardiomyocyte-like cells in vitro. *Huazhong Uni Sci Tech Med Sci.* 2003;23:154-157.
24. Henning RJ, Abu-Ali H, et al. Human umbilical cord blood mononuclear cells for the treatment of acute myocardial infarction. *Cell Transplant.* 2004;13:729-739.
25. Cheng-Heng HU, Gui-Fu WU, et al. Transplanted human umbilical cord blood mononuclear cells improve left ventricular function through angiogenesis in myocardial infarction. *Chin Med J.* 2006;119:1499-1506.
26. Ma N, Stamm C, et al. Human cord blood cells induce angiogenesis following myocardial infarction in NOD/scid-mice. *Cardiovasc Res.* 2005;66:45-54.
27. Kucia M, Ratajczak J, et al. Tissue-specific muscle, neural and liver stem/progenitor cells reside in the bone marrow, respond to an SDF-1 gradient and are mobilized into peripheral blood during stress and tissue injury. *Blood Cells Mol Dis.* 2004;32:52-57.
28. Askari AT, Unzek S, et al. Effect of stromal derived factor 1 on stem cell homing and tissue regeneration in ischaemic cardiomyopathy. *Lancet.* 2003;362:697-703.
29. Otani A, et al. Rescue of retinal degeneration by intravitreally injected adult bone marrow-derived lineage negative hematopoietic stem cells. *J Clin Invest.* 2004;114:765-774.
30. Aoki M, Yasutake M, et al. Derivation of functional endothelial progenitor cells from human umbilical cord

blood mononuclear cells isolated by a novel cell filtration device. *Stem Cells*. 2004;22:994-1002.
31. Hildbrand P, et al. The role of angiopoietins in the development of endothelial cells from cord blood CD34+ progenitors. *Blood*. 2004;104:2010-2019.
32. Otani A, et al. Bone marrow derived stem cells target retinal astrocytes and can promote or inhibit retinal angiogenesis. *Nat Med*. 2002;8:1004-1010.
33. Grant KG, et al. The contribution of adult hematopoietic stem cells to retinal neovascularization. *Adv Exp Med Biol*. 2003;522:37-45.
34. Csaky KG, et al. Recruitment of marrow-derived endothelial cells to experimental choroidal neovascularization by local expression of vascular endothelial growth factor. *Exp Eye Res*. 2004;78:1107-1116.
35. Espinosa-Heidmann DG, et al. Bone marrow derived progenitor cells contribute to experimental choroidal neovascularization. *Invest Ophthalmol Vis Sci*. 2003;44: 4914-4919.
36. Sengupta N, et al. The role of adult bone marrow-derived stem cells in choroidal neovascularization. *Invest Ophthalmol Vis Sci*. 2003;44:4908-4913.
37. Grunewald M, et al. VEGF-induced adult neovascularization: recruitment, retention, and role of accessory cells. *Cell*. 2006;124:175-189.
38. Jin DK, et al. Cytokine-mediated deployment of SDF-1 induces revascularization through recruitment of CXCR4+ hemangiocytes. *Nat Med*. 2006;12:557-567.
39. Ritter ML, et al. Myeloid progenitors differentiate into microglia and promote vascular repair in a model of ischemic retinopathy. *J Clin Invest*. 2006;116:3266-3276.
40. Checchin D, et al. Potential role of microglia in retinal blood vessel formation. *Invest Ophthalmol*. 2006;47: 3595-3602.

Endothelial Progenitor Cells from Cord Blood: Magic Bullets Against Ischemia? 21

Maurizio Pesce, Giulio Pompilio, and Maurizio C. Capogrossi

21.1 EPCs: A Concept, a Marker, or an Identity?

21.1.1 1997: "The Year of the Contact"; the First Definition of EPCs

About 11 years ago, the first report describing the presence of clonogenic cells in the human peripheral blood giving rise to endothelial cells in culture, and participating to angiogenesis *in vivo*, appeared in the Science Journal by J.M. Isner Laboratory.[1,2] The original discovery of these cells was made using a culture method allowing cells obtained by positive selection of $CD34^+$ cells from the peripheral blood mononuclear (PBMNCs) fraction, to adhere to fibronectin-coated dishes in high serum culture medium. The potential clinical relevance of these progenitors was suggested by the demonstration of the ability to give rise to blood vessels *in vivo*, and expression of mature endothelial markers such as CD31 and Tie-2. The EPC definition refers originally to cells expressing a defined set of cellular markers (CD34, VEGF-R2/Flk-1/KDR) in common with hematopoietic stem cells and hematopoietic progenitors.[1-3] EPC *in vitro* clonogenic activity was shown under specific culture conditions such as differential adherence onto extracellular matrix components (fibronectin, collagen) and the presence in the culture medium of high serum levels and/or the presence of *pro*-angiogenic cytokines such as the VEGF and the bFGF.

Today, EPC antigenic or functional quantification in the peripheral circulation has acquired the value of a diagnostic and prognostic "marker" for cardiovascular disease (CVD) and CVD risk factors. For example, it has been shown that the number of cells expressing EPC markers or showing EPC *in vitro* clonogenic activity was correlated with the occurrence of acute ischemic events[4] and vascular trauma.[5] In addition, patients suffering from cardiovascular risk conditions such as old age, male gender, hypertension, diabetes, cigarette smoking, family history of coronary artery disease (CAD), and high LDL cholesterol levels were shown to have significantly reduced levels of circulating EPCs and lower numbers of *in vitro* clonogenic cells.[6-8] These results, together with evidences coming from animal studies, suggest that ischemia or vascular damage represent a major "mobilization" trigger for EPCs from the BM to the peripheral circulation, and that the EPC progenitor function is impaired in patients prone to develop CVD.

21.1.2 Revisiting the Concept and the Definition: The Heterogeneous EPCs' Nature

The heterogeneity of EPCs appeared evident when new methods for obtaining them in culture were devised. For example, two types of EPCs were described based on the timing of appearance in culture by seeding PBMNCs under *pro*-angiogenic conditions.[5,9] Furthermore, an EPC activity has been found in PBMNCs expressing monocyte, hematopoietic, and T-cells specific markers.[10-16]

Both antigenic characterization and the definition of specific culture conditions for human EPCs have been

---

M. Pesce (✉)
Laboratorio di Biologia Vascolare e Terapia Genica, Centro Cardiologico Monzino, Milan, Italy and
Laboratorio di Biologia Vascolare e Terapia Genica, Centro Cardiologico Monzino IRCCS, Via Parea 4, Milan, Italy
e-mail: maurizio.pesce@ccfm.it

N. Bhattacharya and P. Stubblefield (eds.), *Regenerative Medicine Using Pregnancy-Specific Biological Substances*,
DOI: 10.1007/978-1-84882-718-9_21,

revisited by recent studies showing an EPC behavior of different cellular subsets expressing various cellular markers, and obtained at different times in culture by using different media and plating conditions, compared to the originally defined protocol. In a study dissecting the lineage relationships between hematopoietic progenitor cells (HPCs) and EPCs, M. Yoder and colleagues have been able to clearly distinguish between cells in peripheral blood and in cord blood that form endothelial-like colonies (the so-called Colony Forming Units-Endothelial Cells, CFU-EC) and cells that form definitive highly expandable EPC colonies (the so-called Endothelial Colony Forming Cells, ECFC). In their paradigm, also supported by observations raised by other groups,[17,18] Yoder and colleagues have proposed that cells having an EPC potential in PB, CB, and BM are likely derived from CD34$^+$, CD133$^+$, or KDR$^+$ cells that do not express the *pan*-hematopoietic lineage marker CD45.[19,20]

21.1.3 The Search for a Common Progenitor of EPCs and HPCs. But Does It Really Exist?

Figure 21.1 represents the current established lineage relationships of HPCs and EPCs in the BM and the CB. The essential novelty of this current understanding might imply a revisitation of the "hemangioblast" concept in adult life. The hemangioblast is a stem cell type having a double differentiation potency into endothelial and hematopoietic cells and is considered the most undifferentiated cell type in the adult stem cell niche. The existence of the hemangioblast in adults has been theorized based on the expression of the KDR and the CD34 markers in the most undifferentiated CB and BM stem cells.[21,22] However, since in the original identification of the hemangioblast, CD45 was not considered in the panel of cellular markers identifying cells with endothelial

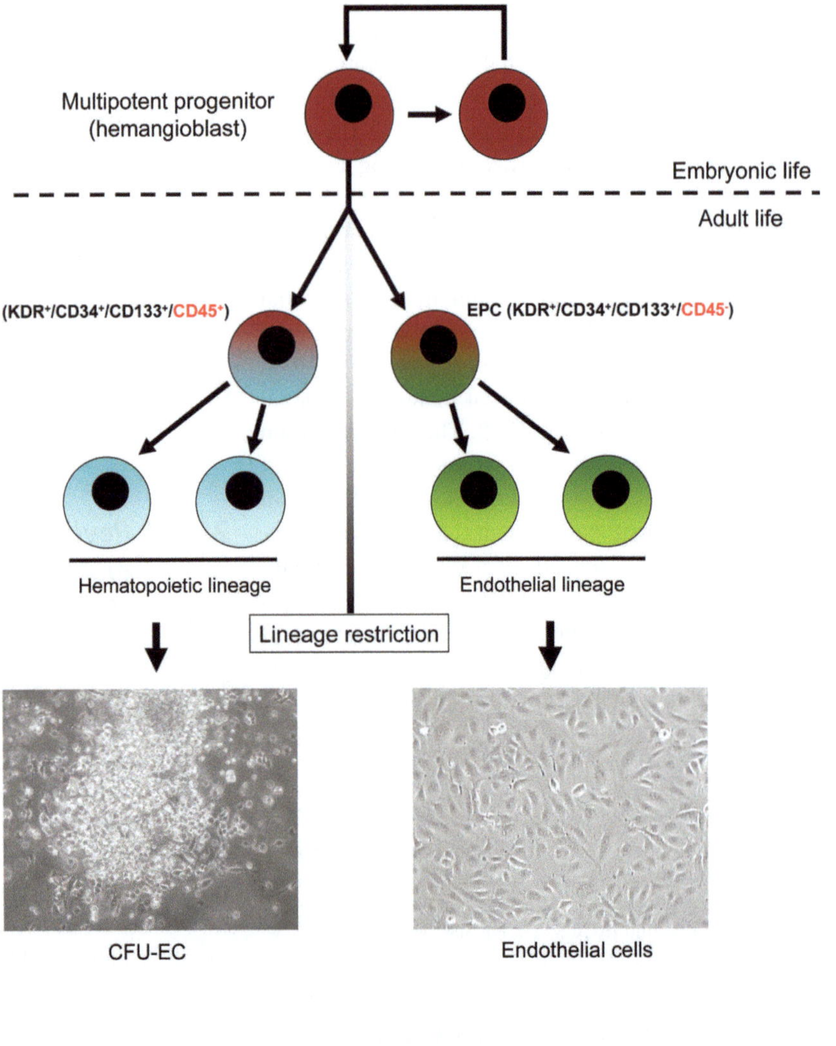

Fig. 21.1 The possible lineage relationships between HPCs and EPCs. Two distinct stem cell populations resident in CB, BM, and PB may coexist in adult life. The expression of CD45 represents the distinctive feature of HPCs (CD45$^+$) vs. EPCs (CD45$^-$). While under *pro*-angiogenic conditions CD45$^+$ HPCs can give rise to CFU-EC cells, characterized by a heterogeneous myelo/monocytic-endothelial mixed phenotype, CD45$^-$ EPCs give rise to ECFCs, representing the most clonogenic EPC type. Both cell types contribute to neo-vascularization *in vivo*

differentiation potency, these recent findings[17,18,23] raise the possibility that CB and BM represent reservoirs of two distinct lineages sharing markers in common (CD34, CD133, KDR) and distinguished by the expression of CD45. These lineages may be related or unrelated, and may be separately established during the embryonic development. Thus, while during mammalian embryonic life, the most primitive hematopoietic stem cells (pHSCs) behave as hemangioblasts and appear to be unrestricted in their ability to promote angiogenesis and hematopoiesis,[24] it is possible that a similar plasticity is not maintained during postnatal life. Sub-fractionation of distinct cellular populations expressing sets of these and other mature endothelial markers (CD144, CD146, CD31, CD105), associated to differential global gene expression and *in vivo* analyses will help unravel the fascinating and intriguing question of the EPC ancestors and their existence in adulthood.

21.2 EPCs "At Work": Multiple Ways to Promote Blood Vessel Formation

21.2.1 Some Important Definitions

The restoration of the blood supply into an ischemic tissue is linked to formation of collateral blood vessels that function as biological "bypasses." During embryonic development, blood vessel formation is called "primary" angiogenesis. By contrast, in adult life, formation of blood vessels in ischemic tissues is distinct through three separate events called angiogenesis, vasculogenesis, and arteriogenesis.[25] Vasculogenesis is referred to the event involving recruitment of circulating progenitor cells, while angiogenesis is a process mainly dictated by sprouting and proliferation of preexisting endothelial cells and arteriogenesis involves remodeling of high-resistance collateral arterioles.

To what extent do EPCs participate in the different processes of adult vasculogenesis? Is it possible to clearly discriminate between the function of EPCs and that of other cell types (inflammatory cells, preexisting endothelial cells) in induction of neo-vascularization? Is there any relationship between the EPC contribution and that of local inflammation and hypoxia in the angiogenic process?

21.2.2 EPCs as "Builders" of New Blood Vessels into Ischemic Tissues

Several experiments performed in animal models of heart and hind-limb ischemia have shown that different cohorts of PB, BM, and CB-derived EPCs become incorporated into newly formed vessels.[2,26-30] Different strategies to recognize EPC-derived cells into ischemic tissues include the adoption of techniques to label injected PB, CB, or BM-derived progenitor cells using living dyes or quantum dots, fluorescent proteins by viral vector infection, BM transplantation, and analysis of the tissue chimerism in human subjects undergoing sex-mismatched BM transplantation. These techniques all allow to follow, although at different extents and with different efficiency/reliability, the trafficking and the fate of progenitor cells into ischemic tissues at a single-cell level, and thus to reveal the acquisition of differentiated phenotypes. The literature, however, is not unequivocally in agreement about the direct participation of EPCs or EPC-derived cells as one of the major mechanisms involved in *neo*-vascularization. For example, Ziegelhoffer and colleagues,[31] using a BM transplantation model in mice, showed that circulating progenitors, although efficiently recruited into the ischemic limb tissues, failed to adopt a vascular phenotype and to be incorporated into newly formed blood vessels. This conclusion, although surprising, raises the hypothesis that a major EPC's role in ischemic tissues may be that of "priming" neo-vascularization, by activating endogenous endothelial cells via a paracrine mechanism (thereby enhancing angiogenesis), and by cross-talking with inflammatory cells.

21.2.3 EPCs as "Organizers" of Novel Blood Vessel Development

A potent paracrine function by recruited EPCs has been suggested as a second important way of action to induce *neo*-angiogenesis in hypoxic sites. In a transcriptomic screen for inflammatory and angiogenic cytokines expressed by EPCs, Urbich and colleagues have shown that these cells express a set of *pro*-angiogenic and *pro*-inflammatory cytokines[32] that is surprisingly overlapping with the array of factors expressed by $CD14^+$ monocytes.

Moreover, our unpublished observations and other reports showing the mixed CFU-ECs endothelial/myeloid composition, suggest that cells from CFU-EC clusters likely enhance angiogenesis through secretion of an array of inflammatory mediators and angiogenic factors that might represent the main way of action into ischemic tissues.[11,12] This hypothesis has an analogy to the angiogenic effects of infiltrating monocyte/macrophage cells into sites of solid tumor development.[33,34] These evidences lead us to conclude that although the recent studies on EPCs identity characterization have offered a new way of conceiving EPCs, they have not led to significant advancements on the issue of their *in vivo* angiogenic activity and, thus, the identification of the best-suited cell type for tissue repair. In fact, according to the above report, CFU-ECs should be excluded due to their prevalent myeloid/immune cells composition and the lack of clonogenic and self-renewal abilities that might reflect in the lack of long-term engraftment onto host tissues, compared to ECFCs.[17,20] On the other hand, the presence of T-cells myeloid progenitors into cell clusters representing CFU-ECs might promote angiogenesis owing to secretion of *pro*-inflammatory and *pro*-angiogenic cytokines that elicit a potent angiogenic response into ischemic tissues, and "prime" regeneration of the vascular bed. This activity might be the prevalent way of action of injected progenitors in preclinical models as well as in clinical studies where the secretion of cytokines might have beneficial effects in terms of collateral vessel formation[35,36] or "cardioprotective" effects.[37] A possible escape to achieve the maximum efficiency in ischemic tissue repair has been indicated in a study by Yoon and colleagues showing the synergistic effect on angiogenesis when early and late EPCs (similar to CFU-EC and ECFC, respectively) were co-injected in a subcutaneous Matrigel plug assay.[38] The hypothesis of these authors is that the presence of both EPC "types" is important for a reciprocal stimulation that results in an enhanced angiogenesis.

Hypoxia is reported to promote recruitment of circulating EPCs through the formation of chemotactic gradients that guide their invasion from the adjacent tissues. The most studied factors that are involved in the establishment of such gradients are VEGF and the chemokine SDF-1. They are both upregulated by hypoxia through the HIF-1α transcription factor activation pathway[39,40] and they both promote progenitor cells migration into the ischemic sites.[27,41-43] In addition, it has been suggested that hypoxia triggers EPCs' migratory activity by upregulating the expression of SDF-1 receptor, the CXCR4 transmembrane protein, through an HIF-1α-dependent mechanism.[44] Other possible mechanisms include an effect of SDF-1 on the VLA-4 integrin receptors[45-47] and the expression of MMP-9 metalloprotease,[48] thus reflecting in enhanced EPC invasive and migratory activities. Finally, SDF-1 and/or VEGF may promote survival of progenitor cells into hypoxic environments through the activation of the PI3K/AKT pathway.[49,50]

21.3 Cord-Blood-Derived EPCs to Promote Angiogenesis: Why and How?

21.3.1 *Newborn Versus Adult EPCs: Advantages, Disadvantages, and Possible Escapes*

Patients with CVD are, in general, individuals suffering from several concomitant conditions predisposing them to the disease and depressing the cellular functions, especially endothelial and myocardial cells. This represents an outstanding clinical problem, but also a biological issue when planning the use of autologous progenitors for induction of neo-vascularization. In fact, the use of "defective" EPCs may lead to only incremental effects on *neo*-vascularization. For example, EPCs obtained from diabetic patients or EPCs experimentally exposed to high glucose levels in culture exhibit lower proliferation potential, increased apoptosis, and enhanced oxidative stress.[51-61] EPC preconditioning has thus been suggested as a possible strategy to be adopted in culture to "rejuvenate" or reactivate EPC function prior to injection into patients. Preconditioning includes treatment with statins, inhibitors of the p38-MAPK, transduction with *pro*-survival factors, treatment by NO-donors (discussed in[62,63]) or treatment with SDF-1 chemokine to enhance EPC adhesion and differentiation potential.[64]

Cord-blood-derived EPCs represent an alternative option to enhance neo-vascularization. The choice on CB to derive EPCs for ischemic tissue repair may be justified by a number of reasons including (1) the higher progenitor cells frequency compared to BM or PB, (2) their wider self-renewal/proliferation potential, (3) the lack of exposure to risk factors that may reduce their *in vivo* engraftment and angiogenic potential.

Limitations to the use of CB cells for ischemic tissues treatment may derive from (1) the risk for chromosome aberrations due to storage or culturing for long periods of time and (2) from possible immune responses due to the allogenic nature of transplanted cells, thus requiring the adoption of host immunosuppression.

In a comparison between the clonogenic activity of long-term ECFC formation from PB and the CB, it was shown that CB cells are significantly more efficient in terms of clonogenic expansion, have a significantly lower ECFC colony generation time and can be expanded for up to 100 population doublings with no signs of senescence.[9] These findings may be explained by a more immature phenotype of endothelial progenitors in the CB compared to PB and the BM, or by the lack of exposure to conditions (i.e., high glucose concentration) that are known to reduce EPCs clonogenicity.[55] Given the extremely low efficiency of ECFC colony formation from PB, CB may be thus an ideal substrate to derive long-term and highly clonogenic cells to repopulate ischemic tissues. However, in the absence of tests clearly comparing animal models of heart or peripheral ischemia, the *pro*-angiogenic and the *pro*-vasculogenic activity of ECFCs selected from CB using different culture methods[18,23] or different markers (i.e., CD146 expression[17,65];), it remains to be determined the best suited cell phenotype to proceed with clinical applications in ischemic diseases.

Finally, problem that may result from ex vivo amplification of EPCs in culture is the potential risk of genetic instability. For example, in a recent report describing the karyotype of CB-derived ECFCs during expansion, it was shown that a significant number of ECFC clones (five out of seven total CB samples analyzed) displayed aberrant chromosomes, even at early times during the course of *in vitro* expansion.[66] Thus, although the recent identification of ECFCs has led to an important advancement for the definition of the EPC identity, it is possible that such cells may not be used for clinical application due the risk for tumorigenesis.

21.3.2 In Vivo Studies

Experience from preclinical studies by us and by others have suggested that CB cells purified from the total CB mononuclear fraction using magnetic separation methods such as the MACS technology and antibodies raised against CD34 or CD133 have a high *in vivo* angiogenic potential that might be of clinical interest. In 2003, we used CB CD34$^+$ cells to enhance angiogenesis and induce muscle repair in a mouse model of hind-limb ischemia.[67] In that report, by using a retroviral vector system to introduce EGFP protein into CD34$^+$ cells and be able to recognize injected cells into recipient tissues, we described the direct incorporation of CD34$^+$ cells into newly formed vascular structures and their participation to enhancement of muscle regeneration in ischemic limbs. We also showed that CD34$^+$ cells expanded using a serum-free medium containing mitogenic cytokines maintained the ability to enhance arteriole length density and regenerate muscle fibers.

In our study, the ability of CD34$^+$ cells to enhance muscle regeneration was particularly intriguing in the view of the so-called stem cell "plasticity" that determined apparent *trans*-differentiation of EPCs into myogenic cells. For this reason we further analyzed the ability of CD34$^+$ cells to directly differentiate into myogenic cells by *in vitro* culture or *in vivo* studies (our unpublished observations). Unfortunately, we never found evidences for cell-autonomous differentiation into myogenic cells by using *pro*-myogenic conditions such as myoblast conditioned media, Wnt-activated signaling[68] or IL-4.[69] However, we observed the presence of myogenic cells derived from the EGFP$^+$/CD34$^+$ cells injected into ischemic limbs co-purifying with the satellite cell fraction (Fig. 21.2). These results suggested that environmental cues such as those acting in ischemic conditions may trigger CB-derived CD34$^+$ cells to adopt a muscle in addition to vascular phenotype.

The evidence that CB-derived CD34$^+$ or CD133$^+$ cells promote angiogenesis and are involved in repair of ischemic tissues, both in the heart and limbs, has been provided by numerous other studies.[70-76] The majority of these investigations, however, have been performed using genetically immunosuppressed animals such as the athymic nude or the NOD/SCID mouse strains that have little or no rejection of human cells. This makes it impossible to assess the potential of cord-blood stem cells to be used for allogenic transplantation into ischemic patients. In our study,[67] we used a pharmacologic-induced immunosuppression by Cyclosporin-A in normal CD-1 mice. We found that only 9% at day 7 ($n=2$) and 35.7% at day 14 ($n=14$) of mice that were treated with CD34$^+$, CD34$^-$, or expanded CD34$^+$ CB cells showed histological evidence of rejection, such as massive leukocyte infiltration and tissue

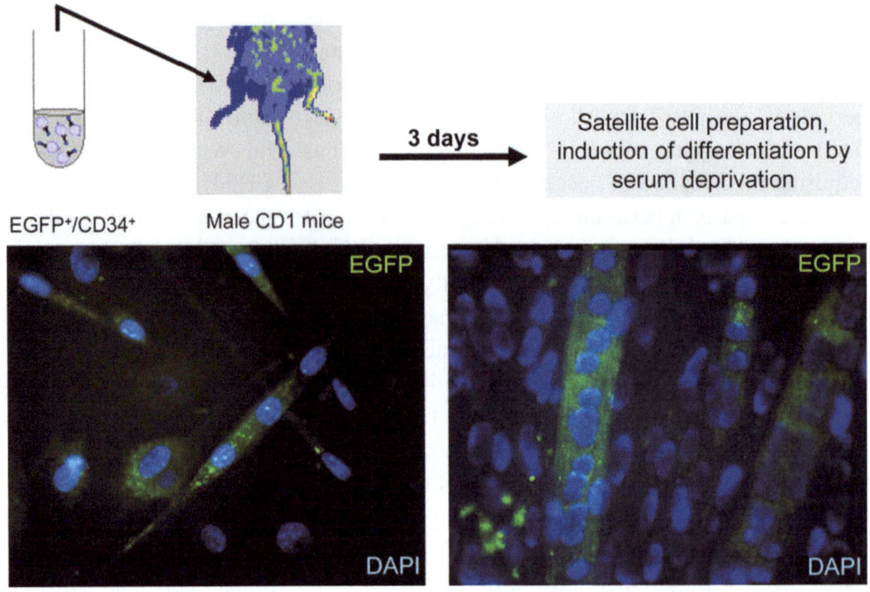

Fig. 21.2 Contribution of injected EGFP+/CD34+ cells to myogenic lineage in vivo. Retrovirus-mediated EGFP-labeled CD34+ cells were injected into ischemic limbs in immunosuppressed mice. After three days, ischemic adductor muscles were removed and enzymatically dissociated to obtain the myogenic satellite cells fraction. Some EGFP+ satellite cells appeared in culture. These cells formed large EGFP+ myotubes upon induction of differentiation by serum deprivation

necrosis. This suggests that, at least in cases in which transplantation of autologous cells is not feasible, the choice of an optimal matching between the donor and recipient HLA types, and the adoption of immunosuppression protocols may help patients to well-tolerate the treatment with CB-derived EPCs.

21.3.3 The Perspectives

Cord-blood banking is becoming an increasingly common and widespread procedure that ensures the maintenance of CB-derived stem cells in native conditions.[77] Originally conceived to represent a useful resource to combat hematologic disorders and malignancies, CB blood banking may be today an important source of stem cells for allogenic transplantation to treat other diseases such as peripheral and cardiac ischemia.[78,79] We believe that CB-derived EPCs may be advantageous in terms of optimal clonogenic activity and repopulation ability of ischemic tissues compared to other stem cell sources such as the PB or the BM. However, preclinical testing should be done in order to clearly assess the tolerance of these cells when proceeding with allogenic transplantation. Future studies will be necessary to address this specific issue and understand the real potential of cord-blood "magic bullets" to combat ischemia diseases.

Acknowledgments The Authors wish to thank Dr. Ilaria Burba and Dr. Anita Gianella for critically reading the manuscript.

References

1. Asahara T, Murohara T, Sullivan A, et al. Isolation of putative progenitor endothelial cells for angiogenesis. *Science*. 1997;275:964-967.
2. Takahashi T, Kalka C, Masuda H, et al. Ischemia- and cytokine-induced mobilization of bone marrow-derived endothelial progenitor cells for neovascularization. *Nat Med*. 1999;5:434-438.
3. Shi Q, Rafii S, Wu MH, et al. Evidence for circulating bone marrow-derived endothelial cells. *Blood*. 1998;92: 362-367.
4. Shintani S, Murohara T, Ikeda H, et al. Mobilization of endothelial progenitor cells in patients with acute myocardial infarction. *Circulation*. 2001;103:2776-2779.
5. Gill M, Dias S, Hattori K, et al. Vascular trauma induces rapid but transient mobilization of VEGFR2(+)AC133(+) endothelial precursor cells. *Circ Res*. 2001;88:167-174.
6. Hill JM, Zalos G, Halcox JP, et al. Circulating endothelial progenitor cells, vascular function, and cardiovascular risk. *N Engl J Med*. 2003;348:593-600.
7. Liew A, Barry F, O'Brien T. Endothelial progenitor cells: diagnostic and therapeutic considerations. *Bioessays*. 2006; 28:261-270.
8. Vasa M, Fichtlscherer S, Aicher A, et al. Number and migratory activity of circulating endothelial progenitor cells inversely correlate with risk factors for coronary artery disease. *Circ Res*. 2001;89:E1-E7.
9. Hur J, Yoon CH, Kim HS, et al. Characterization of two types of endothelial progenitor cells and their different

contributions to neovasculogenesis. *Arterioscler Thromb Vasc Biol*. 2004;24:288-293.
10. Hur J, Yang HM, Yoon CH, et al. Identification of a novel role of T cells in postnatal vasculogenesis: characterization of endothelial progenitor cell colonies. *Circulation*. 2007;116:1671-1682.
11. Kuwana M, Okazaki Y, Kodama H, Satoh T, Kawakami Y, Ikeda Y. Endothelial differentiation potential of human monocyte-derived multipotential cells. *Stem Cells*. 2006;24: 2733-2743.
12. Rehman J, Li J, Orschell CM, March KL. Peripheral blood "endothelial progenitor cells" are derived from monocyte/macrophages and secrete angiogenic growth factors. *Circulation*. 2003;107:1164-1169.
13. Rohde E, Bartmann C, Schallmoser K, et al. Immune cells mimic the morphology of endothelial progenitor colonies *in vitro*. *Stem Cells*. 2007;25:1746-1752.
14. Rohde E, Malischnik C, Thaler D, et al. Blood monocytes mimic endothelial progenitor cells. *Stem Cells*. 2006;24: 357-367.
15. Romagnani P, Annunziato F, Liotta F, et al. CD14+CD34low cells with stem cell phenotypic and functional features are the major source of circulating endothelial progenitors. *Circ Res*. 2005;97:314-322.
16. Urbich C, Heeschen C, Aicher A, Dernbach E, Zeiher AM, Dimmeler S. Relevance of monocytic features for neovascularization capacity of circulating endothelial progenitor cells. *Circulation*. 2003;108:2511-2516. Epub 2003 Oct 27.
17. Nagano M, Yamashita T, Hamada H, et al. Identification of functional endothelial progenitor cells suitable for the treatment of ischemic tissue using human umbilical cord blood. *Blood*. 2007;110:151-160.
18. Timmermans F, Van Hauwermeiren F, De Smedt M, et al. Endothelial outgrowth cells are not derived from CD133+ cells or CD45+ hematopoietic precursors. *Arterioscler Thromb Vasc Biol*. 2007;27:1572-1579.
19. Case J, Mead LE, Bessler WK, et al. Human CD34+ AC133+VEGFR-2+ cells are not endothelial progenitor cells but distinct, primitive hematopoietic progenitors. *Exp Hematol*. 2007;35:1109-1118.
20. Prater DN, Case J, Ingram DA, Yoder MC. Working hypothesis to redefine endothelial progenitor cells. *Leukemia*. 2007;21:1141-1149.
21. Pelosi E, Valtieri M, Coppola S, et al. Identification of the hemangioblast in postnatal life. *Blood*. 2002;100:3203-3208.
22. Ziegler BL, Valtieri M, Porada GA, et al. KDR receptor: a key marker defining hematopoietic stem cells. *Science*. 1999;285:1553-1558.
23. Yoder MC, Mead LE, Prater D, et al. Redefining endothelial progenitor cells via clonal analysis and hematopoietic stem/progenitor cell principals. *Blood*. 2007;109:1801-1809.
24. Takakura N, Watanabe T, Suenobu S, et al. A role for hematopoietic stem cells in promoting angiogenesis. *Cell*. 2000;102:199-209.
25. Jones WS, Annex BH. Growth factors for therapeutic angiogenesis in peripheral arterial disease. *Curr Opin Cardiol*. 2007;22:458-463.
26. Asahara T, Masuda H, Takahashi T, et al. Bone marrow origin of endothelial progenitor cells responsible for postnatal vasculogenesis in physiological and pathological neovascularization. *Circ Res*. 1999;85:221-228.
27. Asahara T, Takahashi T, Masuda H, et al. VEGF contributes to postnatal neovascularization by mobilizing bone marrow-derived endothelial progenitor cells. *Embo J*. 1999;18: 3964-3972.
28. Kamihata H, Matsubara H, Nishiue T, et al. Implantation of bone marrow mononuclear cells into ischemic myocardium enhances collateral perfusion and regional function via side supply of angioblasts, angiogenic ligands, and cytokines. *Circulation*. 2001;104:1046-1052.
29. Murayama T, Tepper OM, Silver M, et al. Determination of bone marrow-derived endothelial progenitor cell significance in angiogenic growth factor-induced neovascularization *in vivo*. *Exp Hematol*. 2002;30:967-972.
30. Tepper OM, Capla JM, Galiano RD, et al. Adult vasculogenesis occurs through in situ recruitment, proliferation, and tubulization of circulating bone marrow-derived cells. *Blood*. 2005;105:1068-1077.
31. Ziegelhoeffer T, Fernandez B, Kostin S, et al. Bone marrow-derived cells do not incorporate into the adult growing vasculature. *Circ Res*. 2004;94:230-238.
32. Urbich C, Aicher A, Heeschen C, et al. Soluble factors released by endothelial progenitor cells promote migration of endothelial cells and cardiac resident progenitor cells. *J Mol Cell Cardiol*. 2005;39:733-742.
33. Dirkx AE, Oude Egbrink MG, Wagstaff J, Griffioen AW. Monocyte/macrophage infiltration in tumors: modulators of angiogenesis. *J Leukoc Biol*. 2006;80:1183-1196.
34. Sica A, Bronte V. Altered macrophage differentiation and immune dysfunction in tumor development. *J Clin Invest*. 2007;117:1155-1166.
35. Kajiguchi M, Kondo T, Izawa H, et al. Safety and efficacy of autologous progenitor cell transplantation for therapeutic angiogenesis in patients with critical limb ischemia. *Circ J*. 2007;71:196-201.
36. Tateishi-Yuyama E, Matsubara H, Murohara T, et al. Therapeutic angiogenesis for patients with limb ischaemia by autologous transplantation of bone-marrow cells: a pilot study and a randomised controlled trial. *Lancet*. 2002; 360:427.
37. Fazel S, Cimini M, Chen L, et al. Cardioprotective c-kit+cells are from the bone marrow and regulate the myocardial balance of angiogenic cytokines. *J Clin Invest*. 2006;116: 1865-1877.
38. Yoon CH, Hur J, Park KW, et al. Synergistic neovascularization by mixed transplantation of early endothelial progenitor cells and late outgrowth endothelial cells: the role of angiogenic cytokines and matrix metalloproteinases. *Circulation*. 2005;112:1618-1627.
39. Carmeliet P, Dor Y, Herbert JM, et al. Role of HIF-1alpha in hypoxia-mediated apoptosis, cell proliferation and tumour angiogenesis. *Nature*. 1998;394:485-490.
40. Ceradini DJ, Kulkarni AR, Callaghan MJ, et al. Progenitor cell trafficking is regulated by hypoxic gradients through HIF-1 induction of SDF-1. *Nat Med*. 2004;10:858-864.
41. Abbott JD, Huang Y, Liu D, Hickey R, Krause DS, Giordano FJ. Stromal cell-derived factor-1alpha plays a critical role in stem cell recruitment to the heart after myocardial infarction but is not sufficient to induce homing in the absence of injury. *Circulation*. 2004;110:3300-3305.
42. Askari AT, Unzek S, Popovic ZB, et al. Effect of stromal-cell-derived factor 1 on stem-cell homing and tissue

regeneration in ischaemic cardiomyopathy. *Lancet*. 2003; 362:697-703.
43. Ceradini DJ, Gurtner GC. Homing to hypoxia: HIF-1 as a mediator of progenitor cell recruitment to injured tissue. *Trends Cardiovasc Med*. 2005;15:57-63.
44. Staller P, Sulitkova J, Lisztwan J, Moch H, Oakeley EJ, Krek W. Chemokine receptor CXCR4 downregulated by von Hippel-Lindau tumour suppressor pVHL. *Nature*. 2003; 425:307-311.
45. De Falco E, Porcelli D, Torella AR, et al. SDF-1 involvement in endothelial phenotype and ischemia-induced recruitment of bone marrow progenitor cells. *Blood*. 2004;104: 3472-3482.
46. Hidalgo A, Sanz-Rodriguez F, Rodriguez-Fernandez JL, et al. Chemokine stromal cell-derived factor-1alpha modulates VLA-4 integrin-dependent adhesion to fibronectin and VCAM-1 on bone marrow hematopoietic progenitor cells. *Exp Hematol*. 2001;29:345-355.
47. Peled A, Kollet O, Ponomaryov T, et al. The chemokine SDF-1 activates the integrins LFA-1, VLA-4, and VLA-5 on immature human CD34(+) cells: role in transendothelial/stromal migration and engraftment of NOD/SCID mice. *Blood*. 2000;95:3289-3296.
48. Petit I, Goichberg P, Spiegel A, et al. Atypical PKC-zeta regulates SDF-1-mediated migration and development of human CD34+ progenitor cells. *J Clin Invest*. 2005;115: 168-176.
49. Hiasa KI, Ishibashi M, Ohtani K, et al. Gene transfer of stromal cell-derived factor-1{alpha} enhances ischemic vasculogenesis and angiogenesis via vascular endothelial growth factor/endothelial nitric oxide synthase-related pathway. Next-generation chemokine therapy for therapeutic neovascularization. *Circulation*. 2004;17:17.
50. Lataillade JJ, Clay D, Bourin P, et al. Stromal cell-derived factor 1 regulates primitive hematopoiesis by suppressing apoptosis and by promoting G(0)/G(1) transition in CD34(+) cells: evidence for an autocrine/paracrine mechanism. *Blood*. 2002;99:1117-1129.
51. Callaghan MJ, Ceradini DJ, Gurtner GC. Hyperglycemia-induced reactive oxygen species and impaired endothelial progenitor cell function. *Antioxid Redox Signal*. 2005;7: 1476-1482.
52. Chen YH, Lin SJ, Lin FY, Wu TC, Tsao CR, Huang PH, Liu PL, Chen YL, Chen JW. High glucose impairs early and late endothelial progenitor cells by modifying nitric oxide-related but not oxidative stress-mediated mechanisms. *Diabetes*. 2007;56(6):1559-1568.
53. Fadini GP, Miorin M, Facco M, et al. Circulating endothelial progenitor cells are reduced in peripheral vascular complications of type 2 diabetes mellitus. *J Am Coll Cardiol*. 2005; 45:1449-1457.
54. Fadini GP, Sartore S, Schiavon M, et al. Diabetes impairs progenitor cell mobilisation after hindlimb ischaemia-reperfusion injury in rats. *Diabetologia*. 2006;49:3075-3084.
55. Ingram DA, Lien IZ, Mead LE, et al. *In vitro* hyperglycemia or a diabetic intrauterine environment reduces neonatal endothelial colony-forming cell numbers and function. *Diabetes*. 2008;57:724-731.
56. Krankel N, Adams V, Linke A, et al. Hyperglycemia reduces survival and impairs function of circulating blood-derived progenitor cells. *Arterioscler Thromb Vasc Biol*. 2005;25: 698-703.
57. Loomans CJ, de Koning EJ, Staal FJ, et al. Endothelial progenitor cell dysfunction: a novel concept in the pathogenesis of vascular complications of type 1 diabetes. *Diabetes*. 2004;53:195-199.
58. Marchetti V, Menghini R, Rizza S, et al. Benfotiamine counteracts glucose toxicity effects on endothelial progenitor cell differentiation via Akt/FoxO signaling. *Diabetes*. 2006;55: 2231-2237.
59. Seeger FH, Haendeler J, Walter DH, et al. p38 mitogen-activated protein kinase downregulates endothelial progenitor cells. *Circulation*. 2005;111:1184-1191.
60. Tepper OM, Galiano RD, Capla JM, et al. Human endothelial progenitor cells from type II diabetics exhibit impaired proliferation, adhesion, and incorporation into vascular structures. *Circulation*. 2002;106:2781-2786.
61. Thum T, Fraccarollo D, Schultheiss M, et al. Endothelial nitric oxide synthase uncoupling impairs endothelial progenitor cell mobilization and function in diabetes. *Diabetes*. 2007;56:66674.
62. Chavakis E, Urbich C, Dimmeler S. Homing and engraftment of progenitor cells: a prerequisite for cell therapy. *J Mol Cell Cardiol*. 2008;45(4):514-522.
63. Seeger FH, Zeiher AM, Dimmeler S. Cell-enhancement strategies for the treatment of ischemic heart disease. *Nat Clin Pract Cardiovasc Med*. 2007;4(suppl 1):S110-S113.
64. Zemani F, Silvestre JS, Fauvel-Lafeve F, et al. Ex vivo priming of endothelial progenitor cells with SDF-1 before transplantation could increase their proangiogenic potential. *Arterioscler Thromb Vasc Biol*. 2008;28:644-650.
65. Delorme B, Basire A, Gentile C, et al. Presence of endothelial progenitor cells, distinct from mature endothelial cells, within human CD146+ blood cells. *Thromb Haemost*. 2005;94:1270-1279.
66. Corselli M, Parodi A, Mogni M, et al. Clinical scale ex vivo expansion of cord blood-derived outgrowth endothelial progenitor cells is associated with high incidence of karyotype aberrations. *Exp Hematol*. 2008;36:340-349.
67. Pesce M, Orlandi A, Iachininoto MG, et al. Myoendothelial differentiation of human umbilical cord blood-derived stem cells in ischemic limb tissues. *Circ Res*. 2003;93:e51-e62.
68. Torrente Y, Belicchi M, Sampaolesi M, et al. Human circulating AC133(+) stem cells restore dystrophin expression and ameliorate function in dystrophic skeletal muscle. *J Clin Invest*. 2004;114:182-195.
69. Horsley V, Jansen KM, Mills ST, Pavlath GK. IL-4 acts as a myoblast recruitment factor during mammalian muscle growth. *Cell*. 2003;113:483-494.
70. Au P, Daheron LM, Duda DG, et al. Differential *in vivo* potential of endothelial progenitor cells from human umbilical cord blood and adult peripheral blood to form functional long-lasting vessels. *Blood*. 2008;111:1302-1305.
71. Botta R, Gao E, Stassi G, et al. Heart infarct in NOD-SCID mice: therapeutic vasculogenesis by transplantation of human CD34+ cells and low dose CD34+KDR+cells. *Faseb J*. 2004;18:1392-1394.
72. Leor J, Guetta E, Feinberg MS, et al. Human umbilical cord blood-derived CD133+ cells enhance function and repair of the infarcted myocardium. *Stem Cells*. 2006;24:772-780.

73. Ma N, Ladilov Y, Moebius JM, et al. Intramyocardial delivery of human CD133+ cells in a SCID mouse cryoinjury model: bone marrow vs. cord blood-derived cells. *Cardiovasc Res.* 2006;71:158-169.
74. Melero-Martin JM, Khan ZA, Picard A, Wu X, Paruchuri S, Bischoff J. In vivo vasculogenic potential of human blood-derived endothelial progenitor cells. *Blood.* 2007; 109:4761-4768.
75. Murohara T, Ikeda H, Duan J, et al. Transplanted cord blood-derived endothelial precursor cells augment postnatal neovascularization. *J Clin Invest.* 2000;105:1527-1536.
76. Ott I, Keller U, Knoedler M, et al. Endothelial-like cells expanded from CD34+ blood cells improve left ventricular function after experimental myocardial infarction. *Faseb J.* 2005;19:992-994.
77. Lemarie C, Esterni B, Calmels B, et al. CD34(+) progenitors are reproducibly recovered in thawed umbilical grafts, and positively influence haematopoietic reconstitution after transplantation. *Bone Marrow Transplant.* 2007;39:453-460.
78. Bonanno G, Mariotti A, Procoli A, et al. Human cord blood CD133+ cells immunoselected by a clinical-grade apparatus differentiate *in vitro* into endothelial- and cardiomyocyte-like cells. *Transfusion.* 2007;47:280-289.
79. Jang JH, Kim SK, Choi JE, et al. Endothelial progenitor cell differentiation using cryopreserved, umbilical cord blood-derived mononuclear cells. *Acta Pharmacol Sin.* 2007;28:367-374.

Therapeutic Potential of Placental Umbilical Cord Blood in Cardiology

22

Shunichio Miyoshi, Nobuhiro Nishiyama, Naoko Hida, Akihiro Umezawa, and Satoshi Ogawa

22.1 Introduction

22.1.1 Possible Cellular Source for Cardiac Stem Cell Therapy

Cardiomyocytes do not undergo cell division after birth. Once cardiomyocytes become necrotic by myocardial infarction, residual cardiomyocytes do not undergo cell division and cannot restore damaged heart tissue. Therefore, in order to restore severely damaged heart function, heart transplantation from a living donor is performed, which, however, is restricted by a shortage of donors. Embryonic stem cells and somatic stem cells, which have a potential to transdifferentiate into cardiomyocytes, may be able to supply newly generated cardiac muscle cells and restore a severely impaired heart. Although embryonic stem cells are known to generate cardiomyocytes,[1,2] there is still an ethical problem, and possible tumorigenicity.[3] Therefore, there are many problems to overcome before progressing to clinical applications. On the other hand, some somatic stem cells have been shown to have a potential to transdifferentiate into cardiomyocytes. Our group reported for the first time that murine marrow-derived mesenchymal stem cells (MSCs) can transdifferentiate into cardiomyocytes in vitro by use of 5-azacytidine, which is known to cause nonspecific demethylation of DNA.[4] However, in humans, MSCs could not transdifferentiate into cardiomyocytes by use of 5-azacytidine; cocultivation with murine cardiomyocytes was essential.[5] Moreover, the cardiomyogenic transdifferentiation ratio was extremely low in human marrow-derived MSCs (0.1–0.3%).[5] This result appears reasonable since we believe that the human nucleus is protected from spontaneous mutation of gene and neoplasm formation, because human life is longer than that of popular experimental animals. In fact, spontaneous translocation of genes and immortalization of MSCs in mice are commonly observed[4] in vitro; however, MSCs obtained from humans seldom cause such mutation. This implies that the plasticity of the nucleus in animal cells is higher than in human cells. Therefore, experiments on human cells are an important component of these studies. Concordant with these results, recent clinical trials showed modest efficacy of marrow-derived stem cells in cardiology.[6–10] Although this was disappointing to many researchers, some thought marrow might be an acceptable cellular source for cardiac stem-cell-based therapy in mice, but not the best source in human beings. We have not yet intensively studied human somatic stem cells obtained from other organs.

22.1.2 Allograftable Cellular Source

Since marrow-derived MSCs can be obtained from patients themselves, they can be utilized in an autograft manner requiring no immunosuppressive agent against rejection. On the other hand, patients who require cardiac stem-cell therapy usually have severe coronary artery disease and severe coronary risk factors, i.e., aged patients with diabetes mellitus, etc. In such cases, ironically, adult somatic stem cells may be damaged, and not function well.[11,12] Therefore, we continue to search for allograftable MSC sources. MSCs do not express HLA-DR, which is most important for MHC class matching to suppress rejection in the allotransplantation;

S. Miyoshi (✉)
Department of Cardiology,
Keio University School of Medicine,
Tokyo, Japan
e-mail: smiyoshi@cpnet.med.keio.ac.jp

N. Bhattacharya and P. Stubblefield (eds.), *Regenerative Medicine Using Pregnancy-Specific Biological Substances*,
DOI: 10.1007/978-1-84882-718-9_22,

but MSCs do express HLA-G,[13] which is associated with a reduced incidence of rejection in the heart after transplantation.[14] Concordant with these findings, some reports showed long-term survival of xenografted MSCs in the heart,[15,16] suggesting that we may be able to perform allograft transplantation of MSCs by minimum matching in MHC class. To perform an allograft transplantation, it would be better to match major HLA types, so as to establish a MSC cellular bank system, which should be essential. Taking this situation into account, researchers have focused on medical waste as a cellular source for MSCs[17–19] because of the ability to collect cells from many young volunteers with no additional risk from collecting stem cells. This will allow us to cover all HLA types, and to perform allotransplants without, or with only minimal, use of immunosuppressive agents. In this section, we are focused on the placental umbilical cord blood (UCB) as a stem-cell source in cardiology.[18,20] There is an advantage in utilizing existing UCB banking, which has been playing a major role in hematopoietic stem-cell transplantation for leukemia treatment worldwide.

22.1.3 Lineage of Stem Cells in Placental Umbilical Cord Blood

Placental umbilical cord blood contains numerous hematopoietic stem cells (HSCs), small amounts of MSCs, and may contain other unknown types of stem cells. In the marrow-derived stem cell, CD-34 positive and/or CD133 positive HSCs do not transdifferentiate into cardiomyocytes,[21,22] but contribute to angiogenesis[23,24] or to a favorable paracrine effect on host heart tissue.[25,26] Such paracrine and neovascularization effects are also very important for partial recovery of damaged cardiomyocytes and, to some extent, are expected to have the therapeutic potential of stem-cell therapy. However, in this section, we would like to focus on stem cells that can regenerate and supply new cardiomyocytes. As previously described, MSCs can transdifferentiate into cardiomyocytes.[4,5] Although the proportion of MSCs in fresh blood samples or tissue samples is very low, we can obtain a substantial number of MSCs by ex vivo expansion, since MSCs can multiply markedly under culture conditions. This implies that the character of MSCs in an experiment may be affected strongly by isolation and culture conditions. Isolation, characterization, and differentiation of clonally expanded human MSCs derived from umbilical cord blood have been reported,[27,28] and the MSCs have been found to have multipotentcy. The immunophenotype of the clonally expanded cells is consistent with that reported from marrow mesenchymal stem cells.[20,27,28] Kögler et al.[29] also reported that CD45-negative populations from human cord blood, termed as unrestricted somatic stem cells (USSCs), have a pluripotent differentiation potential, and their phenotype is almost consistent with MSCs. These established mesenchymal phenotype cells from UCB may not have the same potential; however, they may have powerful cardiomyogenic transdifferentiation ability and a favorable effect on cardiac function.

22.1.4 Cardiomyogenic Transdifferentiation Potential In Vitro

Murine MSCs can be transdifferentiated into cardiomyocytes by 5-azacytidine treatment[4]; however, it is difficult for human marrow-derived MSCs to transdifferentiate in vitro.[5] It is believed that MSCs can transdifferentiate toward the surrounding cell, and unknown "environmental factors" derived from the surrounding cells are the key elements for transdifferentiation.[30] In fact, in order to cause in vitro cardiomyogenesis, cocultivation with cardiomyocytes obtained from other animals should be required as an "environmental factor," as the specific factor for cardiomyogenic transdifferentiation is still unclear. Cardiomyogenic transdifferentiation efficiency of human UCB-derived MSCs (UCBMSCs) was significantly higher (45–50%[18]) than that of human marrow-derived MSCs (0.1–0.3%[5]) in vitro. The differentiated UCBMSCs-derived cardiomyocytes showed a clear striation pattern of contractile protein and gap junctions between the cells (Fig. 22.1). Furthermore, they contracted spontaneously, had cardiomyocyte-specific action potential (Fig. 22.2), and caused a physiological response to pharmacological agents.

The in vitro cocultivation system plays an important role in understanding the mechanism of cardiomyogenic transdifferentiation from MSCs, and also in improving the efficiency of transdifferentiation in situ.

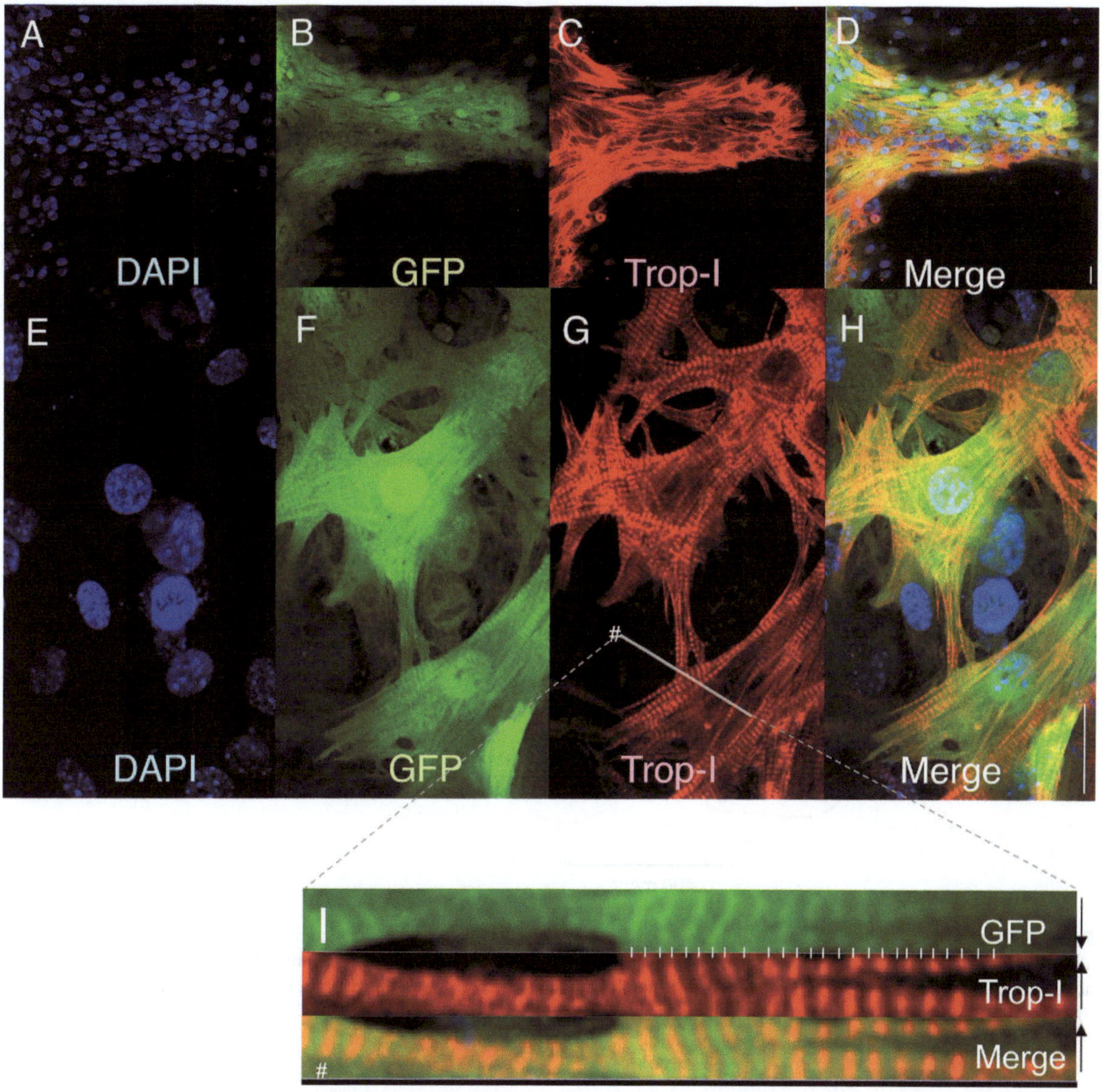

Fig. 22.1 Cardiomyogenic transdifferentiation of umbilical cord blood mesenchymal stem cells. Lower magnification view (**a–d**) and higher magnification view (**e–h**) of laser confocal microscopic view of immunocytochemistry of differentiated umbilical cord blood mesenchymal stem cells (UCBMSCs) with anti-cardiac troponin-I (Trop-I) antibody. Significant numbers of differentiated GFP-positive EMCs (*green*) had troponin-I (*red*) in their cytoplasm (yellow as a result of "merging," **d**, **h**). Nuclei are stained with DAPI (**a**, **e**; *blue*). Clear troponin-I (Trop-I, *red*) staining with striation pattern can be observed. GFP (**b**, *green*) and Trop-I (**c**, *red*) along the white line in (**g**) are magnified in panel I. Interestingly, troponin-I staining and GFP were observed alternately in a striated manner, suggesting that troponin-I is expressed in the GFP-positive cells. Scale bars in the figure denoted 20 μm

However, the cocultivation system also causes some confusion in identifying human MSCs from feeder animals' cardiomyocyte in culture, i.e., selection in cell tracking dye, neglect of possible cell-fusion-dependent acquisition of cardiomyocyte phenotype, and difficulty in direct application for clinical patients. Therefore, defining the key element(s) in the "environmental factor," and utilizing the element(s) in the establishment of a feeder-cell-independent in vitro cardiomyogenic differentiation method should be important to improve the efficacy of cardiac stem-cell therapy.

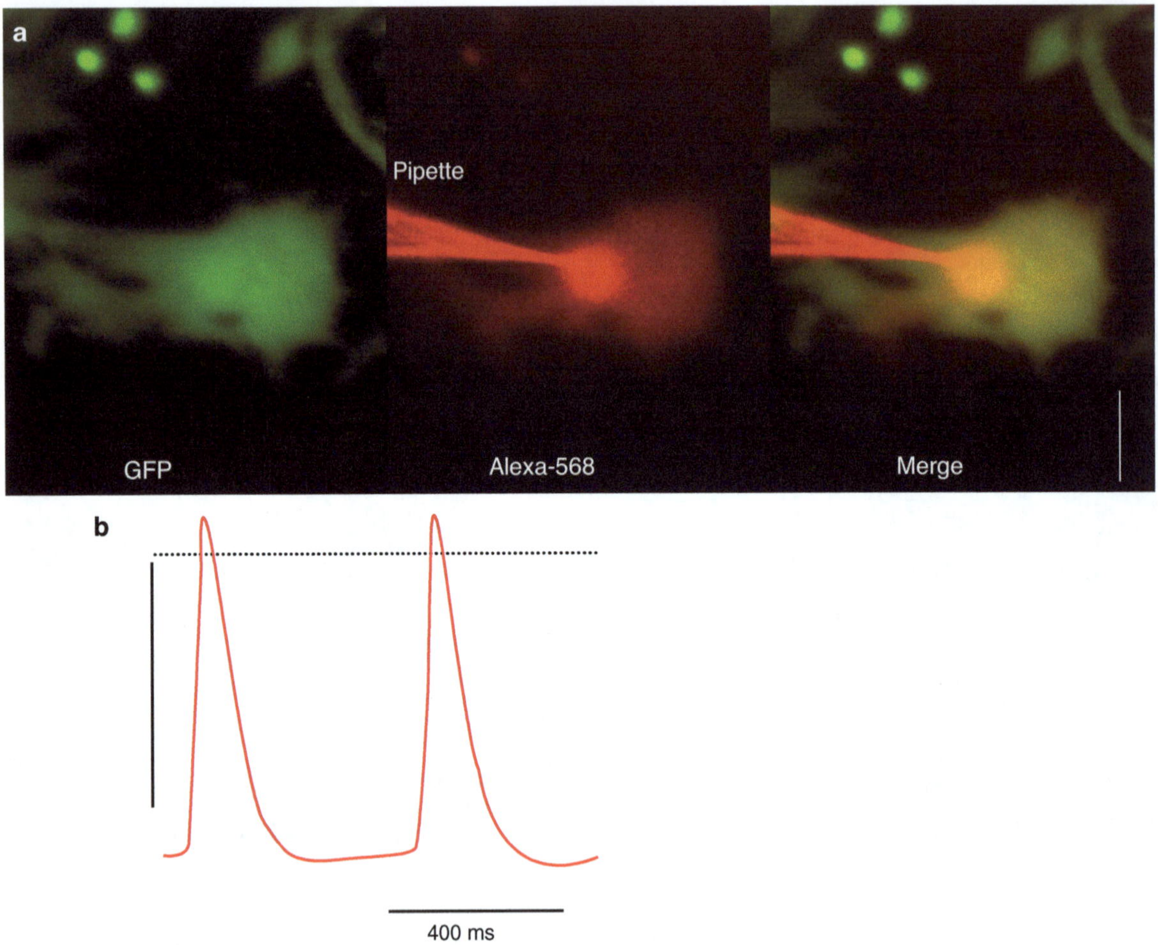

Fig. 22.2 Functional analysis of UCBMSCs and UCBMSCs-TERT. Representative fluorescent microscopic image during action potential (AP) recording is shown in (**a**). Immediately after the AP recordings, alexa-568 dye (*red*) was injected into the cell via the same recording electrode to confirm that the recorded AP was obtained from GFP-positive UCBMSCs. (**b**) Representative APs. The dotted line denotes the 0 mV level and the vertical line denotes 50 mV

22.1.5 Effect of UCB-Derived Stem Cell in Cardiology In Vivo

Marrow-derived stem-cell transplantation was reported to have favorable effects on the cardiac function in both experimental settings[16,23,31,32] and in clinical patients.[6–10,33] Recent reports, however, have shown that the efficacy was modest and the mechanisms of the favorable effects were caused not by newly generated stem-cell-derived cardiomyocytes,[21,22] but by angiogenesis and antiapoptotic effects of stem cells (paracrine effects), since the efficiency of cardiomyogenic transdifferentiation is extremely low in human marrow-derived stem cells.[5] UCB transplantation has been reported to improve cardiac function.[34–36] That study, however, used a fraction of hematopoietic lineage and failed to show any evidence for cardiomyogenesis in vivo. Therefore, the mechanism of the improvement may be due to angiogenesis and/or paracrine effects. Kim et al.[37] showed modest but significant functional recovery of impaired cardiac function by transplantation of human USSCs obtained from umbilical cord blood that expressed mesenchymal cell surface markers; therefore, mesenchymal lineage of the cells obtained from UCB may also have potential therapeutic advantage in cardiac stem-cell therapy. Ahn et al.[38] also showed that UCBMSC transplantation decreases fibrosis, apoptosis, and preserves ventricular function in an ischemia-reperfusion myocardial injury model in vivo. However, no report has shown clear evidence of cardiomyogenic transdifferentiation ability of UCBMSCs in vivo, in spite of significant efficiency of UCBMSCs in vitro. This discrepancy may be caused by differences in an "environmental factor." Since the key element in the "environmental factor" is still unclear, no one knows whether the environmental factor of an in situ heart is enough for UCBMSC to transdifferentiate. Moreover, cardiomyogenic transdifferentiation ability in vivo may be underestimated by the difficulty of devising a tracking system for identifying engrafted cells in situ.

22.1.6 Limitations and Future of UCBMSCs in Cardiology

UCBMSCs are a promising cellular source for cardiac stem-cell therapy, but the number of stem cells containing umbilical cord blood is small. In our previous study,[20] we could establish only a single colony of UCBMSCs from four culture challenges of 10 cc of umbilical cord blood; this is quite low in efficiency when compared to marrow-derived MSCs (usually about 1,000,000 cells can be obtained at the first passage of the culture). Fortunately, the speed of population doubling is faster and the number of times the population doubles is greater in UCBMSCs.[20] Therefore, if we establish UCBMSCs from the placental umbilical cord blood, we may be able to expand the number of cells and finally obtain enough stem cells. Thus, the effort to establish an efficient method to collect UCBMSCs from umbilical cord blood should be put forth.

As described in the previous section, the key element(s) in the "environmental factor" for genesis of cardiomyocytes from MSCs is (are) still unclear. UCBMSCs and our established immortalized UCBMSC line, which contained human TERT gene, was introduced by a retrovirus and called UCBMSC-TERT, along with our cocultivation system in vitro, has extremely high cardiomyogenic differentiation efficiency. Comparison of microarray analysis between UCBMSCs and marrow-derived MSCs at the default state may help us understand the dramatic difference in cardiomyogenic transdifferentiation efficiency. Furthermore, since UCBMSCs showed extremely high cardiomyogenic differentiation efficiency in our system, the key element or elements of the "environmental factor" does exist in our cocultivation system. By analyzing our coculture system and UCBMSC-TERT, we may be able to define the key element(s) in the "environmental factor." There has been no evidence of cardiomyogenesis of UCBMSCs in vivo at the present time; however, by administration with the defined key element(s), we may be able to supply newly generated cardiomyocytes.

References

1. Klug MG, Soonpaa MH, Koh GY, Field LJ. Genetically selected cardiomyocytes from differentiating embronic stem cells form stable intracardiac grafts. *J Clin Invest.* 1996; 98(1):216-224.
2. Min JY, Yang Y, Converso KL, et al. Transplantation of embryonic stem cells improves cardiac function in postinfarcted rats. *J Appl Physiol.* 2002;92(1):288-296.
3. Leor J, Gerecht-Nir S, Cohen S, et al. Undifferentiated human embryonic stem cells are not guided to form new myocardium by transplantation into normal and infarcted heart. *J Am Coll Cardiol.* 2005;45(3 suppl A):151 (abstract).
4. Makino S, Fukuda K, Miyoshi S, et al. Cardiomyocytes can be generated from marrow stromal cells in vitro. *J Clin Invest.* 1999;103(5):697-705.
5. Takeda Y, Mori T, Imabayashi H, et al. Can the life span of human marrow stromal cells be prolonged by bmi-1, E6, E7, and/or telomerase without affecting cardiomyogenic differentiation? *J Gene Med.* 2004;6(8):833-845.
6. Assmus B, Fischer-Rasokat U, Honold J, et al. Transcoronary transplantation of functionally competent BMCs is associated with a decrease in natriuretic peptide serum levels and improved survival of patients with chronic postinfarction heart failure: results of the TOPCARE-CHD Registry. *Circ Res.* 2007;100(8):1234-1241.
7. Assmus B, Schachinger V, Teupe C, et al. Transplantation of progenitor cells and regeneration enhancement in acute myocardial infarction (TOPCARE-AMI). *Circulation.* 2002; 106(24):3009-3017.
8. Wollert KC, Meyer GP, Lotz J, et al. Intracoronary autologous bone-marrow cell transfer after myocardial infarction: the BOOST randomised controlled clinical trial. *Lancet.* 2004;364(9429):141-148.
9. Lunde K, Solheim S, Aakhus S, Arnesen H, Abdelnoor M, Forfang K. Autologous stem cell transplantation in acute myocardial infarction: the ASTAMI randomized controlled trial. Intracoronary transplantation of autologous mononuclear bone marrow cells, study design and safety aspects. *Scand Cardiovasc J.* 2005;39(3):150-158.
10. Schachinger V, Erbs S, Elsasser A, et al. Improved clinical outcome after intracoronary administration of bone-marrow-derived progenitor cells in acute myocardial infarction: final 1-year results of the REPAIR-AMI trial. *Eur Heart J.* 2006; 27(23):2775-2783.
11. Fadini GP, Miorin M, Facco M, et al. Circulating endothelial progenitor cells are reduced in peripheral vascular

complications of type 2 diabetes mellitus. *J Am Coll Cardiol*. 2005;45(9):1449-1457.
12. Heiss C, Keymel S, Niesler U, Ziemann J, Kelm M, Kalka C. Impaired progenitor cell activity in age-related endothelial dysfunction. *J Am Coll Cardiol*. 2005;45(9):1441-1448.
13. Selmani Z, Naji A, Zidi I, et al. Human leukocyte antigen-G5 secretion by human mesenchymal stem cells is required to suppress T lymphocyte and natural killer function and to induce CD4+CD25highFOXP3+ regulatory T cells. *Stem Cells*. 2008;26(1):212-222.
14. Lila N, Amrein C, Guillemain R, et al. Human leukocyte antigen-G expression after heart transplantation is associated with a reduced incidence of rejection. *Circulation*. 2002;105(16):1949-1954.
15. Makkar RR, Price MJ, Lill M, et al. Intramyocardial injection of allogenic bone marrow-derived mesenchymal stem cells without immunosuppression preserves cardiac function in a porcine model of myocardial infarction. *J Cardiovasc Pharmacol Ther*. 2005;10(4):225-233.
16. Shake JG, Gruber PJ, Baumgartner WA, et al. Mesenchymal stem cell implantation in a swine myocardial infarct model: engraftment and functional effects. *Ann Thorac Surg*. 2002;73(6):1919-1925. Discussion 1926.
17. Miyoshi S, Hida N, Nishiyama N, et al. Human menstrual blood is a potential cell source for cardiac stem cell therapy [abstract]. *J Am Coll Cardiol*. 2005;45(3 suppl A):156.
18. Nishiyama N, Miyoshi S, Hida N, et al. The significant cardiomyogenic potential of human umbilical cord blood-derived mesenchymal stem cells in vitro. *Stem Cells*. 2007;25(8):2017-2024.
19. Okamoto K, Miyoshi S, Toyoda M, et al. 'Working' cardiomyocytes exhibiting plateau action potentials from human placenta-derived extraembryonic mesodermal cells. *Exp Cell Res*. 2007;313:2550-2562.
20. Terai M, Uyama T, Sugiki T, Li X, Umezawa A, Kiyono T. Immortalization of human fetal cells: the life span of umbilical cord blood-derived cells can be prolonged without manipulating p16INK4a/RB braking pathway. *Mol Biol Cell*. 2005;16:1491-1499.
21. Balsam L, Wagers A, Christensen J, Kofidis T, Weissman I, Robbins R. Haematopoietic stem cells adopt mature haematopoietic fates in ischaemic myocardium. *Nature*. 2004;428:668-673.
22. Murry C, Soonpaa M, Reinecke H, et al. Haematopoietic stem cells do not transdifferentiate into cardiac myocytes in myocardial infarcts. *Nature*. 2004;428:664-668.
23. Gojo S, Gojo N, Takeda Y, et al. In vivo cardiovasculogenesis by direct injection of isolated adult mesenchymal stem cells. *Exp Cell Res*. 2003;288(1):51-59.
24. Tang YL, Zhao Q, Zhang YC, et al. Autologous mesenchymal stem cell transplantation induce VEGF and neovascularization in ischemic myocardium. *Regul Pept*. 2004;117(1):3-10.
25. Kocher AA, Schuster MD, Szabolcs MJ, et al. Neovascularization of ischemic myocardium by human bone-marrow-derived angioblasts prevents cardiomyocyte apoptosis, reduces remodeling and improves cardiac function. *Nat Med*. 2001;7(4):430-436.
26. Gnecchi M, He H, Liang O, et al. Paracrine action accounts for marked protection of ischemic heart by Akt-modified mesenchymal stem cells. *Nat Med*. 2005;11:367-368.
27. Goodwin HS, Bicknese AR, Chien SN, Bogucki BD, Quinn CO, Wall DA. Multilineage differentiation activity by cells isolated from umbilical cord blood: expression of bone, fat, and neural markers. *Biol Blood Marrow Transplant*. 2001;7(11):581-588.
28. Lee OK, Kuo TK, Chen WM, Lee KD, Hsieh SL, Chen TH. Isolation of multipotent mesenchymal stem cells from umbilical cord blood. *Blood*. 2004;103(5):1669-1675.
29. Kögler G, Sensken S, Airey J, et al. A new human somatic stem cell from placental cord blood with intrinsic pluripotent differentiation potential. *J Exp Med*. 2004;200:123-135.
30. Tomita S, Li RK, Weisel RD, et al. Autologous transplantation of bone marrow cells improves damaged heart function. *Circulation*. 1999;100(19 suppl):II247-II256.
31. Orlic D, Kajstura J, Chimenti S, et al. Bone marrow cells regenerate infarcted myocardium. *Nature*. 2001;410(6829):701-705.
32. Wang JA, Fan YQ, Li CL, He H, Sun Y, Lv BJ. Human bone marrow-derived mesenchymal stem cells transplanted into damaged rabbit heart to improve heart function. *J Zhejiang Univ Sci B*. 2005;6(4):242-248.
33. Chen SL, Fang WW, Ye F, et al. Effect on left ventricular function of intracoronary transplantation of autologous bone marrow mesenchymal stem cell in patients with acute myocardial infarction. *Am J Cardiol*. 2004;94(1):92-95.
34. Hirata Y, Sata M, Motomura N, et al. Human umbilical cord blood cells improve cardiac function after myocardial infarction. *Biochem Biophys Res Commun*. 2005;327:609-614.
35. Leor J, Guetta E, Feinberg M, et al. Human umbilical cord blood-derived CD133+ cells enhance function and repair of the infarcted myocardium. *Stem Cells*. 2006;24:772-780.
36. Ma N, Stamm C, Kaminski A, et al. Human cord blood cells induce angiogenesis following myocardial infarction in NOD/scid-mice. *Cardiovasc Res*. 2005;66:45-54.
37. Kim BO, Tian H, Prasongsukarn K, et al. Cell transplantation improves ventricular function after a myocardial infarction: a preclinical study of human unrestricted somatic stem cells in a porcine model. *Circulation*. 2005;112(9 suppl):I96-I104.
38. Ahn YK, Song CH, Nam KI, et al. Cord blood derived mesenchymal stem cell injection into ischemia-reperfusion myocardial injury decreases fibrosis, apoptosis, and significantly preserves ventricular systolic function. *Circ J*. 2005;69(suppl):396.

Stem Cell Therapy for Heart Failure Using Cord Blood

23

Amit N. Patel, Ramasamy Sakthivel, and Thomas E. Ichim

23.1 Introduction

For the approximately 5 million Americans with heart failure, of which only a small proportion are eligible for transplantation, regenerative medicine is the only therapeutic hope. CHF is caused by many factors such as poor perfusion due to atherosclerotic disease, a previous heart attack, a congenital defect, or previous viral infection, but the end result is usually similar: a self-perpetuating cycle of cardiomyocyte death, inflammatory mediator release, myocardial compensatory hypertrophy, and additional cardiomyocyte death, culminating in a deterioration of ejection fraction. Numerous common themes are associated with the progression to heart failure. We will discuss below how stem cell therapy may act on these factors in a therapeutic sense.

23.2 Inhibition of Inflammatory Cascade by Mesenchymal Stem Cells

Ongoing inflammation is part of the cascade leading to heart failure. Acute inflammation occurs during infarction as a result of tissue damage; however, chronic inflammatory markers are present in post-infarct patients, as well as ischemic heart failure patients, and patients with congenital defects. In general, a positive correlation between advanced heart failure and levels of the inflammatory marker, the pentraxin C-reactive protein (CRP) has been reported.[1,2] While CRP elevation is conventionally seen as a marker of ongoing inflammation, produced by the liver in response to cytokines such as IL-1, IL-6, and TNF-alpha,[3] it also plays an active role in cardiac deterioration through induction of endothelial dysfunction,[4,5] as well as exacerbation of inflammatory processes through activation of complement.[6,7] In addition to CRP, elevated levels of inflammatory cytokines are also noted in CHF patients.[8] Inflammatory mediators are produced not only as a result of cardiomyocyte ischemia, but also stretch injury as a result of hypertrophic accommodation[9,10] and systemic activation of immune cells including T cells[11] and monocytes.[12] Functionally, inflammatory mediators induce direct apoptosis of cardiomyocytes. For example, TNF-alpha is known to induce reduction of bcl-2 gene expression and activate a caspase-dependent apoptosis in cardiac cells at physiological concentrations.[13] Reduction of TNF-alpha activity using soluble receptors has demonstrated beneficial effects in animal models of heart failure.[14]

The importance of inflammatory stimuli in heart failure can be seen in animal models in which activators of inflammatory agents, such as toll-like receptors (TLRs) are knocked out. Generally, TLRs, particularly TLR 2 and 4, recognize endogenous "danger signals" associated with damaged tissue such as extracellular matrix degradation products,[15,16] and heat shock proteins.[17] Doxorubicin-induced heart failure is substantially attenuated in animals lacking TLR-2[18] or TLR-4.[19] TLR-2 knockout mice have a substantially better clinical outcome after experimental infarction, including reduction in remodeling, wall

A.N. Patel (✉)
Associate Professor of Surgery, Director of Cardiovascular Regenerative Medicine, Divison of Cardiothoracic Surgery, Director of Clinical Regenerative Medicine, CTF, 30 N 1900E 3C127, Salt Lake City, UT 84132

N. Bhattacharya and P. Stubblefield (eds.), *Regenerative Medicine Using Pregnancy-Specific Biological Substances*,
DOI: 10.1007/978-1-84882-718-9_23,

thinning, and preservation of LVEF as compared to wild-type controls.[20] Clinically, expression of TLR-4 is associated with poor prognosis in post-infarct patients.[21] Thus, it appears that inflammation is associated with progression of heart failure.

Mesenchymal stem cells (MSCs) were originally identified as "stromal cells," believed to play a role in shaping the bone marrow microenvironment where hematopoiesis occurs.[22] More recently, MSC-like populations have been isolated from a diverse range of sources such as adipose,[23] heart[24], Wharton's Jelly,[25] dental pulp,[26] peripheral blood,[27] cord blood,[28] and more recently menstrual blood.[29-31] In addition to potent regenerative activities of MSCs, which we will describe below, MSCs have potent anti-inflammatory activities which appear to be present regardless of tissue of origin.[32,33] Mechanistically, MSCs appear to suppress inflammation through secretion of anti-inflammatory cytokines such as IL-10,[34] TGF-beta,[35] LIF,[36] soluble HLA-G,[37] and IL-1 receptor antagonist,[38] expression of immune regulatory enzyme such as cycloxygenase[39] and indolamine 2,3 deoxygenase,[40] and ability to induce generation of anti-inflammatory T regulatory cells.[41] The in vivo anti-inflammatory effects of MSCs may be witnessed by success in treating animal models of immune mediate/inflammatory pathologies such as multiple sclerosis,[42] colitis,[43] graft versus host,[44] rheumatoid arthritis,[45] and ischemia/reperfusion injury.[46] In heart failure, administration of MSC post-infarct has been demonstrated to decrease production of TNF-alpha and IL-6, but upregulate generation of the anti-inflammatory cytokine IL-10, which correlated with therapeutic benefit.[47] Clinically, MSCs have demonstrated repeatedly potent therapeutic activity at suppressing graft versus host (GVHD), for which Phase III FDA-registration trials are currently ongoing.[48–53] Thus, one angle in which stem cell therapy may be useful for heart failure is by suppressing ongoing self-perpetuating inflammatory cascade.

23.3 Inhibition of Death/Repair

Cardiomyocyte death, either by apoptosis,[54] or other types of death such as autophagy and programmed necrosis, is part of the self-perpetuating cascade leading to heart failure.[55,56] Thus, the manipulation of these death pathways and upregulation of endogenous repair mechanisms in the heart could be a possible method of decreasing the progression to heart failure. For example, suppression of apoptotic machinery such as the transgenic expression of a dominant negative form of Mammalian sterile 20-like kinase-1 (Mst1) inhibits post-infarct remodeling, through suppression of cardiomyocyte death[57]; similar protective effects can be attained by transfection of anti-apoptotic genes such as IAP-2.[58] Transfection of genes such as hepatocyte growth factor (HGF-1) into cardiac cells has also been demonstrated to reduce progression to heart failure in animal models.[59] ACE inhibitors have been postulated to have some beneficial effects through inhibition of cardiomyocyte apoptosis.[60] Thus, one attractive method of addressing the progression to heart failure would be the identification of methods to prevent ongoing cell death.

Cell death in the heart causes some level of replacement by resident cardiac stem cells (CSCs). These cells are relatively rare and are believed to respond to signals associated with damage to the myocardium. Fransioli et al. generated a transgenic mouse expressing GFP under control of the c-kit promoter. Subsequent to infarct, increased proliferation of c-kit positive cells was seen in the myocardium.[61] In humans, Urbanek et al. examined 20 human hearts from patients who died after acute infarct, 20 hearts with chronic infarct that were transplanted and 12 control hearts. A population of cells expressing c-kit, MDR1, and Sca-1 were seen to enter cell cycle from a basal rate of 1.5% cycling cells in controls to 28% and 14% in acute and chronic infarcts, respectively. The cells expressing the phenotype were demonstrated to be capable of differentiating into myocyte, smooth muscle, and endothelial cell lineages.[62] Isolated CSCs have been successfully expanded ex vivo and administered via the intracoronary route in rats post-infarct. Successful transmigration of the CSC across the endothelium and active regeneration of myocardium was demonstrated.[63] Thus, it appears that a functional population of stem cells exists in the heart that can, to some extent, cause regeneration post-injury.

Both hematopoietic stem cells (HSC) and MSC are capable of secreting factors that on the one hand inhibit apoptosis[64–66] and on the other hand stimulate activation of CSC.[63] For example, it was demonstrated that administration of non-fractionated bone marrow cells containing both cell populations protects against apoptosis in a doxorubin-induced cardiomyopathy model.[67] Furthermore, bone marrow cells are known to produce HGF,[68] and IGF-1[69] that are anti-apoptotic and activate endogenous cardiomyocyte stem cells,[70] have been reported. Interestingly, production of these factors is upregulated in response to inflammatory mediators associated with heart failure such as TNF-alpha.[66] Therefore, it may be possible to believe that MSCs not only migrate to injured tissue but can "sense" inflammatory stimuli such as TNF-alpha and actually try to grade the level of their therapeutic response according to the level of damage sensed.

Another means by which stem cells may repair the heart is through actually differentiating into new heart muscle. Reports exist of both hematopoietic[71] and mesenchymal stem cells[72,73] differentiating into cardiac-like cells, although this is controversial and some groups have reported this to be a product of cell fusion.[74] Additionally, stem cells promote angiogenesis, thus providing nutrients to ischemic areas and potentially allowing regeneration.[75,76]

23.4 Currently Stem Cell Therapy Helps Heart Patients: Just Not That Well

Cell-based regenerative therapy is based on the concept that replacing dead or deficient cardiac muscle with injected therapeutic stem cells might augment contractile function in the heart.[77] Various experimental studies provided evidence that the infusion or injection of stem or progenitor cells may reduce scar formation and fibrosis. The most promising approach for regeneration of necrotic myocardium is cell transplantation; the implantation of muscle cells, progenitor cells, or pluripotent stem cells in infarcted myocardium has been reported by many investigators.[77–81] Initial encouraging results from animal studies[82–85] and several small, uncontrolled clinical trials[86–88] have prompted scientists and clinicians to focus more on critical evaluation of the scientific basis for myocyte regeneration and the efficacy of such proposed cell therapy in clinical practice.

Initial studies on myocardial colonization by contractile cells were demonstrated by injecting cells taken from transgenic mice expressing the gene of β-galactosidase which could then be identified by specific staining.[89] Likewise, the presence of allogeneic dystrophin-positive cells in the heart of dogs suffering from Duchenne muscular dystrophy has brought additional evidence for the capacity of transplanted cells to be inserted into the recipient myocardium and integrated within the cardiac tissue, as demonstrated by the formation of intercalated discs between host and recipient cells.[79,89] Finally, fetal cardiomyocytes were shown to be able to survive in the zone bordering a myocardial infarct, opening the possibility of colonization of an ischemic myocardium.[90]

Moreover, Li et al. (1996) have showed in cryoinjury-induced myocardial infarction in the rat that the intramyocardial injection of fetal cardiomyocytes improved systolic and diastolic function up to two months after transplantation, as assessed by ex vivo Langendorff perfusion studies.[91] In a reperfused infarcted area in rats, fetal cardiomyocytes were intramyocardially implanted and, one month later, function was found to be significantly improved in transplanted animals, compared with controls.[90] Furthermore, in a mouse model of anthracycline-induced toxic cardiomyopathy, it was possible to show that injected fetal cardiomyocytes also improved cardiac function one month after transplantation, as compared with control nontransplanted animals indicating cellular transplantation could effectively improve function of ischemically damaged myocardium.

However, issues relating to availability, ethical problems regarding the fetal source of myocytes and the necessity for immunosuppressive therapy limit the potential clinical application of this allograft technique. In the search of alternative cellular types, investigators have refocused on the clinical use of other cell types such as skeletal myoblasts marrow-derived stromal cells and peripheral blood-derived precursor cells to be used in cell therapy for cardiac ischemia.

The precursors of skeletal muscle fibers, the myoblasts, present in adult animals as quiescent cells and may become activated, proliferate, and differentiate upon muscle injury (in vivo) or following tissue dissociation (in vitro) in culture. Working on primary myoblast transplantations in a cryoinjury model of myocardial infarction in dogs, Chiu et al. (1995) were able to characterize the donor cells in the myocardium 14 weeks after injection.[92] Murry et al. (1996) observed the formation of myotubes and skeletal muscle fibers within the cardiac tissue but they were unable to identify cardiac-specific markers within the tissue formed by the injection of myoblasts.[93] Taylor et al. (1998) assessed transplanted myoblasts in a model of rabbit heart cryoinfarction and observed a functional improvement of the skeletal muscle cells within the scar tissue following transplantation by sonomicrometry.[81] More importantly, in this study, the functional improvement was only seen in those animals in which implanted cells were histologically identified, thereby bringing strong evidence for a causal relationship between the presence of engrafted cells and the functional outcome. On the other hand, Murry et al. (1996) could not identify changes in the myoblast transplanted phenotype.[93] In a myocardial infarction model in rats created by coronary artery ligation, Scorsin et al. (1998, 2000) showed that one month after myoblast transplantation the treated group displayed a significant improvement of function, primarily manifested as a limitation of post-infarction ventricular remodeling, compared with nontransplanted controls.[90,94] No gap junction was detected between transplanted cells as demonstrated by the negative staining for connexin-43. Further experimental studies have shown that when implanted into post-infarction scar tissue, skeletal myoblasts (satellite cells) improved the left ventricular hemodynamic parameters, including contractile function.[94] The results of experimental studies[95,96] have encouraged Menasché et al.[97] to perform the first autologous skeletal myoblast transplantation in a patient, a survivor of myocardial infarction, during CABG. This was shortly followed by a similar procedure by Siminiak et al.[98] and many other investigators.[82,99]

Despite promising results in animal models and initial human studies, skeletal myoblast transplantation into damaged myocardium remains an unproven technology. A number of technical issues remain unresolved, including optimum cell type, ideal number of cells, factors that promote engraftment, surgical delivery method, and patient selection criteria. Several studies are geared toward addressing these issues by many investigators[82–85] and several encouraging small, uncontrolled clinical trials.[86–88] In 2007, Genezyme in association with Medtronic's MAGIC clinical trial on evaluation of skeletal myoblast transplant for treating ischemic heart failure was stopped due to lack of efficacy. The long-term viability and functionality of the transplanted cells has not been proven. Additionally, none of the studies have demonstrated an improvement in patient functional status or survival. Thus, transplantation of skeletal myoblasts remains experimental for the treatment of damaged myocardium. The above studies clearly demonstrate that skeletal myoblasts do not fulfill the major criteria required for a true cardiac regeneration: a coupling of the grafted cells with those of the recipient myocardium and the subsequent generation of a contractile force. These observations provide strong rationale to explore alternative cell populations, among which marrow-derived stem cells are particularly attractive.

Cardiomyocytes formation from circulating BM cells has been first demonstrated by Bittner et al.[100] Goodel's group then demonstrated that in infarcted hearts of adult mice, cardiomyocytes and vascular cells can be generated from stem cell population isolated from mouse BM cells, termed side population (SP) cells.[101] The BM-derived SP cells that express a hematopoietic stem cell antigen Sca-1 and a VEGF receptor Flt-1 have a capacity to form cardiomyocytes and vascular cells in damaged hearts, whereas the exact origin for these cells in the SP compartment has not been confirmed yet.[102] Cardiomyocytes are also formed after direct injection of BM-derived cells such as hematopoietic lineage marker-negative (Lin^-) and c-kit^+ (the receptor for stem cell factor) similar to HSCs[103] or MSCs[104] into damaged heart tissue. Recent work documented that there is a close relationship between hematopoietic stem cells (HSCs) and Endothelial Progenitor Cells (EPC),[105] as well as between mesenchymal stem cells (MSCs) and EPCs.[106,107] Recently, implantation of BM mononuclear cells was shown to provide angiogenesis and enhance regional function of the ischemic myocardium in the pig. Despite the very interesting and promising option to restore myocardial viability, the use of BM stem cells raises two important questions. First, they have not yet shown the potential to multiply and produce a very large number of differentiated cells in vitro, the first condition to fully recolonize diseased myocardium and thus improving ventricular function. Second, there is a risk of developing other types of tissues including severe calcification[108] if undifferentiated BM stem cells are used.

Strauer et al. reported a case in which a 46-year-old patient received autologous bone marrow mononuclear cells by a percutaneous transluminal catheter placed in the infarct-related artery. Ten weeks after administration, the transmural infarct area had been reduced from 24.6% to 15.7% of left ventricular circumference, while ejection fraction, cardiac index, and stroke volume had increased by 20–30%.[109] A subsequent paper in the same year reported administration of similar cells in five patients with advanced ischemia undergoing coronary artery bypass grafting. Cells were administered intramuscularly into areas deemed ungraftable and perfusion was assessed by imaging.

Specific improvement in areas injected was documented in three of the five patients. Perhaps more importantly, no ectopic growths or adverse effects were reported at 1-year follow-up.[110] Since these pioneering studies, cardiac stem cell therapy has been used by numerous groups for numerous conditions causing heart failure. These can be broken down into (a) inhibiting post-acute myocardial infarction remodeling; (b) stimulation of regeneration in chronically injured hearts; and (c) induction of angiogenesis in coronary artery disease. The methods of administering stem cells have included the intracoronary, epicardial, and intravenous routes. Stem cells used to date are bone marrow mononuclear cells, mobilized peripheral blood stem cells, purified CD34 or CD133 cells, autologous mesenchymal stem cells, and allogeneic bone marrow and placental mesenchymal stem cells.

Several meta-analyses of ongoing clinical trials performed indicated that both hematopoietic and mesenchymal cells have promising clinical effects in various types of heart failure. Briefly, Abdel-Latif et al. described 999 patients enrolled in 18 independent controlled cardiac trials in which patients were treated with either unfractionated bone marrow cells, bone marrow mesenchymal, or mobilized peripheral blood.[111] They found that in comparison to controls, there was a statistically significant improvement in ejection fraction, reduction in infarct size, and left ventricular end-systolic volume. Importantly, no safety issues or serious treatment-associated adverse events were noted. In another such comprehensive review, Martin-Rendon et al. focused on bone marrow therapy for post-acute infarction trials. Of 13 randomized studies conducted, encompassing 811 participants, the authors of the review stated that more trials are needed to establish efficacy in terms of clinical endpoints such as death. However, authors of the review did observe a consistent improvement in LVEF, as well as trends for decrease in left ventricular end-systolic and end-diastolic volumes, and infarct size.[112] Two other meta-analyses of randomized trials in the area of bone marrow stem cell infusions also supported the conclusion of safety and mild but statistically significant improvement in LVEF.[113,114] These data suggest that stem cell therapy, both hematopoietic and mesenchymal have clinical effects in various types of heart failure. Theoretically, the leap between these clinical trials and widespread implementation is more of a business than a medical question. In order to postulate on the future of cardiac stem cell therapy, we will discuss several possible means of optimizing existing work.

However, the lack of concordance between emerging animal data showing low or undetectable contributions of BM cells to new cardiomyocyte formation[115,116] and the contrasting clinical benefits derived in early uncontrolled trials[117–119] raises the question of what mechanism underlies contractile-function improvement following human BM cell therapy. Mechanisms of cell benefit in addition to myogenesis might involve intrinsic remodeling of the ventricle in the absence of a direct cellular effect, improvement in contractile function, and myocardial perfusion secondary to the primary revascularization procedure (in the case of stent placement or coronary bypass surgery) or impairment of myocyte death. Additional nonmyogenic contributions to BM cell improvements in cardiac function might include vasculogenic effects of angioblastic, myeloid, mesenchymal or other stem cells resident in the marrow and heart, or paracrine effects of these cells through the release of pro-angiogenic growth and survival factors. Moreover, hematopoietic stem cells characterized by CD133+ and/or CD34+ expression might contribute directly or indirectly to cardiac remodeling, myocyte survival, or long-term improvements in contractile function.

CD133+ EPC are of particular interest in studies directed to therapeutic angiogenesis. Reports by Asahara and other groups indicate that these cells differentiate into endothelial cells after short-term culture.[120] Stamm et al. injected high dose of autologous BM-derived CD133+ cells in six patients in an akinetic infarcted area not amenable to revascularization at the time of CABG.[119] Four patients demonstrated improved ejection fraction (EF), and five patients demonstrated decreased ischemic defect on nuclear scintigraphy.[119] A larger phase I trial with 12 patients demonstrated improvement in ventricular perfusion and dimensions by scintigraphic imaging and no episodes of ventricular arrhythmias.[121] In patients with recent myocardial infarction, Bartunek et al. performed intracoronary administration of enriched CD133+ cells. Among 35 patients with acute myocardial infarction treated with stenting, 19 underwent intracoronary administration of CD133+ progenitor cells. These authors noted that intracoronary infusion of selected CD133+ EPC is associated with improved left ventricular performance paralleled with increased myocardial perfusion and viability.[122] Taken together, these data support the efficacy of patient-derived CD133+ EPC in mediating vasculogenesis in response to ischemia and lay the basis for our Umbilical Cord Blood (UCB-) derived CD133+ EPC or individual patient-derived CD133+ EPC[123-126] cells as a source for cellular therapy for cardiac ischemia.

23.5 How to Increase Stem Cell Efficacy?

Attempts at increasing efficacy of stem cells for cardiac indications have taken several avenues of investigation: increasing trafficking efficacy; enhancing plasticity of administered cells; and increasing growth factor production. Endowment of these features has been performed by gene transfection or modification of culture conditions such as exposure to cytokines or hypoxia. Another interesting approach is addition of chemotactic agents to the area of tissue injury to enhance trafficking. These approaches will be discussed below.

23.6 Making Stem Cells Home Better

Mesenchymal stem cells are known to migrate to injured tissue and hypoxic tissue through expression of receptors such as CD44[127-129] and CXCR-4,[130] respectively. One method to increase efficacy of these cells is to increase their ability to traffic to where they are needed. This has been performed using various approaches. Cheng et al. used retroviral transfection to over-express CXCR-4 on rat bone-marrow-derived MSC. These cells were functionally competent as judged by similar growth profiles and differentiation ability when compared to control-transfected MSC. Intravenous administration of the modified cells in a rat model of myocardial infarction led to a significant improvement in migration to the area of infarct, and LVEF, as well as decreased wall thinning and fibrosis when compared to animals receiving control MSC.[114] Although many fears surrounding genetically modified cells exist, current advances in delivery vectors have increased safety features which may allow such modified MSC to become a clinical reality.[131] An alternative and perhaps easier way of inducing MSC to expression CXCR-4 is by simply "pulsing" them with a brief period of hypoxia,[132] or exposure to cytokines such as SCF, IL-6, Flt-3 ligand, HGF, and IL-3.[133]

Instead of increasing affinity of the stem cells to the chemoattractant, the other way to achieve the same result is to increase the concentration of the chemoattractant. One way is to provide an exogenous depot of angiogenic cytokines in proximity to the area where stem cell migration is desired. Tang et al. administered a SDF-1 expressing plasmid into the ischemic border zone 2 weeks after induction of infarct in BALB/c mice. To determine whether the expressed chemoattractant actually caused stem cell homing, syngeneic labeled bone marrow cells were intravenously injected 3 days after SDF-1 plasmid administration. A significantly increased number of labeled cells were observed in the group receiving the plasmid, in the area where the plasmid was injected.[134] These data suggest that it is feasible to reproduce mobilization induced by infarcts through the administration of homologous cytokines. However, the authors did not describe therapeutic benefit. In another experiment, a more clinically translatable approach was taken. Fibrin glue, fibrinogen and thrombin mixed at the point of care, is used in surgery to control bleeding.[135] Zhang et al. used pegylation technology to covalently bind recombinant SDF-1 to fibrinogen and demonstrated that subsequent to mixing with thrombin, the resultant "patch" could serve as a means of controlled release of SDF-1. The patch was placed on the infarct area of the left ventricle of mice after ligation of the left anterior descending coronary artery. In comparison to control mice receiving a fibrin patch lacking SDF-1, an increase in cells with a stem cell antigen and c-kit positive phenotype was observed in the experimental group. Additionally, at completion of experiment, an increased LVEF was observed in the treatment mice.[136] Since endogenous cardiac stem cells also express a similar phenotype,[137] and cell labeling was not performed, it is difficult to determine whether the therapeutic effect was mediated by mobilization of bone marrow progenitors or cardiac resident stem cells. One interesting way of enhancing activity of such a localization of chemoattractant is to concurrently administer exogenous stem cells, or to mobilize endogenous bone marrow stem cells. In fact, the latter was performed in a study where fibroblasts expressing SDF-1 were injected into the hind limbs of mice after femoral ligation. A synergistic induction of angiogenesis was detected when endogenous bone-marrow-derived stem cells were mobilized with G-CSF.[138] Other clinically used methods may be implemented to enhance stem cell trafficking. For example, erythropoietin (EPO), in addition to its well-known anti-apoptotic effects on cardiomyocytes,[139] has actually been shown to stimulate responsiveness of bone-marrow-derived stem cells to SDF-1 when administered in vivo.[140] Combination therapies of this sort will be interesting to evaluate clinically, especially when the various components are already approved.

23.7 Revitalize Stem Cells

Once we can make sure that stem cells arrive to the site where they are needed to stimulate regeneration, how do we know that they can do this effectively? For example, we do know that in general, stem cell activity diminishes with age,[141] and specifically, in patients with cardiovascular risk factors stem cell activity is additionally suppressed as compared to healthy age-matched controls.[142] There are several issues that must be taken into consideration. Perhaps, most importantly, is how do the stem cells mediate their therapeutic effects? On the one hand, people will state that adult stem cells, such as hematopoietic[143] and even in some cases mesenchymal stem cells,[144] do not differentiate into functional cardiomyocytes, so therefore therapy with these cells is a futile endeavor. As discussed above, efficacy of cardiac stem cell therapy does not rely on cell replacement but could be, and most likely is, mediated by trophic, angiogenic, anti-inflammatory, and anti-apoptotic effects. Regardless of this, the concept of "revitalizing" an adult stem cell so as to be able to actually replace cardiac cells is very exciting.

One method of such "revitalization" involves making the stem cells take a more primitive, embryonic stem-cell-like phenotype. It is known that the more differentiated cells become, the less plasticity they have, and the more restricted epigenetically, they become. Perhaps this was associated with the reason why DNA methyltransferase inhibitors such as 5-azacytidine were initially added to stem cells before implantation into infracted hearts.[80, 145] Other agents that act epigenetically, such as the histone deacetylase inhibitor valproic acid have been demonstrated to enhance hematopoietic stem cell self-renewal capacity in vitro,[146, 147] and have a positive effect on post-infarct remodeling in vivo, although it is not clear whether stem cell activation is implicated.[148] Instead of using agents such as these that upregulate factors associated with pluripotency such as Nanog,[149] an alternative approach is to simply transfect the cells with such genes. For example, Go et al. transfected bone-marrow-derived MSC with Nanog and reported superior expansion potential and ability to differentiate as compared to control transfected cells.[150] Transfection of such "retrodifferentiation" genes is particularly exciting in light of the recent discovery that fibroblasts can be induced to pluripotency through introduction of the pluripotency genes Oct3/4, Sox2, c-Myc, and Klf4 in mice[151] and humans.[152] These "inducible pluripotent stem cells" (iPS) appear to be functional, not only by gene transcription profile, but also by ability to reconstitute animals hematopoietically.[153] Theoretically, it would make sense that retrodifferentiation of an adult stem cell into an iPS would be easier than a skin fibroblast. Indeed, Kim et al. demonstrated that in order to derive iPS cells from neural stem cells, only the factors Oct-4 and klf-4 or c-Myc are needed.[154] Furthermore, newer transfection methods of generating iPS through non-retroviral means have been reported, giving the possibility of generating clinically applicable therapies from these cells.[155] Unfortunately, carcinogenesis associated with the viral vectors is not the main limitation. It is known in general that ES cells are carcinogenic.[156] Additionally, the very transcriptional profile associated with cancer stem cells appears to be related to that of pluripotent cells, regardless if they are generated by iPS or from ES cells.[157]

Thus, one way of increasing potency of MSC-based therapy is through induction of such "rejuvenation"; unfortunately, too much rejuvenation leads to the possibility of carcinogenesis, and additionally may have implications on ability of the cells to evade immune responsiveness and/or migration to the area of injury. For example, it is known that embryonic stem cells are hypoimmunogenic, as seen by weak ability to stimulate allogeneic lymphocyte proliferation.[158] However, it remains an open question whether ES cells can actively suppress ongoing immune responses as is the case with MSC both in animal models[43] and clinically.[159,160] In terms of migratory ability, it is known that functionally various adult stem cells play a protective role in the physiological response to injury. Although the effects in clinical situations are minor, there is suggestive evidence, for example in stroke patients that a correlation between endogenous stem cell mobilization and positive outcome exists.[161,162] While in cardiac infarct cases we do know that mobilization occurs,[163] but correlation with infarct recovery has not been made. Regardless, the question of what stage of differentiation the best cell population is for treatment of cardiac indications remains unclear.

23.8 Use Stem Cell Combinations

Given that we do not know the best stage of differentiation to administer the stem cells, as well as the various drawbacks of transfection and reprogramming approaches, one possible way of advancing efficacy of stem cell therapy would be to combine various stem cell types that we know have trophic activity. One interesting combination would be the use of CD34 cells, which are primarily hematopoietic, but also angiogenic, together with allogeneic mesenchymal stem cells, which have trophic, angiogenic, and potent anti-inflammatory potential. The rationale for combining these two approaches comes from several perspectives: (a) After tissue injury, both mesenchymal[127,164,165] and hematopoietic stem cells[166–168] are mobilized, thus potentially both cells may have therapeutic synergistic activity in a physiological sense; (b) In vivo MSCs provide a microenvironment for CD34 stem cells both embryonically,[169] and postnatally,[170] in vitro MSCs promote expansion of CD34 stem cells[171,172]; and (c) animal models suggest synergy of function.[173]

We have previously published data from an end-stage patient suffering from dilated cardiomyopathy who underwent a profound improvement in ejection fraction after receiving a combination of cord blood expanded CD34 cells and placental-matrix-derived mesenchymal stem cells.[174]

23.9 Cardiovascular Regenerative Cell Therapy Using UCB-Derived HSC

While the approach of utilizing the patient's own HSCs has the advantage of avoiding potential immune response relative to allogeneic cells, it has several disadvantages as well. Patients with acute myocardial infarction may experience significant morbidity with attempted large-volume BM harvest or attempted cytokine-induced BM stem cell mobilization. Moreover, a majority of cardiac patients are of advanced age. Increasing evidence suggests reduced potential and regenerative capacity of marrow-derived circulating EPC with increasing age,[175,176] and compromised capability of responding to inflammatory signals and cytokines released from the ischemic bed.[177] Thus, the ready availability of an "off-the shelf" allogeneic EPC cellular infusion is optimal.

Cord blood stem cells have been used successfully for more than 20 years. Currently, they are used to treat approximately 70 diseases including immunodeficiencies; genetic and neurological disorders; certain types of cancers; and blood disorders, such as leukemia, lymphoma, sickle cell anemia, and aplastic anemia.[178] Today, physicians have performed more than 8,000 cord blood stem cell transplants worldwide. These stem cells hold vast therapeutic promise to address major unmet medical needs and are increasingly being used in medical therapies to improve – and save – lives. Cord blood stem cell treatments differ from bone marrow stem cell treatments in three key areas: increased tolerance of HLA-mismatching, decreased risk of graft-versus-host disease, and enhanced proliferation ability.[179]

Studies in a clinical setting of leukemia have shown that Umbilical Cord Blood Transplants (UCBTs) are more useful than peripheral blood or bone-marrow-derived stem cell transplants.[180] The UCBTs tend to need less HLA matching for graft survival. Lubin and Greene[180] found that a 4/6 and 5/6 HLA match was still useful if the stem cells were UCB-derived. UCBTs are also more advantageous because it is easier and less painful for the donors and is associated with a decreased risk of viral infections after transplantation. UCBT also has a larger number of potential donors, longer telomeres, and is more quickly available.[179–184] Barker et al. found that individuals receiving UCBT received their cord blood transplant an average of 25 days sooner than individuals receiving bone-marrow-derived stem cell transplants.[185]

Umbilical cord blood (UCB) can be utilized as a source of potential therapeutic allogeneic endothelial progenitor cells (EPC). The advantages of cord blood include: collection at no risk to the donor, easy storage, low inherent viral contamination risk, and decreased donor discovery time.[179] However, in the setting of vascular regeneration allogeneic cord blood EPCs are expected to be detected by the immune system of an immune competent patient. Few in vitro studies have shown that UCB EPC cells express HLA class I and II surface molecules and elicit allogeneic T-cell proliferation by immune-competent adult mononuclear cells in mixed lymphocyte cultures (MLR). It is important to address whether the immunogenicity of a UCB EPC allogeneic cellular product is advantageous (i.e., augmenting a vasculogenesis response by recipient cells in situ) or potentially deleterious (i.e., dampening vasculogenesis response or worsening vascular ischemia via allogeneic inflammatory responses). More importantly, preclinical studies by our group and others to date have identified that in vascular injury rodent models, augmentation of murine endogenous microvascular collateralization is not completely due to anatomic incorporation of infused human EPC into the murine vascular endothelium.[186] These observations suggest possible paracrine-mediated angiogenic effects elicited by the infused human EPC responding to inflammatory signals, cytokines, and growth factors released from the ischemic region,[187] suggesting that stimulation of inflammatory signaling molecules by allogeneic cells in the injured area augments growth factor-induced neovascularization.

23.10 Cardiac Angiogenesis Using UCB-Derived HSCs

While various stem cell sources have been studied to induce myogenesis, recent interest has focused on promoting cardiac angiogenesis by proangiogeneic factors.[188] The existence of angiogenic factors such as acidic and basic fibroblast growth factor (FGF1 and 2), VEGF, PDGF, insulin-like growth factor-1 (IGF-1), angiogenin, transforming growth factor (TGF-α and TGF-β), tumor necrosis factor (TNF-α), hepatocyte growth factors (HGF), granulocyte colony-stimulating factor (G-CSF), placental growth factor (PGF), interleukin 8 to be mitogenic for endothelial cells.[189-195] Of the large number of angiogenesis factors that have been described, the FGF and VEGF families have been most intensively studied.[196,197] Although several elements are likely to be involved in the process of angiogenesis, in vivo studies have amply demonstrated that the simple administration of an angiogenic growth factor is sufficient to stimulate the cascade of events that lead to angiogenesis and to the augmentation of blood delivery. However, administration of growth factors in the normal heart does not result in angiogenesis.[198] These techniques retain an important potential as an adjunct to cellular transplantation in inducing angiogenesis in injured myocardium when revascularization cannot be achieved by standard techniques, namely percutaneous angioplasty and coronary artery bypass grafting.

Although a number of angiogenesis studies using gene therapy technique to deliver growth factors are in clinical trials, the resident population of vascular endothelial cells in adults competent to respond to available angiogenic growth factors limits these interventions due to age-related diminution of vascular endothelial cell number. Additionally, vascular endothelial cell function may limit the efficacy of patient-derived progenitor cells in mediating neovascularization.[199-201] These data support the concept that an exogenous source of EPC such as UCB-derived HSCs, rather than autologous patient-derived cells, may be optimal for cellular therapeutics intended to enhance angiogenesis and collateralization around stenosed or occluded vessels to relieve ischemia. UCB-derived HSC offer distinct advantages as a cell source, including greater potential life span and reparative proliferation, relative to existing models of therapeutic angiogenesis derived from patient peripheral blood or marrow.

The available data thus far indicate that there is no proven "off-the-shelf" stem therapy to repair or regenerate heart after acute myocardial infarction (MI) or congestive heart failure (CHI). The limited capacity of heart cells to regenerate and also the available alternative source of stem cells for myocyte regeneration and cardiac revascularization is also limited.[202] Human UCB contains several different types of stem cells including HSC, EPC, and MSC.[72,203-205] These studies suggest human UCB-derived stem cells as a potential alternative source of stem cells for potent cardiomyocyte regeneration and revascularization. Although more studies are needed to prove the full clinical benefit of UCB-derived stem cells, scientists and clinicians believe that the human UCB has the most immediate benefit as ideal source for stem cells.

23.11 Conclusion

Adult stem cell therapy for cardiac conditions has reached the point where new directions are needed to optimize effects. Possibilities of next-generation approaches include the use of "in vitro supercharged" cells, combinations of cells and cytokines, and of course combination of cellular therapies. Currently, the use of cord blood cells as a possible substitute for bone marrow mesenchymal stem cells is being evaluated by our group.

References

1. Sanchez-Lazaro IJ, Almenar L, Reganon E, et al. Inflammatory markers in stable heart failure and their relationship with functional class. *Int J Cardiol.* 2007;129(3):388-393.
2. Alonso-Martinez JL, Llorente-Diez B, Echegaray-Agara M, Olaz-Preciado F, Urbieta-Echezarreta M, Gonzalez-Arencibia C. C-reactive protein as a predictor of improvement and readmission in heart failure. *Eur J Heart Fail.* 2002;4:331-336.
3. Nakou ES, Liberopoulos EN, Milionis HJ, Elisaf MS. The role of C-reactive protein in atherosclerotic cardiovascular disease: an overview. *Curr Vasc Pharmacol.* 2008;6:258-270.
4. Galarraga B, Khan F, Kumar P, Pullar T, Belch JJ. C-reactive protein: the underlying cause of microvascular dysfunction in rheumatoid arthritis. *Rheumatology (Oxford).* 2008; 47(12):1780-1784.
5. Nabata A, Kuroki M, Ueba H, et al. C-reactive protein induces endothelial cell apoptosis and matrix metalloproteinase-9 production in human mononuclear cells: Implications for the destabilization of atherosclerotic plaque. *Atherosclerosis.* 2008;196:129-135.
6. Griselli M, Herbert J, Hutchinson WL, et al. C-reactive protein and complement are important mediators of tissue damage in acute myocardial infarction. *J Exp Med.* 1999; 190:1733-1740.
7. Pepys MB, Hirschfield GM, Tennent GA, et al. Targeting C-reactive protein for the treatment of cardiovascular disease. *Nature.* 2006;440:1217-1221.
8. Satoh M, Minami Y, Takahashi Y, Nakamura M. Immune modulation: role of the inflammatory cytokine cascade in the failing human heart. *Curr Heart Fail Rep.* 2008;5:69-74.
9. Yokoyama T, Sekiguchi K, Tanaka T, et al. Angiotensin II and mechanical stretch induce production of tumor necrosis factor in cardiac fibroblasts. *Am J Physiol.* 1999;276:H1968-H1976.
10. Wang BW, Hung HF, Chang H, Kuan P, Shyu KG. Mechanical stretch enhances the expression of resistin gene in cultured cardiomyocytes via tumor necrosis factor-alpha. *Am J Physiol Heart Circ Physiol.* 2007;293:H2305-H2312.
11. Satoh S, Oyama J, Suematsu N, et al. Increased productivity of tumor necrosis factor-alpha in helper T cells in patients with systolic heart failure. *Int J Cardiol.* 2006;111:405-412.
12. Conraads VM, Bosmans JM, Schuerwegh AJ, et al. Intracellular monocyte cytokine production and CD 14 expression are up-regulated in severe vs mild chronic heart failure. *J Heart Lung Transplant.* 2005;24:854-859.
13. Haudek SB, Taffet GE, Schneider MD, Mann DL. TNF provokes cardiomyocyte apoptosis and cardiac remodeling through activation of multiple cell death pathways. *J Clin Invest.* 2007;117:2692-2701.
14. Kubota T, Bounoutas GS, Miyagishima M, et al. Soluble tumor necrosis factor receptor abrogates myocardial inflammation but not hypertrophy in cytokine-induced cardiomyopathy. *Circulation.* 2000;101:2518-2525.
15. Scheibner KA, Lutz MA, Boodoo S, Fenton MJ, Powell JD, Horton MR. Hyaluronan fragments act as an endogenous danger signal by engaging TLR2. *J Immunol.* 2006; 177:1272-1281.
16. Termeer C, Benedix F, Sleeman J, et al. Oligosaccharides of Hyaluronan activate dendritic cells via toll-like receptor 4. *J Exp Med.* 2002;195:99-111.
17. Asea A. Heat shock proteins and toll-like receptors. *Handb Exp Pharmacol.* 2008;183:111-127.
18. Nozaki N, Shishido T, Takeishi Y, Kubota I. Modulation of doxorubicin-induced cardiac dysfunction in toll-like receptor-2-knockout mice. *Circulation.* 2004;110:2869-2874.
19. Riad A, Bien S, Gratz M, et al. Toll-like receptor-4 deficiency attenuates doxorubicin-induced cardiomyopathy in mice. *Eur J Heart Fail.* 2008;10:233-243.
20. Shishido T, Nozaki N, Yamaguchi S, et al. Toll-like receptor-2 modulates ventricular remodeling after myocardial infarction. *Circulation.* 2003;108:2905-2910.
21. Sheu JJ, Chang LT, Chiang CH, et al. Prognostic value of activated toll-like receptor-4 in monocytes following acute myocardial infarction. *Int Heart J.* 2008;49:1-11.
22. Friedenstein AJ, Petrakova KV, Kurolesova AI, Frolova GP. Heterotopic of bone marrow. Analysis of precursor cells for osteogenic and hematopoietic tissues. *Transplantation.* 1968;6:230-247.
23. Zannettino AC, Paton S, Arthur A, et al. Multipotential human adipose-derived stromal stem cells exhibit a perivascular phenotype in vitro and in vivo. *J Cell Physiol.* 2008; 214:413-421.
24. Hoogduijn MJ, Crop MJ, Peeters AM, et al. Human heart, spleen, and perirenal fat-derived mesenchymal stem cells have immunomodulatory capacities. *Stem Cells Dev.* 2007;16:597-604.
25. Chao KC, Chao KF, Fu YS, Liu SH. Islet-like clusters derived from mesenchymal stem cells in Wharton's Jelly of the human umbilical cord for transplantation to control type 1 diabetes. *PLoS ONE.* 2008;3:e1451.
26. Jo YY, Lee HJ, Kook SY, et al. Isolation and characterization of postnatal stem cells from human dental tissues. *Tissue Eng.* 2007;13:767-773.
27. He Q, Wan C, Li G. Concise review: multipotent mesenchymal stromal cells in blood. *Stem Cells.* 2007;25:69-77.
28. Oh W, Kim DS, Yang YS, Lee JK. Immunological properties of umbilical cord blood-derived mesenchymal stromal cells. *Cell Immunol.* 2008;251(2):116-123.
29. Meng X, Ichim TE, Zhong J, et al. Endometrial regenerative cells: a novel stem cell population. *J Transl Med.* 2007;5:57.
30. Hida N, Nishiyama N, Miyoshi S, et al. Novel cardiac precursor-like cells from human menstrual blood-derived mesenchymal cells. *Stem Cells.* 2008;26(7):1695-1704.
31. Patel AN, Park E, Kuzman M, Benetti F, Silva FJ, Allickson JG. Multipotent menstrual blood stromal stem cells: isolation, characterization, and differentiation. *Cell Transplant.* 2008;17:303-311.
32. Le Blanc K, Ringden O. Immunomodulation by mesenchymal stem cells and clinical experience. *J Intern Med.* 2007;262:509-525.

33. Keyser KA, Beagles KE, Kiem HP. Comparison of mesenchymal stem cells from different tissues to suppress T-cell activation. *Cell Transplant.* 2007;16:555-562.
34. Nasef A, Chapel A, Mazurier C, et al. Identification of IL-10 and TGF-beta transcripts involved in the inhibition of T-lymphocyte proliferation during cell contact with human mesenchymal stem cells. *Gene Expr.* 2007;13:217-226.
35. Ryan JM, Barry F, Murphy JM, Mahon BP. Interferon-gamma does not break, but promotes the immunosuppressive capacity of adult human mesenchymal stem cells. *Clin Exp Immunol.* 2007;149:353-363.
36. Nasef A, Mazurier C, Bouchet S, et al. Leukemia inhibitory factor: role in human mesenchymal stem cells mediated immunosuppression. *Cell Immunol.* 2008;253:16-22.
37. Selmani Z, Naji A, Zidi I, et al. Human leukocyte antigen-G5 secretion by human mesenchymal stem cells is required to suppress T lymphocyte and natural killer function and to induce CD4+CD25highFOXP3+ regulatory T cells. *Stem Cells.* 2008;26:212-222.
38. Ortiz LA, Dutreil M, Fattman C, et al. Interleukin 1 receptor antagonist mediates the antiinflammatory and antifibrotic effect of mesenchymal stem cells during lung injury. *Proc Natl Acad Sci USA.* 2007;104:11002-11007.
39. English K, Barry FP, Field-Corbett CP, Mahon BP. IFN-gamma and TNF-alpha differentially regulate immunomodulation by murine mesenchymal stem cells. *Immunol Lett.* 2007;110:91-100.
40. Jones BJ, Brooke G, Atkinson K, McTaggart SJ. Immunosuppression by placental indoleamine 2, 3-dioxygenase: a role for mesenchymal stem cells. *Placenta.* 2007; 28:1174-1181.
41. Casiraghi F, Azzollini N, Cassis P, et al. Pretransplant infusion of mesenchymal stem cells prolongs the survival of a semiallogeneic heart transplant through the generation of regulatory T cells. *J Immunol.* 2008;181:3933-3946.
42. Kassis I, Grigoriadis N, Gowda-Kurkalli B, et al. Neuropro tection and immunomodulation with mesenchymal stem cells in chronic experimental autoimmune encephalomyelitis. *Arch Neurol.* 2008;65:753-761.
43. Parekkadan B, Tilles AW, Yarmush ML. Bone marrow-derived mesenchymal stem cells ameliorate autoimmune enteropathy independently of regulatory T cells. *Stem Cells.* 2008;26:1913-1919.
44. Li H, Guo Z, Jiang X, Zhu H, Li X, Mao N. Mesenchymal stem cells alter migratory property of T and dendritic cells to delay the development of murine lethal acute graft-versus-host disease. *Stem Cells.* 2008;26(10):2531-2541.
45. Augello A, Tasso R, Negrini SM, Cancedda R, Pennesi G. Cell therapy using allogeneic bone marrow mesenchymal stem cells prevents tissue damage in collagen-induced arthritis. *Arthritis Rheum.* 2007;56:1175-1186.
46. Semedo P, Wang PM, Andreucci TH, et al. Mesenchymal stem cells ameliorate tissue damages triggered by renal ischemia and reperfusion injury. *Transplant Proc.* 2007;39:421-423.
47. Du YY, Zhou SH, Zhou T, et al. Immuno-inflammatory regulation effect of mesenchymal stem cell transplantation in a rat model of myocardial infarction. *Cytotherapy.* 2008; 10:469-478.
48. Le Blanc K, Frassoni F, Ball L, et al. Mesenchymal stem cells for treatment of steroid-resistant, severe, acute graft-versus-host disease: a phase II study. *Lancet.* 2008;371:1579-1586.
49. Ning H, Yang F, Jiang M, et al. The correlation between cotransplantation of mesenchymal stem cells and higher recurrence rate in hematologic malignancy patients: outcome of a pilot clinical study. *Leukemia.* 2008;22:593-599.
50. Ball L, Bredius R, Lankester A, et al. Third party mesenchymal stromal cell infusions fail to induce tissue repair despite successful control of severe grade IV acute graft-versus-host disease in a child with juvenile myelo-monocytic leukemia. *Leukemia.* 2008;22:1256-1257.
51. Ringden O, Uzunel M, Rasmusson I, et al. Mesenchymal stem cells for treatment of therapy-resistant graft-versus-host disease. *Transplantation.* 2006;81:1390-1397.
52. Le Blanc K, Rasmusson I, Sundberg B, et al. Treatment of severe acute graft-versus-host disease with third party haploidentical mesenchymal stem cells. *Lancet.* 2004;363: 1439-1441.
53. Muller I, Kordowich S, Holzwarth C, et al. Application of multipotent mesenchymal stromal cells in pediatric patients following allogeneic stem cell transplantation. *Blood Cells Mol Dis.* 2008;40:25-32.
54. Narula J, Haider N, Virmani R, et al. Apoptosis in myocytes in end-stage heart failure. *N Engl J Med.* 1996;335:1182-1189.
55. Dorn GW 2nd. Apoptotic and non-apoptotic programmed cardiomyocyte death in ventricular remodelling. *Cardiovasc Res.* 2008;81(3):465-473.
56. Rodriguez M, Lucchesi BR, Schaper J. Apoptosis in myocardial infarction. *Ann Med.* 2002;34:470-479.
57. Odashima M, Usui S, Takagi H, et al. Inhibition of endogenous Mst1 prevents apoptosis and cardiac dysfunction without affecting cardiac hypertrophy after myocardial infarction. *Circ Res.* 2007;100:1344-1352.
58. Chua CC, Gao J, Ho YS, et al. Overexpression of IAP-2 attenuates apoptosis and protects against myocardial ischemia/reperfusion injury in transgenic mice. *Biochim Biophys Acta.* 2007;1773:577-583.
59. Jayasankar V, Woo YJ, Bish LT, et al. Gene transfer of hepatocyte growth factor attenuates postinfarction heart failure. *Circulation.* 2003;108(suppl 1):II230-II236.
60. Filippatos G, Uhal BD. Blockade of apoptosis by ACE inhibitors and angiotensin receptor antagonists. *Curr Pharm Des.* 2003;9:707-714.
61. Fransioli J, Bailey B, Gude NA, et al. Evolution of the c-kit-positive cell response to pathological challenge in the myocardium. *Stem Cells.* 2008;26:1315-1324.
62. Urbanek K, Torella D, Sheikh F, et al. Myocardial regeneration by activation of multipotent cardiac stem cells in ischemic heart failure. *Proc Natl Acad Sci.* 2005;102:8692-8697.
63. Dawn B, Stein AB, Urbanek K, et al. Cardiac stem cells delivered intravascularly traverse the vessel barrier, regenerate infarcted myocardium, and improve cardiac function. *Proc Natl Acad Sci USA.* 2005;102:3766-3771.
64. Raffaghello L, Bianchi G, Bertolotto M, et al. Human mesenchymal stem cells inhibit neutrophil apoptosis: a model for neutrophil preservation in the bone marrow niche. *Stem Cells.* 2008;26:151-162.
65. Mirotsou M, Zhang Z, Deb A, et al. Secreted frizzled related protein 2 (Sfrp2) is the key Akt-mesenchymal stem cell-released paracrine factor mediating myocardial survival and repair. *Proc Natl Acad Sci USA.* 2007;104:1643-1648.
66. Wang M, Crisostomo PR, Herring C, Meldrum KK, Meldrum DR. Human progenitor cells from bone marrow or adipose

tissue produce VEGF, HGF, and IGF-I in response to TNF by a p38 MAPK-dependent mechanism. *Am J Physiol Regul Integr Comp Physiol*. 2006;291:R880-R884.
67. Li TS, Takahashi M, Ohshima M, et al. Myocardial repair achieved by the intramyocardial implantation of adult cardiomyocytes in combination with bone marrow cells. *Cell Transplant*. 2008;17:695-703.
68. Crisostomo PR, Wang Y, Markel TA, Wang M, Lahm T, Meldrum DR. Human mesenchymal stem cells stimulated by TNF-alpha, LPS, or hypoxia produce growth factors by an NF kappa B- but not JNK-dependent mechanism. *Am J Physiol Cell Physiol*. 2008;294:C675-C682.
69. Fu X, He Y, Xie C, Liu W. Bone marrow mesenchymal stem cell transplantation improves ovarian function and structure in rats with chemotherapy-induced ovarian damage. *Cytotherapy*. 2008;10:353-363.
70. Urbanek K, Rota M, Cascapera S, et al. Cardiac stem cells possess growth factor-receptor systems that after activation regenerate the infarcted myocardium, improving ventricular function and long-term survival. *Circ Res*. 2005;97:663-673.
71. Zeng F, Chen MJ, Baldwin DA, et al. Multiorgan engraftment and differentiation of human cord blood CD34+ Lin- cells in goats assessed by gene expression profiling. *Proc Natl Acad Sci USA*. 2006;103:7801-7806.
72. Nishiyama N, Miyoshi S, Hida N, et al. The significant cardiomyogenic potential of human umbilical cord blood-derived mesenchymal stem cells in vitro. *Stem Cells*. 2007; 25:2017-2024.
73. Kawada H, Fujita J, Kinjo K, et al. Nonhematopoietic mesenchymal stem cells can be mobilized and differentiate into cardiomyocytes after myocardial infarction. *Blood*. 2004;104:3581-3587.
74. Vieyra DS, Jackson KA, Goodell MA. Plasticity and tissue regenerative potential of bone marrow-derived cells. *Stem Cell Rev*. 2005;1:65-69.
75. Fazel S, Cimini M, Chen L, et al. Cardioprotective c-kit+cells are from the bone marrow and regulate the myocardial balance of angiogenic cytokines. *J Clin Invest*. 2006;116:1865-1877.
76. Kocher AA, Schuster MD, Szabolcs MJ, et al. Neovascularization of ischemic myocardium by human bone-marrow-derived angioblasts prevents cardiomyocyte apoptosis, reduces remodeling and improves cardiac function. *Nat Med*. 2001;7:430-436.
77. Li RK, Jia ZQ, Weisel RD, et al. Cardiomyocyte transplantation improves heart function. *Ann Thorac Surg*. 1996; 62:654-661.
78. El Oakley RM, Ooi OC, Bongso A, Yacoub MH. Myocyte transplantation for myocardial repair: a few good cells can mend a broken heart. *Ann Thorac Surg*. 2001;71:1724-1733.
79. Klug MG, Soonapaa MH, Koh GY, Field LJ. Genetically selected cardiomyocytes from differentiating embryonic stem cells form stable intracardiac grafts. *J Clin Invest*. 1996;98:216-224.
80. Tomita S, Li RK, Weisel RD, et al. Autologous transplantation of bone marrow cells improves damaged heart function. *Circulation*. 1999;100:II247-II256.
81. Taylor DA, Atkins BZ, Hungspreugs P, et al. Regenerating functional myocardium: improved performance after skeletal myoblast transplantation. *Nat Med*. 1998;4:929-933.
82. Dib N, McCarthy P, Campbell A, et al. Feasibility and safety of autologous myoblast transplantation in patients with ischemic cardiomyopathy. *Cell Transplant*. 2005;14:1-9.
83. Dib N, Michler RE, Pagani FD, et al. Safety and feasibility of autologous myoblast transplantation in patients with ischemic cardiomyopathy: four-year follow-up. *Circulation*. 2005;112:1748-1755.
84. Dimarakis I, Habib NA, Gordon MYA. Adult bone marrow-derived stem cells and the injured heart: just the beginning? *Eur J Cardiothorac Surg*. 2005;28:665-676.
85. Dec GW. Management of heart failure: crossing boundary over to the surgical country. *Surg Clin North Am*. 2004;84:1-25.
86. Erbs S, Linke A, Adams V, et al. Transplantation of blood-derived progenitor cells after recanalization of chronic coronary artery occlusion: first randomized and placebo-controlled study. *Cir Res*. 2005;97:756-762.
87. Evers BM, Weissman IL, Flake AW, Tabar V, Weisel RD. Stem cells in clinical practice. *J Am Coll Surg*. 2003; 197:458-478.
88. Forrester JS, Price MJ, Makkar RR. Stem cell repair of infarcted myocardium: an overview for clinicians. *Circulation*. 2003;108:1139-1145.
89. Soonpaa MH, Koh GY, Klug MG, Field LJ. Formation of nascent intercalated disks between grafted fetal cardiomyocytes and host myocardium. *Science*. 1994;264:98-101.
90. Scorsin M, Marotte F, Sabri A, et al. Can grafted cardiomyocytes colonize periinfarction myocardial areas? *Circulation*. 1996;94:337-340.
91. Li RK, Mickle D, Weisel RD, et al. In vivo survival and function of transplanted rat cardiomyocytes. *Cir Res*. 1996;78:283-288.
92. Chiu RC, Zibaitis A, Kao RL. Cellular cardiomyoplasty: myocardial regeneration with satellite cell implantation. *Ann Thorac Surg*. 1995;60:12-18.
93. Murry CE, Wiseman RW, Schwartz SM, Hauschka SD. Skeletal myoblast transplantation for repair of myocardial necrosis. *J Clin Invest*. 1996;98:2512-2523.
94. Scorsin M, Hagege A, Vilguin J-T, et al. Comparison of the effects of fetal cardiomyocyte and skeletal myoblast transplantation on postinfarction left ventricular function. *J Thorac Cardiovasc Surg*. 2000;119:1169-1175.
95. Taylor DA, Atkins BZ, Hungspreugs P, et al. Regenerating functional myocardium: improved performance after skeletal myoblast transplantation. *Nat Med*. 1998;4:929-933.
96. Atkins BZ, Lewis CW, Kraus WE, et al. Intracardiac transplantation of skeletal myoblasts yields two populations of striated cells in situ. *Ann Thorac Surg*. 1999;67:124-129.
97. Menasché P, Hagege AA, Scorsin M, et al. Myoblast transplantation for heart failure. *Lancet*. 2001;357:279-280.
98. Siminiak T, Kalawski R, Kurpisz M. Myoblast transplantation in the treatment of postinfarction myocardial contractility impairment. *Kardiol Pol*. 2002;56:131-137.
99. Herreros J, Prósper F, Perey A, et al. Autologous intramyocardial injection of cultured skeletal muscle-derived stem cells in patients with non-acute myocardial infarction. *Eur Heart J*. 2003;24:2012.
100. Bittner RE, Schofer C, Weipoltshammer K, et al. Recruitment of bone-marrow-derived cells by skeletal and cardiac muscle in adult dystrophic mdx mice. *Anat Embryol (Berl)*. 1999;199:391-396.
101. Goodell MA, Brose K, Paradis G, Conner AS, Mulligan RC. Isolation and functional properties of murine hematopoietic stem cells that are replicating in vivo. *J Exp Med*. 1996;183:1797-1806.

102. Zhou S, Schuetz JD, Bunting KD, et al. The ABC transporter Bcrp1/ABCG2 is expressed in a wide variety of stem cells and is a molecular determinant of the side-population phenotype. *Nat Med.* 2001;7:1028-1304.
103. Jackson KA, Majka SM, Wang H, et al. Regeneration of ischemic cardiac muscle and vascular endothelium by adult stem cells. *J Clin Invest.* 2001;107:1395-1402.
104. Toma C, Pittenger MF, Cahill KS, Byrne BJ, Kessler PD. Human mesenchymal stem cells differentiate to a cardiomyocyte phenotype in the adult murine heart. *Circulation.* 2002;105:93-98.
105. Nishikawa S. A complex linkage in the developmental pathway of endothelial and hematopoietic. *Curr Opin Cell.* 2001;13:673-678.
106. Reyes M, Dudek A, Jahagirdar B, Koodie L, Marker PH, Verfaillie CM. Origin of endothelial progenitors in human postnatal bone marrow. *J Clin Invest.* 2002;109:337-346.
107. Majka SM, Jackson KA, Kienstra KA, Majesky MW, Goodell MA, Hirschi KK. Distinct progenitor populations in skeletal muscle are bone marrow derived and exhibit different cell fates during vascular regeneration. *J Clin Invest.* 2003;111:71-79.
108. Yoon YS, Park JS, Tkebuchava T, Luedeman C, Losordo DW. Unexpected severe calcification after transplantation of bone marrow cells in acute myocardial infarction. *Circulation.* 2004;109:3154-3157.
109. Strauer BE, Brehm M, Zeus T, et al. Intracoronary, human autologous stem cell transplantation for myocardial regeneration following myocardial infarction. *Dtsch Med Wochenschr.* 2001;126:932-938.
110. Hamano K, Nishida M, Hirata K, et al. Local implantation of autologous bone marrow cells for therapeutic angiogenesis in patients with ischemic heart disease: clinical trial and preliminary results. *Jpn Circ J.* 2001;65:845-847.
111. Abdel-Latif A, Bolli R, Tleyjeh IM, et al. Adult bone marrow-derived cells for cardiac repair: a systematic review and meta-analysis. *Arch Intern Med.* 2007;167:989-997.
112. Martin-Rendon E, Brunskill S, Doree C, Hyde C, Watt S, Mathur A, Stanworth, S. Stem cell treatment for acute myocardial infarction. *Cochrane Database Syst Rev.* 2008;(4):CD006536.
113. Kang S, Yang YJ, Li CJ, Gao RL. Effects of intracoronary autologous bone marrow cells on left ventricular function in acute myocardial infarction: a systematic review and meta-analysis for randomized controlled trials. *Coron Artery Dis.* 2008;19:327-335.
114. Cheng Z, Ou L, Zhou X, et al. Targeted migration of mesenchymal stem cells modified with CXCR4 gene to infarcted myocardium improves cardiac performance. *Mol Ther.* 2008;16:571-579.
115. Murry CE et al. Haematopoietic stem cells do not transdifferentiate into cardiac myocytes in myocardial infarcts. *Nature.* 2004;429:664-668.
116. Balsam LB et al. Haematopoietic stem cells adopt mature haematopoietic fates in ischaemic myocardium. *Nature.* 2004;428:668-673.
117. Assmus B, Schachinger V, Teupe C, et al. Transplantation of cells and regeneration enhancement in acute myocardial infarction (TOPCARE-AMI). *Circulation.* 2002;106: 3009-3017.
118. Strauer BE, Brehm M, Zeus T, Kostering M, Hernandez A, Sorg RV. Repair of infarcted myocardium by autologous intracoronary mononuclear bone marrow cell transplantation in humans. *Circulation.* 2002;106:1913-1918.
119. Stamm C, Westphal B, Kleine HD. Autologous bone-marrow stem-cell transplantation for myocardial regeneration. *Lancet.* 2003;361:45-46.
120. Gehling UM, Ergun S, Schumacher U, et al. In vitro differentiation of endothelial cells from AC133-positive progenitor cells. *Blood.* 2000;95:3106-3112.
121. Stamm C et al. CABG and bone marrow stem cell transplantation after myocardial infarction. *Thorac Cardiovasc Surg.* 2004;52:152-158.
122. Bartunek J, Vanderheyden M, Vandekerckhove B, et al. Intracoronary injection of CD133-positive enriched bone marrow progenitor cells promotes cardiac recovery after recent myocardial infarction: feasibility and safety. *Circulation.* 2005;112:178-183.
123. Pierelli L, Bonanno G, Rutella S, Marone M, Scambia G, Leone G. CD105 (endoglin) expression on hematopoietic stem/progenitor cells. *Leuk Lymphoma.* 2001;42: 1195-1206.
124. Barry FP, Boynton R, Haynesworth S, Murphy JM, Zaia J. The monoclonal antibody SH-2, raised against human mesenchymal stem cells, recognizes an epitope on endoglin (CD105). *Biochem Biophys Res Commun.* 1999;265: 134-139.
125. Cheng T, Scadden D. Cell cycle entry of hematopoietic stem and progenitor cells controlled by distinct cyclin-dependent kinase inhibitors. *Int J Hematol.* 2002;75: 460-465.
126. Wang Z, Miura N, Bonelli A, et al. Receptor tyrosine kinase, EphB4 (HTK), accelerates differentiation of select human hematopoietic cells. *Blood.* 2002;99:2740-2747.
127. Herrera MB, Bussolati B, Bruno S, et al. Exogenous mesenchymal stem cells localize to the kidney by means of CD44 following acute tubular injury. *Kidney Int.* 2007; 72:430-441.
128. Sackstein R, Merzaban JS, Cain DW, et al. Ex vivo glycan engineering of CD44 programs human multipotent mesenchymal stromal cell trafficking to bone. *Nat Med.* 2008; 14:181-187.
129. Zhu H, Mitsuhashi N, Klein A, et al. The role of the hyaluronan receptor CD44 in mesenchymal stem cell migration in the extracellular matrix. *Stem Cells.* 2006; 24:928-935.
130. Wang Y, Deng Y, Zhou GQ. SDF-1alpha/CXCR4-mediated migration of systemically transplanted bone marrow stromal cells towards ischemic brain lesion in a rat model. *Brain Res.* 2008;1195:104-112.
131. Neschadim A, McCart JA, Keating A, Medin JA. A roadmap to safe, efficient, and stable lentivirus-mediated gene therapy with hematopoietic cell transplantation. *Biol Blood Marrow Transplant.* 2007;13:1407-1416.
132. Hung SC, Pochampally RR, Hsu SC, et al. Short-term exposure of multipotent stromal cells to low oxygen increases their expression of CX3CR1 and CXCR4 and their engraftment in vivo. *PLoS ONE.* 2007;2:e416.
133. Shi M, Li J, Liao L, et al. Regulation of CXCR4 expression in human mesenchymal stem cells by cytokine treatment: role in homing efficiency in NOD/SCID mice. *Haematologica.* 2007;92:897-904.
134. Tang YL, Qian K, Zhang YC, Shen L, Phillips MI. Mobilizing of haematopoietic stem cells to ischemic

myocardium by plasmid mediated stromal-cell-derived factor-1alpha (SDF-1alpha) treatment. *Regul Pept*. 2005; 125:1-8.
135. Gibble JW, Ness PM. Fibrin glue: the perfect operative sealant? *Transfusion*. 1990;30:741-747.
136. Zhang G, Nakamura Y, Wang X, Hu Q, Suggs LJ, Zhang J. Controlled release of stromal cell-derived factor-1 alpha in situ increases c-kit+cell homing to the infarcted heart. *Tissue Eng*. 2007;13:2063-2071.
137. Barile L, Messina E, Giacomello A, Marban E. Endogenous cardiac stem cells. *Prog Cardiovasc Dis*. 2007;50:31-48.
138. Tan Y, Shao H, Eton D, et al. Stromal cell-derived factor-1 enhances pro-angiogenic effect of granulocyte-colony stimulating factor. *Cardiovasc Res*. 2007;73:823-832.
139. Latini R, Brines M, Fiordaliso F. Do non-hemopoietic effects of erythropoietin play a beneficial role in heart failure? *Heart Fail Rev*. 2008;13:415-423.
140. Brunner S, Winogradow J, Huber BC, et al. Erythropoietin administration after myocardial infarction in mice attenuates ischemic cardiomyopathy associated with enhanced homing of bone marrow-derived progenitor cells via the CXCR-4/SDF-1 axis. *Faseb J*. 2009;23(2):351-361.
141. Chambers SM, Shaw CA, Gatza C, Fisk CJ, Donehower LA, Goodell MA. Aging hematopoietic stem cells decline in function and exhibit epigenetic dysregulation. *PLoS Biol*. 2007;5:e201.
142. Schmidt-Lucke C, Rossig L, Fichtlscherer S, et al. Reduced number of circulating endothelial progenitor cells predicts future cardiovascular events: proof of concept for the clinical importance of endogenous vascular repair. *Circulation*. 2005;111:2981-2987.
143. Wagers AJ, Sherwood RI, Christensen JL, Weissman IL. Little evidence for developmental plasticity of adult hematopoietic stem cells. *Science*. 2002;297:2256-2259.
144. Rose RA, Jiang H, Wang X, et al. Bone marrow-derived mesenchymal stromal cells express cardiac-specific markers, retain the stromal phenotype and do not become functional cardiomyocytes in vitro. *Stem Cells*. 2008;26(11):2884-2892.
145. Makino S, Fukuda K, Miyoshi S, et al. Cardiomyocytes can be generated from marrow stromal cells in vitro. *J Clin Invest*. 1999;103:697-705.
146. De Felice L, Tatarelli C, Mascolo MG, et al. Histone deacetylase inhibitor valproic acid enhances the cytokine-induced expansion of human hematopoietic stem cells. *Cancer Res*. 2005;65:1505-1513.
147. Bug G, Gul H, Schwarz K, et al. Valproic acid stimulates proliferation and self-renewal of hematopoietic stem cells. *Cancer Res*. 2005;65:2537-2541.
148. Lee TM, Lin MS, Chang NC. Inhibition of histone deacetylase on ventricular remodeling in infarcted rats. *Am J Physiol Heart Circ Physiol*. 2007;293:H968-H977.
149. Hattori N, Imao Y, Nishino K, et al. Epigenetic regulation of Nanog gene in embryonic stem and trophoblast stem cells. *Genes Cells*. 2007;12:387-396.
150. Go MJ, Takenaka C, Ohgushi H. Forced expression of Sox2 or Nanog in human bone marrow derived mesenchymal stem cells maintains their expansion and differentiation capabilities. *Exp Cell Res*. 2008;314:1147-1154.
151. Takahashi K, Yamanaka S. Induction of pluripotent stem cells from mouse embryonic and adult fibroblast cultures by defined factors. *Cell*. 2006;126:663-676.
152. Takahashi K, Tanabe K, Ohnuki M, et al. Induction of pluripotent stem cells from adult human fibroblasts by defined factors. *Cell*. 2007;131:861-872.
153. Hanna J, Wernig M, Markoulaki S, et al. Treatment of sickle cell anemia mouse model with iPS cells generated from autologous skin. *Science*. 2007;318:1920-1923.
154. Kim JB, Zaehres H, Wu G, et al. Pluripotent stem cells induced from adult neural stem cells by reprogramming with two factors. *Nature*. 2008;454:646-650.
155. Okita K, Nakagawa M, Hyenjong H, Ichisaka T, Yamanaka S. Generation of mouse induced pluripotent stem cells without viral vectors. *Science*. 2008;322(5903):949-953.
156. Yang S, Lin G, Tan YQ, et al. Tumor progression of culture-adapted human embryonic stem cells during long-term culture. *Genes Chromosomes Cancer*. 2008;47:665-679.
157. Ben-Porath I, Thomson MW, Carey VJ, et al. An embryonic stem cell-like gene expression signature in poorly differentiated aggressive human tumors. *Nat Genet*. 2008;40:499-507.
158. Li L, Baroja ML, Majumdar A, et al. Human embryonic stem cells possess immune-privileged properties. *Stem Cells*. 2004;22:448-456.
159. Slavin S, Kurkalli BG, Karussis D. The potential use of adult stem cells for the treatment of multiple sclerosis and other neurodegenerative disorders. *Clin Neurol Neurosurg*. 2008;110(9):943-946.
160. von Bonin M, Stolzel F, Goedecke A, et al. Treatment of refractory acute GVHD with third-party MSC expanded in platelet lysate-containing medium. *Bone Marrow Transplant*. 2008;43(3):245-251.
161. Dunac A, Frelin C, Popolo-Blondeau M, Chatel M, Mahagne MH, Philip PJ. Neurological and functional recovery in human stroke are associated with peripheral blood CD34+ cell mobilization. *J Neurol*. 2007;254:327-332.
162. Taguchi A, Nakagomi N, Matsuyama T, et al. Circulating CD34-positive cells have prognostic value for neurologic function in patients with past cerebral infarction. *J Cereb Blood Flow Metab*. 2008;29(1):34-38.
163. Grundmann F, Scheid C, Braun D, et al. Differential increase of CD34, KDR/CD34, CD133/CD34 and CD117/CD34 positive cells in peripheral blood of patients with acute myocardial infarction. *Clin Res Cardiol*. 2007;96:621-627.
164. Herrera MB, Bussolati B, Bruno S, Fonsato V, Romanazzi GM, Camussi G. Mesenchymal stem cells contribute to the renal repair of acute tubular epithelial injury. *Int J Mol Med*. 2004;14:1035-1041.
165. Seebach C, Henrich D, Tewksbury R, Wilhelm K, Marzi I. Number and proliferative capacity of human mesenchymal stem cells are modulated positively in multiple trauma patients and negatively in atrophic nonunions. *Calcif Tissue Int*. 2007;80:294-300.
166. Ciulla MM, Ferrero S, Gianelli U, Paliotti R, Magrini F, Braidotti P. Direct visualization of neo-vessel formation following peripheral injection of bone marrow derived CD34+ cells in experimental myocardial damage. *Micron*. 2007;38:321-322.
167. Wojakowski W, Tendera M. Mobilization of bone marrow-derived progenitor cells in acute coronary syndromes. *Folia Histochem Cytobiol*. 2005;43:229-232.
168. Paczkowska E, Larysz B, Rzeuski R, et al. Human hematopoietic stem/progenitor-enriched CD34(+) cells

are mobilized into peripheral blood during stress related to ischemic stroke or acute myocardial infarction. *Eur J Haematol*. 2005;75:461-467.
169. Wang XY, Lan Y, He WY, et al. Identification of mesenchymal stem cells in aorta-gonad-mesonephros and yolk sac of human embryos. *Blood*. 2008;111:2436-2443.
170. Dexter TM. Stromal cell associated haemopoiesis. *J Cell Physiol Suppl*. 1982;1:87-94.
171. Bakhshi T, Zabriskie RC, Bodie S, et al. Mesenchymal stem cells from the Wharton's jelly of umbilical cord segments provide stromal support for the maintenance of cord blood hematopoietic stem cells during long-term ex vivo culture. *Transfusion*. 2008;48(12):2638-2644.
172. Huang GP, Pan ZJ, Jia BB, et al. Ex vivo expansion and transplantation of hematopoietic stem/progenitor cells supported by mesenchymal stem cells from human umbilical cord blood. *Cell Transplant*. 2007;16:579-585.
173. Urban VS, Kiss J, Kovacs J, et al. Mesenchymal stem cells cooperate with bone marrow cells in therapy of diabetes. *Stem Cells*. 2008;26:244-253.
174. Ichim TE, Solano F, Brenes R, et al. Placental mesenchymal and cord blood stem cell therapy for dilated cardiomyopathy. *Reprod Biomed Online*. 2008;16:898-905.
175. Hill JM, Zalos G, Halcox JP, et al. Circulating endothelial progenitor cells, vascular function, and cardiovascular risk. *N Engl J Med*. 2003;348:593-600.
176. Van Zant G, Liang Y. The role of stem cells in aging. *Exp Hematol*. 2003;31:659-672.
177. Wojakowski W, Tendera M, Michalowska A, et al. Mobilization of CD34/CXCR4+, CD34/CD117+, c-met+stem cells, and mononuclear cells expressing early cardiac, muscle, and endothelial markers into peripheral blood in patients with acute myocardial infarction. *Circulation*. 2004;110: 3213-3220.
178. Locatelli F, Rocha V, Reed W, et al. Related umbilical cord blood transplantation in patients with thalassemia and sickle cell disease. *Blood*. 2003;101:2137-2143.
179. Cornetta K, Laughlin M, Carter S, et al. Umbilical cord blood transplantation in adults: results of the prospective Cord Blood Transplantation (COBLT). *Biol Blood Marrow Transplant*. 2005;11:149-160.
180. Lubin BH and Shearer WT. Cord blood banking for potential future transplantation. *Pediatrics*. 2007;119:165-170.
181. Gluckman E, Rocha V, Boyer-Chammard A, et al. Outcome of cord-blood transplantation from related and unrelated donors. Eurocord Transplant Group and the European Blood and Marrow Transplantation Group. *N Engl J Med*. 1997;337:373-381.
182. Laughlin MJ, Barker J, Bambach B, et al. Hematopoietic engraftment and survival in adult recipients of umbilical-cord blood from unrelated donors. *N Engl J Med*. 2001;344:1815-1822.
183. Matsumura T, Narimatsu H, Kami M, et al. Cytomegalovirus infections following umbilical cord blood transplantation using reduced intensity conditioning regimens for adult patients. *Biol Blood Marrow Transplant*. 2007;13:577-583.
184. Sasazuki T, Juji T, Morishima Y, et al. Effect of matching of class I HLA alleles on clinical outcome after transplantation of hematopoietic stem cells from an unrelated donor. Japan Marrow Donor Program. *N Engl J Med*. 1998;339:1177-1185.
185. Barker JN, Krepski TP, DeFor TE, Davies SM, Wagner JE, Weisdorf DJ. Searching for unrelated donor hematopoietic stem cells: availability and speed of umbilical cord blood versus bone marrow. *Biol Blood Marrow Transplant*. 2002;8:257-260.
186. Limbourg F, Ringes-Lichtenberg S, Schaefer A, et al. Haematopoietic stem cells improve cardiac function after infarction without permanent cardiac engraftment. *Eur J Heart Fail*. 2005;7:722-729.
187. Asahara T. Stem cell biology for vascular regeneration. *Ernst Schering Res Found Workshop*. 2005;54:111-129.
188. Distler JH, Hirth A, Kurowska-Stolarska M, Gay RE, Gay S, Distler O. Angiogenic and angiostatic factors in the molecular control of angiogenesis. *Q J Nucl Med*. 2003;47:149-161.
189. Brogi E, Wu T, Namiki A, Isner JM. Indirect angiogenic cytokines upregulate VEGF and bFGF gene expression in vascular smooth muscle cells, whereas hypoxia upregulates VEGF expression only. *Circulation*. 1994;90: 649-652.
190. Nicosia RF, Nicosia SV, Smith M. Vascular endothelial growth factor, platelet-derived growth factor, and insulin-like growth factor-1 promote rat aortic angiogenesis in vitro. *Am J Pathol*. 1994;145:1023-1029.
191. Bos R, van Diest PJ, de Jong JS, van der Groep P, van der Valk P, van der Wall E. Hypoxia-inducible factor-1alpha is associated with angiogenesis, and expression of bFGF, PDGF-BB, and EGFR in invasive breast cancer. *Histopathology*. 2005;46:31-36.
192. Schlingemann RO. Role of growth factors and the wound healing response in age-related macular degeneration. *Graefes Arch Clin Exp Ophthalmol*. 2004;242:91-101.
193. Mentlein R, Held-Feindt J. Angiogenesis factors in gliomas: a new key to tumour therapy? *Naturwissenschaften*. 2003;90:385-394.
194. Zhao L, Eghbali-Webb M. Release of pro- and anti-angiogenic factors by human cardiac fibroblasts: effects on DNA synthesis and protection under hypoxia in human endothelial cells. *Biochim Biophys Acta*. 2001;1538:273-282.
195. Dunn IF, Heese O, Black PM. Growth factors in glioma angiogenesis: FGFs, PDGF, EGF, and TGFs. *J Neurooncol*. 2000;50:121-137.
196. Kano MR, Morishita Y, Iwata C, et al. VEGF-A and FGF-2 synergistically promote neoangiogenesis through enhancement of endogenous PDGF-B-PDGFRbeta signaling. *J Cell Sci*. 2005;118:3759-3768.
197. Laschke MW, Elitzsch A, Vollmar B, Vajkoczy P, Menger MD. Combined inhibition of vascular endothelial growth factor (VEGF), fibroblast growth factor and platelet-derived growth factor, but not inhibition of VEGF alone, effectively suppresses angiogenesis and vessel maturation in endometriotic lesions. *Hum Reprod*. 2006;21:262-268.
198. Kelly BD, Hackett SF, Hirota K, et al. Cell type-specific regulation of angiogenic growth factor gene expression and induction of angiogenesis in nonischemic tissue by a constitutively active form of hypoxia-inducible factor 1. *Circ Res*. 2003;93:1074-1081.
199. Chauhan AMR, Mullins PA, Taylor G, Petch C, Schofield PM. Aging-associated endothelial dysfunction in humans is reversed by L-arginine. *J Am Coll Cardiol*. 1996;28: 1796-1804.

200. Tschudi MR, Barton M, Bersinger NA, et al. Effect of age on kinetics of nitric oxide release in rat aorta and pulmonary artery. *J Clin Invest*. 1996;98:899-905.
201. Hill JM, Syed MA, Arai AE, et al. Outcomes and risks of granulocyte colony-stimulating factor in patients with coronary artery disease. *J Am Coll Cardiol*. 2005;46:1643-1648.
202. Charles A. Goldthwaite, J. Mending a broken heart: stem cells and cardiac repair. In: *Regenerative Medicine*. Bethesda, MD: (National Institute of Health. Department of Health and Human Services; 2006.
203. Bonanno G, Mariotti A, Procoli A, et al. Human cord blood CD133+ cells immunoselected by a clinical-grade apparatus differentiate in vitro into endothelial- and cardiomyocyte-like cells. *Transfusion*. 2007;47(2):280-289.
204. Cheng F, Zou P, Yang H, et al. Induced differentiation of human cord blood mesenchymal stem/progenitor cells into cardiomyocyte-like cells in vitro. *J Huazong Univ Sci Technol Med Sci*. 2003;23:154-157.
205. Yamada Y, Yokoyama S, Fukuda N, et al. A novel approach for myocardial regeneration with educated cord blood cells cocultured with cells from brown adipose tissue. *Biochem Biophys Res Commun*. 2007;353:182-188.

Human Umbilical Cord Blood Mononuclear Cells in the Treatment of Acute Myocardial Infarction

24

Robert J. Henning

24.1 Introduction

Acute coronary occlusion with myocardial infarction is the leading cause of morbidity and mortality in the Western world and, according to the World Health Organization, it will be the major cause of death in the world by the year 2020.[1] Each year more than one million Americans experience an acute myocardial infarction and approximately 400,000 die from acute complications of myocardial infarction.[2] In addition, every year more than 400,000 Americans develop new onset congestive heart failure.[2] A critical determinant of the prognosis of every patient with ischemic heart disease is the size of their myocardial infarction, which directly determines the magnitude of heart dilation, the degree of impairment of heart pump function, the development of heart failure, and, ultimately, the prognosis of the patient.

In order to limit myocardial infarction size and minimize or prevent heart failure, cardiovascular investigators have recently begun to transplant cells into infarcted hearts. Embryonic stem cells, skeletal myoblasts, and bone marrow stem cells have been transplanted into hearts.[3]

Embryonic stem cells can be derived from the inner cell mass of the human blastocyst and have the capacity to differentiate into cells from all three primary germ layers. When injected into hearts, embryonic cells can express cardiac myocyte actin, myosin heavy chain, and troponin proteins, form intercalated disks, sinus nodal and atrial cells, and induce new blood vessel formation in the host ventricle.[4–7] In addition, embryonic stem cells can attenuate thinning of the infarcted cardiac ventricular wall, left ventricular (LV) dilation, and myocardial dysfunction.[4,6] However, ethical, legal, and societal issues involving the procurement and use of human embryonic stem cells for therapeutic purposes have significantly limited the availability and use of these cells. In addition, human embryonic stem cells in culture are most often maintained on mouse feeder fibroblast cells, which raise concerns about transmission of rodent viruses and prions to humans. Moreover, long-term culture and propagation of human embryonic stem cells may result in genetic mutations that limit their usefulness. These important issues have spurred researchers to pursue the use of skeletal myoblasts, isolated from the basal lamina propria of skeletal muscle, as alternative cells for repair of damaged hearts. Approximately 4% of mammalian skeletal muscle cells are skeletal myoblasts, which are capable of cellular division, formation of skeletal myocytes, and muscle repair. Skeletal myoblasts have been directly injected into infarcted hearts or injected into the coronary arteries for implantation into the myocardium.[8,9] These cells can replicate in the myocardium, form multinucleated myotubules, differentiate into mature skeletal myofibers, and can contract when stimulated.[9] Approximately 20% of myofibers from skeletal myoblasts develop characteristics of slow twitch cardiac muscle as the myoblasts mature in the myocardium. Transplantation of skeletal myoblasts into infarcted hearts limits left ventricular remodeling and prevents significant deterioration of the left ventricular ejection fraction.[10,11] Moreover, the new muscle is more resistant to ischemic injury than normal cardiac muscle. However, the number of skeletal myoblasts present in skeletal muscles decreases with age. Consequently, as much as 10 g of

R.J. Henning
Center for Cardiovascular Research,
James A. Haley Medical Center/University of South Florida,
13000 Bruce B. Downs Blvd, 33612, Tampa, FL, USA
e-mail: robert.henning@va.gov

N. Bhattacharya and P. Stubblefield (eds.), *Regenerative Medicine Using Pregnancy-Specific Biological Substances*,
DOI: 10.1007/978-1-84882-718-9_24,

skeletal muscle are necessary for myoblast isolation for transplantation into the hearts of large research animals or humans. Propagation of myoblasts in cultures is mandatory to obtain adequate numbers of autologous or allogeneic myoblasts for successful transplantation, and this may require ≥ 12 days. The histological features of the majority of the mature muscle cells in the myocardium are those of well-differentiated skeletal muscle, not cardiac muscle.[11] Moreover, skeletal myocyte communication with host cardiac myocytes via gap junction proteins is poor because the skeletal myoblast transplants do not form gap junctions with cardiac myocytes and are often insulated from the myocardium by scar tissue.[11] Consequently, cardiac arrhythmias can occur because of electrical re-entry in skeletal muscle "islands" in the myocardium. Therefore, the clinical trials of skeletal myoblasts in patients with ischemic cardiomyopathies frequently require that cardiac patients who receive these cells have an automatic cardiac defibrillator in place.

Recently, investigators have discovered that bone marrow, which contains both mesenchymal and hematopoietic stem cells, has the capacity to colonize different tissues, proliferate, and transdifferentiate into cell lineages of the host organ. Mesenchymal bone marrow stem cells can serve as precursors for cardiac myocytes and hematopoietic stem cells can serve as precursors for vascular endothelial cells and vessels.[12-15] Moreover, bone marrow mesenchymal cells express class I human leukocyte antigens but do not express class II antigens, which significantly limits the possibility of immune rejection by the host.[13]

Bone marrow mesenchymal stem cells (MSCs) implanted into the left ventricle can express the myocardial proteins α-actinin, tropomyosin, troponin, myosin heavy chain, and phospholamban.[14,15] After culture with 5 azacytidine, transplanted bone marrow cells may also induce angiogenesis in myocardial infarctions.[16] Enriched hematopoietic bone marrow progenitor cells can contribute to capillary endothelial cells in newly formed blood vessels in the "at risk" myocardium adjacent to the infarction and may limit inflammation and possibly regenerate myocardium.[16-19] The differentiation process for bone-marrow-derived cells appears to require specific paracrine growth signals from host cardiomyocytes and electromechanical stimulation in the adult heart.[18]

Implantation of bone marrow stem cells attenuates ventricular infarction area and ventricular wall thinning.[16,17,20] In addition, bone marrow stem cells transplanted into infarcted hearts can augment ventricular systolic wall thickening with variable effects on left ventricular pressure and changes in pressure per unit change in time [dP/dt].[16,17] However, the number and viability of stem cells in bone marrow decreases with the increasing age of the donor and/or the presence of associated chronic diseases such as diabetes mellitus or ischemic cardiomyopathy. Consequently, days or weeks may be required for the preparation and expansion of autologous bone marrow stem cells in culture in order to obtain adequate numbers of stem cells for injection into a patient.[18,19] Moreover, the use of allogeneic bone marrow cells frequently requires the use of immunosuppressant drugs in the host, which may contribute to an immature bone marrow transplant cellular phenotype.

Each cell type for transplantation previously discussed is associated with significant ethical, biological, or technical limitations that must be overcome in order to be generally applicable to patients with heart disease. Furthermore, there is currently no scientific consensus about the optimal number of embryonic, skeletal, or adult bone marrow stem cells or the optimal time for stem cell transplantation in patients after myocardial infarction.

For these reasons we have recently begun to investigate the use of human umbilical cord blood mononuclear cells (HUCBCs), isolated from human umbilical cord blood, for the treatment of acute myocardial infarction in research animals. HUCBCs are potentially available in unlimited quantities, can be cryopreserved for periods of 15 or more years with recovery of 60% to 100% viable cells[21] and, therefore, can be readily available for treatment of patients with acute myocardial infarction. In the past, the umbilical cord, cord blood, and placenta were considered waste products that were frequently incinerated after the birth of an infant. Currently, however, umbilical cord blood mononuclear cells are recognized as a rich source of hematopoietic endothelial and mesenchymal stem cells.[21,22] Umbilical cord blood mononuclear cells are currently used for repopulating cells in patients who are being treated for acute lymphoid leukemia, acute myeloid leukemia, chronic myeloid leukemia, myelodysplastic syndrome, neuroblastoma and nonmalignant diseases such as Fanconi's anemia and aplastic anemia.[21,23-26] A distinct advantage of HUCBCs is the immature immunogenicity of these cells, which is similar to embryonic cells, and which significantly reduces the risk of rejection by the host.[21,27]

HUCBC mesenchymal and hematopoietic cells can be propagated in tissue culture to facilitate transplantation.[28,29] The total content of hematopoietic progenitor cells in umbilical cord blood equals or exceeds that of bone marrow but the highly proliferative hematopoietic stem cells are reported to be eightfold higher in HUCBC than in bone marrow.[30,31] These cells express hematopoietic markers such as CD34 and CD14 and can be enriched by as much as 77–95%.[30,31] Cord CD34 positive hematopoietic stem cells contain thrombopoietin and interleukin that cause proliferation of hematopoietic cells and suppress cell apoptosis.[21,30] HUCBCs with a mesenchymal phenotype express the surface protein markers SH2, SH3, SH4, α-smooth muscle actin, MAB 1470, CD13, CD29, and CD49.[28] Cell cycle analysis indicates that >85% of the mesenchymal cells are in the G0/G1 stage of the cell cycle. However, these cells are capable of proliferating with a population-doubling time of 48 h.[28] The immunotype and functional properties displayed by human cord blood mesenchymal cells closely resemble the characteristics assigned to bone-marrow-derived mesenchymal progenitor cells that have been transplanted into myocardium.[28]

In our initial studies with HUCBC, we created acute myocardial infarctions in research animals (rats) by permanent ligation of the left coronary artery.[32] Group I consisted of 24 rats with no interventions in order to determine heart changes due to changes in age and body weight. Group 2 consisted of 33 rats with anterior wall myocardial infarctions in which we directly injected Isolyte, an electrolyte solution, into the ischemic border zones of the infarction 1 h after permanent ligation of the left coronary artery. Group 3 consisted of 38 rats with anterior wall myocardial infarctions in which we injected one million HUCBCs in Isolyte into the ischemic borders of the infarction 1 h after permanent coronary ligation. We did not give immunosuppressive therapy to any rat in any of our studies. All the rats were monitored with echocardiograms that were performed prior to and at 1, 2, 3, and 4 months after myocardial infarction in order to determine the temporal response of ventricular remodeling. Randomized rats from each group also underwent hemodynamic studies at 1, 2, 3, and 4 months prior to being euthanatized.

Neither clinical nor histologic signs of immunorejection occurred in Group 3 HUCBC-treated rats. In Group 3, the left ventricular ejection fractions (EF) decreased from a baseline of 88±3% to a mean of 63.0±3% at 1 month, but progressively increased to

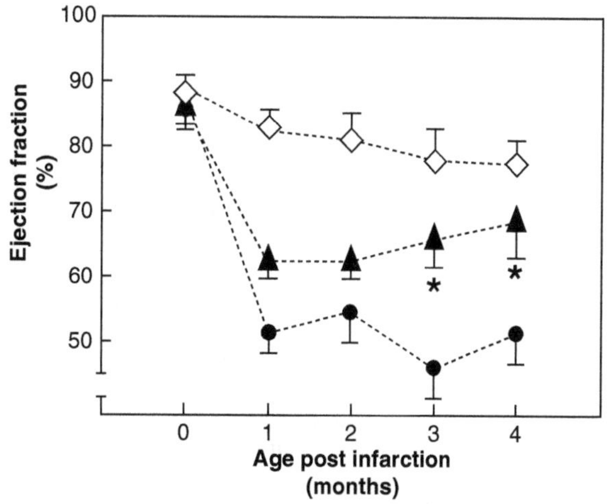

Fig. 24.1 Mean left ventricular ejection fraction (%) determined by echocardiograms for controls (◊), infarct+Isolyte group (•), and infarct+HUCBC group (▲) at baseline (0) and at 1, 2, 3, and 4 months. The ejection fraction in the infarct + HUCBC group was significantly greater (*$p<0.02$) than in the infarct+Isolyte group at 3 and 4 months (From Henning et al.[32] With permission from the publisher)

71.0±6% at 3 and 4 months (See Fig. 24.1). These values significantly differed from Group 2 Isolyte-treated infarcted rats ($p<0.02$) in which the ejection fractions decreased to and remained at 51±3% between 1 and 4 months after infarction.[32] The ejection fractions in the HUCBC-treated rats approximated the ejection fractions in the control rats at 4 months (See Fig. 24.1). At 4 months after myocardial infarction, the myocardial anteroseptal wall thickening in the HUCBC group was 57.9±11.6% and nearly approximated the myocardial anteroseptal wall thickening of 59.2±8.9% in the control rats but significantly exceeded the anteroseptal wall thickening in infarct+Isolyte-treated rats of 27.8±7% ($p<0.02$).[32]

The infarct sizes at 4 months after coronary artery ligation averaged 9.2±2.0% for HUCBC-treated Group 3 rats versus 40.0±9.2% for the Isolyte-treated Group 2 rats ($p<0.01$)[32] (See Fig. 24.2). Hematoxylin and eosin and trichrome staining of myocardial tissue from Groups 1–3 performed at 1, 2, 3, and 4 months after the infarctions, and examined by two pathologists unaware of the rat treatment, showed no histological evidence of immunorejection in the rat hearts treated with HUCBC. We conclude from these experiments that HUCBC can significantly limit myocardial infarct size and improve LV function in infarcted rat hearts without the requirements for host immunosuppression treatment.

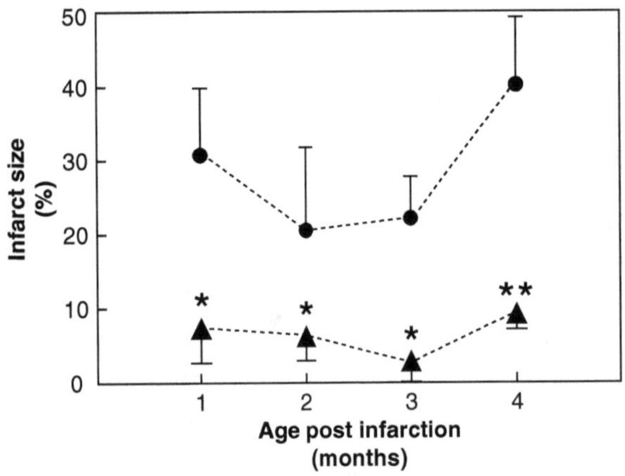

Fig. 24.2 Mean infarct size (% of the left ventricle) in the infarct+Isolyte group (•) and the infarct+HUCBC group (of the left ventricle ▲) hearts at 1 (n=5), 2 (n=6), 3 (n=14), and 4 (n = 15) months post-infarction. The infarct+HUCBC group infarctions were smaller than the infarct+Isolyte group infarctions at each time point (*p<0.01, **p < 0.001) (From Henning et al.[32] with permission from the publisher)

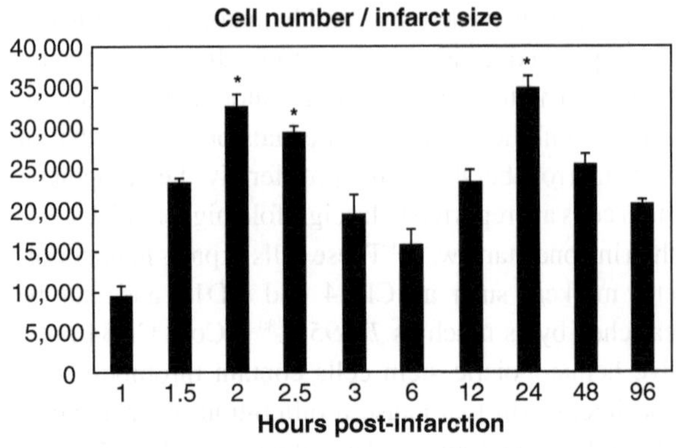

Fig. 24.3 Number of HUCBCs that migrated to infarct tissues at different times after coronary ligation and acute myocardial infarction. The HUCBC number is corrected for infarct size. The 2–2.5 and 24 h post-infarction groups are significantly different from the 1 h group (p<0.01) (From Henning et al.[33] With permission from the publisher)

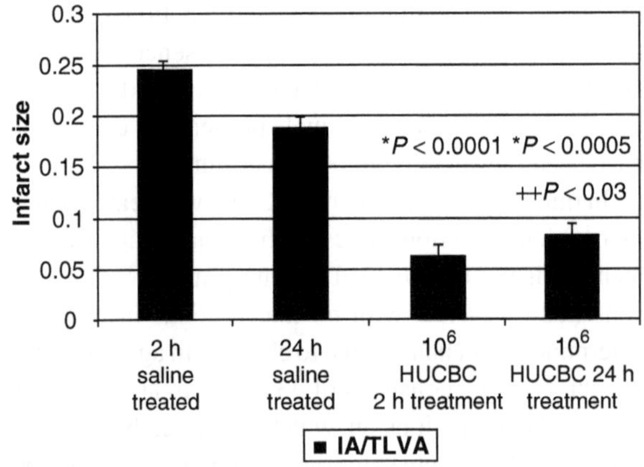

Fig. 24.4 Measured infarction sizes in saline-treated hearts and hearts treated with HUCBCs. The hearts treated with HUCBCs within 2 and 24 h of acute myocardial infarction are significantly smaller than the saline-treated hearts. IA/TLVA: infarction area divided by total left ventricular area. *Compared with 2 h saline. ++Compared with 24 h saline (From Henning et al.[33] With permission from the publisher)

In our next series of investigations,[33] we determined the optimal time to inject HUCBCs after myocardial infarction. We examined the migration of HUCBCs labeled with fluorescent 4′,6-diamidino-2-phenylindole dihydrochloride (DAPI) from separate wells in the upper part of a Boyden Double Chamber Apparatus across polycarbonate membranes with pores of 4 μm, which are significantly less than the 12 μm diameter of HUCBC, to separate wells in the lower part of the apparatus containing left ventricular infarct tissue obtained at 1, 2, 2.5, 3, 6, 12, 24, 48, and 96 h after acute myocardial infarction in rats. In these investigations, we determined that infarcted myocardium has the greatest chemoattraction for HUCBC at 2–2.5 h and at 24 h after acute myocardial infarction[33] (See Fig. 24.3). We then injected 10^6 HUCBCs into infarcted rat hearts at 2 h and into separate rat hearts at 24 h after permanent coronary ligation/acute myocardial infarction and measured the infarction size 1 month later with Tetrazolium staining of the myocardium and computer-assisted infarct sizing (See Fig. 24.4). Injection of HUCBC within 2 h of coronary occlusion resulted in infarction sizes at 1 month of 6.4±0.01% of the LV muscle area (N=10) in comparison with infarct sizes of 24.5±0.02% of the LV area in saline-treated infarcted rat hearts (p<0.0001). Injection of HUCBC at 24 h after coronary occlusion in HUCBC-treated rats resulted in infarction sizes at 1 month of 8.4±0.02% that were significantly less than the saline-treated infarctions at either 2 or 24 h (p<0.005 and p < 0.03, respectively) (See Fig. 24.4). These experiments demonstrate that HUCBCs can be injected early (i.e., 2 h) or late (i.e., 24 h) after the onset of myocardial infarction and have significant beneficial effects in limiting myocardial infarction size.

We then determined the optimal number of HUCBCs to inject in order to most significantly reduce myocardial infarction size.[34] We also defined the optimal technique to inject the HUCBCs.[34] Prior to our study, no consensus existed regarding the optimal dose of stem cells or the optimal route of administration of stem cells for the treatment of acute myocardial infarction. In this study, we administered HUCBC to groups of four to five rats, 1 to 2 h after myocardial infarction, in doses of 0.5×10^6, 1×10^6, 2×10^6, 4×10^6, 8×10^6, 16×10^6 or 32×10^6 in 0.3–0.5 ml of Isolyte S pH 7.4. A second group of rats with infarcted hearts was given only 0.3–0.5 ml of Isolyte pH 7.4. In order to determine the optimal technique for stem cell administration, we injected the HUCBCs directly into the myocardium (IM), or into the coronary arteries (intracoronary artery, *IA*), or intravenously (IV). One month later, we determined the infarct sizes by Tetrazolium staining in the HUCBC-treated group and in the Isolyte-treated group. In these experiments, the optimal dose of HUCBC for infarct size reduction was 4×10^6 HUCBC for IM and IA injections and 16×10^6 HUCBC for IV injection[34] (See Fig. 24.5). The infarct size in the control rat hearts treated with Isolyte averaged $23.7\pm1.7\%$ of the LV muscle area. Intramyocardial injection of HUCBC reduced the infarct size by 93% with 4×10^6 HUCBCs to a value of $1.7\pm1.3\%$ of the LV muscle area ($p<0.001$). Intra-coronary artery injection of HUCBCs reduced the infarction size by 80% with 4×10^6 HUCBC ($p<0.001$) and IV HUCBCs reduced the infarct size by 75% with 16×10^6 HUCBC ($p<0.001$). With 4×10^6 HUCBCs, infarction size was 65% smaller with IM HUCBCs than with IA HUCBCs and 78% smaller than with IV HUCBC ($p<0.03$). However, the intra-coronary artery injection of $\geq32\times10^6$ HUCBC was associated with decreases in oxygen saturation, cardiac arrhythmias, and sudden death in four rats, which suggested that large doses of stem cells given into the coronary arteries can cause microemboli and myocardial infarctions. Nevertheless, HUCBC, when administered directly into the myocardium or into the coronary arteries in doses of 4×10^6, or intravenously in doses of 16×10^6 can produce significant reductions in infarct size in comparison with Isolyte-treated infarcted hearts, without the necessity of host immune suppression.[34] The largest reductions in infarct size occur with the direct intra-myocardial injection of HUCBCs.[34]

In order to determine the effects of HUCBC on inflammation in acute myocardial infarctions, we have measured inflammatory cytokines and chemokines in left ventricular myocardium taken from rat hearts at 2, 6, 12, 24 and 72 h after myocardial infarction[35] (See Fig. 24.6). In the infarcted rat hearts treated with only Isolyte after myocardial infarction, the myocardial concentration of TNFα increased from $6.7\pm0.9\%$ to $52.3\pm4.7\%$, monocyte chemoattraction protein (MCP) increased from $9.5\pm1.2\%$ to $39.8\pm2.1\%$, fractalkine increased from $11.5\pm1.5\%$ to $28.1\pm1.3\%$, ciliary neurotrophic protein, a member of the interleukin-6 (IL-6) family, increased from $12.1\pm0.02\%$ to $25.9\pm1.1\%$, macrophage inflammatory protein (MIP) from $10.3\pm1.5\%$ to $23.9\pm1.4\%$ and interferon-gamma (IFN-γ) increased from $8.7\pm0.4\%$ to $26.0\pm1.6\%$ in comparison with known cytokine controls in non-infarcted myocardium.[35] In contrast, these inflammatory cytokines/chemokines did not significantly change in the rat infarctions that were treated with 4×10^6 HUCBC[35] (See Fig. 24.6).

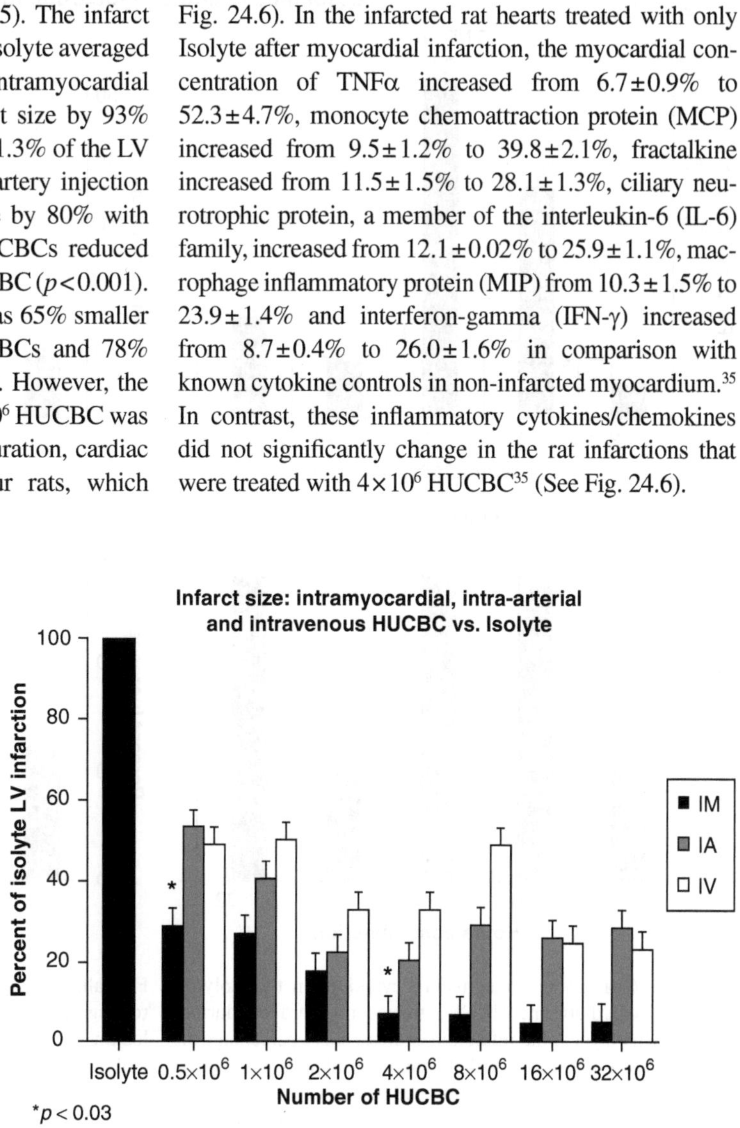

Fig. 24.5 The infarctions in the Isolyte-treated rat hearts are normalized to 100% and the HUCBC-treated intra-myocardial (IM), intra-coronary artery (IA), and intravenous (IV) infarctions are expressed as a percent of the Isolyte-treated left ventricular infarctions for comparison. IM HUCBC produced the largest reductions in infarct size. With 0.5×10^6 HUCBC, the IM technique resulted in infarction sizes that were 46% and 41% smaller than the IA technique and the IV technique, respectively. With 4×10^6 HUCBC, IM-HUCBC-treated infarctions were 65% and 78% smaller than the IA- and IV-HUCBC-treated infarctions, respectively. *$p<0.03$ (From Henning et al.[34] With permission from the publisher)

Using fluorescence-activated cell sorting (FACS) we then measured neutrophils, lymphocytes, and macrophages in left ventricular myocardium at 12, 24, and 72 h after myocardial infarctions in rats treated with either Isolyte or 4×10^6 HUCBC in Isolyte that was directly injected into the peri-infarction myocardium. The percentage of neutrophils significantly increased from $0.04 \pm 0.2\%/50{,}000$ heart cells in the control non-infarcted myocardium to $5.3 \pm 1.2\%/50{,}000$ heart cells($p < 0.001$) 12 h after infarction in Isolyte treated myocardial infarctions.[35] In contrast, the percentage of neutrophils averaged only $1.3 \pm 0.7\%/50{,}000$ heart cells in HUCBC-treated myocardial infarctions ($p < 0.02$ compared with Isolyte)[35] (See Fig. 24.7). Thereafter, the percentages of neutrophils rapidly decreased in the myocardium at 24 h and at 72 h after infarction. At 72 h after the acute infarctions, the percentages of neutrophils averaged $0.6 \pm 0.2\%/50{,}000$ heart cells in Isolyte-treated hearts in contrast to $0.2 \pm 0.1\%/50{,}000$ cells in HUCBC-treated hearts ($p < 0.05$). Moreover, the percentages of neutrophils at 24 and 72 h in HUCBC-infarcted hearts were not significantly different from the controls.[35]

At 24 h post-infarction, the percentage of CD3 and CD4 lymphocytes in infarcted myocardium were $10.7 \pm 1.4\%$ and $6.3 \pm 1.1\%/50{,}000$ cells, respectively,

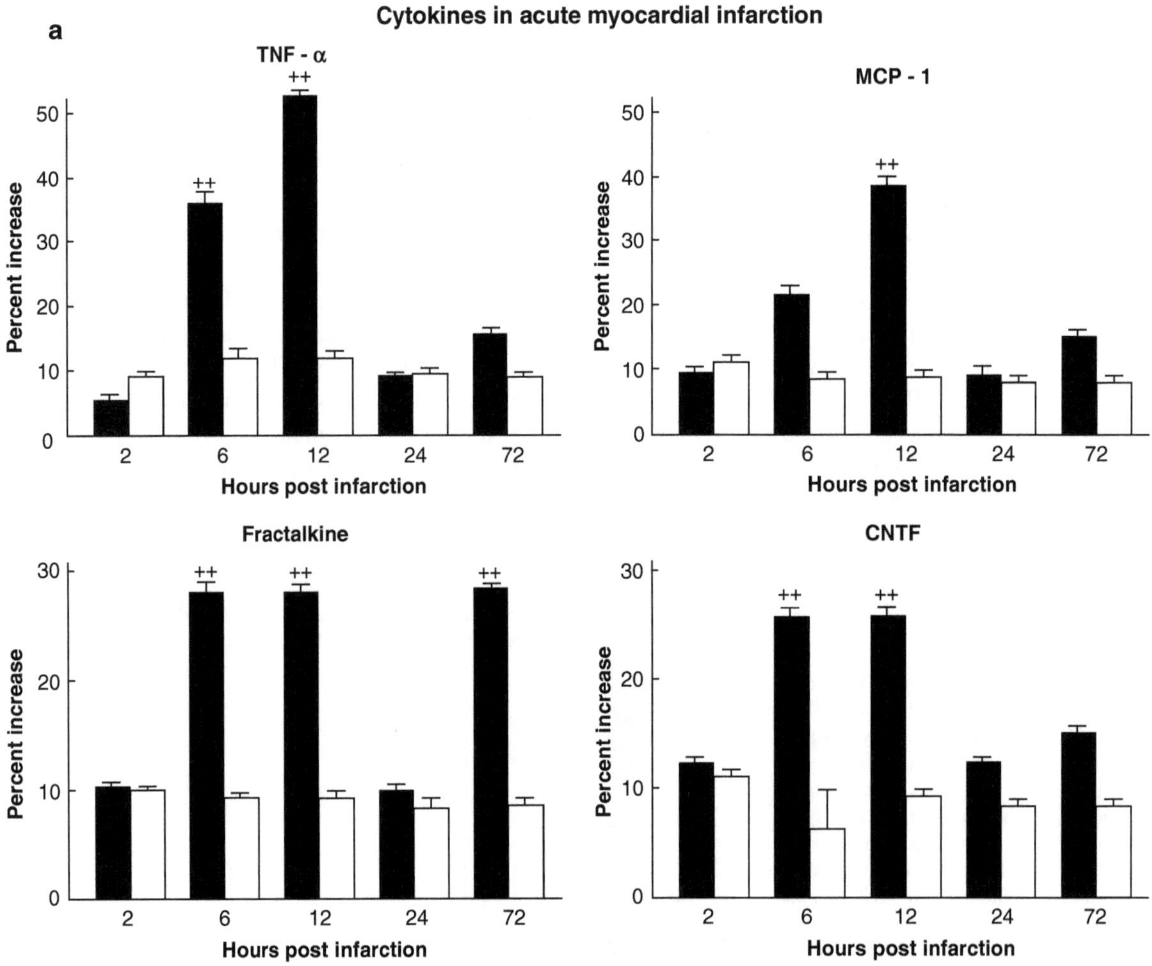

Fig. 24.6 Changes in inflammatory cytokines in the Isolyte-treated myocardium and HUCBC-treated myocardium during the 72 h after myocardial infarction expressed as a percentage of control myocardium. Myocardial cytokines did not significantly change in HUCBC-plus Isolyte-treated hearts. (**a**) *TNFα* tumor necrosis factor α, *MCP-1* monocyte chemoattraction protein-1, Fractalkine, *CNTF* ciliary neurotrophic factor. (**b**) *MIP* macrophage inflammatory protein, *INF-γ* interferon-γ, *IL-1β* interleukin-1β, *IL-4* interleukin-4. $^{++}p < 0.001$, $^{*}p < 0.05$. *Black bars*, Isolyte-treated infarctions. *White open bars*, HUCBC + Isolyte-treated infarctions (From Henning et al.[35] With permission from the publisher)

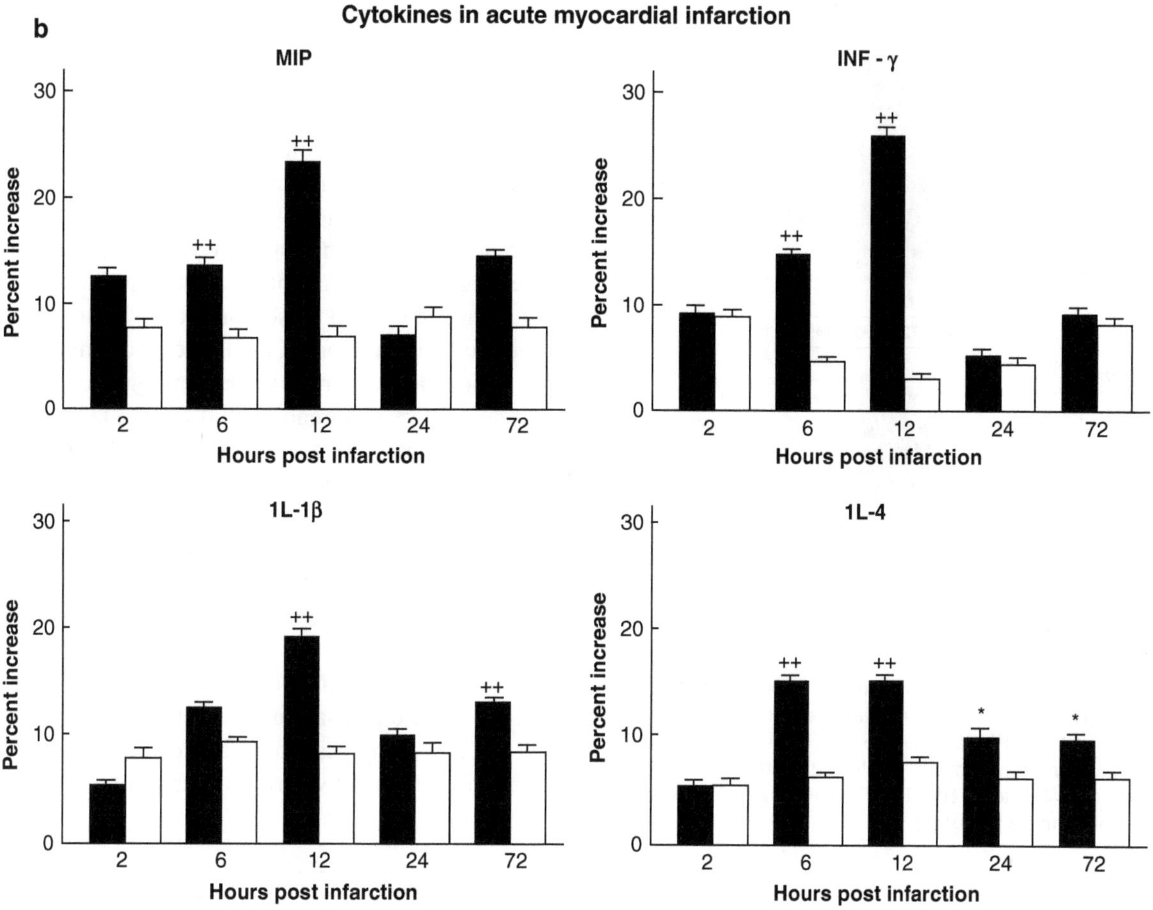

Fig. 24.6 (continued)

in Isolyte-treated infarcted hearts in comparison with only 4.9 ± 0.8% for CD3 and 2.9 ± 0.5% for CD4 in HUCBC-treated infarcted hearts ($p < 0.005$ HUCBC compared with Isolyte)[35] (See Fig. 24.8). The percentage of CD11b macrophages at 24 h post-infarction was significantly larger in Isolyte-treated hearts and averaged 2.8 ± 0.3% in contrast to HUCBC-treated hearts that averaged 1.9 ± 0.2% ($p < 0.05$). At 72 h after infarction, the percentage of CD3 and CD4 lymphocytes averaged 8.0 ± 1.1% and 5.1 ± 0.8%/50,000 heart cells, respectively, in Isolyte-treated myocardial infarctions in comparison with only 4.1 ± 0.5% and 2.3 ± 0.4%/50,000 heart cells, respectively, in the HUCBC-treated infarctions ($p < 0.005$).[35] In these investigations, the left ventricular infarct sizes were more than 40% smaller in HUCBC-treated infarctions in comparison with Isolyte-treated infarctions.[35] Moreover, in rats followed for 2 months post-infarction, the left ventricular ejection fractions in Isolyte-treated hearts decreased to 65.4 ± 1.9% and 69.1 ± 1.9% at one and two months after infarction and were significantly lower than the left ventricular ejection fractions in HUCBC-treated hearts that averaged 72.1 ± 1.3% and 75.7 ± 1.4% (both $p < 0.02$).[35]

Our investigations suggest that HUCBCs are beneficial in infarcted myocardium and do not require host immunosuppression. HUCBCs significantly decrease inflammatory cytokines in hearts with myocardial infarctions, and this decrease in inflammatory cytokines is associated with significant decreases in the percentages of myocardial neutrophils and CD3 and CD4 T lymphocytes in the infarcted myocardium. As a consequence, HUCBC can produce a substantial reduction in acute myocardial infarction size in comparison with

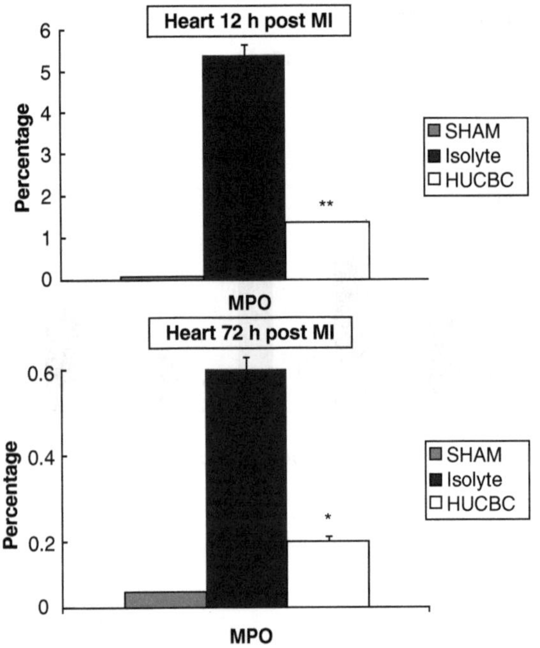

Fig. 24.7 Percentages of neutrophils in 50,000 myocardial heart cells at 12 and 72 h after myocardial infarction and treatment with Isolyte or 4×10^6 HUCBC+Isolyte. HUCBC significantly reduced the number of neutrophils in the infarcted hearts. $*p<0.05$, $**p < 0.02$: Isolyte vs. HUCBC (From Henning et al.[35] With permission from the publisher)

untreated infarcted hearts when these cells are directly injected into the myocardium, or into the coronary arteries, or given intravenously. The reduction in left ventricular infarction size is associated with left ventricular ejection fractions that are significantly greater in HUCBC-treated infarcted hearts than in hearts treated with only Isolyte.

An additional mechanism whereby HUCBC may be beneficial in ischemic/infarcted myocardium is by stimulating new blood vessel formation. Endothelial progenitor cells are normal components of umbilical cord blood that can release pro-angiogenic molecules such as vascular endothelial growth factor.[36–38] These cells can also express KDR, Tie2/Tek, and VE-cadherin, which are expressed by endothelial cells during new blood vessel formation.[36–38] In addition, $CD34^+$ HUCBC can integrate into the walls of blood vessels in the periphery of injured tissue and can increase capillary density in ischemic/infarcted muscles.[39,40]

Investigations are ongoing in our Center to determine the precise mechanisms whereby HUCBCs can limit the size of myocardial infarctions and improve the function of the left ventricle of the heart.

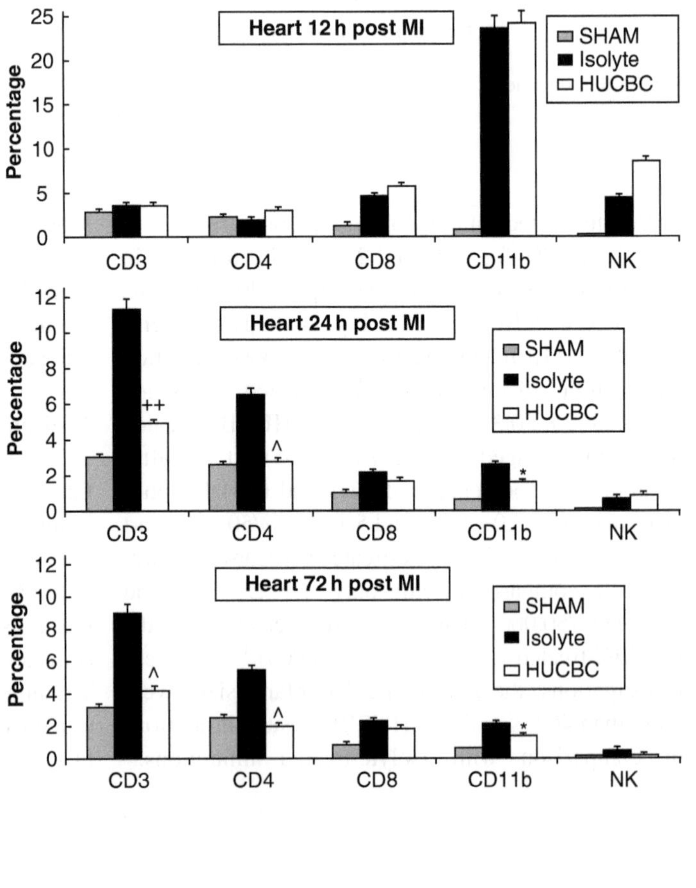

Fig. 24.8 Percentages of inflammatory CD3 lymphocytes, CD4 lymphocytes, CD8 lymphocytes, CD11b macrophages, and CD 161 (natural killer cells) at 12, 24, and 72 h in Isolyte-treated infarcted hearts and in infarcted hearts treated with HUCBC. HUCBC reduced the CD3 and CD4 cells in the infarcted hearts at 24 and 72 h such that the percentages of these cells were not significantly different from the controls. $^{++}p<0.001$, $^p < 0.005$, $*p < 0.05$: Isolyte vs. HUCBC (From Henning et al.[35] With permission from the publisher)

References

1. Lopez AD, Murray C. The global burden of disease, 1990–2020. *Nat Med.* 1998;4:1241-1243.
2. Rosamond W, Flegal K. Heart disease statistics, 2007. *Circulation.* 2007;115:e69-e171.
3. Murry CE, Field LJ, Menasche P. Cell-based cardiac repair: reflections at the 10 year point. *Circulation.* 2005;112: 3174-3183.
4. Kehat I, Kenyagin-Karsenti D, Snir M, Segev H. Human embryonic stem cells can differentiate into myocytes with structural and functional properties of cardiomyocytes. *J Clin Invest.* 2001;108:407-414.
5. McDevitt TC, Laflamme MA, Murry CE. Proliferation of cardiomyocytes derived from human embryonic stem cells is mediated via the IGF/PI3-kinase/Akt signaling pathway. *J Mol Cell Cardiol.* 2005;39:865-873.
6. Laflamme MA, Chen KY, Naumova AV, et al. Cardiomyocytes derived from human embryonic stem cells in pro-survival factors enhance function of infarcted rat hearts. *Nat Biotechnol.* 2007;25:1015-1024.
7. Soonpaa MH, Koh GY, Klug MG, Field LJ. Formation of nascent intercalated disks between grafted fetal cardiomyocytes and host myocardium. *Science.* 1994;264: 98-101.
8. Murry CE, Wiseman RW, Schwartz SM, Hauschka SD. Skeletal myoblast transplantation for repair of myocardial necrosis. *J Clin Investig.* 1996;98:2512-2523.
9. Taylor DA, Atkins BZ, Hungspreugs P. Regenerating functional myocardium: improved performance after skeletal myoblast transplantation. *Nat Med.* 1998;4:929-933.
10. Scorsin M, Hagege A, Vilquin J. Comparison of fetal cardiomyocytes and skeletal myoblast transplantation on postinfarction left ventricular function. *J Thorac Cardiovasc Surg.* 2000;119:1169-1175.
11. Ghostine S, Carrion C, Souza L, et al. Long-term efficacy of myoblast transplantation on regional structure and function after myocardial infarction. *Circulation.* 2002;106: I-131-I-137.
12. Liechty KW, MacKenzie TC, Shaaban AF. Human mesenchymal stem cells engraft and demonstrate site-specific differentiation after in-utero transplantation in sheep. *Nat Med.* 2000;6(11):1282-1286.
13. Pittenger MF, Mackay AM, Beck S, et al. Multilineage potential of adult human mesenchymal stem cells. *Science.* 1999;284:143-147.
14. Shake JG, Gruber PJ, Baumgartner WA. Mesenchymal stem cell implantation in a swine myocardial infarct model: Engraftment and function effects. *Ann Thorac Surg.* 2002;73: 1919-1926.
15. Strauer BE, Brehm M, Zeus T, Kostering M. Repair of infarcted myocardium by autologous intracoronary mononuclear bone marrow cell transplantation in humans. *Circulation.* 2002;106:1913-1918.
16. Tomita S, Li R, Weisel RD, et al. Autologous transplantation of bone marrow cells improve damaged heart function. *Circulation.* 1999;110:II247-II256.
17. Orlic D, Kajstura J, Chimenti S, et al. Bone marrow cells regenerate infarcted myocardium. *Nature.* 2001;410:70-75.
18. Perin EC, Dohmann HFR, Borojevic R, Silva S. Transendocardial, autologous bone marrow cell transplantation for severe, chronic ischemic heart failure. *Circulation.* 2003;107: 2294-2302.
19. Prockop DJ. Marrow stromal cells as stem cells for nonhematopoietic tissues. *Science.* 1997;276:71-74.
20. Toma C, Pittenger MF, Cahill KS, Byrne BJ, Kessler PD. Human mesenchymal stem cells differentiate to a cardiomyocyte phenotype in adult murine heart. *Circulation.* 2002; 105:93-98.
21. Broxmeyer HE. *Cellular Characteristics of Cord Blood and Cord Blood Transplantation.* Bethesda, MD: AABB Press; 1998.
22. Broxmeyer HE, Hangoc G, Cooper S, et al. Growth characteristics and expansion of human umbilical cord blood. *Proc Natl Acad Sci USA.* 1992;89:4109-4113.
23. Kohli-Kumar M, Shahidi NT, Broxmeyer HE. Haemopoietic stem/progenitor cell transplantation in Fanconi anemia using HLA matched umbilical cord blood cells. *Br J Haematol.* 1993;85:419-422.
24. Wagner JE, Broxmeyer HE, Byrd RL, Zehnbauer B. Transplantation of umbilical cord blood after myeloablative therapy: analysis of engraftment. *Blood.* 1992;79:1874-1881.
25. Lu M, Shen RN, Broxmeyer HE. Stem cells from bone marrow, umbilical cord and peripheral blood for clinical application. *Crit Rev Oncol Hematol.* 1996;22:61-78.
26. Lu L, Ge Y, Li Z-H, Breie B, Clapp DW. CD34 stem/progenitor cells purified from cryopreserved normal cord blood can be transduced with high efficiency. *Cell Transplant.* 1995;4:493-503.
27. Traylor S, Bryson YJ. Impaired production of gamma-interferon by newborn cells in-vitro due to a functionally immature macrophage. *J Immunol.* 1985;134:1493-1497.
28. Erices A, Conget P, Minguell JJ. Mesenchymal progenitor cells in human umbilical cord blood. *Br J Haematol.* 2000;109:235-242.
29. Nieda M, Nicol AN, Denning-Kendall P, Sweetenham J. Endothelial cell precursors are normal components of human umbilical cord blood. *Br J Haematol.* 1997;98:775-777.
30. Broxmeyer HE, Douglas GW, Hangoc G, Cooper S. Human umbilical cord blood as a source of transplantable hematopoietic stem/progenitor cells. *Proc Natl Acad Sci USA.* 1989; 86:3828-3832.
31. Piacibello W, Sanavio F, Garretto L, et al. Extensive amplification and sell-renewal of human primitive hematopoietic stem cells from cord blood. *Blood.* 1997;89: 2644-2653.
32. Henning RJ, Abu-Ali H, Balis J, Morgan MB, Willing AE, Sanberg PR. Human umbilical cord blood mononuclear cells for the treatment of acute myocardial infarction. *Cell Transplant.* 2004;13:729-739.
33. Henning RJ, Burgos JD, Ondrovic L, Sanberg P, Morgan MB. Human umbilical cord blood progenitor cells are attracted to infarcted myocardium and significantly reduce myocardial infarction size. *Cell Transplant.* 2006;15: 647-658.
34. Henning RJ, Burgos JD, Vasko M, et al. Human cord blood cells and myocardial infarction: effect of dose and route of administration on infarct size. *Cell Transplant.* 2007; 16:907-917.
35. Henning RJ, Shariff M, Eadula U, et al. Human cord blood mononuclear cells decrease cytokines and inflammatory cells in acute myocardial infarction. *Stem Cells Dev.* 2008; 17(6):1207-1219.

36. Murohara T, Ikeda H, Duan J, et al. Transplanted cord blood-derived endothelial precursor cells augment postnatal neovascularization. *J Clin Investig*. 2000;105:1527-1536.
37. Pomyje J, Zivny J. Expression of genes regulating angiogenesis in human circulating hematopoietic cord blood CD 34+/CD133+ cells. *Eur J Haematol*. 2003;70:143-150.
38. Abboud M, Xu F, LaVia M. Study of early hematopoietic precursors in human cord blood. *Exp Hematol*. 1992;20:10119-10122.
39. Pesce M, Orland A, Iachininoto MG, et al. Myoendothelial differentiation of human umbilical cord blood-derived stem cells in ischemic limb tissues. *Circ Res*. 2003;92:e51-e362.
40. Ma N, Stamm C, Kaminski A, et al. Human cord blood cells induce angiogenesis following myocardial infarction in NOD/scid-mice. *Cardiovasc Res*. 2005;66:45-54.

Part X

Use of Placental Umbilical Cord Blood in Other Subspecialities of Regeneration Medicine

Umbilical Cord-Derived Mesenchymal Stem Cells

25

Jose J. Minguell

Hematopoiesis is a time- and site-dependent event during ontogeny of vertebrates. The first wave of hematopoietic activity appears in the ventral blood islands of the yolk sac, where primitive nucleated erythrocytes are formed to control the oxygen demand of the growing embryo. A second wave of primitive hematopoiesis (aorta-gonads-mesonephros, AGM region) is continued by a shift to the fetal liver where production of all hematopoietic cells is initiated. From near birth until the end of life, hematopoiesis resides in the bone marrow. Thus, embryonic, fetal, and adult hematopoieses are associated with a common stem cell, which, during each migratory event, reaches the specific milieu that is permissive for the programmed and sequential expression of the different forms of hematopoiesis.[1]

Hematopoiesis is sustained by a subset of stem cells, which although present as committed progenitors on the yolk sac are unable to reconstitute the entire hematopoietic system. The multipotent hematopoietic stem cell (HSC) emerges in the AGM region just before the establishment of the hematopoietic liver, where it subsequently expands and colonizes the hepatic tissue and, finally, the newly formed bone marrow. Concomitantly, with the full expression of the self-renewal and differentiation potential of HSCs, a fetal liver microenvironment (niche) is formed, which plays an instructive role with regard to the HSC. Apparently, one of the first events dealing with the interaction of the duplex stem/progenitor/mature cell and hepatic niche is the expression of cell adhesion proteins (α4, α5, and β1 integrins), on the surface of the primitive erythroid cell, which migrate into the fetal liver (FL) and interact with macrophages within the erythroblastic islands in a stage-specific and VCAM-1-dependent process.[2]

Although HSCs can be mobilized into the blood stream, they do not perform their program in the blood; it is performed only after they interact again with the hematopoietic microenvironment of the bone marrow. This implies that stem cells are not autonomous units of development; rather, tissue-specific niches control their destiny.

In placental mammals, the umbilical cord connects the developing embryo or fetus to the placenta. The umbilical cord normally contains two umbilical arteries and the umbilical vein, which supply the fetus with oxygenated, nutrient-rich blood from the placenta. The umbilical vein continues toward the liver, where it splits into two. One of these branches joins with the hepatic portal vein, which carries blood into the liver. The second branch (known as the ductus venosus) allows the majority of the incoming blood (approximately 80%) to bypass the liver and flow directly into the inferior vena cava, which carries blood toward the heart.

Almost 20 years ago, researchers found that UCB contains a relatively high content of HSCs, which, in comparison to adult HSCs, are more primitive and have high proliferative potential. In turn, in 2000, the presence of MSC in UCB was reported.[3] Therefore, it became clear that in placental mammals, the umbilical cord blood contains not only blood, but stem cells supposedly present only in the bone marrow.

Now we know that the regulated movement of these stem cells is critical for the establishment of the hematopoietic activity in the bone marrow, and for homeostasis and, probably, repairs in adulthood. HSCs homing to, and seeding of, the fetal hematopoietic

J.J. Minguell
Scientific Director, TCA Cellular Therapy, LLC,
101 E. Fairway Dr., Suite 514, Covington,
LA, 70433, USA
e-mail: drminguell@tcaway.com

N. Bhattacharya and P. Stubblefield (eds.), *Regenerative Medicine Using Pregnancy-Specific Biological Substances*,
DOI: 10.1007/978-1-84882-718-9_25,

niche is a highly regulated process and is dependent on the synergistic effect of stroma-cell-derived factor-1α (SDF-1α), a main regulator of cell trafficking, and Steel factor (SLF).[4,5]

Seeding in the maternal circulation, fetal progenitor cells can be detected in the circulation of pregnant women during most pregnancies. Decades after delivery, fetal CD34+ or mesenchymal stem cells are still detectable in maternal circulation or bone marrow. Recent data have suggested that fetal progenitor cells that persist after pregnancy, may home maternal injured tissues and various phenotypes (pregnancy-associated progenitor cells).[6]

Thus, the presence of HSCs in the UCB during the last weeks of pregnancy is not an unpredicted issue as it may be the consequence of a programmed change in the production of fetal SLF and SDF-1α, to "offer" fetal liver HSCs a chance to colonize and interact with a more "sophisticated" microenvironment,[7] capable of supporting self-renewal of HSCs.

The presence of relatively mature hematopoietic progenitor cells (HPCs) in human umbilical cord blood (UCB) was demonstrated in 1974 by Knudtzon, and later reports documented the main characteristic of these primitive HPC in UCB.[8–10] The migratory nature of these cells is documented by the observation that from 17 to 41 weeks of gestation, the frequency of CD34+ cells in cord blood decreases. This decline is coincident with the transition from hepatic to bone marrow hematopoiesis.[9,11]

The shifting nature of hematopoiesis during development and the presence of hematopoietic progenitor cells in umbilical cord blood anticipated the notion that other components of the hematopoietic process, like stromal cells may also be traveling, from early fetal sites to the newly formed bone marrow, to establish a proper hematopoietic microenvironment. Whether there is a "common" (and circulating) stromal-like stem cell that gives rise to the various supportive niches for hematopoiesis during fetal development has been a question for speculation. Answers to these questions have been comprehensively undertaken by a model of microenvironmental regulation of fetal hematopoietic niche, postulated on the basis of the existence of circulating precursors that migrate to the various hematopoietic organs and give rise to specialized stromal niches.[12]

Therefore, the reaction of the scientific community to the description of MSC in UCB was not a surprise; rather, it was the acceptance of a "sooner or later" issue. In other words, it somewhat resembled "Cronicas de una muerte anunciada".[13] Thus, the hypothesis that both hematopoietic and stromal progenitors are traveling, via cord blood, from early hematopoietic fetal liver microenvironmental sites[14] to the newly formed bone marrow was proved to be correct.[1,15]

It is well known that the organization of a cellular niche(s) plays a key role in regulating normal hematopoietic stem cell differentiation, maintenance, and regeneration. The hematopoietic niche is composed of stromal cells, including mesenchymal stem cells, which, either through direct cell-to-cell contact or via release of soluble factors, maintain the typical features of stem cells, mainly stem cell quiescence, maintenance, commitment, and maturation.[16–18]

A main characteristic of fetal hematopoiesis is the higher frequency of cycling HSCs undergoing self-renewal, a condition that does not occur in adult bone marrow, where HSCs are largely quiescent and undergo limited self-renewal. This difference seems to be the consequence of the dissimilar cellular and molecular nature of the microenvironment under both developmental stages.[16,19] Fetal liver stromal cells as compared to adult bone marrow cells exhibit a greater capacity to modulate HSC expansion, which seems to be linked to the higher expression of regulators of the Wnt signaling pathway in fetal liver stroma. In the case of bone marrow stroma, the expression of the Notch signaling pathway is predominant.[14]

Thus, the hematopoietic niche is cellular and molecular controlled at cellular and molecular levels during human development to produce conditions for the expansion (in the fetal liver) and for quiescence and limited self-renewal (in the adult bone marrow) of hematopoietic progenitors.

25.1 Characteristics of UCB-Derived Mesenchymal Stem Cells

The first report on the presence of MSC in UCB[3] described that when UCB-derived mononuclear cells were cultured in the presence of fetal bovine serum, a layer of adherent mesenchymal-like cells was formed.

These cells (UCB-MSCs) consisted of colonies of bipolar fibroblastoid cells which, after subcultivation, proliferated with a population-doubling time of 48 h and reached a confluent growth-arrested condition. Cell cycle analysis indicated that 85% of the cells were in the Go/G1 phases, of which 5% displayed a pattern of RNA and DNA staining distinctive for Go, which are quiescent cells. UCB-MSCs were strongly positive for a-naphthyl acetate esterase and periodic acid Schiff, and the immunophenotype, as assessed by flow cytometry, disclosed the homogeneous expression of antigens, including three mesenchymal-related antigens (SH2, SH3, SH4), and α-smooth muscle actin.[20,21] In addition, UCB-MSCs express neither typical myeloid antigens (CD14, CD34, CD45, CD49d, CD106) nor endothelial-related antigens, such as CD31 and vWF.

Further characterization of UCB-MSC has been strengthened by data from several studies.[22–26] The examination of these studies confirmed a striking similarity between the biological properties of UCB-MSC and adult bone-marrow-derived MSC.[27]

In addition to the characterization of UCB-MSC, the work by Erices demonstrated the following: (a) Not all UCB-derived MNCs develop into MSC-like cells when set in culture. In many cases, adherent cells in primary cultures displayed the morphology and characteristics of multinucleated osteoclasts, as evidenced by a strong tartrate-resistant acid phosphatase activity and the expression of antigens CD45 and CD51/CD61.These results confirmed results of previous studies, showing that mature osteoclasts or their progenitors circulate in umbilical cord blood.,[28] (b) Preterm as compared with term cord blood is richer in mesenchymal progenitors. Similar results have been reported for stromal cells isolated from mid-trimester fetal blood.[29] These observations are coincident with results showing that the concentration of UCB-circulating hematopoietic progenitors is higher at preterm conditions than at full-term conditions.[11,30]

UCB-MSCs exhibit the potential to differentiate into osteoblasts, adipocytes, chondrogenic-like cells,[3,31] and neural-like cells, as evidenced by immunofluorescence and RT-PCR analysis to detect expression of neuronal cell-specific protein markers.[32,33] Recent reports have shown that UCB-MSCs also exhibit the potential to support the ex vivo self-renewal, proliferation, and differentiation of $CD34^+$ hematopoietic stem cells. The significant role played by UCB-MSC in maintaining the "stemness" of hematopoietic progenitor cells was demonstrated by cell-to-cell contact studies as well as by indirect interactions (Transwell systems).[34] These studies suggested that not only production of growth factors (SCF, IL-6, TNF-α), but regulation of cell-cell junctions are mediated by UCB-MSC.[31,34,35]

25.2 Another Source of Fetal MSC: Placental Tissues and Umbilical Cord

Placental tissues are considered an attractive source of cells with considerable phenotypic plasticity as well as immunomodulatory properties, including mesenchymal stem cells and amniotic membrane-derived epithelial cells.

Mesenchymal stem cells, or mesenchymal stromal cells as denominated by authors involved in the field,[36] have been isolated from the amniotic and chorionic regions of the fetal placenta, as well as from the umbilical cord. The use of the term "stromal cell" instead of "stem cell" is based on the fact that the hallmarks of stem cells, such as self renewal, stemness, and hierarchy, have not been demonstrated in MSC derived from placental tissues.

Amniotic-, chorionic- and umbilical cord MSC are usually isolated by mechanical peeling or removal of the tissue followed by enzymatic digestion. After primary culture and subcultivation of fibroblast colony-forming units, evolved adherent cells exhibit a pattern of antigen expression (including $CD105^+$, $CD90^+$, $CD73^+$, $CD34^-$, $CD45^-$, Oct-4^+), which is not different from that expressed by UCB-MSC, with the exception of the embryonic marker, Oct-4.

Placental MSC differentiate into cells of the adipogenic, chondrogenic, osteogenic and skeletal myogenic lineages after exposal to the appropriate stimuli. These cells also exhibit a robust hematopoiesis-supportive function.[36–39] In addition, UCB-MSC exhibit a hematopoietic-supportive function.[40]

25.3 Prospects for the Clinical Utilization of UCB-MSC

The last five years have seen a substantial improvement in the understanding of the biology and the potential clinical utilization of adult bone-marrow-derived mesenchymal progenitors, both in the context of cell and gene therapy strategies.

It has been speculated whether MSC transplantation alone or in conjunction with hematopoietic progenitors would facilitate the engraftment of the hematopoietic stem cells after myeloablative therapy. Simultaneously, as precursors of several mesenchymal lineages, MSCs are envisioned as attractive candidates for cell therapy. Their utilization converges to attenuate or correct disorders/damage to mesenchymal tissues, among them, bone, cartilage, and skeletal muscle.[41] Preclinical studies that revealed that ex vivo expanded bone-marrow-derived MSC differentiate into cells exhibiting features of cardiomyocytes (for a review, see[42]), have been decisive in the initiation of clinical trials in patients with cardiovascular diseases.[43]

Although conclusive evidence of the utilization of cord-blood-derived MSCs in regenerative medicine has not been known, it is fully accepted that these cells represent an appealing population of progenitors derived from a tissue considered as a "nontraditional" source of stem cells.[44]

The above prospect is sustained by several cellular, molecular, and preclinical studies, including the following:

(a) Due to their high proliferation rate, less culture time will be required to get a fixed number of ex vivo expanded UCB-MSC. This will result in less subcultivated cells and, thus, fewer chances of expressing apoptotic features.[20]
(b) It is known that fetal cells and, probably, UCB-MSCs have an increased transendothelial migration capacity,[45] which should be important for delivery to damaged tissues.
(c) In xenogeneic transplantation studies, it was demonstrated the presence of human UCB-MSCs in several tissues of immunodeficient mice, several months after systemic infusion. These findings demonstrate homing and survival capability of transplanted UCB-MSC.[46,47]
(d) In terms of immunogenicity, UCB-MSCs express class I human leucocyte antigens (HLA antigens), whereas class II HLA antigens are expressed only after prolonged exposure to interferon-γ.[48,49] Accordingly, UCB-MSC may be used for allogeneic transplantation[50] and/or during ex vivo protocols to improve expansion of adult human CD34+ peripheral blood progenitor cells, as it occurs with adult bone-marrow-derived MSC.[34]
(e) The formulation of optimized ex vivo culture conditions to propagate UCB-MSC, including alternatives to replace fetal bovine serum for clinical-scale MSC expansion[51], and
(f) The definition of critical parameters related to the cellularity and volume of the UCB harvests to prepare UCB-MSC [22].

The attractive differentiation pattern, and the immunologic and other distinctive attributes exhibited by UCB-MSC give support to the contention that, very soon, this up till now "wasted product" in many places, will be used in new clinical protocols to treat diverse diseases.

References

1. Tavassoli M. Embryonic and fetal hemopoiesis: an overview. *Blood Cells*. 1991;17:269-281.
2. Isern J, Fraser ST, He Z, Baron MH. The fetal liver is a niche for maturation of primitive erythroid cells. *PNAS*. 2008; 105:6662-6667.
3. Erices A, Conget P, Minguell JJ. Mesenchymal progenitor cells in human umbilical cord blood. *Br J Haemat*. 2000;109: 235-242.
4. Christensen JL, Wright DE, Wagers AJ, Weissman IL. Circulation and chemotaxis of fetal hematopoietic stem cells. *PLoS Biol*. 2004;2:e75.
5. Kucia M, Reca R, Miekus K, et al. Trafficking of normal stem cells and metastasis of cancer stem cells involve similar mechanisms: pivotal role of the SDF-1–CXCR4 axis. *Stem Cells*. 2005;23:879-894.
6. Nguyen Huu S, Dubernard G, Aractingi S, Khosrotehrani K. Feto-maternal cell trafficking: a transfer of pregnancy associated progenitor cells. *Stem Cell Rev*. 2006;2:111-116.
7. Ara T, Tokoyoda K, Sugiyama T, Egawa T, Kawabata K, Nagasawa T. Long-term hematopoietic stem cells require stromal cell-derived factor-1 for colonizing bone marrow during ontogeny. *Immunity*. 2003;19:257-267.
8. Knudtzon S. In vitro growth of granulocyte colonies from circulating cells in human cord blood. *Blood*. 1974;43:357-361.
9. Leary AG, Ogawa M. Blast cell colony assay for umbilical cord blood and adult bone marrow progenitors. *Blood*. 1987;69:953-956.
10. Nakahata T, Ogawa M. Hemopoietic colony-forming cells in umbilical cord blood with extensive capability to generate mono- and multipotential hemopoietic progenitors. *J Clin Invest*. 1982;70:1324.
11. Shields LE, Andrews RG. Gestational age changes in circulating CD34+ hematopoietic stem/progenitor cells in fetal cord blood. *Am J Obstet Gynecol*. 1998;178:931-937.
12. Badillo AT, Flake AW. The regulatory role of stromal microenvironments in fetal hematopoietic ontogeny. *Stem Cell Rev*. 2006;2:241-246.
13. Garcia Marquez G. *Chronicle of a Death Foretold* (A masterpiece novel by the Nobel Laureate in Literature). New York: Alfred A. Knopf; 1982.
14. Martin MA, Bhatia M. Analysis of the human fetal liver hematopoietic. *Stem Cells Dev*. 2005;14:493-504.
15. Charbord P, Tavian M, Humeau L, Péault B. Early ontogeny of the human marrow from long bones: an immunohistochemical study of hematopoiesis and its microenvironment. *Blood*. 1996;88:4072-4078.
16. Heissig B, Ohki Y, Sato Y, Rafii S, Werb Z, Hattori K. A role for niches in hematopoietic cell development. *Hematology*. 2005;10(3):247-253.

17. McGrath K, Palis J. Ontogeny of erythropoiesis in the mammalian embryo. *Curr Top Dev Biol.* 2008;82:1-22.
18. Tavassoli M, Minguell JJ. Homing of hemopoietic rogenitor cells to the marrow. *Proc Soc Exp Biol Med.* 1991;196: 367-373.
19. Weisel KC, Gao Y, Shieh JH, Moore MA. Stromal cell lines from the aorta-gonado-mesonephros region are potent supporters of murine and human hematopoiesis. *Exp Hematol.* 2006;34:1505-1516.
20. Conget P, Minguell JJ. Phenotypical and functional properties of human bone marrow mesenchymal progenitor cells. *J Cell Physiol.* 1999;181:67-73.
21. Pittenger MF, Mackay AM, Beck CB, et al. Multilineage potential of adult human mesenchymal stem cells. *Science.* 1999;284:143-147.
22. Bieback K, Kern S, Klüter H, Eichler H. Critical parameters for the isolation of mesenchymal stem cells from umbilical cord blood. *Stem Cells.* 2004;22:625-634.
23. Hutson EL, Boyer S, Genever PG. Rapid isolation, expansion, and differentiation of osteoprogenitors from full-term umbilical cord blood. *Tissue Eng.* 2005;11:1407-1420.
24. Kim JW, Kim SY, Park SY, et al. Mesenchymal progenitor cells in the human umbilical cord. *Ann Hematol.* 2004; 83:733.
25. Markov V, Kusumi K, Tadesse MG, et al. Identification of cord blood-derived mesenchymal stem/stromal cell populations with distinct growth kinetics, differentiation potentials, and gene expression profiles. *Stem Cells Dev.* 2007;16:53-73.
26. Parekkadan B, Sethu P, van Poll D, Yarmush ML, Toner M. Osmotic selection of human mesenchymal stem/progenitor cells from umbilical cord blood. *Tissue Eng.* 2007; 13:2465-2473.
27. Minguell JJ, Erices A, Conget P. Mesenchymal stem cells. *Exp Biol Med.* 2001;226:507-520.
28. Roux S, Quinn J, Pichaud F, et al. Human cord blood monocytes undergo terminal osteoclast differentiation in vitro in the presence of culture medium conditioned by giant cell tumor of bone. *J Cell Physiol.* 1996;168:489-498.
29. Yu M, Xiao Z, Shen L, Li L. Mid-trimester fetal blood-derived adherent cells share characteristics similar to mesenchymal stem cells but full-term umbilical cord blood does not. *Br J Haematol.* 2004;124:666-675.
30. Wyrsch A, dalle Carbonare V, Jansen W, et al. Umbilical cord blood from preterm human fetuses is rich in committed and primitive hematopoietic progenitors with high proliferative and self-renewal capacity. *Exp Hematol.* 1999;27: 1338-1345.
31. Wang JF, Wang LJ, Wu YF, et al. Mesenchymal stem/progenitor cells in human umbilical cord blood as support for ex vivo expansion of CD34(+) hematopoietic stem cells and for chondrogenic differentiation. *Haematologica.* 2004;89:837-844.
32. Jeong JA, Gang EJ, Hong SH, et al. Rapid neural differentiation of human cord blood-derived mesenchymal stem cells. *Neuroreport.* 2004;15:1731-1734.
33. Kang JH, Lee CK, Kim JR, et al. Estrogen stimulates the neuronal differentiation of human umbilical cord blood mesenchymal stem cells. *Neuroreport.* 2007;18:35-38.
34. Li N, Feugier P, Serrurrier B, Latger-Cannard V, Lesesve JF, Stoltz JF. Human mesenchymal stem cells improve ex vivo expansion of adult human CD34+ peripheral blood progenitor cells and decrease their allostimulatory capacity. *Exp Hematol.* 2007;35:507-515.
35. Wagner W, Wein F, Roderburg C, et al. Adhesion of hematopoietic progenitor cells to human mesenchymal stem cells as a model for cell–cell interaction. *Exp Hematol.* 2007;35:314-332.
36. Parolini O, Alviano F, Bagnara GP, et al. Isolation and characterization of cells from human term placenta. *Stem Cells.* 2008;26:300-311.
37. Alviano F, Fossati V, Marchionni C, et al. Term Amniotic membrane is a high throughput source for multipotent mesenchymal stem cells with the ability to differentiate into endothelial cells in vitro. *BMC Dev Biol.* 2007;7:11.
38. Portmann-Lanz CB, Schoeberlein A, Huber A, et al. Placental mesenchymal stem cells as potential autologous graft for pre- and perinatal neuroregeneration. *Am J Obstet Gynecol.* 2006;194:664-673.
39. Soncini M, Vertua E, Gibelli L, et al. Isolation and characterization of mesenchymal cells from human fetal membranes. *J Tissue Eng Regen Med.* 2007;1:296-305.
40. Lu LL, Liu YJ, Yang SG, et al. Isolation and characterization of human umbilical cord mesenchymal stem cells with hematopoiesis – supportive function and other potentials. *Haematologica.* 2006;91:1017-1026.
41. Kolf CM, Cho E, Tuan RS. Biology of adult mesenchymal stem cells: regulation of niche, self-renewal and differentiation. *Arthritis Res Ther.* 2007;9:204.
42. Minguell JJ, Erices A. Mesenchymal stem cells and the treatment of cardiac disease. *Exp Biol Med.* 2006;231: 39-49.
43. Clinical trials, 2008. www.clinicaltrials.gov. Identifier NCT # 00555828, 00548613, 00587990.
44. Feldmann RE Jr, Bieback K, Maurer MH, et al. Stem cell proteomes: a profile of human mesenchymal stem cells derived from umbilical cord blood. *Electrophoresis.* 2005;26:2749-2758.
45. Yong KL, Fahey A, Pahal G, et al. Fetal haemopoietic cells display enhanced migration across endothelium. *Br J Haematol.* 2002;116:392-400.
46. Erices EA, Allers CI, Conget PA, Rojas CV, Minguell JJ. Human cord blood-derived mesenchymal stem cells home and survive in the marrow of immunodeficient mice after systemic infusion. *Cell Transplant.* 2003;12:555-561.
47. Jäger M, Degistirici O, Knipper A, Fischer J, Sager M, Krauspe R. Bone healing and migration of cord blood-derived stem cells into a critical size femoral defect after xenotransplantation. *J Bone Miner Res.* 2007;22:1224-1233.
48. Gotherstrom C, Ringden O, Westgren M, Tammik C, Le Blanc K. Immunomodulatory effects of human foetal liver-derived mesenchymal stem cells. *Bone Marrow Transplant.* 2003;32:265-272.
49. Gotherstrom C, Ringden O, Tammik C, Zetterberg E, Westgren M, Le Blanc K. Immunologic properties of human fetal mesenchymal stem cells. *Am J Obstet Gynecol.* 2004; 190:239-245.
50. Tisato V, Naresh K, Girdlestone J, Navarrete C, Dazzi F. Mesenchymal stem cells of cord blood origin are effective at preventing but not treating graft-versus-host disease. *Leukemia.* 2007;21:1992-1999.
51. Reinisch A, Bartmann C, Rohde E, et al. Humanized system to propagate cord blood-derived multipotent mesenchymal stromal cells for clinical application. *Regen Med.* 2007;2:371-382.

26 Cord Blood Stem Cell Expansion Ex Vivo: Current Status and Future Strategies

Jian-Xin Gao and Quansheng Zhou

26.1 Introduction

Allogeneic hematopoietic stem cell (HSC) transplantation has been used to reconstitute immune system for patients with hematopoietic malignancies and advanced autoimmune diseases.[1–3] HSCs can be procured from three sources: bone marrow (BM), mobilized peripheral blood, and cord blood (CB). Since the rarity of HSCs in BM and peripheral blood limits their therapeutic application, CB has emerged as frontline for the source of HSC transplantation.[1–3] The first cord blood transplant was performed in1988 on a 5-year-old Parisian boy who was suffering from Fanconi's anemia, using his newborn sister's HLA-matched umbilical cord blood. Since then, cord blood bank has been established in many institutions to cope with the increasing demand in clinics.[1] Clinical practices indicate that multiple transplantations are required for patients with relapses of malignancy,[3] and thus a large amount of HSCs are required from the same donor.[4] The shortage of HSCs in clinical practices has elicited the research on large ex vivo expansion of CB HSCs.

J.-X. Gao (✉)
Department of Pathology and Comprehensive Cancer Center, Ohio State University, Columbus, OH 43210, USA
e-mail: Jian-Xin.Gao@osumc.edu

26.1.1 Types of Stem Cells

Stem cells are characterized as the cells that can maintain self-renewal and multipotency of differentiation into various cell types. Embryonic stem (ES) cells are pluripotent, which can differentiate into cells of all three germ layers: endoderm, ectoderm, and mesoderm. Unlike ES cells, the stem cells that are isolated from various tissues in fetal and adult animals are multipotent and can differentiate into tissue-specific cell types, such as HSCs and neuronal stem cells.[5,6] A self-renewed daughter stem cell may retain as a parent stem cell, either differentiating into lineage-specific progenitor cells or undergoing apoptosis (programmed cell death). The fate of a stem cell is determined by environmental niches,[7,8] which may direct a stem cell to divide symmetrically or asymmetrically.[5,6,9]

26.1.2 Hematopoietic Stem Cells

HSCs are required for the continuous supply of blood cells and lymphoid precursors throughout adult life. HSCs sustain the capacity of self-renewal and the multipotency of differentiation into all types of blood cells. Increasing data have shown that HSCs might be more than the precursors of blood cells. They might transdifferentiate into non-HSCs, such as brain cells,[10–12] hepatocytes,[13,14] skeletal muscles,[15–17] and cardiomyocytes,[18–20] although these are highly controversial.[21–24] Regardless of whether HSCs are fused with or transdifferentiate into tissue cells, they appear to play an important role in tissue repair as well.[19,25] Thus, not

N. Bhattacharya and P. Stubblefield (eds.), *Regenerative Medicine Using Pregnancy-Specific Biological Substances*,
DOI: 10.1007/978-1-84882-718-9_26,

only can HSCs be used to reconstitute hematopoietic system, but they also have the potential to be used for regenerative medicine.

26.1.3 Cord Blood Hematopoietic Stem Cells (CB HSCs)

CB HSCs have great clinical potentialities in the treatment of multiple diseases, including leukemia, ischemia, neuronal degenerative disorders, and diabetes. However, one unit of CB is not sufficient to achieve robust engraftment in allogeneic transplants.[26] The number of HSCs obtained from a unit of CB is only about one-tenth of a unit of BM, and the limitation has become a major barrier to CB transplant. However, CB HSCs display better ex vivo expansion capability than BM HSCs and have a marked proliferative response to hematopoietic supporting growth factors and cytokines. Thus, CB HSCs are better than BM HSCs used for ex vivo expansion of HSCs.[27]

Obviously, ex vivo expansion of CB-derived HSCs would be of choice to meet the requirements for HSC transplant.[27] However, the methods currently available for HSC expansion are inevitably limited in the magnitude of expansion, because of our limited understanding of cellular and molecular mechanisms underlying stem cell expansion. In this chapter, we will review the progress in stem cell expansion and current understandings of the cellular and molecular bases for HSC expansion.

26.2 Strategy for Ex Vivo CB HSC Expansion

Ex vivo expansion of CB HSCs needs an optimal cell culture system. Such a culture system should not alter the property of expanded HSCs, and provide safe, cost-effective transplantable CB HSCs, which ensure the success for reconstitution of hematopoietic system. Thus, preservation of stem cell property of expanded HSCs is a critical issue for large ex vivo expansion. First, the expanded HSCs should have the capacity of long-term (LT) repopulation without losing the capability of self-renewal and the potency to differentiate into all types of blood cells; second, the expanded HSCs are free of contaminations of feeder cells, nonhuman serum proteins or other microbial agents from the culture system, and third, the expanded HSCs themselves must not be genetically modified to avoid potential transformation caused by the modification.[28, 29]

26.2.1 The Problems for Large Ex Vivo Expansion of HSCs

Currently, although a variety of culture systems for HSC expansion have been developed'[30–34] none of the systems can generate HSCs qualified for the criteria described above. The problems with these culture systems include, (a) HSCs are only expanded in a short period with limited numbers, as they may progressively lose their capability of self-renewal; (b) feeder cells are required for maintenance of stem cell property, but these cells might contaminate the expanded HSCs; and (c) genetic modification of either feeder cells or HSCs is required for large expansion, which might cause transformation of HSCs after transplantation.

Ideally, the expanded HSCs should be derived from a culture system without feeder cells and xenogeneic serum. However, feeder cells can provide growth factors, some of which are not yet identified but critical for the maintaining self-renewal of expanded HSCs.[35] In the absence of feeder cells, HSCs may rapidly lose self-renewal capacity and undergo lineage differentiations. To cope with the issue, some laboratories have genetically modified HSC to promote the magnitude of HSC expansion. This approach may be useful for the researches to reveal molecular mechanisms underlying the stem cell expansion, but also raises a hurdle for clinical application because of the potential genetical and microbial contaminations of HSCs. Thus, the development of a culture system without the needs of feeder cells and genetic modification is a serious challenge for large ex vivo expansion of CB HSCs.

26.2.2 Culture Systems for Stem Cell Expansion

Establishment of a culture system for long-term large ex vivo CB HSC expansion is one of the important goals for stem cell research. However, all culture systems for HSC expansion reported so far are unable to expand HSCs more than 2–3 weeks. In these culture systems, feeder cells are required for maintaining the stem cell property of expanded HSCs. The attempts have been made to develop a 3-D culture system to increase quantity and quality for expanded HSCs.[36–38] The 3-D coculture system constructed with nonwoven fibrous matrices and stromal cells is obviously superior to the 2-D culture system, although it is not as ideal as we expected.[36, 37] The advantage of a 3-D culture system is that it provides microenvironments more closely to stem cell niche than the 2-D culture system does. With the progress in biomaterial research, nanometer polymeric biomaterials may be used as 3-D culture scaffold, providing precisely controlled scaffold architecture that, coupled with the feeder cells and/or small molecules immobilized in the 3-D culture system, regulates the spatio-temporal release of growth factors and morphogens, mimicking in vivo stem cell niches.[39]

To improve the quality and quantity of expanded HSCs, exogenous factors may be added into cultures. Currently, thrombopoietin (TPO), flt3/flk2 ligand (FL), and stem cell factor (SCF) are mostly used for ex vivo expansion of HSCs.[34, 35, 40, 41] These cytokines may support HSC expansion ex vivo in the absence of serum.[41] However, cytokines, growth factors, and/or extracellular matrix proteins added into cultures have not given rise to satisfactory outcomes for HSC expansion,[34, 37, 38, 40] reflecting our limited understanding of cellular and molecular mechanisms that determine stem cell property and constitute stem cell niches.

Since primary stromal cells are limited in resource, the genetically modified stromal cell lines have been widely used for ex vivo expansion of HSCs in research settings. To obtain safe, stable stromal cell lines, telomerase catalytic subunit has been used to transduce BM stromal cells, which can support HSC expansion ex vivo.[34, 35] Although the telomerase transduced stromal cells appears not to be tumorigenic,[34, 35] the safety issue for clinical application remains a concern.

The major barrier for ex vivo expansion of HSCs is that we have not developed a safe, cost-effective and efficient culture system, which can expand HSCs in a large magnitude without sacrificing stem cell property and without the needs of feeder cells and serum. Recently, we have found that a novel condition medium, called XLCM™, can expand hematopoietic progenitors from adult mouse spleen (unpublished data) and BM stem/progenitor cells.[42] The XLCM™ is derived from the supernatant of mitogen-activated human CB mononuclear cells (MNCs), and has demonstrated the ability to expand human CB T cells a millionfold.[43, 44] The XLCM™ consists of many growth factors, cytokines, and chemokines with variable concentrations. After modification, XLCM™ might have the potential to be used to create a novel culture system for CB HSC expansion without the needs of feeder cells and xenogeneic serum.

26.3 Cellular Basis for CB HSC Expansion

Stem cell fate is dependent on cell intrinsic factors and surrounding microenvironments, which are commonly referred to as a stem cell niche. In vivo, HSCs are maintained to be homeostatic through interaction with stem cell niches.[7, 8] The stem cell niche consists of vascular network and cell adhesion components, mesenchymal stromal cells (MSCs), soluble factors, extracellular matrix, neural inputs, and the stem cell itself, providing an environment for stem cell adhesion, mechanical inputs, spatial cues, and homing or engraftment. The MSCs are a main component of stem cell niche,[7, 8, 35] and are usually used as feeder cells for ex vivo expansion of HSCs.[34, 35] The feeder cells may provide receptors and/or ligands to interact with the counterparts on HSCs, producing hematopoiesis-supporting growth factors to collaboratively expand HSCs without impairing self-renewal capacity of HSCs.[35] Some factors produced by feeder cells are largely unknown and thus cannot be replaced by currently known exogenous hematopoietic supporting growth factors.

CB stem/progenitor cells are heterogeneous and consist of long-term repopulation (LT)-HSCs (LT-HSCs), and their hierarchical progenies including short-term repopulation (ST)-HSCs (ST-HSCs), multipotent progenitors (MPPs) and differentiated hematopoietic lineages,[26] although precise origin of HSCs is not clear yet. LT-HSCs are equivalent to, or more primitive than the long-term culture-initiating cell (LTC-IC), and are identified as the SCID-repopulating cells (SRCs) in nonobese diabetic (NOD) mice with severe combined immunodeficiency disease (NOD/SCID mice).[45, 46] The ultimate goal for

CB HSC expansion is to obtain sufficient numbers of LT-HSCs for robust SRC-mediated engraftment. Thus, the choice of HSC subsets may also be critical for the success of ex vivo HSC expansion.

26.3.1 The Mechanisms Underlying Feeder Cells Supporting HSC Expansion Ex Vivo

Feeder cells derived from marrow or CB stromal cells may provide an in vitro mimic of in vivo stem cell niches. Under physiological conditions, the status – self-renewing, differentiating, or remaining static – of an HSC is determined by the outcome of interactions with environmental niches. In culture system, the BM-derived MSCs are a mixture of osteoblasts, adipocytes, and endothelial cells.[47] Osteoblasts are supposed to be the main component of MSCs supporting hematopoiesis, especially for maintaining the self-renewal capacity of HSCs.[35] In addition, endothelial cells may also play a role similar to osteoblasts in stem cell niches, while fibroblasts might provide a niche effecting lineage commitment. In addition, stem cell niches also produce many hematopoiesis-supporting cytokines and growth factors as well as inhibitors, which regulate stem cell proliferation, differentiation, and apoptosis.[48]

In hematopoietic system, a stem cell niche consists of osteoblastic niche and vascular niche.[49] The number of LT-HSCs is closely related to osteoblastic cell population in BM. Increasing osteoblastic cell number causes a parallel increase in LT-HSCs, but not other progenitors. HSCs adhere to osteoblastic cells and keep the cells in quiescent state. Indeed, approximately 75% LT-HSCs are quiescent in G0 phase.[50,51] The mechanisms underlying osteoblasts supporting HSC self-renewal have not been completely understood. Our recent study suggests that surface molecules expressed on osteoblasts such as CD29 can promote HSC expansion, probably through enhancing Wnt5A and stem cell factor (SCF) production.[35] In addition, osteoblastic cells may produce angiopoietins to promote HSC self-renewal.[50,51]

Stem cell vascular niche also plays a role in regulation of stem cell renewal and differentiation. LT-HSCs are associated with sinusoidal endothelial cells to form a vascular niche, which is thought to promote HSCs asymmetric division, differentiation, and mobilization. Thus, vascular niche favors proliferation and differentiation of HSCs, while osteoblastic niche provides a microenvironment for the maintenance of quiescent HSCs.[49] In the BM, these two niches coordinate to maintain stem cell pool and to meet the demands for stem cell self-renewal and differentiation after HSC mobilization and/or hematopoietic insults.

A dividing stem cell faces three different fates: self-renewal, differentiation, and apoptosis. It is important to know the factors that determine HSC fates in order to design new protocols specific for ex vivo expansion of CB HSCs. As indicated above, a self-renewing stem cell can undergo symmetrical and asymmetrical division. As a result of the symmetric division, a stem cell divides into two identical daughter cells with the exactly same property as their parent cells, while asymmetric division leads to the differentiation or programmed cell death of one of two divided daughter cells.[9,52] For large ex vivo expansion of CB HSCs, a culture system should be designed to (a) promote symmetric division or suppress asymmetric division of HSCs; (b) suppress lineage commitment; and (c) ensure the survival of symmetrically divided cells.

The mechanisms of a stem cell undergoing symmetric or asymmetric divisions are largely unknown, although stem cell niche is believed to play a pivotal role in the control of stem cell fates. Initially, the niche effects on stem cell division are observed in *Drosophila*.[53] For symmetric division, a stem cell is in close contact with the stem cell niche and divides parallel to the niche, generating two stem cells, resulting in an increase of stem cells in number. This process happens when a significant number of HSCs are lost in the body and compensation is needed. In the case of asymmetric division, the regulators of self-renewal in a stem cell are asymmetrically located during mitosis of the cells and orient its mitotic spindle to the niche surface; in this way, only one daughter cell can closely contact with stem cell niche to maintain the property of self-renewal, the other cell that does not contact with the niche enters the phase of differentiation,[54] or undergoes apoptosis.[55] Thus, it is possible that CB HSC expansion can be greatly promoted through enhancing symmetric division and/or diminishing stem cell lineage commitment by manipulating stem cell niche ex vivo.

Apoptosis is another fate of stem cells and crucial to ex vivo expansion of CB HSCs. HSCs are sensitive to many factors which induce apoptosis, including growth factor withdrawal or deficiency, exposure to radiation, reactive oxygen species (ROS), heavy metals, and a variety of chemotherapeutic agents. BCL-2 plays an important role in regulation of HSC apoptosis.[56] Over-expression of BCL-2 protected HSCs from apoptosis and increased numbers of HSCs in BM.[57] Another anti-apoptotic protein, Mcl-1, plays a crucial role in maintenance of HSC pool. Mcl-1 is highly expressed in LT-HSCs and declined in all progenitor populations. Inducible knockout of Mcl-1 gene results in a rapid loss of HSCs and a failure of multi-lineage hematopoiesis. SCF can up-regulate Mcl-1 expression, which contributes to maintaining HSC population in vivo and in vitro.[58] Hence, addition of anti-apoptotic agents into CB culture system might be a novel strategy to prevent the loss of ex vivo expanded HSCs.

26.3.2 Manipulation of Feeder Cells for Ex Vivo HSC Expansion

As indicated above, marrow MSCs are the main components of stem cell niches in vivo.[7,8,35] Currently, murine stromal cells or genetically immortalized human BM stromal cells are usually used for ex vitro CB HSC expansion.[30,32,33,35] The advantage of the systems is that the expanded HSCs are less prone to lineage commitment. However, the stem cells expanded in such systems are not appropriate for clinical application because of the potential contamination by xenogeneic or viral-modified stromal cells. In addition, the stromal cells derived from CB and placenta also have the potential to support HSC expansion ex vivo.[59,60]

The ideal approach for ex vivo CB HSC expansion is to create an artificial niche in 3-D culture system, using the cells that function in the BM niches. However, it is still an open question what kind of niche cells is the best for ex vivo expansion of CB HSCs.[36,39]

Human MSCs have displayed the potentials for both large ex vivo expansion of CB HSCs and robust engraftment of expanded HSCs in xeno-recipients. Physiologically, MSCs play an important role in normal hematopoiesis by providing a cell-to-cell interaction and producing hematopoietic-supporting growth factors and cytokines. However, the folds of CB HSC expansion in the co-culture with MSCs are highly variable, probably related to the cellular constitution of MSCs used.[34,61] The CB HSCs expanded in the presence of MSCs showed an enhanced engraftment in NOD/SCID mice as compared with those expanded in the absence of MSCs.[61,62] Thus, co-culture of CB HSCs with tissue HLA-matched human MSCs, such as MSCs from human umbilical cord blood or placenta, is an ideal combination to expand transplantable HSCs.

As mentioned above, osteoblast niche favors the symmetric division and self-renewal of HSCs. A human fetal BM-derived osteoblast cell line with human telomerase catalytic unit (hTERT) has been established.[35] The hTERT-transduced fetal BM-derived osteoblastic cells (FBMOB-hTERT) can support CB HSC expansion ex vivo, actively maintaining their capacity of self-renewal and multipotency of differentiation. As feeder cells, the FBMOB-hTERT are activated to produce more growth factors such as SCF and Wnt-5A while co-cultured with CB HSCs, supporting ex vivo expansion of LTC-IC, LT-HSCs or SRC. The increased production of SCF and Wnt-5A appeared to be mediated by the integrin CD29.[35]

In addition to MSCs and osteoblasts, several other types of feeder cells have been used to promote ex vivo CB HSC expansion, such as human yolk sac endothelial cells, fibroblasts, fetal liver cells, and human aorta-gonad-mesonephros (AGM)-derived stromal cells,[63–65] indicating that multiple cell types may act as feeder cells for ex vivo CB HSC expansion.

Although feeder cells can promote ex vivo CB HSC expansion, the strategy has several pitfalls. For example, some growth factors and cytokines produced by feeder cells, such as IL-3, G-CSF or GM-CSF, may promote lineage commitment of CB HSCs, decreasing the number of SRCs in the expanded HSCs and thus reducing the engraftment or long-term repopulation capacity of expanded HSCs after transplantation. On the other hand, feeder cells may produce inhibitors for stem cell self-renewal (our unpublished observations). Moreover, co-culture of immortalized or engineered feeder cells with CB HSCs usually yields a large expansion of CB HSCs, but also raises a safety concern, because it is difficult to completely remove the contaminated immortalized feeder cells in the expanded HSCs. Technically, this approach is limited in clinical application.

To overcome the shortcomings of living feeder cells, the human BM stromal cells fixed with glutaraldehyde have been used to serve as a feeder layer for ex vivo expansion of CB HSCs.[66] The fixed stromal cells retained the ability to support CD34+ cell ex vivo expansion and increased the numbers of CD34+ cells by 1.8-fold as compared with the control without fixed stromal cells, suggesting that direct contact between CB HSCs and stromal cell surface plays an important role in ex vivo CB HSC expansion, and that stromal cell surface molecules can promote the ex vivo expansion.[66] The molecules on the surface of feeder cells which regulate stem cell self-renewal are unknown; once the molecules are identified, we might use them to replace feeder cells for ex vivo expansion of CB HSCs. However, a disadvantage is that the fixed stromal cells have lost their capacity to produce hematopoietic supporting factors as we have reported.[35] The best strategy for ex vivo HSC expansion might come from the research by comparison of expansion-promoting molecules between living and fixed stromal cells.

Recently, several research groups have used nanofiber to replace feeder cells for ex vivo expansion of CB HSCs. Nanofiber scaffolds with amino groups were conjugated to fiber surface through different spacers. The aminated nanofiber scaffolds in the presence of cytokines markedly promote expansion of HSCs/HPCs (hematopoietic progenitor cells) in CB. Within 10 days, the un-fractionated cells and CD34+CD45+ cells increased by 773 ~ 850-fold and 200 ~ 235-fold, respectively. In addition, the expanded CB HSCs and progenitors significantly enhanced engraftment in NOD/SCID mice.[67] These manmade new materials are expected to have high impact on CB HSC transplantation. Moreover, growing feeder cells in 3-D nanofiber culture system might further promote CB HSC expansion.

26.3.3 The Choice of HSC Subsets for Ex Vivo Expansion

The ultimate goal for ex vivo HSC expansion is to obtain sufficient transplantable LT-HSCs or SRCs. Since HSCs are hierarchically heterogeneous, the choice of a right HSC subset is critical for the success of ex vivo expansion. In mice, hematopoiesis is initiated by LT-HSCs, residing in the Thy-1.1^{lo}Lin$^-$Sca-1$^+$ population.[68–70] CD34 was used as a marker to determine primitive of HSCs in mouse and human.[71] In human, LT-HSCs reside in Lin$^-$CD34$^+$CD38$^-$ population.[45] However, the LT-HSCs have not been uniquely isolated based on a definitive phenotype. Thus, it remains difficult to selectively expand LT-HSCs or SRCs ex vivo.

Recently, the Lin$^-$CD34$^+$CD38$^-$ fraction of cord blood and bone marrow has been subdivided into three subpopulations: CD90$^+$CD45RA$^-$, CD90$^-$CD45RA$^-$, and CD90$^-$CD45RA$^+$. Among them, the Lin$^-$CD34$^+$CD38$^-$CD90$^+$CD45RA$^-$ fraction contains HSC activity demonstrated by as few as 10 purified cells; while the Lin$^-$CD34$^+$CD38$^-$CD90$^-$CD45RA$^-$ fraction exhibits the activity of MPPs.[72] However, Lin$^-$CD34$^-$CD38$^-$ population seems more primitive than Lin$^-$CD34$^+$CD38$^-$ population, the latter appear to be the progenies of the former.[71,73] Indeed, adult human CD34$^+$CD38$^-$ HSCs only contain about 2% transplantable stem cells, which appear to be identical to a population referred to as LTC-ICs.[74] LTC-ICs, which are supposed to contain LT-HSCs or SRCs, have the ability to produce colony-forming cell (CFC) progenies for $\geq$5 weeks when co-cultured with stromal fibroblasts in the presence of cytokines.[74] Therefore, CB Lin$^-$CD34$^+$ cells that are currently used as the source for ex vivo HSC expansion seem not appropriate for optimal ex vivo HSC expansion.

26.3.4 Cell Culture Condition and CB HSC Expansion

Cell culture conditions, such as medium components, culture dimensions, static or stirred cultures, and removal of metabolisms, have been reported to affect CB HSC ex vivo expansion. By comparison of CD34+ HSCs grown in static with stirred culture system, The engraftment of the HSCs expanded in stirred cultures was higher than that of HSCs expanded in static cultures, although static cultures is better than stirred cultures in HSC expansion. If the engraftment capacity of HSCs is reduced during culture, the expanded HSCs are less or not valuable in clinical application. Therefore, based on the capability of engraftment and lineage reconstitution, stirred culture system might be better than static culture system for the maintenance of stem cell property of expanded HSCs.[75] In a 3-D perfusion scaffold culture system, CD34+ HSC retention and proliferation are better than those cells in static culture, and cells from the perfusion scaffold cultures showed higher numbers of primitive progenitor and produced more colony-forming-units than the cells from static culture, indicating that 3-D perfusion culture system has advantages in ex vivo expansion of CB HSCs.[61]

In addition, microencapsulated feeder cells have been used in ex vivo expansion of CB HSCs. In this system, immortalized stromal cells or MSCs are encapsulated and bioactive substances are continuously released from the capsules to effectively support CB HSC proliferation. The expanded CB HSCs are easily separated from the feeder cells in the capsules because of different size and density between CB HSCs and the microcapsules.[76] Thus, microencapsulated feeder cells may be a good system for ex vivo expansion of CB HSCs.

In short, cell-based studies have provided new insights into the mechanisms underlying HSC self-renewal, differentiation, and apoptosis, and collected valuable data and information for further designing new protocols for ex vivo expansion of CB HSCs. These achievements together with the findings in molecular mechanisms, which will be reviewed in the next section, will change the existing views and landscapes for ex vivo CB HSC expansion and stem cell transplantation.

26.3.5 The Factors That Determine the Success and Failure of CB HSC Transplantation

Clearly, quality and quantity of the expanded CB HSCs determine the success or failure of CB HSC transplantation. Current CB transplantation has only achieved limited success in the treatment of child leukemia, because most adults are not permeable to the treatment. Several reasons have been recognized causing the failure of CB transplantation. First, the volume of a single CB unit is limited, usually only 75–150 ml, and the number of nucleated cells in a single CB unit is approximate 2×10^7/kg for an adult, which is only one-tenth of a unit of BM. It has been demonstrated in clinic CB transplantation that HSC dose is a major determinant of engraftment and survival after CB transplantation, notably in adults and older adolescents. In short, cell numbers are a matter. Recently, several groups have tried to transplant two units of partial HLA-matched CB into a receptor and got better clinic outcomes than single unit of CB, indicating that the number of CB HSCs could affect the consequence of CB transplantation.[77] Since an individual donor only provides one unit of CB, the use of multiple CB units from different donors for a single recipient of CB transplant has a big problem in finding major HLA-matched donors. Thus, this strategy has only limited usage.

Second, the quality of ex vivo expanded CB HSCs is not as good as primary CB HSCs with regard to engraftment. Clinical trials have failed to improve engraftment and survival rate using ex vivo expanded CB HSCs, because most of the expanded $CD34^+$ CB HSCs have differentiated into MMPs or lineage-committed cells with limited numbers of LT-HSCs/SRCs before transplantation.

Third, lymphocytic accessory cells usually lack in the expanded CB HSCs, which might be important in facilitating engraftment of LT-HSCs. Recent studies suggest that co-transplantation of lymphocytes with HSCs can promote the engraftment of HSCs in the patient with leukemia, especially those with chronic myelogenous leukemia (CML).[78]

Taken together, not only is quantity, but also quality of CB HSCs, crucial for the success of CB HSC transplantation. New strategies and methods are being extensively explored to improve the quantity and quality of ex vivo expanded CB HSC as measured by SRC.

26.4 Molecular Basis for CB HSC Expansion

The factors that contribute to the stemness (self-renewal and multipotency of differentiation) of HSCs are not completely understood. However, considerable progress has been made in exploring and delineating intrinsic and extrinsic factors that positively or negatively regulate stemness of expanded HSCs ex vivo.[5,30,31,79] Most of the hematopoietic cytokines or growth factors studied promote either survival or differentiation or both of HSCs expanded ex vivo, whereas morphogens-mediated pathways such as Wnt, Notch, and Hedgehog may support HSC expansion by a combination of survival and induced self-renewal.[40,80] Here we briefly review the stemness of transcription factors, signaling transduction pathways, and growth factors, which may positively or negatively regulate HSC self-renewal and multipotency of differentiation and might be operational in ex vivo expansion.

26.4.1 Transcription Factors That Contribute to Stem Cell Expansion

Although both embryonic and adult stem cells have the capability of self-renewal and multipotency of differentiation, the transcription factors that control the stemness vary between them. Recent studies have revealed that several key transcription factors rule the pluripotency of embryonic stem (ES) cells. The transcription factors, such as Oct4, Sox2, and Nanog, appear to be most important in governing the pluripotency of ES cells.[79,81,82] Oct4 and Nanog are essential for the ES cell pluripotency, and disruption of Oct4 and Nanog results in loss of ES cell pluripotency.[83–85] Sox2 binds to Oct4 and contributes to ES pluripotency. Oct4 gene expression is rapidly and completely shut down during early ES cell differentiation.[84,86] Oct4, Sox2, and Nanog work coordinately to regulate expression of hundreds of genes which contribute to stem cell self-renewal or differentiation, and they often bind together at the promoter regions of their target genes to promote or inhibit ES cell either self-renewal or differentiation.[87,88] Indeed, these transcription factors not only play pivotal role in ES cell pluripotency, but also can trigger retrodifferentiation of somatic cells into ES-like cells, i.e., induced pluripotent stem cells (iPSCs), when they are ectopically expressed in somatic cells.[88–91] Ectopic expression of Oct4, and Sox2 in combination with c-Myc and Klf4[89] or with Nanog and LIN28[91] in fibroblasts through virus infection results in reprogramming of the somatic cells into a pluripotent state, which retro-differentiate into ES-like cells within 3 to 4 weeks.[88] Since the iPSCs induced by the ES cell-associated genes has the potential to develop into tetratomas[89–91] and the tumor development is associated with the subversion of ES cell-associated gene programs,[9,42] the possibility of ectopic expression of these genes in the ex vivo expanded HSCs would raise a safety concern for transplantation. Thus, the ex vivo expanded HSCs should be excluded from expression of ES cell-associated genes.[9,42]

The transcription factors that regulate self-renewal and differentiation of pluripotent HSCs are distinct from ES cells.[92] The transcription factors Runx-1, Scl/tail-1, Lomo-2, Mil, Tel, Bmi-1, Dfi-1, and GATA-2 are currently best understood in association with self-renewal and differentiation of HSCs.[93] No blood cells are generated if Runx-1, Scl/tal-1, and LMO2 are disrupted, respectively.[93] However, the levels of the transcription factors in HSCs may determine their physiological or pathological states. Abnormal high activity of these transcription factors due to chromosomal translocation and subsequent formation of fusion proteins may contribute to leukemogenesis.[94] For an example, Bmi-1 expression is stringently required for the development of both normal and leukemic stem cells.[95,96]

On the other hand, differentiation of pluripotent HSCs to multipotent progenitors and committed precursors is triggered by several distinct sets of transcription factors. GATA-1, GATA-2, and FOG-1b promote HSC differentiation into the progenitors of megakaryoctyes and erythrocytes; PU.1 and C/EBPα can drive HSC differentiation into granulocyte and macrophage progenitors; and Ikaros and PU.1 trigger HSC differentiation into common lymphoid progenitors.[97] Therefore, these transcription factors may be an important target for screening novel agents from chemical and peptide libraries to promote HSC self-renewal and inhibit lineage commitment as well during ex vivo expansion.

26.4.2 Signal Transduction Pathways That Govern Stem Cell Fate and Expansion

Although signal transduction pathways that regulate the self-renewal and differentiation of HSCs are not completely understood, several pathways, including Notch, Wnt, Sonic Hedgehog (shh), Smad, and FGF/Erk, have been found to play an important role in control of HSC fate and expansion.[98]

Activation of morphogens pathways may lead to significant expansion of HSCs. The embryonic or early developmental growth factors, such as bone morphogenetic protein (BMP), shh, ligands for notch receptors and Wnt proteins have growth-promoting activity on HSCs.[80,99–106] Activation of Notch-1 by soluble

Jagged-1 added to culture media resulted in expansion of multipotent hematopoietic stem/progenitor cells via activation of BMP pathway.[104] Soluble shh can also promote human HSC expansion via BMP regulation.[99] However, these expansions still remain modest.

Recently, it has been demonstrated that the Wnt signaling pathway is crucial for normal HSC homoeostasis in vitro and in vivo.[101–103,106,107] Activation of Wnt pathway led to increased expression of Notch-1 and HOXB4 genes.[102] HOXB4 is a member of homeobox transcription factor family critical to the control of embryonic development and has emerged as an important member of genes that determines hematopoiesis during early development.[108,109] HOXB4 expression is up-regulated by either TPO[110] or Wnt.[102] In contrast, the function of HOXB4 is restricted by a partner protein PBX1.[111] HOXB4 over-expression mediates HSC regeneration and competitive repopulation.[112] Human HSCs were significantly expanded when cultured in recombinant HOXB4 homeoprotein-conditioned media.[30,31] Thus, Wnt and HOXB4 signaling axis appear to be critical for self-renewal and competitive repopulation of HSCs.

The Notch pathway plays a critical role for HSC fates. Notch ligands, such as Delta 1–3and jagged 1–2, can expand hematopoietic progenitors in vitro. Delta 1–3or Jagged 1–2 bind to Notch receptors, and then the receptor is cut by γ-secretase, resulting in the formation of Notch intracellular domain (NICD). NICD enters into nucleus and forms a complex with the transcription factor CSL and cofactors (MAML) to activate transcription of a variety of target genes.[113,114] Enforced activation of Notch signaling has been shown to elevate self-renewal and long-term in vivo repopulation of HSCs, whereas inhibition of Notch signaling pathway diminishes HSC self-renewal.[115] Thus, Notch signaling pathway may be a good target for manipulation of ex vivo expansion of CB HSCs.

The Smad-mediated signaling pathway also plays a crucial role in the determination of HSC fates.[116] TGF-β family of ligands, such as TGF-β, BMPs, and activin/nodal, deliver signals through Smad pathway and regulate HSC self-renewal. TGF-β binds its receptor 2 and activates ALK5 and ALK1, and consequently activates Smad4. Activated Smad4 and co-Smad translocate into nucleus, bind to target genes, and up-regulate expression of cyclin-dependent kinase inhibitors, p21 and p57, leading to inhibition of HSC growth. TGF-β is a strong inhibitor of HSC growth in vitro. BMP4 contributes to the maintenance of HSCs in vitro and induces both proliferation and differentiation of human hematopoietic progenitors. Smad has eight redundant family members. The roles and the mechanisms of each Samd member in HSC self-renewal and differentiation need to be further elucidated.[116]

Recent studies have also uncovered that FGF/Erk-mediated signaling pathway plays very important roles in stem cell self-renewal and differentiation. Over-expression of FGF4 leads to loss of ES cell self-renewal capability and propels ES cell differentiation.[117] Lack of FGF4 or deficiency in Erk1/2 signaling in ES cells impairs cell differentiation. Genetic disruption of FGF4/Erk and the inhibitors of FGF4/Erk have been used to maintain ES self-renewal and avoid ES cell differentiation. It is interesting to know whether the FGF/Erk pathways also negatively regulate HSC self-renewal.

In addition, STAT3, one of the JAK-STATs signaling proteins, can change the balance of ES cells between self-renewal and differentiation. Leukemia inhibitor factor (LIF) stimulates STAT3 and blocks ES cell differentiation. LIF has been widely used to diminish HSC differentiation and ensure stem cell ex vivo expansion.[118] Furthermore, glycogen synthetase kinase-3 (GSK-3) also contributes to ES cell differentiation; and inhibition of GSK-3 leads to suppress ES cell differentiation.[119] Importantly, GSK-3 inhibitor has been shown to block mouse HSC differentiation in vivo and promote their repopulation activity through modulation of gene targets of Wnt, Hedgehog, and Notch pathways.[120] Thus, blocking GSK-3 pathway might promote expansion of LT-HSCs or SRC ex vivo.

26.4.3 Growth Factors and Cytokines That Regulate CB HSC Expansion

It is well known that hematopoietic growth factors and cytokines play pivotal roles in ex vivo CB HSC expansion.[40] Several cocktails consisting of growth factors and cytokines, such as SCF, TPO, FL/Flt3L, interleukin-3 (IL-3), IL-6, IL-11, erythropoietin (EPO), and GM-CSF, have been used to expand CB HSCs.[5,34,35,66,121] Several protocols using the combinations of these growth factors and cytokines usually expand nucleated CB cells from 20 to several hundred folds within 10–14 days but with limited expansion of LT-HSCs or SRCs. Most of the expanded cells are lineage-committed hematopoietic progenitors, precursors, and mature cells.

In vivo CB HSC transplantation studies have indicated that only LT-HSCs or SRCs can maintain long-term hematopoiesis in mice and human.[122–124] LT-HSCs express c-kit and c-mpl, the receptors for SCF and TPO, respectively, but not Flt3, a receptor for FL. It is well known that SCF is the most important growth factor for stem cell self-renewal and crucial to CB HSC expansion ex vivo. However, a recent study indicates that pretreatment of $CD34^+$ CB HSCs with SCF led to reduced long-term cell engraftment in NOD/SCID mice.[125] TPO has been reported to inhibit apoptosis of HSCs and support HSC growth.[126] Mice with genetic mutation of TPO or c-mpl exhibit decreased numbers of HSCs.[127] Mechanistic studies indicated that the TPO can suppress apoptosis of HSCs rather than enhance their expansion.[128] Thus, although SCF and TPO are important growth factors for ex vivo CB HSC expansion, their definitive function on LT-HSCs or SRCs needs to be more precisely defined.

26.4.4 Proteins That Promote CB HSC Expansion

The most exciting recent development in ex vivo expansion of CB HSCs is that several proteins can effectively promote LT-HSC self-renewal from human CB, including angiopoietin-like proteins (Angptls) and IGF-binding protein 2 (IGFBP2). Angptls, such as Angptl-2, 3, and 5, are expressed in stromal cells and promote ex vivo expansion of HSCs. IGFBP2 was found to enhance ex vivo expansion of HSCs. These two proteins have been used to expand CB HSCs in vitro in combination with other hematopoietic growth factors and cytokines. In a serum-free culture medium, a cocktail of SCF, TPO, FGF-1, Angptl 5, and IGFBP2 expanded human LT-HSCs/SRCs in CB $CD133^+$ cells by 20-fold after 10 days of culture; in contrast, the cocktail in the absence of Angptl 5 and IGFBP2 did not significantly expand LT-HSCs, despite promoting nucleated cell expansion approximately by 260-folds.[129] Thus, Angptl 5 and IGFBP2 favor CB LT-HSC expansion ex vivo and engraftment in vivo.

A number of other soluble proteins that are related to activation of stemness-related transcriptional and signal transduction pathways have been reported to enhance HSC self-renewal ex vivo, including BMP4, HOXB4, sonic Hedgehog (Shh), Wnt-3, notch ligands, proliferin-2, and FGF-1.[129–133] For examples, BMP4 secreted from AFT04 stromal cell line supported ex vivo CB HSC expansion[130]; forced expression of the polycomb group gene Bmi-1 increased ex vivo LT-HSC expansion and engraftment in NOD/SCID mice, and preserved colony-forming activity of expanded CB HSCs by 20 weeks in the stromal cell-free, cytokine-dependent medium[131]; HOX proteins interact with PBX1 to regulate expression of target genes, and HOX decoy peptide has been reported to enhance ex vivo CB HSC expansion, resulting in a twofold increase in HSC engraftment in NOD/SCID mice[132]; a plant mannose-binding lectin NTL can promote CB HSC expansion ex vivo and cell engraftment in BM of NOD/SCID mice after transplantation.[134] Thus, identification of novel proteins enabling LT-HSC/SRC expansion ex vivo will have important clinical impact on CB HSC transplantation.

26.4.5 Small Molecules That Enhance CB HSC Expansion

Cellular copper (Cu) is involved in the regulation of proliferation, differentiation, and apoptosis of hematopoietic progenitor cells through generation of oxidative stress in cells. Copper chelator tetraethylenepentamine (TEPA) has been used to reduce the level of cellular copper and found to promote ex vivo expansion of hematopoietic progenitors and attenuate cell differentiation.[135] CB-derived purified $CD34^+$ cells were grown in liquid medium supplemented with the SCF, TPO, FL, IL-6, and TEPA for 3 to 11 weeks. Control cultures were supplemented with the same growth factors and cytokines in the absence of TEPA for entire culture duration. TEPA significantly increased the proliferation of early progenitors over the controls. Transplantation of the expanded cells into sub-lethally irradiated NOD/SCID mice showed that the engraftment potential of the ex vivo expanded $CD133^+$ cells was near threefold higher than that of unexpanded cells. A phase I/II clinic trial using ex vivo expanded CB HSCs in the presence of TEPA showed that nine out of ten patients with advanced hematopoietic malignancies were engrafted at 30 days for neutrophils and 48 days for platelets; the 100 days survival was 90%, and there were no acute graft-versus-host disease (GVHD) in these cases.[136] These data suggest that TEPA has significant benefits for ex vivo expansion of transplantable CB HSCs. Hence, a series of search for small molecules from chemical and herbal libraries might lead to discovery of novel molecules potential for promoting ex vivo expansion of transplantable CB HSCs.

26.5 Concluding Remarks

Stem cells have brought new hopes for curative therapy of advanced blood malignancies and autoimmune diseases, such as leukemia, diabetes, cardiovascular disorders, Parkinson's and Alzheimer's diseases. However, stem cell research from the bench to clinic is still facing enormous challenges. Shortage of transplantable HSCs is a major obstacle. The expected breakthrough in large ex vivo expansion of transplantable CB HSCs could overcome the difficulty.

Since the frequency of transplantable HSCs (or LT-HSCs/SRCs) is low in normal marrow and multiple HSC transplants are required for reconstitution of HSC compartment in patients,[137] CB HSCs are chosen for large ex vivo expansion to obtained sufficient number of LT-HSCs. The technologies for expanding CB HSCs ex vivo have been extensively explored but remain modest and transient. Thus, there is still a long distance from bench to clinic. Although HSCs can proliferate in the cultures in responding to particular cytokines and growth factors, such as SCF, FL, TPO, IL-3, and IL-11, either alone or in combination, they are inevitably prone to differentiation into lineage-committed hematopoietic cells, programmed cell death, or loss of the capacity for long-term repopulation.[56,121,122,138–143] On the other hand, since stromal (feeder) cells are required for ex vivo HSC expansion, the factors that regulate stem cell expansion have not been defined well. In particular, the interactions between stromal cells and HSCs mediated by receptors and ligands, such as the families of adhesion molecules and matrix proteins, may confound HSC expansion ex vivo.[35,55,144] Thus, the experimental conditions designed for ex vivo expansion of HSCs must meet the requirements for therapeutic application.

Although much effort has been made to expand HSCs ex vivo by using retrovirus to transduce self-renewal-associated genes, such as HOXB4 and Bmi-1, into HSCs[112,145] or using immortalized stromal cells,[34,35] the deleterious effects of the approaches become overt with regard to the risk for cell transformation or tumorigenesis after transplantation.[28,29,146] An alternative approach currently explored is to use soluble gene products of interest such as HOXB4, Wnt, Angptl 5, and IGFBP2 to expand HSCs ex vivo.[30,31,103,129] In addition, discovery of small molecules that target signal transduction pathways required for the maintenance of stem cell property would improve the culture system for ex vivo CB HSC expansion.

To meet the requirements for clinical applications, expanded HSCs should be safe and cost-effective without the risk for cell transformation, preserve the capacity of self-renewal and multipotency of differentiation, and be competent for long-term repopulation in acceptors. To meet these criteria, an appropriate culture system needs to be developed. Based on the current researches, the prospective culture systems for the ex vivo expansion of transplantable HSCs should be serum- and feeder-cell-free, with a 3-D nanofiber scaffold immobilized with the small molecules that can promote expansion but prevent lineage differentiation of proliferating HSCs. These molecules, such as integrins, antibody, and matrix proteins, create a microenvironment in the nanofiber scaffolds, mimicking in vivo stem cell niche. To reach the goal, better understandings of stem cell biology is a must.

In short, recent studies on HSCs have provided new insight on CB HSC self-renewal, differentiation, and apoptosis at both cellular and molecular bases, which are valuable for ex vivo CB HSC expansion. These achievements will change the previous views on ex vivo CB HSC expansion and transplantation, leading to invention of novel strategies and technologies for ex vivo CB HSC expansion and transplantation.

References

1. Gluckman EG, Roch VV, Chastang C. Use of cord blood cells for banking and transplant. *Oncologist.* 1997;2:340-343.
2. Tse W, Bunting KD, Laughlin MJ. New insights into cord blood stem cell transplantation. *Curr Opin Hematol.* 2008;15:279-284.
3. Brown JA, Boussiotis VA. Umbilical cord blood transplantation: basic biology and clinical challenges to immune reconstitution. *Clin Immunol.* 2008;127:286-297.

4. Gluckman E, Rocha V, Boyer-Chammard A, et al. Outcome of cord-blood transplantation from related and unrelated donors. Eurocord Transplant Group and the European Blood and Marrow Transplantation Group. *N Engl J Med.* 1997;337:373-381.
5. Kondo M, Wagers AJ, Manz MG, et al. Biology of hematopoietic stem cells and progenitors: implications for clinical application. *Annu Rev Immunol.* 2003;21:759-806.
6. Morrison SJ, Uchida N, Weissman IL. The biology of hematopoietic stem cells. *Annu Rev Cell Dev Biol.* 1995;11:35-71.
7. Li L, Xie T. Stem cell niche: structure and function. *Annu Rev Cell Dev Biol.* 2005;21:605-631.
8. Moore KA, Lemischka IR. Stem cells and their niches. *Science.* 2006;311:1880-1885.
9. Gao JX. Cancer stem cells: the lessons from precancerous stem cells. *J Cell Mol Med.* 2008;12:67-96.
10. Brazelton TR, Rossi FM, Keshet GI, et al. From marrow to brain: expression of neuronal phenotypes in adult mice. *Science.* 2000;290:1775-1779.
11. Kopen GC, Prockop DJ, Phinney DG. Marrow stromal cells migrate throughout forebrain and cerebellum, and they differentiate into astrocytes after injection into neonatal mouse brains. *Proc Natl Acad Sci USA.* 1999;96:10711-10716.
12. Zhao LR, Duan WM, Reyes M, et al. Human bone marrow stem cells exhibit neural phenotypes and ameliorate neurological deficits after grafting into the ischemic brain of rats. *Exp Neurol.* 2002;174:11-20.
13. Theise ND, Badve S, Saxena R, et al. Derivation of hepatocytes from bone marrow cells in mice after radiation-induced myeloablation. *Hepatology.* 2000;31:235-240.
14. Tang XP, Zhang M, Yang X, et al. Differentiation of human umbilical cord blood stem cells into hepatocytes in vivo and in vitro. *World J Gastroenterol.* 2006;12:4014-4019.
15. Jackson KA, Mi T, Goodell MA. Hematopoietic potential of stem cells isolated from murine skeletal muscle. *Proc Natl Acad Sci USA.* 1999;96:14482-14486.
16. Camargo FD, Green R, Capetanaki Y, et al. Single hematopoietic stem cells generate skeletal muscle through myeloid intermediates. *Nat Med.* 2003;9:1520-1527.
17. Abedi M, Foster BM, Wood KD, et al. Haematopoietic stem cells participate in muscle regeneration. *Br J Haematol.* 2007;138:792-801.
18. Jackson KA, Majka SM, Wang H, et al. Regeneration of ischemic cardiac muscle and vascular endothelium by adult stem cells. *J Clin Invest.* 2001;107:1395-1402.
19. Orlic D, Kajstura J, Chimenti S, et al. Mobilized bone marrow cells repair the infarcted heart, improving function and survival. *Proc Natl Acad Sci USA.* 2001;98:10344-10349.
20. Kajstura J, Rota M, Whang B, et al. Bone marrow cells differentiate in cardiac cell lineages after infarction independently of cell fusion. *Circ Res.* 2005;96:127-137.
21. Nygren JM, Jovinge S, Breitbach M, et al. Bone marrow-derived hematopoietic cells generate cardiomyocytes at a low frequency through cell fusion, but not transdifferentiation. *Nat Med.* 2004;10:494-501.
22. Balsam LB, Wagers AJ, Christensen JL, et al. Haematopoietic stem cells adopt mature haematopoietic fates in ischaemic myocardium. *Nature.* 2004;428:668-673.
23. Rizvi AZ, Swain JR, Davies PS, et al. Bone marrow-derived cells fuse with normal and transformed intestinal stem cells. *Proc Natl Acad Sci USA.* 2006;103:6321-6325.
24. Murry CE, Soonpaa MH, Reinecke H, et al. Haematopoietic stem cells do not transdifferentiate into cardiac myocytes in myocardial infarcts. *Nature.* 2004;428:664-668.
25. Orlic D. BM stem cells and cardiac repair: where do we stand in 2004? *Cytotherapy.* 2005;7:3-15.
26. Brown JM, Weissman IL. Progress and prospects in hematopoietic stem cell expansion and transplantation. *Exp Hematol.* 2004;32:693-695.
27. Mazurier F, Doedens M, Gan OI, et al. Characterization of cord blood hematopoietic stem cells. *Ann NY Acad Sci.* 2003;996:67-71.
28. Hacein-Bey-Abina S, von Kalle C, Schmidt M, et al. A serious adverse event after successful gene therapy for X-linked severe combined immunodeficiency. *N Engl J Med.* 2003;348:255-256.
29. Hacein-Bey-Abina S, Von Kalle C, Schmidt M, et al. LMO2-associated clonal T cell proliferation in two patients after gene therapy for SCID-X1. *Science.* 2003;302:415-419.
30. Amsellem S, Pflumio F, Bardinet D, et al. Ex vivo expansion of human hematopoietic stem cells by direct delivery of the HOXB4 homeoprotein. *Nat Med.* 2003;9:1423-1427.
31. Krosl J, Austin P, Beslu N, et al. In vitro expansion of hematopoietic stem cells by recombinant TAT-HOXB4 protein. *Nat Med.* 2003;9:1428-1432.
32. Breems DA, Blokland EA, Siebel KE, et al. Stroma-contact prevents loss of hematopoietic stem cell quality during ex vivo expansion of CD34+ mobilized peripheral blood stem cells. *Blood.* 1998;91:111-117.
33. Bennaceur-Griscelli A, Tourino C, Izac B, et al. Murine stromal cells counteract the loss of long-term culture-initiating cell potential induced by cytokines in CD34(+)CD38(low/neg) human bone marrow cells. *Blood.* 1999;94:529-538.
34. Kawano Y, Kobune M, Yamaguchi M, et al. Ex vivo expansion of human umbilical cord hematopoietic progenitor cells using a coculture system with human telomerase catalytic subunit (hTERT)-transfected human stromal cells. *Blood.* 2003;101:532-540.
35. Yang YX, Miao ZC, Zhang HJ, et al. Establishment and characterization of a human telomerase catalytic subunit-transduced fetal bone marrow-derived osteoblastic cell line. *Differentiation.* 2007;75:24-34.
36. Li Y, Ma T, Kniss DA, et al. Human cord cell hematopoiesis in three-dimensional nonwoven fibrous matrices: in vitro simulation of the marrow microenvironment. *J Hematother Stem Cell Res.* 2001;10:355-368.
37. Okamoto T, Takagi M, Soma T, et al. Effect of heparin addition on expansion of cord blood hematopoietic progenitor cells in three-dimensional coculture with stromal cells in nonwoven fabrics. *J Artif Organs.* 2004;7:194-202.
38. Takagi M. Cell processing engineering for ex-vivo expansion of hematopoietic cells. *J Biosci Bioeng.* 2005;99:189-196.
39. Sands RW, Mooney DJ. Polymers to direct cell fate by controlling the microenvironment. *Curr Opin Biotechnol.* 2007;18:448-453.
40. Zhang CC, Lodish HF. Cytokines regulating hematopoietic stem cell function. *Curr Opin Hematol.* 2008;15:307-311.
41. Yamaguchi M, Hirayama F, Kanai M, et al. Serum-free coculture system for ex vivo expansion of human cord blood primitive progenitors and SCID mouse-reconstituting cells using human bone marrow primary stromal cells. *Exp Hematol.* 2001;29:174-182.

42. Chen L, Shen R, Ye Y, et al. Precancerous stem cells have the potential for both benign and malignant differentiation. *PLoS ONE*. 2007;2:e293.
43. Skea D, Chang NH, Hedge R, et al. Large ex vivo expansion of human umbilical cord blood CD4+ and CD8+ T cells. *J Hematother*. 1999;8:129-139.
44. Skea D, Hedge R, Dabek B, et al. The selective expansion of functional T cell subsets. *J Hematother Stem Cell Res*. 1999;8:525-538.
45. Larochelle A, Vormoor J, Hanenberg H, et al. Identification of primitive human hematopoietic cells capable of repopulating NOD/SCID mouse bone marrow: implications for gene therapy. *Nat Med*. 1996;2:1329-1337.
46. Dick JE, Bhatia M, Gan O, et al. Assay of human stem cells by repopulation of NOD/SCID mice. *Stem Cells*. 1997;15 (Suppl 1):199-203. discussion 204–197.
47. Keating A. Mesenchymal stromal cells. *Curr Opin Hematol*. 2006;13:419-425.
48. Jones DL, Wagers AJ. No place like home: anatomy and function of the stem cell niche. *Nat Rev Mol Cell Biol*. 2008;9:11-21.
49. Yin T, Li L. The stem cell niches in bone. *J Clin Invest*. 2006;116:1195-1201.
50. Arai F, Suda T. Maintenance of quiescent hematopoietic stem cells in the osteoblastic niche. *Ann NY Acad Sci*. 2007;1106:41-53.
51. Martinez-Agosto JA, Mikkola HK, Hartenstein V, et al. The hematopoietic stem cell and its niche: a comparative view. *Genes Dev*. 2007;21:3044-3060.
52. Metcalf D. On hematopoietic stem cell fate. *Immunity*. 2007;26:669-673.
53. Deng W, Lin H. Asymmetric germ cell division and oocyte determination during *Drosophila* oogenesis. *Int Rev Cytol*. 2001;203:93-138.
54. Knoblich JA. Mechanisms of asymmetric stem cell division. *Cell*. 2008;132:583-597.
55. Seoh JY, Woo SY, Im SA, et al. Distinct patterns of apoptosis in association with modulation of CD44 induced by thrombopoietin and granulocyte-colony stimulating factor during ex vivo expansion of human cord blood CD34+ cells. *Br J Haematol*. 1999;107:176-185.
56. Domen J, Weissman IL. Hematopoietic stem cells need two signals to prevent apoptosis; BCL-2 can provide one of these, Kitl/c-Kit signaling the other. *J Exp Med*. 2000;192:1707-1718.
57. Domen J, Cheshier SH, Weissman IL. The role of apoptosis in the regulation of hematopoietic stem cells: overexpression of Bcl-2 increases both their number and repopulation potential. *J Exp Med*. 2000;191:253-264.
58. Oguro H, Iwama A. Life and death in hematopoietic stem cells. *Curr Opin Immunol*. 2007;19:503-509.
59. Bieback K, Kern S, Kocaomer A, et al. Comparing mesenchymal stromal cells from different human tissues: bone marrow, adipose tissue and umbilical cord blood. *Biomed Mater Eng*. 2008;18:S71-S76.
60. Chen CP, Lee MY, Huang JP, et al. Trafficking of multipotent mesenchymal stromal cells from maternal circulation through the placenta involves vascular endothelial growth factor receptor-1 and integrins. *Stem Cells*. 2008;26:550-561.
61. Zhang Y, Chai C, Jiang XS, et al. Co-culture of umbilical cord blood CD34+ cells with human mesenchymal stem cells. *Tissue Eng*. 2006;12:2161-2170.
62. Friedman R, Betancur M, Boissel L, et al. Umbilical cord mesenchymal stem cells: adjuvants for human cell transplantation. *Biol Blood Marrow Transplant*. 2007;13:1477-1486.
63. Hu L, Cheng L, Wang J, et al. Effects of human yolk sac endothelial cells on supporting expansion of hematopoietic stem/progenitor cells from cord blood. *Cell Biol Int*. 2006;30:879-884.
64. Ma LJ, Gao L, Zhou H, et al. Effects of human mesenchymal stem cells and fibroblastoid cell line as feeder layers on expansion of umbilical cord blood CD34(+) cells in vitro. *Zhongguo Shi Yan Xue Ye Xue Za Zhi*. 2006;14:949-954.
65. Chen HQ, Zhang XC, Tang XY, et al. Hematopoietic growth factors expressed in human aorta-gonad-mesonephros (AGM)-derived stromal cells in vitro. *Zhongguo Shi Yan Xue Ye Xue Za Zhi*. 2006;14:999-1003.
66. Ito Y, Hasauda H, Kitajima T, et al. Ex vivo expansion of human cord blood hematopoietic progenitor cells using glutaraldehyde-fixed human bone marrow stromal cells. *J Biosci Bioeng*. 2006;102:467-469.
67. Chua KN, Chai C, Lee PC, et al. Functional nanofiber scaffolds with different spacers modulate adhesion and expansion of cryopreserved umbilical cord blood hematopoietic stem/progenitor cells. *Exp Hematol*. 2007;35:771-781.
68. Uchida N, Aguila HL, Fleming WH, et al. Rapid and sustained hematopoietic recovery in lethally irradiated mice transplanted with purified Thy-1.1lo Lin-Sca-1+ hematopoietic stem cells. *Blood*. 1994;83:3758-3779.
69. Uchida N, Jerabek L, Weissman IL. Searching for hematopoietic stem cells. II. The heterogeneity of Thy-1.1(lo)Lin(-/lo) Sca-1+ mouse hematopoietic stem cells separated by counterflow centrifugal elutriation. *Exp Hematol*. 1996;24:649-659.
70. Uchida N, Weissman IL. Searching for hematopoietic stem cells: evidence that Thy-1.1lo Lin- Sca-1+ cells are the only stem cells in C57BL/Ka-Thy-1.1 bone marrow. *J Exp Med*. 1992;175:175-184.
71. Engelhardt M, Lubbert M, Guo Y. CD34(+) or CD34(-): which is the more primitive? *Leukemia*. 2002;16:1603-1608.
72. Majeti R, Park CY, Weissman IL. Identification of a hierarchy of multipotent hematopoietic progenitors in human cord blood. *Cell Stem Cell*. 2007;1:635-645.
73. Hess DA, Karanu FN, Levac K, et al. Coculture and transplant of purified CD34(+)Lin(-) and CD34(-)Lin(-) cells reveals functional interaction between repopulating hematopoietic stem cells. *Leukemia*. 2003;17:1613-1625.
74. Petzer AL, Hogge DE, Landsdorp PM, et al. Self-renewal of primitive human hematopoietic cells (long-term-culture-initiating cells) in vitro and their expansion in defined medium. *Proc Natl Acad Sci USA*. 1996;93:1470-1474.
75. Yang S, Cai H, Jin H, et al. Hematopoietic reconstitution of CD34+ cells grown in static and stirred culture systems in NOD/SCID mice. *Biotechnol Lett*. 2008;30:61-65.
76. Fujimoto N, Fujita S, Tsuji T, et al. Microencapsulated feeder cells as a source of soluble factors for expansion of CD34+ hematopoietic stem cells. *Biomaterials*. 2007;28:4795-4805.
77. Brunstein CG, Setubal DC, Wagner JE. Expanding the role of umbilical cord blood transplantation. *Br J Haematol*. 2007;137:20-35.
78. Tanaka J, Sugita J, Kato N, et al. Expansion of natural killer cell receptor (CD94/NKG2A)-expressing cytolytic CD8 T cells and CD4+CD25+ regulatory T cells from the same cord blood unit. *Exp Hematol*. 2007;35:1562-1566.

79. Constantinescu S. Stemness, fusion and renewal of hematopoietic and embryonic stem cells. *J Cell Mol Med.* 2003;7:103-112.
80. Campbell C, Risueno RM, Salati S, et al. Signal control of hematopoietic stem cell fate: Wnt, Notch, and Hedgehog as the usual suspects. *Curr Opin Hematol.* 2008;15:319-325.
81. Chickarmane V, Troein C, Nuber UA, et al. Transcriptional dynamics of the embryonic stem cell switch. *PLoS Comput Biol.* 2006;2:e123.
82. Imamura M, Miura K, Iwabuchi K, et al. Transcriptional repression and DNA hypermethylation of a small set of ES cell marker genes in male germline stem cells. *BMC Dev Biol.* 2006;6:34.
83. Guo Y, Mantel C, Hromas RA, et al. Oct-4 is critical for survival/antiapoptosis of murine embryonic stem cells subjected to stress: effects associated with Stat3/Survivin. *Stem Cells.* 2008;26:30-34.
84. Chambers I, Colby D, Robertson M, et al. Functional expression cloning of Nanog, a pluripotency sustaining factor in embryonic stem cells. *Cell.* 2003;113:643-655.
85. Chambers I, Silva J, Colby D, et al. Nanog safeguards pluripotency and mediates germline development. *Nature.* 2007;450:1230-1234.
86. Nichols J, Zevnik B, Anastassiadis K, et al. Formation of pluripotent stem cells in the mammalian embryo depends on the POU transcription factor Oct4. *Cell.* 1998;95:379-391.
87. Silva J, Smith A. Capturing pluripotency. *Cell.* 2008;132:532-536.
88. Jaenisch R, Young R. Stem cells, the molecular circuitry of pluripotency and nuclear reprogramming. *Cell.* 2008;132:567-582.
89. Takahashi K, Tanabe K, Ohnuki M, et al. Induction of pluripotent stem cells from adult human fibroblasts by defined factors. *Cell.* 2007;131:861-872.
90. Takahashi K, Yamanaka S. Induction of pluripotent stem cells from mouse embryonic and adult fibroblast cultures by defined factors. *Cell.* 2006;126:663-676.
91. Yu J, Vodyanik MA, Smuga-Otto K, et al. Induced pluripotent stem cell lines derived from human somatic cells. *Science.* 2007;318:1917-1920.
92. Orkin SH, Zon LI. Hematopoiesis: an evolving paradigm for stem cell biology. *Cell.* 2008;132:631-644.
93. Teitell MA, Mikkola HK. Transcriptional activators, repressors, and epigenetic modifiers controlling hematopoietic stem cell development. *Pediatr Res.* 2006;59:33R-39R.
94. Rosenbauer F, Tenen DG. Transcription factors in myeloid development: balancing differentiation with transformation. *Nat Rev Immunol.* 2007;7:105-117.
95. Liu S, Dontu G, Mantle ID, et al. Hedgehog signaling and Bmi-1 regulate self-renewal of normal and malignant human mammary stem cells. *Cancer Res.* 2006;66:6063-6071.
96. Raaphorst FM. Self-renewal of hematopoietic and leukemic stem cells: a central role for the Polycomb-group gene Bmi-1. *Trends Immunol.* 2003;24:522-524.
97. Orkin SH. Diversification of haematopoietic stem cells to specific lineages. *Nat Rev Genet.* 2000;1:57-64.
98. Blank U, Karlsson G, Karlsson S. Signaling pathways governing stem-cell fate. *Blood.* 2008;111:492-503.
99. Bhardwaj G, Murdoch B, Wu D, et al. Sonic hedgehog induces the proliferation of primitive human hematopoietic cells via BMP regulation. *Nat Immunol.* 2001;2:172-180.
100. Bhatia M, Bonnet D, Wu D, et al. Bone morphogenetic proteins regulate the developmental program of human hematopoietic stem cells. *J Exp Med.* 1999;189:1139-1148.
101. Eaves CJ. Manipulating hematopoietic stem cell amplification with Wnt. *Nat Immunol.* 2003;4:511-512.
102. Reya T, Duncan AW, Ailles L, et al. A role for Wnt signalling in self-renewal of haematopoietic stem cells. *Nature.* 2003;423:409-414.
103. Willert K, Brown JD, Danenberg E, et al. Wnt proteins are lipid-modified and can act as stem cell growth factors. *Nature.* 2003;423:448-452.
104. Karanu FN, Murdoch B, Gallacher L, et al. The notch ligand jagged-1 represents a novel growth factor of human hematopoietic stem cells. *J Exp Med.* 2000;192:1365-1372.
105. Karanu FN, Yuefei L, Gallacher L, et al. Differential response of primitive human CD34- and CD34+ hematopoietic cells to the Notch ligand Jagged-1. *Leukemia.* 2003;17:1366-1374.
106. Murdoch B, Chadwick K, Martin M, et al. Wnt-5A augments repopulating capacity and primitive hematopoietic development of human blood stem cells in vivo. *Proc Natl Acad Sci USA.* 2003;100:3422-3427.
107. Reya T. Regulation of hematopoietic stem cell self-renewal. *Recent Prog Horm Res.* 2003;58:283-295.
108. Look AT. Oncogenic transcription factors in the human acute leukemias. *Science.* 1997;278:1059-1064.
109. Buske C, Humphries RK. Homeobox genes in leukemogenesis. *Int J Hematol.* 2000;71:301-308.
110. Kirito K, Fox N, Kaushansky K. Thrombopoietin stimulates Hoxb4 expression: an explanation for the favorable effects of TPO on hematopoietic stem cells. *Blood.* 2003;102:3172-3178.
111. Krosl J, Beslu N, Mayotte N, et al. The competitive nature of HOXB4-transduced HSC is limited by PBX1: the generation of ultra-competitive stem cells retaining full differentiation potential. *Immunity.* 2003;18:561-571.
112. Antonchuk J, Sauvageau G, Humphries RK. HOXB4 overexpression mediates very rapid stem cell regeneration and competitive hematopoietic repopulation. *Exp Hematol.* 2001;29:1125-1134.
113. Kopan R. Notch: a membrane-bound transcription factor. *J Cell Sci.* 2002;115:1095-1097.
114. Hurlbut GD, Kankel MW, Lake RJ, et al. Crossing paths with Notch in the hyper-network. *Curr Opin Cell Biol.* 2007;19:166-175.
115. Varnum-Finney B, Xu L, Brashem-Stein C, et al. Pluripotent, cytokine-dependent, hematopoietic stem cells are immortalized by constitutive Notch1 signaling. *Nat Med.* 2000;6:1278-1281.
116. Larsson J, Karlsson S. The role of Smad signaling in hematopoiesis. *Oncogene.* 2005;24:5676-5692.
117. Kunath T, Saba-El-Leil MK, Almousailleakh M, et al. FGF stimulation of the Erk1/2 signalling cascade triggers transition of pluripotent embryonic stem cells from self-renewal to lineage commitment. *Development.* 2007;134:2895-2902.
118. Sekkai D, Gruel G, Herry M, et al. Microarray analysis of LIF/Stat3 transcriptional targets in embryonic stem cells. *Stem Cells.* 2005;23:1634-1642.
119. Sato N, Meijer L, Skaltsounis L, et al. Maintenance of pluripotency in human and mouse embryonic stem cells through activation of Wnt signaling by a pharmacological GSK-3-specific inhibitor. *Nat Med.* 2004;10:55-63.

120. Trowbridge JJ, Xenocostas A, Moon RT, et al. Glycogen synthase kinase-3 is an in vivo regulator of hematopoietic stem cell repopulation. *Nat Med.* 2006;12:89-98.
121. Audet J, Miller CL, Eaves CJ, et al. Common and distinct features of cytokine effects on hematopoietic stem and progenitor cells revealed by dose-response surface analysis. *Biotechnol Bioeng.* 2002;80:393-404.
122. Conneally E, Cashman J, Petzer A, et al. Expansion in vitro of transplantable human cord blood stem cells demonstrated using a quantitative assay of their lympho-myeloid repopulating activity in nonobese diabetic-scid/scid mice. *Proc Natl Acad Sci USA.* 1997;94:9836-9841.
123. Bhatia M, Bonnet D, Murdoch B, et al. A newly discovered class of human hematopoietic cells with SCID-repopulating activity. *Nat Med.* 1998;4:1038-1045.
124. Bhatia M, Wang JC, Kapp U, et al. Purification of primitive human hematopoietic cells capable of repopulating immune-deficient mice. *Proc Natl Acad Sci USA.* 1997;94:5320-5325.
125. Wulf-Goldenberg A, Eckert K, Fichtner I. Cytokine-pretreatment of CD34+ cord blood stem cells in vitro reduces long-term cell engraftment in NOD/SCID mice. *Eur J Cell Biol.* 2008;87:69-80.
126. Jacobsen SE, Borge OJ, Ramsfjell V, et al. Thrombopoietin, a direct stimulator of viability and multilineage growth of primitive bone marrow progenitor cells. *Stem Cells.* 1996;14(suppl 1):173-180.
127. Kimura S, Roberts AW, Metcalf D, et al. Hematopoietic stem cell deficiencies in mice lacking c-Mpl, the receptor for thrombopoietin. *Proc Natl Acad Sci USA.* 1998;95:1195-1200.
128. Borge OJ, Ramsfjell V, Veiby OP, et al. Thrombopoietin, but not erythropoietin promotes viability and inhibits apoptosis of multipotent murine hematopoietic progenitor cells in vitro. *Blood.* 1996;88:2859-2870.
129. Zhang CC, Kaba M, Iizuka S, et al. Angiopoietin-like 5 and IGFBP2 stimulate ex vivo expansion of human cord blood hematopoietic stem cells as assayed by NOD/SCID transplantation. *Blood.* 2008;111:3415-3423.
130. Hutton JF, Rozenkov V, Khor FSL, et al. Bone morphogenetic protein 4 contributes to the maintenance of primitive cord blood hematopoietic progenitors in an ex vivo stroma-noncontact co-culture system. *Stem Cells Dev.* 2006;15:805-813.
131. Rizo A, Dontje B, Vellenga E, et al. Long-term maintenance of human hematopoietic stem/progenitor cells by expression of BMI1. *Blood.* 2008;111:2621-2630.
132. Tanaka H, Matsumura I, Itoh K, et al. HOX decoy peptide enhances the ex vivo expansion of human umbilical cord blood CD34+ hematopoietic stem cells/hematopoietic progenitor cells. *Stem Cells.* 2006;24:2592-2602.
133. Nikopoulos GN, Duarte M, Kubu CJ, et al. Soluble jagged1 attenuates lateral inhibition, allowing for the clonal expansion of neural crest stem cells. *Stem Cells.* 2007;25:3133-3142.
134. Li K, Ooi VE, Chuen CK, et al. The plant mannose-binding lectin NTL preserves cord blood haematopoietic stem/progenitor cells in long-term culture and enhances their ex vivo expansion. *Br J Haematol.* 2008;140:90-98.
135. Prus E, Fibach E. The effect of the copper chelator tetraethylenepentamine on reactive oxygen species generation by human hematopoietic progenitor cells. *Stem Cells Dev.* 2007;16:1053-1056.
136. de Lima M, McMannis J, Gee A, et al. Transplantation of ex vivo expanded cord blood cells using the copper chelator tetraethylenepentamine: a phase I/II clinical trial. *Bone Marrow Transplant.* 2008;41:771-778.
137. Harrison DE, Stone M, Astle CM. Effects of transplantation on the primitive immunohematopoietic stem cell. *J Exp Med.* 1990;172:431-437.
138. Fraser CC, Eaves CJ, Szilvassy SJ, et al. Expansion in vitro of retrovirally marked totipotent hematopoietic stem cells. *Blood.* 1990;76:1071-1076.
139. Glimm H, Eaves CJ. Direct evidence for multiple self-renewal divisions of human in vivo repopulating hematopoietic cells in short-term culture. *Blood.* 1999;94:2161-2168.
140. Miller CL, Eaves CJ. Expansion in vitro of adult murine hematopoietic stem cells with transplantable lympho-myeloid reconstituting ability. *Proc Natl Acad Sci USA.* 1997;94:13648-13653.
141. Ema H, Takano H, Sudo K, et al. In vitro self-renewal division of hematopoietic stem cells. *J Exp Med.* 2000;192:1281-1288.
142. Yagi M, Ritchie KA, Sitnicka E, et al. Sustained ex vivo expansion of hematopoietic stem cells mediated by thrombopoietin. *Proc Natl Acad Sci USA.* 1999;96:8126-8131.
143. Bryder D, Jacobsen SE. Interleukin-3 supports expansion of long-term multilineage repopulating activity after multiple stem cell divisions in vitro. *Blood.* 2000;96:1748-1755.
144. Jiang X-S, Chai C, Zhang Y, et al. Surface-immobilization of adhesion peptides on substrate for ex vivo expansion of cryopreserved umbilical cord blood CD34+ cells. *Biomaterials.* 2006;27:2723-2732.
145. Iwama A, Oguro H, Negishi M, et al. Enhanced self-renewal of hematopoietic stem cells mediated by the polycomb gene product Bmi-1. *Immunity.* 2004;21:843-851.
146. Williams DA, Baum C. Medicine. Gene therapy – new challenges ahead. *Science.* 2003;302:400-401.

Embryonic-Like Stem Cells and the Importance of Human Umbilical Cord Blood for Regenerative Medicine

27

Colin P. McGuckin and Nicolas Forraz

27.1 Introduction

"Ethics" is one of the first words to apply in the world of stem cells. If one cannot apply ethical guides to our patients it is very hard to justify putting any medical research forward. Unfortunately, stem cells raise more ethical questions and debate than anyone ever could have expected. Despite this, there are good options for ethical and appropriate interventions with stem cells, using umbilical cord-derived tissues. Politicians and the scientific community, industrialists, religious groups, and last but not least, patients have the same common fascination with "Stem Cells" and "Regenerative Medicine." However, these terms are often confused as the same thing. Stem cells can be defined as relatively unspecialized cells at the origin of all tissues and organs in the body. In humans, stem cells can be categorized into three main sources:

1. *Embryonic stem cells*, usually derived from unwanted donated human embryos following in vitro fertilization[1]
2. *Adult stem cells,* which can be isolated from adult tissues, including bone marrow, muscles, or even fat[2]
3. *Umbilical cord-derived stem cells,* which are at an intermediate point between embryonic and adult stem cells as they are harvested from the blood present in the umbilical cord that links the baby to the placenta or from the Wharton's Jelly or placental tissues

In the laboratory, stem cells can be instructed to specialize or differentiate into many tissue types. This incredible potential led scientists and clinicians to come up with the concept of Regenerative Medicine, which aims at the repair of tissues and organs damaged accidentally or by a disease, by using stem cell or drug-assisted therapies.[3] Many ethical debates are going on in the world of stem cells, particularly with human embryonic stem cells, which requires the destruction of a human embryo in order to be propagated in a laboratory. However, our work focuses on the umbilical cord-derived stem cells, which are acceptable to every major religion and community.

27.2 The Rise of Umbilical Cord Stem Cells to Prominence

The concept of Regenerative Medicine may be relatively new, but stem cells derived from bone marrow and cord blood have a long track record in successful clinical treatments, including blood (hematopoietic) and immune system reconstitution, through what is commonly known as "bone marrow transplantation," especially for cancer patients whose bone marrow has been destroyed by chemotherapy and radiotherapy. These treatments emerged in the 1950s to become successful clinical procedures in the late 1970s with better clinical and scientific understanding of "HLA-typing" (human histocompatibility system), and led to the general understanding that the cells transplanted from a human volunteer had to be "matched" immunologically to the human recipient.[4]

However, as early as 1939, cord blood was proposed as a potential cell source for blood transfusion by Dr Halbrecht in the Lancet.[5] Of course, adult-to-adult transfusions proved to be more realistic, but it was the first record to propose that such a blood source might be usable. Then, in 1972, for the first time in the world,

C.P. McGuckin (✉)
University of New Castle Upon, Tyne, UK
e-mail: c.mcguckin@conoworld.com

N. Bhattacharya and P. Stubblefield (eds.), *Regenerative Medicine Using Pregnancy-Specific Biological Substances*,
DOI: 10.1007/978-1-84882-718-9_27,

Drs Ende and Ende reported the transfusion of eight umbilical cord blood units to repair the blood system of a 16-year-old child with leukemia, although long-term follow-up of the patient was not available.[6] This amazing treatment may not have been recommended by immunologists, but the double and triple cord blood transplants that are now carried out do show that multiple cord blood transplantations can be achieved.

Further, several research groups around the world confirmed that cord blood contained blood progenitor and stem cells that could be used for HLA-matched bone marrow transplantation.[7,8] In 1988, in a joint collaboration between Europe and the USA, a 6-year-old child from North Carolina who was suffering from Fanconi's anemia [a genetic blood disorder] received the first related HLA-matched cord blood transplantation from his baby sister at Hopital St Louis, in Paris.[9] Not only is this patient still alive and well nearly 20 years later, but his entire blood and immune system also was reconstituted by this transplant. Needless to say that many physicians remained skeptical about the potential of what was considered a waste product of labor. Nevertheless, in the few years following this case, over 60 HLA-matched related cord blood transplantations took place, encouraging Dr Wagner and colleagues at University of Minnesota, USA, to create a volunteer registry, collecting reports that demonstrated that cord blood not only reconstituted the blood system of the patients (all children), but also involved fewer complications than transplantations using bone marrow.[10] Such encouraging clinical observations, led to the potential for unrelated donor transplant and rapidly led to the creation of the world's first public cord blood bank at the New York Blood Center in 1991, which now holds one of the largest inventory of cord blood units worldwide. This bank notably provided the cord blood unit for the first unrelated HLA-matched cord blood transplant performed on a 3-year-old child treated for leukemia,[11] and started the exponential use of cord blood in therapeutic applications.

27.3 Over 20,000 Transplants Achieved

Cord blood transplant is theoretically applicable for conditions that require reconstruction of the blood and immune systems and, since the 1990s, many diseases, including cancer (e.g., lymphoma, leukemia), inherited genetic disorders such as hemoglobinophathies (e.g., sickle cell anemia, beta-thalassemia), immune deficiencies (e.g., severe combined immunodeficiency), bone marrow failure syndromes (e.g., Fanconi's or aplastic anemias) and some metabolic disorders (Krabbe's disease) have been using this therapy[12] (Table 27.1).

Hematologists, having pioneered cord blood, also demonstrated their highly conservative protectionist side, resisting change, and for a long time, cord blood transplantations mostly applied to infants, children, and adolescents because transplanters correlated transplant cell dose to the success of the transplant (e.g., survival, time to engraftment), and cord blood samples were limited in terms of harvestable volumes and cell content. Bone marrow and mobilized peripheral blood (where stem cells are collected following stimulation into the peripheral circulation) have long been preferred to cord blood to treat adults. However, data gathered in the last 15 years suggest that although cord blood causes delayed engraftment, it performs just as well as bone marrow, with the added advantage of allowing for a degree of HLA mismatching.[13] The reality of the situation, however, is that no transplant requiring myeloablation is perfect. Recently, several clinical reports have reported the successful engraftment of cord blood into adults. When considering cell dose limitations in cord blood, two options are available: (1) expansion (i.e., augmenting the cell numbers artificially), which to date has shown no proven advantage over non-expanded units for hematological conditions[14]; or (2) use of several cord blood units. Double cord blood transplants (and even triple) are increasingly becoming standard practice for adults.[15] In our previous center, in Newcastle upon Tyne, the inspiring Professor Steve Proctor and his team were left with no choice but to transplant seven cord blood units into an adult patient with relapsed acute lymphoblastic leukemia and resistant to chemotherapy, when a suitable bone marrow donor could not be found. One cord blood unit closely matched to the patient was co-supported by the other six mismatched units. Unfortunately, this patient was in remission for 8 months post-successful transplant before "minimal residual disease" occurred, causing a relapse in leukemia.[16] Current data on the application of double and triple cord blood transplants indicate that one donor cord blood eventually takes precedence in the bone marrow while the

Table 27.1 Diseases in which cord blood has been used to support or treat

Oncologic Disorders	
Acute lymphoblastic leukemia	Hemaphagocytic lymphohistiocytosis
Acute myeloid leukemia	Hodgkin's disease
Autoimmune lymphoprolifera-tive syndromes	Juvenile myelomonocytic leukemia
Burkitt lymphoma	Langerhans cell histocytosis
Chronic myeloid leukemia	Myelodysplastic syndromes
Cytopenia related to monosomy	Non-Hodgkin's lymphoma
Familial histocytosis	
Immune Deficiencies	
Ataxia telangectasia	Mucolipidosis type II
Cartilage-hair hypoplasia	Myelokathesis
Chronic granulomatous disease	Severe combined immunodeficiency
DiGeorge syndrome	Wiscott–Aldrich syndrome
Hypogammaglobulinemia	x-linked agammaglobulinemia
IKK gamma deficiency	x-linked immunodeficiency
Immune dysregulation polyendocrinopathy	x-linked lymphoprolifera-tive syndrome
Hematological Disorders	
Amegarakarocytic thrombocytopenia	Pancytopenia
Autoimmune neutropenia	Red cell aplasia
Congenital dyserythropoietic anemia	Refractory anemia
Congenital sideroblastic anemia	Severe aplastic anemia
Cyclic neutropenia	Shwachman syndrome
Diamond Blackfan anemia	Severe neonatal thrombocytopenia
Evan's syndrome	Sickle cell disorders
Fanconi anemia	Severe neonatal thrombocytopenia
Glanzmann's disease	Sickle cell disorders
Hypoproliferative anemia	Systemic mastocytosis
Juvenile dermatomyositis	Thalassemia
Juvenile xanthogranulomas	Thrombocytopenias with absent radius
Kostmann's syndrome	
Metabolic Disorders	
Adrenoleukodystrophy	Krabbe's disease
Alpha mannosidosis	Maroteau-lamy syndrome
Diabetes Mellitus, Type 1	Metachromatic leukodystrophy
Gaucher's disease (infantile)	Mucolipidosis, types II, III
Globoid cell leukodystrophy	Neimann Pick syndrome, types A and B
Gunther disease	Osteopetrosis
Hermansky–Pudlak syndrome	Sandoff syndrome
Hurler syndrome	Sanfilippo syndrome
Hurler–Scheie syndrome	Tay Sachs disease

other one/two provide support. This possibility, therefore, gives much hope that small cord blood units may be usable.

Therefore, the major milestone of transplanting cord blood was achieved, leading to more than 20,000 transplants in hematological and immunological conditions.

27.4 Cord-Derived Transplants Leave the Hematology Clinic

Although the early successes were in hematology/oncology applications, recent years have raised our hopes that cord-derived cells could be useful in other regenerative strategies. The possibility that cord blood could be just as, or perhaps more, useful for autologous therapies as allogeneic, has been a controversial subject for hematologists with a limited perspective. Such hematologists have often argued that cord-derived cells can only be used for allogeneic therapies, but the increase in sibling-related transplants within the same family have made the prospect of HLA identical or near identical transplants a reality. So now the near and longer-term future suggests that cord blood and cord stem cells can be useful to help develop novel therapeutics. Many hypotheses have been made, but only in recent years has the research and clinical paths started to come together to make possibilities a reality.

An example of new developments is the trial in Florida, USA, which has revolutionized the management of Type I juvenile diabetes (autoimmune disease destroying pancreatic beta cells, which provide insulin, the regulator of blood sugar). A team led by Drs Atkinson, Shatz, and Haller at the University of Florida in Gainesville, USA were approached by a family who had stored their child's umbilical cord, asking if the blood could be a therapeutic option for their child who had been diagnosed with Type I diabetes. The clinicians reflected on this question and figured that since they would be infusing the child's own cells (a process called autologous therapy), there was a limited risk in the procedure since no chemotherapy or immunosuppressive drugs would be necessary. This led to a full clinical evaluation registered under the number # NCT00305344 with the Food and Drug Administration, enabling the clinical team to recruit 23 more children diagnosed with Type I diabetes who had their cord blood stored in a private family bank. The preliminary analysis on the first cohort of patients is interesting. Not only were there no adverse reactions observed in the children, but the diabetes also did not progress as expected, with some patients gaining an improvement in their insulin production (thereby reducing the daily insulin injections needed).[17] Although important conclusions will be drawn once the trial is completed, these initial results fit our own observations that cord blood contains not only stem cells but also a range of other cells, including primitive immune cells, which may have a role in cell therapy.

Cord blood has also been applied to treatments where there are few or limited options, such as osteopetrosis, a rare disease of infants and young children, characterized by excessive accumulation of mineralized bone and abnormal blood cell formation. Not only did cord blood cells restore blood cell formation, but also as early as 9 months post-transplant, the patients saw a reduction in the excess bone density.[18]

A further report presented at the 2007 American Association of Blood Banks annual meeting, reported that a private family cord blood bank had released six cord blood units for clinical infusion to treat six distinct cases of neurological conditions, including cerebral palsy, anoxic brain injury, and traumatic brain injury.[19] Six samples were released to treat neurological conditions, including cerebral palsy (four samples), anoxic brain injury (one sample), and traumatic brain injury (one sample) at Chicago Children's Memorial Hospital and Duke University in the USA. These clinical interventions were not part of a specific trial, but anecdotal evidence by physicians involved with these cases suggests the treatments were safe with possible improvement in quality of life.

While these trials are exciting, it is clear that more structured trials will be necessary to confirm the preliminary investigations. However, a further debate is also ongoing in cell therapy centers as to what the appropriate placebo should be. Many centers are against the use of normal placebo-based trials where cellular intervention is concerned, although this has taken place in a heart-related intervention in Germany where patients were given no therapy though they were told that they were being trialed with cells and still had bone marrow harvests.

In 2005, the injection of HLA-matched umbilical cord blood stem cells directly at the site of a spinal cord injury in a 37-year-old female patient with functional evidence of sensory perception, hip movements, and neural regeneration in the spine was reported from Seoul National University.[20] Beyond this clinical case study, Professor Young and colleagues (Rutgers University, USA) will lead a series of phase II/phase III clinical trials to assess the potential of cord blood stem cells for spinal cord regeneration.

Scientific research is the impetus behind these clinical innovations, so it is worth demonstrating why research is so important in the field of cord blood.

27.5 Research Leads the Way in Developing Cord-Related Stem Cells

In 2005, our research group made a very significant breakthrough which led to much interest in the scientific community and the media and put cord blood at the forefront of the stem cell debate. We were the first in the world to show that it was possible to isolate, from human umbilical cord blood, pluripotent stem cells expressing markers and characteristics similar to those observed with embryonic stem cells (Fig. 27.1). This rare population of cells called Cord Blood Embryonic-LIKE stem cells had the additional advantage of being expandable in a laboratory.[21] In achieving this, we demonstrated for the first time that the so-called embryonic

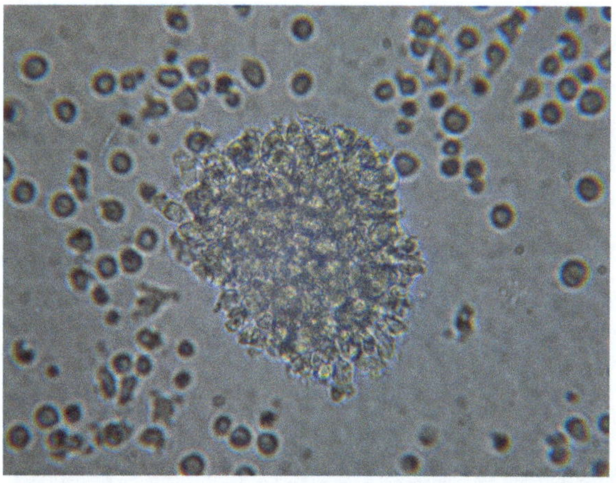

Fig. 27.1 Cord blood embryonic-like stem cells. This colony represents a standard CBE colony growing in two-dimensional culture. Positive for Oct 4, and a range of other ES markers, the cells prefer to grow in groups, but can also be grown by circulating bioreactor

stem cell markers were in fact not so, as they were also expressed in a non-embryonic cell group. Our work and observations have now been confirmed by other research groups around the world, even, amusingly, by a group in the USA, who went on to attempt to patent our invention. Further to this, just like embryonic stem cells, if not better and in a controlled fashion, we were able to turn CBEs into cells and tissues from all three germ layers. In doing so, we were also the first in the world to create hepatocyte-like cells from cord blood, something which we can also achieve from Wharton's Jelly.

Our strategy is to develop something useful for humanity, either by directly using our cells for transplant, or by developing in vitro based systems for drug development and analysis. We have characterized many different groups of cells in cord blood and cord, by applying tissue-engineering principles, combining bio- or chemo-engineered scaffolds and peptides. In doing so, we produce three-dimensional models of organ development for functional testing and analysis. Some of the tissues, including hepatic and neural, can be efficiently grown in "bioreactors" including Rotating Cell Culture Systems originally developed by NASA and produced by Synthecon in the USA, which optimize our process by reducing the effect of gravity.[22]

We are focusing our efforts on key cell types in which we see a faster route to the clinic. For instance, we were the first group in the world to produce pancreatic-like insulin-secreting cells from cord blood stem cell in vitro.[23] Such cells may be either transplanted directly in the future or used in extracorporeal devices, in the future.

The liver hepatocyte-like tissues have also been joined by our ability to produce biliary epithelium-related cells. The liver-like cells not only express liver-specific markers (e.g., cytokeratin 7, 18, 19) but they have also proved to have functional activity (e.g., cytochrome P450, albumin production). We believe this work may yield significant clinical applications because liver is a key organ in detoxification and metabolism of drugs and toxins in the body (80% of the drugs we ingest are processed by the Liver). Liver diseases affect a significant patient population and are caused by excessive alcohol/drug consumption, accidental poisoning, fibrosis, or infections. As many as 500,000 patients worldwide suffer from hepatitis and as many as 70% of patients awaiting a liver transplant never find a match. If enough cord blood was stored in banks, then it might provide the necessary tissue types to assist these patients.

The area where cord blood and cord-derived cells may make a faster impact is likely to be the neurological domain. For a number of years, we have been able to produce neural cells from cord blood stem cells.[24] We and our collaborators have shown that cord blood stem-cell-derived neural cells could be implanted successfully in the brain of rat models. These cells were also capable of spontaneously generating electrical activity known as "Action Potentials" in the laboratory, which are a key requirement to indicate capability to function[25] (Fig. 27.2).

Many other pioneering studies are taking place worldwide using cord blood stem cells as a universal vector for regenerative medicine research. Dr Lazzari and colleagues in Italy are investigating cord blood stem cells as a source for cell repair in acute renal failure.[26] Building on current clinical trials has demonstrated the potential of bone marrow to play a role in the treatment of myocardial infarct; Dr Stamm and colleagues in Germany are now involved in preclinical investigation of cord blood stem cells for cardiovascular disease.[27] Professor Revoltella and colleagues have tested in mice the potential of human cord blood stem cells to correct injury-led deafness.[28]

Such emerging studies lead to one conclusion. Cord-related therapies will increase and to cover all such therapies in the clinic we are going to need readily accessible cord blood.

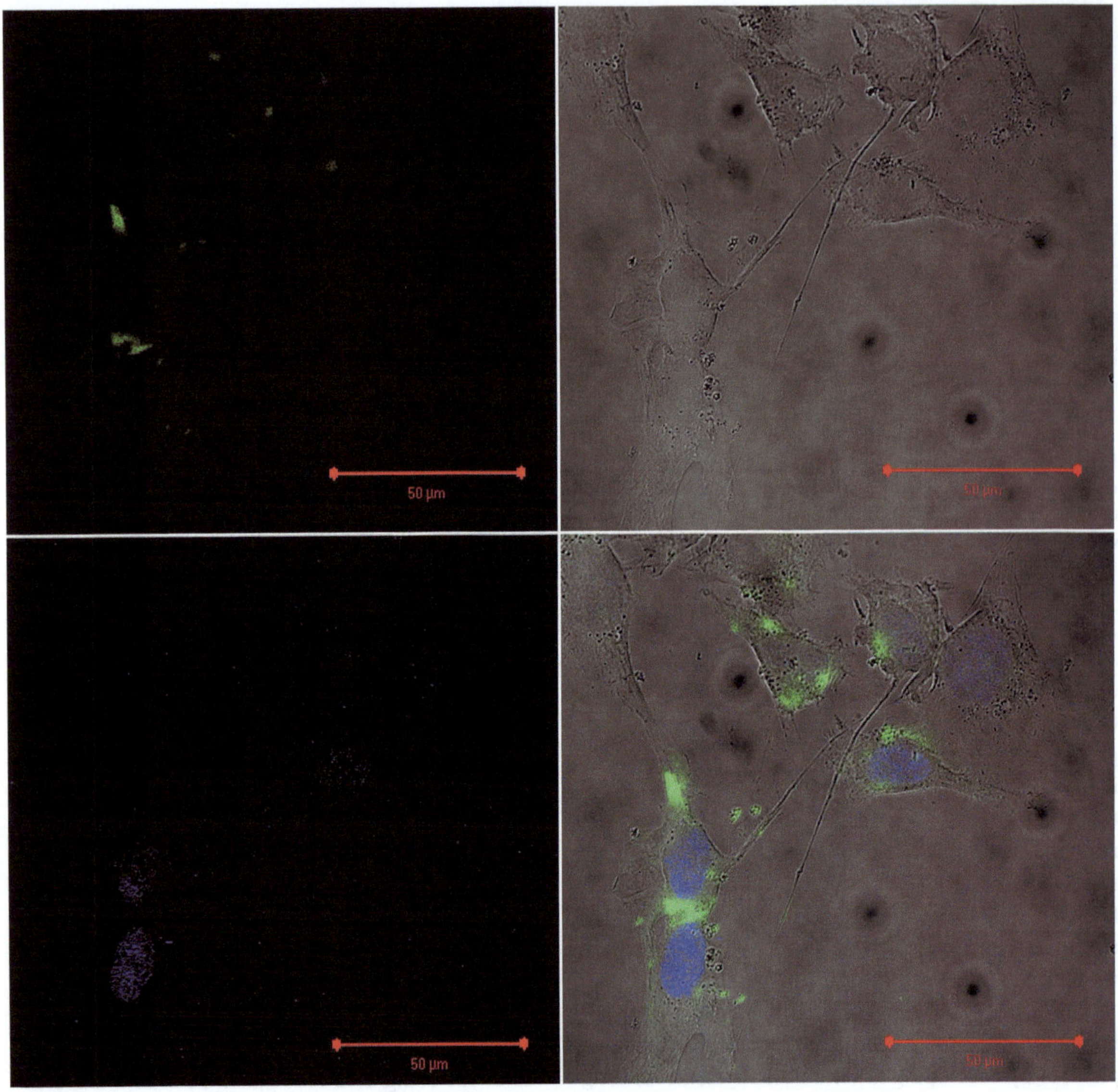

Fig. 27.2 Neural cell outgrowths in a neural cell culture developed from cord blood. A two- and three-dimensional tissue-engineering strategy for serum-free and xeno-free culture has been established in our laboratory allowing GMP-driven cell growth

27.6 Cord Blood Banking – A Controversial but Necessary Choice

The fact that over 130 million children are born every year around the world leaves no doubt that umbilical cord blood is the largest source of easy accessible stem cells, potentially offering unlimited ethnic and genetic diversity for current and future cell-based therapies. Where the law and ethics start to collide with cord blood is in terms of collection, banking, and research. Unfortunately again, some hematologists' inability to keep up with their colleagues' clinics and requirements has left a gap in what is required and what is available and with the increased number of treatments, this is set to get worse (Fig. 27.3).

Cord blood can be collected after birth by a simple puncture of the umbilical cord vein which can be performed in utero (during third stage of labor, prior to the placenta being delivered) or ex utero (by clamping and

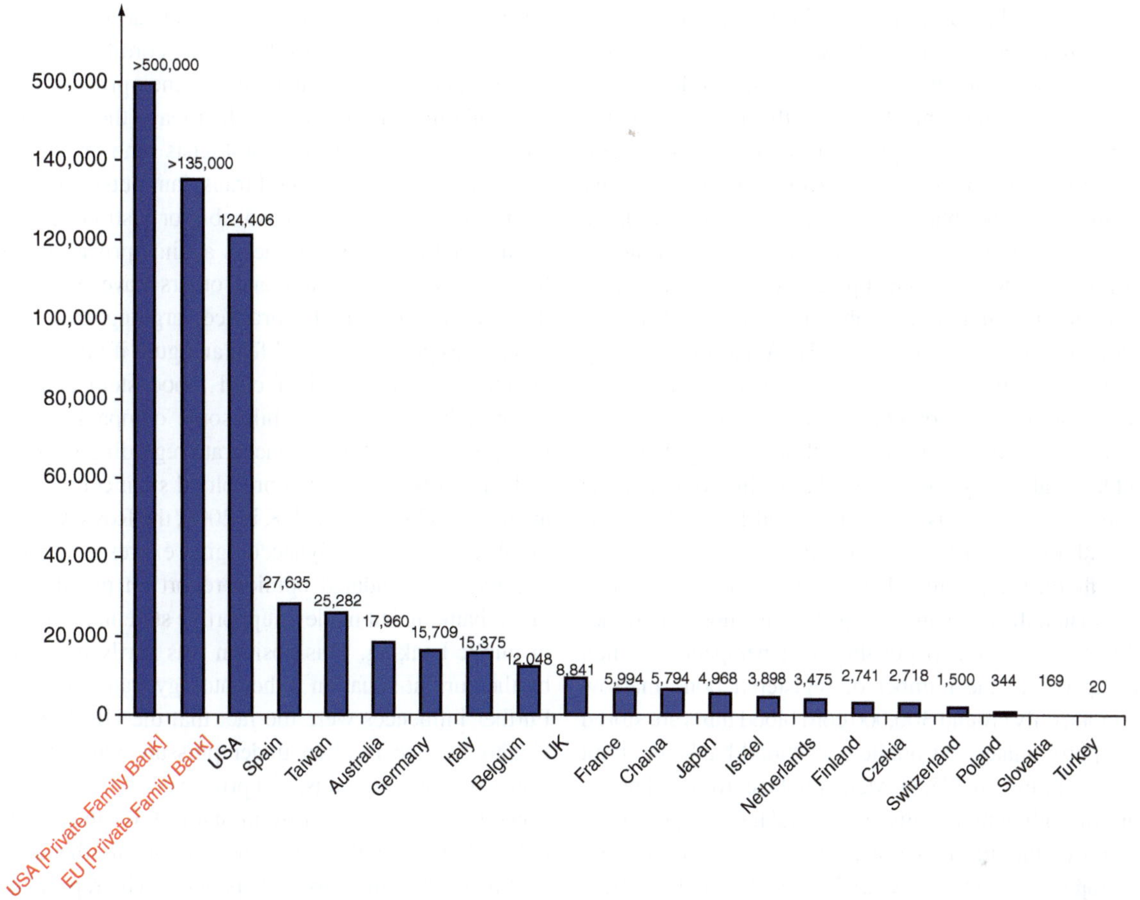

Fig. 27.3 Graphical representation of the cord blood units stored internationally (Source NMDP and personal research)

cutting the umbilical cord after birth and collecting the blood in a separate room). After appropriate training, both collections can yield similar amounts of cord blood (on average between 70–120 mL/unit).

Cord blood is processed in a laboratory through various methods and means, which consist of separating the white blood cells from the red blood cells. The white blood cell fraction contains, in very low quantity, the so-called stem cells necessary for research, but, more importantly, for sustenance of engraftment during a transplant. Processed cord blood cells are then usually cryopreserved in a controlled rate freezer and stored at −196°C in a liquid nitrogen tank with a cryopreserving mixture. This overview of cord blood processing is the basis of cord blood banking. However, this process is relatively little undertaken around the world, leaving many countries with no access to stem cells. France spent over €5 million buying cord blood from other countries in 2007 due to the lack of stored cord blood.

There are, so far, two main types of cord blood banks around the world:

1. Public cord blood banks storing anonymized units that can be released to any patients with a partial or total HLA match
2. Private "family" banks, which store samples against a fee for exclusive use by the family who pay for the service

Public cord blood banks have been developed in many countries since the 1990s, particularly after the progress of unrelated HLA-matched cord blood transplantation as more donor samples were needed to answer the demand. Public cord blood banks sometimes meet strict quality control criteria and national/international accreditation. The rationale behind public cord blood banking is that despite a world registry exceeding 11 million bone marrow donors, as many as 50% of the patients awaiting a bone marrow transplant do not find a donor, although this fact is hotly denied by some hematologists who imply that mobilized peripheral blood is a suitable alternative. However, the transplant of mobilized peripheral blood stem cells is a variable business that relies on timing of collection and is not as straightforward as it first appears. Although the lack of bone marrow donors available at the required time is partly due to the complexity of HLA and immunology matching, other considerations prevent the timely localization of a potential bone marrow donor. For instance, donors often move without leaving their new address, also they have to still be willing to donate the bone marrow but also be healthy and fit for the donation. Public cord blood banks offer the advantage of having the sample already stored in a bank, and if there is a match the unit can be shipped to the hospital immediately. However, to put things in perspective, when compared to the number of registered bone marrow donors, only around 300,000 cord blood units are stored in approximately 34 public cord blood banks in about 20 countries (Fig. 27.3). Many public cord blood banks are struggling financially and several have stopped their activity due to inconsistent financial support from national governments. Apart from public funding, these "public" banks also recover a fee (between "US $20,000 and $50,000") every time they release a unit for transplantation, which is a significant burden to the transplant hospital requiring the stem cells. In some countries, such as the UK, this is done under the name of "recouping costs," as selling human tissues for profit is not legally allowed. Unfortunately, public banks with low activity do not recover enough costs to make the bank financially stable, nor sustainable and many public banks over the years have stopped operating.

However, where debate really rages is in the ethics of private family cord blood banking.[29] Private banks charge on average between "US $1,500" and $3,000 to store umbilical cord blood samples for about 20–25 years. It is estimated that approximately 700,000 units are stored in private cord blood banks worldwide, including 500,000 in the USA and over 135,000 in Europe. Some of the criticisms are that not enough private cord blood banks are complying with regulations and accreditations for processing quality control although the regulatory framework in Europe and in the USA in particular will evolve to better prevent unreliable services. To be fair, accrediting organizations have not always been open to consider private cord blood banks who wanted to validate and improve their process, often not realizing that they would then leave those companies in a legal tangle. Hematologists who are against any form of the use of blood transplant outside of their clinics, often with little reasonable or justifiable arguments, lobby with governments against private family banking. Some clinicians and others have expressed their reservation on the practice, arguing that these companies play on parents' fear and guilt if missing out on the future potential of cord blood stem cells for regenerative medicine. While some companies' marketing strategy is highly inaccurate regarding what can and cannot be done with cord blood stem cells, others are more realistic. In the UK in 2006, the Royal College of Obstetricians and Gynaecologists expressed a rather negative and outdated opinion report on private cord blood banking but made a supportive statement in favor of public banking. This position was partly motivated by the current situation in hematology transplantation. Further influences were the fact that the UK National Health Service is often under pressure with staffing issues in maternity units, and private cord blood collection could be seen as an additional task for the personnel on duty: it conflicts with the duty of care that a free healthcare system owes to its patients. The report covered many important points, including ethics, but scandalously failed to sufficiently acknowledge emerging developments in clinical trials (particularly non-hematology trials) and regenerative medicine research and also failed in two other key areas. First, they did not invite true experts in the cord blood regenerative medicine field who might have balanced the position. Second, they did not take sufficient account of the European Law, particularly Human Rights laws because if a hospital were to refuse a cord blood collection, they would be leaving themselves open to be sued for breaking these laws and going against the parent's wishes, who are the guardians of the child's blood in the first place.

In the UK, more than once new fathers have been refused cord blood collections and they have offered to do it themselves, only for the hospital to rapidly change their mind, knowing full well the liability they would fall under. In Ireland, the author's own brother was refused upon the birth of his daughter and ultimately collected the cord blood himself.

The freedom of choice involved is clear. But also from the perspective of governments preparing for the future, it is necessary to start increased cord blood collections. Thinking long-term, as cord blood-based regenerative medicine applications are emerging, what will happen in the future when parents and children realize they were not properly informed and supported in this endeavor? Hospitals who refuse to help with collections may currently be putting themselves at risk of future liability. In the USA, this situation has happened, with the American Medical Association in 2007 now ruling in favor of informing pregnant women of the different options available to them in terms of cord blood donations, including public and private banking as well as donation to research.[30] Further to this, it is now the law in 12 American states that obstetricians and healthcare professionals have a duty to inform all pregnant women on their options with regard to cord blood banking.[31] The RCOG UK report, however, also did not sufficiently recognize that currently public cord blood banking is underfunded and underdeveloped in the UK and, indeed, most international countries provide no solutions to the problem.

One of the final and most difficult legal questions in the cord blood debate is: "Who owns the stem cells?" Because in reality the parents are the custodians of the cord blood, refusing parental wishes to store the cord blood, may well be putting the child's legal human rights in jeopardy. What is clearly called for is a true legal review of this area to evaluate whether the variable laws in different countries should be taking this into consideration.

But the question is, if cord blood therapies are indeed useful for patients, who will cover the costs of storage? In our opinion, it is totally unrealistic to expect governments to pay for public cord blood banking, in a free healthcare system. There is however, an alternative: public–private partnerships.

We strongly believe that both public cord blood banking and private family banking should be promoted in a strictly regulated environment. Currently, there are not enough cord blood units stored around the world to satisfy present clinical needs, let alone future ones. For example, if 10 fully funded public cord blood banks were to open in the UK today, they would not be able to fulfill the current requirements before 2020.

An interesting additional pattern is the slow emergence of "mixed" or "hybrid" or "public–private partnership" cord blood banks, where both autologous and allogeneic storage is promoted in a financially stable manner. This might, then, be the "happy medium" solution. Indeed, in Spain, where the cord blood of the King's grandchild was stored privately, Royal Decree has now, with governmental support, allowed public–private hybrid banking to take place.

Each country, therefore, must consider the best method of storing cord blood for its population, and with the clear knowledge that patients are waiting for it today.

27.7 Conclusion

While some governments have and are pouring millions into embryonic stem cell research with no cures, no new drugs, and no clinical trials to show for the money, cord blood therapies have already helped over 20,000. Given that the new clinical trials, not least with Type 1 Diabetes, show that there is a growing need, many people alive today could have been treated or supported if a cord blood had been stored for them. Many urgently needed public cord blood banks around the world are struggling to stay afloat despite proven and established therapies waiting desperately for samples to help patients now! To date, not one patient has been treated or cured with human embryonic stem cells. Beyond the important ethical debate on whether human embryos should or should not become a tool when developing cell therapies, it is worth pointing out to governments the patients who are waiting without hope for a therapy that they could have had, had cord blood been stored. Our research with hepatocyte-like cells, pancreatic, and also nervous tissues demonstrates the great leap forward that has been made in regenerative medicine. However, it will be difficult to put this research into practice if cord blood stem cells are not stored and we have to wait for patients when the therapy is ready to go. This is the great paradox in the use of cord blood. Some hematologists had predicted that autologous therapies with cord blood would never happen. However, they did happen, and they are happening, and many patients are now not in a position to receive the therapy because their cord blood was not stored. This is not ethical. It is time to remove cord blood from only the hematology centers and for the obstetrics units to take a greater role in the decision-making process.

Twenty years ago, cord blood could treat only one or two diseases, 10 years ago only a handful. Now, nearly 80 conditions are treatable or supportable with cord blood stem cells. Governments need to put in more effort, support, and funding into this field of medicine. The development of cord blood stem cells for therapy has been the greatest achievement of stem cell research and now patients are waiting for the therapies. We predict that in 10 years' time, liver, pancreatic, and nervous diseases will have prevention strategies related to cord blood therapy. We dream of a day when there will be cord blood banks in every metropolitan city: a dream which is in the making, and a revolution which will continue to grow.

Acknowledgments We are grateful to our staff in the Cell Therapy Research Institute, Lyon, France, and in Newcastle upon Tyne, UK, for their hard work in developing cord blood research and therapies, particularly our team who focuses on the neural cell project pictured here. We also thank the countless obstetric, gynecology, and neonatal centers that have and continue to support our work, and, last but not least, the parents who have donated cord blood to our research program.

References

1. Gluckman E, Broxmeyer HE, Auerbach AD, et al. Hematopoietic reconstitution in a patient with Fanconi's anemia by means of umbilical-cord blood from a HLA-identical sibling. *N Engl J Med*. 1989;321:1174-1178.
2. Prasad VK, Mendizabal A, Parikh SH, et al. Unrelated donor umbilical cord blood transplantation for inherited metabolic disorders in 159 pediatric patients from a single center: influence of cellular composition of the graft on transplantation outcomes. *Blood*. 2008;112(7):2979-2989.
3. Gluckman E, Rocha V. Donor selection for unrelated cord blood transplants. *Curr Opin Immunol*. 2006;18(5):565-570.
4. Laughlin MJ, Barker J, Bambach B, et al. Hematopoietic engraftment and survival in adult recipients of umbilical-cord blood from unrelated donors. *N Engl J Med*. 2001;344(24):1815-1822.
5. Migliaccio AR, Adamson JW, Stevens CE, et al. Cell dose and speed of engraftment in placental/umbilical cord blood transplantation: graft progenitor cell content is a better predictor than nucleated cell quantity. *Blood*. 2000;96(8):2717-2722.
6. Takahashi S, Ooi J, Tomonari A, et al. Post transplantation engraftment and safety of cord blood transplantation with grafts containing relatively low cell doses in adults. *Int J Hematol*. 2006;84(4):359-362.
7. Gluckman E, Rocha V, Boyer-Chammard A, et al. Outcome of cord-blood transplantation from related and unrelated donors. *N Engl J Med*. 1997;337(6):373.
8. Hogan CJ, Shpall EJ, McNiece I, Keller G. Multilineage engraftment in NOD/LtSz-scis/scid mice from mobilized human CD34+ peripheral blood progenitor cells. *Biol Blood Marrow Transplant*. 1997;3:236-246.
9. Hogan CJ, Shpall EJ, McNulty O, et al. Engraftment and development of human CD34(+)-enriched cells from umbilical cord blood in NOD/LtSz-scid/scid mice. *Blood*. 1997;90(1):85-96.
10. Kogler G, Sensken S, Airey JA, et al. A new human somatic stem cell from placental cord blood with intrinsic pluripotent differentiation potential. *J Exp Med*. 2004;200: 123-135.
11. Weiss L. The hematopoietic microenvironment of the bone marrow: an ultrastructural study of the stroma in rats. *Anatom Rec*. 1976;186:161.
12. Westen H, Bainton DF. Association of alkaline phosphatase-positive reticulum cells in bone marrow with granulocyte precursors. *J Exp Med*. 1979;150:919.
13. Lichtman MA. The ultrastructure of the hematopoietic microenvironment of the marrow: a review. *Exp Hematol*. 1981;9:391.
14. Bianco P, Riminucci M. The bone marrow stroma in vivo: ontogeny, structure, cellular composition and changes in disease. In: Beresford JN, Owens ME, eds. *Marrow Stromal Cell Culture. Handbooks in Practival Animal Cell Biology*. Cambridge, UK: Cambridge University Press; 1998:1025.
15. Dexter TM. Stromal cell associated haemopoiesis. *J Cell Physiol Suppl*. 1982;1:87.
16. Dominici M, Le Blanc K, Mueller I, et al. Minimal criteria for defining multipotent mesenchymal stromal cell. The International Society of Cellular Therapy position statement. *Cytotherapy*. 2006;8(4):315.
17. Friedenstein AJ. Precursor cells of mechanocyte. *Int Rev Cytol*. 1976;47:327.
18. Quirici N, Soligo D, Bossolasco P, et al. Isolation of bone marrow mesenchymal stem cells by anti-nerve growth factor receptor antibodies. *Exp Hematol*. 2002;30(7):783.
19. Jones E, English A, Kinsey SE, et al. Optimization of a flow cytometry-based protocol for detection and phenotypic characterization of multipotent mesenchymal stromal cells from human bone marrow. *Cytometry B Clin Cytom*. 2006;70: 391-399.
20. Zannettino A, Paton S, Kortesidis A, et al. Human multipotential mesenchymal/stromal stem cell are derived from a discrete subpopulation of STRO-1^{bright}/CD34-/CD45-/glycophorin-A- bone marrow cells. *Haematogica*. 2007;92(12): 1707.
21. Decker C, Greggs R, Duggan K, et al. Adhesive multiplicity in the interaction of embryonic fibroblasts and myoblasts with extracellular matrices. *J Cell Biol*. 1984;99:1398.
22. Choy M, Oltjen SL, Otani YS, et al. Fibroblast growth factor-2 stimulates embryonic cardiac mesenchymal cell proliferation. *Dev Dyn*. 1996;206:193.
23. Baudino TA, Carver W, Giles W, Borg TK. Cardiac fibroblasts: friend or foe? *Am J Physiol Heart Circ Physiol*. 2006;291:H1015.
24. Seshareddy K, Troyer D, Weiss ML. Method to isolate mesenchymal-like cells from Wharton's Jelly of umbilical cord. *Methods Cell Biol*. 2008;86:101-119.
25. Troyer DL, Weiss ML. Wharton's jelly-derived cells are a primitive stromal cell population. *Stem Cells*. 2008;26(3): 591-599.
26. Urbanek K, Ceselli D, Rota M, et al. Stem cell niches in the adult mouse heart. *Proc Natl Acad Sci USA*. 2006;103: 9226-9231.

27. Mazhari R, Hare JM. Mechanisms of action of mesenchymal stem cells in cardiac repair: potential influences on the cardiac stem cell niche. *Nat Clin Pract Cardiovasc Med.* 2007;4(suppl 1):S21-S26.
28. Le Blanc K, Tammik C, Rosendahl K, et al. HLA expression and immunologic properties of differentiated and undifferentiated mesenchymal stem cells. *Exp Hematol.* 2003;31(10):890-896.
29. Klyushnenkova E, Shustova V, Mosca J, et al. Human mesenchymal stem cells induce unresponsiveness in preactivated but not naïve alloantigen specific T cells. *Exp Hematol.* 1999;27:abstract 122.
30. Klyushnenkova E, Mosa JD, Zernetkina V, et al. T cell responses to allogeneic human mesenchymal stem cells: immunogenicity, tolerance, and suppression. *J Biomed Sci.* 2005;12(1):47-57.
31. Le Blanc K, Ringden O. Immunobiology of human mesenchymal stem cells and future use in hematopoietic stem cell transplantation. *Biol Blood Marrow Transplant.* 2005;11(5):321-334.

Suggested Readings

1. Le Blanc K, Rasmusson I, Gotherstrom C, et al. Mesenchymal stem cells inhibit the expression of CD25 (interleukin-2 receptor) and CD38 on phytohaemagglutinin-activated lymphocytes. *Scand J Immunol.* 2004;60(3):307-315.
2. Bartholomew A, Sturgeon C, Siatskas M, et al. Mesenchymal stem cells suppress lymphocyte proliferation in vitro and prolong skin graft survival in vivo. *Exp Hematol.* 2002;30(1):42-48.
3. Lazarus HM, Koc ON, Devine SM, et al. Cotransplantation of HLA-identical sibling culture-expanded mesenchymal stem cells and hematopoietic stem cells in hematologic malignancy patients. *Biol Blood Marrow Transplant.* 2005;11(5):389-398.
4. Aggarwal S, Pittenger MF. Human mesenchymal stem cells modulate allogeneic immune cell responses. *Blood.* 2005;105(4):1815-1822.
5. Lietchy KW, MacKenzie TC, Shaaban AF, et al. Human mesenchymal stem cells engraft and demonstrate site-specific differentiation after in utero transplantation in sheep. *Nat Med.* 2000;6(11):1282-1286.
6. Grinnemo KH, Mansson A, Dellgren G, et al. Xenoreactivity and engraftment of human mesenchymal stem cells transplanted into infracted rat myocardium. *J Thorac Cardiovasc Surg.* 2004;127(5):1293-1300.
7. Arinzeh TL, Peter SL, Archambault MP, et al. Allogeneic mesenchymal stem cells regenerate bone in a critical-sized canine segmental defect. *J Bone Joint Surg Am.* 2003;85-A(10):1927-1935.
8. Murphy JM, Fink DJ, Hunziker EB, Barry FP. Stem cells therapy in a caprine model of osteoarthritis. *Arthritis Rheum.* 2003;48(12):3464-3474.
9. Mahmud N, Pang W, Cobbs C, et al. Studies on the route of administration and role of conditioning with radiation on unrelated allogeneic mismatched mesenchymal stem cell engraftment in a nonhuman primate model. *Exp Hematol.* 2004;32(5):494-501.
10. Haylock DN, Nilsson SK. Stem cell regulation by the hematopoietic stem cell niche. *Cell Cycle.* 2005;4(10):1353-1355.
11. Briddell R, Kern BP, Zilm KL, et al. Purification of CD34+ cells is essential for optimal ex vivo expansion of umbilical cord blood cells. *J Hematotherapy.* 1997;6:145-150.
12. McNiece I, Briddell R, Stoney G, et al. Large scale isolation of CD34+ cells using the Amgen cell selection device results in high levels of purity and recovery. *J Hematotherapy.* 1997;6(1):5-11.
13. Majhail NS, Brunstein CG, Wagner JE. Double umbilical cord blood transplantation. *Curr Opin Immunol.* 2006;18(5):571-575.

Use of Non-hematopoietic Stem Cells of Fetal Origin from Cord Blood, Umbilical Cord, and Placenta in Regeneration Medicine

28

Zygmunt Pojda

28.1 Introduction

During consecutive fetal development stages, many specific stem cell populations arise, expand, migrate, and differentiate, participating in the formation of the fetus, umbilical cord, and placenta. Some of them may be found only in the organs of developing fetus; others are present in cord blood or reside in umbilical cord and placenta. Collection of cells from fetuses is ethically questionable, and the material from aborted fetuses is heterogeneous both for its biological (poor cell viability, presence of pathogens) and genetic qualities. The only sources of fetal stem and progenitor cells, not causing technical or ethical problems, are cord blood, umbilical cord, and placenta, being the "biological waste material" after newborn's delivery. The developmental age of cells present in cord blood, and even more cells residing in umbilical cord and placenta tissues, may vary – some of them may be the remnants of the cells migrating at the early stages of fetus development, residing in the tissues in dormant, noncycling state.

The history of the research on the fetal stem cells resembles the history of stem cell research and clinical applications in adults: the fetal hematopoietic stem cells were those which were earliest discovered and introduced into practical clinical applications. The development of the methods used for the morphological and functional characterization of tissue-specific stem cells allowed for the detection of fetal cells which did not fit the profile of hematopoietic stem cells. The newly detected cells were characterized by different than hematopoietic stem cells surface marker profile, different morphology, in vitro culture requirements, and the multipotential differentiation capability. Since it is unclear, if some categories of cells fulfill all criteria obligatory for "stemness" it is reasonable the caution in using the term "stem" when characterizing the new, not fully yet characterized cell populations capable of multipotential differentiation. It is now also impossible to systematize all stem and progenitor cells described by various authors in cord blood, umbilical cord, and placenta. Some criteria, as surface marker composition or cell morphology in comparable in vitro culture conditions may be used as relatively objective parameters, but the others, like the differentiation potential, may vary depending on the method of research. As a result, the cell described for example as the "neural stem cell" by authors using techniques favoring the neuropoietic stimulation, may be the same as characterized as multipotential stem cell by the others, who use the tests allowing for differentiation into several other tissue lineages.

Fetal stem cells, like their adult counterparts, are able to differentiate into several tissues, and migrate to the site of tissue injury or to the location of cancer cells. They may form the new cells replacing those destroyed by injury or illness, and also modify the healing process by regulating the activities of local cells. Some stem and progenitor cell populations have strong immunomodulatory capabilities. When compared to adult stem cells, fetal cells have greater proliferation and differentiation potential (longer telomeres and telomerase activity), accumulate lesser amount of DNA lesions, and cause lower risk of transmission of pathogens during transplantation. Both the good availability and the biological characteristics of fetal cells from cord blood,

Z. Pojda
Department of Experimental Hematology, Maria Sklodowska-Curie Memorial Cancer Center and Department of Regenerative Medicine, WIHiE Institute of Hygiene and Epidemiology, Warsaw, Poland
e-mail: zpojda@coi.waw.pl

N. Bhattacharya and P. Stubblefield (eds.), *Regenerative Medicine Using Pregnancy-Specific Biological Substances*,
DOI: 10.1007/978-1-84882-718-9_28,

umbilical cord, and placenta make them the prospective tool for the future clinical applications.

28.2 Non-hematopoietic Cord Blood Stem Cells

Early phase of cord blood stem cell research, started by Broxmeyer,[8-10] and subsequently culminated with the first clinical allogeneic cord blood transplantation,[39] was concentrated on hematopoietic stem cells. Early attempts of identification of non-hematopoietic (mesenchymal stromal cells, MSC) stem and progenitor cells in cord blood collected during mature deliveries were unsuccessful.[42,92,148] Demonstration of the presence of MSC in the first trimester cord blood[12,13] augmented the hypothesis that cord blood MSC may play the physiological role at the early stages of fetus development, being nonexistent at the age of birth. Although the early experiments have shown the beneficial effects of in vivo transplantation of mononuclear cord blood cells into the site of injury (brain ischemia),[20] the results could be easily explained by the plasticity of hematopoietic cells. The progress in the identification and characterization of the other than hematopoietic stem cells in fetal tissues resulted in characterization of multiple stem and progenitor cells which may be translated into clinical applications.

28.2.1 Tissue Specific Monopotent Stem Cells in Cord Blood?

Several in vitro/in vivo experiments allowed for the assumption, that there exist the monopotential, tissue-specific cord blood stem or progenitor cells, capable of differentiating into single lineage – the most numerous studies concerned the neural progenitor cells.[5,11,47,54,125] Other studies revealed the existence of cord blood cells capable of differentiation to hepatocytes,[45,58,63] skeletal muscles,[36,72] myocardium,[79] dendritic cells,[40] cartilage,[34] or bone.[49,121] Basing on the more contemporary results it is, however, not possible to eliminate the probability, that these so-called tissue-specific cells were in fact the multipotential cells (MSC?) given no alternative than to differentiate into the single lineage, according to the specific experimental conditions.[136]

28.2.2 Unrestricted Somatic Stem Cells (USSC)

In 2004, novel, pluripotent stem cells from cord blood have been characterized.[70] These cells are present in a low frequency in cord blood, but after successful collection can be expanded to 10^{15} cells in in vitro culture. USSC, similarly as MSC, are spindle-shaped, plastic-adherent cells, which can be cultured in FCS-supplemented media without any growth factor addition. Telomere length, longer than of adult bone marrow-derived MSC, decreases along the cell divisions. Pluripotentiality of USSC was confirmed both in in vitro and in vivo experiments. In vitro they showed differentiation ability into osteoblasts, chondroblasts, adipocytes, and hematopoietic and neural cells. In vivo experiments, based on the implantation of USSC into various animal species, confirmed the differentiation ability into bone, cartilage, neural cells, and, additionally, capability of formation of parenchymal hepatic cells and cardiomyocytes. When transplanted in a noninjury model into preimmune fetal sheep, USSC participated in 5% of hematopoiesis. Animal experiments did not show any tendency for tumor production by transplanted cells. The expression of surface markers by USSC is, in general, similar to this which is characteristic for MSC. Cells are negative for hematopoietic cell-specific markers, and positive for MSC and pluripotent cell markers (CD14–, CD33–, CD34–, CD45–, CD49b–, CD49c–, CD49f–, CD50–, CD62E– CD62L–, CD62P–, CD106–, CD117–, glycophorin A–, HLA-DR-, and CD13+, CD29+, CD44+, CD49e+, CD90+, CD105+, vimentin+, cytokeratin 8 and 18+, low CD10 and HLA-A,B,C. USSC differs from adult bone marrow MSC in the expression of CD50, CD62L, CD106, and HAS1 – all these markers are present on MSC. In contrary, epithelial markers cytokeratin 8 and 18, and the endothelial marker KDR, are expressed on USSC and absent on adult MSC. USSC may be cultured in vitro throughout over 20 passages. The major disadvantage of USSC, as the pluripotent cells for clinical applications, is their relatively (35% of cord blood collections) low frequency in cord blood from mature deliveries.[71,73]

28.2.3 Cord Blood Mesenchymal Stromal Cells (MSC)

The preliminary attempts to identify the cells fulfilling the officially accepted[25,46] minimal criteria for defining mesenchymal stem cell (MSC) (in vitro plastic-adherent, expressing CD105, CD73, CD90, and lacking the expression of CD45, CD34, CD14 or CD11b, CD79a or CD19, HLA-DR surface molecules, and differentiating to osteoblasts, adipocytes, and chondroblasts) were disappointing – the authors were unable to identify any cord blood MSCs,[42,92,148] or were finding them only in the first trimester[12,13,41] or at latest in mid-trimester (up to 26 weeks)[159] of fetal age. First trimester MSC can be cultured in vitro for at least 20 passages (mean cumulative population doubling of 50.3 ± 4.5). Cells express CD29, CD44, SH2, SH3, and SH4, produce prolyl–4-hydroxylase, α-smooth muscle actin, fibronectin, laminin, and vimentin, and are CD45–, CD34–, CD14–, CD68–, vWF-, and HLA-DR-. In in vitro culture conditions they express the pluripotency stem cell markers Oct-4, Nanog, Rex-1, SSEA-3, SSEA-4, Tra-1-60, and Tra-1-81. In comparison with adult bone marrow MSC, fetal MSC have longer telomeres, greater telomerase activity, express more TERT, faster expand, and later senesce.[41] First trimester fetal blood MSC are able to differentiate into chondrocytes, osteocytes, and adipocytes, and are able to support the cord blood CD34+ cells in long-term culture.[41] Similarly, the mid-trimester fetal blood MSC can also differentiate to adipo- chondro- and osteogenesis, and their phenotype is CD34–, CD45–, CD44+, CD71+, CD90+, and CD105+.[159] The discrepancy between the early findings of MSC presence only in early fetal age,[12,13,41] and lately published data on the MSC presence in mature cord blood[6,73,91,111] can be easily explained by the improvement in techniques of MSC detection and in vitro culture. On the other hand, assuming, that the more discriminating MSC analysis techniques were used formerly, it may be suggested that there exists the quantitative or qualitative difference in first trimester fetal blood MSC and cord blood MSC resulting in the identification of MSCs in first trimester blood with the same research tool which fails to detect them in mature delivery cord blood.

MCSs can be isolated from the term delivery cord blood, albeit the efficiency of their successful isolation is relatively low (20–50% of collections), and the probability of successful in vitro expansion is even much lower.[91,155] The protocols for collection and expansion of cord blood MSC may probably be further optimized,[6,76,77,111] but it seems that the reduction in numbers of fetal blood MSC along with fetal age is the real biological phenomenon, and the frequency of MSC in term delivery cord blood may remain the main obstacle for clinical application of cord blood mesenchymal stromal cells. Cord blood MSCs are spindle-shaped cells in vitro, grow plastic-adherent, and express the surface markers typical for MSC cell family (Table 28.1).

Mesenchymal stromal cells of cord blood origin express several cytokines, both interleukins and growth factors. Under appropriate stimulation, MSC secrete ENA-78, GM-CSF, GRO, IL-1β, IL-6, IL-8, MCP-1, OSM, VEGF, FGF-9, GCP-2, IGFBP-1, IGFBP-2, IGFBP-3, IGFBP-4, IP-10, LIF, MIF, MIP-3α, osteoprotegerin, PARC, PIGF, TGF-β2, TGF-β3, TIMP-1, and TIMP-2. Expression of IL-4, IL-5, IL-7, IL-13, TGF-β1, TNF-α, and TNF-β was not observed under normal culture conditions.[83,104] The ability of reaction to multiple environmental signals, and response by expression and secretion the large repertoire of

Table 28.1 Expression of cellular markers during *in vitro* cord blood mesenchymal stromal cell culture

Marker category	Surface antigens	References
Mesenchymal/hematopoietic	CD29+, CD44+, CD49b+, CD49e+, CD51+, CD54+, CD51/61+, CD58+, CD71+, CD73+, CD90+, CD105+, CD106+, HLA-ABC+, SH-2+, SH-3+, SH-4+	
	CD3–, CD7–, CD19–, CD31–, CD33–, CD34–, CD45–, CD14–, CD11a–, CD62L, CD62P–, CD102–, CD117–, CD133±, CD135–, CD166–,	Gang et al.,[35] Lee et al.,[78] Manca et al.,[91] Tondreau et al.[139]
Embryonic/pluripotency/others	SSEA-3+, SSEA-4+, Tra-1-60+, Tra-1-81+	
	Oct-4+, Nanog+, Rex-1+	

intercellular signaling factors, may explain the role of MSC in tissue repair and immunomodulation even in such situations, when MSC is not repairing tissues by itself by production of the progeny directly replacing the affected cells.

Cord blood MSCs differentiate into three basic lineages: adipo-, chondro-, and osteogenesis, and, depending on experimental design, are able to differentiate into neuronal/glial cells,[26,61,63,112,139] hepatocytes,[63,78] skeletal muscle,[107,112] myocardium,[98] and hematopoietic microenvironment.[53]

28.2.4 Cord Blood Endothelial Progenitors

Endothelial progenitor cell (EPC), characterized by the expression of CD34, KDR (VEGFR-2), and CD133 markers, is an important tool for treatment of diseases caused by insufficient vascularization (limb ischemia, myocardial infarction, etc.).[101,117] EPCs can be isolated both from adult peripheral blood and cord blood.[100] Number of EPCs per equivalent blood volume, measured as the number of EPC-formed plastic adherent colonies, is increased 15-fold in cord blood compared with adult samples. Cord blood cell colonies emerge in culture 1 week earlier than adult-derived, and are consistently larger.[51] This observation leads to the conclusion that adult blood-derived EPCs differ from their cord blood-derived counterparts, belonging to the different cell population, or the differences may fit to the general picture, where fetal stem and progenitor cells have greater proliferative potential than their adult counterparts. Cord blood EPC, similarly as adult cells, express endothelial cell surface antigens CD31, CD141, CD105, CD146, CD144, VWF, and flk-1,[51] but not hematopoietic markers CD45 and CD14. Early observations[117] suggested that endothelial progenitors express the antigens CD34, AC133, or CD117, which are common with hematopoietic progenitor cells. Ingram,[51] however, observed only small percentage of EPC expressing these markers, and did not find any hematopoietic activity of these cells in in vitro assays. Contrary to the adult EPS, cord blood EPCs retain high levels of telomerase activity along the serial passages in culture. In an attempt to characterize the subpopulation of EPC having the greater proliferative potential, Nagano et al.[103] subdivided EPC into two populations according to their aldehyde dehydrogenase (ALDH) activity. They found that EPCs with low ALDH activity (Alde-Low) possess a greater ability to proliferate and migrate, when compared to cells with high ALDH activity (Alde-High). Also hypoxia-inducible factor proteins are up-regulated, and VEGF, CXCR4, and GLUT-1 mRNAs are increased in Alde-Low under hypoxic conditions, while the response of Alde-High is insignificant. In animal experiment, they found that Alde-Low EPC significantly reduced tissue damage caused by ischemia in a mouse flap model.

Endothelial progenitor cells from cord blood are the promising tool for potential clinical applications, but before then, they need to be better characterized, and the techniques of their isolation and in vitro expansion have to be optimized.

28.2.5 Embryonic-Like Stem Cells in Cord Blood

The concept of "embryonic-like" stem cells is older than the search for such cell population in cord blood. In 2001, Vacanti et al.[140] published the description of so-called spore-like cells, which had to be present in all tissues in the dormant state until activated by injury or disease to regenerate the affected tissue. These cells had to survive extremely unfavorable conditions (extreme temperatures, oxygen deprivation), being very small (<5 μm) and containing predominantly nucleus with scanty cytoplasm and few mitochondria. Such cells had to survive dormant until old age, but the logical hypothesis was that they should be more numerous in fetal tissues. Some groups indeed reported the cells of phenotype resembling those of "spore-like cells." First, McGuckin et al.[94-96] described the pluripotent cells of 2–3 μm diameter, lacking the expression of CD45, CD33, CD7, and CD235a, positive for TRA-1-60, TRA-1-81, SSEA-4, SSEA-3, Oct-4, but not SSEA-1. Subsequently, Ratajczak et al.[74,75] demonstrated the presence in cord blood of very small cells (CB-VSEL), 3–5 μm diameter, of phenotype CXCR4+, Oct-4+, SSEA-1+, Sca-1+, lin-, CD45–, expressing transcription factors Oct-4 and Nanog, and several markers for skeletal muscles, heart, neural tissue, liver, pancreas, epidermis, and intestinal epithelium.

The presence and characteristics of the embryonic-like stem cells in cord blood need further research before concluding about their possible clinical utility[113]. Since, however, these cells are present dormant and unchanged also in adult tissues, it seems very probable that even if they find the way to clinics, they will be collected from adult patients, so it will be possible to use them both for autologous and allogeneic applications.

28.2.6 Cord Blood Stem Cells – Many Categories of Stem Cells with Different Characteristics or Different Approaches to Characterize the Same Cells?

Both, recent and more historical literature on cord blood stem cells, do not allow for the "Carl Linnaeus" task of systematization and organization of the hierarchy of cells. When constructing any "Systema Naturae," there is a need of information obtained according to the same criteria. The stem cell characterization, however, was done using various laboratory designs, differing in the conditions allowing for cell differentiation. As a result, it is impossible to compare the cells of similar morphology on the basis of their pluripotentiality – the same cell in different culture conditions may differentiate into varying numbers of lineages. An example may serve the elegant comparison of USSC and cord blood MSC published by Kogler et al.[71] The authors suggest that USSC can differentiate to more lineages than "classical" MSC (not only adipo-, chondro-, and osteopoietic, but also neuro-, cardio-, and hematopoietic). There are, however, numerous papers[73,78,139,160] describing also the wider spectrum of differentiation of cells called by the authors "cord blood MSC," and probably the only difference would be production by USSC of hematopoietic cell progeny. This is only the example of common situation, when it seems to be preferred to describe the new and newly named cell (see also the number of names for umbilical cord MSC) – maybe some strategy of "Occkham razor" would simplify the categorization of cord blood stem and progenitor cells?

28.3 Umbilical Cord Stromal Cells

Umbilical cord plays in fetal age a role of structure connecting the fetus with placenta, allowing to transfer gases, metabolites, regulatory factors, and nutrients from mother to fetus and vice versa. All fetal blood circulates through umbilical cord and placenta blood vessels, offering the easy opportunity for homing into cord and placenta tissues of wide variety of cells. All components of the cord are of fetal origin. The main components of umbilical cord are two arteries and one vein, surrounded by connecting tissue called Wharton's jelly. The cord is covered by the epithelium developed from amnion. Umbilical cord stromal cells (UCSCs), called by the various authors umbilical cord-mesenchymal stem cells (UC-MSCs),[110] mesenchymal stem cell-like cells,[120] umbilical fibroblast-like cells,[156] human umbilical cord perivascular cells (HUCPVCs),[3] human umbilical cord matrix cells (HUCMCs),[2] umbilical cord stromal cells (UCSCs),[64] umbilical cord matrix stem cells (UCMS cells),[116] or umbilical cord-derived stem cells (UCDS cells),[150] are the cells of morphological, proliferative, and pluripotential characteristics of fetal mesenchymal stromal cells. UCSCs can be collected from all structures of umbilical cord: Wharton's jelly,[99,146] perivascular regions,[3,128] and subamnion[64,65]; MSC-like cells are also present in the subendothelium of umbilical vein.[120] It has been suggested that cells residing in Wharton's jelly are more primitive and have greater proliferative capacity than perivascular cells[64] whereas umbilical vein MSCs resemble by their differentiation potential rather adult bone-marrow-derived MSCs.[24,67,120] Techniques used for cell isolation may be divided into those which include enzymatic digestion of umbilical cord structures,[64,65] or simple mincing of MSC-containing tissues (Wharton jelly) and allowing for the expansion of cells in plastic-adherent way in in vitro culture.[99] UCSCs, similarly to the other MSCs, grow in plastic-adherent manner in serum-supplemented medium, and do not need any additions of growth factors.[64,99] Contrary to the cord blood, collections of MCS from umbilical cords are close to 100% successful, and large quantities of cells may be collected and expanded in vitro for up to 30 passages, mean doubling time 24 h.[86] The comparison of the efficiency in obtaining MSCs from match-paired cord blood and umbilical cord samples, processed simultaneously and under the same culture conditions,

Table 28.2 Expression of cellular markers during *in vitro* umbilical cord mesenchymal stromal cell culture

Marker category	Surface antigens	References
Mesenchymal/hematopoietic	CD13+, CD29+, CD44+, CD105+, SH2+, CD106+, CD73+, SH3+, CD166+, HLA-ABC+	
	CD14–, CD34–, CD38–, CD45–, CD31–, HLA-DR-	Hoynowski et al.,[48] Lu et al.,[86] Qiao et al.[115]
Embryonic/pluripotency/others	SSEA4+, TRA-1-60+	
	Oct4+, Sox-2+, Nanog+	

resulted in 10% efficiency of MSC cord blood collection, and 100% success of the isolation of MSC from umbilical cords.[132] Although the experiment was based on 10 matched pairs of cord blood/umbilical cord only, the results illustrate the differences in the availability of MSCs routinely collected from both sources. UCSCs express MSC-type antigen repertoire (Table 28.2).

UCSCs express the mRNA, or secrete the cytokines SCF, LIF, M-CSF, Flt-3, IL-6, GM-CSF, G-CSF, SDF-1, and VEGF-1, but not IL-3,[86] and express LIFr, and TERT.[147] In vitro differentiation potential of UCSC, apart of the (obligatory for MSC) osteo-, chondro-, and adipogenesis, include differentiation into heart muscle,[57,144] nerve and glial cells,[88,99,144] hepatocytes,[14] and skeletal muscle.[23,69] Differentiation pattern of umbilical cord cells allowed for successful tissue engineering of artificial blood vessels and heart valves.[7,44,130] Scarce data suggest that umbilical cord may be also the source of progenitor cells, which, after proper stimulation in vitro by VEGF and bFGF, express endothelial-specific proteins, such as PECAM and CD34, and form the cells of endothelial characteristics both in in vitro and in vivo models.[150]

28.4 Placental Stem Cells

The complex structure of placenta includes the tissues both from fetal and maternal origin. Cells isolated from placental tissue are verified by testing their fetal origin by the techniques able to detect less than 1% of maternal contamination. The cell categories isolated from placenta depend on the region of collection: amniotic epithelial cells from amniotic epithelial region, amniotic mesenchymal stromal cells from amniotic mesenchymal region, chorionic mesenchymal stromal cells and chorionic trophoblast cells from chorionic mesenchymal and chorionic trophoblast regions, respectively. Out of these cells, amniotic and chorionic mesenchymal stromal cells represent the characteristics similar to the adult bone marrow mesenchymal stromal cells (MSCs) both in aspect of their in vitro growth characteristics, surface antigen expression, and the differentiation potential. Both cell types are hematopoietic markers-negative (CD34–, CD45–), HLA-DR$^-$, and positive for markers attributed for MSC: CD73, CD90, and CD105. The characteristics of the amniotic epithelial cells is somewhat more complex – they are able to proliferate shorter than MSCs in in vitro culture (two to six passages), proliferate only in higher densities in presence of epidermal growth factor (EGF), and change the expression of selected markers (HLA-A,B,C, CD90) depending on the culture time. This latter phenomenon may suggest that the amniotic epithelial cells are a heterogeneous population being subsequently selected to higher homogeneity by culture conditions.

28.4.1 Mesenchymal Stromal Cells from Amniotic and Chorionic Regions

Amniotic MSCs are isolated from amnions at any gestation stage when placenta is fully developed. Amnion must be carefully dissected from fetal membranes to avoid the presence of maternal cells. The most popular isolation protocol is based on the two-step digestion procedure, first with trypsin, and subsequently with collagenase.[15] Resulting plastic-adherent cells can be expanded in in vitro culture similarly as adult bone marrow MSCs.[1] Cells at second passage stage express mesenchymal, but not hematopoietic markers (Table 28.3).

Chorionic MSCs are isolated from chorions or chorionic villi.[161] The trophoblast layer is removed mechanically and digested with dispase; subsequently,

Table 28.3 Expression of cellular markers during *in vitro* amniotic mesenchymal stromal cell culture (>2nd passage)

Marker category	Surface antigens	References
Mesenchymal/hematopoietic	CD10+, CD13+, CD 29+, D44+, CD49c+, CD49d+, CD49e+, CD54+, CD73+, CD90+, CD105+, CD166+	
	CD3–, CD14–, CD31–, CD34–, CD45–, CD133–, HLA-DR-	
Embryonic/pluripotency/others	SSEA-3+, SSEA-4+, Tra-1-60+, Tra- 1-81+, SSEA-1-, CD140b+, CD349+, D324–	Alviano et al.,[1] In't Anker et al.,[52] Portmann-Lanz et al.,[114] Soncini et al.,[135] Wolbank et al.,[149] Yen et al.[157]
	Oct-4+	

Table 28.4 Expression of cellular markers during *in vitro* chorionic mesenchymal stromal cell culture (>2nd passage)

Marker category	Surface antigens	References
Mesenchymal/hematopoietic	CD10+, CD13+, CD 29+, CD44+, CD49e+, CD73+, CD90+, CD105+, CD14-, CD 34–, CD45–, CD117–, CD133–, HLA-DR-	
Embryonic/pluripotency/others	SSEA-4+, CD349+, CD140B+, CD324–	Battula et al.,[4] Portmann-Lanz et al.,[114] Soncini et al.[135]

chorion is treated with collagenase alone[114] or collagenase supplemented with DNAse.[135] Markers expressed on chorionic MSC in in vitro culture are listed in Table 28.4.

Animal in vivo experiments confirmed that human placental MSCs, transplanted into animals, are able to migrate into various organs: bone marrow, thymus, spleen, lung, liver, spinal cord[152] brain[164], and kidney.[135] In vitro these cells are able to differentiate into cartilage,[1,114,135] bone,[1,52,114,124,135,145,163] fat tissue,[1,114,135] skeletal muscles[50,114,135] heart muscle,[141,163] epithelium,[1] nerve cells,[114] or pancreatic islets.[145]

28.4.2 Amniotic Epithelial Cells

The amniotic epithelial cells are isolated following stripping of amniotic membrane from chorion by trypsin digestion.[97] Such procedure allows for selection of relatively homogeneous cell suspension, which attach to plastic in in vitro culture. Contrary to MSC-type cells, amniotic epithelial cells need the addition of EGF growth factor to DMEM medium supplemented with FSC.[97] Cells grow throughout two to six passages, displaying typical epithelial morphology. Both the expression of CD90 antigen and HLA-A,B,C human leukocyte antigens increase in culture – the initial expression levels are too low for using these antigens as identification/selection markers for freshly isolated epithelial cells. Among the other markers (listed in Table 28.5) cells express molecular markers of pluripotent stem cells (SRY-related HMG-box gene SOX-2, octamer-binding protein 4 Oct-4, and Nanog[97]). Contrary to placental MSCs, epithelial cells do not express Cd49d marker. Both the molecular markers, and differentiation experiments suggest that amniotic epithelial cells are pluripotent, having adipogenic,[50,114] osteogenic,[114,149] chondrogenic,[50] but also myogenic,[114] cardiomyogenic,[60,97,141] neurogenic,[50,60,97] panrceato-,[97,114] and hepatogenic[60,97,137] potential.

Placenta is composed of both maternal and fetal tissues. Isolation of placental stem cells technically does not differ significantly from isolation of umbilical cord (mechanical excision and enzymatic digestion), but demands the scrutiny in avoiding of admixture of maternal cells into the collected material.

28.5 Possible Clinical Applications of Non-hematopoietic Fetal Stem Cells

Non-hematopoietic fetal stem cells have several advantages over adult stem cells when considering their potential clinical use. Those residing in organs and

Table 28.5 Expression of cellular markers during *in vitro* amniotic epithelial cell culture (>2nd passage)

Marker category	Surface antigens	References
Mesenchymal/hematopoietic	CD10+, CD13+, CD29+, CD44+, CD49e+ CD49d-, CD73+, CD90+, CD105+, CD117±	
	CD14−, CD34−, CD45−, HLA-DR-	
	Oct-4+, SOX-2+, Nanog+	Casey and MacDonald,[15] Miki et al.,[97] Portmann-Lanz et al.,[114] Wolbank et al.[149]
Embryonic/pluripotency/others	SSEA-3+, SSEA-4+, Tra 1-60+, Tra 1–81+, SSEA-1-, CD324+, CD140b+, CFC-1+, POU5F1+, DPPA3+,PROM1+,PAX6+,GCTM2+,FOXD3−,GDF3−	

tissues, which are discarded during delivery (cord blood, umbilical cord, and placenta), can be easily isolated without any ethical or medical contraindications. Generally, all fetal cells, when compared to their adult counterparts,[18,66,68,96,102,118,133,142,154] have greater proliferative potential, comparable or better differentiation potential, and, having longer telomeres and persistent telomerase activity, are not so limited in division numbers as adult cells, which makes easier the task of their in vitro expansion. The gene expression profile of fetal MSC is, however, similar to the profile of adult MSC.[55] When comparing the frequency of successful collections of non-hematopoietic fetal cells, their best source seems to be the umbilical cord; cord blood from pair-matched mature deliveries allowed for tenfold lower frequency of successful collections of these cells.[132] Placenta allows for collection of large numbers of cells with better than cord blood success rate, but the protocols of cell isolation need more experienced surgery and multistep enzyme digestion, and the presence in placenta of cells of maternal origin demands additional tests preventing accidental collection of admixture of non-fetal maternal cells.

Cord blood, umbilical cord, and placenta are the sources of cells of very similar characteristics: mesenchymal stromal-like cells, endothelial progenitors, putative monopotent tissue-specific cells, and maybe embryonic-like cells, the latter being until now very poorly characterized.[114] Embryonic-like cells are supposed to be present also in adult tissues throughout all lifetime, so it seems that collecting them at the fetal age is not the optimum solution (much easier is to use them for both autologous and allogeneic clinical applications after isolation from adults). Similarly to the former cells, the endothelial progenitors from umbilical cord[150] seem to need more detailed characterization. It has been documented that the progenitors having endothelial morphology, and responding in vitro to EGF, VEGF, and bFGF stimulation, are able to form endo- or epithelial structures both in vitro and in vivo.[7,105,129-131,150,151] To make the picture more confusing, amniotic epithelial cells display typical epithelial morphology in in vitro cultures, require the addition of EGF to culture medium, but express also the pluripotency markers, and under proper stimulation differentiate into several cell lineages.[108,109,114,122,123,126,127,149]

Probably the most important and promising for future clinical application purposes are the fetal MSCs. These cells are present in all fetal tissues, although in varying quantities. They express the pluripotency markers, selected embryonic markers, and differentiate not only into "obligatory" adipo-, osteo-, and chondrogenesis traits, but are also able to produce nerve and glia cells,[32,33,84,88,143]hepatocytes and liver stroma cells,[21,106,134] pancreatic islet cells,[19,27,28,37] skeletal muscle cells,[23,35] cardiomyocytes,[37,80,98] or hematopoietic stroma, functionally supporting hematopoiesis when co-cultured with CD34+ cells[86,119,153,158,162] or with embryonic cells.[43] The future fetal MSC clinical applications should include the cell replacement therapies in regenerative medicine. Another application, eliminating excessive use of laboratory animals, would be the use of MSC and their progeny for toxicology testing or drug screening.[93] Fetal stem cells are also a promising tool for gene therapy.[85] The other specificity of fetal MSC is their regulatory role in the sites of injury – even, when definitely not forming the new cells of injured tissue, they are able to modify and optimize the process of healing, or slowing the progress of disease – such effects, as observed in heart infarction,[89] ischemia,[56] brain injury/stroke,[30,38] animal model of Parkinson disease,[59,60] or animal model of retina photoreceptors degeneration,[87] may be attributed to the very extensive set of tools (cytokines),[83,104] enabling them to perform the regulatory role through cell-to-cell dialogue. MSCs are able to migrate to the injury site, but also to the areas of tumor growth,[62,116] so they possibly can be used for tumor-targeting technologies. The other, clinically important, is MSC ability to modulate the immune reactions to allogeneic cells, allowing to ameliorate the host-versus-graft or graft-versus-host reactions.[16,17,22,29,31,80–82,90,138]

28.6 Conclusion

The fetal non-hematopoietic stem cells are pluripotent cells of high proliferative potential. Both their biological characteristics and the simplicity of methods used for their isolation, in vitro expansion, and cryostorage, make them the material of choice for clinical application purposes.

Acknowledgments Supported by the European Union through the FP6-LIFESCIHEALTH Project "Magselectofection," Contract No LSHB-CT-2006-019038

References

1. Alviano F, Fossati V, Marchionni C, et al. Term amniotic membrane is a high throughput source for multipotent mesenchymal stem cells with the ability to differentiate into endothelial cells in vitro. *BMC Dev Biol.* 2007;7:11.
2. Bailey MM, Wang L, Bode CJ, et al. A comparison of human umbilical cord matrix stem cells and temporomandibular joint condylar chondrocytes for tissue engineering temporomandibular joint condylar cartilage. *Tissue Eng.* 2007;13: 2003-2010.
3. Baksh D, Yao R, Tuan RS. Comparison of proliferative and multilineage differentiation potential of human mesenchymal stem cells derived from umbilical cord and bone marrow. *Stem Cells.* 2007;25:1384-1392.
4. Battula VL, Bareiss PM, Treml S, et al. Human placenta and bone marrow derived MSC cultured in serum-free, b-FGF-containing medium express cell surface frizzled-9 and SSEA-4 and give rise to multilineage differentiation. *Differentiation.* 2007;75:279-291.
5. Bicknese AR, Goodwin HS, Quinn CO. Human umbilical cord blood cells can be induced to express markers for neurons and glia. *Cell Transplant.* 2002;11:261-264.
6. Bieback K, Kern S, Kluter H, et al. Critical parameters for the isolation of mesenchymal stem cells from umbilical cord blood. *Stem Cells.* 2004;22:625-634.
7. Breymann C, Schmidt D, Hoerstrup SP. Umbilical cord cells as a source of cardiovascular tissue engineering. *Stem Cell Rev.* 2006;2:87-92.
8. Broxmeyer HE. Self-renewal and migration of stem cells during embryonic and fetal hematopoiesis: important, but poorly understood events. *Blood Cells.* 1991;17:282-286.
9. Broxmeyer HE, Douglas GW, Hangoc G, et al. Human umbilical cord blood as a potential source of transplantable hematopoietic stem/progenitor cells. *Proc Natl Acad Sci USA.* 1989;86:3828-3832.
10. Broxmeyer HE, Kurtzberg J, Gluckman E, et al. Umbilical cord blood hematopoietic stem and repopulating cells in human clinical transplantation. *Blood Cells.* 1991;17: 313-329.
11. Buzanska L, Machaj EK, Zablocka B, et al. Human cord blood-derived cells attain neuronal and glial features in vitro. *J Cell Sci.* 2002;115:2131-2138.
12. Campagnoli C, Bellantuono I, Kumar S, et al. High transduction efficiency of circulating first trimester fetal mesenchymal stem cells: potential targets for in utero ex vivo gene therapy. *BJOG.* 2002;109:952-954.
13. Campagnoli C, Roberts IAG, Kumar S, et al. Identification of mesenchymal stem/progenitor cells in human first-trimester fetal blood, liver, and bone marrow. *Blood.* 2001;98:2396-2402.
14. Campard D, Lysy PA, Najimi M. Native umbilical cord matrix stem cells express hepatic markers and differentiate into hepatocyte-like cells. *Gastroenterology.* 2008;134:833-848.
15. Casey ML, MacDonald PC. Interstitial collagen synthesis and processing in human amnion: a property of the mesenchymal cells. *Biol Reprod.* 1999;55:1253-1260.
16. Chan SL, Choi M, Wnendt S, et al. Enhanced in vivo homing of uncultured and selectively amplified cord blood CD34+ cells by cotransplantation with cord blood-derived unrestricted somatic stem cells. *Stem Cells.* 2007;25:529-536.
17. Chang CJ, Yen ML, Chen YC, et al. Placenta-derived multipotent cells exhibit immunosuppressive properties that are enhanced in the presence of interferon-gamma. *Stem Cells.* 2006;24:2466-2477.
18. Chang YJ, Shih DT, Tseng CP, et al. (2006) Disparate mesenchyme-lineage tendencies in mesenchymal stem cells from human bone marrow and umbilical cord blood. *Stem Cells.* 2006;24:679-685.
19. Chao KC, Chao KF, Fu YS, et al. Islet-like clusters derived from mesenchymal stem cells in Wharton's jelly of the human umbilical cord for transplantation to control type 1 diabetes. *PLoS ONE.* 2008;3:e1451.
20. Chen J, Sanberg PR, Li Y, et al. Intravenous administration of human umbilical cord blood reduces behavioral deficits after stroke in rats. *Stroke.* 2001;32:2682-2688.
21. Chien CC, Yen BL, Lee FK, et al. In vitro differentiation of human placenta-derived multipotent cells into hepatocyte-like cells. *Stem Cells.* 2006;24:1759-1768.
22. Cho PS, Messina DJ, Hirsh EL, et al. Immunogenicity of umbilical cord tissue derived cells. *Blood.* 2008;111: 430-438.
23. Conconi MT, Burra P, Di Liddo R, et al. CD105(+) cells from Wharton's jelly show in vitro and in vivo myogenic differentiative potential. *Int J Mol Med.* 2006;18:1089-1096.
24. Covas DT, Siufi JLC, Silva ARL, et al. Isolation and culture of umbilical vein mesenchymal stem cells. *Braz J Med Biol Res.* 2003;36:1179-1183.
25. Dominici M, Le Blanc K, Mueller I, et al. Minimal criteria for defining multipotent mesenchymal stromal cells. The International Society for Cellular Therapy position statement. *Cytotherapy.* 2006;8:315-317.

26. El-Badri NS, Hakki A, Saporta S, et al. Cord blood mesenchymal stem cells: potential use in neurological disorders. *Stem Cells Dev*. 2006;15:497-506.
27. Ende N, Chen R, Reddi AS. Effect of human umbilical cord blood cells on glycemia and insulitis in type 1 diabetic mice. *Biochem Biophys Res Commun*. 2004;325:665-669.
28. Ende N, Chen R, Reddi AS. Transplantation of human umbilical cord blood cells improves glycemia and glomerular hypertrophy in type 2 diabetic mice. *Biochem Biophys Res Commun*. 2004;321:168-171.
29. Ennis J, Gotherstrom C, Le Blanc K, et al. In vitro immunologic properties of human umbilical cord perivascular cells. *Cytotherapy*. 2008;10:174-181.
30. Fan CG, Zhang QJ, Tang FW, et al. Human umbilical cord blood cells express neurotrophic factors. *Neurosci Lett*. 2005;380:322-325.
31. Friedman R, Betancur M, Boissel L, et al. Umbilical cord mesenchymal stem cells: adjuvants for human cell transplantation. *Biol Blood Marrow Transplant*. 2007;13:1477-1486.
32. Fu YS, Shih YT, Cheng YC, et al. Transformation of human umbilical mesenchymal cells into neurons in vitro. *J Biomed Sci*. 2004;11:652-660.
33. Fu YS, Cheng YC, Lin MY, et al. Conversion of human umbilical cord mesenchymal stem cells in Wharton's jelly to dopaminergic neurons in vitro: potential therapeutic application for Parkinsonism. *Stem Cells*. 2006;24:115-124.
34. Fuchs JR, Hannouche D, Terada S, et al. Cartilage engineering from ovine umbilical cord blood mesenchymal progenitor cells. *Stem Cells*. 2005;23:958-964.
35. Gang EJ, Hong SH, Jeong JA, et al. In vitro mesengenic potential of human umbilical cord blood-derived mesenchymal stem cells. *Biochem Bophys Res Commun*. 2004;321:102-108.
36. Gang EJ, Jeong JA, Hong SH, et al. Skeletal myogenic differentiation of mesenchymal stem cells isolated from human umbilical cord blood. *Stem Cells*. 2004;22:617-624.
37. Gao F, Wu DQ, Hu YH, et al. In vitro cultivation of islet-like cell clusters from human umbilical cord blood-derived mesenchymal stem cells. *Transl Res*. 2008;151:293-302.
38. Garbuzova-Davis S, Willing AE, Zigova T, et al. Intravenous administration of human umbilical cord blood cells in a mouse model of amyotropic lateral sclerosis: distribution, migration, and differentiation. *J Hematother Stem Cell Res*. 2003;12:255-270.
39. Gluckman E, Broxmeyer HA, Auerbach AD, et al. Hematopoietic reconstitution in a patient with Fanconi's anemia by means of umbilical-cord blood from an HLA-identical sibling. *N Engl J Med*. 1989;321:1174-1178.
40. Goussetis E, Theodosaki M, Paterakis G, et al. In vitro identification of a cord blood CD133$^-$ CD34$^-$ Lin$^-$ cell subset that gives rise to myeloid dendritic precursors. *Stem Cells*. 2006;24:1137-1140.
41. Guillot PV, Gotherstrom C, Chan J, et al. Human first-trimester fetal MSC express pluripotency markers and grow faster and have longer telomeres than adult MSC. *Stem Cells*. 2007;25:646-654.
42. Gutierrez-Rodriguez M, Reyes-Maldonado E, Mayani H, et al. Characterization of the adherent cells developed in Dexter-type long-term cultures from human umbilical cord blood. *Stem Cells*. 2000;18:46-52.
43. Hiroyama T, Sudo K, Aoki N, et al. Human umbilical cord-derived cells can often serve as feeder cells to maintain primate embryonic stem cells in a state capable of producing hematopoietic cells. *Cell Biol Int*. 2008;31:1-7.
44. Hoerstrup SP, Kadner A, Breymann C, et al. Living, autologous pulmonary artery conduits tissue engineered from human umbilical cord cells. *Ann Thorac Surg*. 2002;74:46-52.
45. Hong SH, Gang EJ, Jeong JA, et al. In vitro differentiation of human umbilical cord blood-derived mesenchymal stem cells into hepatocyte-like cells. *Biochem Biophys Res Commun*. 2005;330:1153-1161.
46. Horwitz EM, Le Blanc K, Dominici M, et al. Position paper. Clarification of the nomenclature for MSC: the International Society for Cellular Therapy position statement. *Cytotherapy*. 2005;7:393-395.
47. Hou L, Cao H, Wang D, et al. Induction of umbilical cord blood meenchymal stem cells into neuron-like cells in vitro. *Int J Hematol*. 2003;78:256-261.
48. Hoynowski SM, Fry MM, Gardner BM, et al. Characterization and differentiation of equine umbilical cord-derived matrix cells. *Biochem Biophys Res Commun*. 2007;362:347-353.
49. Hutson EL, Boyer S, Genever PG. Rapid isolation, expansion, and differentiation of osteoprogenitors from full-term umbilical cord blood. *Tissue Eng*. 2005;11:1407-1420.
50. Ilancheran S, Michalska A, Peh G, et al. Stem cells derived from human fetal membranes display multi-lineage differentiation potential. *Biol Reprod*. 2007;77:577-588.
51. Ingram DA, Mead LE, Tanaka H, et al. Identification of a novel hierarchy of endothelial progenitor cells using human peripheral and umbilical cord blood. *Blood*. 2004;104:2752-2760.
52. In'tAnker P, Scherjon SA, Kleijburg-van der Keur C, et al. Isolation of mesenchymal stem cells of fetal or maternal origin from human placenta. *Stem Cells*. 2004;22:1338-1345.
53. Jang YK, Jung DH, Jung MH, et al. Mesenchymal stem cells feeder layer from human umbilical cord blood for ex vivo expanded growth and proliferation of hematopoietic progenitor cells. *Ann Hematol*. 2006;85:212-225.
54. Jeong JA, Gang EJ, Hong SH, et al. Rapid neural differentiation of human cord blood-derived mesenchymal stem cells. *Neuroreport*. 2004;15:1731-1734.
55. Jeong JA, Hong SH, Gang EJ, et al. Differential gene expression profiling of human umbilical cord blood-derived mesenchymal stem cells by DNA microarray. *Stem Cells*. 2005;23:584-593.
56. Jomura S, Uy M, Mitchell K, et al. Potential treatment of cerebral global ischemia with Oct4+ umbilical cord matrix cells. *Stem Cells*. 2008;25:98-106.
57. Kadivar M, Khatami S, Mortazavi Y, et al. In vitro cardiomyogenic potential of human umbilical vein-derived mesenchymal stem cells. *Biochem Biophys Res Commun*. 2006;340:639-647.
58. Kakinuma S, Tanaka Y, Chinzel R, et al. Human umbilical cord blood as a source of transplantable hepatic progenitor cells. *Stem Cells*. 2003;21:217-227.
59. Kakishita K, Elwan MA, Nakao N, et al. Human amniotic epithelial cells produce dopamine and survive after implantation into the striatum of a rat model of Parkinson's disease: a potential source of donor for transplantation therapy. *Exp Neurol*. 2000;165:27-34.
60. Kakishita K, Nakao N, Sakuragawa N, et al. Implantation of human amniotic epithelial cells prevents the degeneration of

nigral dopamine neurons in rats with 6-hydroxydopamine lesions. *Brain Res*. 2003;980:48-56.
61. Kang JH, Lee CK, Kim JR, et al. Estrogen stimulates the neuronal differentiation of human umbilical cord blood mesenchymal stem cells (CD34-). *Neuroreport*. 2007;18:35-38.
62. Kang SG, Jeun SS, Lim JY, et al. Cytotoxicity of human umbilical cord blood-derived mesenchymal stem cells against human malignant glioma cells. *Child Nerv Syst*. 2008;24:293-302.
63. Kang XQ, Zang WJ, Bao LJ, et al. Fibroblast growth factor-4 and hepatocyte growth factor induce differentiation of human umbilical cord blood-derived mesenchymal stem cells into hepatocytes. *World J Gastroenterol*. 2005;11:7461-7465.
64. Karahuseyinoglu S, Cinar O, Kilic E, et al. Biology of the stem cells in human umbilical cord stroma: in situ and in vitro surveys. *Stem Cells*. 2007;25:319-331.
65. Karahuseyinoglu S, Kocaefe C, Balci D, et al. Functional structure of adipocytes differentiated from human umbilical cord stroma-derived stem cells. *Stem Cells*. 2008;26:682-691.
66. Kern S, Eichler H, Stoeve J, et al. Comparative analysis of mesenchymal stem cells from bone marrow, umbilical cord blood, or adipose tissue. *Stem Cells*. 2006;24:1294-1301.
67. Kim JW, Kim SY, Park SY, et al. Mesenchymal progenitor cells in human umbilical cord. *Ann Hematol*. 2004;83:733-738.
68. Klingemann H. Discarded stem cells with a future? *Expert Opin Biol Ther*. 2006;6:1251-1254.
69. Kobayashi K, Kubota T, Aso T. Study of myofibroblast differentiation in the stromal cells of Wharton's jelly: expression and localization of alpha-smooth muscle actin. *Early Hum Dev*. 1998;51:223-233.
70. Kogler G, Sensken S, Airley JA, et al. A new human somatic stem cell from placental cord blood with intrinsic pluripotent differentiation potential. *J Exp Med*. 2004;200:123-135.
71. Kogler G, Sensken S, Wernet P. Comparative generation and characterization of pluripotent unrestricted somatic stem cells with mesenchymal stem cells from human cord blood. *Exp Hematol*. 2006;34:1589-1595.
72. Kong KY, Ren J, Kraus M al. Human umbilical cord blood cells differentiate into muscle in sjl muscular dystrophy mice. *Stem Cells*. 2004;22:981-993.
73. Korbling M, Robinson S, Estrov Z, et al. Umbilical cord blood-derived cells for tissue repair. *Cytotherapy*. 2005;7:258-261.
74. Kucia M, Oldak K, Ratajczak MZ, et al. Percoll gradient separation of cord blood mononuclear cells reveals the presence of a novel population of CXCR4+ Oct-4+ small embryonic-like stem cells. *Blood*. 2005;106:1069.
75. Kucia M, Halasa M, Wysoczynski M, et al. Morphological and molecular characterization of novel population of CXCR4+ SSEA-4+ Oct-4+ very small embryonic-like cells purified from human cord blood: preliminary report. *Leukemia*. 2007;21:297-303.
76. Lee MW, Choi J, Yang MS, et al. Mesenchymal stem cells from cryopreserved human umbilical cord blood. *Biochem Biophys Res Comm*. 2004;320:273-278.
77. Lee MW, Yang MS, Park JS, et al. Isolation of mesenchymal stem cells from cryopreserved human umbilical cord blood. *Int J Hematol*. 2005;81:126-130.
78. Lee OK, Kuo TK, Chen WM, et al. Isolation of multipotent mesenchymal stem cells from umbilical cord blood. *Blood*. 2004;103:1669-1675.
79. Leor J, Guetta E, Chouraqui P, et al. Human umbilical cord blood cells: a new alternative for myocardial repair? *Cytotherapy*. 2005;7:251-257.
80. Li C, Zhang W, Jiang X, et al. Human-placenta-derived mesenchymal stem cells inhibit proliferation and function of allogeneic immune cells. *Cell Tissue Res*. 2007;330:437-446.
81. Li CD, Zhang WY, Li HL, et al. Mesenchymal stem cells derived from human placenta suppress allogeneic umbilical cord blood lymphocyte proliferation. *Cell Res*. 2005;15:539-547.
82. Li H, Niederkorn JY, Neelam S, et al. Suppressive factors secreted by human amniotic epithelial cells. *Invest Ophthalmol Vis Sci*. 2005;46:900-907.
83. Liu CH, Hwang SM. Cytokine interactions in mesenchymal stem cells from cord blood. *Cytokine*. 2005;32:270-279.
84. Low CB, Liou YC, Tang BL. Neural differentiation and potential use of stem cells from the human umbilical cord for central nervous system transplantation therapy. *J Neurosci Res*. 2008;86:1670-1679.
85. Lu FZ, Fujino M, Kitazawa Y, et al. Characterization and gene transfer in mesenchymal stem cells derived from human umbilical-cord blood. *J Lab Clin Med*. 2005;146:271-278.
86. Lu LL, Liu YJ, Yang SG, et al. Isolation and characterization of human umbilical cord mesenchymal stem cells with hematopoiesis-supportive function and other potentials. *Haematologica*. 2006;91:1017-1026.
87. Lund RD, Wang S, Lu B, et al. Cells isolated from umbilical cord tissue rescue photoreceptors and visual functions in a rodent model of retinal disease. *Stem Cells*. 2007;25:602-611.
88. Ma L, Feng XY, Cui BL, et al. Human umbilical cord Wharton's Jelly-derived mesenchymal stem cells differentiation into nerve-like cells. *Chin Med J (Engl)*. 2005;118:1987-1993.
89. Ma N, Stamm C, Kaminski A, et al. Human cord blood cells induce angiogenesis following myocardial infarction in NOD/scid-mice. *Cardiovasc Res*. 2005;66:4-6.
90. Magatti M, De Munari S, Vertua E, et al. Human amnion mesenchyme harbors cells with allogeneic T-cell suppression and stimulation capabilities. *Stem Cells*. 2008;26:182-192.
91. Manca MF, Zwart I, Beo J, et al. Characterization of mesenchymal stromal cells derived from full-term umbilical cord blood. *Cytotherapy*. 2008;10:54-68.
92. Mareschi K, Biasin E, Piacibello W, et al. Isolation of human mesenchymal stem cells: bone marrow versus umbilical cord blood. *Haematologica*. 2001;86:1099-1100.
93. Matikainen T, Laine J. Placenta-an alternative source of stem cells. *Toxicol Appl Pharmacol*. 2005;207:544-549.
94. McGuckin C, Jurga M, Ali H, et al. Culture of embryonic-like stem cells from human umbilical cord blood and onward differentiation to neural cells in vitro. *Nat Protoc*. 2008;3:1046-1055.
95. McGuckin CP, Forraz N, Baradez MO, et al. Production of stem cells with embryonic characteristics from human umbilical cord blood. *Cell Prolif*. 2005;38:245-255.
96. Miao Z, Jin J, Chen L, et al. Isolation of mesenchymal stem cells from human placenta: comparison with human bone marrow mesenchymal stem cells. *Cell Biol Int*. 2006;30:681-687.
97. Miki T, Lehmann T, Cai H, et al. Stem cell characteristics of amniotic epithelial cells. *Stem Cells*. 2005;23:1549-1559.
98. Min JJ, Ahn Y, Moon S, et al. In vivo bioluminescence imaging of cord blood derived mesenchymal stem cell transplantation into rat myocardium. *Ann Nucl Med*. 2006;20:165-170.

99. Mitchell KE, Weiss ML, Mitchell BM, et al. Matrix cells from Wharton's jelly form neurons and glia. *Stem Cells.* 2003;21:50-60.
100. Murga M, Yao L, Tosato G. Derivation of endothelial cells from CD34- umbilical cord blood. *Stem Cells.* 2004;22: 385-395.
101. Murohara T, Ikeda H, Duan J, et al. Transplanted cord blood-derived endothelial precursor cells augment postnatal neovascularization. *J Clin Invest.* 2000;105:1527-1536.
102. Musina RA, Bekchanova ES, Sukhikh GT. Comparison of mesenchymal stem cells obtained from different human tissues. *Bull Exp Biol Med.* 2005;139:504-509.
103. Nagano M, Yamashita T, Hamada H, et al. Identification of functional endothelial progenitor cells suitable for the treatment of ischemic tissue using human umbilical cord blood. *Blood.* 2007;110:151-160.
104. Neuhoff S, Moers J, Rieks M, et al. Proliferation, differentiation, and cytokine secretion of human umbilical cord blood-derived mononuclear cells in vitro. *Exp Hematol.* 2007;35:1119-1131.
105. Ng W, Nishiyama C, Mizoguchi M, et al. Human umbilical cord epithelial cells express Notch1: implications for its epidermal-like differentiation. *J Dermatol Sci.* 2008;49: 143-152.
106. Nonome K, Li XK, Takahara T, et al. Human umbilical cord blood-derived cells differentiate into hepatocyte-like cells in the Fas-mediated liver injury model. *Am J Physiol Gastrointest Liver Physiol.* 2005;289:1091-1099.
107. Nunes VA, Cavacana N, Canovas M, et al. Stem cells from umbilical cord blood differentiate into myotubes and express dystrophin in vitro only after exposure to in vivo muscle environment. *Biol Cell.* 2007;99:185-196.
108. Ochsenbein-Kolble N, Bilic G, Hall H, et al. Inducing proliferation of human amnion epithelial and mesenchymal cells for prospective engineering of membrane repair. *J Perinat Med.* 2003;31:287-294.
109. Okawa H, Okuda O, Arai H, et al. Amniotic epithelial cells transform into neuron-like cells in the ischemic brain. *Neuroreport.* 2001;12:4003-4007.
110. Panepucci RA, Siufi JLC, Silva WAS Jr, et al. Comparison of gene expression of umbilical cord vein and bone marrow-derived mesenchymal stem cells. *Stem Cells.* 2004;22: 1263-1278.
111. Parekkadan B, Sethu P, van Poll D, et al. Osmotic selection of human mesenchymal stem/progenitor cells from umbilical cord blood. *Tissue Eng.* 2007;13:2465-2473.
112. Park KS, Lee YS, Kang KS. In vitro neuronal and osteogenic differentiation of mesenchymal stem cells from human umbilical cord blood. *J Vet Sci.* 2006;7:343-348.
113. Pipes BL, Ablib RJ. Embryonic stem cell co-transplantation revisited: utility of umbilical cord blood "embryonic-like " stem cells. *Ann Clin Lab Sci.* 2006;36:105-106.
114. Portmann-Lanz CB, Schoeberlein A, Huber A, et al. Placental mesenchymal stem cells as potential autologous graft for pre- and perinatal neuroregeneration. *Am J Obstet Gynecol.* 2006;194:664-673.
115. Qiao C, Xu W, Zhu W, et al. Human mesenchymal stem cells isolated from the umbilical cord. *Cell Biol Int.* 2008;32:8-15.
116. Rachakatla RS, Marini F, Weiss ML, et al. Development of human umbilical cord matrix stem cell-based gene therapy for experimental lung tumors. *Cancer Gene Ther.* 2007;14: 828-835.
117. Rafii S, Lyden D. Therapeutic stem and progenitor cell transplantation for organ vascularization and regeneration. *Nat Med.* 2003;9:702-712.
118. Reinisch A, Bartmann C, Rohde E, et al. Humanized system to propagate cord blood-derived multipotent mesenchymal stromal cells for clinical application. *Regen Med.* 2007;2:371-382.
119. Robinson SN, Ng J, Niu T, et al. Superior ex vivo cord blood expansion following co-culture with bone marrow-derived mesenchymal stem cells. *Bone Marrow Transplant.* 2006;37:359-366.
120. Romanov YA, Svintsitskaya VA, Smirnov VN. Searching for alternative sources of postnatal human mesenchymal stem cells: candidate MSC-like cells from umbilical cord. *Stem Cells.* 2003;21:105-110.
121. Rosada C, Justesen J, Melsvik D, et al. The human umbilical cord blood: a potential source for osteoblast progenitor cells. *Calcif Tissue Int.* 2003;72:135-142.
122. Sakuragawa N, Thangavel R, Mizuguchi M, et al. Expression of markers for both neuronal and glial cells in human amniotic epithelial cells. *Neurosci Lett.* 1996;209: 9-11.
123. Sakuragawa N, Enosawa S, Ishii T, et al. Human amniotic epithelial cells are promising transgene carriers for allogeneic cell transplantation into liver. *J Hum Genet.* 2000;45:171-176.
124. Sakuragawa N, Kakinuma K, Kikuchi A, et al. Human amnion mesenchyme cells express phenotypes of neuroglial progenitor cells. *J Neurosci Res.* 2004;78:208-214.
125. Sanchez-Ramos J, Song S, Kamath SG, et al. Expression of neural markers in human umbilical cord blood. *Exp Neurol.* 2001;171:109-115.
126. Sankar V, Muthusamy R. Role of human amniotic epithelial cell transplantation in spinal cord injury repair research. *Neuroscience.* 2003;118:11-17.
127. Sanmano B, Mizoguchi M, Suga Y, et al. Engraftment of umbilical cord epithelial cells in athymic mice: in an attempt to improve reconstructed skin equivalents used as epithelial composite. *J Dermatol Sci.* 2005;37:29-39.
128. Sarugaser R, Lickorish D, Baksh D, et al. Human umbilical cord perivascular (HUCPV) cells: a source of mesenchymal progenitors. *Stem Cells.* 2005;23:220-229.
129. Schmidt D, Breymann C, Weber A, et al. Umbilical cord blood derived endothelial progenitor cells for tissue engineering of vascular grafts. *Ann Thorac Surg.* 2004;78: 2094-2098.
130. Schmidt D, Mol A, Odermatt B, et al. Engineering of biologically active living heart valve leaflets using human umbilical cord-derived progenitor cells. *Tissue Eng.* 2006;12:3223-3232.
131. Schmidt D, Mol A, Breymann C, et al. Living autologous heart valves engineered from human prenatally harvested progenitors. *Circulation.* 2006;114(suppl 1):I125-I131.
132. Secco M, Zucconi E, Vieira NM, et al. Multipotent stem cells from umbilical cord: cord is richer than blood! *Stem Cells.* 2008;26:146-150.
133. Selesniemi KL, Reedy MA, Gultice AD, et al. Identification of committed placental stem cell lines for studies of differentiation. *Stem Cells Dev.* 2005;14:535-547.

134. Sharma AD, Cantz T, Richter R, et al. Human cord blood stem cells generate human cytokeratin 18-negative hepatocyte-like cells in injured mouse liver. *Am J Pathol.* 2005;167:555-564.
135. Soncini M, Vertua E, Gibelli L, et al. Isolation and characterization of mesenchymal cells from human fetal membranes. *J Tissue Eng Regen Med.* 2007;1:296-305.
136. Sudo K, Kanno M, Miharada K, et al. Mesenchymal progenitors able to differentiate into osteogenic, chondrogenic, and/or adipogenic cells in vitro are present in most primary fibroblast-like cell populations. *Stem Cells.* 2007;25: 1610-1617.
137. Takashima S, Ise H, Zhao P, et al. Human amniotic epithelial cells possess hepatocyte-like characteristics and functions. *Cell Struct Funct.* 2004;29:73-84.
138. Tisato V, Naresh K, Girdlestone J, et al. Mesenchymal stem cells of cord blood origin are effective at preventing but not treating graft-versus-host disease. *Leukemia.* 2007;21: 1992-1999.
139. Tondreau T, Meuelman N, Delforge A, et al. Mesenchymal stem cells derived from CD133-positive cells in mobilized peripheral blood and cord blood: proliferation, Oct4 expression, and plasticity. *Stem Cells.* 2005;23:1105-1112.
140. Vacanti MP, Roy A, Cortiella J, et al. Identification and initial characterization of spore-like cells in adult mammals. *J Cell Biochem.* 2001;80:455-460.
141. Ventura C, Cantoni S, Bianchi F, et al. Hyaluronan mixed esters of butyric and retinoic acid drive cardiac and endothelial fate in term placenta human mesenchymal stem cells and enhance cardiac repair in infarcted rat hearts. *J Biol Chem.* 2007;282:14243-14252.
142. Wagner W, Wein F, Seckinger A, et al. Comparative characteristics of mesenchymal stem cells from human bone marrow, adipose tissue, and umbilical cord blood. *Exp Hematol.* 2005;33:1402-1416.
143. Walczak P, Chen N, Eve D, et al. Long-term cultured human umbilical cord neural-like cells transplanted into the striatum of NOD SCID mice. *Brain Res Bull.* 2007;74: 155-163.
144. Wang HS, Hung SC, Peng ST, et al. Mesenchymal stem cells in the Wharton's jelly of the human umbilical cord. *Stem Cells.* 2004;22:1330-1337.
145. Wei JP, Zhang TS, Kawa S, et al. Human amnion-isolated cells normalize blood glucose in streptozotocin-induced diabetic mice. *Cell Transplant.* 2003;12:545-552.
146. Weiss ML, Mitchell KE, Hix JE, et al. Transplantation of porcine umbilical cord matrix cells into the rat brain. *Exp Neurol.* 2003;182:288-299.
147. Weiss ML, Medicetty S, Bledsoe AR, et al. Human umbilical cord matrix stem cells: preliminary characterization and effect of transplantation in a rodent model of Parkinson's disease. *Stem Cells.* 2006;24:781-792.
148. Wexler SA, Donaldson C, Denning-Kendall P, et al. Adult bone marrow is a rich source of human mesenchymal"stem" cells but umbilical cord and mobilized adult blood are not. *Br J Haematol.* 2003;121:368-374.
149.Wolbank S, Peterbauer A, Fahrner M, et al. Dose-dependent immunomodulatory effect of human stem cells from amniotic membrane: a comparison with human mesenchymal stem cells from adipose tissue. *Tissue Eng.* 2007;13:1173-1183.
150. Wu KH, Zhou B, Lu SH, et al. In vitro and in vivo differentiation of human umbilical cord derived stem cells into endothelial cells. *J Cell Biochem.* 2007;100:608-616.
151. Wu X, Lensch MW, Wylie-Sears J, et al. Hemogenic endothelial progenitor cells isolated from human umbilical cord blood. *Stem Cells.* 2007;25:2770-2776.
152. Wu ZY, Hui GZ, Lu Y. Transplantation of human amniotic epithelial cells improves hindlimb function in rats with spinal cord injury. *Chin Med J (Engl).* 2006;119:2101-2107.
153. Yamamura K, Ohishi K, Masuya M, et al. Ex vivo culture of human cord blood hematopoietic stem/progenitor cells adversely influences their distribution to other bone marrow compartments after intra-bone marrow transplantation. *Stem Cells.* 2008;26:543-549.
154. Yan H, Yu C. Repair of full-thickness cartilage defects with cells of different origin in a rabbit model. *Arthroscopy.* 2007;23:178-187.
155. Yang SE, Ha CW, Jung M, et al. Mesenchymal stem/progenitor cells developed in cultures from UC blood. *Cytotherapy.* 2004;6:476-486.
156. Yarygin KN, Suzdal'tseva YG, Burunova VV, et al. Comparative study of adult human skin fibroblasts and umbilical fibroblast-like cells. *Bull Exp Biol Med.* 2006;141:161-166.
157. Yen BL, Huang HI, Chien CC, et al. Isolation of multipotent cells from human term placenta. *Stem Cells.* 2005;23:3-9.
158. Yoo ES, Lee KE, Seo JW, et al. Adherent cells generated during long-term culture of human umbilical cord blood CD34+ cells have characteristics of endothelial cells and beneficial effect on cord blood ex vivo expansion. *Stem Cells.* 2003;21:228-235.
159. Yu M, Xiao Z, Shen L, et al. Mid-trimester fetal blood-derived adherent cells share characteristics similar to mesenchymal stem cells but full-term umbilical cord blood does not. *Br J Haematol.* 2004;124:666-675.
160. Zang WJ, Bao LJ, Li DL, et al. Differentiating characterization of human umbilical cord blood-derived mesenchymal stem cells in vitro. *Cell Biol Int.* 2006;30:569-575.
161. Zhang X, Mitsuru A, Igura K, et al. Mesenchymal progenitor cells derived from chorionic villi of human placenta for cartilage tissue engineering. *Biochem Biophys Res Commun.* 2006;340:944-952.
162. Zhang Y, Changdong L, Jiang X, et al. Human placenta-derived mesenchymal progenitor cells support culture expansion of long-term culture-initiating cells from cord blood $CD34^+$ cells. *Exp Hematol.* 2004;32:657-664.
163. Zhao P, Ise H, Hongo M, et al. Human amniotic mesenchymal cells have some characteristics of cardiomyocytes. *Transplantation.* 2005;79:528-535.
164. Zigova T, Song S, Willing AE, et al. Human umbilical cord blood express neural antigens after transplantation into the developing rat brain. *Cell Transplant.* 2002;11:265-274.

Animal Studies of Cord Blood and Regeneration

29

Thomas E. Ichim, Michael P. Murphy, and Neil Riordan

29.1 Cord Blood Transplants: History

Cord blood has been used successfully as an alternative stem cell source to marrow, particularly in pediatric patients with hematopoietic malignancies, bone marrow failure, or inborn errors of metabolism, and currently expanding to adults. Cord blood was known since the 1930s to be useful as a substitute for peripheral blood in transfusions.[1] This may have been what prompted the original report of using cord blood as a clinical source of hematopoietic stem cells occurring in 1972 in a paper describing a pediatric acute lymphoblastic leukemia patient under 6-mercaptopurine and prednisone therapy.[2] Although the treatment did not substantially affect clinical outcome, engraftment was demonstrated for 38 days by differentiation based on erythrocyte markers. Supporting the notion that cord blood may be a useful source of stem cells were laboratory reports identifying high concentration of colony-forming cells within this population in vitro in the 1970s and 1980s.[3,4] The first successful use of cord blood transplants was in 1989 by Gluckman et al.[5] who used sibling cord blood to treat a 5-year-old patient with Fanconi anemia who at last report was still in good health 18 years later.[6] After this initial success, cord blood transplantation rapidly became one of the treatments of choice for pediatric patients lacking sibling donors. The limitation of stem cell number in cord blood units is overcome in pediatric patients due to lower body mass. Accordingly, more than approximately 7,000–8,000 transplants have been performed,[7,8] with the general consensus being that in comparison to bone marrow, cord blood possesses several unique advantages and disadvantages. The advantages include less stringent matching requirements, lower graft versus host disease, and lower risk of contamination. The disadvantages include delayed kinetics of engraftment, limited supply of stem cells, and lack of ability to perform donor-lymphocyte infusions.[9]

The use of cord blood transplantation for the treatment of hematological malignancies has become a standard medical procedure in many areas. Cord blood transplants generally involve administration of cord blood mononuclear cells at approximately 1.5–2.5 × 107 cells/kg into patients having undergone either myeloablative conditioning, or non-myeloablative conditioning. Matching requirements are not as strict as in bone marrow or peripheral blood stem cell transplants. Typically a 4/6 HLA loci match is clinically acceptable. Typical protocols for neutralizing host hematopoiesis include components such as total body irradiation (TBI), cyclophosphamide, busulfan, etoposide, other chemotherapeutics, and/or anti-thymocyte globulin. Protocols that are non-myeloablative seek to eradicate host lymphocytes through administration of anti-thymocyte globulin/TBI/busulfan/fludarabine. Although sometimes similar agents that are used for myeloablation are also used for non-myeloablative conditioning, these agents are used at a lower concentration or reduced frequency of administration. The rationale of non-myeloablative conditioning is to allow for graft-versus-tumor effect to occur, without subjecting patient to severe physiological stress of complete myeloablation.[10,11]

In adults there have been numerous reports and publications regarding myeloablative conditioning followed by cord blood transplantation for malignancy.[12-17] Here, we briefly touch upon two well-cited studies that

T.E. Ichim (✉)
Indiana University, Bloomington, IN, USA

N. Bhattacharya and P. Stubblefield (eds.), *Regenerative Medicine Using Pregnancy-Specific Biological Substances*,
DOI: 10.1007/978-1-84882-718-9_29, © Springer-Verlag London Limited 2011

strongly supported this approach as an alternative to patients lacking an HLA-matched sibling donor. The first study was by the Acute Leukemia Working Party of European Blood and Marrow Transplant Group. This study assessed outcomes of 682 patients with acute leukemia that were recipients of stem cells from unrelated donors. Of these patients, 98 had received cord blood and 584 received bone marrow transplants. Bone marrow was HLA-matched at 6/6 loci, whereas cord blood was mismatched up to 4/6 loci. Multivariate analysis revealed that cord blood recipients had a lower risk of grade II-IV GVHD. Transplant-related mortality, relapse, and leukemia-free survival were similar between patients receiving cord blood. Neutrophil engraftment was significantly delayed in the group receiving cord blood. These findings led to the conclusion that unrelated cord blood transplant can be performed in patients with acute leukemia that do not have an HLA-matched bone marrow donor.[18] The second study compared leukemia patients that received cord blood grafts mismatched for one or two HLA loci, with patients who received bone marrow matched at six loci, and with patients who received bone marrow but were mismatched at one loci. Of the patients who received mismatched bone marrow and mismatched cord blood there was no difference in mortality associated with transplant or in leukemic relapse. The authors of the study, members of the International Bone Marrow Transplant Registry, concluded, similarly to the previous study cited, that HLA-mismatched (up to 4/6 loci) cord blood transplant should be recommended as an alternative to adult patients lacking an HLA-matched adult donor.[19]

Non-myeloablative transplantation is also used in some situations for treatment of malignant disease. The rationale being, as previously stated, that graft versus tumor effect is preserved so the need for complete destruction of host hematopoiesis is minimized. Another possible advantage of non-myeloablative conditioning in terms of malignancy is the enhanced ability of T-cells to reconstitute the host due to preservation of peripheral T-cell niches.[20] This may theoretically allow for an enhanced graft versus tumor effect. In a typical study, 13 patients (median age 49) suffering from various advanced hematological malignancies were transplanted with partially matched cord blood with a median nucleated cell dose of 2.07 × 10(7)/kg following non-myeloablative conditioning. Eight of the patients converted to donor chimerism between 4 weeks and 24 weeks. Median survival was 288 days after transplant.[21] In another representative study, 20 patients with advanced malignant lymphoma were conditioned with low-dose fludarabine, melphalan, and TBI prior to infusion with an average of 2.75 × 10(7)/kg cord blood cells matched at 4/6 and 5/6 HLA loci. Neutrophil engraftment occurred in 15 of the patients at an average of 20 days. Ten patients achieved complete response and estimated 1-year probability of progression-free survival was 50%.[22] These and numerous other studies demonstrate that although delayed in engraftment in comparison to allogeneic bone marrow transplants, cord blood is a suitable alternative for an easily accessible stem cells source for allotransplantation in patients with malignancy.[23,24]

Overall, the main obstacle to cord blood transplantation in general, and particularly after myeloablative conditioning regimens, is the low number of donor cells that are available in the graft. Approximately, the number of CD34+ cells in a unit of cord blood is tenfold less than obtained during a bone marrow graft.[9,25] It is known from several trials that the lower number of CD34+ cells in the cord blood graft correlates with extended time until hematopoietic recovery.[26-28] Accordingly, a variety of attempts have been made to enhance the stem cell content of cord blood grafts using ex vivo expansion. A Phase I study using the proprietary Aastrom Replicell system which includes culture in media supplemented with fetal bovine serum, horse serum, PIXY321, flt-3 ligand, and erythropoietin, demonstrated feasibility of achieving a median 2.4 expansion in overall nucleated cells, a 82-fold expansion in CFU-GM, and a 0.5-fold expansion in lineage-negative CD34+ cells. Patients were administered the cells 12 days post cord blood transplant as a "booster." No serious adverse events associated with administration of expanded cells were observed. Unfortunately, the small patient number did not permit significant analysis of efficacy. [29] Other attempts to increase the number of cord blood cells included administration of two units from different donors,[30] administration of third-party mobilized peripheral blood stem cells,[31] as well as administration of third-party mesenchymal stem cells.[32]

Since cord blood is more readily available as compared to bone marrow, its use for treatment of nonmalignant conditions requiring rapid intervention has been pursued. This use of cord blood can range from need to reconstitute the immune system with cells that are immunocompetent, to the need to deliver a functional enzyme to patients who are deficient in the enzyme, to use of cord blood for repair certain tissues. One example of cord blood transplantation for treatment of an abnormal immune system is a report on eight children suffering from a variety of T-cell immunodeficiencies including severe combined immunodeficiency syndrome (SCID), reticular dysgenesis, thymic dysplasia, combined immunodeficiency disease, and Wiskott–Aldrich syndrome. Following a myeloablative conditioning regimen, administration of 3/6 (two children), 4/6 (four children), and 5/6 (two children) HLA-mismatched cord blood was performed. Engraftment occurred in all but one patient (average time to neutrophil engraftment was 12 days). In the patient that did not engraft, a second cord blood transplant was performed and successful donor hematopoiesis was observed. Based on clinical benefit observed in the patients and similar GVHD profile to bone marrow transplantation, the authors concluded that unrelated umbilical donor cord blood is a suitable alternative source of stem cells for children with severe T-cell immune deficiency disorders that lack a suitable HLA-matched bone marrow donor.[33] A similar report evaluated 12 patients who received unrelated cord blood 7 × 10(7) cells/kg for primary immunodeficiency. All patients engrafted with average time to neutrophil reconstitution being 22 days; 11 patients had full donor T-cell and 6 full donor B-cell chimerism with normal IgG levels and specific antibody responses to tetanus and hepatitis B vaccines 1 year after transplant.[34] In terms of bone marrow failure diseases, such as aplastic anemia, in a recently published report, nine patients (average age 25.3) were subjected to unrelated cord blood transplants. Conditioning was performed in a non-myeloablative manner with cyclophosphamide and antithymocyte globulin. Successful hematopoietic engraftment was found in seven patients. At 32.2-month follow-up (range: 4–69), the patients that engrafted were alive and disease-free.[35]

Besides immune disorders, numerous deficiencies in stem cell function can be corrected by introduction of functional cells. For example, beta-thalassemia, is a hematopoietic disorder characterized by mutation in the beta hemoglobin gene, which in the homozygous state (thalassemia major) leads to severe anemia and transfusion dependence. Five pediatric patients with this condition received unrelated, one or two HLA-mismatched cord blood grafts at an average of 8.8 × 10(7) cells/kg. Preconditioning was performed with busulfan, cyclophosphamide, and antithymocyte globulin. Times to neutrophil engraftment, red blood cell transfusion independence, and platelet engraftment were 12, 34, and 46 days after transplantation, respectively. At the average follow-up time of 303 days after transplantation, complete donor chimerism and lack of need for transfusion were observed in all patients.[36]

Congenital metabolic disorders are another area in which cord blood has been successfully used. For example, Krabbe's Disease is a neurodegenerative disorder that causes death before the age of 2, in part by breakdown of myelin sheaths due to a deficiency in activity of the enzyme lysosomal hydrolase galactosylceramide beta-galactosidase (GALC). This enzyme is normally responsible for degradation of galactosylceramide and psychosine. Accumulation of both sphingolipids sets off a series of biological cascades culminating in demyelination and nervous system dysfunction. Due to the hematopoietic derivation of microglia, which normally express the GALG enzyme, Escolar et al. hypothesized that administration of cord blood into pediatric patients with Krabbe's Disease would result in neurological improvements. The investigators treated a total of 25 patients with Krabbe's Disease: 11 were asymptomatic and younger (12–44-days old) and 14 were symptomatic and older (142–352-days old). Following myeloablative conditioning and unrelated cord blood transplantation, the asymptomatic population had 100% engraftment and 100% survival at median follow-up of 3 years. Furthermore, the same population demonstrated progressive central myelination and approximately normalized gain in developmental skills. In contrast, although the population that was treated during the symptomatic phase also achieved 100% donor engraftment, minimal neurological improvement was observed and survival was only 43% at average follow-up of 3.4 years.[37] The importance of this study is the demonstration that cord blood can be used as a type of cellular "gene therapy" that systemically enters the patient circulation and normalizes cellular function in the area of need. It is important to point out that ablation of the defective microglia cells most likely did not occur in the patients since these cells are long-lived and resistant to usual myeloablative protocols. Accordingly, the dominance of the "healing" capacity of cord blood over the enzymatically defective wild-type cells is an interesting point to consider in light of other studies of regeneration.

29.2 Studies in the Regenerative Use of Cord Blood

Numerous investigations have been performed demonstrating that stem cells found in cord blood can differentiate into a variety of tissues. For example, using a variety of chemical agents and modification of culture conditions, it was demonstrated that cord blood mesenchymal cells, as well as freshly purified cells can be differentiated into cardiomyocyte-like cells which were capable of beating in culture.[38,39] The ability of bone-marrow-derived cells to differentiate into cardiomyocytes has been well-established and that the cells within cord blood that differentiate into cardiomyocytes are of a similar phenotype to the ones in bone marrow.[40,41] In bone-marrow-derived cardiomyocyte experiments, electromagnetic coupling and appropriate gap junction formation with cultured, freshly explanted cardiomyocytes were demonstrated.[42] Furthermore, it has been demonstrated that contacting bone-marrow-derived mesenchymal cells with cardiomyocytes induces differentiation into cardiomyocytes.[43] In contrast to in vivo experiments which suggest a positive effect of bone marrow stem cells in heart disease models, some in vitro evidence suggests that bone-marrow-derived cardiomyocytes may be proarrhythmic.[44] It remains to be seen whether cardiomyocytes derived from cord blood have similar properties, since to date, to the authors' knowledge, no side-by-side comparison has been made between bone marrow and cord blood in terms of cardiomyocyte differentiation.

The naturally residing stem cells in the liver, called "oval cells" express hematopoietic stem cell markers such as CD34 and c-kit, and can be repopulated in vivo by bone-marrow-derived cells, supports the notion that populations within cord blood may be capable of differentiating into hepatocytes.[45] Accordingly, investigators have demonstrated that growth factors such as HGF, alone or in combination with FGF-4, are capable of inducing in vitro generation of albumin-secreting hepatic-like cells.[46-48] In some experiments, it was demonstrated that an enhanced rate of hepatic differentiation from cord blood can be induced by mimicking injury in an in vitro system.[49] The differentiation from cord blood cell to hepatocyte-like cell is believed to occur in some systems by the cells passing through a mesenchymal state prior to differentiation.[50]

Numerous studies have also demonstrated differentiation of cord blood cells into various neuronal lineages.[51-56] Whether it is actually stem cells that differentiate into neurons or other cellular intermediaries exist remains to be completely answered. Some studies suggest that as in hepatic differentiation, cord blood cells pass through a mesenchymal phase before becoming neurons,[57] whereas other studies actually describe a monocytic-like intermediary.[58] It is believed that induction of differentiation can be accomplished by exposure to the local neuronal microenvironment, even in the adult brain.[59] Accordingly, these studies support the notion that cord blood cells may be useful for treatment of neurodegenerative diseases.

Numerous animal models have been performed to assess the potential of cord blood transplantation for treatment of degenerative diseases. We will overview some of these studies to provide a sample of the wide array of potential uses that cord blood may have when it is actually translated into a clinical approach.

Genetic and acquired diseases exist in which regeneration of muscle is desired. Particularly relevant are conditions such as Duchenne Muscular Dystrophy in which one essential gene is defective causing muscular degeneration and premature death (patients rarely live beyond 30). While gene therapy would be theoretically useful, practical clinical implementation has yet to occur. An alternative treatment would be supplementing the diseased individual with stem cells containing the appropriate gene. This was originally investigated using bone marrow stem cells. It is known that bone marrow stem cells are capable of differentiating into a wide variety of muscle-like cells. For example, bone marrow transplant with wild-type murine donors into a mouse model of muscular degeneration (laminin-alpha2-deficient (dy) mice) is capable of extending lifespan and enhancing growth rate, muscle strength, and respiratory function as compared to controls.[60] Similarly, in the mouse model of muscular dystrophy, bone marrow transplantation from wild-type donors results in mdx+ cells migrating and having beneficial function on injured muscles.[61] Accordingly, the use of cord blood transplantation was assessed in the dysferlin-deficient mouse, which is a model of muscle degenerative diseases, limb girdle muscular dystrophy type 2B form and Miyoshi myopathy. Systemic administration of human cord blood nucleated cells, or cord blood CD34+, lineage-negative cells under the cover of immune suppression lead to stable integration of human dysferlin positive cells into muscle. The authors did not comment on therapeutic effect, but suggested that increasing the number of cells trafficking to the muscle may be a useful therapy for development.[62] Another study investigated the effect of direct intramuscular administration of nucleated human cord blood cells into immunocompetent mice directly into injured muscles. The authors demonstrated incorporation of the human cells into regenerating muscle.[63] Unfortunately, neither of the two studies demonstrated therapeutic benefit.

In contrast to the relatively early stages of stem cell research for muscular disorders, utilization of stem cells for myocardial infarction is much more advanced. Patients with myocardial infarction are usually treated with stenting and thrombolytic agents; however, the death of existing myocytes, the formation of scar tissue, and pathological remodeling causes the majority of post-infarct patients to develop congestive heart failure. The rationale for stem cell therapy in the post-infarct situation is to supply cells capable of taking over the function of the cells that have died, and/or to increase local perfusion so as to allow cardiomyocytes that are hibernating to become functional. Bone marrow stem cells have demonstrated ability to reduce pathology left ventricular remodeling and restore left ventricular ejection fraction (LVEF) in numerous clinical studies.[64-66] It is believed that, at least in part, the CD34+ fraction of bone marrow is responsible for this effect, since even CD34+ cells from peripheral blood are also beneficial to post-infarct cardiac function.[67] Given the high content of CD34 cells in cord blood, as well as various cells with cardiomyocyte potential residing therein, numerous studies have investigated the use of cord blood in animal models of infarction. For example, Hirata et al. demonstrated that systemic administration of 2 × 10(5) human cord blood CD34(+) cells into Wistar rats suffering from myocardial infarction led to improvement of LVEF. Microscopic analysis demonstrated engraftment of human cells in the myocardial architecture.[68] Utility of CD133 cells derived from cord blood for myocardial regeneration post-infarct. Administration of 1.2–2 × 10(6) CD133+ cells 7 days post-infarct in athymic rats led to improvement in LV contractility by 42% in treated animals, whereas controls had a decrease in contractility of 39 ± 10% at 30 days post-infarct. Additionally, pathological ventricular remodeling as defined by decrease in thickness of the anterior wall was observed only in the control animals.[69] In order to deal with the low number of cells attainable from cord blood, experiments were performed to investigate the possibility of expanding endothelial progenitors ex vivo and using them for post-infarct repair. Culturing of cord blood in endothelium differentiation media allowed up to 40-fold expansion of cell number. These cells were capable of preserving LVEF in an animal model of infarction.[70] Using a large animal model, administration of 10(8) cultured unrestricted somatic stem cells (USCC) from human cord blood was performed in pigs with artificially occluded left anterior descending 4 weeks after occlusion. Improved regional perfusion, wall motion, and LVEF were observed in comparison to controls at 4 weeks post-cell-administration.[71] These and other animal models experiments[72-76] support the potential of cord blood cells for myocardial infarction, administered systemically, or locally.

Stroke is a significant cause of morbidity and mortality being the third cause of death and disability in the USA. Although rehabilitation procedures exist and are clinically implemented, no medical intervention has been approved as of yet. One therapeutic concept is administration of growth factors to either directly stimulate neurogenesis, or to increase perfusion and thereby allow neuronal populations to exit state of cell cycle arrest. This approach was assessed by systemic administration of the growth factor FGF-2. Although some patients demonstrated improvement in the acute stroke setting, the adverse effects, including hypotension associated with this intervention led to the halting of the Phase III trial.[77,78] Other approaches have included stereotactic administration of neurons derived from the human teratocarcinoma cell line NT-2. It was reported that some patients had increased metabolic activity at the grafted site; however, therapeutic results were not significant.[79,80] Given the ability of cord blood cells to secrete numerous neurotrophic factors,[81] as well as to directly differentiate into a variety of neurons,[82] the use of such cells in animal models of stroke was performed by numerous groups with demonstration of efficacy. For widespread clinical utilization, stereotactic implantation of cells is very difficult. Accordingly, a study was performed using the established middle cerebral artery occlusion (MCAO) rat model of stroke, comparing intravenous versus intrastriatal implantation of human cord blood cells under the cover of cyclosporin immune suppression. In contrast to non-transplanted animals, rats receiving cord blood either through the intravenous or intrastriatal route performed significantly better at task learning by the passive avoidance test, as well as overall behavioral recovery. In the step test, significant improvement was observed only in animals having received cells through the intravenous route. This study demonstrated the feasibility of systemic cord blood administration for treatment of stroke.[83] In order to determine whether cord blood administration induces a dose-dependent neurological recovery, the same group administered 10(4) up to 3–5 × 10(7) human cord blood cells into rats subjected to MCAO. The authors observed a dose-dependent recovery in behavioral performance as well as an inverse relationship between HUCBC dose and infarct size.[84] Using a similar MCAO model, it was reported that an umbilical cord population expressing the embryonic markers Oct-4, Rex-1, and Sox-2, but not hematopoietic markers was able to significantly inhibit behavioral defects.[85] Although the neuroprotective/neuroregenerative effects of cord blood cells are well-established by numerous other experiments,[86-90] the mechanisms of this effect are still being debated. For example, it was demonstrated that angiogenesis plays a critical role in cord blood mediated protection from stroke in a study demonstrating that treatment with angiogenic inhibitors can block beneficial effects of cell administration.[91] Such indirect and/or paracrine effects are also supported by observations that it is not necessary for the transplanted cells to enter the brain to mediate beneficial effects.[92]

In addition to the areas of muscular degeneration, cardiac infarction, and stroke, cord blood stem cells have demonstrated therapeutic efficacy in numerous other animal models such as enzymatic deficiencies,[93,94] autoimmune diabetes,[95,96] liver pathologies,[97-102] and even cancer.[103] Given these powerful preclinical observations, as well as the known multitude of stem cell activities found in cord blood, it only is natural that regenerative applications (besides in the area of hematopoiesis) would be pursued. As of yet, there is one Phase I trial being performed in patients with Type I diabetes involving infusion of autologous cord blood cells for restoration of islet function; however, the trial is ongoing and no data have been published.[104] One of the major limitations that are impeding regenerative application of cord blood transplants is the fact that in contrast to bone marrow, peripheral blood, or adipose derived stem cells, most patients do not have autologous cord blood available. This makes it necessary to use allogeneic, HLA-matched cord blood. It is currently dogma in the minds of most investigators that in the absence of immune suppression, administration of an HLA-matched cord blood graft into a non-immune suppressed host will result in rapid clearance of infused cells without therapeutic benefit. It is the purpose of this paper to reexamine this dogma and to discuss methods of making available cord blood transplantation for regenerative uses without the need for major host preconditioning that would normally preclude patients from having access to this technology. In order to begin this part of the discussion, we will start by first overviewing the basic immunology of cord blood.

29.3 Conclusion

Given the unique regenerative capabilities of cord blood, the easy accessibility of HLA-matched donors, and relative inexpensiveness as compared to other cellular therapies, it is of great interest therapeutically to explore its use into nonconditioned recipients. Another attractive feature of cord blood is that for regenerative activities administration can be systemic since in various models of tissue destruction, local administration does not significantly alter efficacy as compared to systemic.[83,105]

References

1. Halbrecht J. Fresh and stored placental blood. *Lancet.* 1939;2:1263.
2. Ende M, Ende N. Hematopoietic transplantation by means of fetal (cord) blood. A new method. *Va Med Mon.* 1918;1972(99):276-280.
3. Fauser AA, Messner HA. Granuloerythropoietic colonies in human bone marrow, peripheral blood, and cord blood. *Blood.* 1978;52:1243-1248.
4. Broxmeyer HE, Douglas GW, Hangoc G, et al. Human umbilical cord blood as a potential source of transplantable hematopoietic stem/progenitor cells. *Proc Natl Acad Sci USA.* 1989;86:3828-3832.
5. Gluckman E, Broxmeyer HA, Auerbach AD, et al. Hematopoietic reconstitution in a patient with Fanconi's anemia by means of umbilical-cord blood from an HLA-identical sibling. *N Engl J Med.* 1989;321:1174-1178.
6. O'Brien TA, Tiedemann K, Vowels MR. No longer a biological waste product: umbilical cord blood. *Med J Aust.* 2006;184:407-410.
7. Tse W, Laughlin MJ. Umbilical cord blood transplantation: a new alternative option. *Hematology (Am Soc Hematol Educ Program)* 2005:377-383.

8. Rubinstein P. Why cord blood? *Hum Immunol.* 2006;67: 398-404.
9. Schoemans H, Theunissen K, Maertens J, Boogaerts M, Verfaillie C, Wagner J. Adult umbilical cord blood transplantation: a comprehensive review. *Bone Marrow Transplant.* 2006;38:83-93.
10. Bradley MB, Cairo MS. Cord blood immunology and stem cell transplantation. *Hum Immunol.* 2005;66:431-446.
11. Koh LP. Unrelated umbilical cord blood transplantation in children and adults. *Ann Acad Med Singapore.* 2004;33: 559-569.
12. Cornetta K, Laughlin M, Carter S, et al. Umbilical cord blood transplantation in adults: results of the prospective Cord Blood Transplantation (COBLT). *Biol Blood Marrow Transplant.* 2005;11:149-160.
13. Schonberger S, Niehues T, Meisel R, et al. Transplantation of haematopoietic stem cells derived from cord blood, bone marrow or peripheral blood: a single centre matched-pair analysis in a heterogeneous risk population. *Klin Padiatr.* 2004;216:356-363.
14. Lekakis L, Giralt S, Couriel D, et al. Phase II study of unrelated cord blood transplantation for adults with high-risk hematologic malignancies. *Bone Marrow Transplant.* 2006;38:421-426.
15. Tomonari A, Takahashi S, Ooi J, et al. Cord blood transplantation for acute myelogenous leukemia using a conditioning regimen consisting of granulocyte colony-stimulating factor-combined high-dose cytarabine, fludarabine, and total body irradiation. *Eur J Haematol.* 2006;77:46-50.
16. Laporte JP, Lesage S, Portnoi MF, et al. Unrelated mismatched cord blood transplantation in patients with hematological malignancies: a single institution experience. *Bone Marrow Transplant.* 1998;22(suppl 1):S76-S77.
17. Sanz GF, Saavedra S, Jimenez C, et al. Unrelated donor cord blood transplantation in adults with chronic myelogenous leukemia: results in nine patients from a single institution. *Bone Marrow Transplant.* 2001;27:693-701.
18. Rocha V, Labopin M, Sanz G, et al. Transplants of umbilical-cord blood or bone marrow from unrelated donors in adults with acute leukemia. *N Engl J Med.* 2004;351:2276-2285.
19. Laughlin MJ, Barker J, Bambach B, et al. Hematopoietic engraftment and survival in adult recipients of umbilical-cord blood from unrelated donors. *N Engl J Med.* 2001;344:1815-1822.
20. Chao NJ, Liu CX, Rooney B, et al. Nonmyeloablative regimen preserves "niches" allowing for peripheral expansion of donor T-cells. *Biol Blood Marrow Transplant.* 2002;8:249-256.
21. Chao NJ, Koh LP, Long GD, et al. Adult recipients of umbilical cord blood transplants after nonmyeloablative preparative regimens. *Biol Blood Marrow Transplant.* 2004;10: 569-575.
22. Yuji K, Miyakoshi S, Kato D, et al. Reduced-intensity unrelated cord blood transplantation for patients with advanced malignant lymphoma. *Biol Blood Marrow Transplant.* 2005;11:314-318.
23. Perillo A, Ferrandina G, Pierelli L, Bonanno G, Scambia G, Mancuso S. Stem cell-based treatments for gynecological solid tumors. *Eur Rev Med Pharmacol Sci.* 2005;9:93-102.
24. Archuleta TD, Devetten MP, Armitage JO. Hematopoietic stem cell transplantation in hematologic malignancy. *Panminerva Med.* 2004;46:61-74.
25. Gluckman E, Rocha V, Arcese W, et al. Factors associated with outcomes of unrelated cord blood transplant: guidelines for donor choice. *Exp Hematol.* 2004;32:397-407.
26. Wagner JE. Umbilical cord transplantation. *Leukemia.* 1998;12(suppl 1):S30-32.
27. Styczynski J, Cheung YK, Garvin J, et al. Outcomes of unrelated cord blood transplantation in pediatric recipients. *Bone Marrow Transplant.* 2004;34:129-136.
28. Wagner JE, Barker JN, DeFor TE, et al. Transplantation of unrelated donor umbilical cord blood in 102 patients with malignant and nonmalignant diseases: influence of CD34 cell dose and HLA disparity on treatment-related mortality and survival. *Blood.* 2002;100:1611-1618.
29. Jaroscak J, Goltry K, Smith A, et al. Augmentation of umbilical cord blood (UCB) transplantation with ex vivo-expanded UCB cells: results of a phase 1 trial using the AastromReplicell System. *Blood.* 2003;101:5061-5067.
30. Barker JN, Weisdorf DJ, DeFor TE, et al. Transplantation of 2 partially HLA-matched umbilical cord blood units to enhance engraftment in adults with hematologic malignancy. *Blood.* 2005;105:1343-1347.
31. Magro E, Regidor C, Cabrera R, et al. Early hematopoietic recovery after single unit unrelated cord blood transplantation in adults supported by co-infusion of mobilized stem cells from a third party donor. *Haematologica.* 2006;91:640-648.
32. Kim DW, Chung YJ, Kim TG, Kim YL, Oh IH. Cotransplantation of third-party mesenchymal stromal cells can alleviate single-donor predominance and increase engraftment from double cord transplantation. *Blood.* 2004;103:1941-1948.
33. Knutsen AP, Wall DA. Umbilical cord blood transplantation in severe T-cell immunodeficiency disorders: two-year experience. *J Clin Immunol.* 2000;20:466-476.
34. Bhattacharya A, Slatter MA, Chapman CE, et al. Single centre experience of umbilical cord stem cell transplantation for primary immunodeficiency. *Bone Marrow Transplant.* 2005; 36:295-299.
35. Mao P, Zhu Z, Wang H, et al. Sustained and stable hematopoietic donor-recipient mixed chimerism after unrelated cord blood transplantation for adult patients with severe aplastic anemia. *Eur J Haematol.* 2005;75:430-435.
36. Jaing TH, Hung IJ, Yang CP, Chen SH, Sun CF, Chow R. Rapid and complete donor chimerism after unrelated mismatched cord blood transplantation in 5 children with beta-thalassemia major. *Biol Blood Marrow Transplant.* 2005;11:349-353.
37. Escolar ML, Poe MD, Provenzale JM, et al. Transplantation of umbilical-cord blood in babies with infantile Krabbe's disease. *N Engl J Med.* 2005;352:2069-2081.
38. Vanelli P, Beltrami S, Cesana E, et al. Cardiac precursors in human bone marrow and cord blood: in vitro cell cardiogenesis. *Ital Heart J.* 2004;5:384-388.
39. Kadivar M, Khatami S, Mortazavi Y, Shokrgozar MA, Taghikhani M, Soleimani M. In vitro cardiomyogenic potential of human umbilical vein-derived mesenchymal stem cells. *Biochem Biophys Res Commun.* 2006;340:639-647.
40. Makino S, Fukuda K, Miyoshi S, et al. Cardiomyocytes can be generated from marrow stromal cells in vitro. *J Clin Invest.* 1999;103:697-705.
41. Hakuno D, Fukuda K, Makino S, et al. Bone marrow-derived regenerated cardiomyocytes (CMG Cells) express functional adrenergic and muscarinic receptors. *Circulation.* 2002; 105:380-386.

42. Pijnappels DA, Schalij MJ, van Tuyn J, et al. Progressive increase in conduction velocity across human mesenchymal stem cells is mediated by enhanced electrical coupling. *Cardiovasc Res.* 2006;72:282-291.
43. Wang T, Xu Z, Jiang W, Ma A. Cell-to-cell contact induces mesenchymal stem cell to differentiate into cardiomyocyte and smooth muscle cell. *Int J Cardiol.* 2006;109: 74-81.
44. Chang MG, Tung L, Sekar RB, et al. Proarrhythmic potential of mesenchymal stem cell transplantation revealed in an in vitro coculture model. *Circulation.* 2006;113: 1832-1841.
45. Petersen BE, Bowen WC, Patrene KD, et al. Bone marrow as a potential source of hepatic oval cells. *Science.* 1999;284:1168-1170.
46. Tang XP, Zhang M, Yang X, Chen LM, Zeng Y. Differentiation of human umbilical cord blood stem cells into hepatocytes in vivo and in vitro. *World J Gastroenterol.* 2006;12: 4014-4019.
47. Kang XQ, Zang WJ, Bao LJ, et al. Fibroblast growth factor-4 and hepatocyte growth factor induce differentiation of human umbilical cord blood-derived mesenchymal stem cells into hepatocytes. *World J Gastroenterol.* 2005;11: 7461-7465.
48. Teramoto K, Asahina K, Kumashiro Y, et al. Hepatocyte differentiation from embryonic stem cells and umbilical cord blood cells. *J Hepatobiliary Pancreat Surg.* 2005;12: 196-202.
49. Yu J, Zhang FT, Wan HJ, Ye J, Long X, Fang JZ. Conversion of human umbilical cord blood-derived cells into hepatocyte-like cells in a culture system mimicking hepatic injury. *Beijing Da Xue Xue Bao.* 2005;37:402-405.
50. Lee KD, Kuo TK, Whang-Peng J, et al. In vitro hepatic differentiation of human mesenchymal stem cells. *Hepatology.* 2004;40:1275-1284.
51. Buzanska L, Jurga M, Domanska-Janik K. Neuronal differentiation of human umbilical cord blood neural stem-like cell line. *Neurodegener Dis.* 2006;3:19-26.
52. Buzanska L, Jurga M, Stachowiak EK, Stachowiak MK, Domanska-Janik K. Neural stem-like cell line derived from a nonhematopoietic population of human umbilical cord blood. *Stem Cells Dev.* 2006;15:391-406.
53. Yan XH, Huang RB. Differentiation of human umbilical cord blood mesenchymal stem cells toward neurons induced by baicalin in vitro. *Zhonghua Er Ke Za Zhi.* 2006;44:214-219.
54. Jurga M, Markiewicz I, Sarnowska A, et al. Neurogenic potential of human umbilical cord blood: neural-like stem cells depend on previous long-term culture conditions. *J Neurosci Res.* 2006;83:627-637.
55. Ortiz-Gonzalez XR, Keene CD, Verfaillie CM, Low WC. Neural induction of adult bone marrow and umbilical cord stem cells. *Curr Neurovasc Res.* 2004;1:207-213.
56. Chen N, Hudson JE, Walczak P, et al. Human umbilical cord blood progenitors: the potential of these hematopoietic cells to become neural. *Stem Cells.* 2005;23:1560-1570.
57. Fu YS, Cheng YC, Lin MY, et al. Conversion of human umbilical cord mesenchymal stem cells in Wharton's jelly to dopaminergic neurons in vitro: potential therapeutic application for Parkinsonism. *Stem Cells.* 2006;24:115-124.
58. Liu TY, Zhang C, Xiao LL, Yao XL, Feng SW, Zeng Y. Induced differentiation of human cord blood monocytes into neuron-like cells in vitro. *Di Yi Jun Yi Da Xue Xue Bao.* 2005;25:152-155.
59. Walczak P, Chen N, Hudson JE, et al. Do hematopoietic cells exposed to a neurogenic environment mimic properties of endogenous neural precursors? *J Neurosci Res.* 2004;76: 244-254.
60. Hagiwara H, Ohsawa Y, Asakura S, Murakami T, Teshima T, Sunada Y. Bone marrow transplantation improves outcome in a mouse model of congenital muscular dystrophy. *FEBS Lett.* 2006;580:4463-4468.
61. Feng SW, Zhang C, Yao XL, et al. Dystrophin expression in mdx mice after bone marrow stem cells transplantation. *Zhongguo Yi Xue Ke Xue Yuan Xue Bao.* 2006;28:178-181.
62. Kong KY, Ren J, Kraus M, Finklestein SP, Brown RH Jr. Human umbilical cord blood cells differentiate into muscle in sjl muscular dystrophy mice. *Stem Cells.* 2004;22: 981-993.
63. Brzoska E, Grabowska I, Hoser G, et al. Participation of stem cells from human cord blood in skeletal muscle regeneration of SCID mice. *Exp Hematol.* 2006;34:1261-1269.
64. Ge J, Li Y, Qian J, et al. Efficacy of emergent transcatheter transplantation of stem cells for treatment of acute myocardial infarction (TCT-STAMI). *Heart.* 2006;92: 1764-1767.
65. Pudil R, Vojadek J, Filip S, et al. Transplantation of bone marrow derived progenitor cells in acute myocardial infarction. The first results. *Acta Medica (Hradec Kralove).* 2005;48:153-155.
66. Janssens S, Theunissen K, Boogaerts M, Van de Werf F. Bone marrow cell transfer in acute myocardial infarction. *Nat Clin Pract Cardiovasc Med.* 2006;3(suppl 1):S69-S72.
67. Archundia A, Aceves JL, Lopez-Hernandez M, Alvarado M, Rodriguez E. Diaz Quiroz G, Paez A, Rojas FM, Montano LF: Direct cardiac injection of G-CSF mobilized bone-marrow stem-cells improves ventricular function in old myocardial infarction. *Life Sci.* 2005;78:279-283.
68. Hirata Y, Sata M, Motomura N, et al. Human umbilical cord blood cells improve cardiac function after myocardial infarction. *Biochem Biophys Res Commun.* 2005;327:609-614.
69. Leor J, Guetta E, Feinberg MS, et al. Human umbilical cord blood-derived CD133+ cells enhance function and repair of the infarcted myocardium. *Stem Cells.* 2006;24:772-780.
70. Ott I, Keller U, Knoedler M, et al. Endothelial-like cells expanded from CD34+ blood cells improve left ventricular function after experimental myocardial infarction. *FASEB J.* 2005;19:992-994.
71. Kim BO, Tian H, Prasongsukarn K, et al. Cell transplantation improves ventricular function after a myocardial infarction: a preclinical study of human unrestricted somatic stem cells in a porcine model. *Circulation.* 2005;112:I96-I104.
72. Ma N, Ladilov Y, Moebius JM, et al. Characterization of two populations of mesenchymal progenitor cells in umbilical cord blood. *Cardiovasc Res.* 2006;71:158-169.
73. Min JJ, Ahn Y, Moon S, et al. In vivo bioluminescence imaging of cord blood derived mesenchymal stem cell transplantation into rat myocardium. *Ann Nucl Med.* 2006;20: 165-170.
74. Ishikawa F, Shimazu H, Shultz LD, et al. Purified human hematopoietic stem cells contribute to the generation of cardiomyocytes through cell fusion. *FASEB J.* 2006;20: 950-952.

75. Ma N, Stamm C, Kaminski A, et al. Human cord blood cells induce angiogenesis following myocardial infarction in NOD/scid-mice. *Cardiovasc Res*. 2005;66:45-54.
76. Botta R, Gao E, Stassi G, et al. Heart infarct in NOD-SCID mice: therapeutic vasculogenesis by transplantation of human CD34+ cells and low dose CD34+KDR+ cells. *FASEB J*. 2004;18:1392-1394.
77. Bogousslavsky J, Victor SJ, Salinas EO, et al. Fiblast (trafermin) in acute stroke: results of the European-Australian phase II/III safety and efficacy trial. *Cerebrovasc Dis*. 2002;14:239-251.
78. http://www.strokecenter.org/trials/TrialDetail.aspx?tid=48.
79. Kondziolka D, Wechsler L, Goldstein S, et al. Transplantation of cultured human neuronal cells for patients with stroke. *Neurology*. 2000;55:565-569.
80. Meltzer CC, Kondziolka D, Villemagne VL, et al. Serial [18F] fluorodeoxyglucose positron emission tomography after human neuronal implantation for stroke. *Neurosurgery*. 2001;49:586-591 (discussion 591-582).
81. Fan CG, Zhang QJ, Tang FW, Han ZB, Wang GS, Han ZC. Human umbilical cord blood cells express neurotrophic factors. *Neurosci Lett*. 2005;380:322-325.
82. Habich A, Jurga M, Markiewicz I, Lukomska B, Bany-Laszewicz U, Domanska-Janik K. Early appearance of stem/progenitor cells with neural-like characteristics in human cord blood mononuclear fraction cultured in vitro. *Exp Hematol*. 2006;34:914-925.
83. Willing AE, Lixian J, Milliken M, et al. Intravenous versus intrastriatal cord blood administration in a rodent model of stroke. *J Neurosci Res*. 2003;73:296-307.
84. Vendrame M, Cassady J, Newcomb J, et al. Infusion of human umbilical cord blood cells in a rat model of stroke dose-dependently rescues behavioral deficits and reduces infarct volume. *Stroke*. 2004;35:2390-2395.
85. Xiao J, Nan Z, Motooka Y, Low WC. Transplantation of a novel cell line population of umbilical cord blood stem cells ameliorates neurological deficits associated with ischemic brain injury. *Stem Cells Dev*. 2005;14:722-733.
86. Ereniev SI, Semchenko VV, Sysheva EV, et al. Effect of allo- and xenotransplantation of embryonic nervous tissue and umbilical cord blood-derived stem cells on structural and functional state of cerebral cortex of albino rats in posttraumatic period. *Bull Exp Biol Med*. 2005;140:612-615.
87. Newman MB, Willing AE, Manresa JJ, Davis-Sanberg C, Sanberg PR. Stroke-induced migration of human umbilical cord blood cells: time course and cytokines. *Stem Cells Dev*. 2005;14:576-586.
88. Chen SH, Chang FM, Tsai YC, Huang KF, Lin CL, Lin MT. Infusion of human umbilical cord blood cells protect against cerebral ischemia and damage during heatstroke in the rat. *Exp Neurol*. 2006;199:67-76.
89. Vendrame M, Gemma C, de Mesquita D, et al. Anti-inflammatory effects of human cord blood cells in a rat model of stroke. *Stem Cells Dev*. 2005;14:595-604.
90. Naruse K, Hamada Y, Nakashima E, et al. Therapeutic neovascularization using cord blood-derived endothelial progenitor cells for diabetic neuropathy. *Diabetes*. 2005;54:1823-1828.
91. Taguchi A, Soma T, Tanaka H, et al. Administration of CD34+ cells after stroke enhances neurogenesis via angiogenesis in a mouse model. *J Clin Invest*. 2004;114:330-338.
92. Borlongan CV, Hadman M, Sanberg CD, Sanberg PR. Central nervous system entry of peripherally injected umbilical cord blood cells is not required for neuroprotection in stroke. *Stroke*. 2004;35:2385-2389.
93. Garbuzova-Davis S, Gografe SJ, Sanberg CD, et al. Maternal transplantation of human umbilical cord blood cells provides prenatal therapy in Sanfilippo type B mouse model. *FASEB J*. 2006;20:485-487.
94. Garbuzova-Davis S, Willing AE, Desjarlais T, Sanberg CD, Sanberg PR. Transplantation of human umbilical cord blood cells benefits an animal model of Sanfilippo syndrome type B. *Stem Cells Dev*. 2005;14:384-394.
95. Yoshida S, Ishikawa F, Kawano N, et al. Human cord blood–derived cells generate insulin-producing cells in vivo. *Stem Cells*. 2005;23:1409-1416.
96. Ende N, Chen R, Reddi AS. Effect of human umbilical cord blood cells on glycemia and insulitis in type 1 diabetic mice. *Biochem Biophys Res Commun*. 2004;325:665-669.
97. Turrini P, Monego G, Gonzalez J, et al. Human hepatocytes in mice receiving pre-immune injection with human cord blood cells. *Biochem Biophys Res Commun*. 2005;326:66-73.
98. Di Campli C, Piscaglia AC, Pierelli L, et al. A human umbilical cord stem cell rescue therapy in a murine model of toxic liver injury. *Digest Liver Dis*. 2004;36:603-613.
99. Almeida-Porada G, Porada CD, Chamberlain J, Torabi A, Zanjani ED. Formation of human hepatocytes by human hematopoietic stem cells in sheep. *Blood*. 2004;104: 2582-2590.
100. Sharma AD, Cantz T, Richter R, et al. Human cord blood stem cells generate human cytokeratin 18-negative hepatocyte-like cells in injured mouse liver. *Am J Pathol*. 2005;167:555-564.
101. Nonome K, Li XK, Takahara T, et al. Human umbilical cord blood-derived cells differentiate into hepatocyte-like cells in the Fas-mediated liver injury model. *Am J Physiol Gastrointest Liver Physiol*. 2005;289:G1091-G1099.
102. Piscaglia AC, Di Campli C, Zocco MA, et al. Human cordonal stem cell intraperitoneal injection can represent a rescue therapy after an acute hepatic damage in immunocompetent rats. *Transplant Proc*. 2005;37:2711-2714.
103. Ende N, Chen R, Reddi AS. Administration of human umbilical cord blood cells delays the onset of prostate cancer and increases the lifespan of the TRAMP mouse. *Cancer Lett*. 2006;231:123-128.
104. http://www.clinicaltrials.gov/ct/show/NCT00305344?order=1.
105. Hofmann M, Wollert KC, Meyer GP, et al. Monitoring of bone marrow cell homing into the infarcted human myocardium. *Circulation*. 2005;111:2198-2202.

30 Immune Privilege of Cord Blood

Neil H. Riordan and Thomas E. Ichim

30.1 Introduction

Cord blood transplants are commonly thought of as equivalent to bone marrow transplants, with neutropenia associated morbidity and mortality and graft versus host disease (GVHD). This makes it easy for us to forget that cord blood transfusion has been used without immune suppression for many decades. One of the first documented cases is a report from 1939 describing various advantages of using cord blood as a substitute for peripheral blood in certain medical conditions, especially when increased oxygen carrying capacity of fetal hemoglobin is of benefit.[1] It was not until the early 1970s that consideration for cord blood as a source of hematopoietic stem cells was made. Ende et al. reported a case of a patient with acute leukemia treated with high-dose chemotherapy in which cord blood mononuclear cells were administered to accelerate granulocytic recovery. This case report demonstrated feasibility of chimerism induction and lack of serious adverse effects.[2] Subsequent studies by the same group quantified hematopoietic colony forming units in cord blood and proposed the use of this stem cell population source for postradiation disasters.[3] In 1989, cord blood from a related sibling was used to treat a 5-year-old patient with Fanconi anemia following a myeloablative conditioning regimen similar to that used in bone marrow transplantation.[4] Donor cell chimerism and remission of clinical disease was noted. From these early days of cord blood transplantation, the practice has grown substantially. Driven by observations that cord blood, in contrast to bone marrow has less stringent matching requirements, lower GVHD, and lower risk of contamination,[5] cord blood has been used for over 8,000 transplants to date.[6,7] In agreement with its widespread implementation, the cord blood banking has grown substantially. As of last year, over 280,000 stored cord blood units are available for allogeneic use from 36 public nonprofit umbilical cord blood banks in 23 countries.[8] Numerous private banks store cord blood for autologous use. Currently accepted indications for cord blood are primarily as a substitute for bone marrow hematopoietic stem cells. This diminishes the significance of storing autologous cord blood since it is useless for metabolic disorders (the autologous cord blood will still have the mutation) and risky for oncological disorders (since the autologous cells may have premalignant mutations).[9]

With few exceptions, the current practice of cord blood transplantation as reported in the medical literature is based on preconditioning of the recipient to ablate endogenous hematopoietic cells. The few exceptions include unpublished reports by Joanne Kurtzberg using autologous cord blood for patients with cerebral palsy, clinical trials using autologous cord blood for type 1 diabetes, and published reports of Ex-US compassionate use cases. By far, the largest number of publications in the area of non conditioned cord blood use have been generated by the editor of this book, Dr. Bhattacharya who has described safety and therapeutic effects in conditions ranging from rheumatoid arthritis, to HIV, to anemia, to cancer.[10] The use of cord blood without immune suppression is currently growing. Medical tourism clinics such as Beike International, Stem Cell Biotherapies, and BioEden Laboratories have treated large numbers of patients using fresh or ex vivo expanded cord blood stem cells. Given the numerous reports both in the published and non-published literature regarding

N.H. Riordan (✉)
Medistem Panama, Inc., City of Knowledge, Republic of Panama

N. Bhattacharya and P. Stubblefield (eds.), *Regenerative Medicine Using Pregnancy-Specific Biological Substances*,
DOI: 10.1007/978-1-84882-718-9_30,

allogeneic cord blood transplantation, we have decided to address this issue in detail. Furthermore, in conditions where hematopoietic ablation is not required, but stem cells are needed for repair/regeneration of injured tissue, cord blood would be an ideal source of stem cells if the issues of allogeneic immunity are addressed. Currently, most stem cell clinical trials use autologous bone marrow/mobilized cells for conditions such as heart failure,[11] liver failure,[12,13] and peripheral artery disease.[14] Unfortunately, there are several drawbacks with autologous cells, including limitation in number, lose of proliferative activity with age, and degenerative conditions.[15-18] On the one hand, cord blood cells possess a higher regenerative activity, on the other hand, they alleviate the need for bone marrow aspiration or mobilization, procedures that are not unassociated with risks. However, the possible use of allogeneic cord blood stem cells would have to address several questions. For example, why is there no GVHD seen in patients receiving cord blood without immune suppression, while it is seen clinically in patients receiving standard conditioned protocols? Does the cord blood graft get cleared? If it does, can there still be some therapeutic effects? Are the stem cells in the cord blood responsible for therapeutic effects or do other cell types participate? In order to address these questions, let us first begin by discussing the various components of cord blood.

30.2 Cord Blood Hematopoietic Cells

The main reason cord blood is used these days is because of its hematopoietic stem cell content. It is known that before birth, there is an increase in circulating stem cell numbers in fetal circulation, which accounts for the high concentration of CD34-positive cells (approximately 1 per 100 nucleated cells). Thus the concentration of CD34 cells is approximately similar between cord blood and bone marrow. However, in contrast to marrow, CD34 cells from cord blood are capable of significantly larger expansion without differentiation in vitro, as well as having increased numbers of colony forming units.[19] In vivo studies have demonstrated that cord blood CD34 cells contain within them higher numbers of long-term culture initiating cells (LTCIC) and severe combined immunodeficiency (SCID) repopulating cells.[20,21] The advantage of cord blood as opposed to bone marrow in terms of hematopoietic reconstitution and ability for serial passage has been ascribed to higher level of telomerase expression.[22] The CD34 cells from the cord blood are not only able to differentiate into hematopoietic lineages but seem to possess a degree of plasticity. It has been reported that cord blood-derived CD34 cells can be induced to differentiate into cardiomyocytes, mature endothelial cells, alveolar cells, renal cells, smooth muscle, hepatocytes, and neurons.[23-27] Another type of hematopoietic stem cell found in the cord blood is the lineage-negative, CD34-negative population.[28] These cells appear to be precursors to CD34-positive hematopoietic cells and have an increased hematopoietic reconstituting activity based on SCID repopulating assays. An interesting fact regarding this population is that after culture with stromal cells, they acquire CD34 expression, suggesting they are progenitor cells.[29] Hematopoietic cells may be purified from cord blood using a novel system being developed by the company Aldagen that purifies for cells expressing high levels of the enzyme aldehyde dehydrogenase.[30] While in vitro and in vivo reconstitution activity correlates with expression of this enzyme, it is interesting that cells selected on this basis comprise CD34-negative stem cells.[31] In summary, the hematopoietic activity of cord blood stem cells is dispersed over several cellular phenotypes. But, how do these cells contribute to immunity?

30.3 Immune Modulation by Hematopoietic Stem Cells

CD34 hematopoietic stem cells are known to autoregulate proliferation through secretion of the cytokine TGF-®. It is widely published that early CD34 hematopoietic progenitors are maintained in a G0 state through this mechanism and that inhibition of the autocrine loop by either antibody or antisense oligonucleotides induces the stem cells to proliferate.[32-35] It is possible that the local secretion of this cytokine may have effects that are not limited to the stem cell itself. This is especially relevant since CD34 cells have been demonstrated to enter lymph nodes and have ability to participate in various interactions with T cells outside of the bone marrow.[36] TGF-® is a potent suppressor of numerous immunological activities including dendritic cell maturation,[37] T cell activation,[38] and NK activity.[39]

Additionally, this cytokine is known to induce generation of T regulatory cells,[40] which have the ability to inhibit other T cells, as well as suppress function of neutrophils,[41] macrophages,[42] and dendritic cells.[43] Therefore, one possibility is that hematopoietic stem cells may have an immune suppressive effect through secretion of TGF-®. This possibility is supported by a variety of experiments. In the early 1980s, a series of experiments described a "natural suppressor" cell which was capable of forming blood cells, like hematopoietic stem cells, and also suppressed T cell activation in an antigen-nonspecific manner.[44] These cells were initially identified as the antigen-nonspecific immune regulatory cell found in the bone marrow. Initial experiments were performed with mixed lymphocyte reactions using bone marrow and peripheral blood mononuclear cells that demonstrated there was a suppressive cell in the bone marrow that would inhibit lymphocyte proliferation not only in response to alloantigens but also to mitogens. These suppressive cells expressed markers of myeloid progenitors such as CD31 and had demonstrated ability to bind to wheat germ agglutinin and expression of IL-3 receptor.[45,46] In vivo, the myeloid suppressor population exerted immune-suppressive properties in response to myelopoietic cytokines such as GM-CSF, which seemed to be related with their ability to generate myeloid colonies.[47,48] More recent studies have demonstrated that myeloid-committed hematopoietic stem cells are involved in immune suppression in situations ranging from cancer,[49-51] to pregnancy,[52] to postinjury immune suppression.[53,54] In cancer, it has been demonstrated that inducing differentiation of tumor-mobilized hematopoietic myeloid suppressor cells is associated with derepression of antitumor immunity.[55] Given that CD34 cells appear to be intrinsically immune suppressive, one would imagine that administration of high numbers of these cells could be used to induce tolerance or at least immune modulation. Indeed, it has been reported that hematopoietic transplant-associated tolerance is correlated with higher CD34 content in the graft.[56-58] A possible means by which CD34 induce immune suppression is through expression of Fas ligand, which kills activated T cells expressing Fas.[59] Furthermore, human mixed lymphocyte reaction responder cells can be specifically induced to undergo apoptosis by stimulator, but not third party CD34 cells obtained from bone marrow.[60] In summary, it appears that the hematopoietic stem cell component of cord blood has immune modulatory activity. To what extent this immune modulatory activity may be used in absence of immune suppression is not known.

30.4 Mesenchymal Stem Cells in Cord Blood

Mesenchymal stem cells (MSC) are classically defined as adherent, nonhematopoietic cells expressing markers such as CD90, CD105, and CD73, while lacking expression of CD14, CD34, and CD45, and being able to differentiate into adipocytes, chondrocytes, and osteocytes in vitro after treatment with differentiation-inducing agents.[61] Although early studies in the late 1960s initially identified MSC in the bone marrow,[62] more recent studies have reported that these cells can be purified from various tissues such as adipose,[63] heart,[64] Wharton's Jelly,[65] dental pulp,[66] peripheral blood,[67] cord blood,[68] and more recently, menstrual blood.[69-71] Although the mesenchymal stem cell component of cord blood is relatively rare, these cells have been noted to expand greatly in vitro,[72] with the suggestion being made that they may be responsible for various regenerative activities associated with cord blood transplantation.[73] The MSC population is very important to our discussion of cord blood immune privilege, since these cells are not only hypoimmunogenic but in some conditions have been demonstrated to be actively immune suppressive,[74] regardless of source.[75] Several mechanisms of immune suppression have been identified, for example, production of PGE-2, interleukin-10 and expression of the tryptophan catabolizing enzyme indoleamine 2,3,-dioxygenase as well as galectin-1.[76, 77] Additionally, MSC also have the ability to nonspecifically modulate the immune response through the suppression of dendritic cell maturation and antigen presenting abilities.[78,79] Immune suppressive activity is not dependent on prolonged culture of mesenchymal stem cells since functional induction of allogeneic T cell apoptosis was also demonstrated using freshly isolated, irradiated, mesenchymal stem cells.[80] Others have also demonstrated that MSC have the ability to preferentially induce expansion of antigen-specific T regulatory cells with the CD4+ CD25+ phenotype.[81] MSC can antigen specifically inhibit immune responses as observed in a murine model of multiple sclerosis, experimental autoimmune encephalomyelitis, in which administration of these cells lead to inhibition of disease onset.[82]

The immune suppressive properties of MSC are believed to account for the ability to achieve therapeutic effects in an allogeneic manner, without need for matching. Allogeneic bone marrow-derived MSC have been used by academic investigators with clinical benefit in the treatment of diseases such as GVHD,[83-88] osteogenesis imperfecta,[89] Hurler syndrome, metachromatic leukodystrophy,[90] and acceleration of hematopoietic stem cell engraftment.[91-93] The company Osiris Therapeutics has successfully completed Phase I safety studies using allogeneic bone marrow MSCs is now in efficacy finding clinical trials (Phase II and Phase III) for Type I Diabetes, Crohn's Disease, and Graft Versus Host Disease using allogeneic bone marrow-derived MSC. Intravenous administration of allogeneic MSCs by Osiris was also reported to induce a statistically significant improvement in cardiac function in a double-blind study.[94] Other companies have entered clinical trials using allogeneic mesenchymal stem cell-based products. Athersys is currently in Phase I trials using its MultiStem technology, which involves ex vivo expanded MAPC for postinfarct heart repair.[95] Angioblast Systems has recently announced initiation of Phase II trials using Mesenchymal Precursor Cells for stimulation of cardiac angiogenesis.[96] Neuronyx is in Phase I clinical trials using allogeneic human adult bone marrow-derived somatic cells (*hABM-SC*) for postinfarct healing.[97] Allogeneic placenta and cord blood-derived MSC have also been used for treatment of heart failure[98] and Buerger's Disease,[99] respectively.

Since MSC can evade the immune system, it would be interesting to find out whether patients receiving cord blood grafts actually have donor MSC chimerism. Although such studies have not been performed clinically, an experiment of nature supports this possibility. It is known that the placenta does not act as a complete barrier to cellular trafficking. For example, maternal cells have been detected in offspring and vice versa.[100,101] Chen et al. demonstrated that MSC can transplacentally traffic via an active VEGFR-1 and integrin-dependent manner.[102] Clinically, fetal-derived cells have been found in mothers decades after pregnancy. These fetal cells have been postulated to play a reparative/regenerative function since they are found in mothers surrounding areas of injury. For example, a recent publication demonstrated Y-chromosome-positive tissue of fetal origin surrounding injured maternal lung tissue when examined by biopsy.[103] Interestingly, the cells trafficking in a fetal to maternal manner have been demonstrated to be mesenchymal in origin.[104] Thus one wonders whether administration of allogeneic cord blood may provide some "seed MSC" that engraft in the recipient tissue, expand, and provide reparative activity.

30.5 Non-Stem Cell Components of Cord Blood Are Hypoimmunogenic

It is known that cord blood consists of similar immunological populations of blood cells as peripheral blood, with the exception of the immature status of these cells. Accordingly, there are numerous studies that suggest cord blood is less immunogenic as a whole in comparison to peripheral blood. For example, the most potent antigen presenting cell, the dendritic cell, possesses unique properties when freshly extracted from cord blood. Specifically, cord blood dendritic cells are poor stimulators of mixed lymphocyte reaction,[105,106] weakly support mitogen-induced T cell proliferation,[107] possess a predominantly lymphoid phenotype, lack costimulatory molecules,[108-111] and are believed to be involved in the noninflammatory Th2 bias of the neonate.[108-110] Cord blood dendritic cell progenitors also exhibit distinct properties such as enhanced susceptibility to natural and artificial immune suppressants.[112] When cord blood versus peripheral blood-derived dendritic cells are assessed for ability to stimulate immune response to apoptotic or necrotic cells, peripheral blood-derived dendritic cells upregulate costimulatory molecules and stimulate T cell proliferation, whereas cord blood-derived dendritic cells do not. However, dendritic cells from both sources are effectively activated by LPS.[113] A property of cord blood dendritic cell progenitors that is of interest is their propensity toward generating tolerogenic cells. It is reported that growth of cord blood progenitors in M-CSF gives rise to a potently suppressive tolerogenic dendritic cell phenotype. These dendritic cells are not only poor allostimulators, but give rise to CD4+ CD25+ T regulatory cells that are capable of inhibiting mixed lymphocyte reactions.[114] Another interesting tolerogenic feature of cord blood dendritic cells is their propensity to secrete large numbers of MHC II-bearing exosomes that lack expression of costimulatory molecules.[115] This type of exosome was used for prevention of autoimmune disease by other authors.[116] Given the immaturity and anti-inflammatory activity of cord blood dendritic cells, it is suggested that cord blood, in general, will be more poorly immunogenic as compared to other sources of nucleated cells. A comparison may be made between cord blood grafts and liver transplants in that HLA-matching for liver transplants does not seem to affect graft survival.[117] Indeed, dendritic cell populations with a primarily lymphoid phenotype, similar to those found in cord blood are known to predominate in the liver.[117] In summary, the immature nature of the various immunological cells in cord blood is believed to be associated with a propensity toward hypoimmunogenicity.

30.6 Immune Effectors in Cord Blood

As with peripheral blood, cord blood has numerous immunological populations. We will concentrate on the most well-characterized cells in the cord blood with effector function, this being the T cells, and conversely the T regulatory cells. The majority of studies examining cell populations such as NK, NKT, and gamma delta T cells in the cord blood have actually used cord blood as a starting population for in vitro expansion and hence are outside the scope of the current discussion.[118-123] T cells from cord blood are known to have a propensity toward an anti-inflammatory phenotype. This is illustrated, for example, in experiments with CD4+ T cells from cord blood which were shown to produce significantly lower IFN-gamma and higher IL-10 upon activation with mature dendritic cells as opposed to control adult blood-derived CD4+ T cells.[124] Other experiments have demonstrated hyporesponsiveness to mitogen and MLR stimulation,[125,126] as well as reduced levels of IL-2 production and IL-2 responsiveness as opposed to adult T cells.[127] This is not to say that cord blood T cells are not capable of mounting inflammatory and Th1 immunological attacks.[128] For example, GVHD, in some cases lethal, is a clinical reality in some cord blood transplant patients. However, it should be noted that cord blood transplantation in the vast majority of cases takes place following ablation of host T cells. This creation of an "empty compartment" naturally allows for homeostatic expansion, which conceptually primes T cells for aggressive immune reactions without a requirement for a second signal.[129] It is known from the transplantation literature that T cells reconstituting a host that has been lymphoablated are resistant to costimulatory blockade and tolerance induction.[130] Furthermore, the pioneering experiments of Rosenberg's group demonstrated that infusion of tumor-specific lymphocytes following ablation of the recipient T cells, using conditions similar to those used in cord blood transplant preconditioning allows for highly aggressive anti-tumor responses that otherwise would not be observed.[131]

In addition to conventional T cells, cord blood is known to contain a population of T regulatory (Treg) cells that possess immune suppressive activity. The role of Tregs in immunological function is to control, in an antigen-specific manner, hyperimmune activation. Treg depletion in animal models is associated with autoimmunity and transplant rejection,[132] whereas, augmented Treg function is found in pregnancy and cancer.[118,133] These Treg cells typically display the phenotype CD4+ CD25+, are resistant to FasL-mediated apoptosis (in contrast to adult peripheral blood Tregs which are sensitive[134]), and inhibit proliferation of CD4+ CD25-T cells with several-fold more potency than Tregs isolated from adult peripheral blood.[135] Additionally, in comparison to adult peripheral blood, cord blood cells are found at a much higher frequency in cord.[136] It has been demonstrated that Tregs are associated with protection from autoimmune disease in animal models, and clinical remission of autoimmunity.[135,137,138] This suggests that the high Treg content and suppressive activity of cord blood may not only be one of the reasons for lower GVHD as compared to other stem cell sources, but also that cord blood-derived cells may have therapeutic applications of immune dysregulation diseases.

30.7 Clinical Safety of Cord Blood Transplants Without Immune Suppression

While we have presented various scientific perspectives as to why cord blood possesses various nonimmunogenic features, here, we will discuss the clinical reality that allogeneic cord blood infusions appear to have an excellent safety record. Hassal et al. reported in a Lancet paper transfusion of more than a 100 patients with non-HLA matched cord blood. There were no safety issues, nor was GVHD observed in any of the pediatric recipients.[139] The editor of this book has extensively published on the use of allogeneic cord blood without immune suppression. For example, he reported 129 patients (cancer, lupus, anemia, rheumatological disorders, and other diseases) transplanted with a total of 413 Units of cord blood, none having treatment-associated adverse effects.[10,140-143] The same author also reported transient rises in CD34 cells after administration of allogenic cord blood.[144,145] These reports strongly support the safety of cord blood administration without the need for immune suppression or fear of GVHD.

30.8 Why No GVHD?

It is widely known in hematological practice that despite the reduced incidence of GVHD when cord blood is used as a source of hematopoietic stem cells as opposed to bone marrow, GVHD still occurs. So the question is, why would there not be GVHD when patients are receiving cord blood without immune suppression? One explanation could be that the actual process of myeloablation is what predisposes to GVHD. The conditioning regimens used in clinical practice to deplete the recipient's endogenous stem cells also destroy the T and B cell compartments. As a result of this, two things occur. First, when hematopoiesis is being restored by virtue of the stem cell graft, the T cell compartment has to be remade through the thymus. Unfortunately, the thymus atrophies rapidly after the neonatal period. Since positive and negative immunological selection depends on a healthy and functioning thymus, it is likely that a reduction in thymic function causes abnormal generation of T cells that are predisposed to autoreactivity. Another explanation involves the concept of homeostatic expansion. It has been shown that introduction of T cells into a host that is lymphodepleted causes a rapid T cell receptor-independent expansion of T cells. During this process, T cells lose dependence on costimulatory signals for activation and become resistant to tolerance induction.[146,147] In fact, numerous autoimmune diseases are associated with temporary reduction in T cell numbers, after which the rebound homeostatic expansion allows for activation of self-reactive T cells.[129] Therefore, it is likely that GVHD is actually caused by administration of contaminating T cells into an "empty compartment" and not strictly due to allogeneic differences. This is validated by the fact that no evidence of GVHD was seen after administration of paternal lymphocyte doses of up to 2×10^9 in multiple clinical trials of women with immunologically mediated abortions.[148-151] These are higher than the 1.5–3 × 10^7 nucleated cells/kg administered during a cord blood transplant.

30.9 Therapeutic Efficacy

Given that numerous studies and reports have described safety of cord blood administration in absence of immunological reactions or adverse effects, the next question is: how could there be therapeutic effects of allogeneic stem cells if immune suppression is not used? As discussed above, various components of cord blood are hypoimmunogenic, thus suggesting that they may survive the recipient alloimmune pressure. However, is there any evidence of therapeutic benefit? In a case report published by our group, we describe a 50-year-old heart failure patient suffering from nonischemic cardiomyopathy who was treated with intravenous cord blood-expanded CD34 cells and mesenchymal stem cells. Six months after treatment, the patient reported a substantial improvement in quality of life and an increase in left ventricular ejection fraction from 30–40% to 50–55%.[98] A study in four Buerger's disease patients receiving allogeneic cord blood-derived mesenchymal cells reported clinical improvement in pain-free walking distance and increased angiogenesis in previously ischemic areas.[99] A more extensive clinical investigation was reported in 27 patients with ALS treated with allogeneic cord blood, on average 20 units per recipient. Statistically significant improvements in lung function as well as Karnofsky score were reported.[152] Given that there appears to be some clinical improvements in various conditions, the next question is how could such improvements be occurring?

On the one hand, we have described that mesenchymal stem cells may evade immune attack and mediate therapeutic benefits. On the other hand, it may be possible that the actual allogeneic immune response mediated against the implanted cells may be responsible for at least part of the therapeutic effects. It is known that in pregnancy, the size of the placenta formed depends on the genetic mismatch between parents. For example, studies from the early 1970s in inbred mouse strains demonstrated increased litter and placental weight in strains bred that were genetically disparate.[153] More recent studies have demonstrated that allogeneic responses between the maternal immune system and fetal antigens are associated with increased production of angiogenic cytokines and placental growth factors.[154,155] The possibility of an allogeneic immune response having anti-inflammatory effects may be seen in clinical studies in which peripheral blood mononuclear cells were administered into women suffering from spontaneous immunologically mediated abortions. As described above, while these women did not have a GVHD effect, there was suppression of inflammatory cytokines observed systemically.[156] Recently, a study by Yang et al. examined 25 women with unexplained recurrent abortions who were administered paternal or third party leukocyte infusions, at concentration similar to those found in cord blood. An increased number of T cells expressing the T regulatory phenotype were detected. A correlation between increased Treg cells and ability to have children after the therapy was detected.[157] Thus the question becomes, can some of the therapeutic effects of cord blood be related not to stem cell therapy but actually to an effect that could be achieved by administration of simple allogeneic leukocytes? In fact, a study in 11 patients with rheumatoid arthritis demonstrated significant improvement in pain and quality of life, as well as objective measurements after administration of 30–250 million allogeneic leukocytes once every 6 weeks.[158]

30.10 Conclusions

The field of allogeneic cord blood therapy in absence of immune suppression is still in its infancy. While there is growing acceptance that cord blood can be used for nonhematopoietic indications, the majority of hematologists fear exploration into this area due to the belief of graft versus host or host versus graft immune reactions. We have provided scientific rationale and clinical reports that allogeneic cord blood is safe from the perspective of not inducing such immunological reactions. Efficacy data are very preliminary; however, increasing numbers of studies will be performed in the upcoming years. Possible areas for cord blood regenerative trials include use in stroke, as suggested by the editor of this book, in which the injured tissue produces a chemotactic gradient "calling in" hematopoietic stem cells.[159] Support for use in stroke also comes from numerous animal experiments demonstrating stimulation of neurogenesis, reduction in infarct size, and restoration of behavioral abnormalities in animal models.[160]

Additionally, efficacy is seen in stroke models regardless of whether the cells are implanted intracranially or intravenously.[161] Given the profound role of inflammatory mediators in causing pathology in rheumatoid arthritis, this could be a great indication for clinical trials, especially given that allogeneic peripheral blood alone has some level of therapeutic efficacy.[158] On the other hand, given that rheumatoid arthritis is an angiogenesis-driven disease,[162] the angiogenic components of cord blood may be theoretically detrimental. Nevertheless, clinical use of cord blood has demonstrated some therapeutic benefit in RA patients without exacerbation of pathological angiogenesis.[142]

Future studies should explore combination therapies using cord blood and other agents, particularly for conditions such as autoimmunity. For example, it is known that cytokines that are already in clinical use, such as G-CSF, on the one hand, would promote expansion of cord blood CD34 cells, but on the other hand would cause a transient immune modulation. This effect may be useful for conditions such as autoimmunity. For example, it has been demonstrated that administration of G-CSF inhibits onset of diabetes in the nonobese diabetic (NOD) model,[163] Interestingly, prevention of diabetes was associated with upregulation of Treg activity mediated by an interaction between mobilized stem cells and the T cell compartment.[164] Other studies have shown that G- CSF administration can prevent pathological immune responses in models of GVHD,[165] lupus,[166] and multiple sclerosis.[167] In fact, the tolerogenic activities of G-CSF are summarized in a recent review.[168] Other hematopoietic stimulatory cytokines such as M-CSF[114,169] and GM-CSF[170,171] have shown immune suppressive/tolerogenic activities in a variety of settings. Therefore, combinations of cord blood therapy with such agents may be a promising method of expanding activities of already existing methods for maximal therapeutic gain.

References

1. Halbrecht J. Fresh and stored placental blood. *Lancet.* 1939;2:1263.
2. Ende M, Ende N. Hematopoietic transplantation by means of fetal (cord) blood. A new method. *Va Med Mon (1918).* 1972;99:276-280.
3. Ende N, Rameshwar P, Ende M. Fetal cord blood's potential for bone marrow transplantation. *Life Sci.* 1989;44:1987-1990.
4. Gluckman E, Broxmeyer HA, Auerbach AD, et al. Hematopoietic reconstitution in a patient with Fanconi's anemia by means of umbilical-cord blood from an HLA-identical sibling. *N Engl J Med.* 1989;321:1174-1178.
5. Schoemans H, Theunissen K, Maertens J, Boogaerts M, Verfaillie C, Wagner J. Adult umbilical cord blood transplantation: a comprehensive review. *Bone Marrow Transplant.* 2006;38:83-93.
6. Tse W, Laughlin MJ. Umbilical cord blood transplantation: a new alternative option. *Hematology.* 2005;1:377-383.
7. Rubinstein P. Why cord blood? *Hum Immunol.* 2006;67: 398-404.
8. Sullivan MJ. Banking on cord blood stem cells. *Nat Rev Cancer.* 2008;8:555-563.
9. Fisk NM, Roberts IA, Markwald R, Mironov V. Can routine commercial cord blood banking be scientifically and ethically justified? *PLoS Med.* 2005;2:e44.
10. Bhattacharya N. Placental umbilical cord whole blood transfusion: a safe and genuine blood substitute for patients of the under-resourced world at emergency. *J Am Coll Surg.* 2005;200:557-563.
11. Kang S, Yang YJ, Li CJ, Gao RL. Effects of intracoronary autologous bone marrow cells on left ventricular function in acute myocardial infarction: a systematic review and meta-analysis for randomized controlled trials. *Coron Artery Dis.* 2008;19:327-335.
12. Terai S, Ishikawa T, Omori K, et al. Improved liver function in patients with liver cirrhosis after autologous bone marrow cell infusion therapy. *Stem Cells.* 2006;24:2292-2298.
13. Levicar N, Pai M, Habib NA, et al. Long-term clinical results of autologous infusion of mobilized adult bone marrow derived CD34+ cells in patients with chronic liver disease. *Cell Prolif.* 2008;41(suppl 1):115-125.
14. Cobellis G, Silvestroni A, Lillo S, Sica G, Botti C, Maione C, Schiavone V, Rocco S, Brando G, Sica V. Long-term effects of repeated autologous transplantation of bone marrow cells in patients affected by peripheral arterial disease. *Bone Marrow Transplant.* 2008;42(10):667-672.
15. Zhu J, Wang X, Chen J, Sun J, Zhang F. Reduced number and activity of circulating endothelial progenitor cells from patients with hyperhomocysteinemia. *Arch Med Res.* 2006;37:484-489.
16. Fulle S, Di Donna S, Puglielli C, et al. Age-dependent imbalance of the antioxidative system in human satellite cells. *Exp Gerontol.* 2005;40:189-197.
17. Porta C, Caporali R, Epis O, et al. Impaired bone marrow hematopoietic progenitor cell function in rheumatoid arthritis patients candidated to autologous hematopoietic stem cell transplantation. *Bone Marrow Transplant.* 2004;33:721-728.
18. D'Ippolito G, Schiller PC, Ricordi C, Roos BA, Howard GA. Age-related osteogenic potential of mesenchymal stromal stem cells from human vertebral bone marrow. *J Bone Miner Res.* 1999;14:1115-1122.
19. Theunissen K, Verfaillie CM. A multifactorial analysis of umbilical cord blood, adult bone marrow and mobilized peripheral blood progenitors using the improved ML-IC assay. *Exp Hematol.* 2005;33:165-172.
20. Ng YY, van Kessel B, Lokhorst HM, et al. Gene-expression profiling of CD34+ cells from various hematopoietic stem-cell sources reveals functional differences in stem-cell activity. *J Leukoc Biol.* 2004;75:314-323.
21. Hogan CJ, Shpall EJ, McNulty O, et al. Engraftment and development of human CD34(+)-enriched cells from umbilical cord blood in NOD/LtSz-scid/scid mice. *Blood.* 1997;90:85-96.
22. Sakabe H, Yahata N, Kimura T, et al. Human cord blood-derived primitive progenitors are enriched in CD34+c-kit-cells: correlation between long-term culture-initiating cells and telomerase expression. *Leukemia.* 1998;12:728-734.
23. Yeh ET, Zhang S, Wu HD, Korbling M, Willerson JT, Estrov Z. Transdifferentiation of human peripheral blood CD34+ -enriched cell population into cardiomyocytes, endothelial cells, and smooth muscle cells in vivo. *Circulation.* 2003; 108:2070-2073.
24. Sigurjonsson OE, Perreault MC, Egeland T, Glover JC. Adult human hematopoietic stem cells produce neurons efficiently in the regenerating chicken embryo spinal cord. *Proc Natl Acad Sci USA.* 2005;102:5227-5232.
25. Ryu KH, Cho SJ, Jung YJ, et al. In vitro generation of functional dendritic cells from human umbilical cord blood CD34+ cells by a 2-step culture method. *Int J Hematol.* 2004;80:281-286.
26. Liu F, Pan X, Chen G, et al. Hematopoietic stem cells mobilized by granulocyte colony-stimulating factor partly contribute to liver graft regeneration after partial orthotopic liver transplantation. *Liver Transpl.* 2006;12:1129-1137.
27. Zeng F, Chen MJ, Baldwin DA, et al. Multiorgan engraftment and differentiation of human cord blood CD34+ Lin-cells in goats assessed by gene expression profiling. *Proc Natl Acad Sci USA.* 2006;103:7801-7806.
28. Bhatia M, Bonnet D, Murdoch B, Gan OI, Dick JE. A newly discovered class of human hematopoietic cells with SCID-repopulating activity. *Nat Med.* 1998;4:1038-1045.
29. Nakamura Y, Ando K, Chargui J, et al. Ex vivo generation of CD34(+) cells from CD34(-) hematopoietic cells. *Blood.* 1999;94:4053-4059.
30. Gentry T, Deibert E, Foster SJ, Haley R, Kurtzberg J, Balber AE. Isolation of early hematopoietic cells, including megakaryocyte progenitors, in the ALDH-bright cell population of cryopreserved, banked UC blood. *Cytotherapy.* 2007;9: 569-576.
31. Gotze KS, Schiemann M, Marz S, et al. CD133-enriched CD34(-) (CD33/CD38/CD71)(-) cord blood cells acquire CD34 prior to cell division and hematopoietic activity is exclusively associated with CD34 expression. *Exp Hematol.* 2007;35:1408-1414.
32. Hatzfeld J, Batard P, Cardoso AA, et al. Purification and release from quiescence of umbilical cord blood early progenitors reveal their potential to engraft adults. *Blood Cells.* 1994;20:430-434. discussion 434-435.
33. Hatzfeld J, Li ML, Brown EL, et al. Release of early human hematopoietic progenitors from quiescence by antisense

transforming growth factor beta 1 or Rb oligonucleotides. *J Exp Med.* 1991;174:925-929.
34. Akel S, Petrow-Sadowski C, Laughlin MJ, Ruscetti FW. Neutralization of autocrine transforming growth factor-beta in human cord blood CD34(+)CD38(-)Lin(-) cells promotes stem-cell-factor-mediated erythropoietin-independent early erythroid progenitor development and reduces terminal differentiation. *Stem Cells.* 2003;21:557-567.
35. Batard P, Monier MN, Fortunel N, et al. TGF-(beta)1 maintains hematopoietic immaturity by a reversible negative control of cell cycle and induces CD34 antigen up- modulation. *J Cell Sci.* 2000;113(pt 3):383-390.
36. Massberg S, Schaerli P, Knezevic-Maramica I, et al. Immunosurveillance by hematopoietic progenitor cells trafficking through blood, lymph, and peripheral tissues. *Cell.* 2007;131:994-1008.
37. Strobl H, Knapp W. TGF-beta1 regulation of dendritic cells. *Microbes Infect.* 1999;1:1283-1290.
38. Tiemessen MM, Kunzmann S, Schmidt-Weber CB, et al. Transforming growth factor-beta inhibits human antigen-specific CD4+ T cell proliferation without modulating the cytokine response. *Int Immunol.* 2003;15:1495-1504.
39. Malygin AM, Meri S, Timonen T. Regulation of natural killer cell activity by transforming growth factor-beta and prostaglandin E2. *Scand J Immunol.* 1993;37:71-76.
40. Huber S, Schramm C. TGF-beta and CD4+CD25+ regulatory T cells. *Front Biosci.* 2006;11:1014-1023.
41. Lewkowicz P, Lewkowicz N, Sasiak A, Tchorzewski H. Lipopolysaccharide- activated CD4+CD25+ T regulatory cells inhibit neutrophil function and promote their apoptosis and death. *J Immunol.* 2006;177:7155-7163.
42. Mahajan D, Wang Y, Qin X, et al. CD4+CD25+ regulatory T cells protect against injury in an innate murine model of chronic kidney disease. *J Am Soc Nephrol.* 2006;17:2731-2741.
43. Larmonier N, Marron M, Zeng Y, et al. Tumor-derived CD4(+)CD25(+) regulatory T cell suppression of dendritic cell function involves TGF-beta and IL-10. *Cancer Immunol Immunother.* 2007;56:48-59.
44. Mortari F, Bains MA, Singhal SK. Immunoregulatory activity of human bone marrow. Identification of suppressor cells possessing OKM1, SSEA-1, and HNK-1 antigens. *J Immunol.* 1986;137:1133-1137.
45. Sugiura K, Ikehara S, Gengozian N, et al. Enrichment of natural suppressor activity in a wheat germ agglutinin positive hematopoietic progenitor-enriched fraction of monkey bone marrow. *Blood.* 1990;75:1125-1131.
46. Seledtsov VI, Taraban VY, Seledtsova GV, et al. Tumor growth inhibitory and natural suppressor activities of murine bone marrow cells: a comparative study. *Cell Immunol.* 1997;182:12-19.
47. Wright MA, Wiers K, Vellody K, Djordjevic D, Young MR. Stimulation of immune suppressive CD34+ cells from normal bone marrow by Lewis lung carcinoma tumors. *Cancer Immunol Immunother.* 1998;46:253-260.
48. Moore SC, Theus SA, Barnett JB, Soderberg LS. Cytokine regulation of bone marrow natural suppressor cell activity in the suppression of lymphocyte function. *Cell Immunol.* 1992;141:398-408.
49. Gallina G, Dolcetti L, Serafini P, et al. Tumors induce a subset of inflammatory monocytes with immunosuppressive activity on CD8+ T cells. *J Clin Invest.* 2006;116:2777-2790.
50. Huang B, Lei Z, Zhao J, et al. CCL2/CCR2 pathway mediates recruitment of myeloid suppressor cells to cancers. *Cancer Lett.* 2007;252(1):86-92.
51. Huang B, Pan PY, Li Q, et al. Gr-1+CD115+ immature myeloid suppressor cells mediate the development of tumor-induced T regulatory cells and T-cell anergy in tumor-bearing host. *Cancer Res.* 2006;66:1123-1131.
52. Segerson EC, Beetham PK. Suppressor activity of bone marrow cells and localization of fluorescent-labeled bone marrow cells within ovine endometrial tissue. *J Anim Sci.* 2000;78:709-717.
53. Suzuki F, Pollard RB. Mechanism for the suppression of gamma-interferon responsiveness in mice after thermal injury. *J Immunol.* 1982;129:1811-1815.
54. Yang L, Hsu B. The roles of macrophage (M phi) and PGE-2 in postburn immunosuppression. *Burns.* 1992;18:132-136.
55. Wiers KM, Lathers DM, Wright MA, Young MR. Vitamin D3 treatment to diminish the levels of immune suppressive CD34+ cells increases the effectiveness of adoptive immunotherapy. *J Immunother.* 2000;23:115-124.
56. Gur H, Krauthgamer R, Berrebi A, et al. Tolerance induction by megadose hematopoietic progenitor cells: expansion of veto cells by short-term culture of purified human CD34(+) cells. *Blood.* 2002;99:4174-4181.
57. Reisner Y, Gur H, Reich-Zeliger S, Martelli MF, Bachar-Lustig E. Hematopoietic stem cell transplantation across major genetic barriers: tolerance induction by megadose CD34 cells and other veto cells. *Ann NY Acad Sci.* 2003;996:72-79.
58. Donckier V, Troisi R, Toungouz M, et al. Donor stem cell infusion after non-myeloablative conditioning for tolerance induction to HLA mismatched adult living-donor liver graft. *Transpl Immunol.* 2004;13:139-146.
59. George JF, Sweeney SD, Kirklin JK, Simpson EM, Goldstein DR, Thomas JM. An essential role for Fas ligand in transplantation tolerance induced by donor bone marrow. *Nat Med.* 1998;4:333-335.
60. Reisner Y, Gur H, Reich-Zeliger S, Martelli MF, Bachar-Lustig E. Crossing the HLA barriers. *Blood Cells Mol Dis.* 2004;33:206-210.
61. Prockop DJ. Marrow stromal cells as stem cells for nonhematopoietic tissues. *Science.* 1997;276:71-74.
62. Friedenstein AJ, Petrakova KV, Kurolesova AI, Frolova GP. Heterotopic of bone marrow. Analysis of precursor cells for osteogenic and hematopoietic tissues. *Transplantation.* 1968;6:230-247.
63. Zannettino AC, Paton S, Arthur A, et al. Multipotential human adipose-derived stromal stem cells exhibit a perivascular phenotype in vitro and in vivo. *J Cell Physiol.* 2008;214:413-421.
64. Hoogduijn MJ, Crop MJ, Peeters AM, et al. Human heart, spleen, and perirenal fat-derived mesenchymal stem cells have immunomodulatory capacities. *Stem Cells Dev.* 2007;16:597-604.
65. Chao KC, Chao KF, Fu YS, Liu SH. Islet-like clusters derived from mesenchymal stem cells in Wharton's Jelly of the human umbilical cord for transplantation to control type 1 diabetes. *PLoS ONE.* 2008;3:e1451.

66. Jo YY, Lee HJ, Kook SY, et al. Isolation and characterization of postnatal stem cells from human dental tissues. *Tissue Eng*. 2007;13:767-773.
67. He Q, Wan C, Li G. Concise review: multipotent mesenchymal stromal cells in blood. *Stem Cells*. 2007;25:69-77.
68. Oh W, Kim DS, Yang YS, Lee JK. Immunological properties of umbilical cord blood-derived mesenchymal stromal cells. *Cell Immunol*. 2008;251:116-123.
69. Meng X, Ichim TE, Zhong J, et al. Endometrial regenerative cells: a novel stem cell population. *J Transl Med*. 2007; 5:57.
70. Hida N, Nishiyama N, Miyoshi S, et al. Novel cardiac precursor-like cells from human menstrual blood-derived mesenchymal cells. *Stem Cells*. 2008;26:1695-1704.
71. Patel AN, Park E, Kuzman M, Benetti F, Silva FJ, Allickson JG. Multipotent menstrual blood stromal stem cells: isolation, characterization, and differentiation. *Cell Transplant*. 2008;17:303-311.
72. Bieback K, Kern S, Kocaomer A, Ferlik K, Bugert P. Comparing mesenchymal stromal cells from different human tissues: bone marrow, adipose tissue and umbilical cord blood. *Biomed Mater Eng*. 2008;18:S71-S76.
73. Bieback K, Kluter H. Mesenchymal stromal cells from umbilical cord blood. *Curr Stem Cell Res Ther*. 2007;2: 310-323.
74. Le Blanc K, Ringden O. Immunomodulation by mesenchymal stem cells and clinical experience. *J Intern Med*. 2007;262:509-525.
75. Keyser KA, Beagles KE, Kiem HP. Comparison of mesenchymal stem cells from different tissues to suppress T-cell activation. *Cell Transplant*. 2007;16:555-562.
76. Kadri T, Lataillade JJ, Doucet C, et al. Proteomic study of Galectin-1 expression in human mesenchymal stem cells. *Stem Cells Dev*. 2005;14:204-212.
77. Ryan JM, Barry FP, Murphy JM, Mahon BP. Mesenchymal stem cells avoid allogeneic rejection. *J Inflamm (Lond)*. 2005;2:8.
78. Beyth S, Borovsky Z, Mevorach D, et al. Human mesenchymal stem cells alter antigen-presenting cell maturation and induce T-cell unresponsiveness. *Blood*. 2005;105: 2214-2219.
79. Aggarwal S, Pittenger MF. Human mesenchymal stem cells modulate allogeneic immune cell responses. *Blood*. 2005; 105:1815-1822.
80. Plumas J, Chaperot L, Richard MJ, Molens JP, Bensa JC, Favrot MC. Mesenchymal stem cells induce apoptosis of activated T cells. *Leukemia*. 2005;19:1597-1604.
81. Maccario R, Podesta M, Moretta A, et al. Interaction of human mesenchymal stem cells with cells involved in alloantigen-specific immune response favors the differentiation of CD4+ T-cell subsets expressing a regulatory/suppressive phenotype. *Haematologica*. 2005;90:516-525.
82. Zappia E, Casazza S, Pedemonte E, et al. Mesenchymal stem cells ameliorate experimental autoimmune encephalomyelitis inducing T-cell anergy. *Blood*. 2005;106:1755-1761.
83. Le Blanc K, Frassoni F, Ball L, et al. Mesenchymal stem cells for treatment of steroid-resistant, severe, acute graft-versus-host disease: a phase II study. *Lancet*. 2008;371: 1579-1586.
84. Ning H, Yang F, Jiang M, et al. The correlation between cotransplantation of mesenchymal stem cells and higher recurrence rate in hematologic malignancy patients: outcome of a pilot clinical study. *Leukemia*. 2008;22:593-599.
85. Ball L, Bredius R, Lankester A, et al. Third party mesenchymal stromal cell infusions fail to induce tissue repair despite successful control of severe grade IV acute graft-versus-host disease in a child with juvenile myelomonocytic leukemia. *Leukemia*. 2008;22:1256-1257.
86. Ringden O, Uzunel M, Rasmusson I, et al. Mesenchymal stem cells for treatment of therapy-resistant graft-versus-host disease. *Transplantation*. 2006;81:1390-1397.
87. Le Blanc K, Rasmusson I, Sundberg B, et al. Treatment of severe acute graft-versus-host disease with third party haploidentical mesenchymal stem cells. *Lancet*. 2004;363: 1439-1441.
88. Muller I, Kordowich S, Holzwarth C, et al. Application of multipotent mesenchymal stromal cells in pediatric patients following allogeneic stem cell transplantation. *Blood Cells Mol Dis*. 2008;40:25-32.
89. Horwitz EM, Gordon PL, Koo WK, et al. Isolated allogeneic bone marrow-derived mesenchymal cells engraft and stimulate growth in children with osteogenesis imperfecta: Implications for cell therapy of bone. *Proc Natl Acad Sci USA*. 2002;99:8932-8937.
90. Koc ON, Day J, Nieder M, Gerson SL, Lazarus HM, Krivit W. Allogeneic mesenchymal stem cell infusion for treatment of metachromatic leukodystrophy (MLD) and Hurler syndrome (MPS-IH). *Bone Marrow Transplant*. 2002;30:215-222.
91. Le Blanc K, Samuelsson H, Gustafsson B, et al. Transplantation of mesenchymal stem cells to enhance engraftment of hematopoietic stem cells. *Leukemia*. 2007; 21:1733-1738.
92. Lazarus HM, Koc ON, Devine SM, et al. Cotransplantation of HLA- identical sibling culture-expanded mesenchymal stem cells and hematopoietic stem cells in hematologic malignancy patients. *Biol Blood Marrow Transplant*. 2005; 11:389-398.
93. Ball LM, Bernardo ME, Roelofs H, et al. Cotransplantation of ex vivo expanded mesenchymal stem cells accelerates lymphocyte recovery and may reduce the risk of graft failure in haploidentical hematopoietic stem-cell transplantation. *Blood*. 2007;110:2764-2767.
94. http://www.osiristx.com/pdf/PR%2039%2025Mar07%20Provacel%20Pos itive%20Results.pdf
95. http://ir.athersys.com/releasedetail.cfm?ReleaseID=276046
96. http://www.angioblast.com/news/pressrelease10.pdf
97. http://www.neuronyx.com/therapeutic_targets.php
98. Ichim TE, Solano F, Brenes R, et al. Placental mesenchymal and cord blood stem cell therapy for dilated cardiomyopathy. *Reprod Biomed Online*. 2008;16:898-905.
99. Kim SW, Han H, Chae GT, et al. Successful stem cell therapy using umbilical cord blood-derived multipotent stem cells for Buerger's disease and ischemic limb disease animal model. *Stem Cells*. 2006;24:1620-1626.
100. Marleau AM, Greenwood JD, Wei Q, Singh B, Croy BA. Chimerism of murine fetal bone marrow by maternal cells occurs in late gestation and persists into adulthood. *Lab Invest*. 2003;83:673-681.

101. Bianchi DW, Fisk NM. Fetomaternal cell trafficking and the stem cell debate: gender matters. *JAMA*. 2007;297: 1489-1491.
102. Chen CP, Lee MY, Huang JP, et al. Trafficking of multipotent mesenchymal stromal cells from maternal circulation through the placenta involves vascular endothelial growth factor receptor-1 and integrins. *Stem Cells*. 2008;26: 550-561.
103. O'Donoghue K, Sultan HA, Al-Allaf FA, Anderson JR, Wyatt-Ashmead J, Fisk NM. Microchimeric fetal cells cluster at sites of tissue injury in lung decades after pregnancy. *Reprod Biomed Online*. 2008;16:382-390.
104. O'Donoghue K, Chan J. Human fetal mesenchymal stem cells. *Curr Stem Cell Res Ther*. 2006;1:371-386.
105. Borras FE, Matthews NC, Lowdell MW, Navarrete CV. Identification of both myeloid CD11c+and lymphoid CD11c- dendritic cell subsets in cord blood. *Br J Haematol*. 2001;113:925-931.
106. Kawano Y, Noma T, Yata J. Analysis of decreased autologous mixed lymphocyte reaction of cord blood lymphocytes: with special reference to production of and response to interleukin-2 (IL-2). *Asian Pac J Allergy Immunol*. 1984;2:49-55.
107. Petty RE, Hunt DW. Neonatal dendritic cells. *Vaccine*. 1998;16:1378-1382.
108. Sorg RV, Kogler G, Wernet P. Identification of cord blood dendritic cells as an immature CD11c- population. *Blood*. 1999;93:2302-2307.
109. Han P, McDonald T, Hodge G. Potential immaturity of the T-cell and antigen-presenting cell interaction in cord blood with particular emphasis on the CD40-CD40 ligand costimulatory pathway. *Immunology*. 2004;113:26-34.
110. Drohan L, Harding JJ, Holm B, et al. Selective developmental defects of cord blood antigen-presenting cell subsets. *Hum Immunol*. 2004;65:1356-1369.
111. De Wit D, Tonon S, Olislagers V, et al. Impaired responses to toll-like receptor 4 and toll-like receptor 3 ligands in human cord blood. *J Autoimmun*. 2003;21:277-281.
112. Mainali ES, Tew JG. Dexamethasone selectively inhibits differentiation of cord blood stem cell derived-dendritic cell (DC) precursors into immature DCs. *Cell Immunol*. 2004;232:127-136.
113. Wong OH, Huang FP, Chiang AK. Differential responses of cord and adult blood-derived dendritic cells to dying cells. *Immunology*. 2005;116:13-20.
114. Li G, Kim YJ, Broxmeyer HE. Macrophage colony-stimulating factor drives cord blood monocyte differentiation into IL-10(high)IL-12absent dendritic cells with tolerogenic potential. *J Immunol*. 2005;174:4706-4717.
115. Gansuvd B, Hagihara M, Higuchi A, et al. Umbilical cord blood dendritic cells are a rich source of soluble HLA-DR: synergistic effect of exosomes and dendritic cells on autologous or allogeneic T-Cell proliferation. *Hum Immunol*. 2003;64:427-439.
116. Kim SH, Lechman ER, Bianco N, et al. Exosomes derived from IL-10-treated dendritic cells can suppress inflammation and collagen-induced arthritis. *J Immunol*. 2005; 174:6440-6448.
117. Navarro V, Herrine S, Katopes C, Colombe B, Spain CV. The effect of HLA class I (A and B) and class II (DR) compatibility on liver transplantation outcomes: an analysis of the OPTN database. *Liver Transpl*. 2006;12:652-658.
118. Frey O, Brauer R. Regulatory T cells: magic bullets for immunotherapy? *Arch Immunol Ther Exp (Warsz)*. 2006; 54:33-43.
119. Satwani P, Ayello J, Ven C, et al. Immaturity of IL-18 gene expression and protein production in cord blood (CB) versus peripheral blood (PB) mononuclear cells and differential effects in natural killer (NK) cell development and function. *Br J Haematol*. 2005;130:284-292.
120. Wei YM, Cao Q, Zhou HY, et al. Ex vivo expansion of T, NK and CD34+ cells from umbilical cord blood. *Zhongguo Shi Yan Xue Ye Xue Za Zhi*. 2005;13:1076-1081.
121. Okada H, Nagamura-Inoue T, Mori Y, Takahashi TA. Expansion of Valpha24(+)Vbeta11(+) NKT cells from cord blood mononuclear cells using IL-15, IL-7 and Flt3-L depends on monocytes. *Eur J Immunol*. 2006;36:236-244.
122. Kadowaki N, Antonenko S, Ho S, et al. Distinct cytokine profiles of neonatal natural killer T cells after expansion with subsets of dendritic cells. *J Exp Med*. 2001;193: 1221-1226.
123. Subauste CS, Chung JY, Do D, et al. Preferential activation and expansion of human peripheral blood gamma delta T cells in response to Toxoplasma gondii in vitro and their cytokine production and cytotoxic activity against T. gondii-infected cells. *J Clin Invest*. 1995;96:610-619.
124. Liu E, Law HK, Lau YL. Tolerance associated with cord blood transplantation may depend on the state of host dendritic cells. *Br J Haematol*. 2004;126:517-526.
125. Papadogiannakis N, Johnsen SA, Olding LB. Monocyte-regulated hyporesponsiveness of human cord blood lymphocytes to OKT3-monoclonal- antibody-induced mitogenesis. *Scand J Immunol*. 1986;23:91-99.
126. Chen L, Cohen AC, Lewis DB. Impaired allogeneic activation and T-helper 1 differentiation of human cord blood naive CD4 T cells. *Biol Blood Marrow Transplant*. 2006;12:160-171.
127. Hodge S, Hodge G, Flower R, Han P. Cord blood leucocyte expression of functionally significant molecules involved in the regulation of cellular immunity. *Scand J Immunol*. 2001;53:72-78.
128. Canto E, Rodriguez-Sanchez JL, Vidal S. Naive CD4+ cells from cord blood can generate competent Th effector cells. *Transplantation*. 2005;80:850-858.
129. Marleau AM, Sarvetnick N. T cell homeostasis in tolerance and immunity. *J Leukoc Biol*. 2005;78:575-584.
130. Hickman SP, Turka LA. Homeostatic T cell proliferation as a barrier to T cell tolerance. *Philos Trans R Soc Lond B Biol Sci*. 2005;360:1713-1721.
131. Rosenberg SA, Dudley ME. Cancer regression in patients with metastatic melanoma after the transfer of autologous antitumor lymphocytes. *Proc Natl Acad Sci USA*. 2004; 101(suppl 2):14639-14645.
132. Ichim TE, Zhong R, Min WP. Prevention of allograft rejection by in vitro generated tolerogenic dendritic cells. *Transpl Immunol*. 2003;11:295-306.
133. Zenclussen AC. Regulatory T cells in pregnancy. *Springer Semin Immunopathol*. 2006;28:31-39.

134. Fritzsching B, Oberle N, Pauly E, et al. Naive regulatory T cells: a novel subpopulation defined by resistance towards CD95L-mediated cell death. *Blood.* 2006;108: 3371-3378.
135. Godfrey WR, Spoden DJ, Ge YG, et al. Cord blood CD4(+) CD25(+)-derived T regulatory cell lines express FoxP3 protein and manifest potent suppressor function. *Blood.* 2005;105:750-758.
136. Takahata Y, Nomura A, Takada H, et al. CD25+CD4+ T cells in human cord blood: an immunoregulatory subset with naive phenotype and specific expression of forkhead box p3 (Foxp3) gene. *Exp Hematol.* 2004;32:622-629.
137. Sakaguchi S, Ono M, Setoguchi R, et al. Foxp3CD25CD4 natural regulatory T cells in dominant self-tolerance and autoimmune disease. *Immunol Rev.* 2006;212:8-27.
138. Torgerson TR. Regulatory T cells in human autoimmune diseases. *Springer Semin Immunopathol.* 2006;28:63-76.
139. Hassall O, Bedu-Addo G, Adarkwa M, Danso K, Bates I. Umbilical-cord blood for transfusion in children with severe anaemia in under-resourced countries. *Lancet.* 2003;361:678-679.
140. Bhattacharya N. Placental umbilical cord whole blood transfusion to combat anemia in the background of tuberculosis and emaciation and its potential role as an immunoadjuvant therapy for the under-resourced people of the world. *Clin Exp Obstet Gynecol.* 2006;33:99-104.
141. Bhattacharya N. Placental umbilical cord blood transfusion: A novel method of treatment of patients with malaria in the background of anemia. *Clin Exp Obstet Gynecol.* 2006;33:39-43.
142. Bhattacharya N. Placental umbilical cord whole blood transfusion to combat anemia in the background of advanced rheumatoid arthritis and emaciation and its potential role as immunoadjuvant therapy. *Clin Exp Obstet Gynecol.* 2006; 33:28-33.
143. Bhattacharya N. A preliminary study of placental umbilical cord whole blood transfusion in under resourced patients with malaria in the background of anaemia. *Malar J.* 2006; 5:20.
144. Bhattacharya N. A preliminary report of 123 units of placental umbilical cord whole blood transfusion in HIV-positive patients with anemia and emaciation. *Clin Exp Obstet Gynecol.* 2006;33:117-121.
145. Bhattacharya N. Spontaneous transient rise of CD34 cells in peripheral blood after 72 hours in patients suffering from advanced malignancy with anemia: effect and prognostic implications of treatment with placental umbilical cord whole blood transfusion. *Eur J Gynaecol Oncol.* 2006;27:286-290.
146. Prlic M, Blazar BR, Khoruts A, Zell T, Jameson SC. Homeostatic expansion occurs independently of costimulatory signals. *J Immunol.* 2001;167:5664-5668.
147. Wu Z, Bensinger SJ, Zhang J, et al. Homeostatic proliferation is a barrier to transplantation tolerance. *Nat Med.* 2004;10:87-92.
148. Ito K, Tanaka T, Tsutsumi N, Obata F, Kashiwagi N. Possible mechanisms of immunotherapy for maintaining pregnancy in recurrent spontaneous aborters: analysis of anti-idiotypic antibodies directed against autologous T- cell receptors. *Hum Reprod.* 1999;14:650-655.
149. Smith JB, Cowchock FS, Lata JA, Hankinson BT. The number of cells used for immunotherapy of repeated spontaneous abortion influences pregnancy outcome. *J Reprod Immunol.* 1992;22:217-224.
150. Szpakowski A, Malinowski A, Glowacka E, et al. The influence of paternal lymphocyte immunization on the balance of Th1/Th2 type reactivity in women with unexplained recurrent spontaneous abortion. *Ginekol Pol.* 2000;71: 586-592.
151. Hayakawa S, Karasaki-Suzuki M, Itoh T, et al. Effects of paternal lymphocyte immunization on peripheral Th1/Th2 balance and TCR V beta and V gamma repertoire usage of patients with recurrent spontaneous abortions. *Am J Reprod Immunol.* 2000;43:107-115.
152. Ghen MJ, Roshan R, Roshan RO, et al. Potential clinical applications using stem cells derived from human umbilical cord blood. *Reprod Biomed Online.* 2006;13: 562-572.
153. Beer AE, Scott JR, Billingham RE. Histoincompatibility and maternal immunological status as determinants of fetoplacental weight and litter size in rodents. *J Exp Med.* 1975;142: 180-196.
154. Wegmann TG. Placental immunotrophism: maternal T cells enhance placental growth and function. *Am J Reprod Immunol Microbiol.* 1987;15:67-69.
155. Rukavina D, Kapovic M, Radojcic A. Immunoregulatory factors contributing to fetal allograft survival. *Int J Dev Biol.* 1991;35:275-278.
156. Clark DA. Shall we properly re-examine the status of allogeneic lymphocyte therapy for recurrent early pregnancy failure? *Am J Reprod Immunol.* 2004;51:7-15.
157. Yang H, Qiu L, Di W, et al. Proportional change of CD4(+) CD25(+) regulatory T cells after lymphocyte therapy in unexplained recurrent spontaneous abortion patients. *Fertil Steril.* 2008;89:656-661.
158. Smith JB, Fort JG. Treatment of rheumatoid arthritis by immunization with mononuclear white blood cells: results of a preliminary trial. *J Rheumatol.* 1996;23:220-225.
159. Chaudhuri A, Hollands P, Bhattacharya N. Placental umbilical cord blood transfusion in acute ischaemic stroke. *Med Hypotheses.* 2007;69:1267-1271.
160. Vendrame M, Gemma C, Pennypacker KR, Bickford PC. Davis Sanberg C, Sanberg PR, Willing AE: Cord blood rescues stroke-induced changes in splenocyte phenotype and function. *Exp Neurol.* 2006;199:191-200.
161. Chen J, Sanberg PR, Li Y, et al. Intravenous administration of human umbilical cord blood reduces behavioral deficits after stroke in rats. *Stroke.* 2001;32:2682-2688.
162. Khong TL, Larsen H, Raatz Y, Paleolog E. Angiogenesis as a therapeutic target in arthritis: learning the lessons of the colorectal cancer experience. *Angiogenesis.* 2007;10: 243-258.
163. Kared H, Masson A, Adle-Biassette H, Bach JF, Chatenoud L, Zavala F. Treatment with granulocyte colony-stimulating factor prevents diabetes in NOD mice by recruiting plasmacytoid dendritic cells and functional CD4(+)CD25(+) regulatory T-cells. *Diabetes.* 2005;54:78-84.
164. Kared H, Adle-Biassette H, Fois E, et al. Jagged2-expressing hematopoietic progenitors promote regulatory T cell expansion in the periphery through notch signaling. *Immunity.* 2006;25:823-834.
165. Pan L, Delmonte J Jr, Jalonen CK, Ferrara JL. Pretreatment of donor mice with granulocyte colony-stimulating factor

polarizes donor T lymphocytes toward type-2 cytokine production and reduces severity of experimental graft-versus-host disease. *Blood*. 1995;86:4422-4429.
166. Zavala F, Masson A, Hadaya K, et al. Granulocyte-colony stimulating factor treatment of lupus autoimmune disease in MRL-lpr/lpr mice. *J Immunol*. 1999; 163:5125-5132.
167. Zavala F, Abad S, Ezine S, Taupin V, Masson A, Bach JF. G-CSF therapy of ongoing experimental allergic encephalomyelitis via chemokine- and cytokine-based immune deviation. *J Immunol*. 2002;168:2011-2019.
168. Rutella S, Zavala F, Danese S, Kared H, Leone G. Granulocyte colony-stimulating factor: a novel mediator of T cell tolerance. *J Immunol*. 2005;175:7085-7091.
169. Wing EJ, Magee DM, Pearson AC, Waheed A, Shadduck RK. Peritoneal macrophages exposed to purified macrophage colony-stimulating factor (M- CSF) suppress mitogen- and antigen-stimulated lymphocyte proliferation. *J Immunol*. 1986;137:2768-2773.
170. Fu YX, Watson G, Jimenez JJ, Wang Y, Lopez DM. Expansion of immunoregulatory macrophages by granulocyte-macrophage colony- stimulating factor derived from a murine mammary tumor. *Cancer Res*. 1990;50:227-234.
171. Serafini P, Carbley R, Noonan KA, Tan G, Bronte V, Borrello I. High-dose granulocyte-macrophage colony-stimulating factor-producing vaccines impair the immune response through the recruitment of myeloid suppressor cells. *Cancer Res*. 2004;64:6337-6343.

Combination Cellular Therapy for Regenerative Medicine: The Stem Cell Niche

31

Ian K. McNiece

31.1 Introduction

In 1988, Gluckman and colleagues performed the first successful transplantation of umbilical cord blood (CB) cells.[1] The patient was suffering from Fanconi's anemia, complicated by severe aplastic anemia. The mother was shown to be carrying a sibling that was unaffected by the disease and was human leukocyte antigen (HLA) identical to the patient. The CB was collected and stored frozen in liquid nitrogen. After conditioning with low-dose cyclophosphamide and thoraco-abdominal irradiation, the patient received the thawed CB product. Twenty years later this patient is alive and cured of the anemia.

Since the performance of this first CB transplant, more than 10,000 unrelated CB transplantations (CBT) performed in children and adults. The major application of CBT has been to support high-dose chemotherapy treatment for malignant disease. However, a number of other nonmalignant diseases have been successfully treated with CBT, including, lysosomal disorders (Krabbe's disease or Hurler's syndrome), hemoglobinopathies, apalstic anemia, and Fanconi's anemia.[2]

Due to the clinical success of CBT, many CB banks have been established around the world. Several hundred thousand CB units are stored frozen in these banks, making this a unique source of stem cells for clinical applications. Therefore, CB has been the focus of an extensive number of research studies aimed at identifying sources of stem cells for a range of diseases. This review will present an overview of the biology of stem cells in CB products and the potential use of these cells for treatment of a range of diseases.

31.2 Cellular Content of CB Products

The number of nucleated cells collected in a CB product varies markedly, depending upon the experience of the collection staff and the efficiency of collection. The cellular content of CB products has been evaluated in relation to the time to hematopoietic engraftment following high-dose chemotherapy; the total nucleated cell (TNC) dose, CD34+ cell content, and myeloid progenitor content (CFU-GM) have been proposed as predictive of outcome.[3-7]

CB products contain similar cell populations to bone marrow (BM) and mobilized peripheral blood progenitor cell products (PBPC), including hematopoietic stem cells (HSC), primitive progenitor cells, mature progenitor cells, and mature functional cells. However, the total cell number of progenitor cells is much lower in CB compared to BM and PBPC. For example, BM and PBPC contain approx 10^8 CD34+ cells, while CB contains approximately 5×10^6 CD34+ cells. In contrast, the frequency of HSC, as determined by nonobese diabetic/severe combined immunodeficiency (NOD/SCID) engraftment is enriched in the CD34+ cell population of CB compared to BM or PBPC. As few as 100,000 CB CD34+ cells can engraft NOD/SCID mice, while ~1 million BM CD34+ cells and 5 million PBPC CD34+ cells are required for the same levels of engraftment of human cells.[8, 9]

These numbers would suggest that CB products contain similar levels of HSC to BM and PBPC but significantly lower levels of committed progenitor cells. Clinical data are consistent with this theory. Patients transplanted

I.K. McNiece
Interdisciplinary Stem Cell Institute, University of Miami, 1120 NW, 14th Street, Room 1113 Miami, FL, USA
e-mail: imcniece@med.miami.edu

N. Bhattacharya and P. Stubblefield (eds.), *Regenerative Medicine Using Pregnancy-Specific Biological Substances*,
DOI: 10.1007/978-1-84882-718-9_31, © Springer-Verlag London Limited 2011

with CB grafts have delayed neutrophil and platelet engraftment compared to patients transplanted with BM or PBPC products. Patients transplanted with CB products do not have any long-term engraftment problems with long-term donor chimerism persisting in CB recipients. This suggests that CB products contain sufficient long-term engrafting cells (HSC), but minimal short-term engrafting cells (mature progenitor cells).

The total nucleated cell (TNC) dose (per kilogram of body weight) transplanted into recipients following ablative chemotherapy has been shown to be the major predictor of time to neutrophil and platelet engraftment. The minimum dose proposed for adult patients is 1×10^7 TNC/kg, while 3.7×10^7 TNC/kg has been proposed for pediatric patients.

CB products have also been reported to contain stem cells of non-hematopoietic lineages. A range of studies have been reported suggesting that there are multiple types of stem cells in CB products or that CB-derived stem cells are pluripotent. Further research will be necessary to clarify this issue. One stem cell population that has been identified in CB products is the unrestricted somatic stem cell (USSC) defined by Kogler and colleagues.[10] The USSC have some properties overlapping with mesenchymal stem cells (MSCs), but additional tissue potentials that would suggest these cells are more pluripotent.

31.3 Mesenchymal Stem Cells (MSCs)

The bone marrow (BM) from a wide range of mammalian species contains precursor cells that generate adherent colonies of stromal cells in vitro. The BM stroma represents the non-hematopoietic connective tissue elements that provide a system of structural support for developing hematopoietic cells. The complex cellular composition of marrow stromal tissue comprises a heterogeneous population of cells including reticular cells, adipocytes, osteogenic cells near bone surfaces, vascular endothelial cells, smooth muscle cells in vessel walls, and macrophages.[11-14]

The concept that adult hematopoiesis occurs in a stromal microenvironment within the BM was first proposed by Dexter and colleagues, leading to the establishment of the long-term BM culture (LTMC). These studies demonstrated that an adherent stromal-like culture could support maintenance of hematopoietic stem cells (HSC).[15]

MSC are typically isolated, based upon adherence to standard tissue culture flasks. Low-density BM mononuclear cells (MNC) are placed into culture in basal media plus FCS (typically 20%); after 2 to 3 days adherent cells can be visualized on the surface of the flask. The nonadherent cells are removed at this time and fresh media added until a confluent adherent layer forms. The MSC are harvested by treatment with trypsin and further passaged expanding the number of MSC. A number of different cell populations have been isolated using different culture conditions; however, the morphology of these cells is very similar. Phenotypical characterization of MSC has been performed by many groups and standard criteria have been proposed by the International Society of Cellular Therapy (ISCT).[16] The minimal criteria proposed to define human MSC by the Mesenchymal and Tissue Stem Cell Committee of the ISCT consists of the following: (1) the MSC must be plastic-adherent when maintained in standard culture conditions; (2) MSC must express CD105, CD73, and CD90, and lack expression of CD45, CD34, CD14 or CD11b, CD79 alpha or CD19, and HLA-DR surface molecules; and (3) MSC must differentiate into osteoblasts, adipocytes, and chondrocytes in vitro.[16]

A standard in vitro assay for MSC is the colony-forming unit fibroblast (CFU-F) assay.[17] BM mononuclear cells (MNC) are plated at low density and colonies of fibroblasts get attached on the surface of the culture dish. Based upon the results of this assay, the frequency of MSC precursor cells is 1 in 10^4–10^5 BM MNC. The frequency is highly variable between individuals, and the number of MSC has been shown to decrease in older individuals. Other studies have demonstrated that MSC precursors can be isolated based upon surface antigen expression. Antibodies to CD271 and Stro-1 have been used to enrich MSC precursors. CD271, also known as low-affinity nerve growth factor receptor (LNGFR) or p75NTR, belongs to the low-affinity neurotrophin receptor and the tumor necrosis factor receptor superfamily. Selection of CD271+ cells from human BM enriches CFU-F, and MSCs are preferentially selected in the CD271+ fraction compared to the CD271-fraction.[18, 19] Similarly, isolation of Stro-1+ cells from BM MNC results in enrichment of CFU-F in the Stro-1+ fraction compared to the Stro-1- fraction.[20]

The extracellular matrix (ECM) of cardiac tissue provides elasticity and mechanical strength. The cardiac ECM is composed of a number of cells including cardiac fibroblasts, mesenchymal cells, fibronectin, and other matrix proteins.[21-23] We have isolated several stromal cell populations from human fetal heart that are positive for CD105, CD90, and CD73 but negative for CD34 and CD45, which is consistent with the phenotype of BM-derived MSC (McNiece I, unpublished data). Given the homeostatic role of MSC in regulation of HSC, it is highly likely that cardiac stromal cells play a regulatory role in the control of proliferation and differentiation of cardiac stem and progenitor cells (CSC and CPC). This role could be performed through the secretion of a range of growth factors and cytokines.

MI results in ischemic damage that results in cell death of not only cardiomyocytes but also fibroblasts and MSC. Even with migration of viable CSCs and CPCs to the ischemic tissue, the lack of stromal elements would result in the failure of the CSCs and CPCs to proliferate and differentiate, hence failure of remodeling. Therefore optimal tissue regeneration will require regeneration of both cardiomyocytes and stromal cells.

MSC-like cells have been isolated from CB products and from Wharton's jelly.[10, 24, 25] The levels of MSC or MSC precursors are extremely low in CB products with studies demonstrating the isolation of MSC from only 30% of CB products. MSC derived from bone marrow cells have been evaluated for cardiac regenerative therapy and offer advantages over other sources of stem cells because of their availability, immunologic properties, and record of safety and efficacy. Studies of MSC engraftment in rodent and swine models of myocardial infarction (MI) demonstrate (1) functional benefit in post-myocardial infarction (MI) recovery with administration; (2) evidence of neoangiogenesis at the site of the infarct; (3) decrease in collagen deposition in the region of the scar; and (4) some evidence of cells expressing contractile and sarcomeric proteins, but lacking true sarcomeric functional organization[26, 27] Administration of autologous or allogeneic human MSCs to cardiovascular patients has been performed in several clinical studies to date, all in the post-myocardial infarction (MI) setting. The MSC have been administered via the intracoronary route (IC), via peripheral intravenous (IV) injection or direct injection into the cardiac tissue with surgery.

31.3.1 Immunologic Properties of MSCs

MSCs are ideal candidates for allogeneic transplantation because they show minimal MHC class II and ICAM expression and lack B-7 costimulatory molecules necessary for T-cell-mediated immune responses.[28-30] Indeed MSCs do not stimulate a proliferative response from alloreactive T-cells even when the MSCs have differentiated into other lineages or are exposed to proinflammatory cytokines.[30] As previously reviewed,[31] MSCs have significant immunomodulatory effects, inhibiting T-cell proliferation,[32] prolonging skin allograft survival,[33] and decreasing graft-versus-host disease (GVHD).[34] Recently human MSCs were shown to alter the cytokine secretion profile of dendritic cells, T-cells, and natural killer cells in vitro, inhibiting secretion of proinflammatory cytokines (e.g., TNF-a, IFN-γ) and increasing expression of suppressive cytokines (e.g., IL-10), possibly via a prostaglandin E2-mediated pathway.[35] In vivo studies of the fate of MSCs have shown that when transplanted into fetal sheep, human MSCs engraft; undergo site-specific differentiation into various cell types, including myocytes and cardiomyocytes; and persist in multiple tissues for as long as 13 months after transplantation in non-immunosuppressed and immunocompetent hosts.[36] Further, in vivo studies using rodents, dogs, goats, and baboons demonstrate that allogeneic MSCs can be engrafted into these species without stimulating systemic alloantibody production or eliciting a proliferative response from recipient lymphocytes.[37-40] These findings, coupled with the demonstration of efficacy of MSCs for cardiac repair, solidify the notion of using MSCs as an allograft or even xenograft for successful tissue regeneration.

The role of CB-derived MSC in clinical therapies still remains to be defined. The focus on banking both CB products and Wharton's jelly-derived cells may offer options for obtaining MSC from Wharton's jelly cells and stem cells from CB products with the capacity to differentiate to tissue-specific cells. As presented below, the role of MSC may be to regenerate the stem cell niche to facilitate stem cell survival and differentiation. Again the availability of sources of both stem cell populations may offer unique therapeutic options.

31.4 The Stem Cell Niche

The control of proliferation and differentiation of a number of types of stem cells (SC) occurs in the microenvironmental or the stem cell niche. Hematopoietic stem cells (HSC) have been studied in detail and shown to reside in the bone marrow in association with stromal cells, which make up the hematopoietic microenvironment.[41] The stroma consists of several cell populations including mesenchymal stem cells (MSC), fibroblasts, and adventicular reticulocytes.[42] HSC exist in a quiescent state in close relationship with the stromal cells in the bone marrow. These stromal cells produce a number of cytokines and growth factors that are either secreted or expressed as membrane-bound proteins and these cytokines and growth factors control the differentiation and proliferation of the HSC. In vitro, MSC have been shown to support the proliferation and differentiation of HSC, generating committed hematopoietic progenitor cells over a 6-week period. If the microenvironment is compromised, such as in patients who receive multiple rounds of high-dose chemotherapy regimens, normal homeostasis is disrupted and deficiencies in blood cells occur.

We have grown stromal cells with similar properties to MSC from cardiac tissue. Patients with an MI have ischemic tissue that fails to regenerate, and we propose that this is in part due to destruction of cardiac stromal cells. In addition, studies delivering MSC to ischemic cardiac tissue have demonstrated repair with regeneration of cardiomyocytes. Several studies have suggested that paracrine factors were responsible for repair by resident cardiac stem/progenitor cells. Other studies have demonstrated generation of cardiomyocytes from the MSC. Based upon these data we propose that optimal repair of ischemic tissue requires regeneration of both stromal elements and cardiomyocytes. Delivery of MSC to the ischemic tissue can regenerate the stroma and delivery of cardiac stem/progenitor cells can regenerate cardiomyocytes. We further propose that the combination cellular therapy is necessary for optimal repair as delivery of cardiac stem/progenitor cells will result in minimal repair due to the lack of a niche and the absence of appropriate growth factors and cytokines for these cells to proliferate and differentiate.

31.5 Ex Vivo Expansion of Cord Blood Cells on MSC

A number of approaches have been explored for ex vivo expansion of CB products from liquid culture in bags to bioreactors. A number of groups have demonstrated that selection of CD34+ cells or CD133+ cells was necessary for optimal expansion. In 1997 we reported that culture of CB mono nuclear cells (MNC) in a HGF cocktail of stem cell factor (SCF) plus granulocyte colony-stimulating factor (G-CSF) and thrombopoietin (Tpo) resulted in only a 1.4-fold expansion of total cells, 0.8-fold in mature progenitor cells (GM-CFC), and 0.3-fold in erythroid progenitors (BFU-E).[43] In contrast, CD34+ selected CB cells resulted in 113-fold expansion of total cells, 73-fold expansion of GM-CFC, and 49-fold expansion of BFU-E. Based upon these results we initiated expansion cultures in clinical trials with CD34-selected CB cells. Processing of clinical products has led us to two conclusions:

a. Although we can expand significantly the total nucleated cells (TNC) and committed progenitor cells from CD34+ cells, we rarely reached preselection TNC numbers. For a typical CB product starting with a cell dose of 1×10^9 TNC and containing 0.5% CD34+ cells, we would obtain a maximum of 5×10^6 CD34+ cells postselection. Therefore, after culture for 10–14 days we would require a minimum of 200-fold expansion of TNC to obtain preprocessing levels.
b. The performance of clinical trials using CB grafts in the unrelated setting requires the use of frozen CB products. Selection of frozen CB products results in significant losses of CD34+ cells (50% or greater loss of CD34+ cells) and often results in low purities.[44] With a 50% recovery of CD34+ cells after selection we now require at least a 400-fold cell expansion to obtain equivalent TNC as we started with. Again in our experience with clinical studies the purity of the CD34-selected product impacted the level of expansion achieved. The median purity of CD34+ cells was 47.5% (range 14–81%) and the median expansion was 56-fold of TNC.[44] Products with a purity greater than 50% resulted in a median of 139-fold, while products with a purity less than 50% resulted in only 32-fold

expansion. Therefore the use of CD34 selected products has rarely resulted in increased cell doses of ex vivo-expanded cells compared to the starting unmanipulated product.

These studies suggest that the culture conditions currently being undertaken are not capable of expanding the appropriate cell population or that insufficient numbers are being generated to impact the time to recovery of neutrophils or platelets. Our conclusion is that the requirement for selection of CD34+ cells or CD133+ cells from frozen CB products greatly minimizes the potential of generating a suitable expanded CB product to enhance the rate of engraftment. Therefore, in recent studies we have evaluated methods for expanding CB products without an initial CD34- or CD133-selection.

Based upon the ability of MSC to support hematopoietic cells, we have developed a co-culture system that is capable of expanding CB MNC by culturing the CB MCN on confluent MSC layers. As presented above, the literature contains many reports of the ability of MSC to support the growth of hematopoietic cells. It has been demonstrated that MSC produces a number of HGFs and adhesion molecules that may stimulate growth of hematopoietic cells. Our data reproducibly demonstrated a 10- to 20-fold expansion of TNC with 18-fold expansion of GM-CFC and 16- to 37-fold expansion of CD34+ cells.

In previous studies we have evaluated the potential of ex vivo expansion of frozen CB products using the co-culture on MSC. CB products were thawed and washed, resulting in a median of 3.3×10^8 TNC (range $1.4–3.6 \times 10^8$, $n = 5$). For a 50 kg recipient, these CB products would provide only 0.73×10^7 TNC/kg with zero of five products reaching the minimal target dose of 1×10^7 TNC.

Each product was expanded by culturing the MNC fraction from each product on preformed layers of MSC. Ten T162 cm^2 flasks were used for each product such that each flask contained 10% of the CB MNC. After ex vivo culture for 14 days, in the cocktail of SCF, G-CSF, and Tpo in Stemline II media, a median of ninefold expansion of TNC was obtained with a range of 6.5- to 24-fold. The median TNC post expansion was 21.6×10^8 (range $11–79 \times 10^8$ TNC). A median expansion of mature progenitor cells (GM-CFC) of 46-fold was also obtained in the co-culture. For a 50 kg recipient, the expanded CB product would be equivalent to 4.3×10^7 TNC/kg (range $2.2–16 \times 10^7$), with all five expanded products reaching the minimal target of 1×10^7 TNC/kg. In fact all expanded products contained more the 1×10^7 TNC/kg based upon a 100 kg recipient.

We would propose two potential advantages to the use of co-culture for expansion based upon these results. Firstly the possible enhanced engraftment and secondly the ability to use better-matched CB products that may have a low cell dose. Wagner and colleagues have described the use of two CB products to provide an increased cell dose, however, the majority of patients receive a two-antigen mismatched CB unit.[45] Better-matched CB units are routinely identified but are not suitable due to low cell doses. The expansion of the better-matched CB units could potentially decrease the graft-versus-host disease that can result.

A clinical trial to evaluate the potential of ex vivo expanded cells generated using this co-culture approach is currently being conducted at MD Anderson by Dr E. J. Shapll.

31.6 Summary

Stem cells have the potential to repair damaged tissue in a number of organs and tissues of the body; however, these cells require an appropriate niche for survival and proliferation. Inflammation associated with cell death and damage results in further damage to cells and the microenvironment. We propose that Combination Cell Therapy will be necessary for optimal tissue repair. MSCs are a unique cell population that exhibit immune properties that minimize graft rejection or immune activation. The use of MSCs to reestablish the stem cell niche will be critical in several diseases to ensure that stem cells used for regenerative approaches have an appropriate niche. In particular, we are focused on using Combination Cell Therapy for heart disease using MSCs to repair the microenvironment of ischemic tissue and cardiac stem cells for regeneration of cardiomyocytes. We plan to move forward with large animal studies combining MSC with c-kit+ cardiac stem cells in animals with an MI to move toward clinical trials.

References

1. Gluckman E, Broxmeyer HE, Auerbach AD, et al. Hematopoietic reconstitution in a patient with Fanconi's anemia by means of umbilical-cord blood from a HLA-identical sibling. *N Engl J Med*. 1989;321:1174-1178.
2. Prasad VK, Mendizabal A, Parikh SH, et al. Unrelated donor umbilical cord blood transplantation for inherited metabolic disorders in 159 pediatric patients from a single center: influence of cellular composition of the graft on transplantation outcomes. *Blood*. 2008;112(7):2979-2989.
3. Gluckman E, Rocha V. Donor selection for unrelated cord blood transplants. *Curr Opin Immunol*. 2006;18(5): 565-570.
4. Laughlin MJ, Barker J, Bambach B, et al. Hematopoietic engraftment and survival in adult recipients of umbilical-cord blood from unrelated donors. *N Engl J Med*. 2001; 344(24):1815-1822.
5. Migliaccio AR, Adamson JW, Stevens CE, et al. Cell dose and speed of engraftment in placental/umbilical cord blood transplantation: graft progenitor cell content is a better predictor than nucleated cell quantity. *Blood*. 2000;96(8): 2717-2722.
6. Takahashi S, Ooi J, Tomonari A, et al. Post transplantation engraftment and safety of cord blood transplantation with grafts containing relatively low cell doses in adults. *Int J Hematol*. 2006;84(4):359-362.
7. Gluckman E, Rocha V, Boyer-Chammard A, et al. Outcome of cord-blood transplantation from related and unrelated donors. *N Engl J Med*. 1997;337(6):373.
8. Hogan CJ, Shpall EJ, McNiece I, Keller G. Multilineage engraftment in NOD/LtSz-scis/scid mice from mobilized human CD34+ peripheral blood progenitor cells. *Biol Blood Marrow Transplant*. 1997;3:236-246.
9. Hogan CJ, Shpall EJ, McNulty O, et al. Engraftment and development of human CD34(+)-enriched cells from umbilical cord blood in NOD/LtSz-scid/scid mice. *Blood*. 1997;90(1):85-96.
10. Kogler G, Sensken S, Airey JA, et al. A new human somatic stem cell from placental cord blood with intrinsic pluripotent differentiation potential. *J Exp Med*. 2004;200: 123-135.
11. Weiss L. The hematopoietic microenvironment of the bone marrow: an ultrastructural study of the stroma in rats. *Anatom Rec*. 1976;186:161.
12. Westen H, Bainton DF. Association of alkaline phosphatase-positive reticulum cells in bone marrow with granulocyte precursors. *J Exp Med*. 1979;150:919.
13. Lichtman MA. The ultrastructure of the hematopoietic microenvironment of the marrow: a review. *Exp Hematol*. 1981;9:391.
14. Bianco P, Riminucci M. The bone marrow stroma in vivo: ontogeny, structure, cellular composition and changes in disease. In: Beresford JN, Owens ME, eds. *Marrow Stromal Cell Culture. Handbooks in Practical Animal Cell Biology*. Cambridge, UK: Cambridge University Press; 1998:1025.
15. Dexter TM. Stromal cell associated haemopoiesis. *J Cell Physiol Suppl*. 1982;1:87-94.
16. Dominici M, Le Blanc K, Mueller I, et al. Minimal criteria for defining multipotent mesenchymal stromal cell. The International Society of Cellular Therapy position statement. *Cytotherapy*. 2006;8(4):315.
17. Friedenstein AJ. Precursor cells of mechanocyte. *Int Rev Cytol*. 1976;47:327.
18. Quirici N, Soligo D, Bossolasco P, et al. Isolation of bone marrow mesenchymal stem cells by anti-nerve growth factor receptor antibodies. *Exp Hematol*. 2002;30(7):783.
19. Jones E, English A, Kinsey SE, et al. Optimization of a flow cytometry-based protocol for detection and phenotypic characterization of multipotent mesenchymal stromal cells from human bone marrow. *Cytometry B Clin Cytom*. 2006;70:391-399.
20. Zannettino A, Paton S, Kortesidis A, et al. Human multipotential mesenchymal/stromal stem cell are derived from a discrete subpopulation of STRO-1^{bright}/CD34-/CD45-/glycophorin-A-bone marrow cells. *Haematogica*. 2007;92(12):1707.
21. Decker C, Greggs R, Duggan K, et al. Adhesive multiplicity in the interaction of embryonic fibroblasts and myoblasts with extracellular matrices. *J Cell Biol*. 1984;99:1398.
22. Choy M, Oltjen SL, Otani YS, et al. Fibroblast growth factor-2 stimulates embryonic cardiac mesenchymal cell proliferation. *Dev Dyn*. 1996;206:193.
23. Baudino TA, Carver W, Giles W, Borg TK. Cardiac fibroblasts: friend or foe? *Am J Physiol Heart Circ Physiol*. 2006;291:H1015.
24. Seshareddy K, Troyer D, Weiss ML. Method to isolate mesenchymal-like cells from Wharton's Jelly of umbilical cord. *Methods Cell Biol*. 2008;86:101-119.
25. Troyer DL, Weiss ML. Wharton's jelly-derived cells are a primitive stromal cell population. *Stem Cells*. 2008;26(3):591-599.
26. Urbanek K, Ceselli D, Rota M, et al. Stem cell niches in the adult mouse heart. *Proc Natl Acad Sci*. 2006;103:9226-9231.
27. Mazhari R, Hare JM. Mechanisms of action of mesenchymal stem cells in cardiac repair: potential influences on the cardiac stem cell niche. *Nat Clin Pract Cardiovasc Med*. 2007;4(suppl 1):S21-S26.
28. Le Blanc K, Tammik C, Rosendahl K, et al. HLA expression and immunologic properties of differentiated and undifferentiated mesenchymal stem cells. *Exp Hematol*. 2003;31(10):890-896.
29. Klyushnenkova E, Shustova V, Mosca J, et al. Human mesenchymal stem cells induce unresponsiveness in preactivated but not naïve alloantigen specific T cells [abstract]. *Exp Hematol*. 1999;27:122.
30. Klyushnenkova E, Mosa JD, Zernetkina V, et al. T cell responses to allogeneic human mesenchymal stem cells: immunogenicity, tolerance, and suppression. *J Biomed Sci*. 2005;12(1):47-57.
31. Le Blanc K, Ringden O. Immunobiology of human mesenchymal stem cells and future use in hematopoietic stem cell transplantation. *Biol Blood Marrow Transplant*. 2005;11(5):321-334.
32. Le Blanc K, Rasmusson I, Gotherstrom C, et al. Mesenchymal stem cells inhibit the expression of CD25 (interleukin-2 receptor) and CD38 on phytohaemagglutinin-activated lymphocytes. *Scand J Immunol*. 2004;60(3):307-315.
33. Bartholomew A, Sturgeon C, Siatskas M, et al. Mesenchymal stem cells suppress lymphocyte proliferation in vitro and prolong skin graft survival in vivo. *Exp Hematol*. 2002;30(1):42-48.
34. Lazarus HM, Koc ON, Devine SM, et al. Cotransplantation of HLA-identical sibling culture-expanded mesenchymal

stem cells and hematopoietic stem cells in hematologic malignancy patients. *Biol Blood Marrow Transplant.* 2005;11(5):389-398.
35. Aggarwal S, Pittenger MF. Human mesenchymal stem cells modulate allogeneic immune cell responses. *Blood.* 2005;105(4):1815-1822.
36. Lietchy KW, MacKenzie TC, Shaaban AF, et al. Human mesenchymal stem cells engraft and demonstrate site-specific differentiation after in utero transplantation in sheep. *Nat Med.* 2000;6(11):1282-1286.
37. Grinnemo KH, Mansson A, Dellgren G, et al. Xenoreactivity and engraftment of human mesenchymal stem cells transplanted into infracted rat myocardium. *J Thorac Cardiovasc Surg.* 2004;127(5):1293-1300.
38. Arinzeh TL, Peter SL, Archambault MP, et al. Allogeneic mesenchymal stem cells regenerate bone in a critical-sized canine segmental defect. *J Bone Joint Surg Am.* 2003;85-A(10):1927-1935.
39. Murphy JM, Fink DJ, Hunziker EB, Barry FP. Stem cells therapy in a caprine model of osteoarthritis. *Arthritis Rheum.* 2003;48(12):3464-3474.
40. Mahmud N, Pang W, Cobbs C, et al. Studies on the route of administration and role of conditioning with radiation on unrelated allogeneic mismatched mesenchymal stem cell engraftment in a nonhuman primate model. *Exp Hematol.* 2004;32(5):494-501.
41. Haylock DN, Nilsson SK. Stem cell regulation by the hematopoietic stem cell niche. *Cell Cycle.* 2005;4(10):1353-1355.
42. Weiss L. The hematopoietic microenvironment of the bone marrow: an ultrastructural study of the stroma in rats. *Anatom Rec.* 1976;186:161.
43. Briddell R, Kern BP, Zilm KL, et al. Purification of CD34+ cells is essential for optimal ex vivo expansion of umbilical cord blood cells. *J Hematotherapy.* 1997; 6:145-150.
44. McNiece I, Briddell R, Stoney G, et al. Large scale isolation of CD34+ cells using the Amgen cell selection device results in high levels of purity and recovery. *J Hematotherapy.* 1997;6(1):5-11.
45. Majhail NS, Brunstein CG, Wagner JE. Double umbilical cord blood transplantation. *Curr Opin Immunol.* 2006;18(5): 571-575.

Use of Cord Blood in Regenerative Medicine

32

David T. Harris

Abbreviations

CB	Cord blood
CP	Cerebral palsy
T1D	Type I diabetes.

32.1 Introduction

Work that began in the early 1980s revealed that umbilical cord blood (CB; i.e., the leftover blood in the umbilical cord and placenta after the birth of a child) was comparable to bone marrow in terms of its utility in stem cell transplantation.[1-8] CB offered a number of advantages over bone marrow,[8,9] including a lower incidence of graft-versus-host disease (GVHD) and less strict HLA-matching requirements. Over the past 16 years, clinical use of CB has grown to more than 14,000 transplants worldwide.[10] Fortunately, stem cell transplants are still rare when considered at the population level. That is, approximately 30,000–40,000 such transplants are performed annually in the USA for a population base of 300 million; or a rate of 0.01–0.13%.

However, it is estimated that as many as one in three individuals in the USA, or 128 million people, could benefit over their lifetime from the applications of stem cells to regenerative medicine, including therapies for cardiovascular disease, endocrine, and orthopedic treatments (http://www.dhhs.gov/reference/newfuture.shtml). Previously untreatable diseases, such as type 1 diabetes (T1D), myocardial infarction, stroke, and spinal cord injury could potentially be cured. In addition to its use as a stem cell source to regenerate the blood and immune system, CB has recently been utilized in a variety of regenerative medicine applications. Work done by McGuckin et al.,[11,12] Rogers et al.,[13] Kucia et al.,[14] and Harris et al.[15] has shown that CB contains a mixture of multipotent stem cells capable of giving rise to cells derived from the endodermal, mesodermal, and ectodermal lineages. Thus, CB appears to be a practical source of stem cells for use in tissue engineering and regenerative medicine. Recently, clinical trials have begun using CB stem cells to treat type 1 diabetes, cerebral palsy, and peripheral vascular disease among others.[16,17]

In this review, I will summarize the current state of these newly started regenerative medicine clinical trials as well as future uses of CB stem cells.

32.2 Regenerative Medicine Applications

Over the past decade, scientists have demonstrated that CB stem cells can not only differentiate into many cell types in vitro, but these stem cells can also engraft and improve function in a variety of animal disease models. The presence of multiple primitive and multipotential stem cell populations in CB helps to explain the mechanisms for the effects behind these observations.[11-14] This research has paved the way for several clinical trials in human patients (see below). The body of data is extensive for many disease states, including the following preclinical examples.

D.T. Harris
Department of Immunobiology, The University of Arizona, 1656 E. Mabel, MRB 221, Tucson, AZ, 85724, USA
e-mail: davidh@email.arizona.edu.

N. Bhattacharya and P. Stubblefield (eds.), *Regenerative Medicine Using Pregnancy-Specific Biological Substances*,
DOI: 10.1007/978-1-84882-718-9_32, © Springer-Verlag London Limited 2011

32.3 Preclinical Studies

32.3.1 Heart Disease

Cardiovascular disease is the leading cause of morbidity and mortality for both men and women in the USA. Approximately 1 million people die of heart disease each year despite medical intervention. Coronary artery disease comprises approximately half of these deaths. Heart cells have a limited capacity to regenerate after myocardial infarction (MI). Thus, the application of exogenous stem cells seems a logical alternative for therapy. Recently, numerous preclinical and clinical studies have examined the use of adult autologous hematopoietic stem cell sources (see ref.[18]) for details and additional references). There have been no clinical trials using CB stem cells for cardiovascular disease, as most CB donors are not of an age to be at risk for cardiovascular disease, and clinical trials have generally required the use of autologous stem cells. However, although no clinical trials utilizing CB stem cells for heart failure have been conducted to date, a number of preclinical animal studies have been performed.[19-25] Several common observations have been noted in these studies regardless of the protocols utilized including selective migration of the CB stem cells to the injured heart tissue, increased capillary density at the site of injury, decreased infarct size/volume, improved heart function, and a general lack of myogenesis. These observations are thought to be due to the production of angiogenic factors and induction of angiogenesis/vasculogenesis.[19,26,27] That is, the therapeutic effects are not due to stem cell differentiation but rather are trophic in nature. In fact, work done by Gaballa et al.[18,19] in myocardial infracted rats showed that CD34+ CB stem cells induced blood vessel formation, reduced infarct size, and restored heart function. The effects were thought to be due to the release of angiogenic and growth factors (e.g., VEGF, EGF, and Angiopoietin-1, -2) induced by hypoxia as shown by gene array analyses. This work demonstrated that CB stem cells could be induced to become/differentiate into endothelial-like cells. Finally, work from numerous groups seems to indicate that more than one population of pluripotent cells contained in CB is capable of mediating this effect as shown by the ability of CD34(+), CD133(+), and CD45(−) cells to induce cardiac repair after MI.[19,20,24,25,28] Even more importantly, the numbers and potency of these cells found in CB seem sufficient for adult human applications as shown by work performed in a porcine model[24] of approximate human weight.

The ability of CB stem cells to exert a beneficial effect via its angiogenesis-inducing capability appears to be useful for the treatment of various ischemic diseases. Many investigators have demonstrated that not only does CB contain cells displaying the phenotypic characteristics of endothelial precursors that are responsible for blood vessel formation, but that these cells are also capable of differentiating into endothelial cells and becoming blood vessels.[29-36] When placed in animal models, CB stem cells were able to significantly reverse the effects of ischemia in several model systems.[30,33,36] In models of hind-limb ischemia, transplantation of CB stem cells or endothelial cells derived from CB stem cells appeared able to reverse surgery-induced ischemia resulting in limb salvage.[37-40] These observations have led to the recent announcement of a similar trial using CB stem cells would begin in late 2010 at the University of Toronto in conjunction with the University of Minnesota for patients with peripheral vascular disease.

32.3.2 Juvenile Diabetes

Type 1 diabetes is expected to affect 1 in every 300 births in the USA (http://www.diabetes.niddk.nih.gov/dm/pubs/statistics/#youngpeople). Approximately 5–10% of all diabetics will display the type 1 diabetic phenotype (i.e., immune-mediated). That is, approximately 2 million individuals in the USA currently have type 1 diabetes (T1D). Type 1 diabetes results from destruction by the immune system of the beta cells in the pancreatic islets responsible for insulin production. The end result is uncontrolled blood glucose levels. Diabetic complications include cardiomyopathy, coronary artery disease, peripheral vascular disease, and neurological complications. In an effort to treat T1D, surgical procedures have been developed to transplant islets across histocompatibility barriers with limited success due to immune rejection and the lack of cadaver donors. Investigators have tried to address the issue of T1D through the use of stem cells and regenerative medicine.[41] Currently, autologous CB mononuclear (stem) cells are being evaluated in a clinical trial to treat T1D in children.[42] The trial recently completed enrollment of 23 subjects. Preliminary analyses of the first eight subjects to reach 1 year post infusion showed evidence that the treatment is safe and provides some slowing to the decline of endogenous insulin production.[43] The rationale for the clinical trial was established in studies which showed that in animals with T1D, those treated with CB stem cells had lower blood glucose levels, reduced insulitis, and increased lifespan compared to control diabetic animals.[44-46] Similar stem cell trials are being proposed at other centers as well.[42] Although the mechanism(s) of action behind CB stem cell therapy for T1D are not known, it is postulated that once in vivo the infused CB stem cells differentiate into new islet cells *and* mediate an immune tolerance to the new derived islet cells (See ref.[43] and personal communications from M. Haller, University of Florida and unpublished data, D. Harris, University of Arizona). In fact, recent results have indicated that in vitro CB stem cells can indeed be driven to become insulin-secreting islet cells as indicated by the production of C-peptide, an offshoot of the de novo secretion of insulin.[47,48] In both instances, the islet cell differentiation was attributed to the presence of the ES-like stem cells found in CB.

32.3.3 Neurological Diseases and Injuries

In animal models of stroke, amyotrophic lateral sclerosis, Parkinson's disease, cerebral palsy, and spinal cord injury, CB stem cell infusions have resulted in observable behavioral improvement compared to control animals.[12,49-68] The same beneficial effect has also been observed for animals with neurological injury due to traumatic brain injury.[65]

Several investigators have demonstrated that it was possible to derive neurological-like cells utilizing CB stem cells in vitro. McGuckin et al.[12] and Rogers et al.[13] demonstrated that CB stem cells could be differentiated in culture to express neuronal cell morphology as well as expressing typical neuronal markers (GFAP, nestin, musashi-1, and nectin). These neuronal-like cells also released glial-derived neurotrophic factor (GDNF) into the cultures. Jang et al.[49] also showed that CB CD133+ stem cells upon exposure to retinoic acid differentiated into neuronal (astrocytes and oligodendrocytes) and glial cells with neuronal markers (including tubulin βIII, neuron-specific enolase, NeuN, microtubule-associated protein-2 (MAP2), and the astrocyte-specific marker glial fibrillary acidic protein). Further, non-hematopoietic stem cells found in CB (i.e., mesenchymal stem cells or MSC) also could become neural-like cells in vitro (i.e., astrocytes and oligodendrocytes).[50] The utility of these progenitor cell populations for use in cell-based treatments of brain injuries and neurological diseases has recently been reviewed by Chen et al.[51] In confirmation of these reports, work from Harris and Ahmad[52] has also shown that CD133+ and Lin– populations isolated from CB could become glial cells, astrocytes, and oligodendrocytes in vitro. These findings, and those presented below, have led to the initiation of multiple clinical trials for a variety of neurological conditions (see below).

In one study, infusion of CB stem cells into rats with the commonly used middle carotid artery occlusion (MCAO) model of stoke could ameliorate many of the physical and behavioral deficits associated with this disease.[53] Additional studies demonstrated that direct injection of the stem cells into the brain was not required,[54] and in fact, beneficial effects could be observed even if the stem cells did not actually make their way into the target organ (most probably due to trophic effects via the release of growth and repair factors triggered by anoxia).[55,56] The beneficial effects seemed to be dose-dependent and could reduce the size/volume of the infarcted tissue.[57] Again, it appeared that multiple progenitor populations might be capable of mediating these effects.[58] Significantly, CB stem cell therapies were effective up to 48 h after the thrombotic event.[59] In fact, administration of the stem cells within the first 24 h of the stroke was not beneficial, as it was thought that inflammatory events occurring at this time might induce stem cell death. Early studies have also shown benefit in animal models of hemorrhagic (as opposed to embolic) stroke.[60] For additional information one is referred to the recent review in reference.[61]

The observation that CB stem cells can become different types of nervous cells extends its application to other areas of neurological damage, including spinal cord injury. In one study, spinal cord injured rats infused with CB stem cells showed significant improvements 5 days posttreatment compared to untreated animals. The CB stem cells were observed at the site of injured nervous tissue, but not at uninjured regions of the spinal cord.[62] This finding is supported by studies showing that CB stem cells transplanted into spinal-cord-injured animals differentiated into various neural cells, thereby improving axonal regeneration and motor function.[63] Significantly, in a recently reported clinical use of CB stem cells to treat a patient with a spinal cord injury[64] it was stated that transplantation of CB cells improved her sensory perception and mobility in the hip and thigh regions. Both CT and MRI studies revealed regeneration of the spinal cord at the injury site. To date however, this application has not been attempted in other patients.

Based on the success of CB stem cells in treating animals with stroke, similar therapies have been investigated for treating other forms of neurological damage. Along those lines, Lu et al.[65] have demonstrated that intravenous administration of CB mononuclear cells could be used to treat traumatic brain injury in a rat model. Once again the CB cells were observed to enter the brain, selectively migrate to the damaged region of the brain, and reduce the neurological damage. Similarly, CB stem cell infusions could also alleviate symptoms of newborn cerebral palsy in a rat model, with improved neurological effects.[66] These observations have now been turned into clinical therapies. Cord Blood Registry has released over 50 CB stem cell samples for autologous use in the treatment of cerebral palsy, and anoxic and traumatic brain injury (www.cordblood.com and H. Brown, personal communication). Early, albeit anecdotal, reports have indicated beneficial effects from the CB mononuclear cells infusions. Several investigators have begun planning clinical trials to treat such children utilizing autologous CB stem cell infusions (e.g., Duke University; UT-Health Sciences Center, Houston, TX; Medical College Georgia).

32.3.4 Epithelial Tissue Applications

Cord blood also contains stem cell populations capable of giving rise to epithelial tissue, making CB amenable for use in regenerative medicine applications for the eye (cornea), skin (wound healing), and other such tissues (e.g., gut and lung). In terms of the eye, the cornea appears to have the most direct and routine clinical application. The outer layer of the eye is made up of the central cornea, the limbus, and the sclera. The cornea epithelium is a rapidly self-renewing tissue; implicated to have its own source of stem cells (the limbus) specialized for this purpose. Corneal epithelial stem cell deficiency is an important cause of visual disability, resulting from alkali injury, Stevens–Johnson syndrome, ocular cicatricial pemphigoid, aniridia, chronic rosacea keratoconjunctivitis, and iatrogenic causes. Autologous corneal epithelial stem cell grafts have been successful for patients with unilateral disease. However, harvesting cells from the functional eye places the healthy eye at risk for vision loss. Additionally, in bilateral conditions, autologous grafts are not available. The best current solution for bilateral disease is a corneal epithelial stem cell allograft. Allografts require chronic antirejection therapy with possible systemic side effects. In addition, the average survival of allografted corneal stem cells is 2 years. Severe corneal wounds requiring intervention are not uncommon. In fact, corneal wounds make up 37% of all visual disabilities and almost a quarter of all medical visits for ocular problems in North America.[69,70] Work from Nichols et al.[15,71,72] has used CB stem cells as a viable therapeutic alternative for ocular surface disease, as human CB stem cells could represent an unlimited source of tissue for ocular surface reconstruction. Preliminary laboratory and animal data is supportive of this hypothesis. Histology and immunohistochemistry of in vitro differentiated CB stem cells revealed that the resultant cell sheet was morphologically indistinguishable from corneal epithelial cells. Cord blood stem cells were capable of expressing the corneal epithelial-specific cytokeratin, k3. Further, when New Zealand white rabbits were transplanted with the cell sheets, they were able to reconstitute the cornea, forming an optically clear surface. Other investigators have demonstrated that MSC are also capable of reconstituting the cornea in a rat model.[72] In a rat model of chemically induced loss of the cornea, human bone marrow-derived MSC could replace it, and appeared to function like limbal stem cells. As CB also contains MSC populations, this observation may partially explain the mechanism of action of CB stem cells in the previous report.

As CB stem cells can become corneal epithelial cells, it is not far-fetched that these stem cells can also differentiate into other types of epithelial cells, such as those found in the skin, which would be useful in facilitating wound repair (e.g., for diabetic ulcers). Work from the Harris and Ablin laboratories has begun to investigate this premise, knowing that previous studies have demonstrated a bone marrow stem cell contribution to wound healing in mice.[73] In support of this hypothesis, in 2004 there was an initial report of the use of allogeneic CB CD34+ progenitor cells in two patients to promote skin wound/lesion repair.[74] In this application, the progenitor cells were admixed with an autologous fibrin matrix (which acts as a form of glue to keep the stem cells in place) and the stem cells were injected in a volume of 3 ml into the margins of the lesions. At 3–7-months follow-up there was no sign of graft-versus-host disease, and most importantly, there were evidence of significant healing in both patients.

32.4 Current Clinical Trials with CB Stem Cells

Information gained from the laboratory and preclinical animal studies discussed above are beginning to be translated into clinical trials, which are rapidly growing in size and in scope. We will review the largest of the trials below.

32.5 Diabetes

Investigators have tried to address the issue of T1D through the use of stem cells and regenerative medicine.[42] Currently, autologous CB mononuclear (stem) cells are being evaluated in a clinical trial to treat type 1 diabetes in children.[43] Enrollment has recently completed and as described earlier preliminary analyses showed significant improvement in slowing the decline of endogenous insulin production.[43] The protocol for the clinical trial was established in studies that showed that in animals with type 1 diabetes, those treated with CB stem cells had lower blood glucose levels, reduced insulitis, and increased lifespan compared to control diabetic animals.[44-46] Similar stem cell trials are being proposed at other centers as well.[42] Although the mechanism(s) of action behind CB stem cell therapy for T1D are not known, it is postulated that once in vivo the infused CB stem cells differentiate into new islet cells *and* due to the large number of regulatory immune cells mediate an immune tolerance to the new derived islet cells[44] and personal communications from M. Haller, University of Florida and unpublished data, D. Harris, University of Arizona).

32.6 Cerebral Palsy

Cerebral palsy (CP) is a devastating brain disorder that affects many children worldwide (http://www.ucp.org/uploads/cp_fact_sheet.pdf), and stem cells ultimately have the capacity to generate new cells to replace those lost through injury or disease. Umbilical CB stem cells have shown promise in the treatment of CP. Recently, considerable excitement has been generated by patient outcomes observed after umbilical CB stem cell infusions in children treated in a clinical trial at Duke University (see www.cordblood.com). Although not a randomized trial, this treatment has been used to treat more than 100 children with cerebral palsy and has shown functional improvement in many of the subjects (see www.cordblood.com and www.viacord.com). Preliminary results have been significant and encouraging (see http://www.msnbc.msn.com/id/23572206/), and many additional patients are being enrolled. Similar results for children with cerebral palsy have been reported recently by investigators treating children in Europe and Asia (personal communication, Novussanguis Foundation, Paris, France, May 2008). It should be noted that not all children have benefited to the same extent, and it appears that the younger the patient, the more significant the benefits that have been observed. However, the optimal therapeutic regime and the mechanism(s) behind any beneficial effects have yet to be determined.

32.7 Traumatic Brain Injury

The University of Texas at Houston is to begin a Food and Drug Administration (FDA)-submitted clinical trial to treat children with traumatic brain injury utilizing autologous cord blood stem cell infusions (Cox and Baumgartner, 2008, UT Health Sciences Center, Houston, Texas), based on the successful results obtained with a similar autologous bone marrow stem cell study and numerous animal studies demonstrating the efficacy of stem cell treatments in models of traumatic brain injury.[75] The investigators have chosen to focus their efforts on autologous cord blood in part due to the immediate availability and the noninvasive harvest.

32.8 Hearing Loss

A recent animal study demonstrated that the CB stem cells may have clinical utility to repair inner-ear damage and restore hearing.[76] Human CB stem cells were intravenously injected into immunodeficient mice made deaf by exposure to kanamycin, high-intensity noise, or a combination of these insults. The study showed that the CB stem cells migrated and engrafted into the cochlea of the deaf mice and that the levels of engraftment correlated with both the severity of damage and the treatment dose. Analysis at 60 days post-treatment showed that the mice in the CB treatment group had well-repaired cochlea with dramatic hair cell regrowth, while control mice showed no sign of repair or hair cell regeneration. This study has led to discussion and enthusiasm to translate these promising findings into a clinical trial to investigate autologous CB infusions for childhood hearing damage.

32.9 Conclusions

Regenerative medicine has the ability to treat many of the conditions discussed above by replacing or repairing malfunctioning tissues. Because regenerative medicine focuses on functional restoration of damaged tissues, not just the abatement or moderation of symptoms, this field has the potential to cut health-care costs significantly. However, in order for the promise of regenerative medicine to be realized, it is necessary to identify optimal stem cell sources for particular disease states, and make efforts to inform the lay and medical communities as to their options.

Already, in examples of type 1 diabetes and neurological (cerebral palsy and brain injury) applications, cord blood stem cells have transitioned from the laboratory to the clinic and numerous patients are currently being treated in clinical trials. Other trials will surely rapidly follow, including therapies for the eyes and joints, wound healing, and spinal cord injuries. The key to these advances lies in the multipotency of CB stem cells and their ability to be used in many instances under the practice of medicine, since it appears in many instances that it is possible to merely infuse the stem cells directly without timely and costly in vitro culture and differentiation. Whether the beneficial effects are due to stem cell differentiation into new tissues, or due to release of trophic factors, is not yet known. However, the assertion that cord blood stem cells are amenable to regenerative medicine applications today is supported by the clinical trials for type 1 diabetes and cerebral palsy discussed above. We believe that research and clinical trials conducted now and over the next several years will demonstrate that CB stem cells are ideal for use in many diverse regenerative medicine applications

Acknowledgments The funding by The Jerome Lejeune Foundation is gratefully acknowledged.

References

1. Broxmeyer HE, Gluckman E, Auerbach A, et al. Human umbilical cord blood: a clinically useful source of transplantable hematopoietic stem/progenitor cells. *Int J Cell Cloning*. 1990;8(supp 1):76.
2. Gluckman E, Broxmeyer HE, Auerbach A, et al. Hematopoietic reconstitution in a patient with Fanconi's anemia by means of umbilical cord blood from an HLA-identical sibling. *N Eng J Med*. 1989;321:1174-1178.
3. Gluckman E. Stem cell harvesting from cord blood: a new perspective. In: Henon PR, Wunder EW, eds. *Peripheral Blood Stem Cell Autografts*. Berlin: Springer; 1990.
4. Broxmeyer HE, Kurtzburg J, Gluckman E, et al. Umbilical cord blood hematopoietic stem and repopulating cells in human clinical transplantation: an expanded role for cord blood transplantation. *Blood Cells*. 1991;17:330-337.
5. Broxmeyer HE, Kurtzburg J, Gluckman E, et al. Umbilical cord blood hematopoietic stem and repopulating cells in human clinical transplantation. *Blood Cells*. 1991;17: 313-330.
6. Broxmeyer HE, Douglas GW, Hangoc G, et al. Human umbilical cord blood as a potential source of transplantable hematopoietic stem/progenitor cells. *Proc Natl Acad Sci USA*. 1989;86:3828-3832.
7. Vilmer E, Sterkers G, Rahimy C, et al. HLA-mismatched cord blood transplantation in a patient with advanced leukemia. *Transplantation*. 1992;53:1155-1157.
8. Wagner JE, Kernan NA, Steinbuch M, et al. Allogeneic sibling umbilical cord blood transplantation in children with malignant and nonmalignant disease. *Lancet*. 1995;346:214-219.
9. Rubinstein P, Rosenfield RE, Adamson JW, Stevens CE. Stored placental blood for unrelated bone marrow reconstitution. *Blood*. 1993;81:1679-1690.
10. Loper K. AABB Advancements in Cord Blood Transplantation. Available at: http://www.aabb.org/Content/Meetings_and_Events/Annual_Meeting_and_TXPO/61amonline/sunct1.htm. Accessed October 15, 2008.
11. McGuckin C, Forraz N, Baradez MO, et al. Production of stem cells with embryonic characteristics from human umbilical cord blood. *Cell Prolif*. 2005;38:245-255.
12. McGuckin CP, Forraz N, Allouard Q, Pettengell R. Umbilical cord blood stem cells can expand hematopoietic and neuroglial progenitors in vitro. *Exp Cell Res*. 2004; 295:350-359.
13. Rogers I, Yamanaka N, Bielecki R, et al. Identification and analysis of in vitro cultured CD45-positive cells capable of multi-lineage differentiation. *Exp Cell Res*. 2007;313: 1839-1852.

14. Kucia M, Halasa M, Wysoczynski M, et al. Morphological and molecular characterization of novel population of CXCR4+ SSEA-4+ Oct-4+ very small embryonic-like cells purified from human umbilical cord blood-preliminary report. *Leukemia*. 2007;21:297-303.
15. Harris DT, He X, Badowski M, Nicols JC. In: Levicar N, Habib NA, Dimarakis I, Gordon MY, eds. Regenerative Medicine of the Eye: A Short Review. Stem Cell Repair & Regeneration. Vol. 3. London: Imperial College Press; 2008:211–225.
16. Harris DT, Badowski M, Ahmad N, Gaballa M. The potential of cord blood stem cells for use in regenerative medicine. *Expert Opin Biol Ther*. 2007;7(9):1311-1322.
17. Harris DT, Rogers I. Umbilical cord blood: a unique source of pluripotent stem cells for regenerative medicine. *Curr Stem Cell Res Ther*. 2007;2:301-309.
18. Furfaro MEK, Gaballa MA. Do adult stem cells ameliorate the damaged myocardium? Is human cord blood a potential source of stem cells? *Curr Vasc Pharm*. 2007;5:27-44.
19. Sunkomat JNE, Goldman S, Harris DT, et al. Cord blood-derived MNCs delivered intracoronary contribute differently to vascularization compared to CD34+ cells in the rat model of acute ischemia. *J Mol Cell Cardiol*. 2007;42(6 Suppl 1):S97.
20. Botta R, Gao E, Stassi G, et al. Heart infarct in NOD-SCID mice: therapeutic vasculogenesis by transplantation of human CD34+ cells ad low dose CD34+KDR+cells. *FASEB J*. 2004;18:1392-1394.
21. Henning RJ, Abu-Ali H, Balis JU, et al. Human umbilical cord blood mononuclear cells for treatment of acute myocardial infarction. *Cell Transplant*. 2004;13:729-739.
22. Chen HK, Hung HF, Shyu KG, et al. Combined cord blood cells and gene therapy enhances angiogenesis and improves cardiac performance in mouse after acute myocardial infarction. *Eur J Clin Invest*. 2005;35:677-686.
23. Hirata Y, Sata M, Motomura N, et al. Human umbilical cord blood cells improve cardiac function after myocardial infarction. *Biochem Biophys Res Commun*. 2005;327:609-614.
24. Kim BO, Tian H, Prasongsukarn K, et al. Cell transplantation improves ventricular function after a myocardial infarction: a preclinical study of human unrestricted somatic stem cells in a porcine model. *Circulation*. 2006;112 (9 Suppl): 196-204.
25. Leor J, Guetta E, Feinberg MS, et al. Human umbilical cord blood-derived CD133+ cells enhance function and repair of the infracted myocardium. *Stem Cells*. 2006;24(3):772-780.
26. Ma N, Stamm C, Kaminski A, et al. Human cord blood cells induce angiogenesis following myocardial infarction in NOD/scid mice. *Cardiovasc Res*. 2005;66:45-54.
27. Amado LC, Saliaris AP, Schuleri KH, et al. Cardiac repair with intramyocardial injection of mesenchymal stem cells after myocardial infarction. *Proc Natl Acad Sci USA*. 2005;102:11474-11479.
28. Bonnano G, Mariotti A, Procoli A, et al. Human cord blood CD133+ cells imunoselected by a clinical-grade apparatus differentiate in vitro into endothelial- and cardiomyocyte-like cells. *Transfusion*. 2007;47:280-289.
29. Schmidt D, Breymann Y, Weber A, et al. Umbilical cord blood derived endothelial progenitor cells for tissue engineering of vascular grafts. *Soc Thorac Surg*. 2004;78:2094-2098.
30. Murga M, Yao L, Tosato G. Derivation of endothelial cells from CD34⁻ umbilical cord blood. *Stem Cells*. 2004;22: 385-395.
31. Hoerstrup SP, Kadner A, Breymann CI, et al. Living, autologous pulmonary artery conduits tissue engineered from human umbilical cord cells. *Ann Thorac Surg*. 2002;74:46-52.
32. Schmidt D, Mol A, Neuenschwander S, et al. Living patches engineered from human umbilical cord derived fibroblasts and endothelial progenitor cells. *Eur J Cardiothorac Surg*. 2005;27:795-800.
33. Murohara T, Ikeda H, Duan A, et al. Transplanted cord blood-derived endothelial precursor cells augment postnatal neovascularization. *J Clin Invest*. 2000;105:1527-1536.
34. Goldberg JL, Laughlin MJ. UC blood hematopoietic stem cells and therapeutic angiogenesis. *Cytotherapy*. 2007;9(1):4-13.
35. Nieda M, Nicol A, Denning-Kendall P, et al. Endothelial cell precursors are normal components of human umbilical cord blood. *Br J Hematol*. 1997;98:775-777.
36. Murohara T. Therapeutic vasculogenesis using human cord blood-derived endothelial progenitors. *Trends Cardiovasc Med*. 2001;11:303-307.
37. Ikeda Y, Fukada N, Wada M, et al. Development of angiogenic cell and gene therapy by transplantation of umbilical cord blood with vascular endothelial growth factor gene. *Hypertens Res*. 2004;27(2):119-128.
38. Cho S-W, Gwak S-J, Kang S-W, et al. Enhancement of angiogenic efficacy of human cord blood cell transplantation. *Tissue Eng*. 2006;12(6):1651-1661.
39. Finney MR, Greco NJ, Haynesworth SE, et al. Direct comparison of umbilical cord blood versus bone marrow-derived endothelial precursor cells in mediating neovascularization in response to vascular ischemia. *Biol Blood Marrow Transplant*. 2006;12:585-593.
40. Pesce M, Orlandi A, Iachinioto MG, et al. Myoendothelial differentiation of human umbilical cord blood-derived stem cells in ischemic limb tissue. *Circ Res*. 2003;93:51-62.
41. Voltarelli JC, Couri CEB, Stracieri ABP, et al. Autologous nonmyeloablative hematopoietic stem cell transplantation in newly diagnosed type 1 diabetes mellitus. *JAMA*. 2007; 297(14):1568-1576.
42. US National Institutes of Health. Umbilical cord blood infusion to treat type 1 diabetes. Available at: http://www.clinicaltrials.gov/ct/show/NCT00305344?order=1. Accessed September 20, 2006.
43. Haller MJ, Viener HL, Wasserfall C, Brusko T, Atkinson MA, Schatz DA. Autologous umbilical cord blood infusion for type 1 diabetes. *Exp Hematol*. 2008;36(6):710-715.
44. Ende N, Chen R, Reddi AS. Effect of human umbilical cord blood cells on glycemia and insulinitis in type 1 diabetic mice. *Biochem Biophys Res Commun*. 2004;325:665-669.
45. Ende N, Chen R, Mack R. NOD/LtJ type I diabetes in mice and the effect of stem cells (Berashis) derived from human umbilical cord blood. *J Med*. 2002;33:181-187.
46. Harris DT, M Badowski and SM Harman. Treatment of type I diabetes in the NOD mouse with syngeneic cord blood stem cells. Submitted, Open Stem Cell J. 2009;1:62-68, doi: 10.2174/1876893800901010062
47. Sun B, Roh K-H, Lee S-R, Lee Y-S, Kang K-S. Induction of human umbilical cord blood-derived stem cells with embryonic stem cell phenotypes into insulin producing islet-like structures. *Biochem Biophys Res Commun* 2007: doi: 10.1016/j.bbrc, 2007.01.069.
48. Denner L, Bodenburg Y, Zhao JG, et al. Directed engineering of umbilical cord blood stem cells to produce C-peptide and insulin. *Cell Prolif*. 2007;40(3):367-380.

49. Jang YK, Park JJ, Lee MC, et al. Retinoic acid-mediated induction of neurons and glial cells from human umbilical cord-derived hematopoietic stem cells. *J Neurosci Res*. 2004;75:573-584.
50. Buzanska L, Jurga M, Stachowiak EK, Stachowiak MK, Domanska-Janik K. Neural stem-like cell line derived from a nonhematopoietic population of human umbilical cord blood. *Stem Cells Develop*. 2006;15:391-406.
51. Chen N, Hudson JE, Walczak P, et al. Human umbilical cord blood progenitors: the potential of these hematopoietic cells to become neural. *Stem Cells*. 2005;23:1560-1570.
52. Harris DT, Ahmad N, Saxena SK et al. The potential of cord blood stem cells for use in tissue engineering. Abstract, International TESi meeting, Oct 2005 Shanghai, China.
53. Chen J, Sanberg PR, Li Y, et al. Intravenous administration of human umbilical cord blood reduces behavioral deficits after stroke in rats. *Stroke*. 2001;32:2682-2688.
54. Willing AE, Lixian J, Milliken M, et al. Intravenous versus intrastriatal cord blood administration in a rodent model of stroke. *J Neurosci Res*. 2003;73(3):296-307.
55. Borlongan CV, Hadman M, Sanberg CD, Sanberg PR. Central nervous system entry of peripherally injected umbilical cord blood cells is not required for neuroprotection in stroke. *Stroke*. 2004;35:2385-2389.
56. Newman MB, Willing AE, Manressa JJ, Sanberg CD, Sanberg PR. Cytokines produced by cultured human umbilical cord blood (HUCB) cells: implications for brain repair. *Exp Neurol*. 2006;199(1):201-208.
57. Vendrame M, Cassady J, Newcomb J, et al. Infusion of human umbilical cord blood cells in a rat model of stroke dose-dependently rescues behavioral deficits and reduces infarct volume. *Stroke*. 2004;35:2390-2395.
58. Xiao J, Nan Z, Motooka Y, Low WC. Transplantation of a novel cell line population of umbilical cord blood stem cells ameliorates neurological deficits associated with ischemic brain injury. *Stem Cells Dev*. 2005;14:722-733.
59. Newcomb JD, Ajrno CT, Sanberg CD, et al. Timing of cord blood treatment after experimental stroke determines therapeutic efficacy. *Cell Transplant*. 2006;15:213-223.
60. Nan Z, Grande A, Sanberg CD, Sanberg PR, Low WC. Infusion of human umbilical cord blood ameliorates neurologic deficits in rats with hemorrhagic brain injury. *Ann NY Acad Sci*. 2005;1049(1):84-96.
61. Bliss T, Guzman R, Daadi M, Steinberg GK. Cell transplantation therapy for stroke. *Stroke*. 2007;38:817-826.
62. Saporta S, Kim JJ, Willing AE, et al. Human umbilical cord blood stem cells infusion in spinal cord injury: engraftment and beneficial influence on behavior. *J Hematother Stem Cell Res*. 2003;12:271-278.
63. Kuh SU, Cho YE, Yoon DH, et al. Functional recovery after human umbilical cord blood cells transplantation with brain derived-neurotropic factor into the spinal cord injured rats. *Acta Neurochir (Wein)*. 2005;14:985-992.
64. Kang KS, Kim SW, Oh YH, et al. Thirty-seven-year old spinal cord-injured female patient, transplanted of multipotent stem cells from human UC blood with improved sensory perception and mobility, both functionally and morphologically: a case study. *Cytotherapy*. 2005;7:368-373.
65. Lu D, Sanberg PR, Mahmood A, et al. Intravenous administration of human umbilical cord blood reduces neurological deficit in the rat after traumatic brain injury. *Cell Transplant*. 2002;11:275-281.
66. Meier C, Middleanis J, Wasielewski B, et al. Spastic paresis after perinatal brain damage in rats is reduced by human cord blood mononuclear cells. *Ped Res*. 2006;59:244-249.
67. Ende N, Chen R. Parkinson's disease mice and human umbilical cord blood. *J Med*. 2002;33:173-180.
68. Gaebuzova-Davis S, Willing AE, Zigova T. Intravenous administration of human umbilical cord blood cells in a mouse model of amyotrophic lateral sclerosis: distribution, migration, and differentiation. *J Hematother Stem Cell Res*. 2003;12:255-270.
69. Germain L, Auger FA, Grandbois E, et al. Reconstructed human cornea produced in vitro by tissue engineering. *Pathobiology*. 1999;67:140-147.
70. Germain L, Carrier P, Auger FA, Salesse C, Guerin SL. Can we produce a human corneal equivalent by tissue engineering? *Prog Retin Eye Res*. 2000;19(5):497-527.
71. Harris DT, He X, Camacho D, Gonzalez V, Nichols JC. The potential of cord blood stem cells for use in tissue engineering of the eye, Stem Cells & Regenerative Medicine, Jan 23–25, 2006, San Francisco, Abstract
72. Nichols JC, He X, Harris DT. Differentiation of cord blood stem cells into corneal epithelium. *Invest Ophthalmol Vis Sci*. 2005;46:E-Abstract–4772.
73. Badiavas EV, Abedi M, Butmarc J, Falanga V, Quesenberry P. Participation of bone marrow derived cells in cutaneous wound healing. *J Cell Physiol*. 2003;196:245-250.
74. Valbonesi M, Giannini G, Migliori F, Dalla Costa R, Dejana AM. Cord blood (CB) stem cells for wound repair. Preliminary report of 2 cases. *Transfus Apher Sci*. 2004;30(2):153-156.
75. Harting MT, Baumgartner JE, Worth LL, et al. Cell therapies for traumatic brain injury. *Neurosurg Focus*. 2008;24(3–4):E18.
76. Revoltella RP, Papini S, Rosellini A, et al. Cochlear repair by transplantation of human cord blood CD133+ cells to nod-scid mice made deaf with kanamycin and noise. *Cell Transplant*. 2008;17:665-678.

Part XI

Cord Blood Collection Variability and Banking

Comparisons Between Related and Unrelated Cord Blood Collection and/or Banking for Transplantation or Research: The UK NHS Blood and Transplant Experience

33

Suzanne M. Watt, Katherine Coldwell, and Jon Smythe

33.1 Introduction

The first umbilical cord blood (UCB) transplant was carried out approximately 20 years ago for a patient with Fanconi's anemia.[1] Since then, its importance in clinical hematopoietic stem cell (HSC) transplantation, when sourced from both unrelated and related donors, has increased.[2-11] UCB transplants have the advantage of less graft-versus-host (GvH) disease than similarly matched non-T-cell-depleted bone marrow (BM) transplants, although time for engraftment for single UCB unit transplants may be delayed.[2-11] Factors associated with improved UCB engraftment include lower recipient age and weight, closer HLA matching, and cell dose.[12] Low UCB cell doses have often restricted their use for transplantation to children or small adults, where recommended doses range from, or exceed, 2.5 to 5×10^7 total nucleated cells (TNC) per kg.[12] Newer advances have increased cell dosages for specific disease indications. These have included the use of double UCB transplants in adults or for larger recipients, the use of combinations of UCB with selected BM, or mobilized peripheral blood $CD34^+$ cells in an haplo-identical setting, non-myeloablative transplants using UCB in older patients, and genetic testing at the time of in vitro fertilization in the case of designated or directed sibling cord blood donations.[2, 3, 7-17] Indeed, of 8,191 UCB units released for transplant from the NETCORD inventory by June 2008, approximately half the recipients were adults.[18] The NHSBT experience with unrelated UCB donations and transplants is described below.

In the UK, approximately 25% of patients requiring an allogeneic stem cell transplant have an HLA-matched sibling and for the majority of the remainder, if white Caucasian, an appropriate unrelated BM or peripheral blood stem cell (PBSC) harvest or a stored unrelated UCB donation will probably be found.[2, 11] For some children, especially those from ethnic minority groups and those who suffer from inherited disorders such as hemoglobinopathies or immune deficiency, a directed collection of UCB following the birth of a sibling may be the only opportunity for an HLA-matched transplant until the sibling is able to donate BM.[7, 11, 18-21] In addition, such directed or designated UCB (DCB) donations from matched siblings have generally been shown to give better overall results than matched unrelated BM. Pre-implantation genetic testing in the UK to ensure that a pregnancy will result in a child free from a serious inherited disorder known to exist in the family is now accepted practice regulated through the Human Fertilisation and Embryology Authority (HFEA).[22, 23] Combined with pre-implantation HLA typing this practice will increase the chances of finding a matched disease-free DCB collection and so the number of DCB collections is likely to rise.[7, 11, 24, 25] Our current NHSBT DCB experience is described below.

The use of UCB for tissue repair and further studies on UCB leading to more successful hematopoietic stem cell transplantation are dependent on the outcome of

S.M. Watt (✉)
Stem Cell Laboratory, NHS Blood and Transplant, John Radcliffe Hospital, Headington, Oxford, OX3 9BQ, UK and
Nuffield Department of Clinical Laboratory Sciences, University of Oxford, Oxford, UK, OX3 9DU
e-mail: suzanne.watt@nbs.nhs.uk

N. Bhattacharya and P. Stubblefield (eds.), *Regenerative Medicine Using Pregnancy-Specific Biological Substances*,
DOI: 10.1007/978-1-84882-718-9_33,

current research.[2] NHSBT also collects cord blood units for research purposes and a summary of collections and the research being conducted are also outlined below.

33.2 Cord Blood Collection and Banking Practices in NHSBT and Licensing Requirements

As indicated above, NHSBT currently takes responsibility for three integrated types of cord blood donations: (1) altruistic unrelated cord blood collections and banking, (2) designated related cord blood collections and banking, and (3) cord blood donations for research, all on a not-for-profit basis. Progress with each of these activities is described below.

All NHSBT laboratories involved in these activities are licensed with the Human Tissue Authority (HTA) and thus meet the requirements of the UK Human Tissue Act (2004)[26] and the EU Directive for Tissues and Cells (2004/23/EC).[27] The European Union Tissue and Cells Directives set out to establish a harmonized approach to the regulation of tissues and cells across Europe. The Directives set a benchmark for the standards that must be met when carrying out any activity involving tissues and cells for human application (patient treatment). The Directives also require that systems are put in place to ensure that all tissues and cells used in human application are traceable from donor to recipient. The Directives were fully implemented into UK law on the 5th July 2007 by the Human Tissue Authority (HTA) via the Human Tissue (Quality and Safety for Human Application) Regulations 2007 (Q&S Regulations).[28] The HTA's remit was extended, and from July 2008 onward includes the licensing of organizations that procure as well as test, process, distribute, or import/export tissues and cells for human application. Procurement is defined as the processes by which tissues and cells are made available, including the physical act of removing tissue and the donor selection and evaluation.

33.2.1 The NHS Cord Blood Bank in London for Unrelated Donations

The NHS Cord Blood Bank collects altruistic unrelated cord blood units for clinical transplantation and research from maternity units in five NHS Trust hospitals in or close to London, namely, Barnet General, Northwick Park, Luton and Dunstable, St. George's and Watford General. The Luton and Dunstable collections were commenced during 2003 and the Watford General collections during 2007. The fifth collection site at St. George's Hospital in Tooting in 2009. The aim is to bank 20,000 UCB units by 2013, with approximately 40% from ethnic minority groups in England. Further details on banking and testing protocols have been described.[2,, 29-31]

33.2.1.1 Banking and Transplantation of Unrelated Units

Collections ex utero are carried out by trained cord blood bank staff 24 h/day, 5 days/week after taking permission from consenting mothers, and without influencing or interfering with normal routine deliveries and using procedures which minimize medical attention being drawn away from the mother and baby during the collection process. The collections take place in a dedicated area outside the delivery suite so that intrusion on the well-being of the mother and child are minimized. The cord blood collection procedure relies on needle cannulation of the umbilical vein and then gravity feed into a sterile collection bag containing anticoagulant essentially as has been described previously.[2, 29-31] An ISBT-128 bar coding system ensures that the UCB units are traceable from the donation to the therapeutic use of the units.[32, 33] Only those units containing at least 40 mL of blood are currently processed for banking for transplantation if they have appropriate written informed consent, while those not meeting these criteria, and where consent and ethical permission for research have been given, are made available for research. Processing is carried out within 24 h of collection. Further exclusions from banking at present include low total nucleated cell counts medical exclusions (e.g., abnormal blood films, infectious or other disease indicators), processing anomalies (e.g., nucleated cell viability <80% and recoveries after processing of <70%), and language exclusions. Currently, nucleated cell counts $\geq 9\times10^8$/unit for black and ethnic minority donors and $\geq 12\times10^8$/unit for caucasoid donors are banked. Previously, the cut-off for all donations was $\geq 4\times10^8$/unit. Prior to processing and banking, a blood film is made and 3.5 mL of the UCB unit is taken for HLA tissue typing and 0.5 mL for a full blood count.[30, 31] UCB units are volume reduced to 21 mL and depleted of plasma and erythrocytes, previously using the Optipress system[30] and currently using the Biosafe Sepax system.[2, 11, 29, 34] After processing, the erythrocyte fraction is used for blood grouping, plasma is used to screen for mandatory virology (namely, HBsAg, anti-HCV, HIV 1 and 2, syphilis, and HTLV I and II, HBV, HCV, and HIV by polymerase chain reaction (PCR) by the NHSBT at the time of UCB unit reservation), and an aliquot of the buffy coat used for full blood and CD34 counts and viability testing. The UCB buffy coat is then cryopreserved using dimethyl sulfoxide (DMSO) and stored in a BioArchive with the capacity to hold 3,626 UCB volume reduced units.[29]

The NHS Cord Blood Bank was listed as the fourth largest of 31 banks worldwide on the NETCORD inventory (if the banks in Sydney and Melbourne are not combined) as of June 2008 (Table 33.1). It was the second after the New York Blood Centre to obtain FACT–NETCORD accreditation. The percentage of UCB units released for transplantation from these 31 banks was on average 4.35%, almost equally distributed between children and adults if considered overall.[18] There was of course variability depending on the site of the Cord Blood Bank (Table 33.1). Assessment of the inventories of FACT–NETCORD accredited Cord Blood Banks listed on NETCORD revealed 128,933 cord blood units were available for transplantation (June 2008) and usage for transplantation varied from 0.6% to 13.4% per bank, the highest being in France.[18] Although 11,659 unrelated cord blood units were listed on the June 2008 inventory, by August 31st 2008 there were 11,969 unrelated cord blood units banked for transplantation in the NHS Cord Blood Bank (Fig. 33.1). Approximately 40% of banked unrelated UCB units have been sourced from ethnic minority donors. An analysis of the first 8,201 UCB units banked at the NHS Cord Blood Bank has revealed approximately 60% were from donors of northern European Caucasoid origin, 20% from non-European Caucasoid (Asian), 7% African/Afro-Caribbean, 1% oriental, and 10% of mixed ethnic origins, and the ethnicity of 2% was unknown. Of these, HLA alleles that were present at a higher frequency included HLA-A68, A33, A11, B52, B61, B37, Cw7, Cw4, Cw8, DR10, DR14, DR12, and DR8, while haplotypes present at a low frequency or not detected in the bone marrow donor registry included HLA-A33 B44 and HLA A30 B42 DR18.[35] In a further analysis of over 7,000 UCB units, some differences were noted in the collection volume, TNC count, and CD34 cell numbers in UCB units from different ethnic groups[35] and these are illustrated in Table 33.2.

Table 33.1 The June 2008 NETCORD Inventory showing ten cord blood banks with the highest numbers of recorded units and demonstrating the numbers of cord blood units registered and released for transplantation

Cord blood bank	Inventory	Units released for transplant	Ratio of children: adults transplanted
New York	36,638	2,659	1.74:1
Durham	17,378	836	ND
Sydney + Melbourne	15,680	438	1.06:1
Dusseldorf	13,796	508	1.10:1
NHS Cord Blood Bank, London	11,659	213	1.39:1
Malaga	10,921	88	0.57:1
Barcelona	8,487	387	0.83:1
Leuven	8,207	112	1.24:1
Milan	6,981	363	1.14:1
France	6,470	865	0.46:1

ND, no data

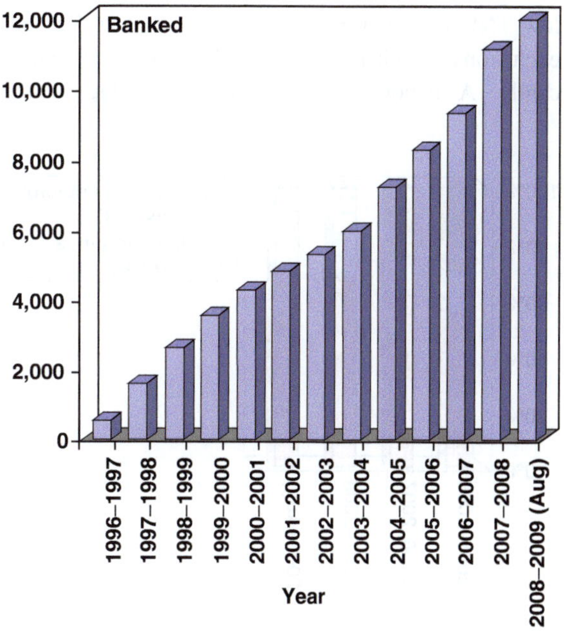

Fig. 33.1 Cumulative increases year on year in the numbers of unrelated cord blood units banked for transplantation until August 31, 2008

Table 33.2 A comparison of TNC, cell volumes, and CD34 numbers in UCB units from different ethnic groups

Ethnic origin	Number of UCB units assessed	Collection volume (mL)[a]	TNC × 10^{7a}	CD34+ cells post-processing (%)
Northern European Caucasoid	4,933	75 ± 25	120 ± 58	3.1 ± 58
Mixed race	781	73 ± 24	113 ± 55	2.8 ± 2.7
Asian	1,534	70 ± 21	104 ± 46	2.4 ± 2.1
Black	554	70 ± 22	97 ± 45	2.6 ± 3.0

33.2.1.2 Cord Blood Units Unsuitable for Banking

A proportion of cord blood units prove unsuitable for banking for transplantation as described above. We have seen this loss in units fall over the period of banking unrelated units, and this is illustrated from 2005–2006 to August 31, 2008 in Fig. 33.2. The number of banked units from those collected had, over this time, risen, from approximately 55% to a current figure of just over 70%. UCB units were unsuitable for transplantation for a variety of reasons, including low volumes, low total nucleated counts (TNC), and medical exclusions. Unrelated UCB units were not at this time banked if volumes were below 40 mL or UCB units contain <4×10^8 TNC. In 2005–2006, low volumes and low TNC accounted for 18.5% of units being unsuitable for transplantation. From April to August 2008, the percentage of UCB units excluded for low volumes and low TNC was approximately the same at 20.5%, but exclusions for other reasons had been reduced significantly. A major reason for this was due to the requirement by the HTA for pre-consent. Those units not suitable for banking, if consented for research, were made available for this purpose, and, if not, were discarded. More recently with the increase in TNC banked 40–50% of UCB units meet the banking criteria.

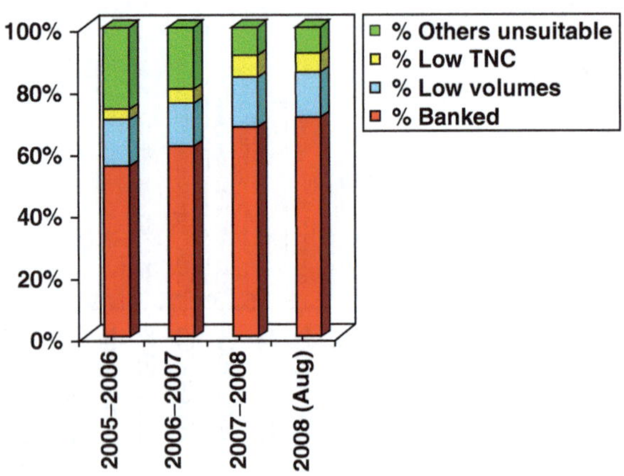

Fig. 33.2 Percentage of units unsuitable for banking compared to those that are banked

33.2.1.3 Transplanted Units, Geographical Distribution, and Transplant Outcomes

By September 30, 2008, the total number of units issued to 22 countries from the NHS Cord Blood Bank had reached 226. There has been a dramatic increase in cord blood units used for transplantation from the NHS Cord Blood Bank in recent years, with almost 50% of units transplanted having been issued over the past 2.5 years, at a time when the numbers of cord blood units banked had exceeded 8,000 (Fig. 33.3). As of August 31, 2010, 15,600 UCB units have been banked and 311 issued for

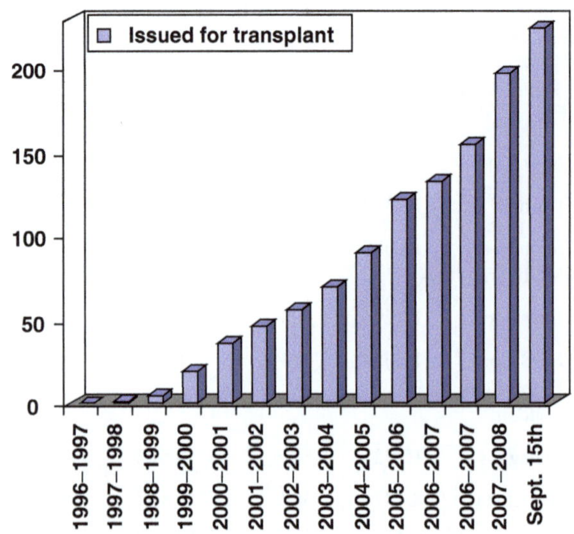

Fig. 33.3 Cumulative numbers of unrelated cord blood units transplanted from the NHS Cord Blood Bank

transplant. Almost 50% of transplanted units have been issued to adult recipients (Table 33.1). Of those issued by 2008, just over 6% were low volume and 55% of these low volume units were issued to non-European Caucasoid (Asian) recipients, where the high degree of HLA matching was the main reason for usage.[35]

A form requesting information on adverse reactions, primary engraftment data, and transplant outcome is issued with the UCB unit for transplant, and additional information is requested if appropriate by telephone. Severe adverse events are reported to the HTA. Approximately one third of the UCB units issued were from ethnic minority donors. Each transplant center determines engraftment using their usual protocols. Neutrophil engraftment is defined as the first of three consecutive days to reach 0.5×10^9 neutrophils per liter and platelet engraftment as the first of three consecutive days to reach 20×10^9 platelets per liter, with at least 7 days since the last platelet transfusion. Transplant outcomes for 144 unrelated UCB units from the NHS Cord Blood Bank had been analyzed up until June 2007. Of these, overall 82.6% engrafted, with the rate being 74% for the first 50 units and 87.2% for the next 94 units. Of 120 units assessed, the 2-year survival rate was 37.5%, while the 5-year survival of 53 UCB units assessed, was 34%.

33.2.2 The Collection and Use of Designated or Directed Cord Blood Units for Transplantation

NHSBT, through its network of licensed and accredited hematopoietic stem cell services laboratories, has taken a leading role in the provision of a national DCB service for high-risk families.[11] This service encompasses liaison with the obstetric delivery unit to arrange collection, processing, testing, storage, and issue for transplantation. This service was developed to operate in a GMP-compliant environment as prescribed by the UK Department of Health Code of Practice for Tissue Banks 2001. The DCB service was also developed to meet the evolving standards of the UK Blood Transfusion Services' "Red Book" quality standards,[36] FACT–NETCORD guidelines,[37] the requirements of the UK Human Tissue Act (2004),[26] and the EU Directive for Tissues and Cells (2004/23/EC).[27,28]

The different aspects of the DCB service are described below, and this includes the quality of the units, practices adopted, regulatory issues, and the outcome of the resulting transplants. We also recommend best practice to ensure a system for DCB collection and storage, which is ethical, robust, high quality, and clinically responsive to patient needs.

33.2.2.1 Eligibility for Designated Cord Blood Donation

NHSBT accepts a referral for a DCB collection if (a) an existing sibling can be treated by allogeneic transplantation, (b) an earlier birth from the same parents indicates an inherited disorder that can result in a future infant needing an allogeneic transplant, (c) the sibling's clinician and hospital are willing to support financially the banking, and (d) the referring clinician takes responsibility for medical review of the mother and potential recipient and obtains informed, written consent, ahead of the delivery. Referring clinicians are those responsible for treating the sibling's condition and therefore specialized in the relevant disorder.

33.2.2.2 Informed Written Consent

Informed written consent is obtained from the donor mother for screening of both the mother and cord blood for mandatory microbiological markers, for tissue typing, and for storage of samples and confidential information. Consent is taken with reference to the UK Department of Health's *Reference Guide to Consent for Examination or Treatment*. An information sheet is provided by NHSBT as an aid to those undertaking consent. Consent is also obtained by the NHSBT from the obstetrician for the delivery hospital to help with the collection.

33.2.2.3 The Human Tissue Authority's Requirements for Designated Cord Blood Donations

The responsibility under the HTA for procurement of directed cord blood donations through the NHSBT is the joint responsibility of both the transplant clinician and the NHSBT. The transplant clinician requires a procurement license for donor selection and counseling and the NHSBT also requires an HTA license for arranging the actual cord blood collection. A service level agreement (SLA) between the clinician and the NHSBT outlines the different responsibilities. If the NHSBT arranges for the physical collection to take place at other premises and/or to be carried out by non-NHSBT staff, then the NHSBT has a third party agreement (TPA) with the midwifery department. NHSBT is responsible for ensuring the midwifery premises and any non-NHSBT collection staff are suitable under the requirements of the HTA regulations. The transplant clinician and the NHSBT will also include arrangements for processing, storage, and distribution by the NHSBT in the SLA where the NHSBT is licensed for and carries out these activities on behalf of the clinician.

33.2.2.4 Mandatory Screening

Samples from mothers are tested in advance for HBsAg, anti-HCV, HIV 1 and 2, syphilis, and more recently HTLV I and II, HCV, HBV and HIV by PCR, by the NHSBT. These same tests are required for unrelated UCB units. Potential DCB collections are not necessarily excluded on the basis of positive maternal virology results, whereas unrelated UCB units that are positive are excluded.

33.2.2.5 Designated Cord Blood Collection and Transport to NHSBT

DCB collections are organized and processed by four NHSBT centers located across England to limit transit times. This means collections can be carried out in the hospital selected by the mother for birth. Initially NHSBT staff used to collect the DCB, but now, for logistical reasons, midwives collect the DCB using detailed instructions provided by NHSBT specialist staff. The instructions cover both collection in utero and ex utero with additional advice from the NHSBT provided by telephone when necessary. NHSBT also offers training for midwives at the sites collecting for the unrelated bank. NHSBT collection kits are sent to a named individual in the delivery hospital in an insulated box, validated for transport of the DCB at 2–8°C using cool-pack inserts. The kit includes a spare collection pack (Macopharma, Twickenham, England) and alcohol and iodine sterile swabs for cleaning the cord. A partially completed label is sent along with line clips, a sealer, and a labeled bottle for a segment of the umbilical cord as a backup source of DNA for HLA typing. Where the mother's history indicates the fetus/infant is at risk of an inherited disorder, the referring hospital liaises directly with the delivery hospital to request neonatal samples for genetic testing. The kit includes details of a courier able to deliver at short notice and shipping records are maintained. The NHSBT centers are staffed 24 h a day and collections are stored at 4°C ahead of cryopreservation within 24 h of collection.

33.2.2.6 HLA Typing, Including Pre-implantation Studies

A blood sample from the potential recipient (and family members if required) is requested for HLA typing. Alternatively, a copy of the HLA typing results is provided by the referring hospital. HLA typing by the NHSBT for HLA-A, B, Cw, DRB1, and DQB1 antigens is carried out by polymerase chain reaction with sequence-specific primers (PCR-SSP) as detailed previously.[38] Before 1997, some HLA typing was done by serology. DNA was stored for any confirmatory HLA typing required subsequently. Pre-natal HLA typing from chorionic villus sampling is as described.[39] Pre-implantation screening[40] was more recently introduced and is the responsibility of the referring clinician and follows UK guidelines issued by the HFEA.[41]

33.2.2.7 Designated Cord Blood Processing and Storage

If the DCB is badly clotted or of negligible volume, the laboratory head discusses with the referring clinician whether to proceed. The default position however is to proceed, with a final decision on storage being made when all results are available. A unique 12-digit barcode number is used for labeling, paperwork, and files in addition to other details. A bleed-line sample is removed for total nucleated cell (TNC) and $CD34^+$ counts, cell viability, and blood grouping. CD34 analysis is performed using a single platform lyse no-wash flow cytometric protocol[29,32,42] which incorporates calibrated beads to quantitate CD34 cell concentration directly and 7-actin-actinomycin (7AAD) to determine viability.

Processing is carried out according to EU GMP standards[43] under grade A air-quality conditions in a clean room. A sample is taken prior to processing for tissue typing and virology screening. Cryoprotectant solution is prepared from DMSO and 10% (w/v) Dextran-40/saline or 4.5% (w/v) Human Albumin Solution (HAS).[29] An equal volume of cooled 20% (v/v) DMSO solution is added to the cells in a controlled manner. To maximize the cell dose stored, volume reduction is not carried out.[11] A small volume is taken post-processing for bacteriological screening and to prepare frozen reference samples before the cells are frozen in cryocyte bags (Baxter Healthcare, Newbury, England). DCB collections are processed individually to avoid inadvertent mixing of bags or samples. A contiguous line sample is left attached to the frozen bags for future confirmatory HLA typing, although this practice has varied in the past. The collections are frozen double-wrapped using a controlled rate freezer and then transferred to liquid-nitrogen-based storage tanks, maintained at less than −150°C and continuously monitored.[11,29] DCB units from mothers known to be positive for microbiology markers or awaiting results are stored in quarantine tanks. Reference samples are cryopreserved with the DCB unit and stored in the same tank.[11]

33.2.2.8 Reporting to the Referring Clinicians

A preliminary report is sent indicating volume, cell counts, and any adverse events with a further report sent once all test results are available. Where HLA typing of the potential recipient and family members is done by the NHSBT, the report includes information on the degree of match. NHSBT histocompatibility staff are available to discuss the HLA typing with the referring hospital's histocompatibility laboratory or transplant medical consultant.

33.2.2.9 Storage Policy

The NHSBT advice on long-term storage of the DCB depends on HLA compatibility, the potential recipient weight, disease progression, and the likelihood and timing of using the DCB unit. Recently, the NHSBT policy has been to store units for patients with at least one haplotype match and cell counts greater than 0.3×10^9 TNC. A written statement from the referring clinician is requested to confirm whether or not long-term storage was required. NHSBT costs (on a not-for-profit basis) are recovered from the referring hospital.

33.2.2.10 Designated Cord Blood Issue for Transplantation and Transplant Outcomes

All final decisions on transplantation remain the responsibility of the transplant unit. All requests to issue the DCB are confirmed in writing. Frozen reference samples are thawed by the NHSBT for cell viability, CFU testing, confirmatory HLA typing, and short tandem repeat (STR) analysis[44] to confirm identity with the sample that has been HLA typed at the time of banking. Where the reference sample is not the unit bleed-line, the sample has been given the same unique barcode number as the unit to avoid incorrect identification. The NHSBT has shown that viability testing on reference samples from both the bleed-line and vials can be used to predict viability in the whole units. CFU assays have been carried out in recent times using a methylcellulose-based medium according to the manufacturer's instructions (Stem Cell Technologies, Vancouver, Canada).[45] DCB units are transported in temperature-monitored dry shipper containers at less than −150°C and were thawed by NHSBT or hospital staff, with instructions provided by the NHSBT. As for unrelated UCB transplants, requests are made for information on adverse reactions, primary engraftment data, and transplant outcome at the time of DCB unit issue and additional information requested by telephone.

33.2.2.11 Breakdown of Collections by Patient Disorder

Over a 13-year period, 412 collections were made, with 80% for an existing sibling and 20% collected where there was a family history of a serious inherited disease. Diagnoses included 179 hematological malignancies, 115 hereditary anemias and 118 immune deficiencies or metabolic disorders. A further 12 requests (3%) were received for collections that were not successful, usually due to a damaged cord/placenta at delivery. The requests for banking have changed over time with a gradual shift from hematological malignancies toward hereditary anemias and immunodeficiencies.[11] Where collections were made from infants who subsequently were found to have inherited the same genetic mutation as the potential recipient they were discarded or used for research.

33.2.2.12 Characteristics of Designated Cord Blood Collections

The median volume minus anticoagulant was 73 mL. The volume exceeded 40 ml for 88% of the collections. The median TNC count was 8.8×10^8 (mean $9.9 \pm 5.5 \times 10^8$; range 1.0 to 31.0) and exceeded 4.0×10^8 for 895 of the collections. Although mean in utero and ex utero collection volumes and nucleated cell counts differed (a median volume of 75 mL and count of 9.5×10^8 versus 67 mL and 7.8×10^8 respectively), this difference was not statistically significant. The median total CD34⁺ cell count per collection was 2.3×10^6 (mean $3.2 \pm 3.4 \times 10^6$; range 0.13–26.67). The median TNC viability prior to freezing for the 50% of collections where this information was collected was 99.0% (mean 98.2 ± 3.7%) with only four less than 90% (73%, 78%, 84%, and 84%).

33.2.2.13 Microbiology Results

Three collections tested positive for anti-HCV, but negative for HCV by PCR, with maternal samples testing positive for HCV. At the request of the referring clinician, these collections were not discarded, but stored in a quarantine vessel pending HLA typing results. Prior to 1999, 7% of collections showed evidence of bacterial contamination. After a change in practice where an iodine swab was used in addition to alcohol wipes, this decreased to 2.4%. The microbiology laboratory indicated that contamination probably took place at the time of collection and provided antibiotic sensitivities. Where the contamination did not preclude use for transplantation, with provision of suitable antibiotics, units were retained in storage.

33.2.2.14 Transplant Characteristics

Of the units collected for existing siblings, 28% were a full 10:10 antigen match at HLA A, B, Cw, DR, and DQ loci. Of the matched units, 20 were issued for transplantation with all but 3 being used for non-malignant disorders. Cord blood was stored on average for 17 months before transplant. None of the units collected from mothers with a family history of inherited disease were transplanted; however the mothers may yet give birth to affected children who are HLA matches. The median time to neutrophil engraftment, where records were available was 18 days (mean 20 ± 8; range 14–28) and for platelets 30 days (mean 30 ± 12; range 11–59). Engraftment rates did not correlate with TNC or $CD34^+$ cell dose, although numbers are small.

33.2.3 *General Observations on the DCB Program and Comparisons with the Unrelated UCB Program*

Since the amalgamation of regional services into the NHSBT in England, the DCB and unrelated UCB procedures have followed a standard practice. The DCB collections are now carried out by midwifery staff in the hospital selected for delivery by the mother. The midwifery departments were willing to help because the collections were for high-risk families rather than private companies, a distinction made by the UK Royal College of Midwives, the Royal College of Obstetricians and Gynaecologists, the American Academy of Pediatrics, and the World Marrow Donor Association (WMDA).[46-49] In contrast, unrelated UCB units are collected by dedicated scientific staff.

We believe our NHSBT infrastructure, with its uniform standards and policies, is an important factor in the success of both unrelated UCB and the DCB programs. It provides the expertise, coordination, and capacity to respond to clinical needs throughout the country and is supported by the extensive NHSBT quality systems. Over a 13-year period, NHSBT has seen an increase in the number of DCB referrals to over 60 per year as awareness amongst clinicians and mothers increases. Despite collections from over 60 hospitals, the vast majority of DCB were suitable for transplantation based on TNC counts. The proportion of unsuccessful DCB collection attempts (3%) is in line with figures for allogeneic UCB banks.[19,20] The median TNC and $CD34^+$ cell counts are comparable to those reported in a multi-center DCB program in the USA.[19,20] The CD34 counts compare very well with the UCB banks that also use this as a measure of suitability.[12,50] A proportion of unrelated UCB banks[2,14,15] adopt a minimum volume of 40 mL and cell count of

4×10^8 TNC, and these criteria were met for 88% of the our DCB collections. It is worth noting that six DCB collections of less than 40 mL had TNC counts over 4.0×10^8 cells. Of further interest, three of the units transplanted had volumes of 14, 18, and 25 mL, yielding TNC counts of 3.5, 2.1, and 7.5×10^8 with doses of 1.6, 0.84, and 4.2×10^7 TNC/kg respectively and all three engrafted within 30 days. These results confirm our policy not to automatically exclude units with less than 40 mL and we agree with Walters et al.[20] that this cutoff is not appropriate for DCB banking because the collection has unique potential. For Walters et al.,[20] the choice not to process units less than 20 mL represented a 4.4% loss of their DCB collections, whereas we collected only 6 (2.2%) units of less than 20 mL. We conclude that collection of a restricted numbers of units by a range of midwifery departments is a viable option for DCB banking.

We recommend that the decision to retain the DCB is best made once the cell counts and HLA typing results have been obtained. Other studies using unrelated UCB units indicate that a single HLA antigen mismatch can be compensated for by higher cell doses and recommend that only HLA-A, -B, -DR-matched units be used when cell doses $<2.5 \times 10^7$ TNC/kg are available.[12] However, even for low-volume DCB collections, there is the possibility of using the DCB unit in conjunction with small volumes of BM from the same donor once the child has reached sufficient age for this to be an option.[17,19,21] There is also the possibility of using donor lymphocytes to counter relapse or infection post-transplantation.[12-14,51,52] Donor lymphocytes were used successfully in this study to combat a rejection episode following one transplant for thalassemia.

Those DCB collections that were less than 40 mL were not associated with one hospital or NHSBT center, but were slightly more likely to have been collected ex utero (57%). We and Wall et al.[53] reported little difference in the volume of UCB units collected using either in utero or ex vivo techniques, while others[54] have observed higher CD34+ cell counts and lower rates of contamination with in utero collections. However the safety of mother and child is paramount and the UK Royal College of Obstetricians and Gynaecologists[47] has suggested all collections should be made ex utero and we know comply with this recommendation.

The median TNC viability prior to cryopreservation of DCB units was 99%, with only four lower than 90%. This suggests that a validated means of transport with a reliable courier is an effective way to avoid loss of viability in transit. Wada et al.[55] have noted lower viabilities of UCB units with increased transit times and lower volumes. However, there was no clear reason (as noted by others[56,57]) to explain the few low viabilities in our studies. We recommend using a single platform CD34 assay that includes viability assessment on fresh samples and combining this with the CFU assay on frozen samples to measure cell numbers accurately and determine viability of the CD34+ cell population.

UCB banks typically volume-reduce to 21 mL,[29-32] whereas we have not reduced the volume of DCB units. The numbers of DCB units are relatively low and so the benefit from not losing as many as 20% of cells through volume reduction[54] or even total loss during processing becomes more important than the larger storage space required. Improving umbilical cord cleaning by using alcohol wipes followed by iodine swabs dropped bacterial contamination of DCB units to 2.4%, which compares very well with other DCB programs[19,20] (3.3%) and is probably at a minimum.[58,59] Bacterial contamination can be addressed at the time of transplant with antibiotics, but units contaminated with dangerous organisms are discarded. Units from an HCV-positive mother could be considered where follow-up testing of the donor child indicates there was no vertical HCV transmission.

All units are now typed at the time of collection using PCR-SSP for HLA-A, -B, -Cw, -DRB1, and -DQB1 to ensure the degree of match can be determined fully and so best inform the decision to store the cord blood. NHSBT histocompatibility and immunogenetics laboratories are all accredited by the European Federation for Immunogenetics. The NHSBT recommends high-resolution (allele) typing for all potential transplant donors and recipients where any uncertainty regarding HLA compatibility is identified. We found that 28% of DCB collections for a sibling were an HLA 10/10 match, in line with Mendelian segregation. An identical matched sibling donor is preferred as the standard of care to an unrelated donor as a clinical option in many transplant settings, as indicated recently by the fourth EBMT report on current practice for transplantation in Europe.[60] Where stem cells from related and unrelated donors are both listed as standard of care options, an unrelated UCB unit may be preferred to a related UCB unit in order to provide a sufficient cell dose. As the practice of pre-implantation

HLA testing becomes more widely accepted, it is likely that the proportion of related matched units and transplants will increase.[61-63]

UCB collections that are not a full match may be used successfully as shown for unrelated UCB transplants where 80% of transplants have one or two mismatches. Indeed, a limited mismatch may be preferable for some malignancies bearing in mind that cell doses of $3–5 \times 10^7$ TNC/kg have been recommended for transplants with a one or two HLA disparity.[12] Where there is a family history of an inherited disorder, the DCB collections are almost always stored for potential future use. Most of the collections for non-malignant disorders were for SCID and matched UCB units have been successfully used for transplant in immunodeficiency from both related and unrelated donors.[12,64] UCB units have been transplanted after many years,[11,12,65] so long-term storage may be appropriate. Continued DCB storage should however be reviewed at least annually to ensure units are not stored unnecessarily. A DCB unit with one HLA haplotype match or less is unlikely to be used and so storage of such a unit over the longer term is most likely not required.

The final decision to store DCB units or otherwise is the responsibility of the referring clinician, but where this ignores current acceptable practice for transplantation[60] the NHSBT may request that the unit is transferred to a private bank. For non-identical DCB units, the decision should be made with regard to the availability of better matched unrelated UCB units of sufficient dose or BM/PBSC being available from UCB banks and BM registries. The initial search against HLA -A, -B, and -DRB1 does not attract a charge but a number of factors should be taken into account when considering the potential availability of unrelated UCB or BM.[2,11,12,14,30,66] First, the majority of potential unrelated donors will be BM or PBSC donors and so the advantages of UCB over these two will not be provided. Second, at least 30% of identified registry donors will not be available.[67] Third, the match will only be for HLA-A, -B, and -DRB1 and the DRB1 typing may have only been of low resolution leading to a reduction in the number and degree of matches once full high-resolution allele-level typing has been performed. Fourth, the availability of an unrelated unit or donor cannot be guaranteed if the transplant is likely to take place sometime later. Fifth, an UCB unit may not contain sufficient cells for the heavier prospective patient but there is no opportunity to "top-up" the cell dose with a BM collection from the same donor.

In our analysis, 20 DCB collections had been transplanted and all were matched for HLA-A, -B, -Cw, -DRB1, and -DQB1 loci. This represents a 5% take-up rate from all collections, a 5.5% transplant rate where a sibling is alive and a 20% transplant rate for matched units. The figures compare very well with DCB programs in several countries.[19,20] Although from low numbers, the figures for transplant engraftment compare very well with those published for unrelated and related cord blood and BM transplants for malignant and non-malignant disorders.[12,13,15,68-71] Our results demonstrate that DCB transplantation is as effective, as unrelated UCB transplantation, albeit in the limited number of cases where a DCB collection is possible.

We are able to offer a cost-effective national service for DCB banking by using the NHSBT's network of stem cell and immunotherapy laboratories that already have the appropriate facilities, staff, and equipment in place for handling PBSC and BM. The costs for DCB cryopreservation are therefore similar to those for PBSC. The laboratories were first accredited by the UK Medicines and Healthcare Products Regulatory Authority (MHRA) and more recently by the Joint Accreditation Committee for ISCT-Europe and EBMT (JACIE) or FACT–NETCORD. Our laboratories have recently also been given licenses by the UK Human Tissue Authority to operate under the EU Directive for Tissues and Cells, which became mandatory in April 2006.

Over recent years, large investments in cord blood banking were announced by US Congress and recommendations were published by the Institute of Medicine on establishing a national cord blood bank program in the USA.[12] The approach that we have taken within the NHSBT encompasses many of these recommendations and demonstrates that a national high-quality program can be maintained despite the logistical difficulties of organizing collections from many hospitals and issuing units for transplantation to diversely situated transplant units.

33.2.4 Cord Blood Units for Research

UCB units collected from related or unrelated sources described above but that are unsuitable for transplant and which have consent and ethical permission for research and development are used for this purpose. We also hold a collection of approximately 300 frozen unrelated UCB donations for research purposes resulting from a Leukemia Research Fund (LRF)–NBS collaboration. These units have been tissue typed and are available for use for defined projects through the LRF cell bank website.[72] Additionally and as private collections for one's own use are not currently recommended,[9,73] we have been collecting UCB units specifically for research and development (CBRD) from the John Radcliffe Hospital in Oxford, where the proportion of Caucasian donors is high, and ethnic minority groups are low, and where we have informed written pre-consent and ethical approval. The NHSBT sites for UCB collections and storage for research are HTA licensed. In order to use UCB for research, units are anonymized and remain traceable via an ISBT-128 barcode. Potential donors are informed of the non-commercial nature of the research being carried out and provided with adequate information to understand the nature and purpose of the research prior to consenting.

Since its inception in March 2004 and over 4 years to 2008, the CBRD team collected over 1,600 UCB donations from mothers giving birth for this purpose. Collections were made ex utero during normal working hours only by our own trained staff. These UCB units for research were predominantly collected from elective cesarean sections (72.8%) supplemented by inductions (0.8%) and an increasing percentage of normal births (initially 5.8% in 2004 rising to an overall average of 26.4%; Fig. 33.4). This increasing rate of UCB donations from normal births corresponds to publicity for the UCB for the research scheme aimed at

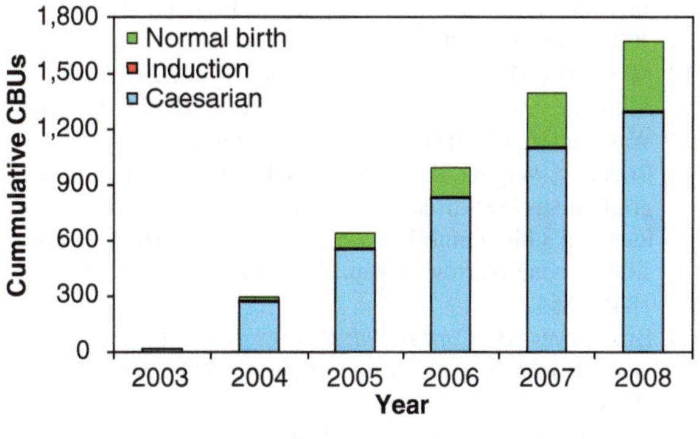

Fig. 33.4 Cumulative numbers of UCB units collected in Oxford for research. The figures show the total numbers and types of delivery. Years refer to total UCB collections since March 2004 based on British financial year (1 April–31 March)

expectant mothers. Originally, cesarean sections presented the best prospect for collecting UCB as the option to donate could be offered to cesarean patients 1 week prior to surgery and written informed consent obtained prior to the section. Unlike normal births and inductions, cesarean sections present a relatively predictable timeframe for delivery and availability of the afterbirth. Cesarean sections are also advantageous due to the lower frequency of damage to afterbirth tissues, which must remain in good condition for collection of UCB (described further below). The mean volume of UCB collected minus anticoagulant was 57 ± 1 mL with a mean total nucleated cell count per UCB unit of $6.0 \pm 0.1 \times 10^8$ TNC/unit (Mean ± SEM, $n = 1{,}594$).

Although over 1,600 UCB collections were made at Oxford from 2004 to 2008, more expectant mothers in Oxford consented to UCB donation for research than the number of UCB units eventually collected. This is in part due to the fact that collection staff were not present 24/7. The following requirements must be satisfied in order to make an UCB collection for research purposes. The mother (aged ≥ 18 years) must have at least 24 h to consider the option to donate for research prior to birth and then give written informed consent. This is usually obtained before the collection takes place, otherwise if this is not possible verbal informed consent is obtained and documented prior to the collection and written consent obtained after the delivery and cord blood collection but before the cord blood leaves the unit and is used for any research. The mother and child must not be affected by any excluding medical conditions, e.g. premature birth, twins, abnormalities of the placenta/cord requiring further investigation, infection with MRSA, HIV, Hepatitis B, Hepatitis C, or other infection. The cord/placenta must be available to the CBRD team quickly after delivery (within about 20 min of birth) and during the working hours of the collection team. Many inductions initiated during the day result in delivery of the baby in the early hours of the morning, hence rarely fall in the collection hours of our CBRD team. By far the most common reason (58% of failed collections) for a failed collection in our experience was poor condition of placental tissue and/or umbilical cord following the delivery. If the umbilical cord sustains damage or is torn (with <10 cm of undamaged cord remaining), it is too short to safely insert a needle and massage the blood from the placenta. If placentas/cords do not reach the collection team reasonably quickly following the delivery, the blood will clot and become unusable or may leak from the placental tissue and be unusable. In order to collect good quality UCB, the placenta must be intact with little or no damage.

The UCB unit undergoes post-collection virology screening by NHSBT to protect the research staff unless the mother does not consent to the testing. UCB units are anonymized before distribution to researchers (identified only by barcode, volume of blood, and TNC/mL) but can be tracked by trained staff not undertaking the research in case of positive virology results or accidental exposure of research staff through needle stick incidents. Since its inception, the CBRD team has operated with two full-time (WTE) collection staff with one 0.1 WTE supervisor. The CBRD team inform and obtain consent from potential donors, as well as collect the UCB units. To maximize the efficiency of CBRD staff, various working time arrangements have been considered or trialed and a cost-benefit analysis was made based on birth rates at the John Radcliffe Hospital and funding available. Given the costs of staffing, funding constraints, and the HTA and EU requirements for consent, we have chosen a daytime collection arrangement that encompasses 38% of births occurring at the John Radcliffe Hospital and covers the scheduling of elective cesarean sections.

UCB units collected were initially funded for R&D projects within NHSBT and for collaborative ventures, but are now provided to other researchers. Projects or groups benefitting from the UCB collected for research at NHSBT are indicated below or by examples of publications from these groups.[74-97] Projects supported include (1) the use of UCB for the development of novel and improved cellular therapies for patients with cancers and degenerative diseases; (2) regulation of the expansion, engraftment, cryopreservation, and function of hematopoietic, endothelial, and mesenchymal stem cells from UCB; (3) studies on leukemia in Down's Syndrome patients; (4) studies on the multipotency of human UCB-derived stem cells; (5) reprogramming of UCB cells to multipotent progenitors; and (6) developing humanized animal models of hematopoiesis.

Acknowledgments The authors would like to thank all those staff within the NHSBT specialist services who provided data for this review, and Ms Milly Lymer for her assistance with references.

Grant Funding: The authors wish to acknowledge the support of NHS Blood and Transplant, The Oxford Biomedical Research Centre, The Experimental Cancer Medicine Centre – Oxford, National Institutes of Research UK, Restore Burns & Wound Healing Trust, the Royal College of Surgeons, the Cord Blood Charity, the British Heart Foundation, the Leukaemia Research Fund, and E.U. Framework VI Thercord and VII Cascade Programme Grants.

References

1. Gluckman E, Broxmeyer HE, Auerbacm AD, et al. Haematopoietic reconstitution in a patient with Fanconi's anaemia by means of umbilical cord from an HLA-identical sibling. *N Engl J Med.* 1989;321:1174-1178.
2. Watt SM, Contreras M. Stem cell medicine: umbilical cord blood and its stem cell potential. *Sem Fetal Neonat Med.* 2005;28:1-12.
3. Ballen KK. New trends in umbilical cord blood transplantation. *Blood.* 2005;105:3786-3792.
4. Rubinstein P. Why cord blood? *Hum Immunol.* 2006;67: 398-404.
5. Brunstein CG, Wagner J. Umbilical cord blood transplantation and banking. *Ann Rev Med.* 2006;57:403-417.
6. Brown JA, Boussiotis VA. Umbilical cord blood transplantation: basic biology and clinical challenges to immune reconstitution. *Clin Immunol.* 2008;127(3):286-297.
7. Pinto FO, Roberts I. Cord blood stem cell transplantation for hematoglobinopathies. *Br J Haematol.* 2008;141(3):309-324.
8. Tse WW, Zang SL, Bunting KD, Laughlin MJ. Umbilical cord blood transplantation in adult myeloid leukemia. *Bone Marrow Transplant.* 2008;41(5):465-472.
9. Sullivan MJ. Banking on cord blood stem cells. *Nat Rev Cancer.* 2008;8(7):555-563.
10. Tse W, Bunting KD, Laughlin MJ. New insights into cord blood stem cell transplantation. *Curr Opin Hematol.* 2008;15(4):279-284.
11. Smythe J, Armitage S, McDonald D, et al. Directed sibling cord blood banking for transplantation: the 10-year experience in the national blood service in England. *Stem Cells.* 2007;25(8):2087-2093.
12. Meyer EA, Hanna K, Gebbie K. *Cord Blood: Establishing a National Hematopoietic Stem Cell Bank Program.* Washington, DC: The National Academies Press; 2005.
13. Rocha V, Wagner JE, Sobocinski K, et al. Comparison of graft versus host disease in children transplanted with HLA identical sibling umbilical cord blood versus HLA identical sibling bone marrow transplant. *N Engl J Med.* 2000;342: 1846-1854.
14. Broxmeyer HE. *Cord Blood: Biology, Immunology, Banking and Clinical Transplantation.* Bethesda, MD: AABB Press; 2004.
15. Schoemans H, Theunissen K, Maertens J, et al. Adult umbilical cord blood transplantation: a comprehensive review. *Bone Marrow Transplant.* 2006;38:83-93.
16. Fernandez MN, Regidor C, Cabrera R, et al. Unrelated umbilical cord blood transplants in adults: early recovery of

neutrophils by supportive co-transplantation of a low number of highly purified peripheral blood CD34+ cells from an HLA-haploidentical donor. *Exp Hematol.* 2003; 31:535-544.
17. Magro E, Regidor C, Cabrera R, et al. Early hematopoietic recovery after single unit unrelated cord blood transplantation in adults supported by co-infusion of mobilized stem cells from a third party. *Haematologica.* 2006;91:640-648.
18. www.NETCORD.org – Virtual Office.
19. Reed W, Smith R, Dekovic F, et al. Comprehensive banking of sibling donor cord blood for children with malignant and nonmalignant disease. *Blood.* 2003;101:351-357.
20. Walters MC, Quirolo L, Trachtenberg ET, et al. Sibling donor cord blood transplantation for thalassemia major: experience of the sibling donor cord blood program. *Ann NY Acad Sci.* 2005;1054:206-213.
21. Cohen Y, Nagler A. Umbilical cord blood transplantation – how, when and for whom? *Blood Rev.* 2004;18:167-179.
22. Dyer C. Couple is given go-ahead to use embryo selection to help existing child. *BMJ.* 2006;332:1114.
23. Dyer C. HFEA widens its criteria for preimplantation genetic diagnosis. *BMJ.* 2006;332:1174.
24. Liao C, Li D, Wei J, Tang X, Li Y, Huang Y. Prenatal HLA-typing in beta-thalassemia before the collection of sibling cord blood. *Prenat Diagn.* 2006;26:89-90.
25. Qureshi N, Foote D, Walters MC, Singer ST, Quirolo K, Vichinsky EP. Outcomes of preimplantation genetic diagnosis therapy in treatment of {beta}-thalassemia: a retrospective analysis. *Ann NY Acad Sci.* 2005;1054:500-503.
26. Human Tissue Authority Codes of Practice. 2006. Department of Health. London, UK.
27. EU Tissues and Cells Directive (2004/23/E). 2004. Official Journal of the European Union.
28. Austin EB, Guttridge M, Pamphilon D, Watt SM. The role of blood services and regulatory bodies in stem cell transplantation. *Vox Sang.* 2008;94(1):6-17.
29. Watt SM, Austin E, Armitage S. Cryopreservation of hematopoietic stem/progenitor cells for therapeutic use. *Methods Mol Biol.* 2007;368:237-259.
30. Davey S, Armitage S, Vanderson R, et al. The London cord blood bank; analysis of banking and transplantation outcome. *Br J Haematol.* 2004;125:358-365.
31. Warwick R, Armitage S. Cord blood banking. *Best Pract Res Clin Obs Gyn.* 2004;18:995-1011.
32. Ashford P, Distler P, Gee A, et al. Terminology and labeling of cellular products: 1. Standards. *Bone Marrow Transplant.* 2007;40(11):1075-1083.
33. Ashford P, Distler P, Gee A, et al. Terminology and labeling of cellular products-2: implementation plan. *Bone Marrow Transplant.* 2007;40(11):1085-1090.
34. Takahashi TA, Rebulla P, Armitage S, et al. Multi-laboratory evaluation of procedures for reducing the volume of cord blood: influence on cell recoveries. *Cytotherapy.* 2006;8(3): 254-264.
35. Brown C, Ellis J, Davey S, Brown J, Green A, Smythe J, Ord J, Regan F, Watt S, Navarrete CV. HLA profile and TNC and CD34 content of the NHS Cord Blood Collected and Issued Units. Abstract S135. *Trans Med.* 2008;18 (suppl. 2): 22.
36. Guidance for the Blood Transfusion Service in the United Kingdom. 7th ed. Norwich, UK: The UK Government Stationary Office; 2005.
37. International Standards for Cord Blood Collection, Processing, Testing, Banking, Selection and Release. 3rd ed. Omaha, NE: Foundation for the Accreditation of Hematopoietic Cell Therapy; 2006.
38. Bunce M, O'Neill CM, Barnardo MC, et al. Phototyping: comprehensive DNA typing for HLA-A, B, C, DRB1, DRB3, DRB4, DRB5 & DQB1 by PCR with 144 primer mixes utilizing sequence-specific primers (PCR-SSP). *Tissue Antigens.* 1995;46:355-367.
39. Forsi L, Brown J, Railton D, et al. HLA typing of chorionic villus samples: assessment of potential for cord blood collection for the treatment of siblings. *Eur J Immunogenetics.* 2002;29:363-368.
40. Verlinsky Y, Rechitsky S, Sharapova T, Morris R, Taranissi M, Kuliev A. Preimplantation HLA testing. *JAMA.* 2004; 291:2079.
41. The Human Fertilisation and Embryology Authority Code of Practice. 6th ed. London, UK: Department of Health; 2003.
42. Allan DS, Keeney M, Howson-Jan K, et al. Number of viable CD34+ cells reinfused predicts engraftment in autologous hematopoietic stem cell transplantation. *Bone Marrow Transplant.* 2002;29:967-972.
43. Rules and Guidance for Pharmaceutical Manufacturers and Distributors. *Annex 1- Manufacture of Sterile Medicinal Products.* Norwich, UK: The UK Department of Health Stationary Office; 2002.
44. Patel R, Davey S, Turner D, Brown C. The use of STR analysis to confirm identity of a cord blood unit prior to transplantation (abstract). *Eur J Immunogenet.* 2003; 30:310.
45. Nissen-Druey C, Tichelli A, Meyer-Monard S. Human haematopoietic colonies in health and disease. *Acta Haematologica.* 2005;113:1-96.
46. Fisk NM, Roberts IA, Markwald R, Mironov V. Can routine commercial cord blood banking be scientifically and ethically justified? *PLoS Med.* 2005;2:87-90.
47. Royal College of Obstetricians and Gynaecologists. Umbilical cord blood banking. Scientific advisory committee opinion paper 2. Revised June 2006.
48. Cord Blood Banking for Potential Future Transplantation. Policy Statement. *Pediatrics.* 2007;119:165-170.
49. Welte K, Costeas P, Brunoehler D et al. The world marrow donor association (WMDA) policy statement for the utility of autologous or family cord blood unit storage. Approved May 2006.
50. Wagner JE, Barker JN, DeFor TE, et al. Transplantation of unrelated donor umbilical cord blood in 102 patients with malignant and nonmalignant diseases: influence of CD34 cell dose and HLA disparity on treatment-related mortality and survival. *Blood.* 2002;100:1611-1618.
51. Hamza NS, Lisgaris M, Yadavalli G, et al. Kinetics of myeloid and lymphocyte recovery and infectious complications after unrelated umbilical cord blood versus HLA-matched unrelated donor allogeneic transplantation in adults. *Br J Haematol.* 2004;124:488-498.
52. Montagna D, Locatelli F, Moretta A, et al. T lymphocytes of recipient origin may contribute to the recovery of specific immune response toward viruses and fungi in children undergoing cord blood transplantation. *Blood.* 2004;103: 4322-4329.

53. Wall DA, Noffsinger JM, Mueckl KA, et al. Feasibility of an obstetrician-based cord blood collection network for unrelated donor umbilical cord blood banking. *J Maternal-Fetal Med.* 1997;6:320-323.
54. Surbek DV, Visca E, Steinmann C, et al. Umbilical cord blood collection before placental delivery during cesarean delivery increases cord blood volume and nucleated cell number available for transplantation. *Am J Obstet Gynecol.* 2000;183:218-221.
55. Wada RK, Bradford A, Moogk M, et al. Cord blood units collected at a remote site: a collaborative endeavor to collect umbilical cord blood through the Hawaii Cord Blood Bank and store the units at the Puget Sound Blood Center. *Transfusion.* 2004;44:111-118.
56. Solves P, Moraga R, Saucedo E, et al. Comparison between two strategies for umbilical cord blood collection. *Bone Marrow Transplant.* 2003;31:269-273.
57. Guttridge M, Sidders C, Booth-Davey E, Pamphilon D, Watt SM. Factors affecting volume reduction and red blood cell depletion of bone marrow using the Cobe Spectra cell separator prior to haematopoietic stem cell transplantation. *Bone Marrow Transplant.* 2006;38:175-181.
58. Rodríguez L, Azqueta C, Azzalin S, García J, Querol S. Washing of cord blood grafts after thawing: high cell recovery using an automated and closed system. *Vox Sang.* 2004; 87:165-172.
59. Stanworth S, Warwick R, Fehily D, et al. An international survey of unrelated umbilical cord blood banking. *Vox Sang.* 2001;80:236-243.
60. Ljungman P, Urbano-Ispizua A, Cavazzana-Calvo et al. Allogeneic and autologous transplantation for haematological diseases, solid tumours and immune disorders: definitions and current practice in Europe. *Bone Marrow Transplant.* 2006;37:439–449
61. Bielorai B, Hughes MR, Auerbach AD, et al. Successful umbilical cord blood transplantation for Fanconi anemia using preimplantation genetic diagnosis for HLA-matched donor. *Am J Hematol.* 2004;77:397-399.
62. Grewal SS, Kahn JP, MacMillan ML, Ramsay NK, Wagner JE. Successful hematopoietic stem cell transplantation for Fanconi anemia from an unaffected HLA-genotype-identical sibling selected using preimplantation genetic diagnosis. *Blood.* 2004;103:1147-1151.
63. Van de Velde H, Georgiou I, De Rycke M, et al. Novel universal approach for preimplantation genetic diagnosis of beta-thalassaemia in combination with HLA matching of embryos. *Hum Reprod.* 2004;19:700-708.
64. Bhattacharya A, Slatter MA, Chapman CE, et al. Single centre experience of umbilical cord stem cell transplantation for primary immunodeficiency. *Bone Marrow Transplant.* 2005; 36:295-299.
65. Broxmeyer HE, Cooper S. High-efficiency recovery of immature haematopoietic progenitor cells with extensive proliferative capacity from human cord blood cryopreserved for 10 years. *Clin Exp Immunol.* 1997;107:45-53.
66. Krishnamurti L, Abel S, Maiers M, Flesch S. Availability of unrelated donors for hematopoietic stem cell transplantation for hemoglobinopathies. *Bone Marrow Transplant.* 2003;31: 547-550.
67. Gluckman E, Rocha V. Cord blood transplantation for children with acute leukaemia: a Eurocord registry analysis. *Blood Cells Mol Dis.* 2004;33:271-273.
68. Locatelli F, Rocha V, Reed W, et al. Related umbilical cord blood transplantation in patients with thalassemia and sickle cell disease. *Blood.* 2003;101:2137-2143.
69. Rocha V, Cornish J, Sievers EL, et al. Comparison of unrelated bone marrow and umbilical cord blood transplants in children with acute leukemia. *Blood.* 2001;97:2962-2971.
70. Bitan M, Or R, Shapira MY, et al. Fludarabine-based reduced intensity conditioning for stem cell transplantation of Fanconi anemia patients from fully matched related and unrelated donors. *Biol Blood Marrow Transplant.* 2006;12:712-718.
71. Rubinstein P, Taylor PE, Scaradavou A, et al. Unrelated placental cord blood for bone marrow reconstitution organisation of the placental blood programme. *Blood Cells.* 2004; 20:587-600.
72. http://www.cellbank.org.uk/
73. Ballen KK, Barker JN, Stewart SK, Greene MF, Lane TA. American Society of Blood and Marrow Transplantation. Collection and preservation of cord blood for personal use. *Biol Blood Marrow Transplant.* 2008;14(3):356-363.
74. Hunt CJ, Pegg DE, Armitage SE. Optimising cryopreservation protocols for haematopoietic progenitor cells: a methodological approach for umbilical cord blood. *Cryo Lett.* 2006;27(2):73-86.
75. Zhang Y, Fisher N, Newey SE, et al. The impact of proliferative potential of umbilical cord-derived endothelial progenitor cells and hypoxia on vascular tubule formation in vitro. *Stem Cells Dev.* 2009;18(2):359-375.
76. Martin-Rendon E, Sweeney D, Lu F, Girdlestone J, Navarrete C, Watt SM. 5-Azacytidine-treated human mesenchymal stem/progenitor cells derived from umbilical cord, cord blood and bone marrow do not generate cardiomyocytes in vitro at high frequencies. *Vox Sang.* 2008;95(2):137-148.
77. Smythe J, Fox A, Fisher N, Frith E, Harris AL, Watt SM. Measuring angiogenic cytokines, circulating endothelial cells, and endothelial progenitor cells in peripheral blood and cord blood: VEGF and CXCL12 correlate with the number of circulating endothelial progenitor cells in peripheral blood. *Tissue Eng Part C Methods.* 2008;14(1):59-67.
78. Fox A, Gal S, Fisher N, et al. Quantification of circulating cell-free plasma DNA and endothelial gene RNA in patients with burns and relation to acute thermal injury. *Burns.* 2008;34(6):809-816.
79. Martin-Rendon E, Hale SJ, Ryan D, et al. Transcriptional profiling of human cord blood CD133+ and cultured bone marrow mesenchymal stem cells in response to hypoxia. *Stem Cells.* 2007;25(4):1003-1012.
80. Cheung JO, Casals-Pascual C, Roberts DJ, Watt SM. A small-scale serum-free liquid cell culture model of erythropoiesis to assess the effects of exogenous factors. *J Immunol Methods.* 2007;319(1–2):104-117.
81. Casals-Pascual C, Kai O, Cheung JO, et al. Suppression of erythropoiesis in malarial anemia is associated with hemozoin in vitro and in vivo. *Blood.* 2006;108(8):2569-2577.
82. Forde S, Tye BJ, Newey SE, et al. Endolyn (CD164) modulates the CXCL12-mediated migration of umbilical cord blood CD133+ cells. *Blood.* 2007;109(5):1825-1833.

83. Jorgensen-Tye B, Levesque JP, Royle L, et al. Epitope recognition of antibodies that define the sialomucin, endolyn (CD164), a negative regulator of haematopoiesis. *Tissue Antigens*. 2005;65(3):220-239.
84. McGuckin CP, Pearce D, Forraz N, Tooze JA, Watt SM, Pettengell R. Multiparametric analysis of immature cell populations in umbilical cord blood and bone marrow. *Eur J Haematol*. 2003;71(5):341-350.
85. McGuckin CP, Forraz N, Baradez MO, et al. Colocalization analysis of sialomucins CD34 and CD164. *Stem Cells*. 2003;21(2):162-170.
86. Zannettino AC, Roubelakis M, Welldon KJ, et al. Novel mesenchymal and haematopoietic cell isoforms of the SHP-2 docking receptor, PZR: identification, molecular cloning and effects on cell migration. *Biochem J*. 2003;370(pt 2):537-549.
87. Connolly NP, Jones M, Watt SM. Human Siglec-5: tissue distribution, novel isoforms and domain specificities for sialic acid-dependent ligand interactions. *Br J Haematol*. 2002;119(1):221-238.
88. Hunt CJ, Armitage SE, Pegg DE. Cryopreservation of umbilical cord blood: 2. Tolerance of CD34(+) cells to multimolar dimethyl sulphoxide and the effect of cooling rate on recovery after freezing and thawing. *Cryobiology*. 2003; 46(1):76-87.
89. Hunt CJ, Armitage SE, Pegg DE. Cryopreservation of umbilical cord blood: 1. Osmotically inactive volume, hydraulic conductivity and permeability of CD34(+) cells to dimethyl sulphoxide. *Cryobiology*. 2003;46(1):61-75.
90. Tunstall-Pedoe O, Roy A, Karadimitris A, et al. Abnormalities in the myeloid progenitor compartment in Down syndrome fetal liver precede acquisition of GATA1 mutations. *Blood*. 2008;112(12):4507-4511.
91. Hamlett I, Draper J, Strouboulis J, Iborra F, Porcher C, Vyas P. Characterization of megakaryocyte GATA1-interacting proteins: the corepressor ETO2 and GATA1 interact to regulate terminal megakaryocyte. *Blood*. 2008;112(7): 2738-2749.
92. Gupta R, Hong D, Iborra F, Sarno S, Enver T. NOV (CCN3) functions as a regulator of human hematopoietic stem or progenitor cells. *Science*. 2007;316(5824):590-593.
93. Hill AJ, Zwart I, Tam HH, et al. Human umbilical cord blood-derived mesenchymal stem cells do not differentiate into neural cells types or integrate into the retina after intravitreal grafting in neonatal rats. *Stem Cells Dev*. 2009; 18(3):399-409.
94. Zwart I, Hill AJ, Girdlestone J, et al. Analysis of neural potential of human umbilical cord blood-derived multipotent mesenchymal stem cells in response to a range of neurogenic stimuli. *J Neurosci Res*. 2008;86(9):1902-1915.
95. Manca MF, Zwart I, Beo J, et al. Characterization of mesenchymal stromal cells derived from full-term umbilical cord blood. *Cytotherapy*. 2008;10(1):54-68.
96. Tisato V, Naresh K, Girdlestone J, Navarrete C, Dazzi F. Mesenchymal stem cells of cord blood origin are effective at preventing but not treating graft-versus-host disease. *Leukemia*. 2007;21(9):1992-1999.
97. Friese MA, Jensen LT, Willcox N, Fugger L. Humanized mouse models for organ-specific autoimmune diseases. *Curr Opin Immunol*. 2006;18(6):704-709.

Donor and Collection-Related Variables Affecting Product Quality in Ex utero Cord Blood Banking

34

Sabeen Askari

The umbilical cord blood transplantation (UCBT) has emerged as a valuable alternative source of hematopoietic stem cells for patients requiring allogeneic transplantation. Recent literature suggests that allogeneic UCBT may potentially emerge as the frontline stem cell source for pediatric patients. Double cord blood units (CBUs) and non-myeloablative engraftment strategies in adults have attracted further attention in clinical practice with the advantages of possible stronger graft-versus-leukemia effect and expanding transplantation indications.[1]

Cord blood has shown several advantages over allogeneic bone marrow for transplantation including immediate availability, absence of risk to the donor, less HLA restriction for donors, lower risk of viral contamination of the graft, and potentially reduced risk of both acute and chronic graft-versus-host disease.[1-9] To date, over 10,000 UCB transplant procedures have been performed worldwide for pediatric[1,3-6,10-12] and adult patients,[6,13-18] and the number is increasing every year. The median delay in platelet engraftment of 59 days, and neutrophil recovery of 25 days following UCBT (compared with 27 and 19 days in marrow recipients, respectively), and their dependence on cell dose, have suggested the possibility of improved engraftment and prolonged patient survival by enhanced product quality.[19-21]

Optimizing product quality is a current focus in cord blood banking and the effect of various variables is being addressed by basic research. The cell dose is considered as the most important factor compared with HLA in donor choice. A minimum cell dose of $>4\times10^7$ NC/kg at collection and 3×10^7 NC/kg at infusion is recommended.[10,14,18] $CD34^+$ cell count correlates with engraftment and a dose of $>2\times10^5$ $CD34^+$ cells/kg is considered optimal; however, it cannot be used for comparative studies between centers due to absence of standardization of counting method.[10,14,18] Colony-forming units of granulocytes-monocytes colonies (CFU-GM) are also used for measuring the stem cell content of CBUs, but there is significant inter-laboratory variability.[10,14,18]

This chapter presents a review of selected donor- and collection-related variables and their effect on total volume, nucleated cell count (TNC), and CD34+ cell count of the CBUs that are collected ex utero.

S. Askari
Department of Laboratory Medicine and Pathology, University of Minnesota, Minneapolis, MN, USA and Department of Pathology and Laboratory Services, Veterans Affairs Medical Center, One Veterans Drive, Minneapolis, MN 55417, USA
e-mail: sabeenaskari@yahoo.com

34.1 Donor-Related Variables

The donor-related variables in ex utero cord blood banking include factors related to both the mother and the newborn (see Table 34.1). Selected donor-related variables include:

1. Maternal age: Controversial results have been reported regarding the effect of maternal age on the collected CBUs. A recent study based on ex utero cord blood collections reported no effect of maternal age on CBU volume, TNC, and CD34+ cell counts.[22] Nakagawa et al.[23] published data from in utero CBU collections and reported higher CD34+ cell count with younger maternal age ($p<0.05$).
2. Maternal race: Superior TNC count has been reported in Caucasians when compared with African Americans.[22]

N. Bhattacharya and P. Stubblefield (eds.), *Regenerative Medicine Using Pregnancy-Specific Biological Substances*,
DOI: 10.1007/978-1-84882-718-9_34, © Springer-Verlag London Limited 2011

Table 34.1 Donor-related variables associated with superior volume, TNC, and/or CD34+ cell counts (based on t-test and analysis of variance) (Askari et al.[22])

Variables	Volume	TNC count	CD34+ count
Primigravida status	NS (n=1,628)	p<0.001 (n=1,628)	NS (n=1,124)
Maternal race other than African American	NS (n=1,623)	p<0.001 (n=1,623)	NS (n=1 123)
Gestational age >40 weeks	p<0.05 (n=1,628)	p<0.001 (n=1,628)	NS (n=1,124)
Placental weight >500 g	p<0.001 (n=428)	p<0.001 (n=428)	p<0.001 (n=273)
Presence of meconium	p<0.001 (n=1,297)	p<0.001 (n=1,297)	p<0.05 (n=872)
Delivery via cesarean section	p<0.001 (n=1,628)	NS (n=1,628)	NS (n=1,124)
Female gender of newborn	NS (n=1,625)	p<0.05 (n=1,625)	NS (n=1,124)
Presence of hemoglobinopathy in the newborn	p<0.05 (n=1,628)	NS (n=1,628)	NS (n=1,124)

Total collected CBUs=2,084. Volume and TNC count available in 1,628; 1,124 of 1,628 CBUs had post-processing CD34+ cell counts. Mean volume, TNC, and CD34+ cell counts were 85.2 mL, 118.9×10^7, and 5.2×10^6, respectively

3. Gravida status: Primigravida status has been associated with a better TNC count.[22]
4. Gestational age: Ballen et al.[24] studied the effects of maternal and neonatal predictors of hematopoietic potential of CBUs. They demonstrated that bigger babies had higher cell counts, more CD34+ cells, and more CFU-GM. Babies of longer gestational age were reported to have higher cell counts, but lower CD34+ cell counts and CFU-GM. Another study by Surbek et al.,[25] evaluating preterm deliveries (22–36 weeks), showed correlation between increasing gestational age and TNC count (p<0.001), and an inverse relation between gestational age and CD34+ cell count (p<0.001); no significant difference in CD34+ cell count was found between early (22–32 week) and late (33–36 week) preterm deliveries (p=0.870). These findings are consistent with another study that demonstrated that while superior volume and TNC count are expected after 40 weeks' gestational age when compared with 36–40 weeks, better CD34+ cell count is not.[22] An in utero cord blood collection study reported a larger product volume associated with larger placenta (p<0.001) and bigger baby (p < 0.001); a higher TNC count in association with longer gestational age; and a higher CD34+ cell count with larger birth weight (p<0.001) and shorter gestational age (p < 0.001).[23]
5. Placental weight: Placental weight >500 g has been shown to be correlated with better volume, TNC, and CD34+ counts.[22]
6. Presence or absence of meconium in the amniotic fluid: Stress during delivery, including prolonged second stage of labor, difficult birth, and lower umbilical arterial pH, has been reported to increase the TNC, the CD34+ cell count, and the total CBU volume.[22,26-29] This is possibly secondary to mobilization of various cell populations by endogenous cytokines. Therefore, intra-labor conditions resulting in fetal distress should not be considered as reasons to defer collection, as on the contrary, the cord blood product in such conditions might prove to be superior, if otherwise safe.
7. Mode of delivery of the newborn (vaginal or cesarean section): Delivery via cesarean section has been shown to produce better volume; however, the data about TNC and CD34+ cell count is variable.[22,30-33] Reboredo et al.[26] reported a 13% higher TNC count following post-vaginal delivery in utero collection when compared with post-cesarean section ex utero collection.

8. Presence or absence of father in the hospital: The volume, TNC, and CD34+ cell counts have been reported to be unaffected by presence of father in the hospital.[22]
9. Gender of the newborn: Both ex utero and post-vaginal delivery in utero CBU collection studies have reported a correlation between female newborn gender and enhanced TNC count.[22,23] However, Reboredo et al.,[26] who evaluated post-cesarean section ex utero and post-vaginal delivery in utero collections, found no correlation with respect to gender of the newborn.
10. Newborn metabolic screening: Presence of hemoglobinopathy in the newborn has been associated with superior volume, but unaltered TNC and CD34+ cell counts.[22]

34.2 Collection-Related Variables

The collection-related variables in ex utero cord blood banking include factors that are unrelated to the donor and mainly involve the collection process (see Table 34.2). Some of these factors may be modified to enhance product quality. Selected collection-related variables include:

1. Time or month of delivery: An ex utero study from a 24 h cord blood collection service reported no effect of the time of delivery or the month of delivery on volume, TNC, or CD34+ cell counts of the collected CBUs.[22]
2. Time interval between delivery of newborn and that of placenta: Equal to or less than 10 min between delivery of newborn and placenta have been reported to predict better volume and CD34+ count; the same effect was not noted in that study for cases with gestational age of >40 weeks.[22] The CD34+ cell count of the CBUs was noted to decline by 10% less if the time interval between the delivery of the newborn and that of the placenta was more than 10 min.
3. Time interval between placental delivery and the beginning of cord blood collection: Commencing collection as soon after the placental delivery as possible has always been desirable and is a potentially modifiable collection-related variable. It has been shown to correlate with increased product volume[22,34] and TNC count (using regression analysis)[22]; however, the same study did not show an effect on the CD34+ cell count when the cord blood collection was started within 5 min of placental delivery.[22]
4. Cord blood collection duration: Superior TNC count, but unaffected CD34+ cell count, has been reported when the cord blood is collected for over 5 min.[22]
5. Two- versus one-person collection: Two-person collection reportedly is more likely to produce a better volume, possibly secondary to enhanced manual placental massage; however, CD34+ cell count or TNC count have not been reportedly significantly different in the two groups.[22]
6. Personnel experience of CBU collection: Ex utero cord blood collection technique is simple and cord blood bank staff can be easily trained in obtaining quality product. An ex utero cord blood collection

Table 34.2 Collection-related variables associated with superior volume, TNC, and/or CD34+ cell counts (based on *t*-test and analysis of variance) (Askari et al.[22])

Variables	Volume	TNC count	CD34+ count
≤10 min interval between delivery of newborn and placenta	$p<0.05$ ($n=1{,}622$)	NS ($n=1{,}622$)	$p<0.05$ ($n=1{,}121$)
≤5 min interval between placental delivery and CBU collection	$p<0.001$ ($n=1{,}628$)	NS ($n=1{,}628$)	NS ($n=1{,}124$)
>5 min CBU collection duration	NS ($n=1{,}627$)	$p<0.05$ ($n=1{,}627$)	NS ($n=1{,}123$)
Two-person versus one-person collection	$p<0.05$ ($n=1{,}624$)	NS ($n=1{,}624$)	NS ($n=1{,}123$)
Personnel experience of <100 CBU collection	$p<0.05$ ($n=1{,}620$)	NS ($n=1{,}620$)	NS ($n=1{,}121$)

study demonstrated that less-experienced staff members were equally successful in collecting a quality product (with respect to TNC and CD34+ cell counts) when compared with staff with history of over 100 CBU collections. Less-experienced staff was in fact reported to have collected CBUs that were more superior in volume, possibly due to prolonged collection duration.[22]

7. Time interval between cord blood collection and processing: Nakagawa et al.[23] have reported a higher CD34+ cell count with shorter time interval between in utero cord blood collection and CBU processing ($p<0.05$).

Ex utero cord blood collection is frequently compared with in utero collection[35-37] which in general, has been considered to be associated with increased product volume.[35] Reboredo et al.[26] published data that demonstrated a normal distribution of concentration of TNCs in cord blood with an average of 10.51×10^6 TNCs obtained per milliliter of cord blood collected; increased CD34+ cell count with increasing product volume was also noted. While these data suggest superiority of in utero UCB collection process in producing better product quality by virtue of providing better product volume, the published data comparing the two collection methods is controversial.[32,36]

A variety of exclusionary factors have previously been proposed for donor selection and designing of a cost-effective collection process.[22,38,39] Recent research has improved our understanding of the variables affecting product quality. CD34+ cell content has been shown to predict the hematopoietic potential of a CBU better than TNC count, because of its better correlation with colony-forming cell content ($p<0.0001$).[40] Given the fact that both volume and TNC count correlate well with CD34+ cell count of the product,[41,42] it indeed appears more practical for cord blood programs with limited resources to cryopreserve and bank CBUs based on TNC count of the product. Full-term pregnant women with placental weights of over 500 g appear to be the most promising donor group. Mothers with history of cesarean section, fetal distress, and female newborn gender should also be considered. An attempt should be made to begin collection within 5 min of placental delivery, and the collection process should be conducted for at least 5 min duration.

This chapter only provides an outline of the topic and it is logical to believe that individual cord blood banks need to identify their respective system variables that might influence the quality of their banked CBUs. Donor selection and collection technique modifications may improve product quality.

Acknowledgments I would like to thank Dr. Jeffrey McCullough, Dr. John Miller, and Gayl R Chrysler for their pertinent support, comments, insight, and advice regarding the process of evaluation of donor and collection-related variables affecting the cord blood quality at our regional cord blood bank.

References

1. Tse W, Bunting KD, Laughlin MJ. New insights into cord blood stem cell transplantation. *Curr Opin Hematol.* 2008; 15(4):279-284.
2. Ballen K, Broxmeyer HE, McCullough J, et al. Current status of cord blood banking and transplantation in the United States and Europe. *Biol Blood Marrow Transplant.* 2001; 7:635-645.
3. Rocha V, Wagner JE Jr, Sobocinski KA, et al. Graft-versus-host disease in children who have received a cord blood or bone marrow transplant from an HLA-identical sibling. *N Engl J Med.* 2000;342:1846-1854.
4. Wagner JE, Rosenthal J, Sweetman R, et al. Successful transplantation of HLA matched and HLA mismatched umbilical cord blood from unrelated donors: analysis of engraftment and acute graft-versus-host disease. *Blood.* 1996;88:795-802.
5. Kurtzberg J, Laughlin M, Graham L, et al. Placental blood as a source of hematopoietic stem cells for transplantation into unrelated recipients. *N Engl J Med.* 1996;335:157-166.
6. Gluckman E, Rocha V, Boyer-Chammard A, Eurocord Transplant Group and the European Blood and marrow Transplantation Group. Outcome of cord blood transplantation from related and unrelated donors. *N Engl J Med.* 1997;337:373-381.
7. Wagner JE, Kurtzberg J. Placental and umbilical cord blood transplantation. In: Hoffman, ed. *Hematology: Basic Principles and Practice.* Philadelphia, PA: Churchill Livingston; 2000.
8. Rubinstein P, Rosenfield RE, Adamson JW, et al. Stored placental blood for unrelated bone marrow reconstruction. *Blood.* 1993;81:679-690.
9. Rubinstein P. Placental blood-derived hematopoietic stem cells for unrelated bone marrow reconstitution. *J Hematother.* 1993;2:207-210.
10. Wagner JE, Barker JN, DeFor TE, et al. Transplantation of unrelated donor umbilical cord blood in 102 patients with malignant and nonmalignant diseases: influence of CD34 cell dose and HLA disparity on treatment-related mortality and survival. *Blood.* 2002;100:1611-1618.
11. Locatelli F, Rocha V, Reed W, et al. Related umbilical cord blood transplantation in patients with thalassemia and sickle cell disease. *Blood.* 2003;101:2137-2143.

12. Rubinstein P, Carrier C, Scaradavou A, et al. Outcomes among 562 recipients of placental-blood transplants from unrelated donors. *N Engl J Med.* 1998;339:1565-1577.
13. Laughlin MJ, Barker J, Bambach B, et al. Hematopoietic engraftment and survival in adult recipients of umbilical-cord blood from unrelated donors. *N Engl J Med.* 2001;344: 1815-1822.
14. Barker JN, Davies SM, DeFor T, et al. Survival after transplantation of unrelated donor umbilical cord blood is comparable to that of human leukocyte antigen-matched unrelated donor bone marrow: results of a matched-pair analysis. *Blood.* 2001;97:2957-2961.
15. Laughlin MJ, Eapen M, Rubinstein P, et al. Outcomes after transplantation of cord blood or bone marrow from unrelated donors in adults with leukemia. *N Engl J Med.* 2004;351: 2265-2275.
16. Rocha V, Labopin M, Sanz G, et al. Transplants of umbilical-cord blood or bone marrow from unrelated donors in adults with acute leukemia. *N Engl J Med.* 2004;351: 2276-2285.
17. Hwang WY, Samuel M, Tan D, et al. A meta-analysis of unrelated donor umbilical cord blood transplantation versus unrelated donor bone marrow transplantation in adult and pediatric patients. *Biol Blood Marrow Transplant.* 2007;13:444-453.
18. Gluckman E, Rocha V. Donor selection for unrelated cord blood transplants. *Curr Opin Immunol.* 2006;18:565-570.
19. Eapen M, Rubinstein P, Zhang MJ, et al. Outcomes of transplantation of unrelated donor umbilical cord blood and bone marrow in children with acute leukaemia: a comparison study. *Lancet.* 2007;369:1947-1954.
20. Gluckman E, Rocha V, Chastang C. Cord blood stem cell transplantation. *Bail Clin Haemat.* 1999;12:279-292.
21. Thomson BG, Robertson KA, Gowan D, et al. Analysis of engraftment, graft-versus-host disease, and immune recovery following unrelated donor cord blood transplantation. *Blood.* 2000;96(8):2703-2711.
22. Askari S, Miller J, Chrysler G, McCullough J. Impact of donor- and collection-related variables on product quality in ex-utero cord blood banking. *Transfusion.* 2005;45(2): 189-194.
23. Nakagawa R, Watanabe T, Kawano Y, et al. Analysis of maternal and neonatal factors that influence the nucleated and CD34+ cell yield for cord blood banking. *Transfusion.* 2004;44:262-267.
24. Ballen K, Wilson M, Wuu J, et al. Bigger is better: maternal and neonatal predictors of hematopoietic potential of umbilical cord blood units. *Bone Marrow Transplant.* 2001;27(1): 7-14.
25. Surbek DV, Holzgreve W, Steinmann C, et al. Preterm birth and the availability of cord blood for HPC transplantation. *Transfusion.* 2000;40(7):817-820.
26. M-Reboredo N, Diaz A, Castro A, Villaescusa RG. Collection, processing and cryopreservation of umbilical cord blood for unrelated transplantation. *Bone Marrow Transplant.* 2000;26: 1263-1270.
27. Lim FT, Scherjon SA, Van Beckhoven JM, et al. Association of stress during delivery with increased numbers of nucleated cells and hematopoietic progenitor cells in umbilical cord blood. *Am J Obstet Gynecol.* 2000;183(5):1144-1152.
28. Lim FT, Van Winsen L, Willemze R, Kanhai HH, Falkenburg JH. Influence of delivery on numbers of leukocytes, leukocyte subpopulations, and hematopoietic progenitor cells in human umbilical cord blood. *Blood Cells.* 1994;20 (2–3):547-558.
29. Aufderhaar U, Holzgreve W, Danzer E, Tichelli A, Troeger C, Surbek DV. The impact of intrapartum factors on umbilical cord blood stem cell banking. *J Perinat Med.* 2003; 31(4):317-322.
30. Yamada T, Okamoto Y, Kasamatsu H, Horie Y, Yamashita N, Matsumoto K. Factors affecting the volume of umbilical cord blood collections. *Acta Obstet Gynecol Scand.* 2000; 79(10):830-833.
31. Vettenranta K, Piirto I, Saarinen-Pihkala UM. The effects of the mode of delivery on the lymphocyte composition of a placental/cord blood graft. *J Hematother.* 1997;6:491-493.
32. Solves P, Moraga R, Saucedo E, et al. Comparison between two strategies for umbilical cord blood collection. *Bone Marrow Transplant.* 2003;31(4):269-273.
33. Sparrow RL, Cauchi JA, Ramadi LT, Waugh CM, Kirkland MA. Influence of mode of birth and collection on WBC yields of umbilical cord blood units. *Transfusion.* 2002;42(2): 210-215.
34. Jones J, Stevens CE, Rubinstein P, Robertazzi RR, Kerr A, Cabbad MF. Obstetric predictors of placental/umbilical cord blood volume for transplantation. *Am J Obstet Gynecol.* 2003;188(2):503-509.
35. Surbek DV, Schönfeld B, Tichelli A, et al. Optimizing cord blood mononuclear cell yield: A randomized comparison of collection before vs. after placental delivery. *Bone Marrow Transplant.* 1998;22:311-312.
36. Lasky LC, Lane TA, Miller JP, et al. In-utero or ex-utero cord blood collection: which is better? *Transfusion.* 2002; 42(10):1261-1267.
37. Rebulla P, Lecchi L, Porretti L, et al. Practical placental blood banking. *Transfus Med Rev.* 1999;13(3):205-226.
38. McCullough J, Clay ME, Wagner JE. Cord blood stem cells. In: Ball ED, Lister J, Law P, eds. *Hematopoietic Stem Cell Therapy.* New York: Churchill Livingstone; 2000:287-297.
39. McCullough J, Herr G, Lennon S, Stroncek D, Clay ME. Factors influencing the availability of umbilical cord blood for banking and transplantation. *Transfusion.* 1998;38(5): 508-510.
40. Aroviita P, Teramo K, Westman P, Hiilesmaa V, Kekomaki R. Associations among nucleated cell, CD34+ cell and colony-forming cell contents in cord blood units obtained through a standardized banking process. *Vox Sang.* 2003; 84(3):219-227.
41. Lim FT, Beckhoven J, Brand A, et al. The number of nucleated cells reflects the hematopoietic content of umbilical cord blood for transplantation. *Bone Marrow Transplant.* 1999;24(9):965-970.
42. Rogers I, Sutherland, Holt D, et al. Human UC-blood banking: impact of blood volume, cell separation and cryopreservation on leukocyte and CD34+ cell recovery. *Cytotherapy.* 2001;3(4):269-276.

Cord Blood as a Source of Hematopoietic Progenitors for Transplantation

35

Pilar Solves, Amando Blanquer, and Vicente Mirabet

35.1 Introduction

Umbilical cord blood (UCB) has become an alternative to bone marrow and peripheral blood as a source of progenitors for hematopoietic stem cell (HSC) transplantation, and its use has increased greatly in recent years.[1-5] Although UCB has broad spectrum of possible uses, allogeneic HSC transplantation is the major indication.[6] Since the first cord blood transplantation was performed in Paris in 1988,[7] knowledge of the biologic characteristics of UCB has improved and the benefits of using UCB stem cells have become apparent. Currently, cord blood is used in the treatment of patients with high-risk or relapsed hematopoietic malignancies, hemoglobinopathies, bone marrow failure syndromes, congenital immunodeficiency syndromes, and inborn errors of metabolism. To date, over 8,000 UCB transplants in children and adults have been performed worldwide using UCB donors.[8-10] Major advantages of UCB as compared to bone marrow include the availability, a low risk of infectious disease transmission such as CMV and EBV, and low risk of graft-versus-host disease (GVHD). The immaturity of UCB cells allows a tolerance from one to two human leukocyte antigen mismatches, which offers the opportunity to extend the donor pool.[11] It is also quicker to perform UCB transplantation from the time of beginning of donor search. However, the main disadvantage consists of the lower number of progenitor cells compared to that of bone marrow and peripheral blood and usually sufficient for smaller recipients, usually children weighing less than 40 kg. Table 35.1 shows the main advantages and disadvantages of UCB as a source of HSC for allogeneic transplantation.

This chapter describes the characteristics of UCB as a source of hematopoietic progenitors for transplantation, and the factors influencing HSC content of UCB units. In addition to this, it contains a description of the cord blood banking activities and the most relevant results of UCB transplantation selected clinical trials.

35.2 Characterization of Cord Blood Hematopoietic Progenitors

UCB quality is basically defined by three parameters: total nucleated cells (TNC), $CD34^+$ cells, and colony-forming units (CFU) content. TNC is a surrogate measure of the stem cell dose in the transplant product and currently the most important factor for donor choice.[9] The most utilized phenotypic marker for stem and progenitor cells is CD34, a glycophosphoprotein. Although it is used as an important clinical marker, it is found also on cells that are not stem or progenitor cells.[12] Colony assays (CFU) determine the in vitro functionality of hematopoietic progenitors. These methods use the growth of cells in semisolid culture media that allow the growth of distinctive colonies. These colonies derive from single cells termed high-proliferative-potential colony-forming cells (HPP-CFCs); multipotential colony-forming units (CFU-GEMM for granulocyte, erythroid, macrophage, and megakaryocite-containing components); and more lineage-restricted progenitors such as CFU-GM (containing granulocyte and macrophage differentiation capacity), CFU-G (with granulocyte differentiation ability), CFU-M (with macrophage differentiation capacity), CFU-Mega (with megakaryocyte differentiation capacity),

P. Solves (✉)
Cord Blood Bank, Valencia Transfusion Center and Valencia Transfusion Center, Avda del Cid 65-A, 46014 Valencia, Spain
e-mail: solves_pil@gva.es

N. Bhattacharya and P. Stubblefield (eds.), *Regenerative Medicine Using Pregnancy-Specific Biological Substances*,
DOI: 10.1007/978-1-84882-718-9_35,

Table 35.1 Advantages and disadvantages of UCB as source of hematopoietic stem cells for allogeneic transplantation as compared to bone marrow/peripheral blood

Advantages	Disadvantages
Easy and safe procurement	Limited number of HPC
Low risk of viral contamination	Impossible to collect additional volumes
Rapid availability	Risk of genetic disease transmission
Immaturity of T-cell immunity	Prolonged immune reconstitution
Reduced severe AGVHD[a]	Prolonged neutrophil/ platelet recoveries
Reduced CGVHD[b]	Increased infectious morbidity
Multiethnic representation	

[a]Acute graft-versus-host disease
[b]Chronic graft-versus-host disease

and BFU-E (burst-forming-unit-erythroid).[13] The content of CFU is based on the number of different colonies formed per number of cells plated. Early studies showed that CFU-GM could be grown in vitro from UCB.[14] Further in vitro studies by Broxmeyer et al. demonstrated that UCB contains sufficient number of HSC to be used for autologous or allogeneic hematopoietic reconstitution.[15] Although these three parameters are well correlated, CD34$^+$ cells and CFU content predicts the hematopoietic potential of a UCB unit better than TNC content.[16,17] UCB characteristics influencing engraftment are total nucleated cells (TNC), CD34$^+$ cell, CFU contents, and degree of donor–recipient HLA-match.[17]

Since the cell dose is the main factor for improved transplantation outcome, the factors influence HSC content of UCB units have been analyzed in many studies.

35.3 Factors Influencing Umbilical Cord Blood Hematopoietic Content

Two important groups of factors have been showed to have any influence on UCB quality: the mode of collection and some obstetric variables.

35.3.1 Influence of Mode of Collection

The collection strategy is the first step for collecting good-quality UCB units and varies among banks and among collection sites for the same UCB bank.[18] There are two main techniques for collecting UCB from the umbilical vein: in the delivery room during the third stage of labor while the placenta is still in uterus by midwives and obstetricians (obstetrician-based) or in an adjacent room after placental delivery by UCB bank-trained technicians (technician-based). In both cases, after the baby delivery the cord is clamped and cut, then wiped with alcohol followed by an iodine swab. The umbilical vein is cannulated with a needle attached to a collection bag containing CPD/CPDA anticoagulant solution. The cord blood is allowed to drain into the bag by gravity, while delivering physician or technician periodically mixes the contents of the collection bag in order to avoid clotting. This closed bag system collection is used by most UCB banks, although open systems have also been used and mostly discarded because they result in excessive microbiological contamination.[19] Both collection strategies have some advantages and some disadvantages. The advantage of the obstetrician-based strategy is that the volume of cells collected is usually higher if the cord is clamped early and the collection is started immediately. However, this can disrupt the normal process of delivery. After delivery, UCB collection is easier and it can be performed by UCB bank-trained personnel. However, fewer cells may be collected and there may be an increase in the risk of bacterial contamination. At present, there is no international consensus on the procedure of UCB collection in the maternity wards.[20] Surbek et al.[21] compared both strategies in vaginal deliveries and concluded that in utero collection yielded a significant higher volume and total number of mononuclear cells. Results from a posterior study demonstrated that the median concentrations of TNC and CFU were significantly lower in CB collected after placental delivery when compared to their counterparts collected before placental delivery. The incidence of macroscopic clots was also higher in ex utero collections. The reduction of stem and progenitor cells is possibly because of clotting activities developed with time.[22] The results of Solves et al. group showed higher numbers of total nucleated cells and CFU for in utero UCB collections.[23] The presence of hemorrhage in the maternal and fetal areas of delivered placenta and the clots formed in fetal placental vessels could explain the loss of hematopoietic progenitor cells in ex utero collections. Time from cord clamping to collection also could contribute to differences between in utero and ex utero methods as shown by some authors for ex utero collections.[24] Other authors have confirmed a larger UCB volume and more TNC content were obtained in the samples collected before placental expulsion also for cesarean deliveries.[25] However results from our group has not found any differences in UCB quality of units collected from cesarean deliveries with the two strategies.[26] Recent studies have not found differences between the two ways of collection for vaginal deliveries, although there was a trend toward a higher cell concentration and collected volume for in utero collections.[27] To date, there is only one study showing a greater volume and CD34+ cells content in UCB units collected after placental expulsion following cesarean sections, but not vaginal deliveries.[28] The difference was thought to be because of the position of the umbilical cord and the infant prior to clamping. Placing the newborn on the maternal abdomen after delivery increases the volume and CD34+ cell content in UCB units.[29,30] Although there is controversy about best collection strategy,[31] most authors agree UCB collection before placental delivery is the best approach for UCB collection that allows optimization of UCB bank methodology.[21-23,25] The feasibility of an obstetrician-based collection network has been demonstrated as having many advantages. First, the close relation that the delivering physician has with the prospective donors facilitates the process of obtaining quality consent for the banking. Second, obstetricians can easily decide not to collect the UCB if there are some maternal-infant problems, and finally this strategy avoids the financial burden that generates the presence of UCB bank personnel in the maternity wards.[32]

35.3.2 Influence of Obstetric Factors

During pregnancy, the frequencies of circulating white blood cells and hematopoietic progenitor cells are relatively constant.[33] However, during delivery the hematopoietic progenitors in UCB increase, showing large variations.[34] To date, many studies have investigated the influence of obstetric and neonatal factors on the volume and hematopoietic content of UCB donations, in order to optimize cord blood donors' selection.[35-39] Some of them have concluded that length of gestation influences volume and cells content of UCB units, supporting the fact that, with increasing gestational age, UCB volume and TNC counts increase, CD34+ cells percentages decrease, and total CD34+ cell content and CFU values are maintained.[36,40-43] For preterm deliveries, the number of CD34+ cells available per UCB unit does not depend on gestational age; therefore preterm delivery should not be a reason to exclude UCB collection.[42] In support of this, NETCORD standards only recommend that UCB collections shall be obtained from infants of at least 34-week gestation.[44] Number of previous gestations also influences UCB quality in the manner that each additional prior birth contributed to a different percentage decrease in volume, TNC, CD34+ cells, and CFU of UCB units.[38] Other studies have focused their attention on stress factors during delivery, demonstrating that prolonged first stage of labor, total length of labor, low venous and arterial pH, and Apgar score resulted in UCB units with higher numbers of hematopoietic progenitors, possibly due to the activation of the cytokine network.[36,37,41,45,46] Solves et al. analyzed the influence of stress factors on UCB quality, showing that the meconium-stained amniotic fluid also increased the hematopoietic content of UCB units.[47] On the contrary, some stress factors that have failed to show any influence in UCB collections quality were rupture membranes for more than 24 h and assisted delivery.[45] In summary, UCB collected from stressful deliveries can be accepted for UCB banking.

Regarding mode of delivery, results from different studies are controversial; while some authors conclude that there are no differences between collections from assisted delivery when compared to spontaneous vaginal delivery or between vaginal and cesarean deliveries,[27,47] others find better UCB quality from vaginal deliveries.[38,39] Only one study found higher volume and CD34+ cells content from cesarean sections and collected after placenta expulsion, as compared to vaginal deliveries.[28] The difference was thought to be due to the position of the umbilical cord and the infant prior to clamping.

Influence of newborn sex on UCB quality is controversial. The higher hematopoietic content observed in UCB units collected from males can be due to the higher weight of males as compared to females.[48] However, some authors have showed that hematopoietic progenitor cell concentration is higher in male infants, even after correcting for birth weight.[49] While UCB form female contains more TNC mainly due to the higher neutrophil concentration, UCB from males have higher CD34+ cell content. Other factors that have been associated to a higher hematopoietic progenitor content in UCB units are younger maternal age[35] and Caucasian race.[40] Table 35.2 shows the obstetric characteristics of

Table 35.2 Obstetric date according to the total nucleated cells (TNC) content of cord blood units. Continuous variables are expressed as mean and SD, and range

	UCB total nucleated cells		
	<80 × 107 *n* = 400	≥80 × 107 *n* = 900	*P*
Maternal age	32.34 ± 5.15 (18–48)	32.84 ± 5.95 (18–47)	ns
Gestational age (weeks)	38.93 ± 1.57 (34–42)	39.41 ± 1.41 (33–42)	<0.05
Gestations number	1.88 ± 1.03 (0–7)	1.97 ± 3.25 (0–9)	ns
Rupture of membranes (hours)	6.26 ± 5.12 (0–24)	7.44 ± 6.49 (0–30)	ns
Duration of delivery (hours)	3.95 ± 2.9 (0–16)	4.36 ± 3.75 (0–21)	ns
Infant weight (g)	3122.22 ± 562.27 (2,000–4,350)	3359.84 ± 399.84 (2,250–4,500)	<0.05
Placental weight (g)	629.41 ± 121.93 (335–1,100)	711.18 ± 242.47 (300–1,130)	<0.05
Mode of delivery			
Cesarean	26	76	
Spontaneous			ns
Vaginal	320	582	
Assisted vaginal	54	242	<0.05
Newborn sex			
Male	225	440	
Female	175	460	<0.05
Mode of collection			
Before placental delivery	203	576	
After placental delivery	197	324	<0.05

UCB donations according to the TNC count was less or more than 8 × 10^8.

The most important and constant obstetric factor influencing UCB quality appears to be the placental and newborn weight. The largest series examined to date has confirmed the observation that the weight of the infant and the placenta are consistent predictors of cord blood volume and cells content.[36-40] In fact, in one study, placental weight was the only variable that independently influenced on volume, TNC, CD34$^+$ cells, and CFU content of collections.[47] Bigger babies have higher cell counts, more CD34$^+$ cells, and CFU. Taking into accounts the direct association between placental and birth weight, in the banks that collect units before placental delivery, birth weight could be used as selection criteria. In fact some authors have suggested that better UCB units could be obtained by selecting first or second babies, babies with a gestational age ≤40 weeks, and weighting >3,600 g.[36] Applying less strict criteria as baby weight ≥3,200 g would optimize UCB collections. On the other hand, it would be possible to establish placental weight as obstetric criteria in those banks that collected UCB units after placental delivery. Some authors have previously estimated that to obtain UCB with ≥80 × 10^8 TNC content, the chosen cut-off points determined by Receiver Operating

Characteristic curves are 695 g for placental weight, and 3,150 for birth weight in our geographic area.[50]

35.4 Unrelated Cord Blood Banking

UCB banking is becoming a practical and feasible solution for providing sufficiently well-matched hematopoietic stem and progenitor cells for all patients within an acceptable timeframe. UCB banks are being established all over the world to support the increasing clinical activity. In 1993, the first three UCB bank programs were started in New York, Milan, and Dusseldorf. Standards for minimal volumes collected, sterile processing, volume reduction, $CD34^+$ measure controls, and other subsets vary among different banks. In spite of this, there continues to be a major effort to standardize banking and regulate all the steps to provide the highest quality for patient use. The American Association of Blood Banks, American Red Cross, American Society of Blood and Marrow Transplantation, European Blood and Marrow Transplantation Society, Eurocord, Foundation for the Accreditation of Hematopoietic Cell therapy, international Society for Hematotherapy and Graft Engineering, Joint Accreditation Committee of ISHAGE-Europe and EBMT, NETCORD, and the National Bone Marrow Donor program are organizations in place to ensure that the quality and standards in CB banking are established and met.[51] UCB banking involves the main following steps: donor selection, collection, volume reduction, cryopreservation and storage, biological controls, and release of the UCB to the transplant center. Table 35.3 summarizes the different activities of UCB banking.

Table 35.3 Activities that are involved in unrelated cord blood banking

Maternity ward	Cord blood bank	Transplant center
Donor selection	UCB reception	Transplantation
UCB collection	Processing – volume reduction Cryopreservation and storage Biological controls Shipment to transplant center Follow-up	Follow-up

35.4.1 UCB Donor Selection

Selection of UCB donors is an important way to improve not only the overall safety but also quality of finally stored UCB units. Because UCB banking is still a relatively new field and despite the quality programs for UCB having been established by AABB, FACT-NETCORD standards, and Council of Europe, there is considerable variation in the proceedings for selection, processing, and quality control of UCB units among different banks worldwide. In fact, the policies regarding donor health history are still in an investigational phase and vary among different UCB banks. Most of them recommend that a family medical history, with particular reference to inherited diseases, must be obtained before donation not only from the mother but also from father. The health history is usually obtained from the mother, because she determines the risk of transfusion-transmitted diseases from the cord blood. The role of the father must be focused on genetic history. In order to apply donor selection criteria established by the cord blood bank, potential donors must answer a questionnaire specially designed to identify specific risk factors for HIV, HCV, HBV, syphilis, and other infectious diseases. The exact questions to be asked, the information to be obtained, and the source of that information continue to be refined and modified. Some authors have reported exclusion rates of 56% of donor pairs, being the presence of sexually transmitted diseases, fever, or medications administered to the mother the most common reasons for deferral of potential donors.[52] Projecting these exclusion factors, other authors have estimated that up 71% of potential donors would be excluded for UCB donation.[53]

35.4.2 UCB Collection

The cord blood can be collected either in utero, before the delivery of the placenta, or ex utero, after placental delivery as shown in Sect. 35.3.1. The St Louis Cord Blood Bank and many others uses an obstetrician-based cord blood collection network, while the UK cord blood bank used trained technicians who collected the cord blood following the delivery of the placenta. Advantages and disadvantages of each method have been described before.

35.4.3 Volume Reduction

According to most standards, CB units must be processed before 48 h from the collection date. Many cord blood banks have set a TNC content ranging from 60 to 100 × 10^7 as minimum required values for processing the units. The proportion of UCB units discarded for cryopreservation ranges from 50% to 80%.[50] UCB units can be cryopreserved as whole blood without any previous processing method. However, reducing the volume of the CB unit, prior to storage, provides a substantial reduction in per unit cost of the cryogenic space, cryopreservation requires smaller volumes of dimethyl sulfoxide (DMSO) and the red cell content is significantly depleted. In addition, the waste products from the reduction process can be used for testing, permitting the bank to maximize the storage of the stem cells collected. Volume reduction provides benefits to the patient such as reducing the risks of ABO incompatibility and DMSO toxicity.[54] Many techniques have been tested for the volume reduction purposes, but to date the most widely used method is the hydroxyethyl starch (HES) sedimentation technique developed by Rubinstein et al. in New York.[55] Concentration of hematopoietic progenitors is achieved through two centrifugation steps. After the addition of 6% HES solution in a ratio of one part HES to five parts blood, the mixture is gently centrifugated at 50 g to produce leukocyte-rich plasma that is transferred to another bag. The leukocyte-rich plasma is further concentrated by a second centrifugation step at 400 g. This is a manual technique that can be performed in a closed system in an efficient way.

The top and bottom process have been used in many banks. This method uses differential centrifugation followed by either manual or automatic expression of excess RC and plasma, reducing the CB to a buffy coat. For this purpose, the CB must be collected in a triple-bag set. This closed-system does not require the addition of exogenous material, and is compatible with large-scale CB banking.[56]

A method using disposable leukocyte filters have been described by several authors.[57,58] This system was designed for the CB to flow through the filter by gravity, NC are trapped on the filter while the RBC and platelets flow through into a drain bag. Although filtration has showed potential for use in CB banking, to date, they have not been introduced into large-scale routine processing.

Two different automatic devices have been developed specifically for CB processing: Sepax and AXP. Sepax consists of a centrifuge and a pneumatic system, with vacuum or pressure capability to fill or empty the separation chamber and lines, using sterile, disposable processing kits with different configurations for dedicated protocols. The system includes two protocols for CB volume reduction suitable for therapeutic banking: the most used based on HES and the second one isolates a buffy coat component without the addition of additives.[59] The AXP™ AutoXpress platform is an automated fully closed system, specifically designed to reduce CB to a precise volume. This system has been developed with integrated sampling capability and a cryoprotectant line incorporating a sterile filter, potentially providing a truly closed system.[54]

35.4.4 Cryopreservation and Storage of the Cord Blood Unit

Cryopreservation is a critical issue for a long-term maintenance of UCB viability and colony-forming capacities. Cryopreservation methods for UCB have been based on techniques established for hematopoietic progenitor cells from bone marrow or peripheral blood.[60] Currently, the most widely used system for UCB cryopreservation is controlled rate freezing (CRF) of 1°C/min in programmed devices.[61] Uncontrolled-rate freezing (URF) in −80°C mechanical freezers is an attractive alternative aimed at reducing costs and ease of routine freezing procedures that has been successfully performed in some cord blood banks.[62]

Recovery of hematopoietic progenitor cells has been studied in cord blood cells frozen for up to 15 years.[63]

35.4.5 Biological Controls

The cord blood units are HLA tested, usually by molecular methods. A nucleated cell count, $CD34^+$ testing, and CFU assays are tested in the cord blood after processing. The lack of standardization among different banks for the $CD34^+$ and CFU assays makes comparison of cord blood units among banks difficult.

Blood from the maternal donor is tested for infectious disease markers, including tests for syphilis, HIV, hepatitis B, hepatitis C, and also for cytomegalovirus antibodies. In some banks the mother and the baby are reevaluated at 6 months after delivery, and a current medical history and repeated infectious disease tests are obtained.[64]

35.4.6 *Release of Cord Blood Unit to Transplant Center*

Possible HLA-matched cord blood units for patients are found via computerized registries. In Europe, NETCORD has 60,000 units listed. UCB units are shipped from the bank to the transplant center using a dry-shipper that allows the unit to remain frozen for at least 48 h and can maintain temperatures below −120°C for 14 days. The unit can be then thawed and infused to the receptor to ensure maximum viability of the cells infused on the day of transplant.

The UCB units can be thawed and directly infused or washed following the technique described by the NY Cord Blood Bank, using dextran and albumin.[55]

35.5 UCB Transplantation Outcomes

As a consequence of the first successful UCB transplant, large-scale clinical trials to evaluate allogeneic UCB transplantation for a variety of diseases were soon established in many centers.

35.5.1 *Allogeneic Transplantation*

Table 35.4 shows the results of selected clinical trials performing unrelated UCB transplantation.

35.5.1.1 Pediatric Studies

Since UCB contains 1–2 logs fewer nucleated cells than a unit of bone marrow, pediatric patients were primarily chosen as the first recipients of UCB transplants.

Table 35.4 Outcomes of UCB transplantation. Results of selected clinical trials

Author (reference)	*n*	Median follow-up (months)	Disease-free survival (%)
Pediatric patients			
Kurtzberg et al.[68]	25	13	48
Gluckman et al.[66]	65	10	29
Wagner et al.[69]	102	32	47
Adult patients			
Laughlin et al.[71]	68	22	26
Sanz et al.[72]	22	8	53
Ooi et al.[73]	18	18	76
Laughlin et al.[77]	116	40	23

Related Donor Transplantation

UCB has been used successfully in related transplants for both malignant and nonmalignant diseases.[65] The Eurocord group reported related and unrelated UCB transplantation in children from 45 centers showing the importance of cell dose as a predictor of neutrophil and platelet engraftment. A nucleated cell dose more than 3.7×10^7/kg correlated with engraftment.[66] The same group analyzed their results for related cord blood transplant for children with sickle cell anemia and thalassemia.[67] The 2-year probability of event-free survival was 70% for thalassemia and 90% for sickle cell anemia.

Unrelated Donor Cord Blood Transplantation

The first unrelated cord blood transplantations were reported in 1996 by Kurtzberg et al.[68] Of these patients, 24 were children. Of 55 patients, 23 engrafted and the event-free survival was 48% with a median follow-up of 12 months. Other studies have shown similar results. Rubinstein et al. from the New York Blood Center reported on 562 cases that underwent transplantation in a variety of centers with different transplantation protocols. Younger age and a higher nucleated cell dose per kilogram infused correlated with improved engraftment and survival. Another study from the University of Minnesota performed transplantations in 102 children with unrelated umbilical cord blood.[69]

The incidence of neutrophil engraftment was 88% and platelet engraftment 65%. The incidence of graft-versus-host disease was low. The results from these studies indicate that cord blood transplantation can be successful, even if the patient and cord blood donor are mismatched at two antigens. The incidence of graft-versus-host disease is low, but engraftment is delayed. The cell dose infused is an important marker for improved engraftment and survival.[70]

35.5.1.2 Adults

In the initial reports of UCB transplantation in adults, the results were poor. In those reports the median UCB TNC dose per kilogram for adult patients was approximately half of the cell doses for pediatric patients. Laughlin et al.[71] analyzed the outcome of 68 adults who received UCB transplants. Transplant-related mortality was very high. A high number of $CD34^+$ cells in the graft correlated with an improved outcome. Better results have been reported by the Spanish and Japanese groups.[72,73] The Spanish group reported 22 patients, median age 29, who received cord blood transplants. Disease-free survival was 53% at 1 year. Younger age was correlated with improved survival.[72] In the Japanese group of 18 patients reported, the 2-year probability of disease-free survival was 76%.[73]

The Eurocord group has recently analyzed the impact of diagnosis, cell dose, and HLA incompatibilities in patients receiving a single UCB.[74] In malignant disease group (925 patients), the cell dose was the most important factor for outcome. The number of HLA mismatches increased the risk of delayed engraftment and led to a higher incidence of transplant-related mortality. Patients with nonmalignant diseases must receive a higher cell dose to obtain engraftment and HLA matches played a major role for engraftment, graft-versus-host disease (GVHD) and survival.

In general, results of unrelated cord blood transplantation in adults show the engraftment can be achieved, but is delayed compared to unrelated bone marrow. Early transplant-related mortality is high, mostly due to infection. The cell dose (NC/kg or $CD34^+$ cells/kg) correlated with outcome in most studies. While a cell dose of 1×10^7/kg was considered to be the lower acceptable limit in 1996,[69] currently 2.5×10^7/kg is considered to be the lower limit.[75] An increasing body of evidence suggests that better HLA-matched UCB grafts positively influence outcomes.[76] Recently a study from the International Bone Marrow Transplant Registry compared survival after unrelated UCB transplantations to survival after bone marrow transplantation.[77] The results of this study suggest that outcomes are improved for patients receiving a matched unrelated bone marrow transplant; recipients of cord blood and 1-antigen-mismatched unrelated bone marrow had similar survivals.

35.5.2 Autologous Transplantation

Use of autologous cord blood when available is challenging and controversial matter. Autologous/private cord blood banks with commercial interests have been established around the world, although some scientific societies encourage the donors to donate the UCB to public UCB banks.[78] Some scientific arguments against commercial UCB banking are that the likelihood that the stored blood will be used is very low (ranging from 1/4,000 and 1/20,000) and that in some malignancies autologous transplantation cannot be the best therapeutic option.[79] Currently in the USA, private UCB collection and storage is performed in less than 3% of all deliveries. To date, only three cord blood transplantations have been reported in one patient with neuroblastoma,[80] one patient with aplastic anemia,[81] and one patient with acute leukemia.[82]

References

1. Gluckman E. Current status of umbilical cord blood hematopoietic stem cell transplantation. *Exp Hematol.* 2000;28:1197-1205.
2. Sanz MA, Sanz GF. Unrelated donor umbilical cord blood transplantation in adults. *Leukemia.* 2002;16:1984-1991.
3. Rocha V, Sanz G, Gluckman E. Umbilical cord blood transplantation. *Curr Opin Hematol.* 2004;33:559-569.
4. Tse W, Laughlin M. Cord blood transplantation in adult patients. *Cytotherapy.* 2005;7:228-242.
5. Eapen M, Rubinstein P, Zhang MJ, et al. Outcomes of transplantation of unrelated donor umbilical cord blood and bone marrow in children with acute leukaemia: a comparison study. *Lancet.* 2007;369:1947-1954.
6. Sanchez-Ramos J. Stem cells from umbilical cord blood. *Semin Reprod Med.* 2006;24:358-369.
7. Gluckman E, Broxmeyer HE, Auerbach AD, et al. Hematopoietic reconstitution in a patient with Fanconi's anemia by means of umbilical-cord blood from an HLA-identical sibling. *N Engl J Med.* 1989;321:1174-1178.

8. Tse W, Laughlin MJ. Umbilical cord blood transplantation: a new alternative option. *Hematology Am Soc Hematol Educ Program*. 2005:377–383.
9. Gluckman E, Rocha V. Donor selection for unrelated cord blood transplants. *Curr Opin Immunol*. 2006;18:565-570.
10. Brunstein CG, Baker KS, Wagner JE. Umbilical cord blood transplantation for myeloid malignancies. *Curr Opin Hematol*. 2007;14:162-169.
11. Barker JN, Wagner JE. Umbilical cord blood transplantation: current state of the art. *Curr Opin Oncol*. 2002;14: 160-164.
12. Broxmeyer HE, Cooper S, Li ZH, et al. Myeloid progenitor cell regulatory effects of vascular endothelial cell growth factor, a ligand for the tyrosine kinase receptor Flk-1. *Int J Hematol*. 1995;62:203-215.
13. Broxmeyer HE. Phenotypic and proliferative characteristics of cord blood hematopoietic stem and progenitor cells and gene transfer. In: *Cellular Characteristics of Cord Blood and Cord Blood Transplantation*. 1st ed. Bethesda, MD: AABB Press;1998.
14. Knudtzon S. In vitro growth of granulocytic colonies from circulating cells in human cord blood. *Blood*. 1974;43: 357-361.
15. Broxmeyer HE, Douglas GW, Hangoc G, et al. Human umbilical cord blood as a potential source of transplantable hematopoietic stem/progenitor cells. *Proc Natl Acad Sci USA*. 1989;86:3828-3832.
16. Aroviita P, Teramo K, Westman P, et al. Associations among nucleated cell, $CD34^+$ cell and colony-forming cell contents in cord blood units obtained through a standardized banking process. *Vox Sang*. 2003;84:219-227.
17. George TJ, Sugrue MW, George SN, et al. Factors associated with parameters of engraftment potential of umbilical cord blood. *Transfusion*. 2006;46:1803-1812.
18. Stanworth S, Warwick R, Fehily D, et al. An international survey of unrelated umbilical cord blood banking. *Vox Sang*. 2001;80:236-243.
19. Elchalal U, Fasouliotis SJ, Shtockheim D, et al. Postpartum umbilical cord blood collection for transplantation: a comparison of three methods. *Am J Obstet Gynecol*. 2000;182:227-332.
20. Surbek DV, Aufderhaar U, Holzgreve W. Umbilical cord blood collection for transplantation: which technique should be preferred? *Am J Obstet Gynecol*. 2000;183:1587-1588.
21. Surbek DV, Schonfeld B, Tichelli A, et al. Optimizing cord blood mononuclear cell yield: a randomized comparison of collection before vs after placenta delivery. *Bone Marrow Transplant*. 1998;22:311-312.
22. Wong A, Yuen PM, Li K, et al. Cord blood collection before and after placental delivery: levels of nucleated cells, haematopoietic progenitor cells, leukocyte subpopulations and macroscopic clots. *Bone Marrow Transplant*. 2001;27: 133-138.
23. Solves P, Moraga R, Saucedo E, et al. Comparison between two strategies for umbilical cord blood collection. *Bone Marrow Transplant*. 2003;31:269-273.
24. Askari S, Miller J, Chrysler G, et al. Impact of donor and collection related variables on product quality in ex utero cord blood banking. *Transfusion*. 2005;45:189-194.
25. Surbek DV, Visca E, Steinmann C, et al. Umbilical cord blood collection before placental delivery during cesarean delivery increases cord blood volume and nucleated cell number available for transplantation. *Am J Obstet Gynecol*. 2000;183(1):218-221.
26. Solves P, Fillol M, Lopez M, et al. Mode of collection does not influence haematopoietic content of umbilical cord blood units from caesarean deliveries. *Gynecol Obstet Invest*. 2006;61:34-39.
27. Sparrow RL, Cauchi JA, Ramadi LT, et al. Influence of mode of birth and collection on WBC yields of umbilical cord blood units. *Transfusion*. 2002;42:210-215.
28. Yamada T, Okamoto Y, Kasamatu H, et al. Factors affecting the volume of umbilical cord blood collections. *Acta Obstet Gynecol Scand*. 2000;79:830-833.
29. Grisaru D, Deutsch V, Pick M. Placing the newborn on the maternal abdomen after delivery increases the volume and CD34 cell content in the umbilical cord blood collected: an old maneuver with new applications. *Am J Obstet Gynecol*. 1999;180:1240-1243.
30. Pafumi C, Zizza G, Russo A, et al. Placing the newborn on the maternal abdomen increases the volume of umbilical cord blood collected. *Gynecol Obstet Invest*. 2001;23:397-399.
31. Lasky LC, Lane TA, Miller JP, et al. In utero or ex utero cord blood collection: which is better? *Transfusion*. 2002;42: 1261-1267.
32. Wall DA, Noffsinger JM, Mueckl KA, et al. Feasibility of an obstetrician-based cord blood collection network for unrelated donor umbilical cord blood banking. *J Matern Fetal Med*. 1997;6:320-323.
33. Migliaccio G, Biaocchi M, Hamel N, et al. Circulating progenitor cells in human ontogenesis: response to growth factors and replating potential. *J Hematother*. 1996;5:161-170.
34. Kawamura T, Toyabe S, Moroda T, et al. Neonatal granulopoiesis is a postpartum event which is seen in the liver as well as in the blood. *Hepatology*. 1997;26:1567-1572.
35. Shlebak AA, Roberts IA, Stevens TA, et al. The impact of antenatal and perinatal variables on cord blood haematopoietic stem/progenitor cell yield available for transplantation. *Br J Haematol*. 1998;103:1167-1171.
36. Ballen KK, Wilson M, Wuu J, et al. Bigger is better: maternal and neonatal predictors of hematopoietic potential of umbilical cord blood units. *Bone Marrow Transplant*. 2001;27:7-14.
37. Aufderhaar U, Holzgreve W, Danzer E, et al. The impact of intrapartum factors on umbilical cord blood stem cell banking. *J Perinat Med*. 2003;31:317-322.
38. Jones J, Stevens CE, Rubinstein P, et al. Obstetric predictors of placental/umbilical cord blood volume for transplantation. *Am J Obstet Gynecol*. 2003;188:503-509.
39. Mohyeddin MA, Alimoghaddam KA, Goliaei ZA, Ghavamzadeh AR, et al. Which factors can affect cord blood variables? *Transfusion*. 2004;5:690-693.
40. Nakagawa R, Watanabe T, Kawano Y, et al. Analysis of maternal and neonatal factors that influence the nucleated and $CD34^+$ cell yield for cord blood banking. *Transfusion*. 2004;44:262-267.
41. Donaldson C, Armitage WJ, Laundy V, et al. Impact of obstetric factors on cord blood donation for transplantation. *B J Haematol*. 1999;106:128-132.
42. Surbek DV, Holzgreve W, Steinmann C, et al. Preterm birth and the availability of cord blood for HPC transplantation. *Transfusion*. 2000;40:817-820.

43. Solves P, Larrea L, Soler MA, et al. Relationship between gestational age and cord blood quality. *Transfusion.* 2001;41:302-303.
44. FACT. *NetCord/FAHCT International Standards for Cord Blood Collection, Processing, Testing, Banking, Selection and Release.* 2nd ed. Omaha, NE: FACT;2001.
45. Lim FT, Scherjon SA, van Beckhoven JM, et al. Association of stress during delivery with increased numbers of nucleated cells and hematopoietic progenitor cells in umbilical cord blood. *Am J Obstet Gynecol.* 2000;183:1144-1151.
46. Ferber A, Grassi A, Akyol D, et al. The association of fetal heart rate patterns with nucleated red blood cell count at birth. *Am J Obstet Gynecol.* 2003;188:1228-1230.
47. Solves P, Perales A, Moraga R, et al. Maternal, neonatal and collection factors influencing the haematopoietic content of cord blood units. *Acta Haematol.* 2005;113(4):241-246.
48. Solves P, Mirabet V, Perales A, et al. Newborns' sex and hematopoietic progenitor cell content of cord blood. *Transfusion.* 2005;45(11):1828.
49. Aroviita P, Teramo K, Hiilesmaa V, et al. Cord blood hematopoietic progenitor cell concentration and infant sex. *Transfusion.* 2005;45:613-621.
50. Solves P, Perales A, Mirabet V, et al. Optimizing donor selection in a cord blood bank. *Eur J Haematol.* 2004;72(2): 107-112.
51. Ballen K, Broxmeyer HE, McCullough J, et al. Current status of cord blood banking and transplantation in the United States and Europe. *Biol Blood Marrow Transplant.* 2001;7:635.
52. Jefferies LC, Albertus M, Morgan MA, et al. High deferral rate for maternal-neonatal donor pairs for an allogeneic umbilical cord blood bank. *Transfusion.* 1999;40:122-123.
53. Lecchi L, Ratti I, Lazzari L, et al. Reasons for discard of umbilical cord blood units before cryopreservation. *Transfusion.* 2000;40:122-123.
54. Armitage S. Cord blood processing: volume reduction. *Cell Preservation Technol.* 2006;4:9-16.
55. Rubinstein P, Dobrilla L, Rosenfield RE, et al. Processing and cryopreservgation of placental/umbilical cord blood for unrelated bone marrow reconstitution. *Proc Natl Acad Sci USA.* 1995;13:533-540.
56. Ademokun JA, Chapman C, Dunn J, et al. Umbilical cord blood collection and separation for haematopoietic progenitor cell banking. *Bone Marrow Transplant.* 1997;19:1023-1028.
57. Eichler H, Kern S, Beck C, et al. Engraftment capacity of umbilical cord blood cells processed by either whole blood preparation or filtration. *Stem Cells.* 2003;21:208-216.
58. Yasutake M, Sumita M, Terashima S, et al. Stem cell collection filter system for human placental/umbilical cord blood processing. *Vox Sang.* 2001;80:101-105.
59. Querol S, Azqueta ,Torrico C, et al. An automated and closed procedure for cord blood processing, freezing and thawing: the Barcelona cord blood bank experience [Abstract]. 30th Annual Meeting of the European Group for Blood and Marrow Transplantation. 2004, Barcelona, Spain.
60. Wagner JE, Broxmeyer HE, Cooper S. Umbilical cord and placental blood hematopoietic stem cells: collection, cryopreservation and storage. *J Hematother.* 1992;1:167-173.
61. Donaldson C, Armitage WJ, Denning-Kendall PA, et al. Optimal cryopreservation of human umbilical cord blood. *Bone Marrow Transplant.* 1996;18:725-731.
62. Itoh T, Minegishi M, Fushimi J, et al. A simple controlled-rate freezing method without a rate-controlled programmed freezer provides optimal conditions for both large-scale and small-scale cryopreservation of umbilical cord blood cells. *Transfusion.* 2003;43:1303-1308.
63. Broxmeyer HE, Srour EF, Hangoc G, et al. High-efficiency recovery of functional hematopoietic progenitor and stem cells from human cord blood cryopreserved for 15 years. *Proc Natl Acad Sci USA.* 2003;100:645-650.
64. Lecchi L, Rebulla P, Ratti I, et al. Outcomes of a program to evaluate mother and baby 6 months after umbilical cord blood donation. *Transfusion.* 2001;41:606-610.
65. Wagner JE, Kernan NA, Steinbuch M, et al. Allogeneic sibling umbilical cord blood transplantation in children with malignant and non malignant disease. *Lancet.* 1995; 346:214-219.
66. Gluckman E, Rocha V, Boyer-Chammard A, et al. Outcome of cord blood transplantation from related and unrelated donors. Eurocord Transplant Group and the European Blood and Marrow Transplantation group. *N Engl J Med.* 1997; 337:373-381.
67. Locatelli F, Rocha V, Reed W, et al. Related umbilical cord blood transplantation in patients with thalassemia and sickle cell disease. *Blood.* 2003;101:2137-2143.
68. Kurtzberg J, Laughlin M, Graham ML, et al. Placental blood as a source of hemnatopoietic stem cells for transplantation into unrelated recipients. *N Engl J Med.* 1996;335:157-166.
69. Wagner JE, Barker JN, DeFor TE, et al. Transplantation of unrelated donor umbilical cord blood in 102 patients with malignant and non malignant diseases: influence of CD34 cell dose and HLA disparity on treatment-related mortality and survival. *Blood.* 2002;100:1611-1618.
70. Ballen KK. New trends in umbilical cord blood transplantation. *Blood.* 2005;105:3786-3792.
71. Laughlin MJ, Barker J, Bambach B, et al. Hematopoietic engraftment and survival in adult recipients of umbilical cord blood from unrelated donors. *N Engl J Med.* 2001; 344:1815-1822.
72. Sanz GF, Saavedra S, Planelles D, et al. Standardized, unrelated donor cord blood transplantation in adults with hematologic malignancies. *Blood.* 2001;98:2332-2338.
73. Ooi J, Iseki T, Takahashi S, et al. Unrelated cord blood transplantation for adult patients with de novo acute myeloid leukaemia. *Blood.* 2004;103:489-491.
74. Gluckman E, Rocha V. Donor selection for unrelated cord blood transplants. *Current Opinion Immunol.* 2006;18: 565-570.
75. Rubinstein P, Stevens CE. Placental blood for bone marrow replacement: The New York Blood Center's program and clinical results. *Bailliere's Best Pract Res Clin Haematol.* 2000;13:565-584.
76. Brunstein CG, Wagner JE. Cord blood transplantation for adults. *Vox Sang.* 2006;91:195-205.
77. Laughlin MJ, Eapen M, Rubinstein P, et al. Outcomes after transplantation of cord blood or bone marrow from unrelated donors in adults with leukemia. *N Engl J Med.* 2004;351:2265-2275.
78. Carpenter RJ Jr. Commercial cord blood banking: public cord blood banking should be more widely adopted. *BMJ.* 2006;333(7574):919.

79. Edozien LC. NHS maternity units should not encourage commercial banking of umbilical cord blood. *BMJ.* 2006;333(7572):801-804.
80. Ferreira E, Pasternak J, Bacal N, et al. Autologous cord blood transplantation. *Bone Marrow Transplant.* 1999;24:1041.
81. Fruchtman SM, Hurlet A, Dracker R, et al. The successful treatment of severe aplastic anemia with autologous cord blood transplantation. *Biol Blood Marrow Transplant.* 2004; 10:741-742.
82. Hayani A, Lampeter E, Viswanatha D, et al. First report of autologous cord blood transplantation in the treatment of a child with leukemia. *Pediatrics.* 2007;119:296-300.

Part XII

Clinical Use of Amniotic Fluid

36 Amniotic Fluid and Placenta Stem Cells

Anthony Atala

36.1 Introduction

Amniotic fluid-derived progenitor cells can be isolated from a small amount of the fluid obtained during amniocentesis, a procedure that is already performed in many pregnancies to screen for congenital abnormalities. Placenta-derived stem cells can be obtained from a small biopsy of the chorionic villi. Cell culture experiments with these two types of cells have provided evidence that they may have the potential to differentiate into various cell types, including adipogenic, osteogenic, myogenic, endothelial, neurogenic, hepatogenic, cardiac, and pancreatic lineages. In this respect, they meet a commonly accepted criterion for pluripotent stem cells, without implying that they can generate every adult tissue. This suggests that amniotic fluid and placenta contain a novel type of pluripotent cell that could one day be used in research and treatment. In particular, neurons, hepatocytes, and osteoblasts are among the cell types for which improved stem cell sources may open up novel therapeutic applications.

36.2 Amniotic Fluid and Placenta in Developmental Biology

Gastrulation is a major milestone in early postimplantation development. At about embryonic day 6.5 (E6.5), gastrulation begins in the posterior region of the embryo. Pluripotent epiblast cells are allocated to the three primary germ layers of the embryo (ectoderm, mesoderm, and endoderm) and germ cells, which are the progenitors of all fetal tissue lineages, as well as to the extraembryonic mesoderm of the yolk sac, amnion, and allantois. The latter forms the umbilical cord as well as the mesenchymal part of the labyrinthine layer in the mature chorioallantoic placenta. The final positions of the fetal membranes result from the process of embryonic turning, which occurs around day 8.5 of gestation and "pulls" the amnion and yolk sac around the embryo. At this time, the specification of tissue lineages is accomplished by the restriction of developmental potency and the activation of lineage-specific gene expression. This process is strongly influenced by cellular interactions and signaling.

The amniotic sac is a tough but thin transparent pair of membranes that holds the developing fetus until shortly before birth. The inner membrane, the amnion, contains the amniotic fluid and the fetus. The outer membrane, the chorion, envelops the amnion and is part of the placenta. The amnion is derived from ectoderm and mesoderm, and as it grows, it begins to fill with fluid. Originally this fluid is isotonic, containing proteins, carbohydrates, lipids, phospholipids, urea, and electrolytes. Later, urine excreted by the fetus increases its volume and changes its composition. The presence of the amniotic fluid ensures symmetrical structure development and growth. In addition, it cushions and protects the embryo, helps maintain consistent pressure and temperature, and permits freedom of fetal movement, which is important for musculoskeletal development and blood flow. The fetus can breathe in the amniotic fluid, allowing normal growth and development of the lungs. The fluid is also swallowed by the fetus, allowing the gastrointestinal tract to

A. Atala
Department of Urology, Wake Forest Institute for Regenerative Medicine, Wake Forest University Health Sciences, Medical Center Boulevard, 5th Floor Watlington Hall, Winston-Salem, NC 27157
e-mail: aatala@wfubmc.edu

N. Bhattacharya and P. Stubblefield (eds.), *Regenerative Medicine Using Pregnancy-Specific Biological Substances*,
DOI: 10.1007/978-1-84882-718-9_36,

develop, and as a result, components of the fluid pass via the fetal blood into the maternal blood.

Amniotic fluid also contains a mixture of different cell types. In 2002, it was reported that a subpopulation of cells cultured from amniotic fluid as well as from placenta have the potential to differentiate into various cell types, suggesting that they may contain potential stem cells. A number of different origins have been suggested for these cells. Cells of both embryonic and fetal origins and cells from all three germ layers have been reported to exist in amniotic fluid. However, studies have shown that the consistent presence of a Y chromosome in cell lines derived from cases in which the amniocentesis donor carried a male child implies that the actual amniotic fluid stem cells originate in the developing fetus. The cells are thought to be sloughed from the fetal amnion and skin, as well as the alimentary, respiratory, and urogenital tracts.

36.3 Isolation and Characterization of Progenitor Cells

Amniotic fluid progenitor cells are isolated by centrifugation of amniotic fluid from amniocentesis. Chorionic villi placental cells are isolated from single villi under light microscopy. Amniotic fluid cells and placental cells are allowed to proliferate in vitro and are maintained in culture for 4 weeks. The culture medium consists of modified Earl's medium, 18% Chang medium B, and 2% Chang medium C with 15% embryonic stem cell certified fetal bovine serum, antibiotics, and L-glutamine.

A pluripotential subpopulation of progenitor cells present in the amniotic fluid and placenta can be isolated through positive selection for cells expressing the membrane receptor c-kit. This receptor binds to the ligand stem cell factor. About 0.8–1.4% of cells present in amniotic fluid and placenta have been shown to be c-kit positive in an analysis by fluorescence-activated cell sorting (FACS). Progenitor cells maintain a round shape for 1 week post-isolation when cultured in nontreated culture dishes. In this state, they demonstrate low proliferative capability. After the first week the cells begin to adhere to the plate and change their morphology, becoming more elongated and proliferating more rapidly, reaching 80% confluence with a need for passage every 48–72 h. No feeder layers are required either for maintenance or expansion. The progenitor cells derived show a high self-renewal capacity with >300 population doublings, far exceeding Hayflick's limit. The doubling time of the undifferentiated cells is noted to be 36 h with little variation with passages.

These cells have been shown to maintain a normal karyotype at late passages and have normal G1 and G2 cell cycle checkpoints. They demonstrate telomere length conservation while in the undifferentiated state as well as telomerase activity even in late passages. Analysis of surface markers shows that progenitor cells from amniotic fluid express human embryonic stage-specific marker SSEA4, and the stem cell marker Oct4, and did not express SSEA1, SSEA3, CD4, CD8, CD34, CD133, C-MET, ABCG2, NCAM, BMP4, TRA1-60, and TRA1-81, to name a few. This expression profile is of interest as it demonstrates expression by the amniotic fluid-derived progenitor cells of some key markers of the embryonic stem cell phenotype, but not the full complement of markers expressed by embryonic stem cells. This may indicate that the amniotic cells are not quite as primitive as embryonic cells, yet maintain greater potential than most adult stem cells. Another behavior showing similarities and differences between these amniotic fluid- and blastocyst-derived cells is that while the amniotic fluid progenitor cells do form embryoid bodies in vitro, which stain positive for markers of all three germ layers, these cells do not form teratomas in vivo when implanted in immunodeficient mice. Last, cells, when expanded from a single cell, maintain similar properties in growth and potential as the original mixed population of the progenitor cells.

36.4 Differentiation of Amniotic Fluid- and Placenta-Derived Progenitor Cells

The progenitor cells derived from amniotic fluid and placenta are pluripotent and have been shown to differentiate into osteogenic, adipogenic, myogenic, neurogenic, endothelial, hepatic, and renal phenotypes in vitro. Each differentiation has been performed through proof of phenotypic and biochemical changes consistent with the differentiated tissue type of interest. We discuss each set of differentiations separately.

36.4.1 Adipocytes

To promote adipogenic differentiation, progenitor cells can be induced in dexamethasone, 3-isobutyl-1-methylxanthine, insulin, and indomethacin. Progenitor cells cultured with adipogenic supplements change their morphology from elongated to round within 8 days. This coincides with the accumulation of intracellular droplets. After 16 days in culture, more than 95% of the cells have their cytoplasm filled with lipid-rich vacuoles. Adipogenic differentiation also demonstrates the expression of peroxisome proliferation-activated receptor γ2 (PPARγ2), a transcription factor that regulates adipogenesis, and of lipoprotein lipase through reverse transcription-polymerase chain reaction (RT-PCR) analysis. Expression of these genes is noted in progenitor cells under adipogenic conditions but not in undifferentiated cells.

36.4.2 Osteocytes

Osteogenic differentiation was induced in progenitor cells with the use of dexamethasone, β-glycerophosphate, and ascorbic acid 2-phosphate. Progenitor cells maintained in this medium demonstrated phenotypic changes within 4 days, with a loss of their spindle-shaped phenotype and development of an osteoblast-like appearance with finger-like excavations into the cytoplasm. At 16 days, the cells aggregated, showing typical lamellar bone-like structures. In terms of functionality, these differentiated cells demonstrate a major feature of osteoblasts, which is to precipitate calcium. Differentiated osteoblasts from the progenitor cells are able to produce alkaline phosphatase (AP) and to deposit calcium, consistent with bone differentiation. Undifferentiated progenitor cells lacked this ability. Progenitor cells in osteogenic medium express specific genes implicated in mammalian bone development (AP, core-binding factor A1 [CBFA1], and osteocalcin) in a pattern consistent with the physiological analog. Progenitor cells grown in osteogenic medium show activation of the AP gene at each time point. Expression of CBFA1, a transcription factor specifically expressed in osteoblasts and hypertrophic chondrocytes and that regulates gene expression of structural proteins of the bone extracellular matrix, is highest in cells grown in osteogenic-inducing medium on day 8 and decreases slightly on days 16, 24, and 32. Osteocalcin is expressed only in progenitor cells under osteogenic conditions at 8 days.

36.4.3 Endothelial Cells

Amniotic fluid progenitor cells can be induced to form endothelial cells by culture in endothelial basal medium on gelatin-coated dishes. Full differentiation is achieved by 1 month in culture; however, phenotypic changes are noticed within 1 week of initiation of the protocol. Human-specific endothelial cell surface marker (P1H12), factor VIII (FVIII), and KDR (kinase insert domain receptor) are specific for differentiated endothelial cells. Differentiated cells stain positively for FVIII, KDR, and P1H12. Progenitor cells do not stain for endothelial-specific markers.

Amniotic fluid progenitor-derived endothelial cells, once differentiated, are able to grow in culture and form capillary-like structures in vitro. These cells also express platelet endothelial cell adhesion molecule 1 (PECAM-1 or CD31) and vascular cell adhesion molecule (VCAM), which are not detected in the progenitor cells on RT-PCR analysis.

36.4.4 Hepatocytes

For hepatic differentiation, progenitor cells are seeded on Matrigel- or collagen-coated dishes at different stages and cultured in the presence of hepatocyte growth factor, insulin, oncostatin M, dexamethasone, fibroblast growth factor 4, and monothioglycerol for 45 days. After 7 days of the differentiation process, cells exhibit morphological changes from an elongated to a cobblestone appearance. The cells show positive staining for albumin on day 45 postdifferentiation and also express the transcription factor HNF4? (hepatocyte nuclear factor 4?), the c-Met receptor, the multidrug resistance (MDR) membrane transporter, albumin, and ?-fetoprotein. RT-PCR analysis further supports albumin production. The maximum rate of urea production for hepatic differentiation-induced cells is upregulated to 1.21×10^3 ng of urea per hour per cell from 50 ng of urea per hour per cell for the control progenitor cell populations.

36.4.5 Myocytes

Myogenic differentiation is induced in amniotic fluid-derived progenitor cells by culture in medium containing horse serum and chick embryo extract on a thin gel coat of Matrigel. To initiate differentiation, the presence of 5-azacytidine in the medium for 24 h is necessary. Phenotypically, the cells can be seen to organize themselves into bundles that fuse to form multinucleated cells. These cells express sarcomeric tropomyosin and desmin, both of which are not expressed in the original progenitor population.

The development profile of cells differentiating into myogenic lineages interestingly mirrors a characteristic pattern of gene expression reflecting that seen with embryonic muscle development. With this protocol, Myf6 is expressed on day 8 and suppressed on day 16. MyoD expression is detectable at 8 days and suppressed at 16 days in progenitor cells. Desmin expression is induced at 8 days and increases by 16 days in progenitor cells cultured in myogenic medium.

36.4.6 Neuronal Cells

For neurogenic induction, amniotic progenitor cells are induced in dimethyl sulfoxide (DMSO), butylated hydroxyanisole, and neuronal growth factor. Progenitor cells cultured under neurogenic conditions change their morphology within the first 24 h. Two different cell populations are apparent: morphologically large flat cells and small bipolar cells. The bipolar cell cytoplasm retracts toward the nucleus, forming contracted multipolar structures. Over subsequent hours, the cells display primary and secondary branches and cone-like terminal expansions. Induced progenitor cells show a characteristic sequence of expression of neural-specific proteins. At an early stage the intermediate filament protein nestin, which is specifically expressed in neuroepithelial stem cells, is highly expressed. Expression of III-tubulin and glial fibrillary acidic protein (GFAP), markers of neuron and glial differentiation, respectively, increases over time and seems to reach a plateau at about 6 days. Progenitor cells cultured under neurogenic conditions show the presence of the neurotransmitter glutamic acid in the collected medium. Glutamic acid is usually secreted in culture by fully differentiated neurons.

36.4.7 Renal Cells

End-stage kidney disease has reached epidemic proportions in the USA. Currently, dialysis and allogenic renal transplant remain the only treatments for this disease, but there are significant drawbacks to each. Dialysis can prolong survival via replacement of filtration functions, but other kidney functions are not replaced, thus leading to long-term consequences such as anemia and malnutrition. Currently, renal transplantation is the only definitive treatment that can restore the entire function of the kidney, including filtration, production of erythropoietin, and 1,25-dihydroxyvitamin D3. However, transplantation presents with several limitations, such as a critical donor shortage, complications due to chronic immunosuppressive therapy, and graft failure.

Over the last decade, stem cells and their possible role in the construction of bioartificial organs such as the kidney have been an area of intense research. Despite their potential in regenerative medicine applications, cells such as embryonic stem cells have ethical concerns associated with their use, and certain types of research with these cells has been banned. Amniotic fluid stem cells, however, do not have these problems, and may represent an exciting new cell source for tissue engineering strategies.

In 2007, Perin et al. showed that AFSC could be induced to differentiate into renal cells when placed into an in vitro embryonic kidney environment. Human AFSCs were obtained from human male amniotic fluid and were labeled with either LacZ or green fluorescent protein (GFP) so that they could be tracked throughout the experiment. These labeled cells were microinjected into murine embryonic kidneys (12.5–18 days gestation) and maintained in a special co-culture system in vitro for 10 days. Using this technique, it was shown that the labeled hAFSCs remained viable throughout the experimental period, and, importantly, they were able to contribute to the development of various primordial kidney structures including the renal vesicle, C- and S-shaped bodies. Studies using reverse transcriptase polymerase chain reaction (RT-PCR) indicated that the implanted hAFSCs began to express early kidney markers such as zona occludens-1 (ZO-1), glial-derived neurotrophic factor, and claudin. Together, these data suggested that hAFSCs have the intrinsic ability to differentiate into a number of different cell types that make up the kidney. Therefore, hAFSCs could represent a potentially limitless, ethically neutral source of cells for tissue engineering and cell therapy applications aimed at regenerating failing renal tissue.

36.5 In Vivo Behavior of Amniotic Fluid Stem Cells

Our group has cultured amniotic fluid stem cells in neuronal differentiation medium for a time and then grafted them into the lateral cerebral ventricles of control mice and the ventricles of the *twitcher* mouse model, in which a progressive loss of oligodendrocytes leads to massive demyelination and neuronal loss. AFS cells integrated into the brains of both strains seamlessly, appeared morphologically indistinguishable from surrounding mouse cells, and survived efficiently for at least 2 months. Interestingly, more of the AFS cells integrated into the injured twitcher brains (70%) than into the normal brains (30%), hinting at the potential for CNS therapies. In this study the phenotypes of the implanted human cells were not assessed. However, the pattern of incorporation and morphologies of cells derived from the AFS cells appeared similar to those obtained previously in the same animal model after implantation of murine neural progenitor and stem-like cells. In that case, the donor-derived cells were identified as astrocytes and oligodendrocytes.

From a tissue engineering perspective, osteogenically differentiated AFS cells were embedded in an alginate/collagen scaffold and implanted subcutaneously into immunodeficient mice. By 18 weeks after implantation, highly mineralized tissues and blocks of bone-like material were observed in the recipient mice using micro CT. These blocks displayed a density somewhat greater than that of mouse femoral bone. This indicates that AFS cells could be used to engineer grafts for the repair of bone defects.

36.6 Amniotic Fluid and Placenta for Cell Therapy

Pluripotent stem cells are ideal for regenerative medicine applications, as they have the capability to differentiate in stages into a huge number of different types of human cells. The discovery of a stem cell population in the amniotic fluid offers a very promising alternative source of stem cells for cellular therapy. The full range of adult somatic cells that AFS cells can produce remains to be determined, but their ability to differentiate into cells of all three embryonic germ layers and their high proliferation rate are two advantages over most adult stem cell sources. AFS cells represent a new class of stem cells with properties somewhere between embryonic and adult stem cell types. However, unlike embryonic stem (ES) cells, AFS cells do not form teratomas, and this low risk of tumorigenicity would be advantageous for eventual therapeutic applications. In addition, these cells are easily obtained without destruction of embryos, and thus their use may avoid some of the ethical concerns surrounding the use of ES cells. Finally, AFS cells could be used for both autologous and allogenic therapy through matching of histocompatible donor cells with recipients.

Amniotic fluid cells can be obtained from a small amount of fluid during amniocentesis at the second trimester, a procedure that is already often performed in many of the pregnancies in which the fetus has a congenital abnormality and to determine characteristics such as sex. Kaviani and coworkers reported that "just 2 mL of amniotic fluid" can provide up to 20,000 cells, 80% of which are viable. Because many pregnant women already undergo amniocentesis to screen for fetal abnormalities, cells can be simply isolated from this fluid and banked for future use.

In addition, while scientists have been able to isolate and differentiate on average only 30% of mesenchymal stem cells (MSCs) extracted from a child's umbilical cord shortly after birth, the success rate for amniotic fluid-derived stem cells is close to 100%. Furthermore, with amniotic fluid cells, it takes 20–24 h to double the number of cells collected, which is faster than for umbilical cord stem cells (28–30 h) and bone marrow stem cells (more than 30 h). This phenomenon is an important feature for urgent medical conditions.

36.7 Conclusion

Pluripotent progenitor cells isolated from amniotic fluid and placenta present an exciting possible contribution to the field of stem cell biology and regenerative medicine. These cells are an excellent source for research and therapeutic applications. The ability to isolate progenitor cells during gestation may also be advantageous for babies born with congenital malformations. Furthermore, progenitor cells can be cryopreserved for future self-use. Compared with embryonic stem cells, progenitor cells isolated from amniotic fluid have many similarities: they can differentiate into all three germ layers, they express common markers, and they preserve their telomere length. However, progenitor cells isolated from amniotic fluid and placenta have considerable advantages. They easily differentiate into specific cell lineages and they avoid the current controversies associated with the use of human embryonic stem cells. The discovery of these cells has been recent, and a considerable amount of work remains to be done on the characterization and use of these cells. In future, cells derived from amniotic fluid and placenta may represent an attractive and abundant, noncontroversial source of cells for regenerative medicine.

Acknowledgment The author wishes to thank Dr. Jennifer Olson for editorial assistance with this manuscript.

References

1. Snow MH, Bennett D. Gastrulation in the mouse: assessment of cell populations in the epiblast of tw18/tw18 embryos. *J Embryol Exp Morphol*. 1978;47:39-52.
2. Downs KM, Harmann C. Developmental potency of the murine allantois. *Development*. 1997;124(14):2769-2780.
3. Downs KM, Hellman ER, McHugh J, Barrickman K, Inman KE. Investigation into a role for the primitive streak in development of the murine allantois. *Development*. 2004;131(1):37-55.
4. Gardner RL, Beddington RS. Multi-lineage 'stem' cells in the mammalian embryo. *J Cell Sci*. 1988;10(suppl):11-27.
5. Loebel DA, Watson CM, De Young RA, Tam PP. Lineage choice and differentiation in mouse embryos and embryonic stem cells. *Dev Biol*. 2003;264(1):1-14.
6. Moser M, Li Y, Vaupel K, et al. Placental failure and impaired vasculogenesis result in embryonic lethality for neuropathy target esterase-deficient mice. *Mol Cell Biol*. 2004;24(4):1667-1679.
7. Smith JL, Gesteland KM, Schoenwolf GC. Prospective fate map of the mouse primitive streak at 7.5 days of gestation. *Dev Dyn*. 1994;201(3):279-289.
8. Kinder SJ, Tsang TE, Quinlan GA, Hadjantonakis AK, Nagy A, Tam PP. The orderly allocation of mesodermal cells to the extraembryonic structures and the anteroposterior axis during gastrulation of the mouse embryo. *Development*. 1999;126(21):4691-4701.
9. Parameswaran M, Tam PP. Regionalisation of cell fate and morphogenetic movement of the mesoderm during mouse gastrulation. *Dev Genet*. 1995;17(1):16-28.
10. Rathjen J, Lake JA, Bettess MD, Washington JM, Chapman G, Rathjen PD. Formation of a primitive ectoderm like cell population, EPL cells, from ES cells in response to biologically derived factors. *J Cell Sci*. 1999;112(Pt 5):601-612.
11. Dang SM, Kyba M, Perlingeiro R, Daley GQ, Zandstra PW. Efficiency of embryoid body formation and hematopoietic development from embryonic stem cells in different culture systems. *Biotechnol Bioeng*. 2002;78(4):442-453.
12. Li L, Arman E, Ekblom P, Edgar D, Murray P, Lonai P. Distinct GATA6- and laminin-dependent mechanisms regulate endodermal and ectodermal embryonic stem cell fates. *Development*. 2004;131(21):5277-5286.
13. Kaviani A, Perry TE, Dzakovic A, Jennings RW, Ziegler MM, Fauza DO. The amniotic fluid as a source of cells for fetal tissue engineering. *J Pediatr Surg*. 2001;36(11):1662-1665.
14. Robinson WP, McFadden DE, Barrett IJ, et al. Origin of amnion and implications for evaluation of the fetal genotype in cases of mosaicism. *Prenat Diagn*. 2002;22(12):1076-1085.
15. Bartha JL, Romero-Carmona R, Comino-Delgado R, Arce F, Arrabal J. Alpha-fetoprotein and hematopoietic growth factors in amniotic fluid. *Obstet Gynecol*. 2000;96(4):588-592.
16. Heidari Z, Isobe K, Goto S, Nakashima I, Kiuchi K, Tomoda Y. Characterization of the growth factor activity of amniotic fluid on cells from hematopoietic and lymphoid organs of different life stages. *Microbiol Immunol*. 1996;40(8):583-589.
17. Sakuragawa N, Elwan MA, Fujii T, Kawashima K. Possible dynamic neurotransmitter metabolism surrounding the fetus. *J Child Neurol*. 1999;14(4):265-266.
18. Srivastava MD, Lippes J, Srivastava BI. Cytokines of the human reproductive tract. *Am J Reprod Immunol*. 1996;36(3):157-166.
19. Baschat AA, Hecher K. Fetal growth restriction due to placental disease. *Semin Perinatol*. 2004;28(1):67-80.
20. Prusa AR, Hengstschlager M. Amniotic fluid cells and human stem cell research: a new connection. *Med Sci Monitor*. 2002;8(11):RA253-RA257.
21. Medina-Gomez P, del Valle M. Cultivo de celas de liquido amniotico. Analisis de colonias, metafases e indice mitotico, con fin de descartar contaminacion de celulas maternas. *Gynecol Obstet Mex*. 1988;56:122-126.

22. In 't Anker PS, Scherjon SA, Kleijburg-van der Keur C, et al. Amniotic fluid as a novel source of mesenchymal stem cells for therapeutic transplantation. *Blood.* 2003;102(4): 1548-1549.
23. Prusa AR, Marton E, Rosner M, et al. Neurogenic cells in human amniotic fluid. *Am J Obstet Gynecol.* 2004;191(1): 309-314.
24. De Coppi P, Bartsch G Jr, Siddiqui MM, et al. Isolation of amniotic stem cell lines with potential for therapy. [see comment]. *Nat Biotechnol.* 2007;25(1):100-106.
25. Siddiqui MJ, Atala A. Aminiotic fluid derived pluripotential cells. In: *Handbook of Stem Cells.* San Diego, CA: Elsevier Academic Press; 2004:175-179.
26. Bryan TM, Englezou A, Dunham MA, Reddel RR. Telomere length dynamics in telomerase-positive immortal human cell populations. *Exp Cell Res.* 1998;239(2):370-378.
27. Cremer M, Schachner M, Cremer T, Schmidt W, Voigtlander T. Demonstration of astrocytes in cultured amniotic fluid cells of three cases with neural-tube defect. *Hum Genet.* 1981;56(3):365-370.
28. Jaiswal N, Haynesworth SE, Caplan AI, Bruder SP. Osteogenic differentiation of purified, culture-expanded human mesenchymal stem cells in vitro. *J Cell Biochem.* 1997;64(2):295-312.
29. Karsenty G. Role of Cbfa1 in osteoblast differentiation and function. *Semin Cell Dev Biol.* 2000;11(5):343-346.
30. Komori T, Yagi H, Nomura S, et al. Targeted disruption of Cbfa1 results in a complete lack of bone formation owing to maturational arrest of osteoblasts [see comment]. *Cell.* 1997;89(5):755-764.
31. Delo DM, De Coppi P, Bartsch G Jr, Atala A. Amniotic fluid and placental stem cells. *Method Enzymol.* 2006;419:426-438.
32. Dunn JC, Yarmush ML, Koebe HG, Tompkins RG. Hepatocyte function and extracellular matrix geometry: long-term culture in a sandwich configuration. *FASEB J.* 1989;3(2):174-177 [erratum appears in *FASEB J* May 1989;3(7):1873].
33. Schwartz RE, Reyes M, Koodie L, et al. Multipotent adult progenitor cells from bone marrow differentiate into functional hepatocyte-like cells. *J Clin Invest.* 2002;109(10): 1291-1302.
34. Hamazaki T, Iiboshi Y, Oka M, et al. Hepatic maturation in differentiating embryonic stem cells in vitro. *FEBS Lett.* 2001;497(1):15-19.
35. Rosenblatt JD, Lunt AI, Parry DJ, Partridge TA. Culturing satellite cells from living single muscle fiber explants. *In Vitro Cell Dev Biol Anim.* 1995;31(10):773-779.
36. Bailey P, Holowacz T, Lassar AB. The origin of skeletal muscle stem cells in the embryo and the adult. *Curr Opin Cell Biol.* 2001;13(6):679-689.
37. Rohwedel J, Maltsev V, Bober E, Arnold HH, Hescheler J, Wobus AM. Muscle cell differentiation of embryonic stem cells reflects myogenesis in vivo: developmentally regulated expression of myogenic determination genes and functional expression of ionic currents. *Dev Biol.* 1994;164(1):87-101.
38. Hinterberger TJ, Sassoon DA, Rhodes SJ, Konieczny SF. Expression of the muscle regulatory factor MRF4 during somite and skeletal myofiber development. *Dev Biol.* 1991; 147(1):144-156.
39. Patapoutian A, Yoon JK, Miner JH, Wang S, Stark K, Wold B. Disruption of the mouse MRF4 gene identifies multiple waves of myogenesis in the myotome. *Development.* 1995;121(10):3347-3358.
40. Black IB, Woodbury D. Adult rat and human bone marrow stromal stem cells differentiate into neurons. *Blood Cell Mol Dis.* 2001;27(3):632-636.
41. Woodbury D, Schwarz EJ, Prockop DJ, Black IB. Adult rat and human bone marrow stromal cells differentiate into neurons. *J Neurosci Res.* 2000;61(4):364-370.
42. Guan K, Chang H, Rolletschek A, Wobus AM. Embryonic stem cell-derived neurogenesis. Retinoic acid induction and lineage selection of neuronal cells. *Cell Tissue Res.* 2001;305(2):171-176.
43. Carpenter MK, Inokuma MS, Denham J, Mujtaba T, Chiu CP, Rao MS. Enrichment of neurons and neural precursors from human embryonic stem cells. *Exp Neurol.* 2001;172(2): 383-397.
44. Chazan JA, Libbey NP, London MR, Pono L, Abuelo JG. The clinical spectrum of renal osteodystrophy in 57 chronic hemodialysis patients: a correlation between biochemical parameters and bone pathology findings. *Clin Nephrol.* 1991;35(2):78-85.
45. Cohen J, Hopkin J, Kurtz J. *Infectious Complications After Renal Transplantation.* Philadelphia, PA: W.B. Saunders; 1994.
46. Ojo AO, Hanson JA, Wolfe RA, Leichtman AB, Agodoa LY, Port FK. Long-term survival in renal transplant recipients with graft function. *Kidney Int.* 2000;57(1):307-313.
47. Ojo AO, Hanson JA, Meier-Kriesche H, et al. Survival in recipients of marginal cadaveric donor kidneys compared with other recipients and wait-listed transplant candidates. *J Am Soc Nephrol.* 2001;12(3):589-597.
48. Perin L, Giuliani S, Jin D, et al. Renal differentiation of amniotic fluid stem cells. *Cell Prolif.* 2007;40(6):936-948.
49. De Coppi P, Bartsch G Jr, Siddiqui MM, et al. Isolation of amniotic stem cell lines with potential for therapy. *Nat Biotechnol.* 2007;25(1):100-106.
50. Hoehn H, Bryant EM, Fantel AG, Martin GM. Cultivated cells from diagnostic amniocentesis in second trimester pregnancies. III. The fetal urine as a potential source of clonable cells. *Humangenetik.* 1975;29(4):285-290.
51. In 't Anker PS, Scherjon SA, Kleijburg-van der Keur C, et al. Amniotic fluid as a novel source of mesenchymal stem cells for therapeutic transplantation 3. *Blood.* 2003; 102(4):1548-1549.
52. Tsai MS, Lee JL, Chang YJ, Hwang SM. Isolation of human multipotent mesenchymal stem cells from second-trimester amniotic fluid using a novel two-stage culture protocol 2. *Hum Reprod.* 2004;19(6):1450-1456.

37 Use of Amniotic Membrane, Amniotic Fluid, and Placental Dressing in Advanced Burn Patients

Niranjan Bhattacharya

37.1 Introduction

Fire-related deaths are a neglected public-health issue in India. These deaths are mostly due to kitchen accidents, self-immolation, domestic violence, and even dowry-related abuses. Death associated with burn injuries primarily depends on the age of the patient, the percentage of the body surface burned, and associated smoke-inhalation injury apart from the role of preexisting diseases such as diabetes, hypothyroid, clinical and subclinical tuberculosis, arthritis, nutrition level, etc.

In a report published in *Lancet*, 163,000 fire-related deaths were reported in 2001 in India, and 106,000 of these were of women, mostly between 15 and 34 years of age. The average ratio of fire-related deaths of young women to young men was 3:1.[1]

Wound sepsis and chronic wound formation are two frequent associates of burn injuries, which enhance the mortality and morbidity of burn patients. Chronic wounds are defined as wounds that have not proceeded through orderly and timely reparation to produce anatomic and functional integrity after 3 months. Burns cause dynamic injuries that may progress over the first 2–3 days to months; therefore, frequent reassessment of the wound is required to ensure optimal management. Many burns are not uniform; the depth of the burn varies from one area to another. Proper surgical management of burn wounds before colonization of the eschar by bacteria and septic liquefaction, which otherwise are inevitable, is very important.

All burn wound types have the potential to become chronic; additional modifying factors such as venous or arterial insufficiency, local-pressure effects, and existence of preexisting diseases as mentioned earlier may retard the process of healing.

Controlling wound infection and covering the burn wound with bio-friendly substances are two essential parts of burn treatment. An ideal skin substitute should be used to cover the wound for dressing purposes. This skin-covering agent for burn wounds should be easily available, nontoxic in nature with elastic and adherence properties, cost little so that it is affordable to all sections of the society, and be universally available. It should act as a semipermeable membrane for essential substances such as oxygen and other micronutrients, and ideally, prevent the entrance of bacteria.

Unfortunately, none of the available artificial skin substitutes, in a truly global perspective, fulfills the ideal requirements for burn injury. The objective of the present study was to examine whether a readily available hospital waste, that is, pregnancy-specific biological substances (PSBS), could be used as a suitable and effective biological dressing in reducing burn wound sepsis in the treatment of thermal burns among all age groups.

37.2 Materials and Methods

Initially, 97 patients were randomly recruited for the present study of the utility of PSBS in the burn management. Of these, 33 patients, who did not agree to the PSBS protocol, were transferred to the burn unit of a tertiary-care hospital for treatment.

The rest, i.e., 64 burn patients with 26–76% of total body surface area burnt calculated on the basis

N. Bhattacharya
Department of General Surgery, Obstertrics and Gynaecology and Clinical Immunology, Advanced Medical Research Institute, Gol Park, B.P. Poddar Hospital and Vidyasagore Hospital, Kolkata 700034, India

N. Bhattacharya and P. Stubblefield (eds.), *Regenerative Medicine Using Pregnancy-Specific Biological Substances*,
DOI: 10.1007/978-1-84882-718-9_37,

of the famous rule of nine, were enrolled in the present study for PSBS treatment (1999–2009).

The patients were asked for a voluntary written consent before carrying out the proposed procedure. The ethical committee of the hospital had approved this research project earlier.

The area affected included both partial to complete thickness and at times the adjacent soft tissue injury in case of thermal burns. Patients suffering from chemical burns or burn injuries in sensitive parts of the body such as the genitals or face were also included in this study from 1999 to 2009.

After admission, all the cases were treated with (1) normal saline for initial removal of the dirt and debris followed by (2) application of freshly collected placenta's raw surface at the site of the wound (Fig. 37.2), after which there was (3) sprinkling of copious amounts of freshly collected clear amniotic fluid at the site of the burn injury (after 5–10 min), and lastly, the application of amniotic membrane (4) at the burned wound site (amniotic or the fetal side of the amniotic membrane in case of superficial or partial thickness skin burn for early epithelialization, and maternal attachment site or the chorionic site in deep burn to improve circulation through angiogenesis-supporting cytokine content of the chorionic site of the membrane) (Figs. 37.3 and 37.4). This amniotic membrane was kept in the amniotic fluid, which was freshly collected from consenting donor mothers who were VDRL, hepatitis B and C, and HIV (1&2)-negative and had undergone cesarean section.

Routine prophylactic antibiotics (ceftriaxone 1 g twice daily intravenously and gentamycin 60–80 mg twice daily through intramuscular route, metronidazole 200 mg thrice daily initially through intravenous route with intravenous isolyte fluid) were routinely given to all patients who consented for the present study 1999–till date. Subsequently, there was switching to oral broad spectrum antibiotics for all cases. Routine multivitamins, minerals, and trace elements were also prescribed to all the patients. Other drugs such as long-acting beta-blocker (Inderal) 40 mg/day in case of adults were also routinely prescribed. Weight is considered as the basic guideline for the dosage calculation of drugs in different age groups. In case of diabetes, or hepatic and renal or cardiac problems, appropriate additional drugs and insulin (Act Rapid variety in standard sliding scale is followed till the healing process is stable) was prescribed in consultation with a senior physician along with stoppage of the aminoglycoside antibiotic in general.

Once the healing process started stabilizing, physiotherapeutic protocol was prescribed by the senior physiotherapist of the hospital. The wounds were inspected regularly to look for exudation, offensive odor, or any other clinical local or systemic sign of infection.

Amniotic membrane dressings, however, have one major drawback, i.e., of rejection of the graft, and hence only serve as a temporary dressing unless the different stem cell components of the freshly collected amniotic fluid, amniotic membrane, placenta, and their intrinsic cytokine network participate; they participate and assist the healing process of the host. All the steps of clinical improvement were meticulously noted with photographic evidences and histological evidences to understand the steps and etiopathogenesis of the reparative process involved with the application of PSBS in burn wounds.

37.3 Results and Analysis

There are many treatment options available for burns and the resulting wounds.[2] In the present study, as mentioned, human pregnancy-specific biological waste materials alone were used; these include the placenta, the amniotic fluid, and the amniotic membrane. These materials are normally discarded and go to the incinerator, but can easily be collected from the obstetric/labor room/operation theater and effectively used in case of burn wound.

In the present series, 64 burn patients (male 24, age 2–96 years, mean 36 ± 5.4 years and female 40, age 7–68 years, mean 32 ± 5.7 years) with 26–76% of total body surface area burned, as mentioned earlier, were enrolled for PSBS treatment (1999–2009).

All the patients were treated with prophylactic antibiotic (splint as and when necessary depending on the site and proximity of the joint) and other support regime including early ambulation and physiotherapy in addition to the specific treatment with the pregnancy-associated biological substances (PABS) as mentioned earlier in the material and method section. The membrane prevents heat and water loss from the wound surface and acts as a barrier against bacterial contamination, thus aiding the healing process and reducing morbidity. Another clinically significant and important property of the amniotic membrane is its ability to offer marked relief from pain.

Among the 64 cases enrolled for the study, 24 (37.5%) were males and the remaining 40 (62.5%) were females. The burnt area varied from 26% to 76% of the surface area as per standard calculation of the rule of 9. The prevalence is noted below. There were six patients in the less than 10 years (9.37%) age group, while 16 cases (25%) were in the range of 11–20 years. There were 22 (34.37%) in the age group of 21–30, 11 (17.18%) in the age group of 31–40, and 4 (6.25%) in the age group of 41–50 years respectively, while 2 (3.12%) were in the 51–60 age range, 2 (3.12%) in the age group of 61–70, and 1(1.56%) in the 71 and above age group.

The thermal injuries were caused by hot water in 20 (31.25%) cases, followed by exposure to direct flame (kerosene or other cooking material), comprising 28 (43.75%) cases. Suicide attempts were the cause in 10 (15.62%) cases. Burn was caused in the other six cases (9.38%) due to accidents involving coal or cigarettes (smoking inside mosquito nets, after the intake of alcohol with the risk increasing with drug addiction), attempts at dowry killing or other attempted murders.

In the present group, no fatality was encountered. Follow-up for (up to) 6 months continued after the total healing of the scar. Problems of keloid and hypertrophic scars were noted in six (9.38%) cases only. Contrary to our expectations, post-burn leucoderma was not encountered in a single case. Some degree of hypopigmentation at the burn scarring site was seen in 14 cases (21.87%), which gradually reverted to the normal skin color and architecture within the 6 months follow-up period. Another important complication was post-burn contracture, which was noted in six cases (9.38%). Again, this problem was treated with appropriate release incision and fresh amniotic membrane application. Partial thickness skin autograft was avoided in the treatment regimen. The most difficult patient, an attempted suicide victim with 76% burns, who was treated with pregnancy specific biological substances only, is presented here with photographic and histological evidences. The clinical photographs of improvement are shown in Figs. 37.1–37.7.

Wound healing is a very complex process that is tightly regulated to achieve wound repair. The process has three important components, i.e., inflammation, proliferation, and maturation. Following the initial tissue injury, inflammatory mediators known as cytokines are released from the injured tissue cells and wound

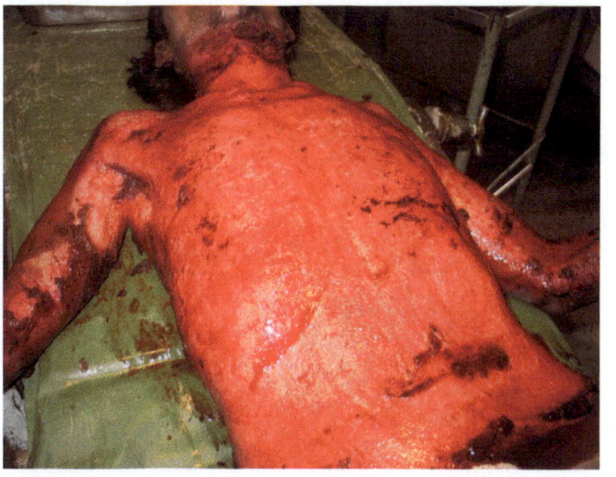

Fig. 37.1 Suicidal attempt with kerosene resulting 76% partial and complete thickness skin burn treated with pregnancy-specific biological substance dressing

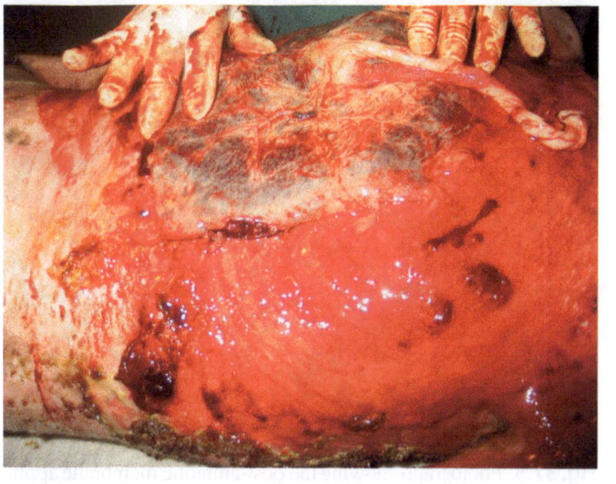

Fig. 37.2 Photograph showing the application of placenta at the burnt site

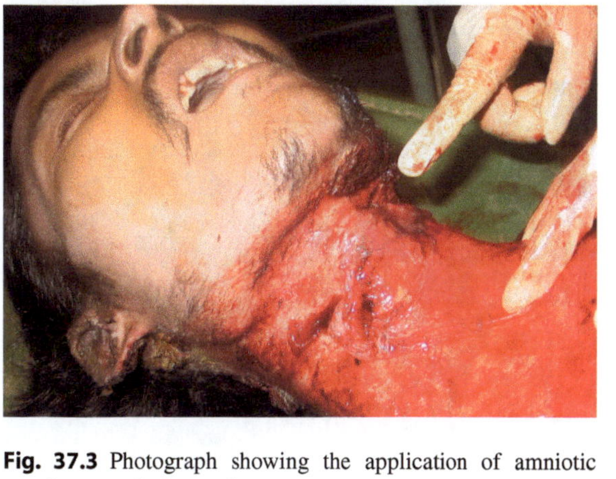

Fig. 37.3 Photograph showing the application of amniotic membrane at the burnt site

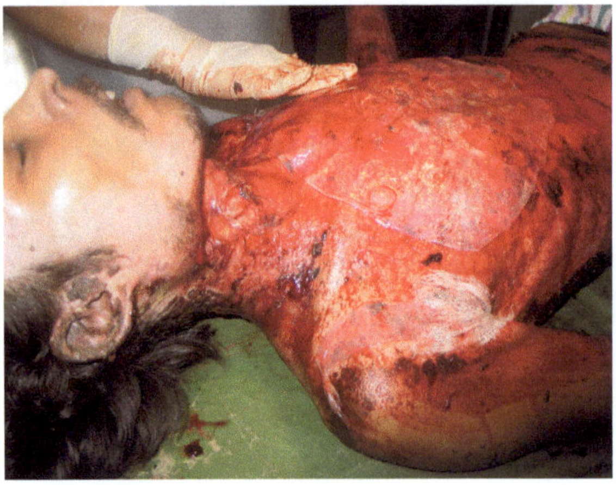

Fig. 37.4 Photograph showing the application of amniotic membrane at the burnt site

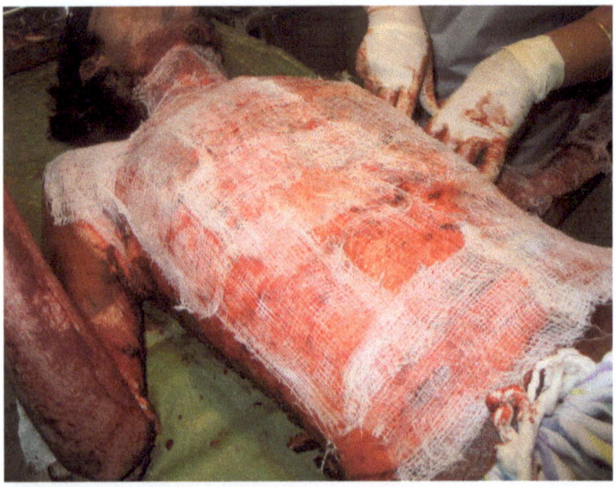

Fig. 37.5 Photograph showing the post-amniotic membrane application of Vaseline gauge to cover the site and to keep it moist

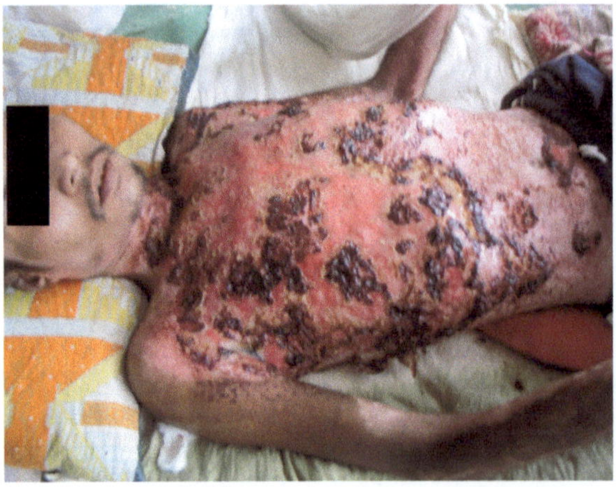

Fig. 37.6 Photograph of the same patient after 6 weeks

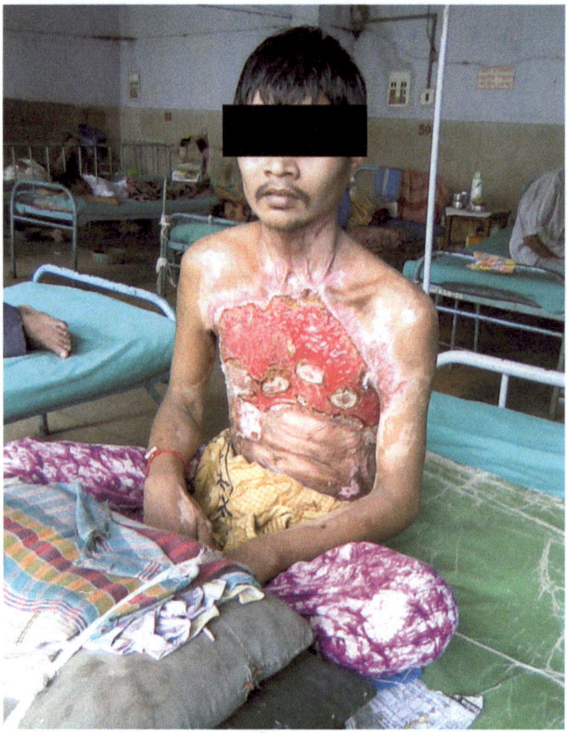

Fig. 37.7 The same patient after 6 months of periodic dressing with the pregnancy-specific biological substances as per text schedule

blood clots, after which the inflammatory phase initiates. Then, the proliferation stage begins several days after injury. In this stage, the platelet degranulation activates the coagulation cascade, and the resultant fibrin clot serves as a scaffold for the proliferation phase of wound healing. During the proliferative phase, fibroblasts in the extracellular matrix increase and synthesize the tissue components, such as proteoglycans, fibronectin, and collagen. New vessels and epithelium are formed as rapidly as possible to maximize the tissue replacement dynamics. All wound cells are maximally active and are sensitive to factors that regulate cell proliferation and protein biosynthesis. All the cells proliferate, and metalloproteinases are simultaneously released into the extracellular fluid to activate a matrix breakdown process. The balance between tissue degradation and biosynthesis permits remodeling of the provisional tissue and its ultimate repair as seen in Figs. 37.8–37.11.

In order to augment the complex process of wound healing, investigators who have used amniotic membrane as a temporary dressing material with the belief

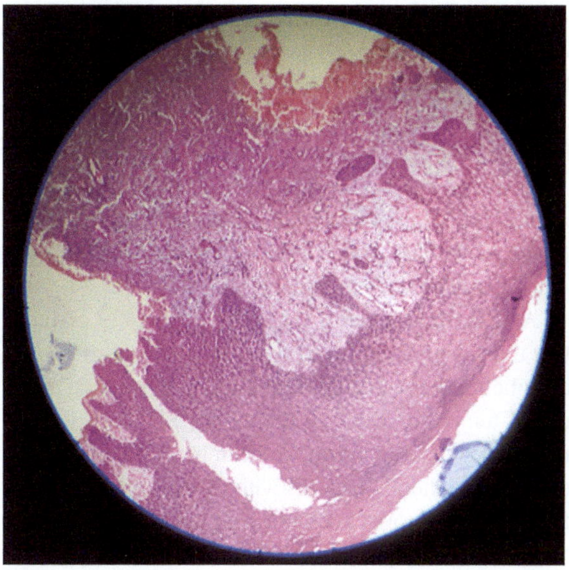

Fig. 37.8 Showing the histological features of the skin stained with hematoxylin and eosin of the same patient under scan power. Wound healing is a very complex process that is tightly regulated to achieve wound repair. The process has three important components, i.e., inflammation, proliferation, and maturation. Following the initial tissue injury, inflammatory mediators known as cytokines are released from the injured tissue cells and wound blood clot, after which the inflammatory phase initiates

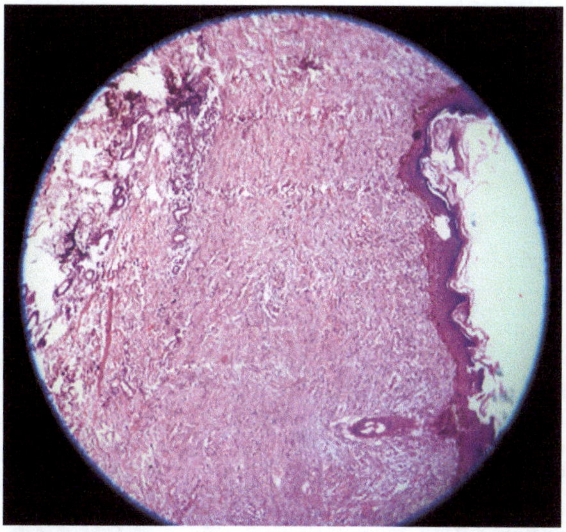

Fig. 37.9 Showing the histological features of the skin stained with hematoxylin and eosin of the same patient under low power suggesting steps of regeneration. During the proliferative phase, fibroblasts in the extracellular matrix increase and synthesize the tissue components, such as proteoglycans, fibronectin, and collagen. New vessels and epithelium are formed as rapidly as possible to maximize the tissue replacement dynamics. All wound cells are maximally active and are sensitive to factors that regulate cell proliferation and protein biosynthesis. All the cells proliferate, and metalloproteinases are simultaneously released into the extracellular fluid to activate a matrix breakdown process. The balance between tissue degradation and biosynthesis permits remodeling of the provisional tissue and its ultimate repair as seen

that human amniotic epithelial cells do not express HLA-A, -B, -C, and -DR antigens, or beta 2-microglobulin on their surfaces.[3] However, there has been some recent controversy regarding this topic. It has been noted that amniotic membrane epithelial cells display some degree of antigenicity and immunogenicity as allografts due to the presence of some (though definitely less than adult) MHC class I and II antigens.

On the other hand, viable human amniotic epithelial cells (HAECs) have been shown to elicit beneficial effects on secretion of anti-inflammatory factors. It has been seen that topical application of culture supernatant from HAECs leads to profound suppression of suture-induced neovascularization in the cornea. In addition, expression of interleukin (IL)-1 beta mRNA was suppressed in cauterized cornea where amniotic membrane was applied. These results suggest that amniotic cells are a source of soluble anti-inflammatory factors that suppress inflammation.[4]

In spite of the scientific controversy, the clinical and practical impression suggests that this dressing is extremely effective. It speeds up the healing process and reduces pain. The relative ease of procurement and preparation and its low cost and easy availability projects the amniotic membrane as an ideal temporary skin substitute for burn wounds. Amniotic membrane has been utilized in various studies to cover the burn wound dressing for less exudation of protein and electrolyte, as well as for its bio-friendly nature and hypoantigenic qualities apart from the cytokine support it lends to wound healing.

This substance when used as a biological dressing not only lowered the hospital's and patients' costs, but also significantly reduced the rate of infection and pain. Sometimes recipients experienced a problem of odor. This was a normal response of amniotic membrane sloughing due to rejection by the host immune system or this may be due to *Pseudomonas aeruginosa* or *Staphylococcus aureus* infection.[5]

The odor becomes offensive at times depending on the type and concentration of the bacteria and the condition of the wound. This is a condition suggesting a change of the dressing with a fresh one. After the first

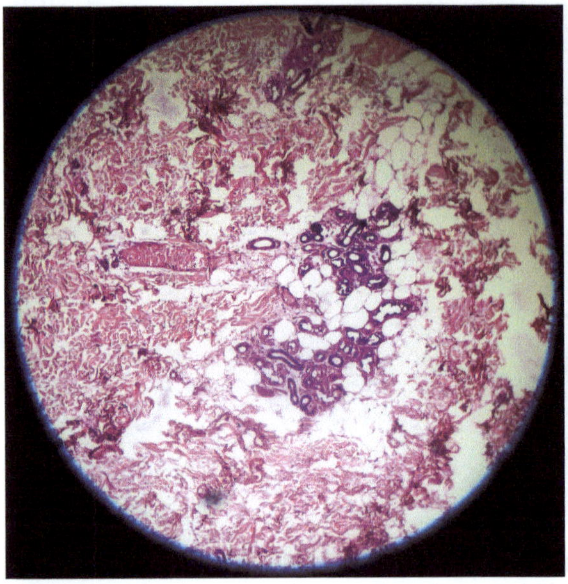

Fig. 37.10 Showing the histological features of the skin-stained examination with hematoxylin and eosin of the same patient under low power suggesting steps of regeneration with cell nest-like structure of fetal origin, which is actively helping the regeneration process

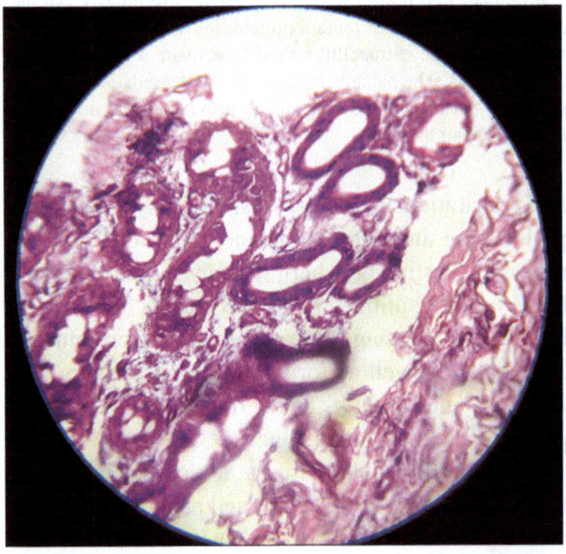

Fig. 37.11 Showing the histological features of the skin stained with hematoxylin and eosin of the same patient under high power. The slide is suggesting steps of regeneration of the skin

dressing on admission till the period when the acute infection of the wound is reasonably controlled, patients were advised to keep the burn wound open and to stay inside mosquito nets. Since the wounds were visible, the amniotic membrane was changed when necessary, i.e., when the integrity of the dressing was lost, which was approximately every fourth day to the seventh day. In infected cases, the same procedures of inspection, debridement, antibiotic treatment, and change of biological dressings were carried out whenever necessary. After recovery, the patients were followed up on a weekly basis for up to 4 weeks, and on a monthly basis for up to 3 years. Any signs of infection or abnormal scar formation were recorded.

Another investigator, Dr. B. Bose, in 1979 claimed that the use of amniotic membranes was cost-effective, even better than other allografts and heterografts, and a useful temporary biological dressing.[6] Further, it does not need special facilities, and can be easily employed even in peripheral hospitals with modest infrastructure facilities. The use of such material has been advocated for use in developing countries because of its promising good results.[7] Amniotic membrane patch has even been found to be effective in the treatment of acute corneal alkali burns.[8] The mechanisms underlying the effectiveness of amniotic membrane as an aid in the treatment of burn wounds have been postulated by a number of researchers. Apart from its physical properties in reducing water and heat loss, Kim et al.[9] have suggested that the mechanism responsible for the rapid healing observed is due to the inhibition of proteinase activity, thus reducing the inflammatory responses by reducing the infiltration of polymorphonuclear leukocytes. Benefits claimed for this procedure of biological dressings include rapid healing with cosmetically satisfactory results (i.e., lack of scarring) and the avoidance of autografting. The success of the procedure is ascribed at least in part to the biological activities of growth factors secreted by the donor fibroblasts that are transiently present in the wounds. The basis for use of all of these biological and artificial skin substitutes is that they create a moist environment, which has been shown to improve the rate of healing under a variety of clinical and experimental conditions. Another renowned plastic surgeon and author, Dr. JB Lynch, strongly justified the use of amniotic membrane as a suitable biological skin dressing in 1979.[10]

Another researcher of repute Dr. Eaglestein reviewed the role of a variety of occlusive and semi-occlusive artificial skin substitutes for control of wound sepsis and early healing.[11] Despite their wide use in many developed countries, none of these artificial skin substitutes serve as ideal dressings besides being very expensive, especially for routine use.

It should be mentioned here that amniotic membrane was actually used as a burn dressing for many years before anyone thought of the stem cell components of the amniotic membrane and other pregnancy-specific biological substances. However, no investigator had ever used freshly collected bacteria-free amniotic fluid, which is a rich source for mesenchymal stem cells and as known, has a bacteriocidal property due to the presence of properdin-like materials embedded in it.

37.4 Discussion

Burn is an injury vis-à-vis inflammation at the site of contact of heat, cold, electricity, chemicals, light, radiation, or friction. The result and the degree of burn depend on the involvement of the underlying skin and the subcutaneous tissue, muscle, bone, blood vessel, and other adjacent tissues.

Pain is severe in case of burn due to exposure of the free nerve endings and also because of the cytokines and other toxins produced at the site of the burn where vascularity and tissue autoregulation are affected. Apart from local metabolites' participation, direct and profound injury of the nerves also plays an important role. In general, thermal injuries to living tissues occur as a function of temperature and duration of exposure to a heat source,[12] which leads to complications such as shock, infection, electrolyte imbalance, and respiratory distress. All these complications cumulatively may at times have a fatal implication.

The survival of a burn patient is largely dependent upon prompt and efficient wound healing and control of infection, which also prevents wound contracture. Sequel due to infection by deadly *Pseudomonas* plays a very negative role in the mortality of burn patients after the first week of the initial burn injury. Appropriate first aid limits progression of the burn depth and influences outcome. Therefore, assessment of area and depth is crucial to formulating a management plan. Burn depth may progress with time, so reevaluation is essential.

Different biological dressings to cover the burn wounds in order to assist the healing process have been attempted to treat burns over the centuries. Biological dressings are classified as Group a: *Biological* (homologous skin) – glycerinized pigskin with or without silver or aldehyde treatment variety, human amniotic membrane (amniotic side for epithelialization and chorionic side for neovascularization). A variety of techniques utilizing different heterografts from lizard to frog skin, then to homografts, and ultimately autografts were used to cover the wound in case of burn and to prevent infection. Further developments made possible by improvement of storage techniques in the 1970s, involved the wider employment of preserved allografts; Group b: *Synthetic skin substitutes* like polyurethane and hydrocolloids, polyurethane film, acrylamide film, and hydroxyethyl methyerylate with polyurethane films; and finally Group c: *Bioengineered skin substitutes*, which are derived from and with the association of biological materials such as collagen and silicone sheets, also used in different centers of the world to cover the burn wound and assist the healing process as a whole. Covering the burn wound prevents loss of water, electrolytes, and proteins and also prevents the dispersion of heat etc.

The Chinese system of traditional medicine has appreciated the wound healing property of the placenta or its different extract preparations, apparently for thousands of years. In a relatively recent published article, investigators have suggested that the application of human placental extract (HPE) in rats both at topical and intramuscular routes (2.5 mL/kg) caused significant lowering of wound size ($P < 0.05$), wound index ($P < 0.05$), and the number of days required for complete healing ($P < 0.01$); significant gain in tensile strength ($P < 0.01$); appreciable increase of tissue DNA, total protein, and collagenesis were observed in the HPE-treated group. The authors concluded that human placental extract systematically helps collagenesis leading to potent healing of wounds.[13] In another article, it was shown that placental-extract gel and cream were effective topical agents for chronic non-healing wounds. In addition, there was less pain and discomfort during dressing change with the placental-extract cream, which the investigators therefore recommended for topical application in chronic non-healing wounds.[14] In a different study, another group of investigators conducted a comparative study on the efficacy and safety of topical application of a purified extract of human placenta (placentrex gel) versus povidone iodine for its wound-healing potential after orthopedic surgeries. The authors concluded that purified placental extract and povidone iodine have comparative wound-healing effects.[15]

Apart from the biological healing property of the placenta and its extract, another pregnancy-specific biological substance, i.e., amniotic fluid, also plays a very important role if used in burn injuries. It has to be remembered in this connection that nature washes the vaginal canal of the mother with the amniotic fluid before the birth of a baby in all species in order to prevent infection to the baby. This is Nature's proof of the sterile and bacteriocidal properties of the amniotic fluid.

In addition, the amniotic fluid possibly possesses a lubricating effect due to its higher viscosity and protein and cellular composition, and may also have a reparative effect due to the progenitor cell/stem cell component in it, i.e., the epithelial and the mesenchymal stem cell population. The stem cells of the amniotic fluid are capable of differentiating into multiple lineages; this may be valuable for therapy. In this context it is relevant to mention that a group of renowned investigators have reported in *Nature Biotechnology* on the isolation of human and rodent amniotic fluid–derived stem (AFS) cells that express embryonic and adult stem cell markers. Undifferentiated AFS cells expand extensively without feeders, double in 36 h and are not tumorigenic. Lines maintained for over 250 population doublings retained long telomeres and a normal karyotype. AFS cells were noted to be broadly multipotent.[16]

Use of amniotic membrane for covering the burn wound decreases oozing from the wound site after debridement and thus decreases the need for blood and albumin transfusion, and causes less electrolyte imbalance. Amniotic membrane also possesses some antibacterial characteristics, namely, bacteriostatic effect on gram-positive bacteria due to the lysozyme content. Amniotic membrane also helps in the epithelialization of the wound.

One important contributor to this book, Prof Andrew Burd of the Chinese University of Hong Kong, a renowned plastic and reconstructive surgeon, has calculated the global production of placentas, amniotic fluid, and the amniotic membrane which are noted in his article in this book. This will convey a picture of the massive global wastage of such materials when we throw them into the dustbin for eventual destruction through incinerators. Before analyzing the utilization potential of pregnancy-specific biological substances in burn wound, it is necessary to give a brief account of the substance-compositions and properties, and the contemporary scientific advances in the field of stem cell research which relate to these important substances.

Placenta:	Weight Volume	50 million kg 60,000 m^3
Amniotic membrane:	Area	15 million m^2
Amniotic fluid:	Volume	60 million L

37.4.1 Amniotic Epithelial Cells

Amniotic epithelial cells are isolated following the stripping of amniotic membrane from the chorion by trypsin digestion.[17] Such procedure allows for selection of relatively homogeneous cell suspension, which attach to plastic in in vitro culture. Contrary to mesenchymal stromal cells (MSC-type cells), amniotic epithelial cells need the addition of epidermal growth factor (EGF) to Dulbecco's Modified Eagle's Medium (DMEM) supplemented with forward scatter (FSC). Cells grow throughout two to six passages, displaying typical epithelial morphology. Both the expression of CD90 antigen and HLA-A, -B, -C (human leukocyte antigens) increase in culture – the initial expression levels are too low for using these antigens as identification/selection markers for freshly isolated epithelial cells. Among the other markers, cells express molecular markers of pluripotent stem cells (SRY-related HMG-box gene SOX-2, octamer-binding protein 4 Oct-4, and Nanog). Contrary to placental MSCs, epithelial cells do not express the Cd49d marker.

Both the molecular markers, and differentiation experiments suggest that amniotic epithelial cells are pluripotent, having not only adipogenic,[18] osteogenic[19] chondrogenic, but also myogenic, and cardiomyogenic, neurogenic,[20] pancreatic,[21] and hepatogenic[22] potentials.

37.4.2 Mesenchymal Stromal Cells from Amniotic and Chorionic Regions

Amniotic MSC are isolated from amnions at any gestation stage when the placenta is fully developed. The amnion must be carefully dissected from fetal membranes to avoid the presence of maternal cells. The most popular isolation protocol is based on the two-step digestion procedure, first with trypsin, and subsequently with collagenase. The resulting plastic-adherent cells can be expanded in in vitro culture similarly as in adult bone marrow MSCs.[23] Cells at the second passage stage (at least) express mesenchymal, but not hematopoietic markers. Animal in vivo experiments confirmed that human placental MSC, transplanted into animals, are able to migrate into various organs: bone marrow, thymus, spleen, lung, liver, brain and kidney. In vitro, these cells are able to differentiate into cartilage, bone,[24] fat tissue,[25] skeletal muscles,[26] heart muscle,[27] epithelium, nerve cells, or pancreatic islets.[28]

37.4.3 Placental Tissue and the Umbilical Cord is an Important Source of Fetal Mesenchymal Stem Cell

Placental tissues are considered an attractive source of cells with considerable phenotypic plasticity as well as immunomodulatory properties, including mesenchymal stem cells and amniotic membrane-derived epithelial cells.

Mesenchymal stem cells or mesenchymal stromal cells have been isolated from the amniotic and chorionic regions of the fetal placenta, as well as from the umbilical cord.

Amniotic-, chorionic-, and umbilical cord MSC are usually isolated by mechanical peeling or removal of the tissue followed by enzymatic digestion. After primary culture and sub-cultivation of fibroblast colony-forming units, evolved adherent cells exhibit a pattern of antigen expression (including CD105+, CD90+, CD73+, CD34–, CD45–, Oct-4+), which is not different from that expressed by UCB-MSC, with the exception of the embryonic marker, Oct-4. Placental MSC has the intrinsic property to differentiate into cells of the adipogenic, chondrogenic, osteogenic, and skeletal myogenic lineages after exposure to the appropriate stimuli. These cells also exhibit a robust hematopoiesis-supportive function.[29,30] Moreover, UCB-MSC exhibit a hematopoietic-supportive function.[31]

Having given the background, it is important to point out that fetal stem cells, like their adult counterparts, are able to differentiate into several tissues, and migrate to the site of tissue injury. They may form the new cells replacing those destroyed by injury or illness, and also modify the healing process. Fetal cells have greater proliferation and differentiation potential (longer telomeres and telomerase activity). The cells collected from the placenta depend on the region of collection: amniotic epithelial cells from the amniotic epithelial region or amniotic mesenchymal stromal cells (MSC) from the amniotic mesenchymal region, to name a few.

Of these cells, amniotic and chorionic mesenchymal stromal cells represent characteristics similar to in vitro growth characteristics, exhibit surface antigen expression, and differentiation potential. Both cell types are hematopoietic markers – negative ($CD34^-$, $CD45^-$), $HLA\text{-}DR^-$, and positive for markers attributed for MSC: CD73, CD90, and CD105. The characteristics of the amniotic epithelial cells are somewhat more complex – they are able to proliferate shorter than MSCs in in vitro culture, proliferate only in higher densities in presence of epidermal growth factor (EGF), and change the expression of selected markers (HLA-A, -B, -C, -CD90) depending on the culture time. This latter phenomenon may suggest that the amniotic epithelial cells are a heterogeneous population being subsequently selected to higher homogeneity by culture conditions.

37.4.4 Feto-Maternal Cell Traffic in Pregnancy and Its Long-Term Consequences

Whether any transplacental cell traffic is a cause of inflammation or is just a bystander in the regeneration of damaged tissue, is a matter of ongoing debate.[32–34] Analyzing murine syngeneic and allogeneic pregnancies, Khoshrotehrani and colleagues showed that fetal microchimerism is a naturally occurring phenomenon leading to detectable levels of mononuclear cells in several maternal tissues, such as lungs, heart, spleen, kidney, and bone marrow.[33,34] It has been shown for human and murine pregnancies that levels decrease after delivery. This phenomenon indicates that the maternal immune effector cells shift back to their normal allo-response. Although the Khoshrotehrani group was unable to detect microchimeric fetal cells in maternal mouse brain during pregnancy, other researchers have found a relevant proportion of fetal progenitors that obviously were able to cross the blood–brain barrier during pregnancy.[35] The cells even increased in number during a period of 4 weeks post partum and adopted a local phenotype such as perivascular macrophage-, neuron-, astrocyte-, and oligodendrocyte-like cell type. The levels of fetal microchimerism were increased in brain injuries suggesting an active role in tissue regeneration. Whether this is just associated with inflammation and phagocytosis of damaged tissue, or the formation of truly new neurons, needs further attention. Similar uncertainties exist for an animal model on murine maternal hepatic injury.[36] Here too, higher levels of microchimerisms were detected after chemical compared to surgical laceration, with increasing levels between 4 and 8 weeks after injury.

It has also been shown that fetal progenitors could persist within the maternal blood and tissue over decades.[37] These CD34+ and CD34+/CD38+ fetal progenitors are capable of differentiating into functional T and B cells, and may help in the regeneration process on the maternal system.

If we come back to our study, it is worth remembering that the skin offers a perfect model system for studying the wound-healing process, which involves a finely tuned interplay between several cell types, pathways, and processes. The dysregulation of these factors may lead to wound-healing disorders, resulting in chronic wounds, as well as abnormal scars such as hypertrophic and keloid scars.

In the present work, the possibility that mesenchymal and epithelial stem cells supplied by the amniotic fluid and placental and membrane sources, are acting as a cell therapy source in combination to augment the process of healing, is strongly postulated. This is possibly the basis for the augmented strategy of healing with the application of pregnancy-specific biological substances as dressing material. The therapeutic effect of the stem cell seems consistent with both the paracrine function and the transdifferentiation. Systemic and micro-milieu factors appear to dictate the fate of implanted stem cells.

Researchers must begin to focus upon a few basic critical issues: the modulation of the systemic and microenvironment for stem cells in order to augment stem cell survival and transdifferentiation; the underlying mechanisms of stem cell therapy and the fate of stem cells; and the differentiation into specific cell types as per the local demand or other terminal cell populations with synchronizing and favorable paracrine functions.

Earlier scientists used stem cells in different routes, namely, subcutaneous or intravenous route. In the present study, the transdermal route, which is seen as deficient, was used for fetal and neonatal-studded PSBS (as burn dressing) with good effect in burn victims. This is the clinical validation of the fetal and neonatal stem cells-studded pregnancy specific biological substances, which is used through the deficient transdermal route in burn victims.

In the earlier paragraphs, we have discussed in brief the major advances in the field of PSBS and its potential regenerative impact through the progenitor or stem cells found in it. This property has made them a unique category on the tissue used for regenerative biology.

The present article is a report on the simultaneous and judicious utilization of the chorionic (for angiogenesis) and amniotic membrane (for epithelization) and is a clinical validation of molecular advances in the field of stem cell biology in case of burn victims. The utilization of bio-friendly amniotic fluid with intrinsic bacteriocidal property for the preservation of the fresh placenta and membrane instead of traditional normal saline, and also dressing with amniotic fluid, not only reduces the unnecessary utilization of costly antibiotic cream in dressing or tulle formation, but also helps in the prevention of the emergence of bacterial resistance and mixed infection due to opportunistic fungi and resistant bacterial infection including anaerobes, bacteriods, and other trouble-making organisms.

Furthermore, it may be noted that there has not been a single case of mortality among all the patients treated with PSBS so far in our study. In the quest to understand the mechanism for this positive outcome, histology and hematoxylin eosin staining was done from the burn patient's skin, which revealed groups of cells that looked as though a cell nest was migrating to the site of burn injury to assist proper epithelization. There was no unusual inflammatory cellular infiltrate or feature suggesting graft versus host reaction.

The present research group has had similar experiences vis-à-vis different sets of experiments on the survival of the fetal human leucocytic antigen (HLA) randomized tissue graft in subcutaneous space of different hosts, without the help of immunosuppressives.[38–43]

This study possibly hints at the transdermal route of attachment, migration, and subsequent assistance on the host's reparative process by the stem cells supplied to the site of burn wound by the freshly collected amniotic membrane, amniotic fluid, or the placenta, collectively known as PSBS for positive outcome.

37.5 Conclusion

The skin offers a perfect model system for studying the wound-healing process, which involves a finely tuned interplay between several cell types, pathways, and processes. As the contribution of stem cells toward tissue regeneration and wound healing is increasingly appreciated, a rising number of stem cell therapies for cutaneous wounds are currently under development, encouraged by emerging preliminary findings in both animal models and human studies.

The present work shows that the proper and judicious use of naturally occurring pregnancy waste material's fluids and membranes not only helps in early recovery and proper epithelization of the wound, but also prevents wound contracture, keloid, and ugly hypertrophic scars in most cases as noted in our preliminary observation.

Acknowledgment The Department of Science and Technology, Government of West Bengal, supported the investigator with a research grant during his tenure at Bijoygarh State Hospital from 1999 to 2006.The author gratefully acknowledges the support of the patients who volunteered for this research work. The guidance of Prof K. L. Mukherjee (Biochemistry) and Prof M. K. Chhetri, former Director of Health Services is also acknowledged.

References

1. Sanghavi P, Bhalla K, Das V. Fire Related Deaths in India in 2001: A Retrospective Analysis of Data. *Lancet*. 2009; 373(9671):1282–1288.
2. Rudolph R, Ballantyne DL Jr. Skin grafts. In: McCarthy JG, ed. *Plastic Surgery*, vol. 1. 1st ed. Philadelphia, PA: W.B. Saunders; 1990:221-274.
3. Akle CA, Adinolfi M, Welsh KI, Leibowitz S, McColl I. Immunogenicity of human amniotic epithelial cells after transplantation into volunteers. *Lancet*. 1981;2(8254): 1003-1005.
4. Hori J, Wang M, Kamiya K, Takahashi H, Sakuragawa N. Immunological characteristics of amniotic epithelium. *Cornea*. 2006;25(10 suppl 1):S53-S58.
5. Pruit BA, Lindberg RB. Pseudomonas aeruginosa infection on burn patients. In: Doggett RG, ed. *Pseudomonas aeruginosa: Clinical Manifestations of Infection and Current Therapy*. 1st ed. New York: Academic; 1979:339-366.
6. Bose B. Burn wound dressing with human amniotic membrane. *Ann R Coll Surg Engl*. 1979;61:447-444.
7. Ramakrishnan KM, Jayaraman V. Management of partial-thickness burn wounds by amniotic membrane: a cost effective treatment in developing countries. *Burns*. 1997; 23(suppl):33-36.
8. Sharma SC, Bagree MM, Bhat AL, et al. Amniotic membrane is an effective burn dressing material. *Jpn J Surg*. 1985;15:140-143.
9. Kim JS, Kim JC, Na BK, et al. Amniotic membrane patching promotes healing and inhibits proteinase activity on wound healing following acute corneal alkali burn. *Exp Eye Res*. 2000;70:327-329.
10. Lynch JB. Thermal burns. In: Smith JW, Aston SJ, eds. *Grabb and Smith Plastic Surgery*. 3rd ed. Boston, MA: Little Brown & Company; 1979:453-575.
11. Eaglestein WH. Experience with biosynthetic dressings. *J Am Acad Dermatol*. 1985;12:434-440.
12. Moritz AR, Henriques FC. Studies of thermal injuries. II. The relative importance of time and surface temperature in the causation of cutaneous burns. *Am J Pathol*. 1947;23: 695-720.
13. Biswas TK, Auddy B, Bhattacharya NP, Bhattacharya S, Mukherjee B. Wound healing property of human placental extracts in rats. *Acta Pharmacol Sin*. 2001;22(12): 1113-1116.
14. Tiwary SK, Shukla D, Tripathi AK, Agrawal S, Singh MK, Shukla VK. Effect of placental-extract gel and cream on non-healing wounds. *J Wound Care*. 2006;15(7): 325-328.
15. Chandanwale A, Langade D, Mohod V, et al. Comparative evaluation of human placental extract for its healing potential in surgical wounds after orthopaedic surgery: an open, randomised, comparative study. *J Indian Med Assoc*. 2008;106(6):405-408.
16. De Coppi P, Bartsch G Jr, Siddiqui MM, et al. Isolation of amniotic stem cell lines with potential for therapy. *Nat Biotechnol*. 2007;25:100-106.
17. Miki T, Lehmann T, Cai H, et al. Stem cell characteristics of amniotic epithelial cells. *Stem Cells*. 2005;23:1549-1559.
18. Ilancheran S, Michalska A, Peh G, et al. Stem cells derived from human fetal membranes display multi-lineage differentiation potential. *Biol Reprod*. 2007;77:577-588.
19. Portmann-Lanz CB, Schoeberlein A, Huber A, et al. Placental mesenchymal stem cells as potential autologous graft for pre- and perinatal neuroregeneration. *Am J Obstet Gynecol*. 2006;194:664-673.
20. Kakishita K, Nakao N, Sakuragawa N, et al. Implantation of human amniotic epithelial cells prevents the degeneration of nigral dopamine neurons in rats with 6-hydroxydopamine lesions. *Brain Res*. 2003;980:48-56.
21. Miki T, Lehmann T, Cai H, et al. Stem cell characteristics of amniotic epithelial cells. *Stem Cells*. 2005;23:1549-1559.
22. Takashima S, Ise H, Zhao P, et al. Human amniotic epithelial cells possess hepatocyte-like characteristics and functions. *Cell Struct Funct*. 2004;29:73-84.
23. Alviano F, Fossati V, Marchionni C, et al. Term amniotic membrane is a high throughput source for multipotent mesenchymal stem cells with the ability to differentiate into endothelial cells in vitro. *BMC Dev Biol*. 2007;7:11.
24. Sakuragawa N, Kakinuma K, Kikuchi A, et al. Human amnion mesenchyme cells express phenotypes of neuroglial progenitor cells. *J Neurosci Res*. 2004;78:208-214.
25. Soncini M, Vertua E, Gibelli L, et al. Isolation and characterization of mesenchymal cells from human fetal membranes. *J Tissue Eng Regen Med*. 2007;1:296-305.
26. ibid 25

27. Zhao P, Ise H, Hongo M, et al. Human amniotic mesenchymal cells have some characteristics of cardiomyocytes. *Transplantation*. 2005;79:528-535.
28. Wei JP, Zhang TS, Kawa S, et al. Human amnion-isolated cells normalize blood glucose in streptozotocin-induced diabetic mice. *Cell Transplant*. 2003;12:545-552.
29. ibid 28
30. Parolini O, Alviano F, Bagnara GP, et al. Isolation and characterization of cells from human term placenta. *Stem Cells*. 2008;26:300-311.
31. Lu LL, Liu YJ, Yang SG, et al. Isolation and characterization of human umbilical cord mesenchymal stem cells with hematopoiesis-supportive function and other potentials. *Haematologica*. 2006;91:1017-1026.
32. Nelson JL. Microchimerism and autoimmune disease. *N Engl J Med*. 1998;338:1224-1225.
33. Khosrotehrani K, Bianchi DW. Fetal cell microchimerism: helpful or harmful to the parous woman? *Curr Opin Obstet Gynecol*. 2003;15:195-199.
34. Khosrotehrani K, Bianchi DW. Multi-lineage potential of fetal cells in maternal tissue: a legacy in reverse. *J Cell Sci*. 2005;118:1559-1563.
35. Khosrotehrani K, Johnson KL, Guégan S, et al. Natural history of fetal cell microchimerism during and following murine pregnancy. *J Reprod Immunol*. 2005;66:1-12.
36. Tan XW, Liao H, Sun L, et al. Fetal microchimerism in the maternal mouse brain: a novel population of fetal progenitor or stem cells able to cross the blood-brain barrier? *Stem Cells*. 2005;23:1443-1452.
37. Khosrotehrani K, Reyes RR, Johnson KL, et al. Fetal cells participate over time in the response to specific types of murine maternal hepatic injury. *Hum Reprod*. 2007;22:654-666.
38. Bianchi DW, Zickwolf GK, Weil GJ, et al. Male fetal progenitor cells persist in maternal blood for as long as 27 years postpartum. *Proc Natl Acad Sci USA*. 1996;93:705-708.
39. Bhattacharya N. Fetal cell/tissue therapy in adult disease: a new horizon in regenerative medicine. *Clin Exp Obstet Gynecol*. 2004;31(3):167-173.
40. Bhattacharya N, Chhetri MK, Mukherjee KL, et al. Can human fetal cortical brain tissue transplant (up to 20 weeks) sustain its metabolic and oxygen requirements in a heterotopic site outside the brain? A study of 12 volunteers with Parkinson's disease. *Clin Exp Obstet Gynecol*. 2002;29(4):259-266.
41. Bhattacharya N, Chhetri MK, Mukherjee KL, et al. Human fetal adrenal transplant: a possible role in relieving intractable pain in advanced rheumatoid arthritis. *Clin Exp Obstet Gynecol*. 2002;29(3):197-206.
42. Bhattacharya N. Fetal tissue/organ transplant in HLA-randomized adult vascular subcutaneous axillary folds: preliminary report of 14 patients. *Clin Exp Obstet Gynecol*. 2001;28(4):233-239.
43. Bhattacharya N, Mukherjee KL, Chettri MK, et al. A unique experience with human pre-immune (12 weeks) and hypoimmune (16 weeks) fetal thymus transplant in a vascular subcutaneous axillary fold in patients with advanced cancer: a report of two cases. *Eur J Gynaecol Oncol*. 2001;22(4):273-277.

Clinical Use of Amniotic Fluid in Osteoarthritis: A Source of Cell Therapy

38

Niranjan Bhattacharya

38.1 Introduction

Nature is the finest physician: it takes care of everyone because its concerns are universal and genuine; however, it always follows certain guidelines and principles and scientists can learn from these. Giving birth is an act of nature through which the realm of creation is opened to the mother. Nature comes from the Latin word *Natus*, "to be born." During birth, if nothing is present, what will prevent infection from affecting the newly born baby? Nature's concern can be seen when the vagina is washed, prior to birth, by a fine baby-friendly liquid containing cell suspension and antibacterial elements, which gives it a disinfectant property; it also possesses lubricant and cell therapy properties that are crucial for the mother and the child at the critical time of birth. This fluid is known as the amniotic fluid. So far, no one appears to have used this fluid for any therapeutic purpose for more than 70 years in the practice of modern medicine. However, there has been some recent awareness about amniotic fluid as a source of mesenchymal stem cells, which can be converted into any cell type given the niche or the environment for its transdifferentiation property, the implication being that it can help in regeneration in a degenerating system.

Knee-joint problem is one of the commonest geriatric problems that makes a person aware that he is aging. Although the exact causes for painful knee joints may be difficult to ascertain in many cases and sometimes remain unknown, it is understood that degenerative damage, especially cartilage damage, plays a central role in the pathogenic mechanism leading to this disorder. Current treatment modalities include pharmacological support, physiotherapy, etc., to palliate the condition. There is growing interest in the development of novel technologies to repair or regenerate the degenerated knee joint.

In 1927, Dr Johnson, a famous American surgeon and investigator first reported on the use of human and bovine amniotic fluid as an agent that stimulates the defense mechanism if injected in a host at the site of the problem or injury. Initially, amniotic fluid collected from mothers undergoing cesarean section was used and later substituted with a bovine amniotic fluid concentrate.[1] Contemporary evidence of other workers suggested that the use of amniotic fluid after abdominal surgery prevents or at least minimizes postoperative adhesions[2–4]. Taking a hint from the available knowledge in the field,[5] Dr Mandell Shimberg, a noted orthopedic surgeon from Kansas, USA, used amniotic fluid in various pathological conditions affecting different joints in the body. He also used the amniotic fluid in a closed reduction attempt in difficult fractures involving or proximal to a joint. Dr Shimberg used the amniotic fluid in an intraarticular route in 46 patients with knee-joint affectations, namely, sympathetic joint effusion, subacute joint infection, atrophic arthritis, gonorrheal joint effusion, and also idiopathic joint effusion cases only to name a few, without facing a single mortality and a very low morbidity.[6]

N. Bhattacharya
Department of General Surgery, Obstertrics and Gynaecology and Clinical Immunology, Advanced Medical Research Institute, Gol Park, B.P. Poddar Hospital and Vidyasagore Hospital, Kolkata 700034, India

N. Bhattacharya and P. Stubblefield (eds.), *Regenerative Medicine Using Pregnancy-Specific Biological Substances*,
DOI: 10.1007/978-1-84882-718-9_38, © Springer-Verlag London Limited 2011

People at that time did not know that there is a group of cells known as stem cells with high telomere content, which possess a unique property of transdifferentiation, depending on the niche provided to them and can migrate to the site of injury, and thus can actively participate in tissue repair and regeneration.

Recent clinical use of stem cell biology includes autologous mesenchymal stem cells application in animal models, which can arrest intervertebral disc degeneration or even partially regenerate it and the effect is suggested to be dependent on the severity of degeneration.[7] It has become abundantly clear to scientific workers in recent times that a stem cell can renew itself through cell division and can be induced to develop into many different specialized cell types. Moreover, stem cells have shown the ability to migrate and engraft within various tissues, as well as exert stimulatory effects on other cell types through various mechanisms (e.g., paracrine effects, cell–cell interactions).[8]

Important investigators in the field, Kaviani and coworkers, have reported that "just 2 milliliters of amniotic fluid" can provide up to 20,000 cells, 80% of which are viable.[9] Actual amniotic fluid stem cells originate in the developing fetus. The cells are thought to be sloughed from the fetal amnion and skin, as well as the alimentary, respiratory, and urogenital tracts. Amniotic fluid also contains a mixture of different cell types and has the potential to differentiate into various cell types.[10] A number of different origins have been suggested for these cells.[11] Cells of both embryonic and fetal origins and cells from all three germ layers have been reported to exist in amniotic fluid.[12, 13]

The focus of the present chapter is on the use of amniotic fluid in osteoarthritis (OA). Osteoarthritis-related knee problem is the commonest cause of disability at older ages.[14] This chapter reports on a study undertaken over 7 years (1999–2006) and its follow-up, which was undertaken to examine the clinical efficacy of freshly collected amniotic fluid from consenting mothers undergoing hysterotomy and ligation, as an effective progenitor or stem cell source for cell therapy procedure to combat the varying stages and grades of degenerative osteoarthritis affectations of the knee. Further, a simple comparison was made with the time-tested palliative procedure of intraarticular, long-acting corticosteroid application aseptically in the O.T. The study got the necessary clearance of the institutional ethical committee.

38.2 Materials and Method

Fresh amniotic fluid was collected from women admitted for hysterotomy and ligation at Bijoygarh State Hospital (1999–2006) and was used for the present study for the treatment of patients with osteoarthritis of the knee joint.

As per the standing direction of the State Family Planning Department, hysterotomy and ligation may be allowed up to 20 weeks of pregnancy, provided the mother has two or more healthy children. For the present study, 10 cm^3 amniotic fluid was collected aseptically in the O.T. from each mother undergoing hysterotomy and ligation, from an intact sac after opening the uterus, when the amniotic membrane containing the amniotic fluid generally herniates outside the uterus. The sac was gently punctured and the amniotic fluid was sucked out aseptically with a wide-bore size 16 needle and syringe. The collection protocol started, after getting the donor's consent and the recipient's informed consent.

Initially, 62 patients volunteered for this project of amniotic fluid cell therapy on degenerative osteoarthritis of the knee joint. Ten cases were discarded due to the association of neurodegenerative diseases such as Parkinsonism, malignancy, dementia of varying etiology and other chronic disease burdens.

The 52 cases that were ultimately enrolled for this trial had earlier not responded to conventional pharmacological or nonpharmacological treatment. The pharmacological treatment had included use of NSAIDs, i.e., naproxen, ibuprofen, etc., as well as the cyclooxygenase-2 inhibitor group of drugs like celecoxib with supporting drugs such as glucosamine, chondroitin, and opiates, only to name a few. The nonpharmacological treatment had included anaerobic exercises, i.e., resistance training, suggestion of weight loss or use of crutch, use of brace for the patella, and correction of tilting or misalignment. Acupuncture for some temporary relief had also been suggested to them but there was either no response or noncompliance.

The patients suffering from osteoarthritis of the knee not responding to oral medication and physiotherapy were given the option of free cell therapy from freshly collected amniotic fluid source, or intraarticular instillation of long-acting steroid. These patients were randomized for age and sex, and eventually divided in two equal groups: Group A (26 patients; 14 male and 12 female, age varying from 39 to 78 years, mean 51.4 ± 4.6 years

SD) and Group B (also contained 26 patients, female 14 and male 12, age varying from 41 to 82 years, mean 49±6.4 years SD).

The patients were asked to mark presence or absence or overall impression of improvement or deterioration with treatment, of some simple clinical parameters like (1) knee pain at rest, (2) little walking is painful, (3) definite increase in walking distance, (4) decrease in flexibility of the joint, (5) swelling of the joint, (6) little power of the joint to move against gradual increasing resistance, (7) difficulty in the initiation of the movement, (8) stiffness of the joint and movement, (9) range of movement is severely restricted. If seven of the nine clinical functional parameters were positive and there was X-ray evidence confirming the osteoarthritis status (standard weight-bearing anteroposterior and lateral knee radiographs) osteoarthritis was confirmed. Each compartment (medial tibiofemoral, lateral tibiofemoral, patellofemoral) was graded 0–3 for overall severity of OA. Clinical assessment of joint effusion (positive bulge sign and patellar tap: present/absent) was documented by a specialist, and knee aspiration was performed via the medial approach, and the volume of aspirated synovial fluid (SF) recorded. Total and differential leukocyte counts were estimated in all SF samples, which were also examined for the presence of calcium pyrophosphate crystals by microscopy.

Following aspiration from the knee, if there was effusion or dryness, each knee would be randomly injected with triamcinolone acetonide (40 mg in 1+9 mL normal saline) for Group A or alternatively with only amniotic fluid 10 ml, as a source of cell therapy for Group B. The amniotic fluid was taken from consenting mothers carrying pregnancy (14 weeks to 20 weeks gestation, calculated from the first day of the last menstrual period [LMP]; mean gestation was 17.6±2.1 weeks in the present study), who were undergoing hysterotomy and ligation as a family planning measure as already mentioned. If both the knees of the patient were affected, they were treated with identical dosages in each knee.

A thorough history of all the patients was taken, i.e., age, sex, height, weight, menstrual history, history of chronic disease like tuberculosis, hypothyroid, frank diabetes, or even altered glucose tolerance, history of diabetes in the family, lipid profile including uric acid level, apart from a history of specific involvement of cancer, systemic lupus erythematosus, ankylosing spondylitis, etc.

Specific rheumatological history with history of oral or intraarticular steroid intake, degree, and pattern of joint involvement with the duration of knee affection were noted. The knee pain was noted on a 100 mm horizontal visual analogue pain scale (VAS). The other parameters that were assessed included the distance walked in 1 min (WD) and also a locally modified and local (Bengali) language-translated Modified Health Assessment Questionnaire[15] was filled up.

Individual features in each compartment (narrowing, sclerosis, osteophytes, cysts, and attrition) were graded 0–3 and presence/absence of chondrocalcinosis was also noted. At follow up visits (1st–6th, 9th, 12th, 18th, and 24th month), a specialist doctor made a subjective assessment of the clinical condition with objective correlation, as much as it was practicable, for all the patients blinded for the type of treatment offered to the patient, to clinically assess the overall status of the treated knee joint (worse, no change, improved). Pain score (VAS), WD, and HAQ were recorded.

Student's paired test (*p* value) was also conducted. Analysis of variance for repeated measures was used to compare differences that were assessed by simple regression analysis. The differences in patient opinion of overall change, and relationship between clinical evidences were calculated by contingency table analysis incorporating mean with standard deviation (SD). Differences that were significant at the 5% confidence interval are quoted in the follow-up chart record.

At the completion of the study, patients who received cell therapy were offered steroid therapy if they voluntarily requested for that procedure, and *vice versa*.

38.3 Result and Analysis

Patient demographic data were more or less similar in both patient groups A and B as noted in Table 38.1. Age varied from 39 to 82 years, and the study included 50% male and 50% female patients; weight varied from 49.8 to 112.6 kg, height varied from 4 ft 11 in. to 6 ft 1 in.; the period of illness varied from 3 to 14 years, with the majority showing involvement of both knees.

As mentioned earlier, out of the 52 patients, we randomize the overall results of the effusion group

Table 38.1 Showing the patients selected for this study (epidemiological profile)

No of patients enrolled for this study: $N=52$
Age of group: 39–82 years
Sex: males 26 and females 26
Weight: 49.8–112.6 kg
Height: 4 ft 11 inches to 6 feet 1 inch
Duration: 3–14 years
Single knee effusion: 16
Both knee affection/effusion: 36
Treatment with analgesic including NSAID and Physiotherapy etc.: All of them

(32 cases) and the noneffusion group (20 cases). The clinical assessment is based arbitrarily on certain easy clinical parameters that the patient could understand irrespective of intelligence or education status.

These parameters are nine in number, namely, subjective appreciable decrease in knee pain at rest, walking without pain for some time, definite increase in walking distance before pain reappeared, etc. If seven of the nine parameters were positive with objective verification, the result was termed as adequate clinical improvement with the therapy in either A or B schedule, whichever was followed by the individual patient. If the result satisfied less than seven clinical parameters out of nine, the result was considered as inadequate clinical improvement.

If the overall impact of treatment in Group A is assessed and compared with the results of Group B as noted in the Table 38.2 and Fig. 38.1, it can be seen that a mean 92.3% patients showed improvement in the steroid-treated group (A) compared to a mean 88.46% of the patients in the amniotic fluid group (B) at the completion of the first month from the procedure ($p < .01$).

At the completion of the second month from the initiation of treatment, mean improvement was reported in 57.69% in the steroid-treated Group A and in 84.61% of the amniotic fluid-treated Group B ($p < .01$). The benefit of treatment was sustained at the end of the third month in Group B particularly, with mean 80.76% in the amniotic fluid-treated Group B and 46.15% in the steroid-treated Group A ($p < .01$) showing continued improvement. Evaluation after completion of the fourth month of treatment showed that a mean 30.76% of Group A and 73.07% of Group B maintained the benefit of the treatment ($p < .02$). The value for the 5th, 6th, 9th, 12th, and 24th months for Group A were noted as decreasing uniformly: mean 26.92%, 23.07%, 19.23%, 15.38%, and 15.38%, respectively. The identical value for the 5th, 6th, 9th, 12th, and 24th months for Group B were mean 65.38%, 57.69%, 53.84%, 50%, and 46.15%, respectively.

Out of 32 patients (61.53% of the patients) who had clinical evidence of joint effusion, which was aspirated before instillation of steroid or the amniotic fluid in the joint space with adequate antiseptic and aseptic precautions, 21(40.38%) patients were treated with amniotic fluid after aspiration of the joint space; the rest, that is, 11 (21.15%) patients were treated identically with intraarticular steroid. The results are worth noting:

(a) In this study, 18 out of the 21 patients (85.71%) with clinical evidence of joint effusion showed benefit with amniotic fluid cell therapy (Group B) as seen after 1 month.
(b) This can be compared to Group A, where seven patients (63.63%) out of the 11 with effusion, treated with intraarticular steroid, showed improvement.
(c) Among patients in the noneffusion group, five patients (B) were treated with amniotic fluid, and four (80%) of them were satisfied with the outcome after completion of 1 month.
(d) Similarly, of the 15 patients of the noneffusion group who were treated with intraarticular steroid (A), eight cases were enjoying the benefits of therapy (53.33%) at the end of the first month period.

This variation of the results in the effusion group and the noneffusion group could be due to the state of the disease. In early stages of osteoarthritis there is irritation of the joint space with the erosion of the joint cartilage. This irritation of the synovial membrane that is responsible for the effusion eventually dries up with the progression of the disease and an initiation of fibrous ankylosis, or in a late stage, bony ankylosis, sets in.

The overall findings of treatment in the effusion and noneffusion group were further supported by the results of VAS, WD, and HAQ study as reported in Table 38.3.

The results are supported by analysis of the VAS (Visual Analogue Pain Scale), WD (walking distance in

Table 38.2 Showing the clinical results of treatment in Groups A and B (subjective and objective improvement of at least seven out of nine clinical parameters)

Group A treated with intra-articular steroid $N=26$	Group B: treated with cell therapy $N=26$	Satisfaction after 1 month	Special comment (p <.01).
Assessment after 1 month showed improvement, i.e., mean subjective and objective assessment of definite relief in 92.3%±3.6% S.D. Lost follow-up (LFU)=Nil	Assessment after 1 month showed improvement, i.e., mean subjective and objective assessment of definite relief in 88.46%±2.8% Lost-follow up (LFU)=Nil	Group A=24 Group B=23	
Assessment after 2 months showed improvement, i.e., subjective and objective assessment of definite relief in mean 57.69%±4.8% (LFU)=Nil	Assessment after 2 months showed improvement, i.e., mean subjective and objective assessment of definite relief mean 84.61%±7.3% (LFU)=Nil	Satisfaction after 2 months Group A=15 Group B=22	(p <.01).
Assessment after 3rd month showed improvement, i.e., subjective and objective assessment of definite relief in Mean 46.15%±3.4% (LFU)=Nil	Assessment after 3rd month showed improvement, i.e., subjective and objective assessment of definite relief in Mean 80.76%±7.4% (LFU)=Nil	Satisfaction after 3 months Group A=12 Group B=21	(p <.01).
Assessment after 4th month showed improvement, i.e., subjective and objective assessment of definite relief in Mean 30.76%±2.9% (LFU)=Nil	Assessment after 4th month showed improvement, i.e., subjective and objective assessment of definite relief in Mean 73.07%±6.8% (LFU)=Nil	Satisfaction after 4 months Group A=8 Group B=19	(p <.02).
Assessment after 5th month showed improvement, i.e., subjective and objective assessment of definite relief in Mean26.92%±2.9% SD (LFU)=Nil	Assessment after 5th month showed improvement, i.e., subjective and objective assessment of definite relief in 65.38%±4.9% SD (LFU)=Nil	Satisfaction after 5 months Group A=7 Group B=17	(p <.01).
Assessment after 6th month showed improvement, i.e., subjective and objective assessment of definite relief in (Mean) 23.07%±2.2% SD (LFU)=Nil	Assessment after 6th month showed improvement, i.e., subjective and objective assessment of definite relief in 57.69%±4.9% SD (LFU)=Nil	Satisfaction after 6 months Group A=6 Group B=15	(p <.01).
Assessment after 9th month showed improvement, i.e., subjective and objective assessment of definite relief in 19.23% (Mean)±2.1% (LFU)=Nil	Assessment after 9th month showed improvement, i.e., subjective and objective assessment of definite relief in Mean 53.84%±4.4% percent (LFU)=Nil	Satisfaction after 9 months Group A=5 Group B=14	(p <.01).
Assessment after 12th month showed improvement, i.e., subjective and objective assessment of definite relief in 15.38%±2.2% (LFU)=Nil	Assessment after 12th month showed improvement, i.e., subjective and objective assessment of definite relief in 50%±4.3% (LFU)=Nil	Satisfaction after 1 year Group A=4 Group B=13	(p <.01).
Assessment after 24th month showed improvement, i.e., subjective and objective assessment of definite relief in 15.38%±2.2% (LFU)=Nil	Assessment after 24th month showed improvement, i.e., subjective and objective assessment of definite relief in 46.15%±5.4% (LFU)=Nil	Satisfaction after 2 years Group A=4 Group B=12	(p <.01).

meters), and HAQ (Health Assessment Questionnaire) assessments. Table 38.3 outlines changes in VAS, WD, and HAQ in steroid (Group A) and cell therapy (Group B) patients at the third- and sixth-month follow-up. The results demonstrated a significant improvement in VAS at third month, which was sustained at the sixth-month interval assessment in both groups, but more so in the cell therapy Group B (p<.001). Again, a better and more positive improvement trend was noted at the third- and sixth-month assessments in WD (walking distance in meters) in case of Group B (cell therapy group with amniotic fluid), when compared to the steroid-treated Group A. The health analysis questionnaire results also supported the VAS and WD results of Group A and B justifying the validity and superiority of cell therapy from steroid therapy in this preliminary report (p<.01).

Fig. 38.1 Showing the overall 24-month follow-up of treatment of Group A (steroid-treated) and Group B (amniotic fluid cell therapy group)

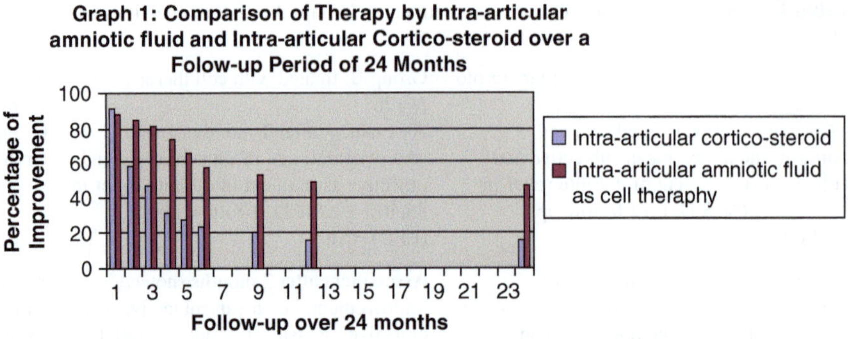

Table 38.3 Shows the value of the VAS, WD, and HAQ in steroid (Group A) and cell therapy (Group B)

(Pretreatment mean ± SD) VAS (mm)	(3rd month mean ± SD) VAS (mm)	(6th month mean ± SD) VAS (mm)	*p* value
Mean Group A values with SD: 56 ± 11.30	21 ± 6.47	32 ± 4.8	(p <.02).
Mean Group B values with SD: 57 ± 10.2	17 ± 3.4	12 ± 4.8	(p <.002).
Walking distance in meters (WD)			
Mean Group A values with SD 38.6 ± 4.8 m	51 ± 4.8 m	42.2 ± 4.8 m	(p <.01)
Mean Group B values with SD 39.8 ± 3.8 m	58.6 ± 6.9 m	61.4 ± 7.2 m	(p <.01)
Local language Modified Health Analysis Questionnaire (1–11)			
Mean Group A values with SD 2.2 ± .2	2.3 ± 0.2	2.2 ± 0.4	(p <. 002)
Mean Group B values with SD 2.4 ± 0.3	2.1 ± 0.12	1.8 ± 0.31	(p <.01)

The *t*-test, one-way analysis of variance (ANOVA) and a form of regression analysis

38.4 Discussion

Randomized clinical trials (RCTs) are regarded as the most reliable method of evaluating the effects of interventions in health care. RCTs are also considered the "golden standard" for providing research evidence for interventions in evidence-based health care.[16]

The validity and reliability of trial results are, however, largely dependent on the study design and the methodology in its conduct. Jadad A.R.[17] has defined the quality of a trial, with emphasis on the methodological quality, as "the confidence that the trial design, conduct, and analysis have minimized or avoided biases in its treatment comparisons." In this paper, the attempt was to follow the basic guideline to minimize investigator or other biases as far as practicable. Our subjective assessment of that scoring in this study is possibly 3 on the Jadad scale.

The present study is the first global report on a clinical comparison of the effect of amniotic fluid cell therapy and the impact of standard intraarticular palliative treatment in case of varying degrees of osteoarthritis-induced degenerated knee joints.

Cell therapy describes the process of introducing new cells into a tissue in order to treat a disease. The material used for cell therapy in this study is freshly collected amniotic fluid from women admitted by the family planning department in a government hospital for hysterotomy and ligation. Under normal circumstances, the fetus and the amniotic fluid contained sac are immediately disposed for eventual clearance through the incinerator of the hospital.

To recapitulate, amniotic fluid is to be found in the amniotic cavity that protects the fetus as a buffer and also helps growth and movement, and prevents adherence to the placenta or the surrounding structures. This clear watery fluid is contributed principally from the maternal blood via the amniotic fluid epithelium but freely intermixes with secretions from the fetal lung, kidney, gastrointestinal tract, and the skin; hence, the properties of this specialized fluid compartment is quite complex with contributions from both the maternal and the fetal side. Toward the outside, the amniotic cavity is delimited by the amniotic epithelium, the chorion laeve, and the decidua capsularis. The main constituents are water and electrolytes (99%) together with glucose, lipids from the fetal lungs, proteins with bactericide properties, and fetal epithelium cells.

As mentioned earlier, pleuripotent progenitor cells isolated from the amniotic fluid and the placenta possibly present an exciting contribution to the field of stem cell biology and regenerative medicine.

Compared with embryonic stem cells, progenitor cells isolated from the amniotic fluid have many similarities: they can differentiate into all three germ layers, they express common markers, and they preserve their telomere length. However, progenitor cells isolated from the amniotic fluid and placenta have considerable advantages. They easily differentiate into specific cell lineages and further, they avoid the current controversies associated with the use of human embryonic stem cells.

Pregnancy results in the acquisition of specialized and unique cells that may have clinical applications and therapeutic potential. Whether the pregnancy-associated progenitor cells (PAPCs) are hematopoietic stem cells (HSCs), mesenchymal stem cells (MSCs), or are a new population of stem cells is an unresolved issue. It is also unknown whether PAPCs respond to all types of maternal injury or only those injuries that recruit stem cells.

It is possible that these cells, since they are fetal in origin, have a higher proliferative capacity or more plasticity than their equivalent adult (maternal) cells. In the current debate over the use of embryonic stem cells for treatment of disease, the discovery of a population of fetal stem cells that apparently differentiate from the ones in adult women, and can be acquired without harming the fetus, may be significant.[18, 19]

The growing fetus in the womb is an eternal source of stem cells. The initial interest in the field started with the use of placental blood-derived hematopoietic stem cells in Fanconi's anemia in 1988 by the legendary Prof Elaine Gluckman. Meanwhile, scientists have been able to isolate and differentiate only 30% of mesenchymal stem cells (MSCs) on an average, extracted from a newborn's umbilical cord jelly-like material, shortly after birth. The success rate for amniotic fluid-derived stem cells, on the other hand, is close to 100%. Analysis of surface markers shows that progenitor cells from amniotic fluid express human embryonic stage-specific marker SSEA4, and the stem cell marker Oct4, and do not express SSEA1, SSEA3, CD4, CD8, CD34, CD133, C-MET, ABCG2, NCAM, BMP4, TRA1-60, and TRA1-81 [51, 52].

38.5 Differentiation of Amniotic Fluid- and Placenta-Derived Progenitor Cells

The progenitor cells derived from amniotic fluid and the placenta are pleuripotent and have been shown to differentiate into osteogenic, adipogenic, myogenic, neurogenic, endothelial, hepatic, and renal phenotypes in vitro. Each differentiation has been performed through proof of phenotypic and biochemical changes consistent with the differentiated tissue type of interest. In 2007, Perin et al. showed that AFSC (amniotic fluid stem cells) could be induced to differentiate into renal cells when placed into an in vitro embryonic kidney environment.[20]

In this preliminary clinical study, freshly collected amniotic fluid has been utilized as a source of cell therapy with the hypothetical assumptions that the mesenchymal cells of the AF (amniotic fluid) will participate in the knee joint repair process, the viscosity of the amniotic fluid will assist lubrication, and the bactericidal property of the amniotic fluid will guard against inadvertent infection. The idea was to match/compare this new therapeutic protocol (cell therapy for regeneration) with the globally accepted standard protocol of intraarticular injection of long-acting steroid triamcinolone.

The problem of knee pain is very common after the age of 50 years. Varying stages and grades of osteoarthritis due to degeneration of the knee joint plays the most important role behind such painful knee problems. The main pharmacological treatments remain analgesics and nonsteroidal antiinflammatory drugs (NSAIDs) although the role of these two treatments in the management of OA has been questioned.[21–24] So far, the most important antiinflammatory drug available in rheumatology, which can give some real relief in osteoarthritis is corticosteroid. In this context, it is first necessary to explain the treatment with corticosteroids to understand the implications of the present study, since intraarticular injection of steroid is a common treatment for osteoarthritis of the knee practiced globally by rheumatologists.

The aim of treatment in patients with osteoarthritis (OA) is to reduce symptoms, minimize disability, and limit the structural changes in the osteoarthritis-affected joint. But it has been observed that in OA a combination of factors, both mechanical and biochemical as well as immunological (and its effects) or cytokine effects, not only cause hyaline cartilage damage with time but also affects the synovial membrane. The subchondral bone, ligaments, and periarticular muscles also show varying degree of involvement and derangement. Synovial membrane inflammation in OA patients is probably related to the destruction of hyaline cartilage and the subsequent release of cartilage breakdown products into the synovial fluid. Clinical evidence suggests that the benefits of even a strong antiinflammatory drug like steroid is short-lived, lasting for usually 1–4 weeks[25] In the present study we have followed the guideline of the use of 40 mg triamcinolone as recommended by the American College of Rheumatologists.[26]

Whether steroid injection flares the pain and deteriorates the joint is a valid question as has been isolated in two cases in this study that were later proved to have tubercular infection. But excepting in cases of undiagnosed tuberculosis of the knee, steroid injection does not appear to have any important adverse effect on the whole. Studies of cartilage damage, however, tend to suggest that changes are more likely due to the underlying disease than the steroid injection.[27,28,29]

38.6 New Horizon for Offering a Cure (Repair) for Osteoarthritis with Simple Cell Therapy

In the developing world, surgical abortion as a method of family planning is practiced widely. Hysterotomy and ligation is a standard surgical method of termination in government hospitals in India. Aseptic collection of the amniotic fluid is not a difficult job for experienced gynecologists and obstetricians who perform this simple surgery with skill and dedication. The aseptically collected amniotic fluid can be easily preserved in special containers in the vapor phase of liquid nitrogen chambers or jars.

This may work as an amniotic fluid bank that can supply amniotic fluid on demand. Amniotic fluid is a unique fluid made by nature; it is a cocktail of mesenchymal stem cells with antibacterial property, which is used in the present study as the cell therapy source for the repair of damaged cartilage, synovial membrane, supporting muscles, and supporting ligaments, as per the niche provided to these specialized stem cells for regeneration purposes, in advanced and degenerative osteoarthritis with satisfying results.

The amniotic fluid, because of its increased viscosity due to protein and other cellular suspension, differs from the steroid-treated fluid (normal saline), and may act as a lubricant that diminishes the irritation at the initial phase; and the mesenchymal cells, which do not express HLA antigens, may possibly help in the repair process of the adjacent structures in the joint space as a whole. Though the epidemiological background (Table 38.1) of Groups A and B are grossly randomized, the result of the therapy (shown in Fig. 38.1 and Table 38.2), strongly supports the potential of this new form of cell therapy in case of advanced osteoarthritis. The present treatment proved to be much superior to, and lasted longer than, the conventional widely practiced therapy with corticosteroid instillation at the joint.

Lastly, it may be noted with interest that in this simple method of cell therapy, Group B maintained superior patient's satisfaction in 12 cases only out of 26 enrolled patients, after completion of the 24-month follow-up period. The corresponding number for the standardized universally practiced protocol of intraarticular long-acting steroid (Group A) therapy for advanced osteoarthritis is a pathetic figure of four cases only (Fig. 38.1). The results are further supported by the VAS, WD, and HAQ assessments as mentioned in Table 38.3, which reiterated a significant improvement in VAS at third month and was sustained at the sixth-month interval assessment in both groups, but more so in the cell therapy Group B ($p < .001$).

38.7 Conclusion

Intraarticular amniotic fluid instillation is a new method of treatment in advanced osteoarthritis when the patient is not getting any relief with conventional analgesic and physiotherapeutic support.

The long-term follow-up result of this type of cell therapy justifies its procedural superiority over conventionally and universally practiced intraarticular long-acting corticosteroid triamcinolone (($p < .001$).

Acknowledgment The Department of Science and Technology, Government of West Bengal supported the investigator with a research grant during his tenure at Bijoygarh State Hospital from 1999 to 2006.The work started in Bijoygarh Government Hospital (1999–2006) and was followed up at Vidyasagore Government Hospital subsequently. The author gratefully acknowledges the support of the patients who volunteered for this research work. Guidance of Prof K. L. Mukherjee of Biochemistry and Prof M. K. Chhetri, former Director of Health Services and Prof B. K. Dutta of Orthopaedics are also acknowledged.

References

1. Johnson HL. Observation on the prevention of post operative peritonitis and abdominal adhesions. *Surg Gynac Obstet.* 1927;XLV:612.
2. Johnson HL. An exposition and the preparation and administration of the amniotic fluid concentrate. *N Engl J Med.* 1935;CCXII:557.
3. Jones ES. Joint lubrication. *Lancet.* 1936;1:1043.
4. Lacey JT. Amniotic fluid - A clinical study. *Ann Surg.* 1935; Cl:529.
5. Collins DH. The pathology of sinovial effusions. *J Pathol Bacteriol.* 1936;XLII:113.
6. Mandell S. The use of amniotic fluid concentrate on orthopedic conditions. *J Bone Joint Surg Am.* 1938;20:167-177.
7. Leung VY, Chan D, Cheung KM. Regeneration of intervertebral disc by mesenchymal stem cells: potentials, limitations, and future direction. *Eur Spine J.* 2006;15(suppl 3): S406-413.
8. Sobajima S, Vadala G, Shimer A, Kim JS, Gilbertson LG, Kang JD. Feasibility of a stem cell therapy for intervertebral disc degeneration. *Spine J.* 2008;8(6):888-896.
9. Kaviani A, Perry TE, Dzakovic A, Jennings RW, Ziegler MM, Fauza DO. The amniotic fluid as a source of cells for fetal tissue engineering. *J Pediatr Surg.* 2001;36(11):1662-1665.
10. Prusa AR, Hengstschlager M. Amniotic fluid cells and human stem cell research: a new connection. *Med Sci Monitor.* 2002;8(11):RA253-257.
11. Medina-Gomez P, del Valle M (1988) Cultivo de celas de liquido amniotico. Analisis de colonias, metafases e indice mitotico, con fin de descartar contaminacion de celulas maternas. *Ginecol Obstet Mex.* 56122–1
12. In't Anker PS, Scherjon SA, Kleijburg-van der Keur C, et al. Amniotic fluid as a novel source of mesenchymal stem cells for therapeutic transplantation. *Blood.* 2003;102(4):1548-1549.
13. Prusa AR, Marton E, Rosner M, et al. Neurogenic cells in human amniotic fluid. *Am J Obstet Gynecol.* 2004;191(1):309-314.
14. Ref: Royal College of General Practitioners. *Morbidity Statistics from General Practice 1991–92.* London, U.K.: HMSO; 1995.
15. McAlindon TE, Cooper C, Kirwan JR, Dieppe P. Determinants of disability in osteoarthritis of the knee. *Ann Rheum Dis.* 1993;52:258-262.
16. Sjögren P, Halling A. Quality of reporting randomised clinical trials in dental and medical research. *Br Dental J.* 2002; 192:100-103.
17. Jadad AR. *Randomised Controlled Trials.* London, U.K.: BMJ Books; 1998:28-36.
18. O'Donoghue K, Choolani M, Chan J, et al. Identification of fetal mesenchymal stem cells in maternal blood: implications for non-invasive prenatal diagnosis. *Mol Hum Reprod.* 2003;9:497-502.
19. O'Donoghue K, Chan J, de La Fuente J, et al. Microchimerism in female bone marrow and bone decades after fetal mesenchymal stem-cell trafficking in pregnancy. *Lancet.* 2004; 364:179-182.
20. Perin L, Giuliani S, Jin D, et al. Renal differentiation of amniotic fluid stem cells. *Cell Prolif.* 2007;40(6):936-948.
21. Friedman DM, Moore ME. The efficacy of intraarticular steroids in osteoarthritis: a double-blind study. *J Rheumatol.* 1980;7:850-856.
22. Mazieres B, Masquelier AM, Capron MH. A French controlled multicenter study of ntraarticular orgotein versus intraarticular corticosteroids in the treatment of knee osteoarthritis: a one-year follow up. *J Rheumatol Suppl.* 1991;27:134-137.
23. Cederlof S, Jonson G. Intraarticular prednisolone injection for osteoarthritis of the knee. A double blind test with placebo. *Acta Chir Scand.* 1966;132:532-537.
24. Dieppe PA, Sathapatayavongs B, Jones HE, Bacon PA, Ring EF. Intra-articular steroids in osteoarthritis. *Rheumatol Rehabil.* 1980;19:212-217.
25. Raynauld J, Buckland-Wright C, Ward R, et al. Safety and efficacy of long-term intraarticular steroid injections in osteoarthritis of the knee. *Arth Rheum.* 2003;48:370-377.
26. American College of Rheumatology subcommittee on osteoarthritis guidelines. Recommendations for the medical management of osteoarthritis of the hip and knee. *Arth Rheum.* 2000;43:1905-1915.
27. Ayral X. Injections in the treatment of osteoarthritis. *Best Pract Res Clin Rehumatol.* 2001;15:609-626.
28. Gaffney K, Ledingham J, Perry JD. Intra-articular triamcinolone hexacetonide in kneeosteoarthritis: factors influencing the clinical response. *Ann Rheumat Dis.* 1995;54: 379-381.

Part XIII
Clinical Issue of Aborted Human Tissue

A Study and Follow-up (1999–2009) of Human Fetal Neuronal Tissue Transplants at a Heterotopic Site Outside the Brain in Cases of Advanced Idiopathic Parkinsonism

39

Niranjan Bhattacharya

39.1 Introduction

During brain development, one of the most important structures is the subventricular zone (SVZ) from which most neurons are generated. In adulthood, the SVZ maintains a pool of progenitor cells that continuously replace neurons in the olfactory bulb. Neurodegenerative diseases induce a substantial upregulation or downregulation of SVZ progenitor cell proliferation, depending on the type of disorder. Far from being a dormant layer, the SVZ responds to neurodegenerative disease in a way that makes it a potential target for therapeutic intervention.[1]

Patients with Parkinson's disease experience tremors, slurred speed, and slowness of movement that eventually progresses to total paralysis. In this progressive, debilitating illness, the cells in a small part of the brain called the substantia nigra are destroyed, depriving the striatum (the part of the brain that controls movement) of a critical molecule called dopamine. Despite devastating loss of motor control, mental faculties in Parkinson's patients remain intact, and while the disease is in itself not fatal, patients often succumb to complications such as injuries from falls or pneumonia. Parkinson's disease (PD) is also an intractable degenerative neurological disorder that affects nerve cells in the part of the brain controlling muscle movement. Age is the single most consistent risk factor, and with the increasing age of the general population, the prevalence of Parkinson's disease is likely to rise steadily in the future. The impact of the disease is indicated by the fact that mortality is two to five times higher among affected persons than among age-matched controls. People with Parkinson's disease often experience trembling, muscle rigidity, difficulty walking, problems with balance, and slowed movements. These symptoms usually develop after age 60, although some people affected by Parkinson's disease are younger than age 50. This disease (PD) is progressive, that is, the signs and symptoms become worse over time. But although Parkinson's disease may eventually be disabling, the disease often progresses gradually, and most people have many years of productive living after a diagnosis. The first line of treatment for Parkinson's disease is drug therapy. Unfortunately, L-dopa, a precursor of dopamine that can be absorbed by the brain, helps only as long as there are some substantia nigra cells still alive to absorb the drug. Once that area of the brain is destroyed, L-dopa becomes ineffective, which until recently meant that patients were left without any available treatment for this disorder.

More than 180 years ago, James Parkinson first described the disorder that bears his name. This is an important disease affecting natural movement and neuromuscular coordination. Other diseases which lack muscular coordination include essential tremor and dystonia. This group of diseases shows varying degrees of abnormalities of the basal ganglia (a large cluster of cells that lie deep in the hemispheres of the brain). The anatomic and biochemical connections from the basal ganglia to other parts of the brain are

N. Bhattacharya
Department of General Surgery, Obstertrics and Gynaecology and Clinical Immunology, Advanced Medical Research Institute, Gol Park, B.P. Poddar Hospital and Vidyasagore Hospital, Kolkata 700034, India

N. Bhattacharya and P. Stubblefield (eds.), *Regenerative Medicine Using Pregnancy-Specific Biological Substances*,
DOI: 10.1007/978-1-84882-718-9_39, © Springer-Verlag London Limited 2011

extremely complex and not fully understood. Although actual weakness can develop in some pathological states involving the basal ganglia, most of these diseases affect an individual's speed, quality, and ease of movement.

The diagnosis of Parkinson's disease in this study is essentially clinical. The primary manifestations are tremor, rigidity, and slowness of movement. At least two of the manifestations along with definite improvement after treatment with levodopa help in the diagnosis of PD. Despite the known benefit of levodopa in reducing the symptoms of Parkinson's disease, concern has been expressed that its use might hasten neuro-degeneration. Though at present the most effective treatment for Parkinsonism is achieved by levodopa, many patients experience undesirable side effects such as nausea, vomiting, cardiac arrhythmias, abnormal involuntary movements, and psychiatric disturbances. In case of advanced PD, where higher doses of agonist/levodopa + carbidopa and other drugs are needed, there may be motor fluctuations with wearing-off effects of levodopa before 4 hours, which can cause a treatment crisis.

There are studies which have assessed the effect of levodopa on the rate of progression of Parkinson's disease. The clinical data suggest that levodopa either slows the progression of Parkinson's disease or has a prolonged effect on the symptoms of the disease. In contrast, neuroimaging data suggest that levodopa either accelerates the loss of nigrostriatal dopamine nerve terminals or that its pharmacologic effects modify the dopamine transporter. The potential long-term effects of levodopa on Parkinson's disease remain uncertain.[2]

Surgical therapy for Parkinson's disease (PD) has been a treatment option for over 100 years. Advances in the knowledge of basal ganglia physiology and in techniques of stereotactic neurosurgery and neuroimaging have allowed more accurate placement of lesions or "brain pacemakers" in the sensorimotor regions of target nuclei. This, in turn, has led to improved efficacy with fewer complications than in the past. Currently, bilateral deep brain stimulation (DBS) of the subthalamic nucleus (STN) or the internal segment of the globus pallidus (GPi) is the preferred option (and is approved by the US Food and Drug Administration) for the surgical treatment of PD. The most important predictors for the outcome of DBS for PD are patient selection and electrode location.[3]The other possibility of treatment includes the stereotactic placement of implanting tissue in the brains of patients with Parkinson's disease to alleviate motor deficiency. This has been a focus of multiple international research efforts for the past several years and has generated much public interest. There was wide enthusiasm and optimism about the clinical benefits of autologous implants in the adrenal medulla.[4]

Fetal tissue transplants, in which organ-specific and nonspecific stem cells live in their natural environment, which are injected into the failing organs of patients, work on the premise that placed in the right environment, the transplanted cells take their cues from their surroundings and develop into the needed tissue. The stem cell components of the fetal tissue, if injected into the brain, become brain cells. If injected into the pancreas, they develop into pancreatic cells. Stem cells seem adaptable to such procedures, growing rapidly after transplantation, and secreting hormones and other chemicals that promote tissue growth. As an added bonus, these "master" cells are too undeveloped to be detected by the recipient's immune system, and thus often avoid the rejection that plagues normal organ transplant procedures.

This chapter focuses on human fetal neuronal tissue transplant in advanced idiopathic Parkinson's disease. It not only attempts to prove the safety of this procedure, but shows that even if the tissue is placed in a heterotopic site such as the axilla, (a) it is not rejected, and (b) there is improvement in the patient, signifying migration (?) of necessary stem cells from the alternate site to the site of injury/ requirement, and/ or secretion of hormones and chemicals to promote tissue growth to replace the damaged ones.

39.2 Fetal Tissue Used to Treat Diseases and Defects

One of the most promising areas in medical research today is fetal tissue transplantation. At stake is the source of stem cells – progenitor cells harvested from human fetuses that can differentiate into any cell in the adult human body. Fetal tissue transplants are being investigated as treatments for a wide range of debilitating human conditions. Researchers hope to cure diabetes by regenerating insulin-producing pancreatic cells in diabetics, and blindness by regrowing retinal tissue in the eye. Scientists hope to develop better treatments for heart attack victims with fetal tissue used to regrow damaged heart muscle. Fetal tissue transplants also look promising for a variety of problems caused by destroyed nerve cells, such as Parkinson's disease, Huntington's chorea, and even spinal cord injuries.

The crux of the method is the therapeutic use of fetal stem cells to replace damaged tissue that the body itself cannot repair. For instance, paralysis is currently incurable because, once destroyed, the nerve cells of the spinal cord are not able to grow back. Researchers hope that stem cells can be used to bridge a spinal cord injury in much the same way as skin cells grow back to cover a cut. Although not ready to be tried in people, procedures that inject fetal tissue cells at spinal cord breaks have shown encouraging results in small animals, as in one study where scientists were able to get partially paralyzed rats to walk again.[5] Similar experiments to regenerate nerve cells of the brain are also being investigated as cures for Huntington's chorea and Parkinson's disease, two diseases caused when specialized nerve cells in the brain begin to die off.

In certain centers of excellence in recent years, pioneering fetal tissue transplants into the brain of Parkinson's patients have shown promise in slowing or even reversing symptoms of the disease. In this treatment, cells from the pre-brain structures of 6- to 8-week-old fetuses are injected into the patient's striatum, where if all goes well they grow into a bundle of nerve cells that produce the needed dopamine. Patients with successful fetal tissue transplants have shown remarkable improvement in the severity of tremors and in their ability to move.

With such exciting results and millions of people in this country alone suffering from Parkinson's and other diseases that may be helped by fetal tissue transplants, patients and their advocates are urging further research into the use of stem cells. However, currently, the only reliable source of fetal stem cells is selectively aborted human fetuses, collected from abortion clinics with the permission of the mother.

The current chapter presents the findings of a study which examined the safety and therapeutic aspects of fetal neuronal tissue transplantation at a subcutaneous heterotopic site under local anesthesia in different patients with severe idiopathic Parkinsonism, not responding to conventional drugs. All the cases went through the voluntary consent protocol and received the approval of the Institutional Ethical Committee of the hospital headed by Prof M. K. Chettri and other senior professors and a sitting Additional District Judge of Alipore Court (who received a no-objection certificate from the Calcutta High Court).

The basic idea of this study is to verify the fundamental property of stem cells, i.e., migration to the site of injury (homing effect). The study examines whether this homing effect operates in case of fetal neuronal tissue transplant, which is a rich source of neuronal progenitor cells, or the stem cell, and helps to restore or repair the neuronal microenvironment with neurocytokines and other essential amines necessary for that action. The idea of selecting the axilla for the transplant of the fetal neuronal tissue came from the fact that it is a convenient site from which the tissue is easily retrievable in both male and female with little local anesthesia; moreover, it has vascularity, leaves less of a scar, and is aesthetically suitable.

The successful development of fetal cell/tissue transplantation in adults has resulted in the possibility of eventual therapeutic solutions in a variety of intractable diseases.[6–16] Umbilical cord blood transfusion in the adult system appears to be safe. Further, an investigator has also reported on the successful transplantation of fetal thymus in HLA randomized adult axilla, to combat leucopenia in the background of non-Hodgkin's lymphoma. One reason that the transplant is not rejected and can be implanted successfully is because, during intrauterine growth, the human fetus passes through the pre-immune (before 15 weeks) and subsequently through the hypo-immune phases of growth and maturation. The expression of hypo-antigenicity of the growing fetus in utero provides an excellent opportunity for fetal tissue/organ transplant, in its pre-HLA state of growth and maturation. The assumption here is that hypo-antigenic naive fetal cells will not be targeted by the hosts' HLA system.

In Calcutta, a group of investigators worked from 1979 on the process through which the human fetus acquires immunocompetence.[7–15] The current investigator was a part of that group and developed concepts on the immunocompetence of fetal tissue, which was perceived to be safe for transplants.

Transplantation of human embryonic dopamine neurons in the brain, to be more precise, the CT-guided stereotactic placement of embryonic mesencephalic tissue in the putamen or putamen + caudate region has shown marked improvement in Parkinson's disease. Even fetal pig neuron cells have shown survival in patients with Parkinsonism, when placed in the caudate + putamen region. Hence, neuro-transplantation has been proposed as a potential treatment for neurodegenerative disorders, from Parkinsonism to Huntington's disease.[17–20]

The science of stem cell biology is based on three cardinal behaviors of stem cells, i.e., stem cells can easily migrate to a site of injury; it has transdifferentiation properties based on its environment; and lastly, stem cells are immortal due to the telomeric reverse transcriptase activity of the stem cells that prevent the shortening impact on the telomeric end after cell division. This study is further aimed at verifying the basic stem cell characteristic in the clinical setting of Parkinsonism. If found effective, this method may prove to be a simple method avoiding the costly stereotactic CT guided basal ganglion placement of the neuronal tissue in patients suffering from Parkinson's disease.

In this work, the fetuses were collected from consenting mothers admitted for hysterotomy and ligation. Fetal tissue was collected from mothers admitted at Bijoygarh State Hospital for transplantation in patients admitted to the same institute. Screening of the cases was done in Bijoygarh State Hospital and sometimes, in case of any confusion, the data were re-verified at A. M. R. I. Hospital, Mahatma Gandhi Research Laboratory, B. P. Poddar Hospital, Roy and Tribedi Laboratory, K. C. M. Clinical Diagnosis and Research Centre, etc.

39.3 Materials and Methods

Initially 56 patients volunteered and were cleared by the Institutional Ethical Committee for enrollment in the fetal brain tissue transplantation protocol; however, this transplantation service could be offered to 48 patients only, who completed the rigorous clinical protocol of diagnosis and treatment. Of the 48 patients, eight were female and 40 were male, and the age varied from 45 to 75 years.

To recapitulate the clinical features of this disease in brief: the disease is named after the English physician, James Parkinson for his famous essay on "An essential on the shaking palsy" (1817). This chronic neurodegenerative disorder of the central nervous system affects the sufferers' motor skills, speech, and other functions. Motor skill affectation includes the symptoms of tremor (akinetic rigid variety), rigidity (cogwheel type), bradykinesia/akinesia (dysrhythmic and detrimental loss of amplitude), and postural instability (loss of reflexes). Apart from gait and postural disturbances including dystonia, speech, and swallowing disturbances, there is often fatigue with musk face, micrographia, loss of motor coordination, akathisia, etc. This disease is often associated with neuropsychiatric disturbances, namely, cognitive disturbances with slow reaction time, executive dysfunction, dementia with short-term memory loss (procedural memory loss more than declarative memory) apart from sleep disturbances, i.e., somnolence-disturbed REM sleep and terminal insomnia. Mood disorders are also associated with this disease, which includes both depression and anxiety.

The 48 patients were included in the study of HLA randomized developing fetal brain tissue transplant protocol, after taking informed consent from the patient's guardian and approval from the multidisciplinary ethical committee, as per the Indian Council of Medical Research (ICMR) guidelines. This group of experts, headed by a senior judge collectively weighed and considered the (a) possible clinical benefits and costs, (b) privacy and safety of the donors and the recipients, (c) utmost care so as to eliminate the possibility of secondary gain in the decision to undergo an abortion, (d) the right of the woman to choose abortion and donate tissue for research.

The criteria for exclusion (by the investigator in coordination with senior physicians and neurologists) included major psychiatric illness and other substantial medical problems such as advanced cancer, uncontrolled diabetes, ulcerative colitis, malignant hypertension, hyperthyroid, and other intractable chronic diseases including Parkinson's-plus syndrome and secondary Parkinson's disease.

Blood was tested for the donor mother who kindly consented to donate her fetus for medical research voluntarily, for HIV (1 and 2), hepatitis B and C, apart from other routine tests as per the anesthetist's guidance for safe surgery, such as, Hb, TC, DC, ESR, urea, creatinine, glucose (fasting and postprandial). Only healthy mothers were chosen for fetal tissue donation.

Similarly blood was drawn from the recipient of the fetal neuronal tissue transplant, for hematological parameters (hemoglobin, total count, differential count, erythrocyte sedimentation rate, platelet count), hepatitis B and C screening, HIV[1, 2] screening and also fasting and postprandial glucose, urea, creatinine, bilirubin, ferritin for eventual comparison of the pre- and post-transplant values. These were studied in the preoperative phase and continued postoperatively at weekly rate up to 4 weeks from the date of transplantation. In addition C-reactive protein (CRP), antinuclear antibody (ANA), Anti-dsDNA were routinely done before and 1 month after the transplantation to see if there was any adverse immunological effect of transplantation.

Following standard antiseptic, aseptic, and premedication protocol, the brain transplant recipients' site was prepared using 1% lignocaine infiltration anesthesia at the proposed site in the axillary fold of the skin. Then, the skin of the recipient was prepared after thoroughly washing with betadine and spirit. A 2 cm long and 2 cm deep tissue space with good vascularity was dissected and prepared in each patient selected for receiving the brain tissue from a developing fetus.

In an adjacent OT, a fetus was collected from a mother undergoing hysterotomy and ligation (up to 20 weeks) under general anesthesia. The fetus and the intact sac were taken out and brought to the transplant recipient's table. The second group of surgeons took out 5–15 g of the fetal subcortical tissue and the adjacent developing frontal subcortical region, i.e., within 1 cm of the frontal opercula of the fetus, under an operative microscope and immediately weighed it on an electronic weighing machine before placing it in the recipient's prepared surgical wound site. After putting the neuronal tissue at the specific site, the wound was then closed by (000 Vicryl) atraumatic interrupted suture. A little tissue was taken for electron microscopical study (SEM) apart from histology.

Three months from the primary operation, some brain tissue was taken out with a small elliptical incision under local anesthesia for serial histological evaluation by hematoxylin and eosin (H&E) stain. In one case, the transplanted tissue was taken out after 10 years.

39.4 Result and Analysis

In the present series, the diagnosis of PD remained clinical,[21] and was conducted by an experienced consultant to exclude early Parkinson's disease, secondary Parkinsonism, and the Parkinson's-plus group of disorders. Only advanced idiopathic Parkinsonism cases were considered as subjects for the study. The problem of treatment of this disease in developing countries is partially the mind-set, which refuses to accept a prolonged treatment which has diminishing returns *vis-à-vis* the cost of treatment. Our referral service provided the options of stereotactic surgery/ablative or deep brain stimulation procedures in various brain nuclei/dopaminergic cell implantation, which could be conducted in private hospitals in our country or abroad at a prohibitive price. All the 48 patients refused to go for these options. They also refused our suggestion to purchase a peripheral apomorphine pump.

The clinical profile of the patients who underwent our transplantation surgery protocol is being followed up including assessment of the mental state (Mini-Mental State Examination), disability assessment, and the mood of the patient to assess depression and prevailing anxiety state. The other studies include assessment of the hematological, immunological, metabolic parameters to see if there is any adverse impact of transplantation of HLA randomized neuronal tissue at the heterotopic site in the axilla on the host system. Lastly, the study of a small amount of retrieved tissue from the axilla (after 3 months from its placement) was done under simple microscopy and scanning electron microscopy, as mentioned earlier, to see if there was graft versus host reaction involving the fetal tissue.

Details regarding the patients enrolled for the fetal subcortical midbrain tissue transplant are given in Table 39.1.

Out of the 48 cases (patients), all were persuaded to allow partial retrieval of the fetal tissue from the axilla under local anesthesia in the operation theater; however, eight patients refused to allow retrieval because they thought that the improvement was due to the transplant, and as such if they continued with the transplant, there would not be any further problem. They were further persuaded and ultimately partial retrieval of the fetal tissue from the last patient was done after more than 10 years (serial no. 47 of Table 39.1) from the date of placement of the transplant.

What is significant is the fact that the 5–15 g of fetal tissue heterotopic subcutaneous graft never caused any graft rejection or any other features of acute, subacute, or chronic graft versus host reaction in a single case after a long follow-up till date, nor were there any apparent changes in the biochemical, i.e., hepatic, renal, or metabolic, parameters in the host system. What is also intriguing is the persistence of the fetal tissue, which was not destroyed by the

Table 39.1 List of patients who took the fetal subcortical fetal brain tissue transplant (number-48) and posttransplant (1 month) follow-up

Case no/ name	Age & sex	Date of transplant	Duration of PD (in years)	Treatment received earlier	Age of the fetal brain (in weeks)	Clinical improvement before graft removal	Date of graft removal	Sequential TC, DC, ESR study
1,G.R	65 M	5.2.2001	5	A-H	14	Mild	5.5.2001	No gross variation
2,A.M.	55 M	8.2.2001	5	-do-	20	Moderate	8.5.2001	-do-
3,K.M.	71 M	19.2.2001	6	-do-	20	Moderate	19.5.2001	-do-
4,C.P.	66 M	19.2.2001	11	-do-	20	Mild	22.5.2001	-do-
5,M.S.	48 M	26.2.2001	7	-do-	10	Mild	27.5.2001	-do-
6,B.P.	45 F	27.3.2001	4	-do-	16	No variation	27.6.2001	-do-
7,K.P.	65 M	27.3.2001	11	-do-	16	Moderate	27.6.2001	-do-
8,T.C.	73 M	27.6.2001	4	-do-	18	No Variation	27.9.2001	-do-
9,B.M.	56 M	4.6.2001	6	-do-	10	Mild	4.9.2001	-do-
10,S.B.	45 M	27.6.2001	13	-do-	18	Mild	27.9.2001	-do-
11,T.P.	75 F	22.5.2001	9	-do-	10	Moderate	22.8.2001	-do-
12,S.T.	55 M	30.6.2001	5	-do-	10	Moderate	30.9.2001	-do-
13,S.M.	61 M	5.2.2001	5	-do-	14	Mild	5.5.2001	-do-
14,P.S.	53 M	8.2.2001	5	-do-	20	Moderate	8.5.2001	-do-
15,R.R.	73 M	19.2.2001	6	-do-	20	Moderate	19.5.2003	-do-
16,P.D.	64 M	19.2.2001	11	-do-	20	Mild	22.5.2001	-do-
17,G.D.	47 M	26.2.2001	7	-do-	10	Mild	27.5.2001	-do-
18,B.S.	48 F	27.3.2001	4	-do-	16	No variation	27.6.2004	-do-
19,K.M.	62M	27.3.2001	11	-do-	16	Moderate	Refused retrieval	-do-
20,R.D.	72 F	22.5.2001	9	-do-	10	Moderate	22.8.2002	-do-
21,G.B.	54 M	4.6.2001	6	-do-	10	Mild	4.9.2002	-do-
22,T.D.	46M	27.6.2001	13	-do-	18	Mild	27.9.2003	-do-
23,G.C.	57 M	27.6.2001	4	-do-	18	No variation	27.9.2001	-do-
24,P.T.	55 M	30.6.2003	5	-do-	10	Moderate	Refused retrieval	-do-

25,B.D.	62 M	5.2.2001	5	-do-	14	Mild	5.5.2003	-do-
26,N.C.	53 M	8.2.2001	5	-do-	20	Moderate	8.5.2006	-do-
27,D.G.	70 M	19.2.2001	6	-do-	20	Moderate	19.5.2003	-do-
28,S.P.	63M	19.4.1999	11	-do-	20	Mild	Refused retrieval	-do-
29,G.M.	49 M	26.2.2001	7	-do-	10	Mild	27.5.2001	-do-
30,N.B.	46 F	27.3.2000	4	-do-	16	No variation	Refused retrieval	-do-
31,S.D.	64 M	27.3.2001	11	-do-	16	Moderate	27.6.2001	-do-
32,G.P.	73 F	22.5.2001	9	-do-	10	Moderate	22.8.2001	-do-
33,R.S.	55M	4.6.2001	6	-do-	10	Mild	4.9.2001	-do-
34,K.C.	49 M	27.6.2000	13	-do-	18	Mild	Refused retrieval	-do-
35,M.D.	71M	27.6.2001	4	-do-	18	No variation	27.9.2001	-do-
36,R.B.	51 M	30.6.2001	5	-do-	10	Moderate	30.9.2001	-do-
37,T.C.	62 M	5.2.2001	5	-do-	14	Mild	5.5.2001	-do-
38,K.R.	53 M	8.2.2002	5	-do-	20	Moderate	Refused retrieval	-do-
39,G.G.	72 M	19.2.2001	6	-do-	20	Moderate	19.5.2001	-do-
40,N.D.	59 M	19.2.2001	11	-do-	20	Mild	22.5.2001	-do-
41,T.R.	42 M	26.2.2001	7	-do-	10	Mild	27.5.2001	-do-
42,G.T.	49 F	27.3.2002	4	-do-	16	No variation	Refused retrieval	-do-
43,S.S.	67M	27.3.2001	11	-do-	16	Moderate	27.6.2001	-do-
44,K.M.	73 F	22.5.2001	9	-do-	10	Moderate	22.8.2001	-do-
45,B.C.	66 M	4.6.2001	6	-do-	10	Mild	4.9.2001	-do-
46,S.D.	71 M	27.6.2001	13	-do-	18	Mild	27.9.2001	-do-
47,A.M.	44 M	27.6.1999	4	-do-	16	No variation	28.7.2009	-do-
48,T.D.	57 M	30.6.2001	5	-do-	10	Moderate	30.9.2006	-do-

All the patients noted in the table were treated with A-H of the available drugs and others forms of treatment from local source. A – levodopa; B – dopamine agonist group of drugs; C – catechol-O-methyltransferase (COMT) inhibitor group of drugs; D – anticholinergics; E – monoamine oxidase-B inhibitors; F – physiotherapy; G – behavioral therapy to combat depression, dementia, psychosis, etc.; H – symptomatic treatment for autonomic dysfunction immunoenzyme network of the host defense and the surveillance system. In the cases, where the graft was removed 10 years after the placement of the tissue, the same survival of the fetal tissue was noted. The conclusion is simple:

1. Fetal neuronal tissue can survive in the adult host where the neuroendocrine, immunoenzymatic, and cytokine regulation is distinctly different from the adult system.
2. There is similarity in the histological and electron microscopic findings of the retrieved fetal tissue from the host adult tissue, from the first month to the tenth year after its placement in the adult host. This is a truly astonishing finding.
3. This observation justifies the hypothesis that there is persistence of the stem cell component of the fetal tissue in the adult host in such a situation.
4. Fetal tissue creates its own microenvironment for its survival.
5. That there is no abnormal growth and differentiation of the fetal tissue justifies the observation that the genetic regulation with its apoptosis mechanism is in full operation in the adult host, which is ineffective in detecting and destroying the primitive hypoantigenic system existing in fetal tissue.
6. One observation/deduction is that it is possible that this may be one of the mechanisms through which pregnancy and neoplasm induce tolerance of the homograft.

The following figures and corresponding analysis are presented to trace the clinical impact of the transplant in the 48 patients who underwent the procedure.

A minimum score of 40 points was required for enrollment in the motor portion of the Parkinson's Disease Unified Rating Scale.[22] When the patient had been without medication, scores in this scale varied from 0 to 108. Clinical improvement is rated individually as mild, moderate, and substantial on the basis of objective assessment by the attending doctor. If the assessment showed a regression or degradation of the pretransplant score of 33.3%, it is shown as mild improvement; if the degradation of the pretransplant score is more than 33.3% and less than 66.6%, it is graded as moderate improvement and further degradation of the pretransplant score would lead to an assessment of substantial improvement. Pretransplant scoring of higher value in the Unified Parkinson's Disease Rating Scale indicates greater severity of symptoms and motor complications that could not be controlled by pharmacological therapy alone.

In the present study, both subjective and objective improvements of 83.3% score from the pretransplant level (to the date of assessment, i.e., 1 month after the installation of the fetal cortical graft), were noted, of which mild improvement was noted in 41.66% and another 41.66% patients showed moderate improvement from the pretransplant Unified Parkinsonism Scoring System; however, 16.66% of the patients did not show any objective positive response[22] (Table 39.1, Fig. 39.1).

In the present series, eight cases (16.66%) were female patients and the rest, i.e., 40 (83.33%) were male (Fig. 39.2).

The age varied from 45 to 75 years; 25.0% patients belonged to the 40+ age group, 27.1% patients belonged to 50+ age group, 25.0% patients belonged to the 60+ age group and the rest, 22.9% patients, belonged to 70+ age group (Fig. 39.3).

Patients with serial number 19, 24, 28, 30, 34, 38, 42 (vide Table 39.1) objected to retrieval of the fetal tissue and continued with the fetal tissue at the heterotopic site, on the basis of their own logic that retrieval may deteriorate the clinical gains achieved. However, they did not register any clinical problems.

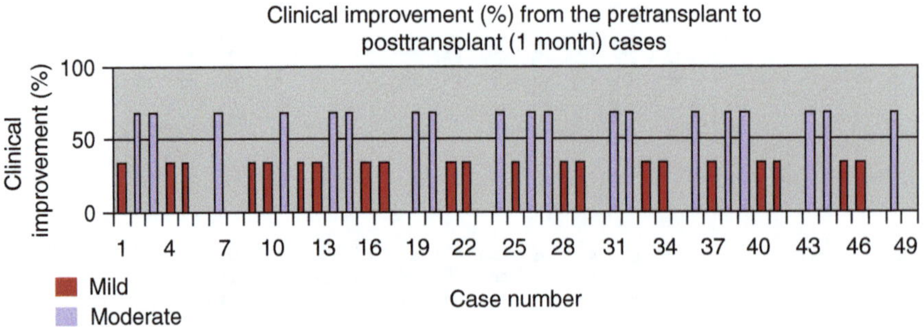

Fig. 39.1 Graphical impact of posttransplant (1 month) clinical improvement

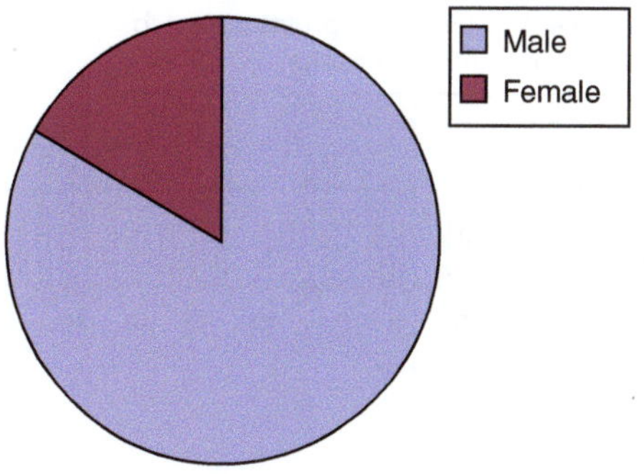

Fig. 39.2 Sex-wise distribution of patients

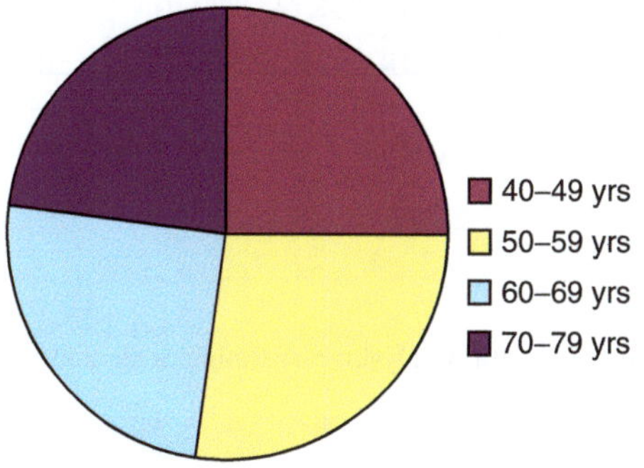

Fig. 39.3 Age group-wise distribution of patients

The same trends toward improvement were seen at the third month follow-up, as noted in Fig. 39.4.

The same improvement trends were seen at the sixth month, first-year, and even in the second- year follow-up, as noted in Figs. 39.5–39.7.

It appears from the study (Table 39.2 and Fig. 39.8) and the follow-up of the 48 cases, where we did subcutaneous heterotopic placement of fetal subcortical tissue that 25% of the cases did not have L-dopa + carbidopa dosage reduction or dyskinesia scoring improvement in subjective and objective assessment. However, in the rest, i.e., 75% of the cases, there was a definite reduction of the L-dopa + carbidopa dosage: from 20% reduction (8.3%), 25% reduction (8.3%), 33% reduction (16.66%), 40% reduction (8.3%), 50% reduction (16.66%), 66% reduction (8.3%), to 75% reduction (8.33%). Similarly, clinical review (Fig. 39.13) revealed partial improvement (33%) of the dyskinetic status in one third of the patients (33%) in the present series in both subjective and objective assessment. While maximum improvement (66.6% from the pretransplant status) was noted in 16.66% cases, there were other shades of improvement in 16.6% cases (50% improvement), and 8.3% cases registered some improvement (25%). There was, however, no apparent objective improvement in 25% of the total cases, which included 16.6% cases where a

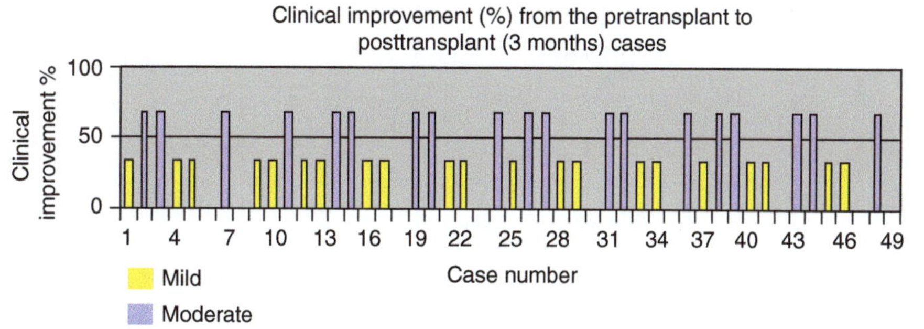

Fig. 39.4 Posttransplant (3 months) clinical improvement

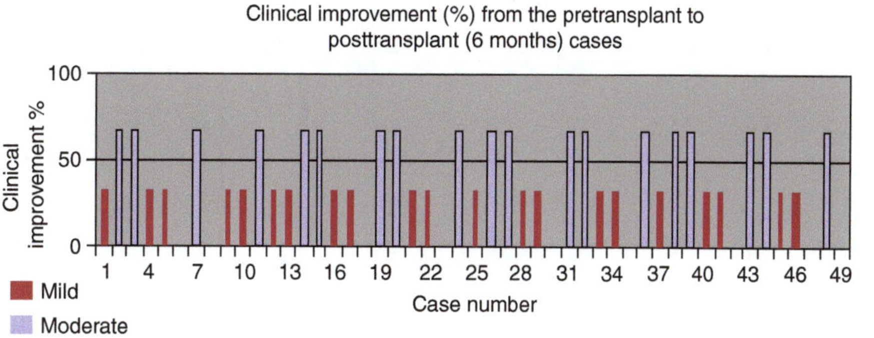

Fig. 39.5 Posttransplant (6 months) clinical improvement

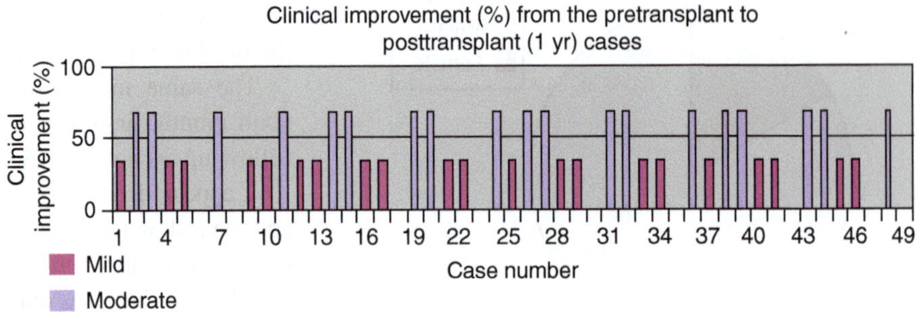

Fig. 39.6 Posttransplant (1 year) clinical improvement

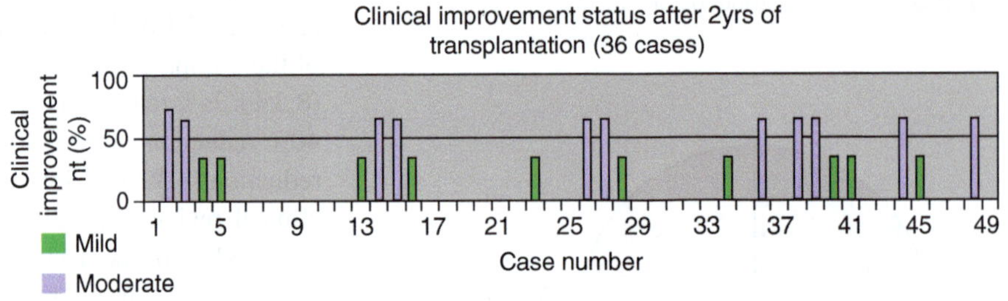

Fig. 39.7 Posttransplant (2 years) clinical improvement (number – 36)

Table 39.2 Clinical impact of fetal subcortical brain tissue transplant after 1 month on Dyskinesia Rating Scale and L-dopa ± carbidopa dosage

Case no	L-dopa + carbidopa dosage before transplant (in tablets/day)	L-dopa + carbidopa dosage 1 month after transplant (in tablets/day)	Dyskinesia rating (0–4)* before transplant	Dyskinesia rating (0–4)* 1 month after transplant	Very troublesome dyskinesia or deterioration	Sense of well-being after the transplant
1	4	2	3	1	Nil	Present
2	5	2	2	1	Nil	Present
3	6	3	4	3	Nil	Present
4	3	3	2	2	Nil	Present
5	4	3	2	2	Nil	Present
6	5	4	3	2	Nil	Present
7	4	1	3	1	Nil	Present
8	3	1	2	2	Nil	Present
9	6	4	4	2	Nil	Present
10	2	2	3	3	Nil	Present
11	3	2	3	2	Nil	Present
12	2	2	2	2	Nil	Present
13	4	2	3	1	Nil	Present
14	5	2	2	1	Nil	Present
15	6	3	4	3	Nil	Present

Table 39.2 (continued)

16	4	4	2	2	Nil	Present
17	4	3	2	2	Nil	Present
18	5	4	3	2	Nil	Present
19	4	1	3	1	Nil	Present
20	3	1	2	2	Nil	Present
21	5	4	4	2	Nil	Present
22	2	2	3	3	Nil	Present
23	3	2	3	2	Nil	Present
24	2	2	2	2	Nil	Present
25	4	2	3	2	Nil	Present
26	5	2	2	1	Nil	Present
27	6	3	4	3	Nil	Present
28	3	3	2	2	Nil	Present
29	4	3	2	2	Nil	Present
30	5	4	3	2	Nil	Present
31	4	1	3	1	Nil	Present
32	3	1	2	2	Nil	Present
33	6	4	4	2	Nil	Present
34	2	2	3	3	Nil	Present
35	3	2	3	2	Nil	Present
36	2	2	2	2	Nil	Present
37	4	2	3	1	Nil	Present
38	5	2	2	1	Nil	Present
39	6	3	4	3	Nil	Present
40	3	3	2	2	Nil	Present
41	4	3	2	2	Nil	Present
42	5	4	3	2	Nil	Present
43	4	1	3	1	Nil	Present
44	3	1	2	2	Nil	Present
45	5	4	4	2	Nil	Present
46	2	2	3	3	Nil	Present
47	3	2	3	2	Nil	Present
48	3	2	2	2	Nil	Present

*0: no dyskinesia; 4: severe dyskinesia on the basis of objective criteria assessment such as walking, putting on a coat, lifting a cup to the lips for drinking, etc. Each tablet of L-dopa contained 250 mg reduction of the L-dopa + carbidopa dosage could be seen without apparent destabilization of the clinical status from the pretransplant level. There was not a single incidence of posttransplant troublesome dyskinesia in this series with the fetal neuronal tissue subcutaneous heterotopic transplant protocol.

It appears from the follow-up that the clinical status was maintained (vide Figs. 39.8–39.12) at the same state in all the patients, excepting in case no. 2, where there was a 20% increase (deterioration) in L-dopa + carbidopa dosage and in case no. 10, where there was 50% reduction in L-dopa + carbidopa dosage (improvement); in case no. 25 there was a 25% increase in L-dopa + carbidopa dosage (deterioration) as graphically shown.

The dyskinesia scale also did not show any gross variation in the clinical status after the first month as noted vide Figs. 39.13–39.17. No intractable or troublesome dyskinesia was noted in any case after the neuronal fetal tissue transplantation, justifying this simple approach as an extremely safe and patient-friendly procedure.

Studying the safety level of this experimental procedure of fetal neuronal tissue transplantation protocol, the essential changes in the leucocytes count and CRP levels before and after the transplant were noted to see if there was any inflammatory impact on the host system due to the fetal neuronal tissue transplant. The overall impression is noted below:

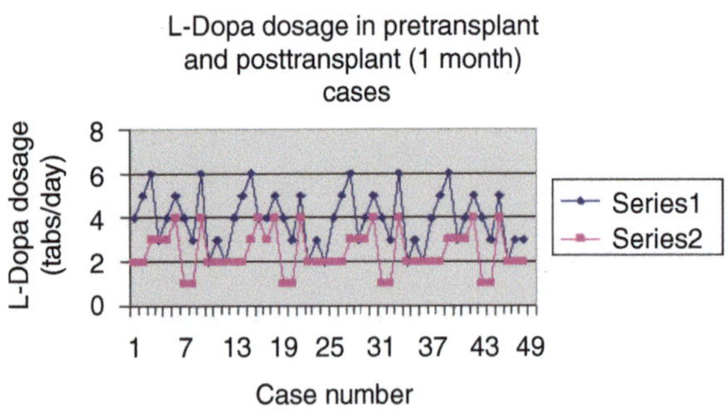

Fig. 39.8 L-dopa ± carbidopa dosage in pretransplant and posttransplant (1 month) casesSeries – 1: pretransplant cases; Series – 2: posttransplant cases

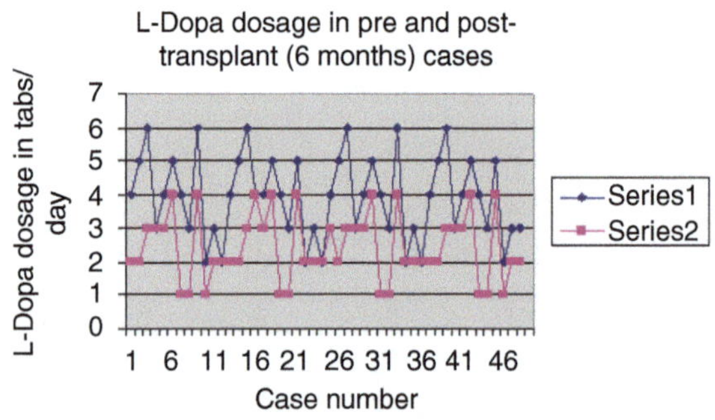

Fig. 39.10 L-dopa ± carbidopa dosage in pretransplant and posttransplant (6 months) CasesSeries – 1: pretransplant cases; Series – 2: posttransplant cases

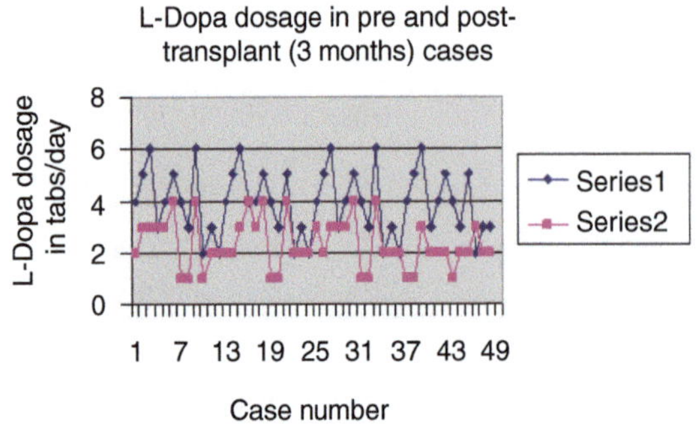

Fig. 39.9 L-dopa± carbidopa dosage in pretransplant and posttransplant (3 months) casesSeries 1: pretransplant cases; Series – 2: posttransplant cases

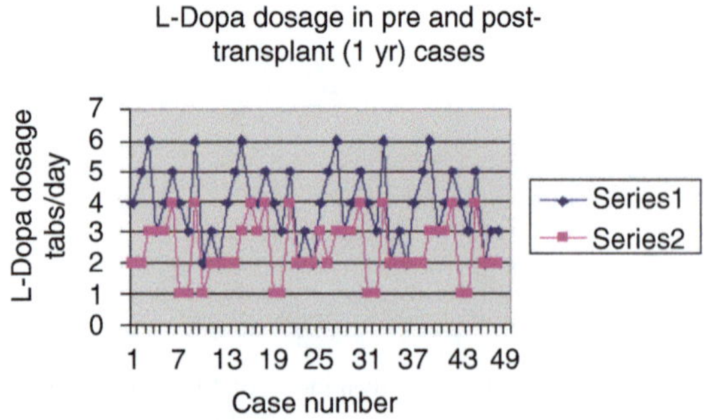

Fig. 39.11 L-dopa ± carbidopa dosage in pretransplant and posttransplant (1 year) casesSeries – 1: pretransplant cases; Series – 2: posttransplant cases

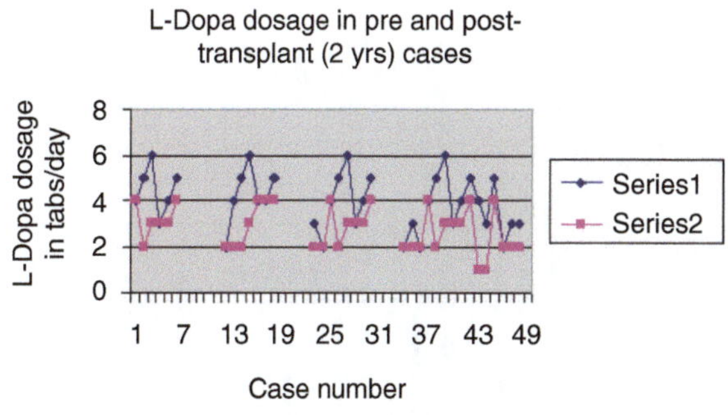

Fig. 39.12 L-dopa± carbidopa dosage in pretransplant and posttransplant (2 years) cases (number 36)Series – 1: pretransplant cases; Series – 2: posttransplant cases

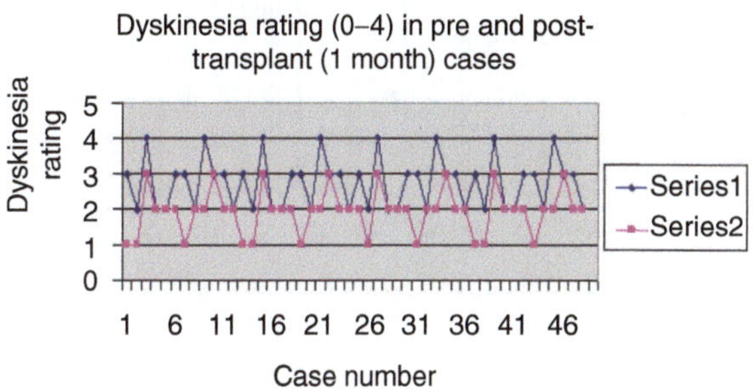

Fig. 39.13 Dyskinesia rating (0–4) in pretransplant and posttransplant (1 month) casesSeries – 1: pretransplant cases; Series – 2 posttransplant Cases

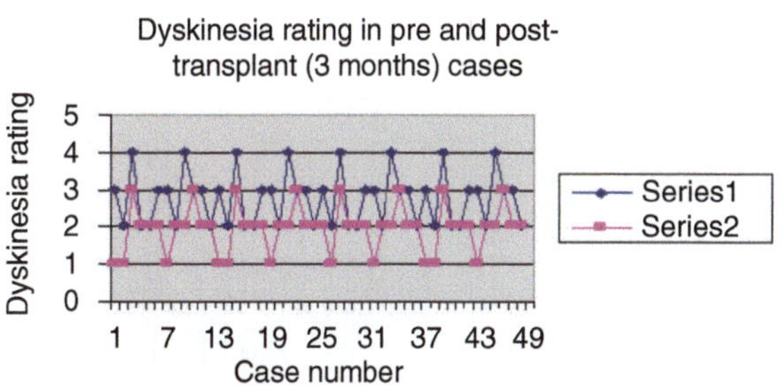

Fig. 39.14 Dyskinesia rating (0–4) in pretransplant and posttransplant (3 months) casesSeries – 1: pretransplant cases; Series – 2: posttransplant cases

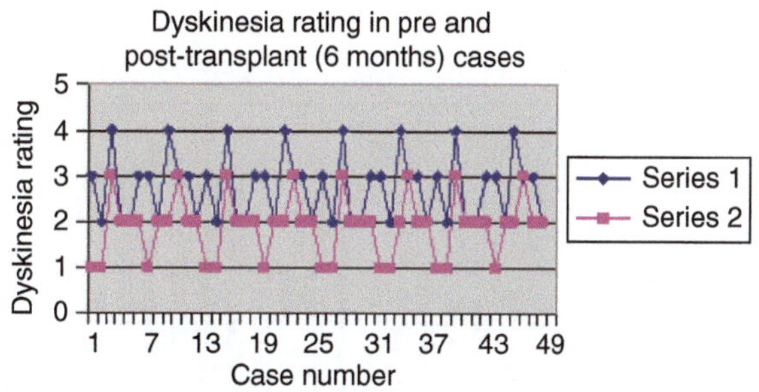

Fig. 39.15 Dyskinesia rating (0–4) in pretransplant and posttransplant (6 months) casesSeries – 1: pretransplant cases; Series – 2: posttransplant cases

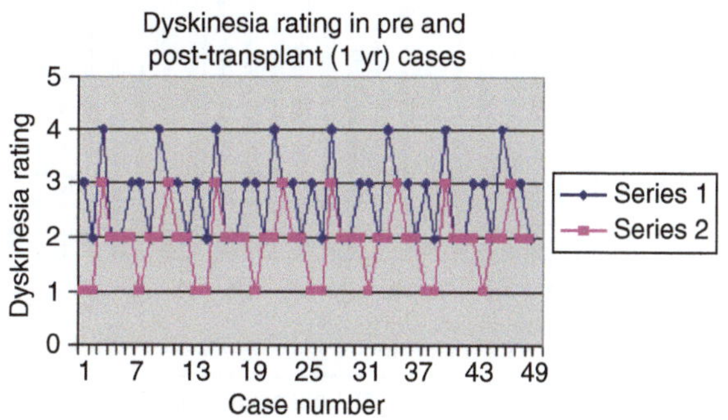

Fig. 39.16 Dyskinesia rating (0–4) in pretransplant and posttransplant (1 year) casesSeries – 1: pretransplant cases; Series – 2: posttransplant cases

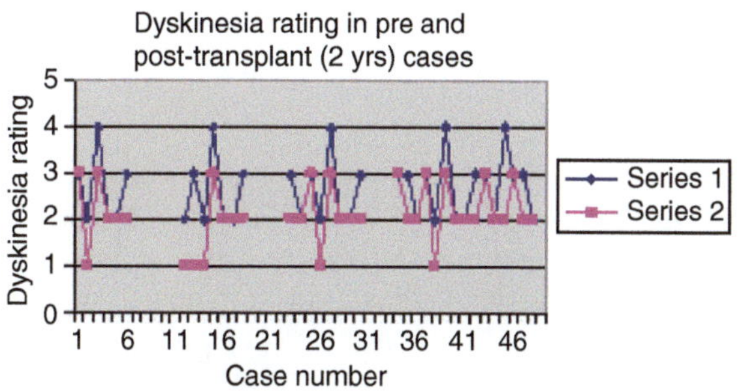

Fig. 39.17 Dyskinesia rating (0–4) in pretransplant and posttransplant (2 years) cases (number-36)Series – 1: Pretransplant cases; Series – 2: posttransplant cases

1. The standard inflammatory marker CRP does not show any appreciable change from the pretransplant to posttransplant value as seen in Fig. 39.18.
2. If the leucocytes value of the pretransplant level is analyzed and compared with the posttransplant level, as seen in Fig. 39.19, practically minimal or no impact can be observed. In this group, frequent respiratory and gastrointestinal bacterial infection, helminthiasis, subclinical tuberculosis, malnutrition, etc., were very common; hence, a slight change in leucocytes count up to 10% of the base level has little clinical-specific significance.
3. Similarly, studying the metabolic, hepatic, and renal functions, it was observed that there was practically no difference in the urea (Fig. 39.20), creatinine (Fig. 39.21), bilirubin (Fig. 39.22), glucose

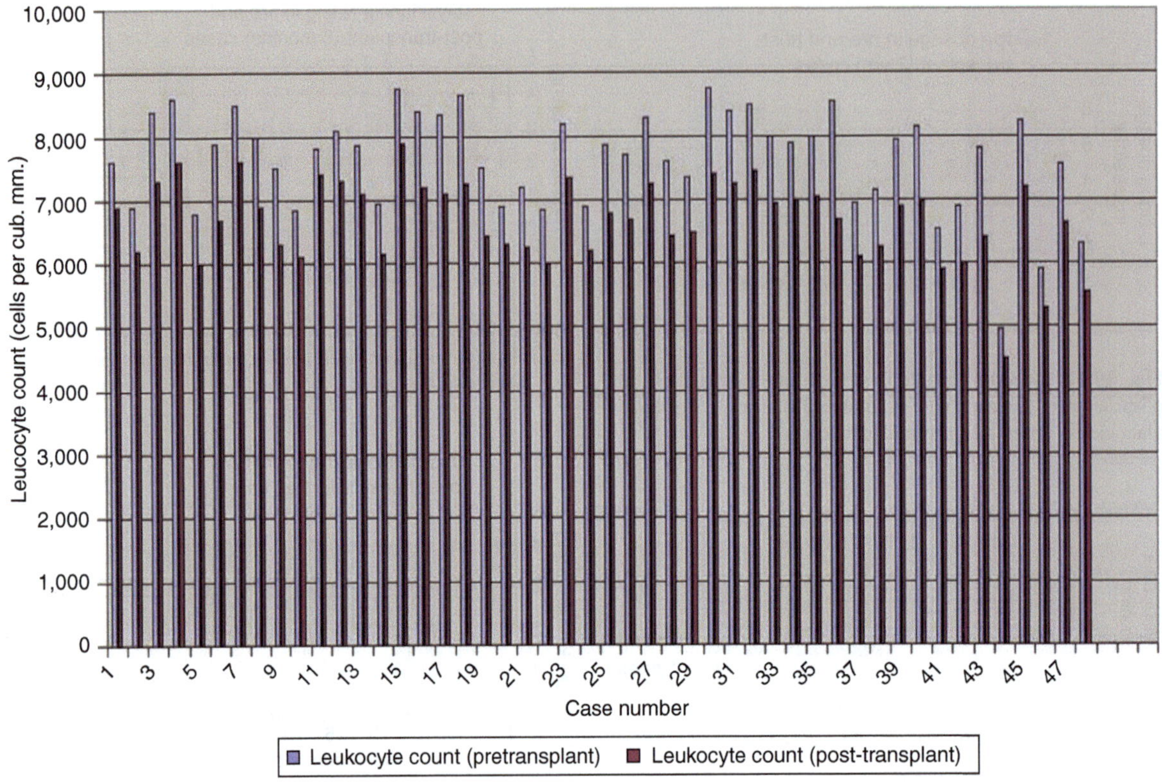

Fig. 39.18 Leucocyte count in pre- and post-transplant (1 month) cases

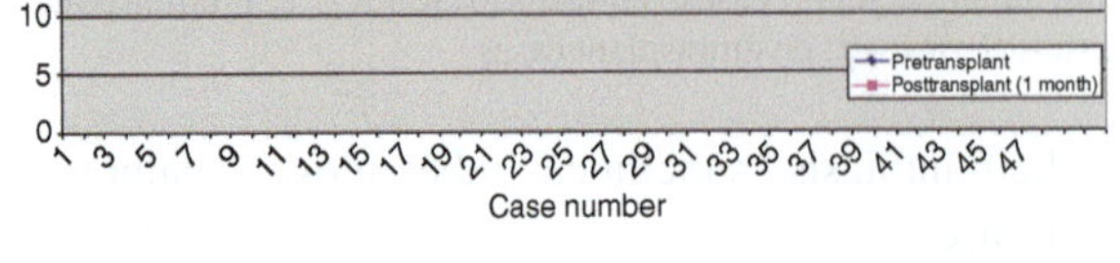

Fig. 39.19 Blood urea level in pre- and post-transplant cases

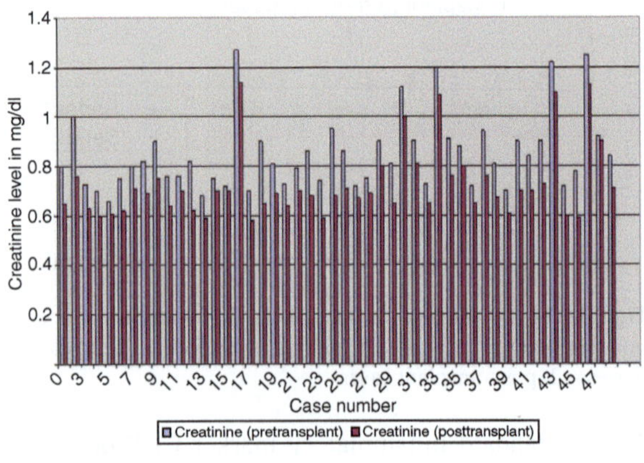

Fig. 39.20 Creatinine level in pre- and post-transplant (1 month) cases

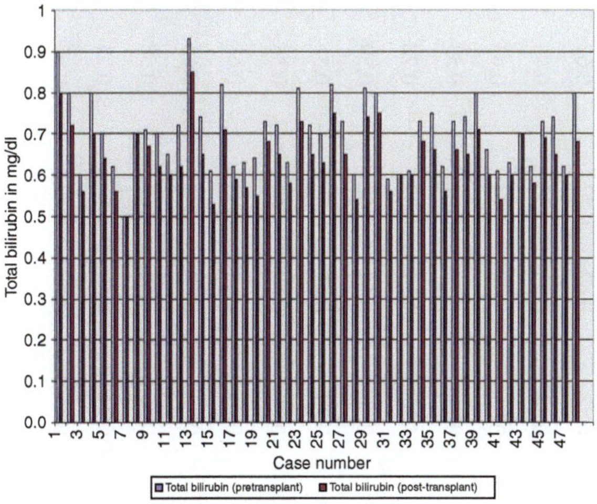

Fig. 39.21 Total bilirubin in pre- and post-transplant (1 month) cases

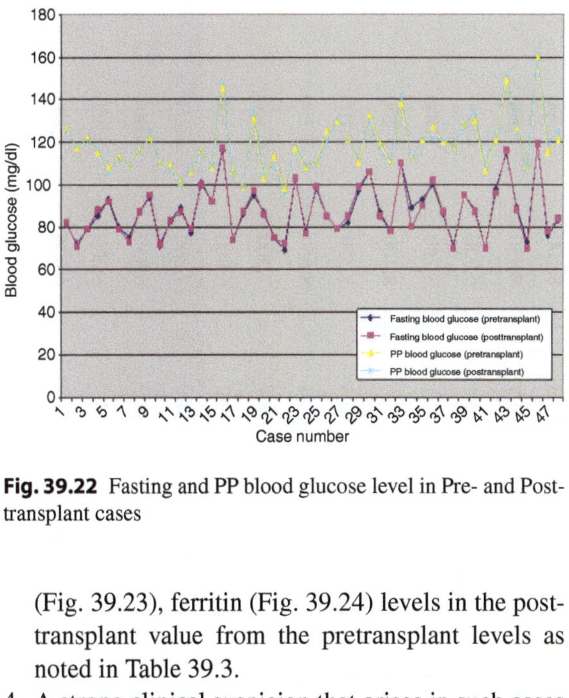

Fig. 39.22 Fasting and PP blood glucose level in Pre- and Post-transplant cases

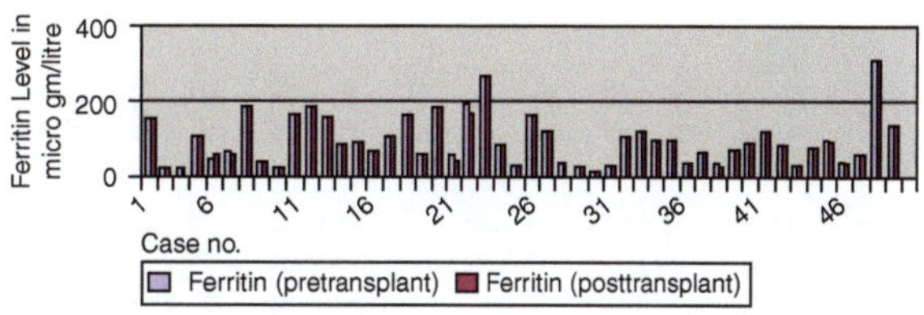

Fig. 39.23 Ferretin level in pre- and post-transplant (1 month) cases

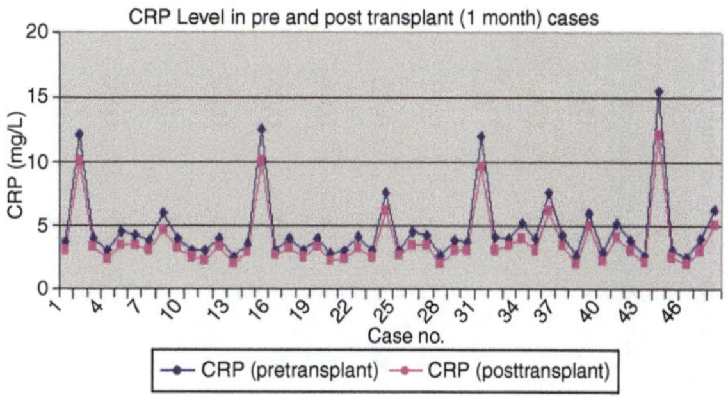

Fig. 39.24 CRP level in pre- and post-transplant (1 month) cases

(Fig. 39.23), ferritin (Fig. 39.24) levels in the post-transplant value from the pretransplant levels as noted in Table 39.3.

4. A strong clinical suspicion that arises in such cases is regarding the immunological sensitization of the host system against the HLA randomized fetal neuronal tissue transplant. But in the present study, no unusual rise in the antinuclear antibody level (Fig. 39.25), or to be more specific, no rise of anti-dsDNA level, was noted in any Parkinsonism patient (vide Fig. 39.26, Table 39.4).
5. Interestingly there was also a rise of hemoglobin (Table 39.3) value from the pretransplant level without any appreciable change in the ferritin value. The cause for the rise of the hemoglobin level could be due to the erythropoietin effect factor of the fetal brain.

After studying the safety parameters of the procedure in order to understand the etiopathogenesis of the clinical improvement in a majority of the patients with the fetal tissue transplantation, the microphotographs need to be examined. Figs. 39.27–39.31 and the scanning electron microscopy photographs (Figs. 39.6 and 39.7) should be interpreted in the same manner.

39.5 Histological Analysis

During fetal life, at around 16 weeks of gestation there is extensive growth of capillaries inside the developing neuronal cells or neuroblasts. There is also, in this age group, development of sulcus in the fetal brain. From the ventricles (neural tubes), the neuroblasts arise, differentiate, and migrate peripherally. Centrally, the fibers predominate and the neural cells become sparse and smaller in size. Here, the background shows the subcutaneous tissue and the skin without any inflammatory or immunological rejection by the adult host site, with

Table 39.3 Body weight, hemoglobin content, total counts, urea, creatinine, total bilirubin content in pre- and posttransplant (1 month) cases (48 cases)

Case no.	Body weight (kg)		Hb content (g/dL)		Total leukocyte count (cells/mm^3)		Urea(mg/dL)		Creatinine (mg/dL)		Total bilirubin (mg/dL)	
	Pre	Post	Pre	Post	Pre	Post	Pre	Post	Pre	Post	Pre	Post
1	54	56	10.3	11.9	7,600	6,900	23	20	0.8	0.65	0.9	0.80
2	59.5	61	10.6	12.0	6,900	6,200	32	26	1.0	0.76	0.8	0.72
3	63	64	10.4	12.6	8,400	7,300	21	18	0.73	0.63	0.6	0.56
4	51	53.5	10.7	12.3	8,600	7,600	18	16	0.70	0.60	0.8	0.7
5	49	51	11.0	13.5	6,800	6,000	17	15	0.66	0.61	0.7	0.64
6	43	44.5	11.2	13.0	7,900	6,700	25	22	0.75	0.62	0.62	0.56
7	55	57	10.5	12.7	8,500	7,600	21	17	0.80	0.71	0.50	0.50
8	54	56	11.2	12.8	8,000	6,900	23	18	0.82	0.69	0.70	0.70
9	56	57.5	11.4	13.8	7,500	6,300	31	26	0.90	0.75	0.71	0.67
10	54	56.5	10.8	12.4	6,850	6,100	19	15	0.76	0.64	0.70	0.62
11	57.5	59	10.9	12.7	7,800	7,400	22	19	0.76	0.70	0.65	0.60
12	58	60	11.1	13.3	8,100	7,300	28	23	0.82	0.62	0.72	0.62
13	57	57	9.9	12.7	7,850	7,100	17	16	0.68	0.59	0.93	0.85
14	52	55	11.0	13.5	6,950	6,150	25	22	0.75	0.70	0.74	0.65
15	61	64.5	12.9	14.1	8,750	7,900	16	16	0.72	0.70	0.61	0.53
16	53	54.5	11.8	13.6	8,400	7,200	43	37	1.27	1.14	0.82	0.78
17	56.5	58.5	11.9	13.4	8,350	7,100	18	17	0.70	0.59	0.62	0.59
18	55	57.5	11.8	13.4	8,650	7,250	25	20	0.90	0.65	0.63	0.57
19	59	61.5	12.3	13.7	7,500	6,450	21	19	0.81	0.69	0.64	0.55
20	53	55.5	10.6	12.2	6,900	6,300	17	16	0.73	0.64	0.73	0.68
21	51.5	55	12.4	14.0	7,200	6,250	29	27	0.79	0.70	0.72	0.65
22	51	54.5	12.8	14.2	6,850	6,000	26	22	0.86	0.68	0.63	0.58
23	65.5	68	12.3	13.9	8,200	7,350	23	20	0.74	0.59	0.81	0.73
24	61	65	12.0	13.8	6,900	6,200	32	28	0.95	0.68	0.72	0.65

25	59.5	59.5	12.3	14.3	7,850	6,800	30	26	0.86	0.71	0.70	0.63
26	58	61	9.9	12.1	7,700	6,700	21	18	0.72	0.67	0.82	0.75
27	61.5	64.5	11.8	13.4	8,300	7,250	19	18	0.75	0.69	0.73	0.65
28	50	52	11.8	13.9	7,600	6,450	26	22	0.90	0.80	0.60	0.54
29	54	56.5	12.1	13.8	7,350	6,500	24	21	0.81	0.65	0.81	0.74
30	53	54.5	9.8	11.4	8,750	7,400	38	33	1.12	1.00	0.80	0.75
31	60.5	62.5	10.2	11.6	8,400	7,250	31	24	0.90	0.81	0.59	0.56
32	63	65	11.9	13.3	8,500	7,450	19	15	0.73	0.65	0.60	0.60
33	56	57	12.6	13.8	8,000	6,950	40	36	1.20	1.09	0.61	0.60
34	58	59	12.9	14.5	7,900	7,000	22	18	0.91	0.76	0.73	0.68
35	61	63.5	12.7	14.1	8,000	7,050	29	24	0.88	0.80	0.75	0.66
36	49	50.5	11.8	13.5	8,550	6,700	20	16	0.72	0.65	0.62	0.56
37	60	60	11.5	12.7	6,950	6,100	33	36	0.94	0.76	0.73	0.66
38	55	56.5	12.2	13.9	7,150	6,250	27	23	0.81	0.67	0.74	0.65
39	64	66.5	12.0	13.8	7,950	6,900	19	16	0.70	0.61	0.80	0.71
40	56	57.5	11.8	12.3	8,150	7,000	30	25	0.90	0.70	0.66	0.60
41	47.5	49.5	11.6	13.2	6,550	5,900	26	22	0.84	0.70	0.61	0.54
42	51	54	10.2	12.1	6,900	6,000	29	28	0.90	0.73	0.63	0.60
43	55	56	12.4	13.5	7,800	6,400	42	36	1.22	1.10	0.70	0.70
44	59	60	10.5	11.8	4,950	4,500	25	23	0.72	0.60	0.62	0.58
45	52	54.5	13.1	14.6	8,250	7,200	27	21	0.78	0.59	0.73	0.69
46	60	61.5	12.0	13.7	5,900	5,300	40	32	1.25	1.13	0.74	0.65
47	52	54	12.8	14.3	7,550	6,650	28	22	0.92	0.90	0.62	0.60
48	60.5	63.5	11.9	13.2	6,330	5,550	19	17	0.84	0.71	0.80	0.68

Pre Pretransplant, *Post* Posttransplant (1 month)

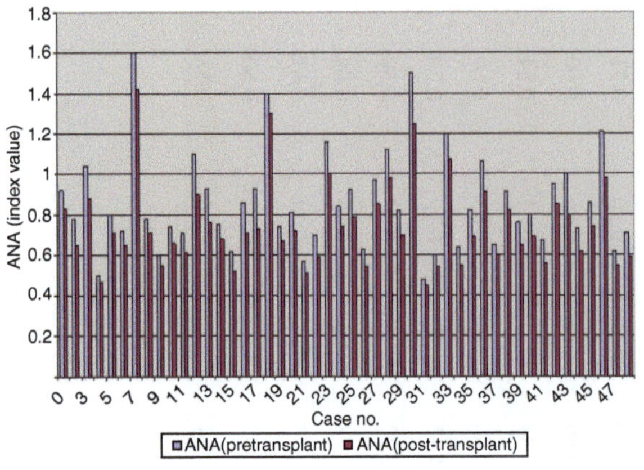

Fig. 39.25 ANA level (index value) in pre- and post-transplant (1 month) cases

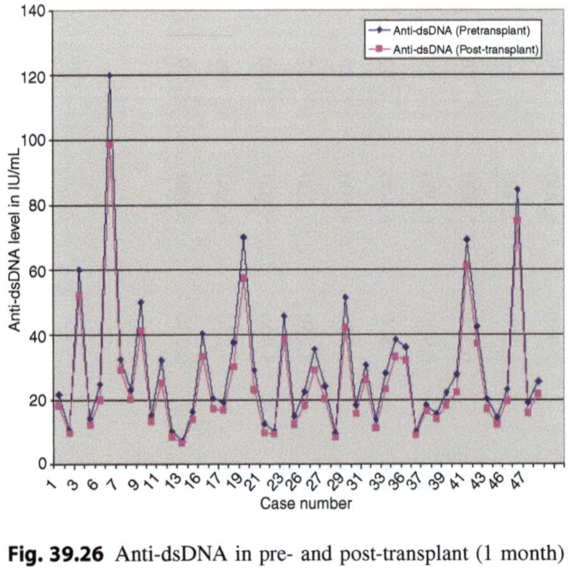

Fig. 39.26 Anti-dsDNA in pre- and post-transplant (1 month) cases

conspicuous absence of the typical features of endarterites, mononuclear invasion, vascular disruption, and thrombosis, even after 3 months (Figs. 39.27–39.29) in the case of Mr. K. P. (serial no. 7). Similar histological findings are noted in the microphotographs (Figs. 39.30 and 39.31) retrieved after the tenth year in the case of Mr. A. M (serial no. 47).

Scanning electron microscopic study of the retrieved tissue from the axilla reaffirmed the presence of fetal neuronal tissue in the background of the host tissue. Investigating further with the electron microscope, the host tissue showed that there were no features of inflammatory cellular reaction when seen in different magnifications, i.e. ×350 (Fig. 39.32) and ×750 (Fig. 39.33). Why there was no obvious inflammatory cellular infiltration or other acute, subacute, or chronic reactions in the host tissue as a result of the fetal tissue transplant remains a scientific mystery to be solved by future researchers. The present study, however, shows a certain degree of inflammatory subcellular cytokine impact on the retrieved tissue. Massive cellular edema can be perceived, which leads to fragmentation and partial loss of collagenous architecture. In fine, the overall impression of the retrieved tissue through scanning electron microscopy suggests noncellular inflammation sequelae justifying subcellular impact at the nano level.

What has been learnt from these histology [microscopy] and electron microscopical results is the fact that there appears to be a kind of insensitivity in the host system to mount any acute, subacute, or chronic inflammatory or immunological reaction. At first, it may appear that the chronic progressive Parkinson's disease itself is causing such an immunological insensitivity in the host system, but in cases of fetal tissue transplant in varying diseases done by the group of researchers in Calcutta, such a condition of host insensitivity or tolerance has been noted and reported on earlier.[7-9]

The present study reaffirms the presence of fetal neuronal tissue in the background of host tissue without any further growth and differentiation. If we try to explain this phenomenon on the basis of existing knowledge in stem cell biology, we can accept its feasibility, as seen in case of adult stem cells, which remains dormant. Further sophisticated molecular marker studies may explain how fetal stem cells become adult stem cells and reside peacefully in the adult immune system; this phenomenon supports the dictum, that the fetal tissue develops its own survival strategy by creating its own microenvironment for its growth and survival.

What is exciting is the fact that neuronal progenitor cell-rich fetal tissue transplant definitely improved the state of clinical disability from the pre- to the post-transplant phase (vide Table 39.5). Of the 43.74% of the patients in the present series who reported with grade 3 clinical disability, 57.14% could be upgraded to the grade-2 level in the 1 month posttransplant phase and the improvement was sustained and could gradually be further upgraded in the 1-year follow-up period. Similarly, of the 29.16% of the patients in the

Table. 39.4 Fasting blood glucose, PP blood glucose, C-reactive protein, antinuclear antibody, anti-dsDNA, ferritin content in pre- and posttransplant (1 month) cases (48 cases)

Case no.	Fasting blood glucose		PP blood glucose (mg/dL)		CRP (mg/L)		ANA (index value)		Anti-dsDNA (IU/mL)		Ferritin (μg/L)	
	Pre	Post	Pre	Post	Pre	Post	Pre	Post	Pre	Post	Pre	Post
1	81	82	126	125	3.7	3.0	0.92	0.83	21.6	18.0	156	154
2	72	71	117	118	12.0	10.0	0.78	0.65	10.5	9.5	27	27
3	79	79	122	120	4.0	3.3	1.04	0.88	60.0	52.0	32	32
4	85	88	115	117	3.0	2.4	0.50	0.47	14.0	12.1	113	111
5	93	92	108	105	4.5	3.5	0.80	0.71	24.7	19.8	52	53
6	80	79	113	112	4.2	3.5	0.72	0.65	120	98.5	64	63
7	75	73	109	109	3.8	3.0	1.60	1.42	32.3	29.0	185	184
8	87	87	116	115	6.0	4.7	0.78	0.71	23.0	20.0	42	41
9	94	95	122	123	4.0	3.2	0.60	0.55	50.0	41.2	21	21
10	71	72	110	110	3.0	2.5	0.74	0.66	15.0	13.3	167	166
11	83	83	109	107	3.0	2.3	0.71	0.61	32.0	25.1	181	182
12	89	87	101	100	4.0	3.3	1.10	0.90	10.0	8.3	162	160
13	77	79	106	107	2.5	2.0	0.93	0.76	7.0	6.4	90	89
14	101	100	116	115	3.6	2.9	0.75	0.68	16.1	13.7	98	99
15	92	92	108	109	12.5	10.0	0.62	0.52	40.2	33.2	63	65
16	116	117	145	148	3.0	2.7	0.86	0.71	20.3	17.0	115	112
17	74	74	106	105	4.0	3.2	0.93	0.73	19.1	16.7	168	165
18	86	88	99	99	3.1	2.5	1.40	1.30	37.5	30.0	59	58
19	95	97	131	134	4.0	3.3	0.74	0.67	70.0	57.4	184	187
20	87	86	103	105	2.8	2.2	0.81	0.72	29.0	23.0	53	52
21	75	75	113	109	3.0	2.4	0.57	0.51	12.3	9.5	194	196
22	69	72	98	95	4.1	3.2	0.70	0.59	10.1	9.2	262	265
23	103	103	117	119	3.0	2.5	1.16	1.00	45.6	38.3	90	91
24	78	77	108	109	7.6	6.1	0.84	0.74	14.8	12.2	32	34

(continued)

Table. 39.4 (continued)

Case no.	Fasting blood glucose		PP blood glucose (mg/dL)		CRP (mg/L)		ANA (index value)		Anti-dsDNA (IU/mL)		Ferritin (µg/L)	
25	98	99	110	110	3.0	2.6	0.92	0.79	22.3	18.1	166	165
26	85	85	125	120	4.5	3.5	0.63	0.54	35.4	29.0	118	118
27	79	79	130	131	4.2	3.4	0.97	0.85	24.0	20.2	38.7	38
28	82	85	121	120	2.6	2.0	1.12	0.98	9.5	8.1	30.6	31
29	97	99	113	110	3.9	3.0	0.82	0.70	51.3	42.0	14.9	14.5
30	106	106	132	130	3.7	3.0	1.50	1.25	18.2	15.5	32.4	32
31	87	85	119	117	11.9	9.6	0.48	0.45	30.5	26.0	114	113
32	78	78	110	109	4.1	3.0	0.60	0.54	13.7	11.0	117	117
33	110	110	138	142	4.0	3.4	1.20	1.07	28.0	23.1	105	104
34	89	80	112	112	5.1	4.0	0.64	0.55	38.4	33.1	96	97
35	93	90	121	122	4.0	3.1	0.82	0.69	36.0	32.0	39	39
36	100	102	127	122	7.5	6.1	1.06	0.91	10.0	8.9	64	65
37	86	87	120	121	4.2	3.4	0.65	0.60	18.2	16.3	37	36
38	71	70	118	118	2.6	2.0	0.91	0.82	15.4	13.9	72	71
39	95	95	128	128	6.0	4.9	0.76	0.65	22.0	18.1	99	98
40	88	87	130	134	2.9	2.3	0.80	0.69	27.6	22.0	121	121
41	70	70	106	109	5.2	4.1	0.67	0.56	69.1	61.2	86	87
42	98	96	121	123	3.9	3.0	0.95	0.85	42.3	37.0	35	37
43	115	116	149	152	2.7	2.1	1.00	0.80	20.1	17.0	78	79
44	89	88	126	124	15.5	12.0	0.73	0.62	14.3	12.2	102	100
45	73	70	108	108	3.1	2.5	0.86	0.72	23.0	19.5	42	40
46	119	118	162	160	2.5	2.0	1.21	0.98	84.5	75.0	56	53
47	76	78	115	119	4.0	3.0	0.62	0.55	19.0	15.6	308	302
48	83	84	121	125	6.2	5.0	0.71	0.59	25.4	21.5	133	135

Pre Pretransplant, *Post* Posttransplant (1 month)

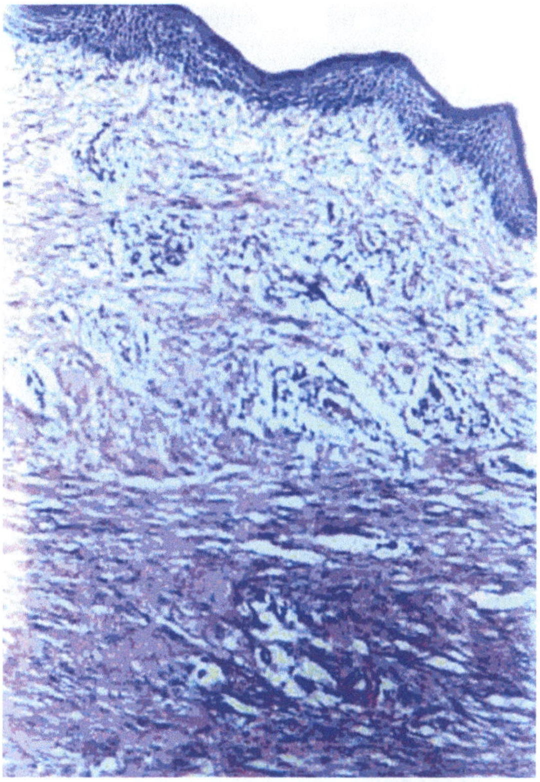

Fig. 39.27 Showing the low-power H&E-stained fetal neuronal tissue after its retrieval from axillary site after 12 weeks

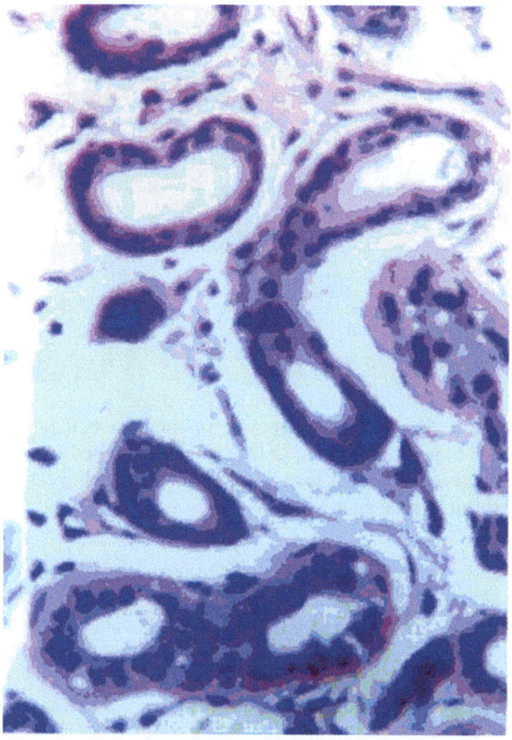

Fig. 39.29 Showing the high power (oil immersion, magnification 375 times) H&E-stained fetal neuronal tissue after its retrieval from the axillary site after 12 weeks

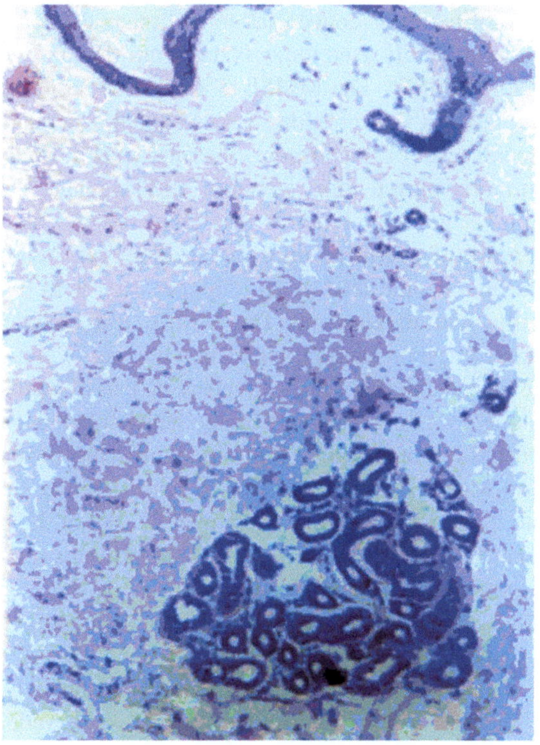

Fig. 39.28 Showing the low power H&E-stained fetal neuronal tissue after its retrieval from axillary site after 12 weeks

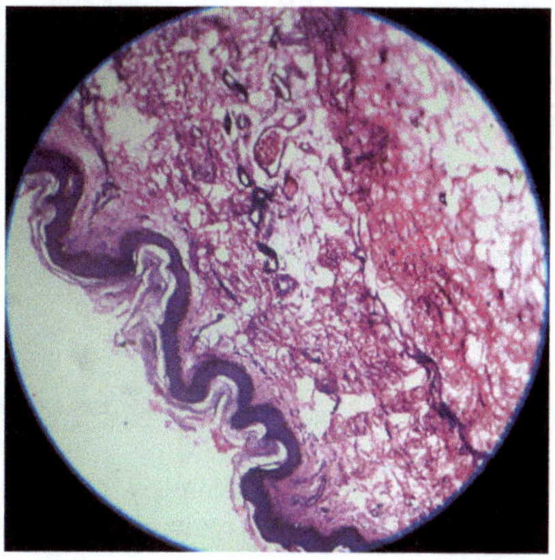

Fig. 39.30 Showing the low-power H&E-stained fetal neuronal tissue after its retrieval on the tenth year from axillary site (serial no. 47)

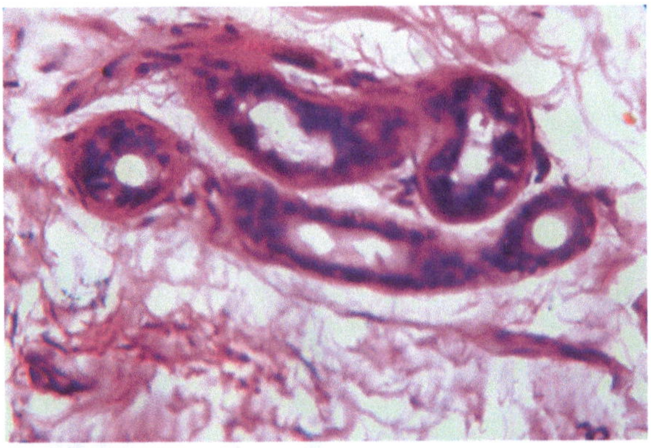

Fig. 39.31 Showing the high power (oil immersion, magnification 375 times) hematoxylin and eosin (H&E) stained 16 weeks fetal neuronal tissue seen after its retrieval on the tenth year from axillary site (serial no. 47)

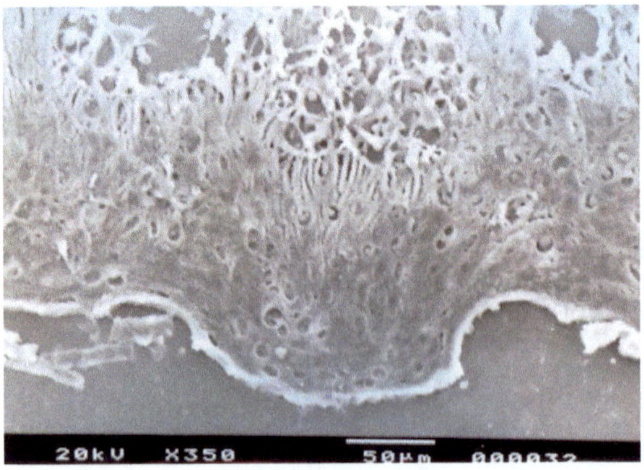

Fig. 39.32 Scanning electron Microscopy (SEM) showing stained fetal neuronal tissue after its retrieval on the third month from axillary site

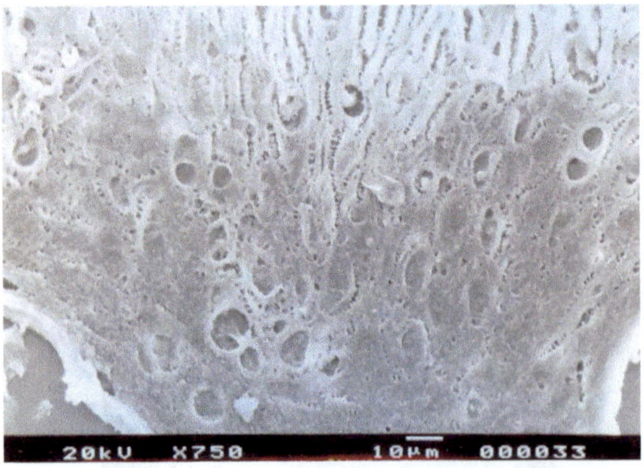

Fig. 39.33 Scanning electron Microscopy (SEM) showing stained fetal neuronal tissue after its retrieval on the third month from axillary site

present series who reported with grade-4 deformity at the pretransplant state, 13.3% could be upgraded to the grade-3 level within 1 month of the progenitor fetal neuronal tissue placement at the axilla. In the present series, 27.1% of the patients reported with grade-5 disability; 23.07% could be upgraded within a year (follow-up). In all the cases there was an overall slow and gradual improvement in the disability status, which is a positive finding.

Any study related to Parkinsonism is grossly incomplete unless we assess cognitive functions during the course of treatment.

In the present study, along with the assessment of the Disability Staging of Parkinson's Disease Table 39.5 (Hoehn and Yahr classifications), the Mini-Mental State Examination (MMSE) has also been assessed vide Table 39.6; further, a study of the mood of the patient using the Hospital Anxiety and Depression Scale (HADS) was also done.

39.6 Mini-Mental State Examination (MMSE)

Of all the mental status examinations for the assessment of patients' concentration and other skills, the most commonly used today is the Mini-Mental State Exam (MMSE). The MMSE is a research-based set of questions that provides a score about a person's general level of impairment covering areas of orientation, recall, retention, attention, and language. The Mini-Mental State Exam is generally a reliable, valid measure of cognitive impairment. However, highly educated people tend to score higher even with disability and impairment. The MMSE test includes simple questions and problems in a number of areas: the time and place of the test, repeating lists of words, simple arithmetic use, and comprehension language, and basic motor skills. The maximum score on the Mini-Mental State Exam is 30. In general, scores fall into four categories. Any score over 27 (out of 30) is effectively normal. Below this, 20–26 indicates some cognitive impairment; 10–19 moderate to severe cognitive impairment, and below 10 very severe cognitive impairment.

If the overall MMSE scoring results of the patients of Parkinsonism in the present series are observed, a

Table 39.5 Showing the disability staging of Parkinson's disease (Hoehn and Yahr classifications) before and after fetal tissue transplants[22]

Patients' total no. = 48	Grading of disability	Patients with different grade (pretransplant) disability	Patients with 1 month different grade (posttransplant) disability	Follow up 3 months posttransplant of disability grade	Follow up 6 months posttransplant of disability grade	Follow up 1 year posttransplant of disability grade
Nil	I	Nil	Nil	Nil	2	2
Nil	II	Nil	9	12	19	16
21	III	21	14	12	10	10
14	IV	14	14	13	9	10
13	V	13	11	11	8	10

Table 39.6 Mini-mental state assessment

Patients' total no = 48 with disability grading	Range with mean with SD scoring of the Mini Mental State exam before transplant	Range and mean with SD scoring of the Mini Mental State exam 1 month after transplant	Range and mean with SD scoring of the Mini Mental State exam 3 months after transplant	Range and mean with SD scoring of the Mini Mental State exam 6 month after transplant	Range and mean with SD scoring of the Mini Mental State exam 1 year after transplant	Lost or irregular or refusal to do the test
21 (III)	10–29 24.6 ± 2.8	14–29 25.8 ± 2.9	15–29 26.1 ± 2.1	17–28 26.3 ± 2.8	18–27 26.6 ± 2.7	
14 (IV)	9–26 20.4 ± 2.2	12–25 23.7 ± 2.1	14–27 25.1 ± 2.8	15–26 25.9 ± 2.1	15–26 24.9 ± 2.8	
13 (V)	9–22 16.6 ± 1.9	9–24 18.6 ± 1.8	9–24 19.2 ± 1.7	10–22 20.1 ± 1.8	11–21 19.8 ± 2.2	

wide fluctuation of cognitive impairment not always related to the physical disability scale can be seen; however, with fetal neuronal tissue transplant there appears to be a definite improvement with up gradation (improvement of the status) of the deformity from grade 3 to grade 2 in 42.85% of the cases. There is also improvement in other grades after the transplantation and the follow-up evaluation of the grade (vide Table 39.5), i.e., in physical disability scoring, as mentioned earlier. Similarly, while assessing the cognitive impairment (Table 39.6), the results showed improvement irrespective of the pretransplant disability grading. This is statistically significant (p value < .03). The procedural sensitivity is 71–92% and the specificity varies from 56% to 96%.[24]

Another widely practiced scale to study the mood of the patient is analysis through the Hospital Anxiety and Depression Scale (HADS). Without this study the overall assessment is incomplete.

39.7 Study of the Mood of the Transplant Patient (e.g., HADS)

There has been considerable controversy regarding the relationship between depression and anxiety. The descriptive, longitudinal, genetic, biological, and treatment response data indicate that there is overlap between depression and anxiety, namely[1] that there are a variety of more or less discrete, but sometimes coexisting, syndromes within the spectrum of anxiety and depression[2]; that symptoms of depression and anxiety represent different external manifestations of a more basic underlying cause[3]; that one condition may predispose to the other[4]; that there could be an association mixed anxiety/depressive disorder. The Hospital Anxiety and Depression Scale (HADS) is a widely used, self-administered questionnaire specifically

Table 39.7 Gain in weight and rise in hemoglobin value at 4 weeks from the date of transplant placement of patients who took the fetal subcortical brain tissue transplant (48 cases)

Case no./ name	Age and sex	Date of transplant	Difference in Hb value from previous value on the date of graft removal (g)	Treatment received earlier	Age of the fetal brain (weeks)	Clinical improvement before graft removal	Specific complication or graft vs. host reaction	Difference in weight on the date of graft removal (kg)
1,G.R.	65 M	5.2.2001	1.6	A-H*	14	Mild	Nil	2
2,A.M.	55 M	8.2.2001	1.4	-do-	20	Moderate	Nil	1.5
3,K.M.	71 M	19.2.2001	2.2	-do-	20	Moderate		1
4,C.P.	66 M	19.2.2001	1.6	-do-	20	Mild	Nil	2.5
5,M.S.	48 M	26.2.2001	2.5	-do-	10	Mild	Nil	2
6,B.P.	45 F	27.3.2001	1.8	-do-	16	No variation	Nil	1.5
7,K.P.	65 M	27.3.2001	2.2	-do-	16	Moderate	Nil	2
8,T.C.	73 M	27.6.2001	1.6	-do-	18	No variation	Nil	2
9,B.M.	56 M	4.6.2001	2.4	-do-	10	Mild	Nil	1.5
10,S.B.	45 M	27.6.2001	1.6	-do-	18	Mild	Nil	2.5
11,T.P.	75 F	22.5.2001	1.8	-do-	10	Moderate	Nil	1.5
12,S.T.	55 M	30.6.2001	2.2	-do-	10	Moderate	Nil	2
13,S.M.	61 M	5.2.2001	2.8	-do-	14	Mild	Nil	No gross variation
14,P.S.	53 M	8.2.2001	2.5	-do-	20	Moderate	Nil	3
15,R.R.	73 M	19.2.2001	1.2	-do-	20	Moderate	Nil	3.5
16,P.D.	64 M	19.2.2001	1.8	-do-	20	Mild	Nil	1.5
17,G.D.	47 M	26.2.2001	1.5	-do-	10	Mild	Nil	2
18,B.S.	48 F	27.3.2001	1.6	-do-	16	No variation	Nil	2.5
19,K.M.	62 M	27.3.2001	1.4	-do-	16	Moderate	Nil	2.5
20,R.D.	72 F	22.5.2001	1.6	-do-	10	Moderate	Nil	2.5
21,G.B.	54 M	4.6.2001	1.6	-do-	10	Mild		3.5
22,T.D.	46 M	27.6.2001	1.4	-do-	18	Mild	Nil	3.5
23,G.C.	57 M	27.6.2001	1.6	-do-	18	No variation	Nil	2.5

24,P.T.	55 M	30.6.2001	1.8	-do-	10	Moderate	Nil	4
25,B.D.	62 M	5.2.2001	2	-do-	14	Mild	Nil	No gross variation
26,N.C.	53 M	8.2.2001	2.2	-do-	20	Moderate	Nil	3
27,D.G.	70 M	19.2.2001	1.6	-do-	20	Moderate	Nil	3
28,S.P.	63M	19.2.2001	2.1	-do-	20	Mild	Nil	2
29,G.M.	49 M	26.2.2001	1.7	-do-	10	Mild	Nil	2.5
30,N.B.	46 F	27.3.2001	1.6	-do-	16	No variation	Nil	1.5
31,S.D.	64 M	27.3.2001	1.4	-do-	16	Moderate	Nil	2
32,G.P.	73 F	22.5.2001	1.4	-do-	10	Moderate	Nil	2
33,R.S.	55M	4.6.2001	1.2	-do-	10	Mild	Nil	2
34,K.C.	49 M	27.6.2001	1.6	-do-	18	Mild	Nil	1
35,M.D.	71 M	27.6.2001	1.4	-do-	18	No variation	Nil	2.5
36,R.B.	51 M	30.6.2001	1.7	-do-	10	Moderate	Nil	1.5
37,T.C.	62 M	5.2.2001	1.2	-do-	14	Mild	Nil	No gross variation
38,K.R.	53 M	8.2.2001	1.7	-do-	20	Moderate		1.5
39,G.G.	72 M	19.2.2001	1.8	-do-	20	Moderate	Nil	2.5
40,N.D.	59 M	19.2.2001	1.2	-do-	20	Mild		1.5
41,T.R.	42 M	26.2.2001	1.6	-do-	10	Mild	Nil	2
42,G.T.	49 F	27.3.2001	1.9	-do-	16	No variation	Nil	3
43,S.S.	67M	27.3.2001	1.1	-do-	16	Moderate	Nil	1
44,K.M.	73 F	22.5.2001	1.3	-do-	10	Moderate	Nil	1
45,B.C.	66 M	4.6.2001	1.5	-do-	10	Mild	Nil	2.5
46,S.D	71 M	27.6.2001	1.7	-do-	18	Mild	Nil	1.5
47,A.M.	44 M	27.6.2001	1.5	-do-	18	No variation	Nil	2
48,T.D	57 M	30.6.2001	1.3	-do-	10	Moderate	Nil	3

*A-H: Details are given in Table 39.1

developed to detect anxiety and depression states in hospital and medical outpatient clinic settings. It is composed of two 7-item scales, one for anxiety and one for depression.

A study was undertaken to explore the usefulness of the HADS scheme in Parkinsonism and to assess the impact of neuronal tissue transplantation on the volunteers of the present study through the HADS scale. All 48 patients were assessed using a socio-demographic and clinical questionnaire and the HADS (individual interview). HADS proved to be easily understood and well accepted by the patients. The Bengali HADS results support the use of the English HADS (which might be a little modified during translation to the Bengali language) as a screening tool for these patients (facial validity, internal consistency, construct validity), though the associations between HADS scores and some demographic and clinical variables question its validity. The study furnished evidence that the Bengali version was a reliable instrument for detecting states of anxiety[25–27] and depression in Bengali-speaking patients in Calcutta. This preliminary validation study of the local Bengali version of the HADS questionnaire showed it as an acceptable and reliable measure of psychological morbidity among Parkinson's disease patients.

Based on Snaith and Zigmond's interpretation of HADS-A and HADS-D scores of 8 or over, patients screened positive for anxiety (72%) and/or depression (85%) before the transplant showed an improvement after the transplant from 1 month onward as noted in HADS-A and HADS-D scoring, eventually reaching a lower value of anxiety (32%) and/or depression (35%)[28] at the end of 1 year.

The result is derived from interviews, and assessment of the pre- and posttransplant questionnaire filled by the patient. This study underscores the importance of assessing patient affective status, as anxiety and depression are common, and provides additional insight into the association of anxiety/depression with certain demographic and clinical variables. The reasons for the inconsistencies between these results and published literature should be explored in future studies by other researchers in the field.

39.8 Secondary Advantages of Neuronal Tissue Transplantation in the Present Study

Some secondary impacts following the fetal cortical brain tissue transplants were also observed, and are reported and analyzed below.

Table 39.7 indicates that the Hb content improved in all the patients following transplantation, increasing between 1.4 and 2.8 g. Why there is improvement of the hemoglobin status with fetal brain tissue transplantation is not yet understood. The possibilities are[1]: the erythropoietin content of the transplanted developing fetal brain tissue, and[2] the erythropoietin receptors in the fetal brain, along with other coordinating neurocytokines, may have a positive impact on the background of anemia[3]; There is also a theoretical possibility that there is a combating catabolic cytokines' impact (TNF) on bone marrow by the anabolic neurocytokines (insulin-like growth factors, IGF) of the developing brain tissue (?) in chronic disease. To understand how erythropoietin may work in this case, it is necessary to give a brief background on anemia in chronic disease, immune responses, and the functions of erythropoietin.

Anemia in chronic disease is a very complex phenomenon of cytokine inter-regulation and belongs to a specific subgroup of anemia and is the second most prevalent cause of anemia. The first important cause is the dietary iron deficiency. In case of anemia in chronic disease, there is acute or chronic immune activation of the specific cytokine system, which helps in shifting of the iron from its normal route. The condition has also been termed as "anemia of inflammation".[29]

This condition is immune-driven; the cytokines and cells of the reticuloendothelial system induce changes in iron homeostasis, proliferation of erythroid progenitor cells, production of erythropoietin, and the lifespan of red cells, all of which contribute to the pathogenesis of anemia. Erythropoiesis can be affected by disease underlying anemia of chronic disease. This can be due to pro-inflammatory cytokines and free radicals that damage erythroid progenitor cells. Bleeding episodes, vitamin deficiencies (e.g., of cobalamin and folic acid), hypersplenism, and autoimmune hemolysis may also contribute to the anemic process affecting diseases such as rheumatoid arthritis, and other granulomatous and non-granulomatous causes. Fetal tissue may have a positive cytokine and growth factor impact on the host's bone marrow to improve the anemic status of the individual.

In the hematopoietic system, the principal function of erythropoietin (Epo) is the regulation of red blood cell production, mediated by its specific cell surface receptor (EpoR). Following the cloning of the Epo gene (EPO) and characterization of the selective hematopoietic action of Epo in erythroid lineage cells, recombinant Epo forms (epoetin-alfa, epoetin-beta, and the long-acting analog darbepoetin-alfa) have been widely used for treatment of anemia in chronic kidney disease, and chemotherapy-induced anemia in cancer patients. Ubiquitous EpoR expression in non-erythroid cells has been associated with the discovery of diverse biological functions for Epo in non-hematopoietic tissues. During development, Epo–EpoR signaling is required not only for fetal liver erythropoiesis, but also for embryonic angiogenesis and brain development. A series of recent studies suggest that endogenous Epo–EpoR signaling contributes to wound-healing responses, physiological and pathological angiogenesis, and the body's innate response to injury in the brain and heart.[30]

The biological effects of Epo in the central nervous system (CNS) involve activation of its specific receptor and corresponding signal transduction pathways. Epo receptor expression is abundant in the developing mammalian brain, and decreases as term approaches. Epo has been identified as a neurotrophic and neuroprotective agent in a wide variety of experimental paradigms, from neuronal cell culture to in vivo models of brain injury.[31]

Epo was once regarded as a cytokine with only hematopoietic effects. It is now clear that the distributions of Epo and Epo-R are more widespread in the developing human system. Epo-R is widely distributed during early fetal development, leading to speculation that Epo acts in concert with other growth factors to optimize growth and development. Areas in which Epo has important recognized effects are on endothelial cells, and in the developing heart, gastrointestinal tract, and brain.[32] This is a precise explanation to justify the uniform rise in hemoglobin in the Parkinson's disease patient after neuronal tissue transplantation.

39.9 Improvement of Aches and Pain with Fetal Neuronal Tissue Transplantation

Another observation following transplant of fetal subcortical developing tissue in 48 patients with PD was that there was a reduction in aches and pain associated with the disease. Why there is improvement in the aches and pain all over the body with varying intensity in different patients of Parkinsonism, especially an improvement in the perception of pain, after fetal brain tissue transplantation, also remains a mystery to date. A specific objective in the present study was to examine the impact from the clinical angle; the possible effect of the endorphin component of the fetal tissue was also considered.

The effect of human fetal neuronal tissue graft (within 20 weeks, i.e., the legal limit of hysterotomy and ligation) transplant in patients with intractable neuropathic pain not relieved with pain-relieving drugs including nonsteroidal inflammatory drugs was evaluated. The impact of fetal tissue transplantation on pain relief was assessed by the specifications and guidelines of the McGill questionnaire.[33] The impact of age differences in the quality of chronic pain was also incorporated.[14]

A short form of the McGill Pain Questionnaire (SF-MPQ) has been used here in patients with Parkinsonism with some pain and aches in the body. The SF-MPQ has shown promise as a useful tool in situations in which the standard MPQ takes too long to administer, yet qualitative information is desired and the PPI and VAS are inadequate.[34] The main component of the SF-MPQ consists of 15 descriptors (11 sensory and 4 affective), which are rated on an intensity scale as 0 = none, 1 = mild, 2 = moderate, or 3 = severe. Three pain scores are derived from the sum of the intensity rank values of the words chosen for sensory, affective, and total descriptors. The SF-MPQ also includes the Present Pain Intensity (PPI) index of the standard MPQ and a visual analogue scale (VAS). The correlations were consistently high and significant. The SF-MPQ was shown to be sufficiently sensitive to demonstrate differences due to treatment at statistical levels comparable to those obtained with the standard form. In the present study of fetal subcortical tissue transplantation in a heterotopic site in patients of Parkinsonism, a definite improvement of pain was observed in 28 patients (58.31%) from the pretransplant state after 1 month following the guidelines of the short-form McGill Pain Questionnaire.

39.10 Weight Gain and Sense of Well-Being with Fetal Tissue Transplantation

Another interesting observation was the sense of well-being in 80% of the patients, with a gain in weight of 1.5 kg or more in 80% of the patients enrolled for this study. The issue here is if the weight gain is an effect of the endorphin, which is found alongside the fetal growing progenitor cells.

Endorphins are endogenous opioid polypeptide compounds.[35] The term "endorphin" implies an activity. It consists of two parts: *endo-* and *-orphin*; these are taken from the words *endogenous* and *morphine*, intended to mean a morphine-like substance originating from within the body. Endorphins are produced by the body during strenuous exercise or orgasm, and they can produce a sense of well-being. They work as "natural pain killers," whose effects may be enhanced by other medications.

Although morphine and other similar drugs of abuse have different acute mechanisms of action, their brain pathways of reward exhibit common functional effects upon both acute and chronic administration. Long known for its analgesic effect, the opioid beta-endorphin is now known to induce euphoria, and to have rewarding and reinforcing properties.

Numerous studies have investigated the behavioral effects of beta-endorphin, both endogenous and exogenously applied. However, the potential for biotransformation of beta-endorphin in the extracellular space of the brain has not been previously directly addressed in vivo.

Beta-endorphin is released into the blood (from the pituitary gland) and into the spinal cord and brain from hypothalamic neurons. The beta-endorphin that is released into the blood cannot enter the brain in large quantities because of the blood–brain barrier. The physiological importance of the beta-endorphin that can be measured in the blood is far from clear. This beta-endorphin is a cleavage product of a precursor hormone for ACTH. The behavioral effects of beta-endorphin are exerted by its actions in the brain and spinal cord, and probably the hypothalamic neurons are the major source of beta-endorphins at these sites. In situations where the level of ACTH is increased, for example, the level of endorphins also increases slightly.

Beta-endorphin has the highest affinity for the μ1-, slightly lower affinity for the μ2-, and δ- and low affinity for the κ1-. μ-receptors which are the main receptors through which it acts. Classically, μ-receptors are presynaptic and inhibit neurotransmitter release; through this mechanism, they inhibit the release of the inhibitory neurotransmitter, and disinhibit the pathways, causing more dopamine to be released. By hijacking this process, exogenous opioids cause inappropriate dopamine release, and lead to the strange reaction that causes addiction.

Opioid receptors, however, have many other and more important roles in the brain and periphery – modulating pain, cardiac, gastric, and vascular functions, and possibly the induction of panic and satiation; receptors are often found at postsynaptic locations as well as presynaptically. In 2008, researchers in Germany reported that the myth of the runner's high was in fact true. Using PET scans combined with recently available chemicals that reveal endorphins in the brain, they were able to compare runners' brains before and after a run. The runners whom the researchers recruited were told that the opioid receptors in their brains were being studied, and did not realize that their endorphin levels were being studied in regard to the runner's high. The participants were scanned and received psychological tests before and after a 2 h run. Data received from the study showed endorphins were produced during the exercise and were attaching themselves to areas of the brain associated with emotions (limbic and prefrontal areas).[36]

Whether fetal tissue with its many unique properties has a growth promoting role is also to be investigated in future. A serial estimation of p21 and p27 levels on the recipients of the transplanted fetal tissue may be helpful.

Unlike starvation, the weight loss seen in chronic illnesses arises from loss of muscle and fat. Muscle damage with a lack of regeneration, manifests itself in several life-threatening diseases, including infection and sepsis. Often misdiagnosed as a condition simply of weight loss, cachexia is actually a highly complex metabolic disorder involving features of anorexia, anemia, lipolysis, and insulin resistance. A significant loss of lean body mass arises from such conditions, resulting in wasting of skeletal muscles.

Indeed, tumor necrosis factor-alpha (TNF-alpha) levels are raised in several animal models of cachectic muscle wasting, whereas the insulin-like growth factor (IGF) system acts potently to regulate muscle development, hypertrophy, and maintenance. This concept of skeletal muscle homeostasis, often viewed as the net balance between two separate processes of protein synthesis and degradation, has, however, changed. More recently, the view is that these two biochemical processes are not occurring independently of each other but in fact are finely coordinated by a web of intricate signaling networks. The mechanisms of degeneration and regeneration, with specific emphasis on the potential and controversial cross-talk, which may exist between anabolic growth factors (e.g., IGF-I) and catabolic cytokines (e.g., TNF-alpha), are not yet settled.[37]

To repeat the cardinal points of this study, the fetal tissue was transplanted in sex- and HLA-randomized adult axillas without concomitant immunosuppressive or radiation support to blunt the hosts' immune response; serial estimation of the peripheral blood of the hosts did not show any gross leukocytosis or lymphocytosis within the first 6 weeks of the observation period. The site of transplantation was also found to be healthy in all cases. The fetal microenvironment is distinctly different from the adult neuroendocrine and metabolic microenvironment.[38] Therefore, it is possible that the transplanted fetal tissue adjusts its own microenvironment to an altered metabolic and immunological situation, using its naive pre-immune or hypo-immune status to prevent immune recognition by the host. Why and how the fetal neuronal tissue in a sex- and HLA-randomized non-primed (no immunosuppressive support or radiation) host escapes the immunological or inflammatory recognition system, and becomes a human homologous chimera, is a mystery to be solved in the future. However, on the basis of the direct experiences of the current study, as mentioned earlier, the present researchers strongly believe that the hypo-antigenicity of the fetal tissue is the most important factor in preventing the hosts' recognition of the fetal tissue, which thus escapes rejection.

39.11 Discussion

If the global experiences of attempted experimental treatments through dopaminergic cell/tissue graft in animal and/or primate model to ameliorate the clinical features of Parkinson's disease are studied, certain interesting facts and trends can be seen, which are noted below:

(a) Most studies suggest that the putamen is the primary transplantation target site as this region exhibits the most marked reductions in dopamine content, and it is the part of the striatal complex that is physiologically most closely linked to motor control. Grafts can reinnervate striatum, release dopamine, and become integrated in the patient's brain. However, the results are widely varied; it should be noted that grafts can give rise to troublesome dyskinesias.[39]

(b) The most important question raised by the clinical transplantation trials performed so far is why the functional outcome has been so variable. Graft efficacy has to be increased and variability reduced. But so far, the improvements after intrastriatal transplantation of fetal DA neurons in patients have not exceeded those found with subthalamic deep-brain stimulation, and there is no convincing evidence that drug-resistant symptoms are reversed by these grafts.

Thus, as mentioned above, these experimentations can give rise to very troublesome dyskinesia even after precisely placing the graft in the putamen or at a specific site inside the brain.

One of the most important observations of the present extensive clinical experiment on Parkinsonism volunteers (from 1999 till date) is that there was survival of some developing neuronal tissue fractions at the site of graft placement at the axillary subcutaneous region without the tacit support of immunosuppressives in HLA randomized patients. Moreover, it should be mentioned, in no medical literature has the axilla ever been claimed as a specialized immune privileged environment, as in case of the brain, cornea, cartilages, and the pregnant uterus. But histological study of the retrieved fetal neuronal tissue even after 10 years from the date of placement revealed its survival at the graft placement site.

If the basic physiological functions of the human brain are examined, it can be seen that the adult brain consumes 15% of the resting cardiac output for its metabolic need and auto-regulation. When most of the tissues can sustain an anaerobic insult from a few to 30 min in the human system, sudden total lack of oxygen supply to the brain cells in an adult will make the person lose consciousness within 5–10 s. A serious complication of abnormal fluid dynamics is the development of brain cell edema, with the possibility of eventual destruction of brain cells as a result of decrease in blood flow due to extra edema fluid compression since the brain is encased in a solid vault. Blood-cerebrospinal fluid and the blood–brain barrier operate simultaneously through the fusion of adjacent endothelial cells to one another, thus decreasing permeability. However, in the case of other tissues in the body, there are extensive slit pores.[40]

Keeping in view the physiological peculiarities of adult brain function, the question may now be asked as to whether the developing fetal brain, up to 20 weeks of gestation (limit of hysterotomy and ligation), has acquired the specialized requirements needed for its survival (as has been delineated for the adult brain system). It was observed from the fetal brain tissue transplants that were conducted that fetal brain cells can live outside their normal position, that is, without the encasement of the skull, and without the presence of cerebrospinal fluid or meningeal support, which are necessary for the survival of adult brain cells. On the basis of the analysis of the histological data (vide Figs. 39.27–39.31), the present researchers have noted that there is no rejection of the transplant in the form of thrombosis, endarteritis, mononuclear invasion, etc. The question is why this is so.

In fine, if the entire medical community combines its wisdom and examines this very complex issue of fetal tissue–adult tissue interaction in health and disease, there may be a major breakthrough in this hitherto vastly unknown and fascinating field, which has many important implications for understanding regeneration biology and engineering with strategies for tissue restoration to combat human miseries in the future.[41]

The newer modality of treatment with exciting therapeutic positive claims, include cell/tissue replacement for the treatment of PD, which is essentially based on two hypotheses: first, the predominant symptoms of PD are dependent on the dysfunction or loss of the dopaminergic neurons in the nigrostriatal pathway; and second, dopaminergic neurons grafted into the dopamine-deficient striatum can replace those neurons lost as a result of the disease process and can reverse, at least in part, the major symptoms of the disease; this is a new approach to the treatment of progressive neurodegenerative diseases.[42]

Certain points come to the fore in the context of the present study:

1. In the present series, the diagnosis of PD was clinical and was conducted by an experienced consultant to exclude early Parkinson's disease with mild to moderate disability and without any motor fluctuations. Only patients with advanced idiopathic Parkinsonism were considered as subjects for the study.
2. Options were suggested to the patients for stereotactic autologous adrenal medullary tissue placement at the selective midbrain region at a specialized Centre of Excellence in India, but they all refused because of the prohibitive cost of this kind of specialized experimental surgery in a private hospital. They even negated a suggestion to purchase a peripheral apomorphine pump. All 48 patients who reported initially, refused to go for such options.
3. The clinical profile of the patients who underwent the transplantation surgery protocol was meticulously followed up. In all the enrolled cases, methodical history-taking revealed that the symptoms were insidious in onset and the common features included tremor, stiffness involving at least one limb, difficulty in walking, fatigue, depression, softness of voice, dysarthria, and poverty of emotional and motor responsiveness.

   More than 80% of the patients who participated in this fetal tissue transplant protocol showed some degree of clinical and also cognitive improvement out of this method of treatment. There is not a single instance of procedure related complication or deterioration of the patients overall clinical condition
4. All 48 patients who underwent the transplant process were persuaded to allow retrieval of the fetal tissue from the axilla under local anesthesia in the operation theater; however, seven of the 48 patients refused to allow retrieval at the last moment, because they thought that the improvement was due to the transplant and as such if they continued to retain the transplant, there would not be any further problem. They continued to be persuaded and finally the last

patient from whom the fetal tissue was retrieved had kept the tissue for more 10 years of the date of placement of the transplant

What was intriguing was the persistence of the fetal tissue: it was not destroyed by the immunoenzyme network of the host defense and surveillance system even in the tenth year of its placement. In the tenth year follow-up visit, one patient (serial no. 47) agreed to repeated persuasion for retrieval of the tissue, which, surprisingly, showed persistence of the fetal tissue in the adult host without any obvious difference from other cases where the tissue was retrieved in the third month posttransplant.

The conclusion is simple:

1. Fetal neuronal tissue can survive in an adult host where the neuroendocrine, immunoenzymatic, and cytokine regulation is distinctly different from the adult system.
2. There are gross similarities in the histological and electron microscopic findings of the retrieved subcortical midbrain fetal neuronal progenitor tissue from the host adult tissue. This is a truly astonishing finding regarding the retrieved tissue, from the first month to the tenth year after its placement in the adult host.
3. This observation justifies the notion that there is a persistence of the stem cell component of the fetal tissue in the adult host in such a situation.
4. Fetal tissue creates its own microenvironment for its survival. There are histological similarities in the different neurological specimens retrieved from the adult host system.
5. That there is no abnormal growth and differentiation of the fetal tissue justifies the idea that while genetic regulation with its apoptosis mechanism is in full operation in the adult host, it is ineffective in detecting and destroying the primitive hypoantigenic system existing in fetal tissue.
6. Studying the disability status, cognitive assessment including mood showed marked improvement from the pretransplant status as mentioned earlier.
7. Another important observation was the sense of well-being and weight gain along with some improvement in the anemic status after the transplant in all the cases.

   It is well known that in a developing fetal brain, there are phase of proliferation, neurogenesis, gliogenesis, angiogenesis, and neovascularization; but all these events vary spatiotemporarily with location and cell types. The present study of the transplanted brain tissue showed features of growth, proliferation, and neurogenesis in small groups as well as in a cluster or cell nest-like formation (Figs. 39.27–39.31).
8. What is interesting is the lack of any leukocytosis or lymphocytosis in the hosts' system even after the graft had survived for 1 month or more in the host's axilla. The present researchers have had similar experiences with thymus[7] and other fetal tissue transplants such as fetal heart, liver, lung, or pancreas.[8]

39.12 Summary and Conclusions

It is of paramount importance to the present group of researchers to determine whether the survival of the fetal brain tissue in heterotopic site and the clinical improvement are directly related to each other, or whether the growth factors and different cytokines of the human developing fetal brain tissue are responsible for the apparent improvement. Another interesting aspect is to examine whether neuro-degeneration in idiopathic Parkinsonism is site-specific or not. If it is, neuro-degeneration could be avoided through a heterotopic transplant site. These are very interesting issues to ponder and are currently under scrutiny.

The present study is a simple option for the treatment of advanced idiopathic Parkinsonism when conventional treatment becomes refractory. The safety and simplicity of the procedure only suggests that it could be an early and valuable option for the treatment of Parkinsonism.

The study also showed long-term survival reports of fetal tissue transplant (10 years after placement) in adult hosts, safety of the fetal tissue in adult host, and also some other nonspecific improvement of fetal tissue transplant on the adult system as manifested with weight gain, rise of hemoglobin, decrease in aches and pains all over the body, which justifies a thorough molecular studies to understand the etiopathogenesis of fetal tissue–adult tissue interaction and its positive ramifications in health and disease.

Figures 39.27–39.29 show the histology of the tissues of a 16-week fetal transplanted midbrain tissue in the background of sex- and HLA-randomized adult axillary tissue, without any immunosuppressive support for its survival. The tissue was taken out surgically after 3 months of its placement. The patient was Mr K. P. (65 years; serial no. 7). Figure 39.27 shows the tissue stained with hematoxylin and eosin stain in low power. Figure 39.28 depicts stain with hematoxylin and eosin in low power. Figure 39.29 shows high power view of the fetal neuronal cells under oil immersion lens (375 times).

Figures 39.30–39.31 depict the histology of the 16 weeks' developing human fetal neuronal tissue in the tenth year after its placement in the axilla in a patient of Parkinsonism. Histology showed identical cellular appearance with Fig. 39.27–39.29, and absence of any immunological and inflammatory reaction from the host's side as perceived in histology. Figure 39.30 showing staining in low power and Fig. 39.31 (patient's serial no= 47) showing the staining in high power in oil immersion lens.

Scanning electron microscopic study of the retrieved tissue from the axilla reaffirmed the presence of fetal neuronal tissue in the background of host tissue. Studying further with electron microscope, the host tissue showed that there was no features of inflammatory cellular reaction as seen in different magnifications (×350), Figs. 39.32 (×750) and 39.33. Why there are no obvious inflammatory cellular infiltration and other acute, subacute, or chronic reactions of the host tissue to the fetal tissue remains a scientific mystery to be solved by future researchers. The present study, however, shows some degree of inflammatory subcellular cytokine impact as seen in the retrieved tissue. There are massive cellular edema, which leads to fragmentation and partial loss of collagenous architecture. In conclusion, the overall impression of the retrieved tissue through scanning electron microscopy suggested noncellular inflammation sequelae justifying subcellular impact at the nano level.

Acknowledgment The author acknowledges with gratitude the support of the patients of advanced Parkinsonism and their families for the support and enthusiasm received for this research program. The author also acknowledges the support of the mothers who donated their aborted fetus for the research project. The author gratefully appreciates the financial support from the Department of Science and Technology, Government of West Bengal, Calcutta.

The author acknowledges with gratitude the support and overall guidance of Prof M. K. Chhetri of AMRI. Prof K. L. Mukherjee of Mahatma Gandhi Clinical Research Laboratory helped in dsDNA, ANA, and Ferritin estimation. The author also thanks the research associates of the project, Dr Bimal Samanta, Dr Mohua Bhattacharya, Dr Asit Ghosh, Dr Ronjit Nandi, Dr Bonya Biswas, and Dr S. P. Das for their material support. Mrs Seema Das, Clinical Psychologist, assisted in the HAD and MMSE scoring of the patients. Prof Sanjukta Bhattacharya, Jadavpur University, Calcutta, is duly acknowledged for her editorial assistance.

References

1. Curtis MA, Faull RLM, Eriksson PS. The effect of neurodegenerative diseases on the subventricular zone. *Nat Rev Neurosci.* 2007;8:712-723.
2. Walton-Hadlock JL, Fahn S, Keiburtz K, Tanner CM. Levodopa and the progression of parkinson's disease. *New Engl J Med.* 2005;352:1386.
3. Obeso JA, Guridi J, Rodriguez-Oroz MC, et al. Deep brain stimulation of the subthalamic nucleus or the pars interna of the globus pallidus in Parkinson's disease. *New Engl J Med.* 2001;345(13):956-963.
4. López-Lozano JJ, Bravo G, Abascal J, Brera B, Millan I. Clinical outcome of cotransplantation of peripheral nerve and adrenal medulla in patients with Parkinson's disease. Clínica Puerta de Hierro Neural Transplantation Group. *J Neurosurg.* 1999 May;90(5):875-882.
5. Grégoire Courtine, Yury Gerasimenko, Rubia van den Brand, et al. Transformation of nonfunctional spinal circuits into functional states after the loss of brain input. *Nat Neurosci.* 2009;12:1333-1342.
6. Bhattacharya N, Mukherjee K, Chhetri MK, Banerjee T, Mani U, Bhattacharya S. A study report of 174 units of placental umbilical whole blood transfusion in 62 patients as a rich source of fetal hemoglobin supply in different indications of blood transfusion. *Clin Exp Obst Gynecol.* 2001; 28(1):47-52.
7. Bhattacharya N, Mukherjee KL, Chhetri MK, et al. An unique experience with human pre-immune (12 weeks) and hypo-immune (16 weeks) fetal thymus transplant in a vascular subcutaneous axillary fold in patients with advanced cancer – a report of two cases. *Eur J Gyn Oncol.* 2001;22(4): 273-277.
8. Bhattacharya N. Fetal tissue / organ transplant in HLA randomized adults' vascular subcutaneous axillary fold – a preliminary report of 14 patients. *Clin Exp Obst Gynecol.* 2001; 28(4):233-239.
9. Bhattacharya N. Fetal cell/tissue therapy in adult disease – A new horizon in regenerative medicine. *Clin Exp Obst Gynecol.* 2004;31:167-173.
10. Thomas M, Yang L, Hornsby PJ. Formation of functional tissue from transplanted adrenocortical cells expressing telomerase reverse transcriptase. *Nat Biotechnol.* 2000;18(1):39-42.
11. Bhattacharya N, Chaudhuri N, Banerjee S, Mukherjee KL. Intraamniotic tetanus toxoid as a safe abortifacient. *Ind J Med Res.* 1979;70:435-439.
12. Bhattacharya N. Letter to the editor. *Clin Exp Obst Gynecol.* 1996;23:272-275.
13. Bhattacharya N (1996) Intraamniotic instillation of tetanus toxoid: A safe, cheap, effective abortifaecient in the light of

our experiences with different intraamniotic instillation of antigens for alteration of pregnancy immunotolerance – A study from 1978 to 1996. In Tambiraja RL, Ho NK, eds. *Relevance and Excellence in Perinatal Care*, Proceedings of the 9th Congress of the Federation of the Asia and Oceania Perinatal Societies, Singapore, Nov. 1996, Monduzzi Editore, Bolonga, Italy, pp 193–200.

14. Bhattacharya N (1996) Dissolution of the fetus: A new experience with intraamniotic BCG instillation. In: Tambiraja RL, Ho NK, eds. *Relevance and Excellence in Perinatal Care*, Proceedings of the 9th Congress of the Federation of the Asia and Oceania Perinatal Societies, Singapore, Nov. 1996, Monduzzi Editore, Bolonga, Italy, pp 201–206.
15. Bhattacharya N (1996) Study of the aborted fetus after intraamniotic instillation of tetanus toxoid. In: Tambiraja RL, Ho NK, eds. *Relevance and Excellence in Perinatal Care*, Proceedings of the 9th Congress of the Federation of the Asia and Oceania Perinatal Societies, Singapore, Nov. 1996, Monduzzi Editore, Bolonga, Italy, pp 187–192.
16. Bhattacharya N (2001) A study on the intraamniotic instillation of tetanus toxoid on a growing human fetus. D. Sc. dissertation, Faculty of Medicine, Calcutta University Kolkata
17. Deacon T, Schumacher J, Dinsmore J, Thomas C, et al. Histological evidence of fetal pig neural cell survival after transplantation into a patient with parkinson's disease. *Nat Med.* 1997;3(3):350-353.
18. Freed CR, Green PE, Breeze RE, Tsai WY. Transplantation of embryonic dopamine neurons for severe Parkinson's disease. *New Engl J Med.* 2001;344:710-719.
19. Wenning GK, Ordin P, Morrish P, Rehncrona S. Short and long term survival and function of unilateral intrastriatal dopaminergic grafts in Parkinson's disease. *Ann Neurol.* 1997;42(1):95-107.
20. Hoffer B, van Horne C. Survival of dopaminergic neurons in fetal tissue graft. *New Engl J Med.* 1995;332(17): 1163-1164.
21. Jankovic J. Parkinson's disease: clinical features and diagnosis. *J Neurol Neurosurg Psychiatr.* 2008;79(4): 368-376.
22. Fahn S, Elton RL. Members of the UPDRS Development Committee. Unified Parkinson's Disease Rating Scale. In: Fahn S, Marden CD, Calne D, Goldstein M, eds. *Recent developments in Parkinson's disease*, vol. 2. Florham Park, NJ: Macmillan Healthcare Information; 1987:153-163.
23. Hoehn MM, Yahr MD. Parkinsonism, onset, progression and mortality. *Neurology.* 1967;17:427-435.
24. Parashos SA, Johnson ML, Erickson-Davis C, Wielinski CL (2000). Assessing cognition in parkinson disease: use of the cognitive linguistic quick test. *J Geriatr Psychiatr Neurol.* 2009;22(4):228-234.
25. Zigmnond AS, Snaith RP. The hospital anxiety and depression-scale. *Acta Psychiatr Scand.* 1983;67:361-370.
26. Snaith RP, Taylor CM (1985) Rating scgles for depression and anxiety:a current perspective. *Br J Clin Pharmacol.* 19:17S-20S.
27. Channer KS, James MA, Papouchado M, Res JR. Anxiety and depression in patients with chest pain referred for exercise testing. *Lancet.* 1985;ii:820-822.
28. Snaith RP, Zigmond AS. *The Hospital Anxiety and Depression Scale Manual.* Windsor: NFER, Nelson; 1994.
29. Weiss G. Pathogenesis and treatment of anaemia of chronic disease. *Blood Rev.* 2002;16:87-96.
30. Acasoy MO. Non haematopoietic biological effect of erythropoietin. *Br J Haematol.* 2008;141(1):14-31.
31. Juul S. Erythropoietin in the central nervous system and its use to prevent hypoxicischaemic brain injury. *Acta Paediatr Suppl.* 2002;91:36-42.
32. Juul S. Nonerythropoietic roles of erythropoietin in the fetus and neonate. *Clin Perinatol.* 2000;27(3):527-541.
33. Melzack R. The McGill pain questionnaire: major properties and scoring method. *Pain.* 1975;1:277-299.
34. Melzack R. The short-form McGill Pain Questionnaire. *Pain.* 1987;30(2):191-197.
35. Goldstein A, Lowry PJ. Effect of the opiate antagonist naloxone on body temperature in rats. *Life Science.* 1975; 17(6):927-931.
36. Catherine Saint Louis, "Exercise Test: Truth or Myth", Published: March 27, 2008, *New York Times.*
37. Waste management – cytokines, growth factors and cachexia, 2006;17(6): 475–486. Epub 2006 Nov 22
38. Miller RK (1991) Fetal drug therapy: Principles and Issues. In: Pitkin MR and Scott JR, eds, *Clin. Obst. Gynecol.* 34(2), 241–250.
39. Olanow CW, Goetz CG, Kordower JH, Stoessl AJ, Sossi V, Brin MF. A double-blind controlled trial of bilateral fetal nigral transplantation in Parkinson's disease. *Ann Neurol.* 2003;54:403-414.
40. Guyton SE, Hall JE. *Text Book of Medical Physiology.* 10th ed. Philadelphia, PA: Harcourt Asia Pvt. Ltd, Saunders Company; 1997:709-714.
41. Stocum DL. Regeneration biology and engineering: Strategies for tissue restoration. *Comment Wound Repair Regen.* 1998; 6(4):273-275.
42. Bjorklund A, Dunnett SB. Neural transplants in Parkinson's disease: do they work? Forum, Neural transplantation for the treatment of Parkinson's disease. *Lancet Neurol.* 2003; 2:437-445.

Part XIV
Ethics

Ethical Issues Surrounding Umbilical Cord Blood Donation and Banking

40

Gabrielle Samuel, Ian Kerridge, and Tracey O'Brien

40.1 Introduction

Hematopoietic stem cell transplantation (HSCT) is curative therapy for many malignant and nonmalignant conditions including leukemia, bone marrow failure syndromes, immunodeficiencies, and inborn errors of metabolisms. Stem cells for HSCTs are typically sourced from compatible bone marrow or peripheral blood donors. However, over the last 2 decades umbilical cord blood (UCB), once considered a biological waste product, has become a routinely used source of hematopoietic stem cells. With this, there has been establishment of UCB banks, both not-for-profit "public" banks and private commercial banks, resulting in a large and growing inventory of this type of stem cell. This has raised a number of important scientific, ethical, legal, and political issues. These include ethical concerns regarding ownership of the blood, the processes for obtaining consent for collection and storage of UCB, issues relating to confidentiality and privacy, questions raised regarding commercial non-altruistic banking, and social justice issues relating to equity of access and equity of care.

40.1.1 Rational for UCB Collection and Storage

HSCT is only an option for patients in need of a transplant if there is a suitably matched human leukocyte antigen (HLA) donor. (Donors and recipients are matched according to the degree to which they share specific HLA's, inherited from both parents, which determine the immunological identity of a cell.) Given the enormous variation in HLA "types" in the community, the best chance of finding a fully matched donor is within one's own family. Unfortunately, as siblings have only a one in four chance of matching with each other, and extended family members a much lower chance of being matched, only 30% of patients needing a donor have a suitably matched related donor. For the majority of patients who do not have a matched sibling or family member, volunteer donor is the only option. However, even with the number of volunteers on international bone marrow donor registries exceeding 10 million, many patients still cannot find a suitably matched donor for transplant. There are three main reasons for this. First, in recent years it has become clear that immunological variation between individuals is a lot greater than first anticipated, meaning that it is harder to identify a fully matched donor. Second, even where a suitable matched donor is identified, donors sometimes are unwilling to proceed with donation or may be unavailable. Third, the probability of finding a donor is directly correlated with the ethnicity of the recipient, with patients from an ethnic group with wide immunogenetic variation and/or from a group underrepresented on donor registries much less likely to find a suitable donor. As most volunteers on the US, European, and Australian registries are of North Caucasian descent there is a powerful ethnic bias that makes transplantation much less of an option for many patients from non-Caucasian, mixed-ethnic, and indigenous populations.

The discovery that UCB was a rich source of hematopoietic progenitor cells, and that these stem cells were immunologically naive, and so could be safely infused even when not completely matched,[1]

G. Samuel (✉)
Centre for Values, Ethics and the Law in Medicine,
University of Sydney NSW, Australia
e-mail: gabbysamuel@gmail.com

N. Bhattacharya and P. Stubblefield (eds.), *Regenerative Medicine Using Pregnancy-Specific Biological Substances*,
DOI: 10.1007/978-1-84882-718-9_40, © Springer-Verlag London Limited 2011

provided hope that UCB banks – analogous to bone marrow donor registries – would expand the immunological "range" of stem cells available for transplantation, increase the number of individuals for whom transplant was an option, increase the number of donors (particularly from ethnic minorities), and address the ethnic underrepresentation evident in stem cell banks worldwide. UCB transplantation also offers a number of other advantages: UCB collection is non-invasive, safe, and painless; it is widely and easily accessible; and it can be cryopreserved, providing a readily accessible source of stem cells, dramatically reducing the time from initiation of search to transplant from 3–4 months to 1–2 weeks[2] – something that becomes particularly important in situations where time to transplant can affect survival and outcome, such as is the case with high-risk malignancies and congenital neurodegenerative disorders.

The main disadvantages of UCB have been the inferior speed of hematopoietic cell recovery and higher graft failure rates compared with peripheral blood and bone marrow. Factors impacting on speed of hematopoietic recovery have varied between reports; however most cite nucleated cell dose, CD 34+ cell dose or colony-forming units–granulocyte–macrophage (CFU-GM) as critical determinants. The average nucleated cell yield in a single cord unit collection provides sufficient cells for hematopoietic reconstitution in recipients less than 40 kg, and is the major factor limiting use in adult patients. However, improved outcomes with single UCB unit transplants and the use of "double cord" blood transplants, where two partially matched UCB units are transplanted simultaneously, have led to increasing use of UCB transplantation in adults and remains the focus of ongoing investigation.[3]

40.1.2 Clinical Evidence Regarding UCB Transplantation

Since the first successful UCB transplant in 1988 over 8,000 UCB transplants have been performed worldwide.[4] Indeed, current estimates indicate that over half of pediatric unrelated transplants, and 20–50% of adult unrelated transplants, now utilize UCB as the stem cell source.[5] And although no randomized trials have been conducted in children comparing the outcomes of allogeneic HSCT using UCB versus bone marrow, numerous uncontrolled trials have found that outcomes of transplantation with well-matched UCB are essentially equivalent and in some cases superior to unrelated donor marrow or peripheral blood stem cells.[6-8] Hwang performed a meta-analysis of pooled data on comparative studies of unrelated donor UCB transplantation and unrelated donor bone marrow transplantation in children requiring allogeneic HSCT and concluded that there were no differences in 2-year overall survival.[9] Likewise, Eapen conducted a retrospective analysis of the outcomes of transplantations in children with acute leukemia between 1995 and 2003, and concluded that 5-year leukemia-free survival was comparable between matched bone marrow and UCB (5/6 or 6/6 with high cell dose) transplants. Matched UCB donor transplants had the best outcome, and 4/6 UCB donor transplants had an inferior outcome, regardless of cell dose.[10] Although most of these studies were for patients with malignant disease, in the series of Rocha, half of the cord blood transplants were for nonmalignant disease.[10]

Further developments, including the use of "double cord" blood transplants,[11] promise to improve the outcomes of UCB transplantation still further. As evidence grows that UCB transplantation is a safe and effective option in children and adults requiring HSCT, and that UCB may provide the only source of stem cells for many people from underrepresented ethnic groups, the need for UCB will become ever more pressing. This need can only be met by the establishment and maintenance of an effective network of UCB banks.

40.1.3 Establishment of UCB Banks

UCB banks fall into two categories: public banks and private, "for-profit" banks. The distinction between these two types of banks is critically important, as these two models of UCB storage differ in relation to their scientific rationale and medical utility, and in relation to the philosophical concepts that may be used to examine their construction and the issues raised by their incorporation into clinical practice.

Government and/or community-funded public UCB banks are facilities where parents can "donate" their child's UCB for use by any individual with a malignant or nonmalignant condition requiring unrelated donor allogeneic hematopoietic stem cells for transplantation (according a strict set of immunological criteria). These banks were established in an effort to increase the likelihood that patients (particularly those from underrepresented ethnic minorities) would be able to find a suitably matched "donor." Many countries have now established their own public UCB banks, with greater than 300,000 UCB units internationally stored for potential future use. Over 45 public UCB banks worldwide report data to the World Marrow Donor Association (WMDA). Other cord blood networks include EuroCord, Auscord, and AsiaCord.

In contrast, private UCB banks, for a fee, store a child's UCB for personal or family use. It is important to note that use of a child's own stem cells, i.e., autologous UCB transplants are not routinely used in the treatment of leukemia and other inherited blood and immune conditions, and the need for autologous transplantation is remote with a lifetime estimated probability in US persons of 0.23%.[12] This is evidenced by the fact that although over 8,000 unrelated cord blood transplants have been reported, there are less than a handful of reported autologous UCB transplants in peer-reviewed publications.[13-15] The reason for this is twofold. First, research shows that the abnormal cells that caused the disease later in life may be present from birth.[16] Second, autologous transplantation for the treatment of childhood leukemia is inferior to allogeneic transplantation as it does not readily allow for the potent-positive immunological reaction of graft versus leukemia (GVL) leading to high relapse rates.[17,18] Despite the low likelihood of autologous UCB use, there has been a rapid emergence of commercial UCB banks. There are now approximately 134 private banks worldwide, with an estimated 780,000 UCB units stored.[19] The driving forces which create such a demand for the private UCB banks include the success of allogeneic HSCT programs using donated UCB, and the promise that UCB stem cells will have therapeutic potential in regenerative medicine. Most private banks charge an upfront fee, plus an annual storage fee (up to 18 years). In the USA, the upfront cost is approximately US$1,600, with annual storage fees of US$125[20]; in Australia, the parents pay Australian $1,950, followed by annual installments of $110,[21] and in the UK the cost is £495 for processing and annual storage of £75.[22]

Because of the marked philosophical, ethical, and clinical distinction between public and private banks, it is important that there is adequate differentiation of the language terms used to describe the two different processes. Terms such as "donation" should be applied only to the system of public UCB "storage," and that storage of UCB within a public institution should be referred to as just that, storage, and not by the term "banking," which should be used only in relation to private UCB storage as it implies an investment – the retention of tissue for one's own benefit at some point in the future.

40.2 Ethical Issues

Although the concept of UCB collection, storage, and use may seem relatively unproblematic on the surface, UCB collection, donation, and banking, all raise a number of important epistemic, social, legal, and ethical concerns. These include questions regarding ownership of the blood, the processes for obtaining consent for collection and storage, issues relating to confidentiality, questions raised regarding commercial non-altruistic banking, and social justice issues relating to equity of access and equity of care.

40.2.1 Ethical Issues Concerning UCB Collection

Ownership of the umbilical cord blood: In Western societies, UCB was traditionally regarded as a waste product, discarded after birth. However, the use of UCB both therapeutically and in research has necessitated a reevaluation of the ownership of this tissue in ethical and legal terms. In an early paper on UCB collection, it was suggested that UCB should have the same status as any donated organ or tissue, which means it should be treated as being owned by the child.[23] This concept has since become broadly accepted in many countries, with decisions regarding a child's UCB collection and use made by the "person responsible" – in this case the child's parents. However, this view of ownership still leaves a number of unresolved questions, for example, issues of consent to donation or banking (see below), and whether children could ever hold their parents liable for not banking the child's UCB.

The question of legal ownership of UCB also remains an area of controversy. The law in most countries has been established to deal with tissue, organ, and blood donation. Some countries, such as the USA, UK, and Australia, draw distinctions between tissue and organ donation, and the use of blood for transfusions. However, because of the unique nature of UCB, and also because there have been no reported cases or laws specifically dealing with UCB, it is still unclear which laws apply to disputes about UCB ownership. Partly this is a reflection of the fact that the blood is taken from extracted tissue rather than through phlebotomy, raising issues of whether UCB is regulated by tissue or blood laws. The lack of clarity is also exacerbated by the uncertainty of whether the umbilicus is legally part of the mother or the child as the laws regulate donation of tissue and blood differently depending on whether the donor is a child or an adult. This leads to uncertainty regarding commodification of UCB, commercialization, donor risks, donor property rights, and the property rights of the organization that receives the UCB through donation or transfer. UCB storage is becoming more and more prevalent, and these issues need to be addressed legally in the near future[24, 25]

Ownership becomes especially uncertain in the context of private UCB banking. For instance, little attention is paid to who owns UCB in the event of company bankruptcy, or if the parents can no longer pay the annual storage fee, or become divorced.[26] In addition, the contracts of some private banks in the USA have extremely controversial language regarding who owns the sample of UCB after a specific contract expires.[26]

Perhaps more challengingly, most discussion of ownership of UCB has tended to adopt a libertarian stance regarding its collection, storage, and use, and few have addressed ownership from a communitarian or feminist perspective, or construed ownership in "nonlegal" terms. This emphasis on bodily autonomy and personal control over a tissue resource that may have a particular market value has inevitably led to a rather impoverished view of the social functions of UCB storage and of the moral significance of transplantation programs. Alternative perspectives of UCB storage that emphasize interdependence, mutual obligations and duties, ontological and cultural security, social relationships and care, rather than autonomy, independence, and noninterference are sorely needed in relation to debates regarding UCB donation and banking, as the latter provide a limited view of altruism, communal obligations, and the values and goals of health care.

Consent: There are a number of possible strategies for UCB collection, including mandated choice, required request, paid donation, altruistic volunteerism, and presumed consent. All, however, are made problematic because of uncertainty as to who should consent, and under what conditions consent should be obtained. Decisions regarding UCB collection are often made by the mother, and based on valid consent. While this would seem uncontroversial, the practical requirement that decisions regarding UCB donation and banking should be "informed," "voluntary," and "competent" has been the focus of intense scrutiny. For example, concerns that women would not be competent to make decisions regarding UCB donation during the course of labor as they may be experiencing physical or emotional distress have led to the introduction of "phased consent protocols," whereby consent is broken down into three phases (before-, during-, and after-labor consent).[27, 28] Similarly, while it is generally accepted that women should be adequately informed regarding UCB donation and banking, there remains considerable uncertainty regarding how much information needs to be disclosed to potential donors and how consent processes for UCB donation should deal with the more complex issues associated with this, such as accessibility of any genetic and infectious disease test results.[29, 30] Further, the coercive nature of private UCB banking campaigns, which often misconstrue the medical necessity of a child ever needing their UCB, can ultimately misinform pregnant women about the medical benefits of UCB donation and banking.

In addition, if UCB is considered to be owned by the child, it would suggest that both parents must exercise their parental power to consent to donation on the child's behalf. If this model is correct, both the mother and the father would have equal rights to consent to the donation and storage. A number of issues then become apparent, such as whether the consent of both parents is required and what should happen when the mother and father disagree about whether the UCB should be donated and stored.

Respect of Privacy and the idea of obligation: Respect for privacy and confidentiality are of fundamental importance in Western society, particularly in the construction of contemporary health services.[31-33] In most industrialized countries the maintenance of donor privacy/confidentiality in UCB banking is controlled in two ways.

First, it is assumed that the results of all investigations routinely performed prior to UCB donation are treated in the same manner as the results of tests done in any other context and in any other health-care setting, and so remain confidential. Second, even though a recipient may have strong grounds for wishing to know the identity of the donor (e.g., if they experienced failure of their initial transplant and required a second hematopoietic stem cell transplant), it is widely accepted that the donor's identity must not be disclosed, as such a disclosure may have harmful physiological, practical, legal, and institutional repercussions.

40.2.2 Social-Justice Concerns for UCB Public Donation

Public UCB banks were established in an effort to increase the likelihood that patients (particularly those from underrepresented ethnic minorities) would be able to find a suitably matched "donor" for HSCT. Although progress in the field of UCB transplantation has been phenomenal, there are still a number of political and structural challenges which impede the public UCB banking system.

Underrepresentation of ethnic minority groups: UCB banks are still characterized by the underrepresentation of many ethnic groups. For example, although in the USA more than 95% of patients are able to find at least one potential 4/6 HLA matched UCB unit,[34] these figures fail to take into account the quality of the donated UCB unit, the amount of cells it contains, and conceals the fact that patients from non-Caucasian ethnic or indigenous populations have a much lower chance of finding a suitably matched unrelated donor. In the USA, African-Americans have the least chance of finding a matched bone marrow, peripheral blood, or UCB donor, followed by Hispanic and Asian patients.[35] Similarly, a number of ethnic groups in Australia are underrepresented in public UCB banks. In particular, Indian patients have the least chance of finding a suitable matched unrelated donor, followed by Asian patients, and Aboriginal Australians.[36] Education and awareness programs have been emphasized in an attempt to try and increase UCB donation rates. However, these educational strategies have still been unable to adequately address the issue of minority group underrepresentation.

Access to UCB donation: Although many developed countries have now established their own public UCB banks (developing countries are now starting to follow this trend, for example in Mexico[37]), the organization of, and access to, UCB collection centers continues to be a social-justice concern. First, public UCB donation in some countries still remains disorganized. For example, while there are a number of UCB banks throughout the USA, not all are affiliated with the National Marrow Donor Registry. And second, UCB public banks in many countries have limited collections centers, which often tend to not be located in culturally diverse areas, meaning that there continues to be low donation/recruitment rates from ethnic minority groups. For instance, Australia only has 11 collection centers nationwide, most of which are in higher socioeconomic urban areas, and the NHS cord blood bank in the UK only collects UCB at four hospitals, all of which are clustered around the north of London. Fortunately, in the USA, although there are also limited UCB collection centers nationally, a mother can donate her child's UCB from any hospital in the USA via Cryobanks international, which will collect the UCB and deposit it in the National Marrow Donation Program (NMDP) registry (Cryobanks international is primarily a private UCB bank, which offers a UCB donation service).

Cost–benefit of UCB donation: Collecting and storing UCB is expensive and requires considerable and continuing government support. This is of particular concern in developing countries, in which UCB public donation systems are beginning to emerge, because the cost of HSCT is extensive, and may be better spent on other, more immediate, public health concerns. However, it is also important to consider the costs of UCB donation programs in western industrialized countries to ensure continued investment in the banks is still providing significant health benefits for patients (in terms of its opportunity cost) and providing increased access to HSCT – especially for those of ethnic groups who are currently underrepresented in bone marrow donor registries. Although reports suggest that UCB public banks are more ethnically diverse than their bone marrow registry counterparts,[36] many groups are still underrepresented and have a much lower likelihood of finding a suitable match. Given the enormity of the challenge of UCB donation, the global nature of the donor registry system and the fact that for many ethnic groups approximately 5,000 UCB units need to be collected before there is a significant increase in the probability that an individual from an underrepresented ethnic minority will find a suitably matched donor (Marcus Vowels, personal communication.), we have argued elsewhere that it may be more valuable and cost-effective to selectively target only a few of the groups that are currently underrepresented on donor registries, such as those populations who originate in countries that do not have large UCB inventories, as it is these populations who will have more difficulties finding a suitably matched donor.[36]

Awareness of UCB donation and banking: To date, there has been limited international research regarding the knowledge, attitudes, and acceptability of UCB donation and banking. A couple of studies have surveyed pregnant women from high socio-demographic areas with very little ethnic diversity (36). Despite the higher educational level of participants, the women still had a very poor understanding of the UCB donation and banking[38, 39] In addition, a study conducted by Perlow demonstrated that ethnic minorities, younger patients, and those with lesser degrees of education were more likely to be poorly informed about UCB donation and banking.[40] These results suggest that few patients receive UCB donation and banking education from health-care providers. In response to this, the Institute of Medicine has provided recommendations endorsing the counseling of pregnant women on this subject.[41] Fox and Perlow have both argued in favor of this, supporting the obstetricians' role as an information-provider.[42, 43] Fox states that it is medically reasonable to recommend public UCB banking where feasible since all citizens have a justice-based obligation to come to the rescue of others when we are in a position to do so, when our efforts are likely to be successful, and when the sacrifice is reasonable. However, Fox goes on to argue that although the provision of information is important, because of the limited uses of autologous UCB transplants, obstetricians have no ethical obligation to offer private banking to those patients who do not request it (although if a patient requests to bank their child's UCB privately this request should be honored).[42] Although Fox's argument may hold true now, as the therapeutic role of UCB begins to expand and move into newer fields such as regenerative medicine (see below), these moral arguments may need reevaluating.

40.2.3 Ethical Issues Concerning UCB Private Banking

The rise of private UCB banks has generated considerable debate in the industrialized world, with several countries, such as Italy and Spain, electing to prohibit private UCB banking.[44, 45] Concerns raised against private banking generally relate to medical necessity, community, benefit, consent, and honesty in advertising.

Medical necessity: The scientific justification for UCB private banking is particularly weak and any claims of medical utility are, at present, enormously contestable. There are a number of medical/scientific problems with claims that banking of one's own UCB stem cells is beneficial. First, the vast majority of people will never develop hematological malignancy or any other indication for HSCT, and so will never require the use of their own hematopoietic stem cells. Second, even where people do have a genetic or immunological disorder or acquire a malignancy that may benefit from transplantation, in the majority of instances the optimal source of stem cells is not the patient's own hematopoietic stem cells, but those from another individual. (This is either because the individuals' stem cells may contain a known or unknown genetic abnormality that may predispose to malignancy, or, more importantly, because immunological/genetic differences between the donor and recipients stem cells are necessary for transplantation to be effective.) And third, even where there is evidence that an individual may benefit from autologous transplantation, such as with lymphoma, multiple myeloma, or relapsed/refractory germ cells tumors, it still does not follow that one should bank one's own stem cells for use in this situation. While autologous transplantation may be beneficial, in the vast majority of cases, autologous progenitors can be "harvested" from an individual following administration of cytokines with or without chemotherapy if and when a transplant is planned. This simple fact makes banking of one's own stem cells for autologous transplantation both unnecessary and inefficient. Given this, it is unsurprising that estimates of the likelihood of ever needing an autologous HSCT are extremely small, ranging from 1/20,000 to 1/200,000.[12] It is noteworthy that one private company with 14 years experience and 200,000 stored UCB units reports in a company document that only 25 UCB units have been used for autologous transplants (and 42 sibling transplants)[46, 47] and that only 3–4 autologous UCB transplants have been reported in peer-reviewed medical literature.[13-15] This lack of scientific justification is often used as an argument against private UCB banking. While it is undoubtedly true that at present there is negligible medical rationale for autologous HSCT, it does not provide a definitive argument against private banking. For it is arguable that as long as the parents have been provided with sufficient and accurate information about the benefits and drawbacks of private banking to allow them to make a decision, they should have the right to choose. It is not appropriate to claim that decisions to bank privately are unethical or wrong simply because the decision may not be one that others agree with. Such decisions become a problem only when they are based on inaccurate information.

Loss of donors: Other concerns expressed about private banking have focused on the potential impact of the loss of donors, and therefore HLA heterogeneity, on public banks.[48] Like arguments about the right of parents to choose private banking, these concerns also seem somewhat overstated. There is already a large source of HLA-types available in public banks, and the fact that transplantation can successfully be performed with HLA-mismatched UCB units extends this range even further. This suggests that private banking would only have an impact on the HLA-variability available in public banks if it became the dominant form of banking.

Social justice concerns: When private UCB banks first started operating, concerns were raised about the inequity of "allowing" only certain individuals to store UCB (i.e., those who can afford it). These concerns were quickly crushed, since, at the time, the medical benefits of private banking were negligible, there was little therapeutic use for autologous HSCT, and the promise of regenerative medicine was still fiction. Although not much has changed since that time, advances in regenerative medicine are moving at an alarming pace. If autologous transplantation does become more of a therapeutic option in the future, when the families concerned are unable to afford private banking, the issue of justice and equity in the health-care system, along with various further ethical issues of state obligation need to be dealt with.

Commercial influences on decision-making: Since UCB donation and banking first became an option for parents, concerns have been raised that the interests of private bank, or the scientific and medical community, would distort the information presented to parents and so reduce the validity of their consent.[49] While this possibility exists for public banking, in the context of private banking, where financial viability is entirely dependent on attracting "donors," this would appear to be a major risk. Indeed, private UCB banks have been criticized for not providing accurate information, and for using powerful advertising campaigns specifically designed to play on the emotions of new parents.[50] These campaigns often appear to sell possible, rather than real applications of UCB, strongly suggest that stem cell therapies (regenerative medicine) derived from UCB are available, or will shortly and inevitably be developed, and tend to overemphasize, or at least misconstrue the likelihood of ever needing the UCB.[50] This misrepresentation of facts seriously impinges on people's right to make an informed decision and ultimately diminishes their personal autonomy. Concerns regarding the coercive nature of information regarding private banking have been so sustained that they have led to recommendations by the European Commission stating, appropriate information should be given to the consumers willing to use their services, including the fact that the likelihood that the sample may be used to treat one's child is currently negligible, that the future therapeutic possibilities are of a very hypothetical nature and that up until now there is no indication that the present research will lead to specific therapeutic applications of one's own UCB cells.[49]

While it may be the case that the potential therapeutic utility of UCB for personal use has been overstated to serve commercial ends, it is important to note that this field of research is extraordinarily vibrant and that there is some (increasing) evidence to support the idea that UCB may have use beyond autologous HSCT (see below). Given the increasing applications for UCB as a therapeutic option, it is likely that the chance of needing one's own UCB in the future may substantially increase.

40.2.4 A Role for Umbilical Cord Blood Stem Cells in Regenerative Medicine

In recent years an enormous amount of attention has been devoted to establishing whether UCB stem cells may have a role in tissue repair (regenerative medicine).

This interest in UCB stem cells has arisen for two reasons. First, as the collection and use of UCB cells does not involve the destruction of an embryo, their use in research and therapy avoids many of the moral concerns raised by embryonic stem cell research. Second, recent research has demonstrated that UCB stem cells can differentiate into numerous cell types in vitro, and repair damaged neurological, cardiovascular, and hepatic tissues.[51, 52] This plasticity (i.e., the ability under the correct conditions to differentiate into a variety of cells) can be attributed to the presence of several cellular populations within UCB, each with varying degrees of pluripotential ability, including hematopoietic, endothelial, epithelial, mesenchymal, unrestricted somatic, and embryonic-like stem cells.[53-57] The expansive range of different UCB stem cells suggests a role for UCB in the treatment of diseases such as diabetes, strokes, and Parkinson disease. At present, UCB stem cells are being evaluated in a clinical trial to treat Type I diabetes (autologous)[58]; they have been used clinically to treat spinal cord injury (allogeneic)[59]; and have been used in a number of preclinical animal studies for heart disease and ischemic disease.[60-62] UCB stem, and have been demonstrated to form bone in vitro when subjected to shockwave induction,[63] suggesting a role in aiding ligament repair. Interestingly, the potential of UCB stem cells to function in bone and ligament repair has driven a number of cottage industries to bank UCB as a "sports industry repair kit" for the future.[64]

While much of the excitement surrounding UCB research is based upon hope, rather than evidence, should UCB stem cells prove to have wide therapeutic application, serious questions will arise regarding the maintenance of social equity in health care when only a small proportion of the population are able to afford UCB storage for personal/family use.[49, 50] And while state-provided storage of all UCB for personal use may, although extremely costly, satisfy social justice concerns,[49] this "solution" will threaten the real and symbolic value attached to altruistic donation of tissues and, in the end, the very existence of public UCB banks.

40.3 Conclusion

Public UCB banks have, for the most part, been very successful in making HSCT transplantation a real option for patients who require a transplant. UCB banks attract considerable public support from the public and the medical community, and are significantly more ethnically diverse than the existing bone marrow registries. In contrast, concerns regarding the legitimacy of storing UCB for personal/family use and the marketing strategies used to "sell" these banks has led many international groups, including the European Group on Ethics in Science and New Technologies,[49] the World Marrow Donor Association,[65] the Royal College of Obstetricians and Gynaecologists in the UK,[66] and the Society of Obstetricians and Gynaecologists of Canada[67] to recommend thatappropriate information should be given to the consumers willing to use their (commercial UCB banks) services, including the fact that the likelihood that the sample may be used to treat one's child is currently negligible, that the future therapeutic possibilities are of a very hypothetical nature and that up until now there is no indication that the present research will lead to specific therapeutic applications of one's own cord blood cells.[49] This advice remains true. However, it must be acknowledged that these recommendations are based on current knowledge of the therapeutic uses of UCB stem cells. However, UCB is moving into a new area of regenerative medicine and should evidence emerge that UCB has value in the treatment of a wide range of degenerative disorders, then the ethical, moral, and legal nature of public and private UCB donation and bank will need to be changed.

References

1. Gluckman E, Koegler G, Rocha V. Human leukocyte antigen matching in cord blood transplantation. *Semin Hematol.* 2005;42(2):85-90.
2. Barker JN, Krepski TP, DeFor TE, Davies SM, Wagner JE, Weisdorf DJ. Searching for unrelated donor hematopoietic stem cells: availability and speed of umbilical cord blood versus bone marrow. *Biol Blood Marrow Transplant.* 2002;8(5):257-260.
3. Brunstein CG, Barker JN, Weisdorf DJ, et al. Umbilical cord blood transplantation after nonmyeloablative conditioning: impact on transplantation outcomes in 110 adults with hematologic disease. *Blood.* 2007;110(8):3064-3070.
4. Gluckman E, Rocha V. Donor selection for unrelated cord blood transplants. *Curr Opin Immunol.* 2006;18(5):565-570.
5. Rubinstein P. Why cord blood? *Human Immunol.* 2006;67(6): 398-404.
6. Rocha V, Wagner JE Jr, Sobocinski KA, et al. Graft-versus-host disease in children who have received a cord-blood or

bone marrow transplant from an HLA-identical sibling. Eurocord and International Bone Marrow Transplant Registry Working Committee on Alternative Donor and Stem Cell Sources. *N Engl J Med*. 2000;342(25):1846-1854.
7. Rocha V, Cornish J, Sievers EL, et al. Comparison of outcomes of unrelated bone marrow and umbilical cord blood transplants in children with acute leukemia. *Blood*. 2001; 97(10):2962-2971.
8. Barker JN, Davies SM, DeFor T, Ramsay NK, Weisdorf DJ, Wagner JE. Survival after transplantation of unrelated donor umbilical cord blood is comparable to that of human leukocyte antigen-matched unrelated donor bone marrow: results of a matched-pair analysis. *Blood*. 2001;97(10):2957-2961.
9. Hwang WY, Samuel M, Tan D, Koh LP, Lim W, Linn YC. A meta-analysis of unrelated donor umbilical cord blood transplantation versus unrelated donor bone marrow transplantation in adult and pediatric patients. *Biol Blood Marrow Transplant*. 2007;13(4):444-453.
10. Eapen M, Rubinstein P, Zhang MJ, et al. Outcomes of transplantation of unrelated donor umbilical cord blood and bone marrow in children with acute leukaemia: a comparison study. *Lancet*. 2007;369(9577):1947-1954.
11. Barker JN, Weisdorf DJ, DeFor TE, et al. Transplantation of 2 partially HLA-matched umbilical cord blood units to enhance engraftment in adults with hematologic malignancy. *Blood*. 2005;105(3):1343-1347.
12. Nietfeld JJ, Pasquini MC, Logan BR, Verter F, Horowitz MM. Lifetime probabilities of hematopoietic stem cell transplantation in the US. *Biol Blood Marrow Transplant*. 2008;14(3):316-322.
13. Hayani A, Lampeter E, Viswanatha D, Morgan D, Salvi SN. First report of autologous cord blood transplantation in the treatment of a child with leukemia. *Pediatrics*. 2007;119(1): e296-e300.
14. Ferreira E, Pasternak J, Bacal N, de Campos Guerra JC, Mitie Watanabe F. Autologous cord blood transplantation. *Bone Marrow Transplant*. 1999;24(9):1041.
15. Fruchtman SM, Hurlet A, Dracker R, et al. The successful treatment of severe aplastic anemia with autologous cord blood transplantation. *Biol Blood Marrow Transplant*. 2004; 10(11):741-742.
16. Gale KB, Ford AM, Repp R, et al. Backtracking leukemia to birth: identification of clonotypic gene fusion sequences in neonatal blood spots. *Proc Natl Acad Sci USA*. 1997;94(25): 13950-13954.
17. Rubinstein P, Carrier C, Scaradavou A, et al. Outcomes among 562 recipients of placental-blood transplants from unrelated donors. *N Engl J Med*. 1998;339(22):1565-1577.
18. Parkman R. The future of placental-blood transplantation. *N Engl J Med*. 1998;339(22):1628-1629.
19. Katiz-Benichou G. Umbilical cord blood banking: economic and therapeutic challenges. *Int J Healthcare Technol Manage*. 2007;8:464-477.
20. CRYO-CELL International Inc. Pricing/Financing – New Clients. Available from: http://www.cryo-cell.com/services/pricing.asp. Accessed March 2008.
21. Biocell Cord blood banking. What's the cost? Available from: http://www.biocell.com.au/whats-the-cost. Accessed March 2008.
22. Rizza J. Boston-based cord blood bank expands to UK. *Transplant News*. 2001;Jan 16.
23. Sugarman J, Reisner EG, Kurtzberg J. Ethical aspects of banking placental blood for transplantation. *JAMA*. 1995; 274(22):1783-1785.
24. Annas GJ. Waste and longing – the legal status of placental-blood banking. *N Engl J Med*. 1999;340(19): 1521-1524.
25. Dame L, Sugarman J. Blood money: ethical and legal implications of treating cord blood as property. *J Pediatr Hematol Oncol*. 2001;23(7):409-410.
26. Saginur M, Kharaboyan L, Knoppers BM. Umbilical cord blood stem cells: issues with private and public banks. *Health Law J*. 2004;12:17-34.
27. Vawter DE, Rogers-Chrysler G, Clay M, et al. A phased consent policy for cord blood donation. *Transfusion*. 2002; 42(10):1268-1274.
28. Askari S, Miller J, Clay M, Moran S, Chrysler G, McCullough J. The role of the paternal health history in cord blood banking. *Transfusion*. 2002;42(10):1275-1278.
29. Fernandez MN. Eurocord position on ethical and legal issues involved in cord blood transplantation. *Bone Marrow Transplant*. 1998;22(suppl 1):S84-S85.
30. Winifred J, Pinch E. Cord blood banking: ethical implications. *Aus J Nursing*. 2001;101(10):55-59.
31. Beauchamp T, Childress JF. *Confidentiality. Principles of Biomedical Ethics*. 5th ed. New York: Oxford University Press; 2001:303-312.
32. Beauchamp T, Childress JF. *Privacy. Principles of Biomedical Ethics*. 5th ed. New York: Oxford University Press; 2001: 293-303.
33. Kerridge I, Lowe M, McPhee J. *Confidentiality and Record Keeping. Ethics and Law for the Health Professions*. 2nd ed. Sydney: The Federation Press; 2005:237.
34. National Marrow Donor Program. Likelihood of Finding an Unrelated Donor or Cord Blood Unit. Available from: http://www.marrow.org/PHYSICIAN/URD_Search_and_Tx/Likelihood_of_Finding_an_URD_o/index.html. Accessed September 2007.
35. National Marrow Donor Program. National Marrow Donor Program-facts and figures. 2005. Available from: http://www.marrow.org/NEWS/MEDIA/Facts_and_Figures/facts_figures.pdf. Accessed March 2008.
36. Samuel G, Kerridge I, Vowels M, Trickett A, Chapman J, Dobbins T. Ethnicity, equity and public benefit: a critical evaluation of public umbilical cord blood banking in Australia. *Bone Marrow Transplant*. 2007;40:729-734.
37. Novelo-Garza B, Limon-Flores A, Guerra-Marquez A, et al. Establishing a cord blood banking and transplantation program in Mexico: a single institution experience. *Transfusion*. 2008;48(2):228-236.
38. Fox NS, Stevens C, Ciubotariu R, Rubinstein P, McCullough LB, Chervenak FA. Umbilical cord blood collection: do patients really understand? *J Perinat Med*. 2007;35(4):314-321.
39. Fernandez CV, Gordon K, Van den Hof M, Taweel S, Baylis F. Knowledge and attitudes of pregnant women with regard to collection, testing and banking of cord blood stem cells. *CMAJ*. 2003;168(6):695-698.
40. Perlow JH. Patients' knowledge of umbilical cord blood banking. *J Reprod Med*. 2006;51(8):642-648.
41. Lubin BH, Shearer WT. Cord blood banking for potential future transplantation. *Pediatrics*. 2007;119(1):165-170.

42. Fox NS, Chervenak FA, McCullough LB. Ethical considerations in umbilical cord blood banking. *Obstet Gynecol.* 2008;111(1):178-182.
43. Perlow JH. Umbilical cord blood banking options and the prenatal patient: an obstetrician's perspective. *Stem Cell Rev.* 2006;2(2):127-132.
44. Fisk NM, Roberts IA, Markwald R, Mironov V. Can routine commercial cord blood banking be scientifically and ethically justified? *PLoS Med.* 2005;2(2):e44.
45. Ecker JL, Greene MF. The case against private umbilical cord blood banking. *Obstet Gynecol.* 2005;105(6): 1282-1284.
46. Cord blood registry. Company review. Available from: http://cordblood.net/pdf/transplant_summary.pdf. Accessed March 2008.
47. Cord Blood Registry. Stem Cell Therapy Data. Available from: http://cordblood.net/pdf/transplant_summary.pdf. Accessed March 2008.
48. Umbilical Cord Blood Banking: The Royal Australian and New Zealand College of Obstetricians and Gynaecologists; 2003. Available from: http://64.233.187.104/search?q=cache:JYxtJls67BoJ:www.ranzcog.edu.au/publications/statements/C-obs18.pdf+200,000+placental+blood+units+stored&hl=en. Accessed February 2006.
49. Opinion on the ethical aspects of umbilical cord blood banking - opinion no. 19. Luxembourg: The European Group on Ethics in Science and New Technologies to the European Commission; 2004. Available from: http://europa.eu.int/comm/european_group_ethics/docs/avis19_en.pdf. Accessed January 2006.
50. Sugarman J, Kaalund V, Kodish E, et al. Ethical issues in umbilical cord blood banking. Working Group on Ethical Issues in Umbilical Cord Blood Banking. *JAMA.* 1997;278(11):938-943.
51. Yamada Y, Yokoyama S, Fukuda N, et al. A novel approach for myocardial regeneration with educated cord blood cells co-cultured with cells from brown adipose tissue. *Biochem Biophys Res Commun.* 2007;353(1):182-188.
52. Sanberg PR, Willing AE, Garbuzova-Davis S, et al. Umbilical cord blood-derived stem cells and brain repair. *Ann NY Acad Sci.* 2005;1049:67-83.
53. Wang HS, Hung SC, Peng ST, et al. Mesenchymal stem cells in the Wharton's jelly of the human umbilical cord. *Stem Cells.* 2004;22(7):1330-1337.
54. Kogler G, Sensken S, Wernet P. Comparative generation and characterization of pluripotent unrestricted somatic stem cells with mesenchymal stem cells from human cord blood. *Exp Hematol.* 2006;34(11):1589-1595.
55. Lee OK, Kuo TK, Chen WM, Lee KD, Hsieh SL, Chen TH. Isolation of multipotent mesenchymal stem cells from umbilical cord blood. *Blood.* 2004;103(5):1669-1675.
56. Zhao Y, Wang H, Mazzone T. Identification of stem cells from human umbilical cord blood with embryonic and hematopoietic characteristics. *Exp Cell Res.* 2006;312(13):2454-2464.
57. Goodwin HS, Bicknese AR, Chien SN, Bogucki BD, Quinn CO, Wall DA. Multilineage differentiation activity by cells isolated from umbilical cord blood: expression of bone, fat, and neural markers. *Biol Blood Marrow Transplant.* 2001;7(11):581-588.
58. Umbilical Cord Blood Infusion to Treat Type 1 Diabetes. Available form: http://www.clinicaltrials.gov/ct/show/NCT00305344?order=1. Accessed March 2008.
59. Kang KS, Kim SW, Oh YH, et al. A 37-year-old spinal cord-injured female patient, transplanted of multipotent stem cells from human UC blood, with improved sensory perception and mobility, both functionally and morphologically: a case study. *Cytotherapy.* 2005;7(4):368-373.
60. Ma N, Stamm C, Kaminski A, et al. Human cord blood cells induce angiogenesis following myocardial infarction in NOD/scid-mice. *Cardiovasc Res.* 2005;66(1):45-54.
61. Henning RJ, Abu-Ali H, Balis JU, Morgan MB, Willing AE, Sanberg PR. Human umbilical cord blood mononuclear cells for the treatment of acute myocardial infarction. *Cell Transplant.* 2004;13(7–8):729-739.
62. Schmidt D, Breymann C, Weber A, et al. Umbilical cord blood derived endothelial progenitor cells for tissue engineering of vascular grafts. *Ann Thorac Surg.* 2004;78(6):2094-2098.
63. Wang FS, Yang KD, Wang CJ, et al. Shockwave stimulates oxygen radical-mediated osteogenesis of the mesenchymal cells from human umbilical cord blood. *J Bone Miner Res.* 2004;19(6):973-982.
64. Pennington B. Sports medicine turns to stem cell 'repair kits'. *International Herald Tribune.* 2007;March 20.
65. World Marrow Donor Association. WMDA Policy Statement on the Utility of Autologous or Family Cord Blood Unit Storage 2006. Available from: http://www.worldmarrow.org/fileadmin/WorkingGroups_Subcommittees/DRWG/Cord_Blood_Registries/WMDA_Policy_Statement_Final_02062006.pdf. Accessed August 2007.
66. Royal College of Obstetricians and Gynaecologists: Umbilical Cord Blood Banking: Scientific Advisory Committee Opinion Paper 2. 2006. Available from: http://www.rcog.org.uk/index.asp?PageID=545. Accessed March 2008.
67. Armson BA. Umbilical cord blood banking: implications for perinatal care providers. *J Obstet Gynaecol Can.* 2005;27(3): 263-290.

Index

V

W

MIX
Papier aus verantwortungsvollen Quellen
Paper from responsible sources
FSC® C105338

If you have any concerns about our products,
you can contact us on
ProductSafety@springernature.com

In case Publisher is established outside the EU,
the EU authorized representative is:
Springer Nature Customer Service Center GmbH
Europaplatz 3, 69115 Heidelberg, Germany

Printed by Libri Plureos GmbH
in Hamburg, Germany